It Felt Love
(A poem by Hafez)

How
Did the rose
Ever open its heart
And give to this world
All its
Beauty?
It felt the encouragement of light
Against its
Being,
Otherwise,
We all remain
Too
Frightened.

Tomb of Hafez, Shiraz, IRAN

Khwāja Shams-ud-Dīn Muḥammad Ḥāfeẓ-e Shīrāzī, known by his pen name **Hafez** "the memorizer; the (safe) keeper," was a 14th century (1315–90) Persian poet. His collected works are regarded as a pinnacle of Persian literature and are often found in the homes of people in the Persian speaking world, who learn his poems by heart and still use them as proverbs and sayings.

Nanoemulsions

Formulation, Applications, and Characterization

Nanoemulsions

Formulation, Applications, and Characterization

Edited by

Seid Mahdi Jafari

David Julian McClements

ELSEVIER

ACADEMIC PRESS

An imprint of Elsevier

Academic Press is an imprint of Elsevier
125 London Wall, London EC2Y 5AS, United Kingdom
525 B Street, Suite 1800, San Diego, CA 92101-4495, United States
50 Hampshire Street, 5th Floor, Cambridge, MA 02139, United States
The Boulevard, Langford Lane, Kidlington, Oxford OX5 1GB, United Kingdom

Library of Congress Cataloging-in-Publication Data
A catalog record for this book is available from the Library of Congress

British Library Cataloguing-in-Publication Data
A catalogue record for this book is available from the British Library

ISBN: 978-0-12-811838-2

Publisher: Andre Gerharc Wolff
Acquisition Editor: Nina Rosa Bandeira
Editorial Project Manager: Amy Clark
Production Project Manager: Vijayaraj Purushothaman
Cover Designer: Miles Hitchen

Typeset by SPi Global, India

Contents

Part III
Production of Nanoemulsions by Mechanical Methods

Part IV
Application of Nanoemulsions

11. Applications of Nanoemulsions in Foods

Francesco Donsì

12. Application of Nanoemulsions in Formulation of Pesticides

Jianguo Feng, Qi Zhang, Qi Liu, Zhengxi Zhu, David J. McClements and Seid Mahdi Jafari

13. Application of Nanoemulsions in Drug Delivery

Kamla Pathak, Satyanarayan Pattnaik and Kalpana Swain

14. Application of Nanoemulsions in Cosmetics

Odile Sonneville-Aubrun, Megumi N. Yukuyama and Aldo Pizzino

15. Application of Nanoemulsions in the Synthesis of Nanoparticles

Rafael Muñoz-Espí and Olaia Álvarez-Bermúdez

Contributors

Numbers in Parentheses indicate the pages on which the author's contributions begin.

Salman Akram (77), University of Strasbourg, Strasbourg, France

Liliana Alamilla-Beltrán (207), National Polytechnic Institute (Instituto Politécnico Nacional), Mexico City, Mexico

Olaia Álvarez-Bermúdez (477), Institut de Ciència dels Materials (ICMUV), Universitat de València, Paterna, Spain; Max Planck Institute for Polymer Research, Mainz, Germany

Nicolas Anton (77), University of Strasbourg, Strasbourg, France

Ana I. Bourbon (519), University of Minho, Braga, Portugal

Cheryl Chung (547), University of Massachusetts, Amherst, MA, United States

Francesco Donsì (349), University of Salerno, Fisciano, Italy

Jianguo Feng (379), Yangzhou University, Yangzhou, China

Seyed M.T. Gharibzahedi (233), Islamic Azad University, Lahijan, Iran

Raquel F.S. Gonçalves (519), University of Minho, Braga, Portugal

Virginia G. Granillo-Guerrero (207), National Polytechnic Institute (Instituto Politécnico Nacional), Mexico City, Mexico

Gustavo F. Gutiérrez-López (207), National Polytechnic Institute (Instituto Politécnico Nacional), Mexico City, Mexico

Andreas Håkansson (103, 175), Kristianstad University, Kristianstad, Sweden

Seid Mahdi Jafari (3, 207, 233, 379, 615), Gorgan University of Agricultural Sciences and Natural Resources, Gorgan, Iran

Heike P. Karbstein (141), Karlsruhe Institute of Technology, Karlsruhe, Germany

Diana E. Leyva-Daniel (207), National Polytechnic Institute (Instituto Politécnico Nacional), Mexico City, Mexico

Qi Liu (379), Yangzhou University, Yangzhou, China

Farooq A. Masoodi (615), University of Kashmir, Srinagar, India

David J. McClements (3, 21, 379, 547, 577, 615), University of Massachusetts, Amherst, MA, United States

Amor Monroy-Villagrana (207), National Polytechnic Institute (Instituto Politécnico Nacional), Mexico City, Mexico

Rafael Muñoz-Espí (477), Institut de Ciència dels Materials (ICMUV), Universitat de València, Paterna, Spain

Kamla Pathak (415), Uttar Pradesh University of Medical Sciences, Etawah, India

Satyanarayan Pattnaik (415), Uttar Pradesh University of Medical Sciences, Etawah, India

Antonio Perazzo (53), Princeton University, Princeton, NJ, United States

Ana C. Pinheiro (519), University of Minho, Braga; Institute of Experimental Biology and Technology, Oeiras, Portugal

Aldo Pizzino (435), L'Oréal, Research and Innovation, St-Ouen, France

Valentina Preziosi (53), University of Naples Federico II, Naples, Italy

Marilyn Rayner (103), Lund University, Lund, Sweden

Odile Sonneville-Aubrun (435), L'Oréal, Research and Innovation, Chevilly-Larue, France

Kalpana Swain (415), Talla Padmavathi College of Pharmacy, Warangal, India

Ulrike S. van der Schaaf (141), Karlsruhe Institute of Technology, Karlsruhe, Germany

Thierry F. Vandamme (77), University of Strasbourg, Strasbourg, France

António A. Vicente (519), University of Minho, Braga, Portugal

Fidel Villalobos-Castillejos (207), National Polytechnic Institute (Instituto Politécnico Nacional), Mexico City, Mexico

Goran T. Vladisavljević (287), Loughborough University, Loughborough, United Kingdom

Touseef A. Wani (615), University of Kashmir, Srinagar, India

Megumi N. Yukuyama (435), University of Sao Paulo, Butanta, Brazil

Zipei Zhang (21), University of Massachusetts, Amherst, MA, United States

Qi Zhang (379), Yangzhou University, Yangzhou, China

Ruojie Zhang (577), University of Massachusetts, Amherst, MA, United States

Zhengxi Zhu (379), Yangzhou University, Yangzhou, China

Preface

There has been a growing interest in the utilization of colloidal systems for various purposes within academy and commerce. These small particles can be used to modify the texture, optical properties, or stability of materials; to act as sensors; or to encapsulate, protect, and deliver active agents. Nanoemulsions are one of the most commonly used types of colloidal systems for these purposes because of their simplicity of preparation and the ability to easily modify their properties, such as particle size, composition, charge, and physical state.

The purpose of this book is to provide an overview of the design, fabrication, physicochemical properties/characterization, and application of nanoemulsions in different fields. The book contains contributions from a number of key researchers in the field of nanoemulsions from a variety of different academic disciplines. After presenting a brief overview of nanoemulsion basics in Section 1 (Chapter 1, general aspects, and Chapter 2, overview of nanoemulsion properties), low-energy and high-energy techniques for production of nanoemulsions have been discussed in Sections 2 (Chapters 3 and 4) and 3 (Chapters 5–10) of the book. In Chapter 3, catastrophic phase inversion techniques and, in Chapter 4, transitional techniques of nanoemulsification have been provided. Chapter 5 presents the general background of high-energy mechanical nanoemulsification techniques. Fabrication of nanoemulsions by rotor-stator devices and high-pressure valve homogenization has been covered in Chapters 6 and 7, respectively. Also, in Chapters 8–10, production of nanoemulsions through microfluidization, ultrasonication, and membrane emulsification has been discussed, respectively. Section 4 (Chapters 11–15) has been devoted to application of nanoemulsions in various fields including foods (Chapter 11), pesticides (Chapter 12), drug delivery (Chapter 13), cosmetics (Chapter 14), and nanoparticle synthesis (Chapter 15). Finally, Section 5 (Chapters 16–19) deals with characterization and analysis of nanoemulsions. Particle properties and physicochemical attributes of nanoemulsions have been explained in Chapters 16 and 17, respectively. Also, gastrointestinal fate and safety of nanoemulsions have been covered in Chapters 18 and 19, respectively.

We hope this book will stimulate further research in this rapidly growing area and will enable scientists to ascertain whether nanoemulsions are the most appropriate colloidal system to solve their particular problems.

We thank all of the authors of the chapters for taking time from their busy schedules to contribute to this project. We also thank all of the editorial staff at

Elsevier for their help and support throughout the project. Finally, we thank all of our families, friends, and colleagues.

Seid Mahdi Jafari
David Julian McClements

Part I

Nanoemulsion Basics

Chapter 1

General Aspects of Nanoemulsions and Their Formulation

David J. McClements* and Seid Mahdi Jafari†
*University of Massachusetts, Amherst, MA, United States, †Gorgan University of Agricultural Sciences and Natural Resources, Gorgan, Iran

Chapter Outline

1.1 INTRODUCTION

There has been a surge of interest in the utilization of nanoparticle dispersions for a variety of medical and industrial applications in the past decade or so, including within the pharmaceutical, food, agrochemical, cosmetics, and personal care industries (Donsi et al., 2011; Hormann and Zimmer, 2016; Kotta et al., 2012; Lu et al., 2012; McClements and Rao, 2011; Mei et al., 2013; Patel et al., 2016; Salvia-Trujillo et al., 2016; Silva et al., 2012; Wu et al., 2013). In particular, nanoparticles have been widely studied for their ability to encapsulate, protect, and release bioactive agents; to modify material's rheological, optical, and stability properties; and to alter the gastrointestinal fate of encapsulated substances (Lu et al., 2012; McClements, 2011; McClements and Xiao, 2012; Jafari, 2017). Nanoparticles have a number of physicochemical and physiological characteristics that make them particularly suitable materials for these applications, which are related to their small particle size and high surface area (Jafari and McClements, 2017). This book focuses on the design,

Nanoemulsions. https://doi.org/10.1016/B978-0-12-811838-2.00001-1

fabrication, properties, characterization, and applications of a particular type of nanoparticle dispersion: *nanoemulsions*. Nanoemulsions can easily be produced on a large scale using common industrial operations, and so they are particularly suitable for commercial applications. In this chapter, we provide a brief overview of some of the most important characteristics of nanoemulsions to help orientate the reader. Many of these subjects are then treated in more detail in the later chapters in this book.

1.2 STRUCTURE OF NANOEMULSIONS

Nanoemulsions consist of a dispersion of small droplets of one immiscible liquid in another immiscible liquid (McClements, 2011; McClements and Rao, 2011). The two immiscible liquids most widely used in commercial applications are oil and water, and so nanoemulsions are typically of either the oil-in-water (O/W) or the water-in-oil (W/O) type (Fig. 1.1). O/W nanoemulsions consist of small oil droplets dispersed in an aqueous medium, whereas W/O nanoemulsions consist of small water droplets dispersed in an oily medium (Jafari et al., 2017). O/W nanoemulsions are much more commonly utilized than W/O ones, and therefore, they will be the major focus of this book. The droplets in O/W nanoemulsions are typically coated by a hydrophilic emulsifier, whereas those in W/O nanoemulsions are coated by a lipophilic emulsifier. The nature of the emulsifier present at the oil-water interface plays a critical role in determining the overall functional attributes of nanoemulsions and should be carefully selected for each specific application.

O/W nanoemulsions are often used as templates to form other types of structured nanoparticle dispersion (Fig. 1.2). Solid lipid nanoparticles (SLNs) and nanostructured lipid carriers (NLCs) consist of fully or partly crystalline lipid

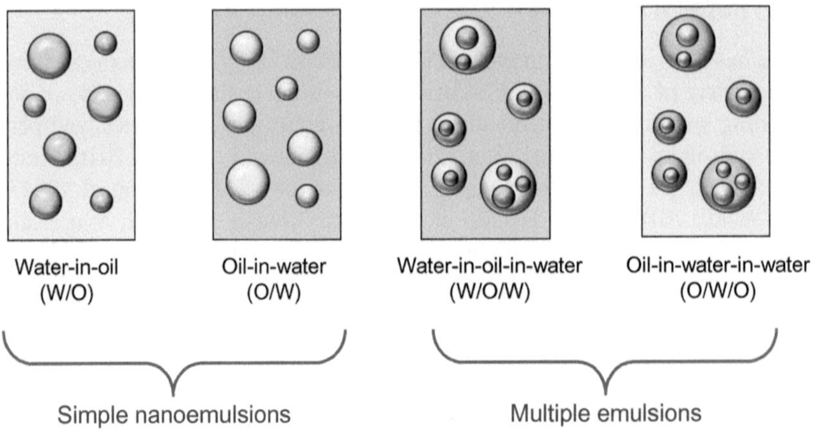

FIG. 1.1 Nanoemulsions may have a number of different structures depending on the relative location of the oil and water phases.

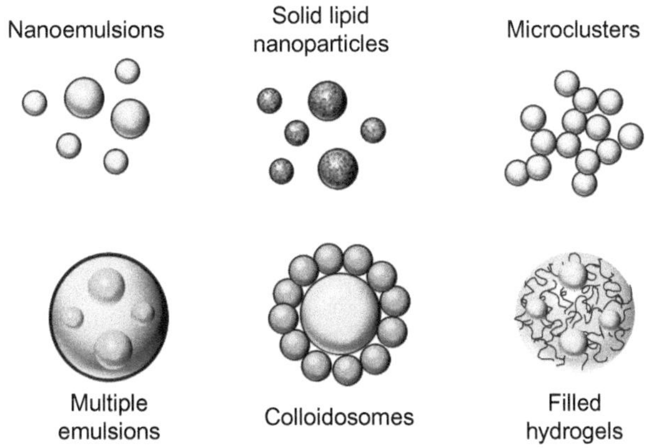

FIG. 1.2 A number of different structures can be created using lipid nanodroplets: nanoemulsions, solid lipid nanoparticles, microclusters, multiple emulsions, colloidosomes, and filled hydrogel particles.

particles dispersed in an aqueous medium, respectively (Muller et al., 2000, 2002; Pyo et al., 2017; Katouzian et al., 2017). Typically, an O/W nanoemulsion is initially formed by homogenizing a high-melting lipid and an aqueous phase containing a hydrophilic surfactant at a temperature above the melting point of the lipid. The system is then cooled below the lipid phase melting point, which results in the crystallization of the oil droplets. The solidified lipid phase in SLNs and NLCs retards molecular diffusion processes, which are useful for inhibiting the chemical degradation or controlling the release of encapsulated substances (Katouzian et al., 2017). The physical state of the lipid droplets may also alter the density and refractive index, which can change the creaming stability and optical properties of nanoemulsions.

Other types of structures can also be produced using nanoemulsions as building blocks, including multiple emulsions of the water-in-oil-in-water (W/O/W) or oil-in-water-in-oil (O/W/O) type (Fig. 1.1). These systems are usually produced using a two-step process (McClements, 2012a). For example, a W/O/W emulsion is produced by initially forming a W/O nanoemulsion by homogenizing a water phase with an oil phase containing a lipophilic surfactant together, and then, this nanoemulsion is homogenized with a water phase containing a hydrophilic surfactant. Multiple emulsions have advantages for certain applications, such as protecting a hydrophilic substance from the external aqueous phase (Assadpour et al., 2016a,b, 2017a,b), controlling the release of a hydrophilic substance (Esfanjani et al., 2015, 2017), reducing the off-flavor of a hydrophilic substance (such as bitterness or astringency), or reducing the overall fat content of the system (McClements, 2015).

Nanoemulsions can also be used as building blocks for other types of structures, such as filled hydrogels (Fig. 1.2). In this case, an O/W nanoemulsion is mixed with a biopolymer solution that is capable of forming a hydrogel, and then, a two-step process is used to form the filled hydrogels, particle formation and particle gelation (McClements, 2012a; Shewan and Stokes, 2013). Initially, a particle is formed that contains small lipid droplets trapped inside a larger biopolymer-rich water droplet, and then the system conditions are changed to cross-link the biopolymers within the water droplet (Mokhtari et al., 2017). Filled hydrogels can be designed to encapsulate, protect, and release bioactive components by altering their dimensions, internal composition, or structure. This can be achieved by altering the fabrication method used and the type and concentration of biopolymer and cross-linking agents used. In addition, the properties of the lipid droplets trapped inside the hydrogels can also be controlled, such as their size, concentration, composition, or charge.

Nanoemulsions can also be used to create other types of structures such as colloidosomes or microclusters (Fig. 1.2). A colloidosome consists of a large central particle with smaller particles adsorbed to its surface, whereas a microcluster consists of a number of smaller particles held together by attractive forces. These kinds of structures may be created from nanoemulsions in order to change the rheological, optical, or stability properties of materials or for controlled release applications.

1.3 NANOEMULSION FABRICATION

Numerous types of fabrication methods have been developed to create nanoemulsions, which can be roughly divided into low-intensity and high-intensity methods (Gupta et al., 2016; McClements, 2011; Shams and Sahari, 2016; Jafari et al., 2015). High-intensity methods are currently the most widely used for producing nanoemulsions in industrial applications. These methods use specially designed mechanical devices to break up and intermingle the oil and water phases by generating intense shear, turbulent, and cavitational flow profiles. The most commonly used mechanical devices for producing nanoemulsions are high-pressure valve homogenization, microfluidization, and sonication (McClements and Rao, 2011; Tadros et al., 2004). Low-intensity methods rely on the spontaneous generation of small droplets in certain surfactant-oil-water mixtures when the system composition or environmental conditions (such as temperature) are changed (Komaiko and McClements, 2016). The most commonly used low-energy methods are the phase inversion temperature, spontaneous emulsification, and emulsion inversion point methods. Membrane emulsification is another special technique that could be applied for producing nanoemulsions (Chapter 10) that are sometimes classified in high-energy emulsification methods, but it is in fact a mechanical technique with low-energy utilization.

The choice of a particular nanoemulsion fabrication method depends on the nature of the materials to be homogenized (particularly the oil and surfactant phases) and the desired physicochemical and functional attributes of the final product (such as optical, rheological, stability, and release characteristics).

The basic principles behind both low- and high-energy fabrication methods will be presented in Parts II and III of this book, and their relative advantages and disadvantages will be highlighted. Knowledge of the different approaches available is important for scientists and technologists to select the most appropriate fabrication method for a particular application.

1.4 NANOEMULSION PARTICLE PROPERTIES

The physicochemical properties and functional attributes of a nanoemulsion are largely determined by particle properties, such as composition, size, electric charge, aggregation state, physical state, and interfacial composition (Fig. 1.3). A researcher or manufacturer can therefore tailor the properties of the particles in a nanoemulsion so as to obtain the physicochemical or physiological properties required for a specific application.

Typically, nanoemulsions contain a range of different droplet sizes, and therefore, their dimensions are characterized in terms of a particle size distribution (Fig. 1.4). In many cases, it is more convenient to report the particle dimensions in terms of a mean droplet diameter and polydispersity index. The mean droplet diameter can be defined in a number of different ways, with the most common being the number-, surface-, and volume-weighted values, that is, d_N (or d_{10}), d_S (or d_{32}), and d_V (or d_{43}). However, other values are sometimes used such as the intensity-weighted (Z-average) diameter that is determined by dynamic light scattering. It is always important to specify the type of mean diameter that is used when reporting particle size data on nanoemulsions since their values can be very different depending on the width of the distribution.

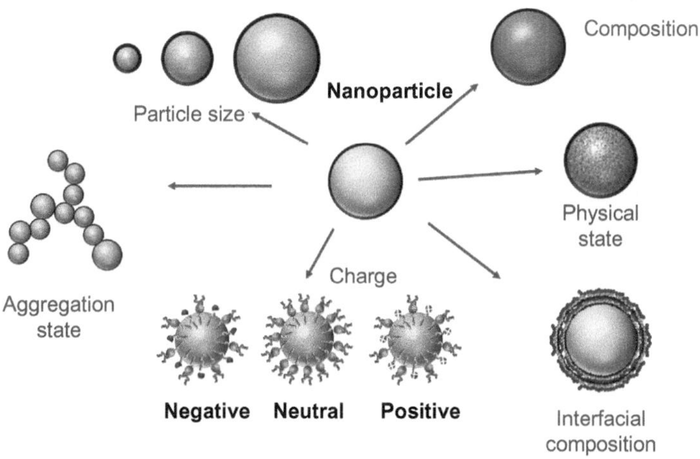

FIG. 1.3 Nanoemulsions may vary in particle characteristics, such as size, composition, and charge.

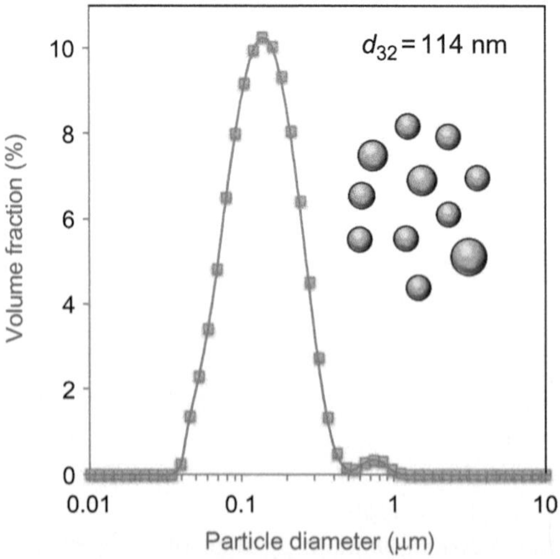

FIG. 1.4 Particle size distribution of a nanoemulsion formed by a low-energy method using MCT as the oil phase and Tween 80 as the surfactant. The measurement was carried out using static light scattering.

An important issue that must be addressed when dealing with nanoemulsions is to distinguish them from both conventional emulsions and microemulsions (McClements, 2012b). The main difference between nanoemulsions and conventional emulsions is the droplet size. Both types of system are thermodynamically unstable, but the droplets in nanoemulsions are smaller than those in conventional emulsions (McClements, 2011). There is some debate and uncertainty about the cutoff size that separates emulsions from nanoemulsions. In this book, we assume that systems with a mean droplet diameter <200 nm can be considered to be nanoemulsions, but systems containing larger droplets can be considered to be emulsions. This definition is somewhat arbitrary since there is no distinct change in the physical, chemical, or biological properties of a system when the droplet diameter falls below 200 nm. Instead, there is usually a gradual change in properties as the droplet size is decreased. Nevertheless, nanoemulsions do have physicochemical characteristics that are distinctly different from those of emulsions:

- **(i)** Nanoemulsions may be optically clear due to the weak light scattering by small droplets.
- **(ii)** Nanoemulsions may be highly stable to gravitational separation due to weak gravitational forces and Brownian motion effects.
- **(iii)** Nanoemulsions may have a higher bioavailability than emulsions due to more rapid digestion.
- **(iv)** Nanoemulsions may have a higher chemical reactivity than emulsions due to the higher oil-water interfacial area.

Typically, the differences between emulsions and nanoemulsions become more pronounced when the droplet diameter falls below about 200nm, and so, this value is a good demarcation point.

The electric characteristics of emulsion droplets are largely determined by the nature of any surface-active substances adsorbed to their surfaces and the ionic composition of the surrounding solution (McClements, 2015). The electric potential may vary from strongly positive, to neutral, to strongly negative depending on the emulsifier type and solution conditions. The electric properties of the oil droplets in a nanoemulsion play a major role in determining its physical and chemical stability, its gastrointestinal fate, and its ability to fabricate more complex structures (Fig. 1.2). This latter effect is because electrostatic interactions are often used to assemble structures from electrically charged colloidal particles and polymers.

As mentioned earlier, the core of the lipid droplets may be liquid (as in nanoemulsions) or solid (as in SLNs or NLCs). The physical state of the droplet interior depends on the nature of the oil phase used (such as its crystallization and melting behavior) and the thermal history of the system. Controlling the physical state of the lipid droplets can be used to control the stability and release characteristics of any encapsulated bioactive agents. Moreover, the physical state of the droplets may also influence their refractive index and density, which will impact the optical properties and creaming stability of the system.

The nature of the emulsifier used to prepare a nanoemulsion largely determines interfacial properties, such as thickness, charge, hydrophobicity, and chemical reactivity (McClements, 2015). Consequently, the selection of an appropriate emulsifier is extremely important in the development of a nanoemulsion for a particular application. Unlike in conventional emulsions, the interfacial layer in nanoemulsions can make up an appreciable volume of the droplets (McClements, 2011):

$$\Phi_S = \frac{(r+\delta_S)^3 - r^3}{(r+\delta_S)^3} \tag{1.1}$$

Here, Φ_S ($=V_S/V_{C+S}$) is the volume of the emulsifier layer (shell) divided by the effective volume of the overall droplet (=core + shell), δ_S is the thickness of the emulsifier layer (shell), and r is the radius of the oil droplet (core). For instance, the value of Φ_S is 25%, 49%, 70%, and 88% for nanoemulsions containing lipid droplets with core radius of 20nm and shell thicknesses of 2, 5, 10, and 20nm, respectively, whereas Φ_S is only 3%, 7%, 14%, and 25%, for emulsions containing lipid droplets with core radius of 200nm and similar shell thicknesses. As a result, many of the properties of a nanoemulsion droplet may be different from those of a corresponding emulsion droplet, such as density, refractive index, and release characteristics.

The droplets in nanoemulsions may exist as separate entities, or they may exist as clusters (flocs) that consist of numerous droplets held together by attractive interactions (McClements, 2015). The nature of the flocs formed has a major impact on the physicochemical properties of nanoemulsions, such as their rheology, optical properties, and stability. Typically, a flocculated nanoemulsion will have a higher viscosity than a nonflocculated one with the same droplet concentration (McClements, 2011). In a dilute nanoemulsion, droplet flocculation usually leads to more rapid gravitational separation due to the increase in particle size. On the other hand, in a concentrated nanoemulsion, droplet flocculation may actually inhibit gravitational separation due to the formation of a three-dimensional network of aggregated droplets that are not free to move. The size, structure, and strength of the flocs formed can be controlled to create flocculated nanoemulsions with different properties.

1.5 NANOEMULSION STABILITY

Nanoemulsions are thermodynamically unstable systems because of the unfavorable molecular interactions that occur at the oil-water interface due to the hydrophobic effect, and therefore, there is a thermodynamic driving force to reduce the contact area between the oil and water phases (McClements, 2011; Tadros et al., 2004). Consequently, nanoemulsions will always break down over time, which may occur through a variety of mechanisms including gravitational separation, flocculation, coalescence, and Ostwald ripening (Fig. 1.5).

Gravitational separation is the process whereby droplets either move upward (creaming) or downward (sedimentation) because their density is lower or higher than that of the surrounding liquid, respectively. The rate of droplet movement due to gravitational separation increases as the droplet size increases, the density contrast increases, and the aqueous phase viscosity decreases. Consequently, gravitational separation can be reduced by decreasing the droplet size, reducing the density contrast by adding weighting agents,

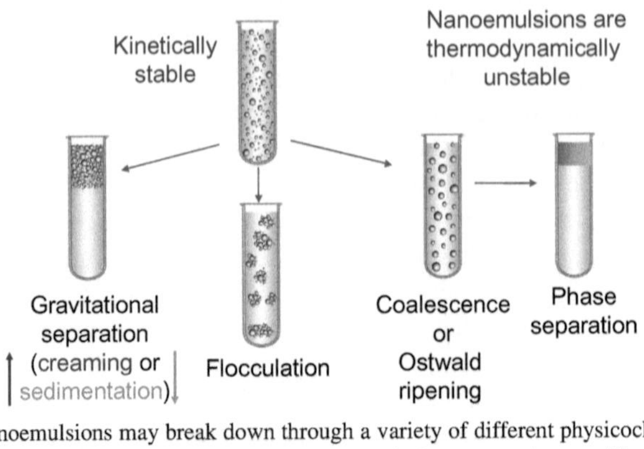

FIG. 1.5 Nanoemulsions may break down through a variety of different physicochemical mechanisms, depending on the composition and structure, and when exposed to specific environmental conditions.

or increasing the aqueous phase viscosity by adding thickening agents. Flocculation and coalescence are two forms of droplet aggregation that depend on the nature of the colloidal interactions between the droplets. Flocculation is the process whereby two or more droplets associate with each other and form clusters due to attractive forces acting between them. Coalescence is the process whereby a number of droplets collide and merge together leading to the formation of a larger droplet. Droplet aggregation can be inhibited by ensuring that the repulsive interactions (e.g., electrostatic and steric) are greater than the attractive interactions (e.g., van der Waals, hydrophobic, and depletion). This can often be achieved by altering the nature of the emulsifier used or the aqueous phase composition. Ostwald ripening is the process whereby smaller droplets shrink and larger droplets grow due to the diffusion of dispersed phase molecules through the continuous phase, which is driven by differences in the curvature of the particles. The solubility of the dispersed phase is higher in small droplets (large curvature) than in large droplets (small curvature), which leads to a concentration gradient that drives droplet growth. Ostwald ripening can be inhibited by using an oil phase that has a low water solubility or by adding ripening inhibitors. Knowledge of the most important instability mechanisms operating within a nanoemulsion is critical to designing systems with appropriate stability characteristics.

Nanoemulsions can be designed to have a kinetic stability that is sufficiently long for commercial applications by carefully controlling their composition and structure. In particular, the choice of a suitable oil and aqueous phase and the selection of the most appropriate additives, such as emulsifiers, texture modifiers, ripening inhibitors, or weighting agents, are critical. The major physicochemical mechanisms responsible for nanoemulsion instability and strategies to improve nanoemulsion stability will be discussed later in this book.

1.6 NANOEMULSION INGREDIENTS

As mentioned earlier, the selection of an appropriate group of ingredients is critical to the formulation of a successful nanoemulsion (McClements, 2015). The oil phase used to formulate a nanoemulsion may be composed of various hydrophobic components, including triglyceride oils, essential oils, flavor oils, oil-soluble vitamins, colors, flavors, preservatives, nutraceuticals, and pharmaceuticals. Oil phases may therefore differ in their physicochemical properties such as viscosity, polarity, refractive index, melting point, and interfacial tension, which alter their ability to both form and stabilize nanoemulsions. Moreover, the physiological response to an ingested nanoemulsion also depends on the oil type, for example, digestibility and bioavailability (McClements and Xiao, 2012; Jafari and McClements, 2017).

The aqueous phase is typically composed primarily of water, but it may also contain other components that play an important role, such as buffers, salts, cosolvents, preservatives, proteins, and carbohydrates. Thus, aqueous phases may also differ considerably in their physicochemical and physiological properties. Nanoemulsions cannot usually be formed by simply homogenizing an oil and aqueous phase together because the system would rapidly separate into its component phases (McClements, 2015). Instead, a number of additional ingredients, often referred to as stabilizers, are typically needed to facilitate the formation and ensure the long-term stability of nanoemulsions, including emulsifiers, texture modifiers, weighting agents, and ripening inhibitors.

An *emulsifier* is a surface-active amphiphilic molecule that adsorbs onto the oil-water interface during homogenization, reduces the interfacial tension thereby facilitating droplet disruption, and creates a protective interfacial layer thereby inhibiting droplet aggregation (McClements and Gumus, 2016). Numerous types of emulsifiers are available to form and stabilize nanoemulsions that vary from industry to industry (Kralova and Sjoblom, 2009; Tadros, 2014). For example, in the food industry, small molecule surfactants, proteins, polysaccharides, and phospholipids are commonly used (McClements and Gumus, 2016). Emulsifiers vary considerably in their ability to form and stabilize nanoemulsions, and it is extremely important to identify the most suitable one for a particular application.

A *texture modifier* is a substance that is typically added to the continuous phase to change its rheological properties and may be a thickening agent or a gelling agent (McClements, 2015). A thickening agent increases the viscosity of a solution due to its ability to alter the fluid flow profile and typically consists of soluble polymers with extended structures. In the food industry, water-soluble polysaccharides, such as xanthan gum, are commonly used as thickening agents because they can be used at low levels to greatly thicken solutions. A gelling agent gives some solid-like characteristics to a solution due to its ability to form physical or chemical cross-links with its neighbors. In the food industry, proteins (such as meat, egg, milk, or plant proteins) and polysaccharides (such as alginate, carrageenan, and pectin) are often used as gelling agents. These texture modifiers improve the stability of nanoemulsions by inhibiting droplet movement, thereby retarding gravitational separation.

A *weighting agent* is a substance that is added to the dispersed phase so that its density becomes similar to that of the continuous phase, thereby reducing the tendency for gravitational separation to occur (McClements, 2015). Weighting agents are most commonly used in O/W nanoemulsions in the beverage industry (Piorkowski and McClements, 2014). In this case, they are hydrophobic substances that have a density that is appreciably higher than that of water, such as brominated vegetable oil (BVO), sucrose acetate isobutyrate (SAIB), or ester gum. Typically, both flavor and triglyceride oils have lower densities than water, and therefore, adding an appropriate amount of weighting agent can increase their density until it matches that of the aqueous phase, thereby inhibiting creaming.

A *ripening inhibitor* is a substance that is added to the dispersed phase of a nanoemulsion to retard droplet growth due to Ostwald ripening

(Wooster et al., 2008). Typically, ripening inhibitors are utilized in O/W nanoemulsions that are prepared from oil phases with a relatively high solubility in water, such as essential or flavor oils (Chang et al., 2012; Rao and McClements, 2012). In the absence of the ripening inhibitors, this type of nanoemulsion rapidly becomes unstable due to droplet growth and accelerated creaming. Ripening inhibitors are typically highly hydrophobic molecules that have a very low solubility in water, such as long-chain triglycerides (e.g., corn, palm, rape seed, or sunflower oil). The presence of a ripening inhibitor in an oil droplet retards droplet growth through an entropy of mixing effect. As the water-soluble oil molecules (flavor or essential oils) diffuse from the small droplets to the large droplets, there is an increase in the concentration of the ripening inhibitor (triglyceride oil) remaining in the smaller droplets. As a result, there is a concentration gradient in the system, which favors the movement of water-soluble oil molecules from large to small droplets, thereby counterbalancing droplet growth through Ostwald ripening.

1.7 PHYSICOCHEMICAL PROPERTIES OF NANOEMULSIONS

The bulk physicochemical characteristics of nanoemulsions, such as their optical and rheological properties, play a major role in determining their suitability for particular applications. Both the optical and rheological properties of nanoemulsions are mainly determined by their composition and structure. The optical properties of nanoemulsions depend on both light scattering and absorption effects (Mason et al., 2006; McClements, 2002; Michels et al., 2008). The turbidity or opacity of a nanoemulsion depends on the light scattering profile, which depends on the concentration, size, and refractive index contrast of the droplets. The color of a nanoemulsion depends on the presence of any chromophores that selectively absorb light at particular wavelengths, and the presence of droplets scatters light differently at different wavelengths. The light scattering from emulsion droplets is typically very weak when their dimensions are much smaller than the wavelength of light, which leads to an optically clear system (Fig. 1.6). As the particle size increases, the efficiency of light scattering increases and reaches a maximum when the dimensions of the droplets are fairly similar to the wavelength of light. It is for this reason that emulsions tend to be either turbid or optically opaque depending on the total droplet concentration.

Typically, the lightness of an emulsion or nanoemulsion increases as the droplet concentration increases (Fig. 1.7). There is usually a steep increase in lightness from 0% to 5% droplets, and then a more gradual change at higher levels. The optical properties of nanoemulsions are important for many commercial applications. For example, the food industry often wants to incorporate lipophilic substances (such as colors, flavors, vitamins, or nutraceuticals) into optically transparent beverages, such as soft drinks or fruit juices (Piorkowski and McClements, 2014).

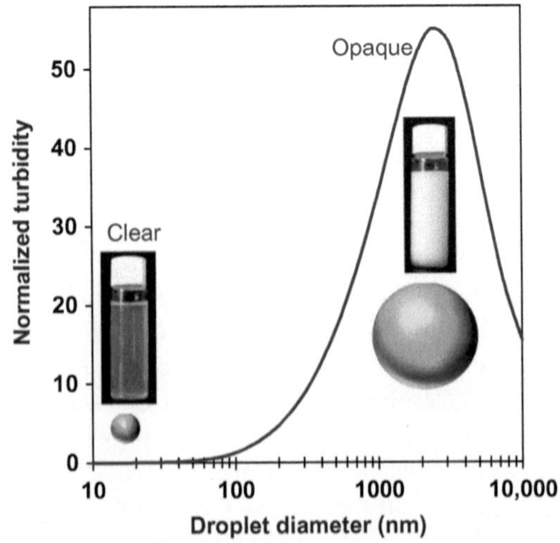

FIG. 1.6 Dependence of the turbidity of a nanoemulsion or emulsion on the droplet size. Nanoemulsions containing very small droplets ($d < 40$ nm) tend to be transparent.

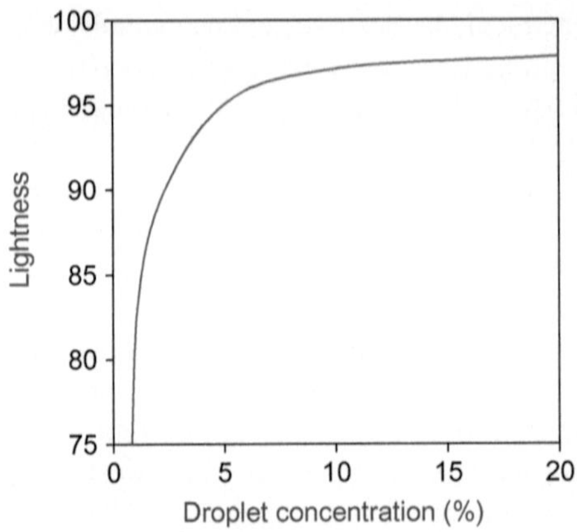

FIG. 1.7 Dependence of the lightness of nanoemulsions on droplet concentration.

The rheology of a nanoemulsion is another important physicochemical property that may influence its practical applications. Typically, the viscosity of a nanoemulsion increases with increasing droplet concentration, gradually at first but then more steeply as the droplets become more closely packed together (Fig. 1.8). Indeed, at high droplet concentrations, a nanoemulsion tends to have viscoelastic properties and may have a yield stress and an elastic modulus (Tadros et al., 2004). The critical droplet concentration where there is a steep rise in viscosity is often lower for nanoemulsions than emulsions because the dimensions of the interfacial layer are comparable with the dimensions of the oil droplet core, and hence, the effective volume of the droplets is greater.

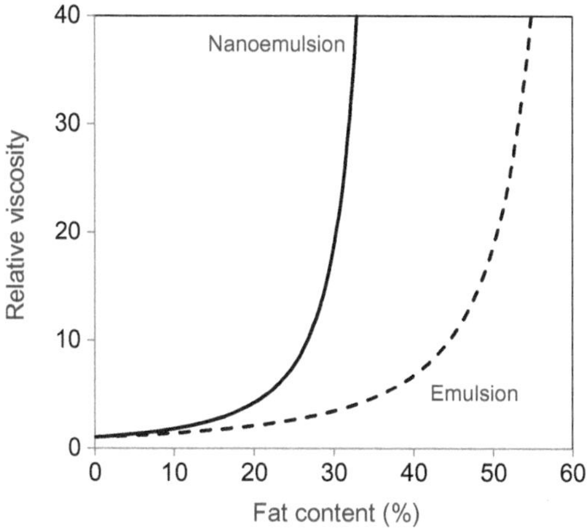

FIG. 1.8 Dependence of the relative viscosity of nanoemulsions on droplet concentration.

The *effective* volume fraction ($\phi_{effective}$) of the emulsifier-coated droplets in a nanoemulsion is given by the following expression:

$$\phi_{effective} = \phi\left(1 + \frac{\delta_S}{r}\right)^3 \tag{1.2}$$

Here, ϕ is the dispersed phase volume fraction (uncoated droplets). In the case of O/W emulsions, $\delta_S \ll r$, and so, the effective volume fraction is close to the oil content ($\phi_{effective} \approx \phi$). However, in nanoemulsions, $\delta_S \approx r$, and so the effective volume fraction may be appreciably larger than the oil content ($\phi_{effective} > \phi$) due to the presence of the emulsifier layer. Nanoemulsions with high viscosities or gel-like properties can therefore be prepared at relatively low oil contents by having a thick interfacial layer or strong electrostatic repulsion between the droplets (McClements, 2011; Tadros et al., 2004). Alternatively, semisolid nanoemulsions can be fabricated by ensuring that there is a strong attractive interaction between the droplets and that there are sufficient droplets present to form a three-dimensional network throughout the system (Mao and McClements, 2013).

The optical and rheological properties of nanoemulsions can therefore be carefully controlled by altering droplet characteristics so as to create materials for particular applications. A more detailed description of the major factors impacting the appearance, rheology, and stability of nanoemulsions is given later in Chapter 2 of this book.

1.8 NANOEMULSION CHARACTERIZATION

The design, fabrication, and testing of nanoemulsions require knowledge of their composition, structure, and properties, and so, appropriate analytic tools are required to provide information about particle characteristics (such as concentration, size, aggregation state, charge, and physical state) and bulk physicochemical properties (such as appearance, rheology, and stability). A wide variety of different analytic tools have been developed to characterize the properties of nanoemulsions (McClements and McClements, 2016). The particle size distribution is typically determined using dynamic or static light scattering methods. The particle charge characteristics are commonly measured using microelectrophoresis methods. The aggregation state and morphology of nanoemulsions are often determined using optical, electron, or atomic force microscopy. The physical state of the droplets in nanoemulsions can be conveniently determined using differential scanning calorimetry or X-ray scattering methods. The different types of analytic instruments and protocols that are commonly used to characterize the properties of nanoemulsions are described in more detail later in Part V of this book.

1.9 APPLICATIONS OF NANOEMULSIONS

Researchers have already examined many different potential applications of nanoemulsions within the pharmaceutical, food, cosmetic, and agrochemical industries. The small size and high surface area of nanoemulsions give them a number of characteristic physicochemical and physiological properties that make them highly suitable for certain applications. A number of the most important applications of nanoemulsions in different industries are discussed later in Part IV of this book. Here, we just provide a few examples to highlight their considerable potentials.

O/W nanoemulsions are particularly suitable for incorporating lipophilic active agents (such as drugs, nutraceuticals, vitamins, colors, flavors, or preservatives) into aqueous environments. The lipophilic active agent is simply mixed with the oil phase prior to nanoemulsion fabrication, so that the final system contains active-loaded lipid droplets dispersed in water. The appearance of the final product can be controlled by altering the oil droplet size. For applications where the final product is turbid or opaque, nanoemulsions with relatively large droplets can be utilized (d from 100 to 200 nm). On the other hand, for applications where the final product should be optically clear, nanoemulsions with relatively small droplets can be used ($d < 50$ nm). Another advantage of using nanoemulsions for this application is that they tend to be highly stable to gravitational separation and aggregation because of the small droplet size, which can increase the shelf life of a product.

Another major advantage of O/W nanoemulsions for certain applications is that they tend to be rapidly digested within the gastrointestinal tract because of their high oil-water surface area (McClements and Xiao, 2012; Jafari and McClements, 2017). As a result, they rapidly release any encapsulated bioactive components and rapidly form mixed micelles that can solubilize and transport these bioactive components to the epithelium cells. Nanoemulsions are therefore particularly suitable for increasing the oral bioavailability of hydrophobic bioactive agents, such as nonpolar drugs or nutraceuticals.

Numerous studies have also shown that O/W nanoemulsions are suitable for increasing the efficacy of antimicrobial essential oils against a broad range of microorganisms, including bacteria, yeast, and molds (Donsi and Ferrari, 2016; Shams and Sahari, 2016). Converting the essential oil into nanosized droplets appears to increase its ability to disrupt the cell membranes of microorganisms. Another advantage of nanoemulsions is that a number of different types of antimicrobial agents can be incorporated into a single delivery system, such as hydrophobic (droplet interior), amphiphilic (droplet surface), and hydrophilic (aqueous phase) agents.

1.10 CONCLUSION

In this chapter, formulation ingredients and properties of nanoemulsions along with their fabrication methods and characterization techniques were described briefly. It is very important to differentiate nanoemulsions from conventional emulsions in terms of droplet size; we introduced a droplet size of 200 nm as the border below which the properties and characteristics of nanoemulsions show different trends such as a higher apparent transparency, more bioavailability, higher stability, and more chemical reactivity. Nanoemulsions can have considerable applications in different fields of food, pharmaceutical, cosmetic, and agrochemical industries. Another crucial aspect is selecting the appropriate ingredients and techniques for formulation and preparation of nanoemulsions so that they can have the optimum applicability in terms of rheological properties, appearance, stability, and efficiency. Therefore, researchers and experts should have an in-depth knowledge of the nanoemulsion formation, and it is very significant to have access to analytic instruments for measuring the nanoemulsion properties such as droplet size, surface tension, viscosity, color, and surface charge.

REFERENCES

Assadpour, E., Maghsoudlou, Y., Jafari, S.-M., Ghorbani, M., Aalami, M., 2016a. Optimization of folic acid nano-emulsification and encapsulation by maltodextrin-whey protein double emulsions. Int. J. Biol. Macromol. 86, 197–207.

Assadpour, E., Maghsoudlou, Y., Jafari, S.-M., Ghorbani, M., Aalami, M., 2016b. Evaluation of folic acid nano-encapsulation by double emulsions. Food Bioprocess Technol. 9 (12), 2024–2032.

Assadpour, E., Jafari, S.-M., Maghsoudlou, Y., 2017a. Evaluation of folic acid release from spray dried powder particles of pectin-whey protein nano-capsules. Int. J. Biol. Macromol. 95, 238–247.

Assadpour, E., Jafari, S.-M., 2017b. Spray drying of folic acid within nano-emulsions: optimization by Taguchi approach. Dry. Technol. 35 (9), 1152–1160.

Chang, Y.H., Mclandsborough, L., McClements, D.J., 2012. Physical properties and antimicrobial efficacy of thyme oil nanoemulsions: influence of ripening inhibitors. J. Agric. Food Chem. 60, 12056–12063.

Donsi, F., Ferrari, G., 2016. Essential oil nanoemulsions as antimicrobial agents in food. J. Biotechnol. 233, 106–120.

Donsi, F., Sessa, M., Mediouni, H., Mgaidi, A., Ferrari, G., 2011. In: Saravacos, G., Taoukis, P., Krokida, M., Karathanos, V., Lazarides, H., Stoforos, N., Tzia, C., Yanniotis, S. (Eds.), Encapsulation of bioactive compounds in nanoemulsion-based delivery systems.11th International Congress on Engineering and Food.

Esfanjani, A.F., Jafari, S.M., Assadpoor, E., Mohammadi, A., 2015. Nano-encapsulation of saffron extract through double-layered multiple emulsions of pectin and whey protein concentrate. J. Food Eng. 165, 149–155.

Esfanjani, A.F., Jafari, S.M., Assadpour, E., 2017. Preparation of a multiple emulsion based on pectin-whey protein complex for encapsulation of saffron extract nanodroplets. Food Chem. 221, 1962–1969.

Gupta, A., Eral, H.B., Hatton, T.A., Doyle, P.S., 2016. Nanoemulsions: formation, properties and applications. Soft Matter 12, 2826–2841.

Hormann, K., Zimmer, A., 2016. Drug delivery and drug targeting with parenteral lipid nanoemulsions—a review. J. Control. Release 223, 85–98.

Jafari, S.M., 2017. An overview of nano-encapsulation techniques and their classification. In: Jafari, S.M., Jafari, S.M. (Eds.), Nano-encapsulation Technologies for the Food and Nutraceutical Industries. Elsevier, San Diego, CA.

Jafari, S.M., Fathi, M., Mandala, I., 2015. Emerging product formation. In: Food Waste Recovery: Processing Technologies and Industrial Techniques. Elsevier Inc, San Diego, CA.

Jafari, S.M., McClements, D.J., 2017. Nanotechnology approaches for increasing nutrient bioavailability. In: Advances in Food and Nutrition Research. Academic Press, San Diego, CA.

Jafari, S.M., Paximada, P., Mandala, I., Assadpour, E., Mehrnia, M.-A., 2017. Encapsulation by nano-emulsions. In: Jafari, S.M. (Ed.), Nano-encapsulation Technologies for the Food and Nutraceutical Industries. Elsevier, San Diego, CA.

Katouzian, I., Esfanjani, A.F., Jafari, S.M., Akhavan, S., 2017. Formulation and application of a new generation of lipid nano-carriers for the food bioactive ingredients. Trends Food Sci. Technol. 68 (Suppl. C), 14–25.

Komaiko, J.S., McClements, D.J., 2016. Formation of food-grade nanoemulsions using low-energy preparation methods: a review of available methods. Compr. Rev. Food Sci. Food Saf. 15, 331–352.

Kotta, S., Khan, A.W., Pramod, K., Ansari, S.H., Sharma, R.K., Ali, J., 2012. Exploring oral nanoemulsions for bioavailability enhancement of poorly water-soluble drugs. Expert Opin. Drug Deliv. 9, 585–598.

Kralova, I., Sjoblom, J., 2009. Surfactants used in food industry: a review. J. Dispers. Sci. Technol. 30, 1363–1383.

Lu, Y., Qi, J.P., Wu, W., 2012. Absorption, disposition and pharmacokinetics of nanoemulsions. Curr. Drug Metab. 13, 396–417.

Mao, Y.Y., McClements, D.J., 2013. Modulation of food texture using controlled heteroaggregation of lipid droplets: principles and applications. J. Appl. Polym. Sci. 130, 3833–3841.

Mason, T.G., Wilking, J.N., Meleson, K., Chang, C.B., Graves, S.M., 2006. Nanoemulsions: formation, structure, and physical properties. J. Phys. Condens. Matter 18, R635–R666.

McClements, D.J., 2002. Theoretical prediction of emulsion color. Adv. Colloid Interf. Sci. 97, 63–89.

McClements, D.J., 2011. Edible nanoemulsions: fabrication, properties, and functional performance. Soft Matter 7, 2297–2316.
McClements, D.J., 2012a. Advances in fabrication of emulsions with enhanced functionality using structural design principles. Curr. Opin. Colloid Interface Sci. 17, 235–245.
McClements, D.J., 2012b. Nanoemulsions versus microemulsions: terminology, differences, and similarities. Soft Matter 8, 1719–1729.
McClements, D.J., 2015. Food Emulsions: Principles, Practices, and Techniques. CRC Press, Boca Raton, FL.
McClements, D.J., Gumus, C.E., 2016. Natural emulsifiers—biosurfactants, phospholipids, biopolymers, and colloidal particles: molecular and physicochemical basis of functional performance. Adv. Colloid Interf. Sci. 234, 3–26.
McClements, D.J., Rao, J., 2011. Food-grade nanoemulsions: formulation, fabrication, properties, performance, biological fate, and potential toxicity. Crit. Rev. Food Sci. Nutr. 51, 285–330.
McClements, D.J., Xiao, H., 2012. Potential biological fate of ingested nanoemulsions: influence of particle characteristics. Food Funct. 3, 202–220.
McClements, J., McClements, D.J., 2016. Standardization of nanoparticle characterization: methods for testing properties, stability, and functionality of edible nanoparticles. Crit. Rev. Food Sci. Nutr. 56, 1334–1362.
Mei, L., Zhang, Z.P., Zhao, L.Y., Huang, L.Q., Yang, X.L., Tang, J.T., Feng, S.S., 2013. Pharmaceutical nanotechnology for oral delivery of anticancer drugs. Adv. Drug Deliv. Rev. 65, 880–890.
Michels, R., Foschum, F., Kienle, A., 2008. Optical properties of fat emulsions. Opt. Express 16, 5907–5925.
Mokhtari, S., Jafari, S.M., Assadpour, E., 2017. Development of a nutraceutical nano-delivery system through emulsification/internal gelation of alginate. Food Chem. 229, 286–295.
Muller, R.H., Mader, K., Gohla, S., 2000. Solid lipid nanoparticles (SLN) for controlled drug delivery—a review of the state of the art. Eur. J. Pharm. Biopharm. 50, 161–177.
Muller, R.H., Radtke, M., Wissing, S.A., 2002. Solid lipid nanoparticles (SLN) and nanostructured lipid carriers (NLC) in cosmetic and dermatological preparations. Adv. Drug Deliv. Rev. 54, S131–S155.
Patel, R.B., Thakore, S.D., Patel, M.R., 2016. Recent survey on patents of nanoemulsions. Curr. Drug Deliv. 13, 857–881.
Piorkowski, D.T., McClements, D.J., 2014. Beverage emulsions: recent developments in formulation, production, and applications. Food Hydrocoll. 42, 5–41.
Pyo, S.M., Müller, R.H., Keck, C.M., 2017. Encapsulation by nano-structured lipid carriers. In: Jafari, S.M. (Ed.), Nano-encapsulation Technologies for the Food and Nutraceutical Industries. Elsevier, San Diego, CA.
Rao, J., McClements, D.J., 2012. Impact of lemon oil composition on formation and stability of model food and beverage emulsions. Food Chem. 134, 749–757.
Salvia-Trujillo, L., Martin-Belloso, O., McClements, D.J., 2016. Excipient nanoemulsions for improving oral bioavailability of bioactives. Nanomaterials 6, 17.
Shams, N., Sahari, M.A., 2016. Nanoemulsions: preparation, structure, functional properties and their antimicrobial effects. Appl. Food Biotechnol. 3, 138–149.
Shewan, H.M., Stokes, J.R., 2013. Review of techniques to manufacture micro-hydrogel particles for the food industry and their applications. J. Food Eng. 119, 781–792.
Silva, H.D., Cerqueira, M.A., Vicente, A.A., 2012. Nanoemulsions for food applications: development and characterization. Food Bioprocess Technol. 5, 854–867.
Tadros, T., 2014. An Introduction to Surfactants. Walter de Gruyter GmbH, Berlin, Germany.

Tadros, T., Izquierdo, R., Esquena, J., Solans, C., 2004. Formation and stability of nano-emulsions. Adv. Colloid Interf. Sci. 108, 303–318.
Wooster, T.J., Golding, M., Sanguansri, P., 2008. Impact of oil type on nanoemulsion formation and ostwald ripening stability. Langmuir 24, 12758–12765.
Wu, Y., Li, Y.H., Gao, X.H., Chen, H.D., 2013. The application of nanoemulsion in dermatology: an overview. J. Drug Target. 21, 321–327.

Chapter 2

Overview of Nanoemulsion Properties: Stability, Rheology, and Appearance

Zipei Zhang and David J. McClements
University of Massachusetts, Amherst, MA, United States

Chapter Outline

2.1 INTRODUCTION

There is a growing interest in the utilization of nanoemulsions in the food, cosmetics, and pharmaceutical industries due to their potential advantages over conventional emulsions (McClements, 2011). Nanoemulsions typically have better stability to droplet aggregation and creaming because of their smaller droplet dimensions (McClements, 2012). Nanoemulsions containing sufficiently small droplets ($d < 50$ nm) scatter light waves very weakly and can therefore be used to create optically transparent products (Mason et al., 2007). Nanoemulsions with highly viscous or gel-like properties can be created at low droplet concentrations because of the overlap of polymer or electrostatic layers surrounding the small droplets (Chen et al., 2011). Nanoemulsions can

Nanoemulsions. https://doi.org/10.1016/B978-0-12-811838-2.00002-3

be designed to promote the bioavailability of bioactive agents trapped inside them or coingested with them (Chen et al., 2011; Zhang and McClements, 2016; McClements et al., 2016). In this chapter, an overview of the physicochemical properties of nanoemulsions is given, with an emphasis on their stability, rheology, and appearance. Moreover, the relationship between the structure and function of nanoemulsions is discussed in terms of the development of nanoemulsion-based products with desirable physicochemical properties and functional attributes.

2.2 IMPORTANCE OF PHYSICOCHEMICAL PROPERTIES

2.2.1 General Physicochemical Properties of Nanoemulsions

Nanoemulsions are colloidal dispersions where one liquid is dispersed as small droplets in another immiscible liquid (McClements, 2011). In most commercial applications, the two immiscible liquids involved are oil and water. According to the relative spatial organization of the oil and water phases, a nanoemulsion can be classified as oil in water (O/W) or water in oil (W/O). Both types of nanoemulsions are being developed for commercial applications, but the O/W type is by far the most commonly used and so will be the main focus of this chapter. The physicochemical and physiological properties of nanoemulsions, such as their appearance, texture, stability, and gastrointestinal fate, are largely determined by the nature of the droplets they contain, such as their composition, concentration, size, and interfacial properties (McClements and Rao, 2011; Jafari et al., 2008). Consequently, there is a great interest in establishing the connection between the microscopic structure and macroscopic properties of nanoemulsions (Mason et al., 2007; McClements, 2011).

2.2.2 Importance of Physicochemical Properties

2.2.2.1 Stability

The term "stability" can be taken to mean the ability of a nanoemulsion to resist changes in its physicochemical properties over time (McClements, 2015). Nanoemulsions are thermodynamically unstable because the separated oil and water phases have a lower free energy than the emulsified ones (McClements, 2011). Consequently, they always tend to break down given sufficient time, with the rate of change depending on the height of any kinetic energy barriers in the system. There are numerous pathways by which nanoemulsions may break down, including flocculation, coalescence, gravitational separation, Ostwald ripening, and phase inversion (McClements, 2015). In many commercial applications, it is important that a nanoemulsion-based product remains both physically and chemically stable when exposed to specific environmental conditions during its manufacture, storage, transportation, and utilization (such as pH, ionic strength, temperature, and mechanical forces). On the other hand, nanoemulsion may have to break down under another specific set of environmental conditions (such as when encountering the small intestine after ingestion). Physical and chemical changes in nanoemulsion properties can lead to changes in their appearance, texture, and release characteristics, which are important for designing them for particular applications.

2.2.2.2 *Appearance*

The optical properties of a nanoemulsion determine its overall appearance that depends on its interactions with light waves, such as reflection, transmission, absorption, and scattering (McClements, 2002). The appearance of a nanoemulsion-based product is normally the first sensory impression that a consumer makes and therefore plays an important role in consumer acceptance (Caivano and del Pilar, 2012). Numerous characteristics contribute to the overall appearance of nanoemulsions, including homogeneity, opacity, and color (McClements, 2015). A nanoemulsion appears homogeneous when the droplets are evenly dispersed throughout it but heterogeneous when extensive droplet aggregation, creaming, or oiling off occurs. Conventional emulsions ($200\,\text{nm} < d < 200\,\mu\text{m}$) tend to have a cloudy or opaque appearance because of strong light scattering by the oil droplets. On the other hand, nanoemulsions may be clear ($d < 50\,\text{nm}$) or cloudy ($50\,\text{nm} < d < 200\,\text{nm}$) depending on the dimensions of the droplets relative to the wavelength of light ($380\,\text{nm} < \lambda < 780\,\text{nm}$) (McClements, 2002). In general, the degree of light scattering depends on the number, size, and refractive index of the droplets in a nanoemulsion. The color of nanoemulsions depends on the presence of any chromophores that selectively adsorb light in the visible region of the electromagnetic spectrum ($380\,\text{nm} < \lambda < 780\,\text{nm}$). The factors influencing the appearance of nanoemulsions are discussed in more detail in Section 2.5.

2.2.2.3 *Rheology*

Rheology is the discipline that deals with the deformation and flow of matter and is an important parameter influencing the manufacture and functional performance of many materials (Van Vliet, 2013; Rao, 2013; Mehrnia et al., 2017). Knowledge of the rheology of nanoemulsions is essential for a number of reasons (Chen and Stokes, 2012; Selway and Stokes, 2014; Helgeson, 2016). First, the efficiency of droplet disruption inside a homogenizer depends on the viscosity of the separate oil and water phases and the rheology of the nanoemulsion produced (Wooster et al., 2008). Second, the shelf life of nanoemulsion-based products depends on the rheology of the individual phases, e.g., the creaming rate of O/W nanoemulsions usually decreases as the viscosity of the aqueous phase increases (McClements, 2015). Third, the design and operation of many important manufacturing processes depends on the way that a nanoemulsion flows, e.g., stirring inside a vessel, flow through a pipe, passage through a heat exchanger, or packaging into product containers (Saguy et al., 2013). Fourth, some of the sensory attributes of nanoemulsion-based products depend on their rheology, e.g., their perceived creaminess, thickness, pourability, or flowability.

Therefore, a manufacturer must carefully design and consistently produce nanoemulsion-based products that have the rheological attributes desired for a particular application. The factors affecting the rheological properties of nanoemulsions are discussed in more detail in Section 2.4.

2.2.2.4 *Release Characteristics*

A potentially important commercial application of nanoemulsions is to encapsulate hydrophobic, hydrophilic, or amphiphilic bioactive agents, such as pharmaceuticals, vitamins, nutraceuticals, flavors, colors, and preservatives (McClements, 2010; Date et al., 2016; Huang et al., 2010). It is therefore important to understand the retention and release characteristics of bioactive components from nanoemulsion-based delivery systems (Assadpour et al., 2017). Nanoemulsions may be designed to retain bioactive compounds during storage within a food product but control their release when they encounter specific environmental conditions, such as the mouth for flavors or the gastrointestinal tract for pharmaceuticals or nutraceuticals (Yang and McClements, 2013; Soukoulis et al., 2017). The main factors impacting the retention and release of bioactive components from nanoemulsions in the gastrointestinal tract are discussed in Chapter 18.

2.2.3 Structure-Function Relationships

The physicochemical characteristics of nanoemulsion-based products are mainly determined by the characteristics of the droplets they contain (Date et al., 2016; Huang et al., 2010). This section therefore provides a brief overview of the different ways that the droplets within a nanoemulsion may vary.

2.2.3.1 *Droplet Composition*

The composition of the droplets in nanoemulsions can be manipulated by careful selection of the ingredients used to fabricate them (McClements, 2011). In the case of O/W nanoemulsions, the droplets consist of a core of hydrophobic material (usually oil) surrounded by a shell of surface-active materials (usually emulsifiers). The composition of nanoemulsion droplets can therefore be controlled by using different types of hydrophobic and surface-active materials to fabricate them. Hydrophobic materials (such as oils) differ in their polarities, densities, refractive indices, viscosities, and phase behavior, which impacts the formation, physicochemical properties, and functional properties of nanoemulsions (McClements, 2015). The overall composition of the droplets within a nanoemulsion is often considerably different from that of the droplets in a conventional emulsion prepared using the same ingredients because of the difference in droplet dimensions (McClements, 2011; McClements and Rao, 2011). In conventional emulsions, the shell-layer thickness (δ_S) is much smaller than the radius (r) of the hydrophobic core ($\delta_S << r$), and so, the droplets mainly

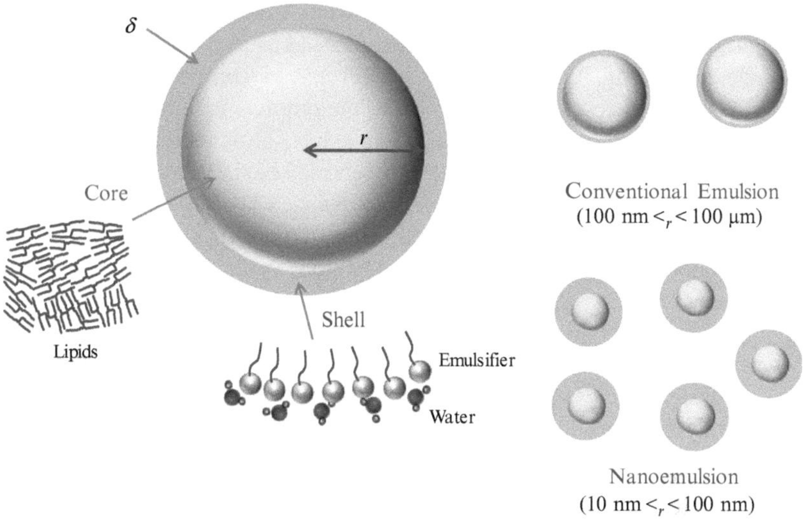

FIG. 2.1 Schematic representation of the core-shell structure of the droplets in nanoemulsions and emulsions. Typically, the droplets consist of a core of oil surrounded by a shell of emulsifier.

consist of oil. Conversely, in nanoemulsions, the shell-layer thickness may have fairly similar dimensions as the hydrophobic core ($\delta_S \approx r$), and so, the shell layer makes an appreciable contribution to the overall droplet composition (Fig. 2.1) (Tadros et al., 2004).

The physicochemical properties of nanoemulsions can be controlled by manipulating the droplet compositions. For example, the creaming stability of the droplets in a nanoemulsion depends on the density contrast between the oil and aqueous phases, whereas the appearance depends on the refractive index contrast between these phases (McClements, 2015). The release characteristics of nanoemulsions can also be controlled by altering droplet composition. For example, the core or shell material can be selected so that the droplet breaks down in a specific location of the human GIT, thereby releasing a bioactive component (McClements and Xiao, 2012; Lesmes and McClements, 2009).

2.2.3.2 *Droplet Concentration*

The droplet concentration in a nanoemulsion can be expressed in a number of ways, including the number, surface area, mass, or volume of droplets per unit volume or mass of a nanoemulsion (McClements, 2011; McClements and Rao, 2011). However, the volume fraction (ϕ), which is the volume of droplets per unit volume of nanoemulsion, is probably the most common means of expressing droplet concentration. The droplet concentration in nanoemulsions can be manipulated by altering the initial amount of oil phase used to fabricate them or by diluting/concentrating them after fabrication. The effective volume fraction

(ϕ_{eff}) of droplets in a nanoemulsion may be appreciably higher than that in a conventional emulsion with the same oil content (ϕ) due to differences in droplet dimensions (Fig. 2.1): $\phi_{eff} = \phi \times (r + \delta_S)^3/r^3$. In conventional emulsions ($\delta_S << r$), the effective volume fraction of the droplets is fairly similar to the volume fraction of the oil phase (ϕ). Conversely, in nanoemulsions ($\delta_S \approx r$), the effective volume fraction of the oil droplets may be substantially higher than that of the oil phase due to the presence of the shell layer.

As mentioned earlier, the droplet concentration can be varied by using different levels of oil to formulate the nanoemulsion. Typically, the oil content in an O/W nanoemulsion is between about 0.1 and 50 wt% during the manufacturing procedure depending on the homogenization method used. The droplet concentration can also be varied after nanoemulsion formation. For instance, it can be decreased by dilution (e.g., by adding more continuous phase), or it can be increased by concentration (e.g., using filtration, centrifugation, or evaporation methods). It should be noted that droplet concentration by some of these methods (particularly filtration and centrifugation) is more difficult for nanoemulsions than for conventional emulsions because of the small droplet dimensions.

2.2.3.3 Droplet Size

Most nanoemulsions contain a huge number of spherical droplets that vary in size over a certain range, and therefore, their dimensions can be described by a particle size distribution (PSD) that represents the fraction of droplets within a range of discrete size classes (McClements, 2015). The PSD of a nanoemulsion is typically controlled by manipulating nanoemulsion homogenization conditions, system composition, and ingredient properties. For example, for high-energy homogenization methods, the PSD depends on the intensity and duration of energy input, emulsifier type and concentration, oil-water interfacial tension, and oil and water-phase viscosities. Similarly, for low-energy homogenization methods, the PSD depends on temperature-time history, stirring speeds, addition rates, surfactant-oil-water ratio, surfactant type, and ionic strength. Different strategies are therefore required to optimize the PSD of a nanoemulsion depending on the preparation method utilized to prepare it.

The effective radius of the droplets within a nanoemulsion depends on the dimensions of the hydrophobic core and surface-active shell: $r_{eff} = r + \delta_S$. Different types of emulsifiers lead to shell layers with different thicknesses, and therefore, the effective particle radius may depend on emulsifier type. Moreover, the thickness of the shell layer may be increased by adsorption of additional materials on to the droplet surfaces, e.g., charged biopolymers or other charged species may adsorb onto the surfaces of oppositely charged emulsifier-coated oil droplets through electrostatic deposition (Guzey and McClements, 2006; Güzey and McClements, 2006).

2.2.3.4 *Droplet Charge*

The droplets in nanoemulsions typically have an electric surface potential due to adsorption of charged materials, such as ionic emulsifiers, biopolymers, or mineral ions (McClements, 2015). The sign and magnitude of the surface potential play an important role in determining the physicochemical properties of nanoemulsions, such as their aggregation stability, chemical stability, ingredient interactions, and surface adhesion. The electric properties of nanoemulsion droplets are usually characterized in terms of their ζ potential, which is typically in the range from −60 to +60 mV depending on emulsifier type and solution conditions (McClements, 2010).

The ζ potential of nanoemulsion droplets can be made to be positive, neutral, or negative by selecting cationic, nonionic, or anionic emulsifiers. The magnitude (and sometimes the sign) of the ζ potential can be manipulated by altering solution pH and ionic strength. For example, the ζ potential on protein-coated droplets changes from highly positive at low pH to neutral at the isoelectric point and highly negative at high pH values. Moreover, the electric characteristics of emulsifier-coated droplets can be modified by adsorption of other charged substances, such as ionic polysaccharides, proteins, or multivalent ions.

2.2.3.5 *Physical State of the Droplets*

The oil droplets in many oil-in-water nanoemulsions are completely liquid, but it is possible to form fully or partially crystalline oil droplets by varying the oil-phase composition and temperature (Weiss et al., 2008). For example, a nanoemulsion containing liquid droplets can be prepared at a temperature above the melting point of the oil phase, and then the system can be cooled below the melting point to promote droplet crystallization (McClements, 2015; Wissing et al., 2004; Muller and Keck, 2004). Two types of lipid nanoparticles can be fabricated using nanoemulsions as a template based on this principle, solid lipid nanoparticles (SLNs) and nanostructured lipid carriers (NLCs) (Müller et al., 2002; Walstra, 2002). SLNs contain oil droplets that are completely crystallized and have a highly ordered crystalline structure, whereas NLCs contain oil droplets that are partly crystallized and/or have a more disordered solid structure. The physical state of the droplets in a nanoemulsion may alter a number of its physicochemical properties and should therefore be carefully controlled.

2.3 STABILITY

Nanoemulsions are thermodynamically unfavorable systems and therefore have a tendency to break down over time due to a variety of physicochemical processes, including creaming, flocculation, coalescence, Ostwald ripening, and phase separation (Fig. 2.2). The rates at which these processes occur in nanoemulsions are often different from those in conventional emulsions due to differences in droplet size. In this section, a brief overview of the

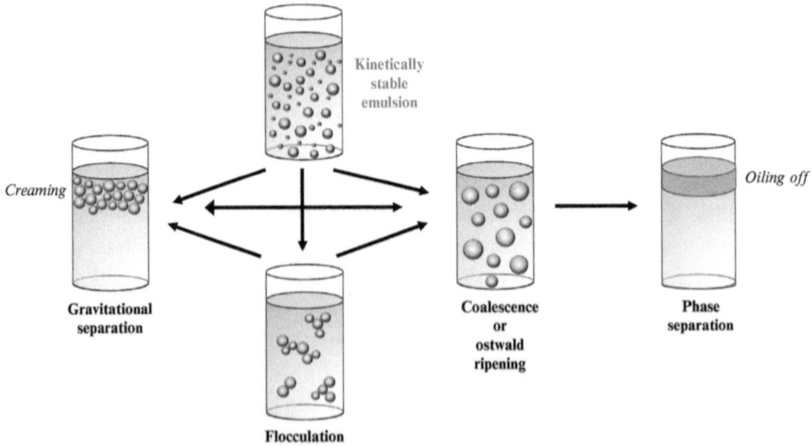

FIG. 2.2 Schematic diagram of instability mechanisms that occur in nanoemulsions: gravitational separation (i.e., creaming or sedimentation), flocculation, coalescence, Ostwald ripening, and phase inversion.

physicochemical basis of each major instability mechanism and a discussion of the impact of droplet size on the physical and chemical stability of nanoemulsions are given.

2.3.1 Gravitational Separation

The droplets in nanoemulsions usually have a different density from the surrounding liquid, and so a gravitational force acts upon them. If the droplets have a lower density than the surrounding liquid, they tend to move upward (creaming), but if they have a higher density, they will tend to move downward (sedimentation) (Fig. 2.3). Most liquid oils have lower densities than liquid water, and so, there is a tendency for oil to move upward and water to move downward. Thus, creaming is more common in O/W nanoemulsions, whereas sedimentation is more common in W/O nanoemulsions. Having said this, the oil droplets in O/W nanoemulsions may sediment under some circumstances: (i) if the oil phase crystallizes and (ii) if they are coated by a thick dense interfacial layer.

The creaming velocity of an isolated rigid homogeneous spherical particle in an ideal (Newtonian) liquid is described by Stokes' law:

$$v_{\text{Stokes}} = -\frac{2gr^2(\rho_2 - \rho_1)}{9\eta_1} \tag{2.1}$$

Here, v_{Stokes} is the creaming velocity, r is the particle radius, g is the acceleration due to gravity, ρ is the density, η is the shear viscosity, and the subscripts 1 and 2 refer to the continuous and dispersed phases, respectively. The sign of

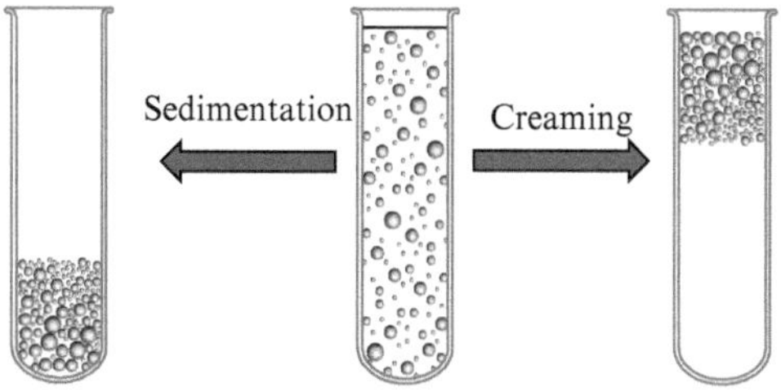

FIG. 2.3 The creaming or sedimentation of the droplets in a nanoemulsion occurs due to the density difference between the oil and water phases.

the creaming velocity determines whether the droplets move upward (+) or downward (−). In the case of nanoemulsions, Stokes' law often has to be modified to take into account the fact that the particles are not isolated rigid homogenous spheres, the surrounding fluid may be nonideal, and Brownian motion effects are important for small droplets (McClements, 2015). Brownian motion favors a random distribution of droplets throughout a nanoemulsion, whereas gravitational forces favor the accumulation of droplets at either the top or bottom of the system. Typically, the vertical distribution of droplets is dominated by gravity for relatively large droplets ($d > 100\,\text{nm}$) but dominated by Brownian motion for relatively small particles ($d < 100\,\text{nm}$).

The relatively small size of the droplets in nanoemulsions may also impact their creaming stability by altering the effective density of the droplets. In nanoemulsions, the volume occupied by the shell layer may be an appreciable fraction of the total volume of the droplets (core + shell). The overall density of the droplets in a nanoemulsion can be described by the following expression (McClements, 2011):

$$\rho_{\text{particle}} = \Phi_S \rho_S + (1 - \Phi_S)\rho_C \tag{2.2}$$

Here, $\Phi_S = (V_S/V_{\text{eff}})$ is the volume of the shell divided by the effective volume of the overall particle (core + shell), and the subscripts S and C refer to the shell and core, respectively. In practice, the shell layer is usually denser than either the oil or aqueous phases and therefore tends to increase the overall droplet density. Thus, the thicker or denser the shell layer, the greater the droplet density. This phenomenon can be utilized to reduce the density contrast between the droplets and the surrounding liquid, and therefore reducing the creaming velocity. This effect is highlighted in Fig. 2.4, which shows that the small droplets in nanoemulsions ($d_{43} = 66\,\text{nm}$) tend to sediment because the overall droplet density is greater than water, whereas the large particles in conventional emulsions ($d_{43} = 325\,\text{nm}$) tend to cream because the overall droplet density is less than that of water. Gravitational separation in nanoemulsions can also be inhibited by adding weighting agents (dense hydrophobic materials) to the oil phase prior to homogenization.

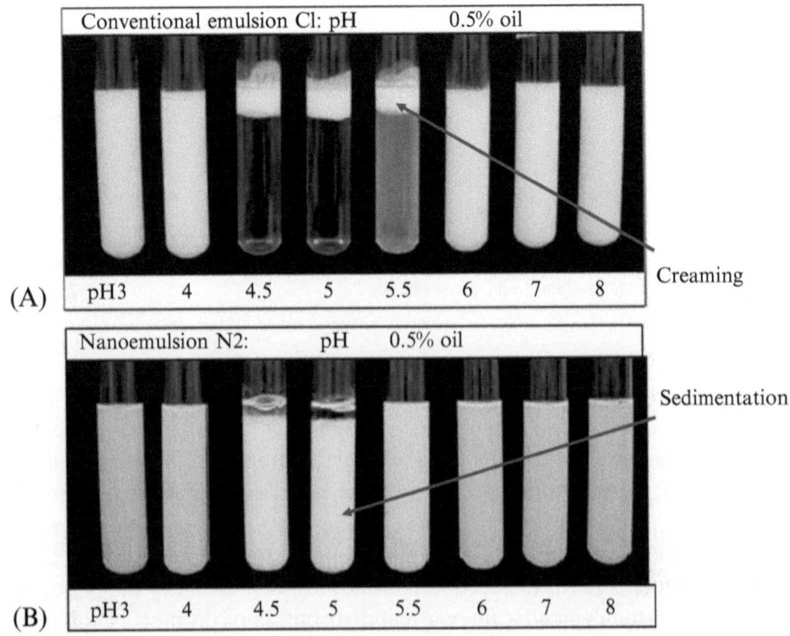

FIG. 2.4 Influence of pH on the gravitational stability of protein-stabilized 0.5 wt% corn oil-in-water (A) conventional emulsions (d_{43} = 325 nm) and (B) nanoemulsions (d_{43} = 66 nm). *(Data from Lee, S.J., McClements, D.J., 2010. Fabrication of protein-stabilized nanoemulsions using a combined homogenization and amphiphilic solvent dissolution/evaporation approach. Food Hydrocoll. 24 (6–7), 560–569.)*

2.3.2 Droplet Aggregation

The droplets in nanoemulsions continually collide with each other as they move around due to Brownian motion, gravity, or mechanical agitation. The droplets may move apart or stick together after a collision, depending on the strength and range of the various attractive and repulsive forces acting between them (McClements, 2011). The two most important types of droplet aggregation in nanoemulsions are flocculation and coalescence. In flocculation, a number of droplets come together and form an aggregate where each of the droplets retains their individual integrity, whereas in coalescence, the droplets merge together to form a single larger droplet (Fig. 2.2). The major types of colloidal interactions operating between the droplets in nanoemulsions are van der Waals, electrostatic, steric, hydrophobic, depletion, and bridging (McClements, 2015). The van der Waals, hydrophobic, depletion, and bridging interactions are usually attractive, whereas the steric and electrostatic interactions are usually repulsive. In general, the strength of both the attractive and repulsive colloidal interactions tends to increase when increasing droplet size.

A typical dependence of the interaction potential between a pair of emulsifier-coated oil droplets with droplet separation is shown in Fig. 2.5. The droplets in a nanoemulsion may exist in a number of different aggregation

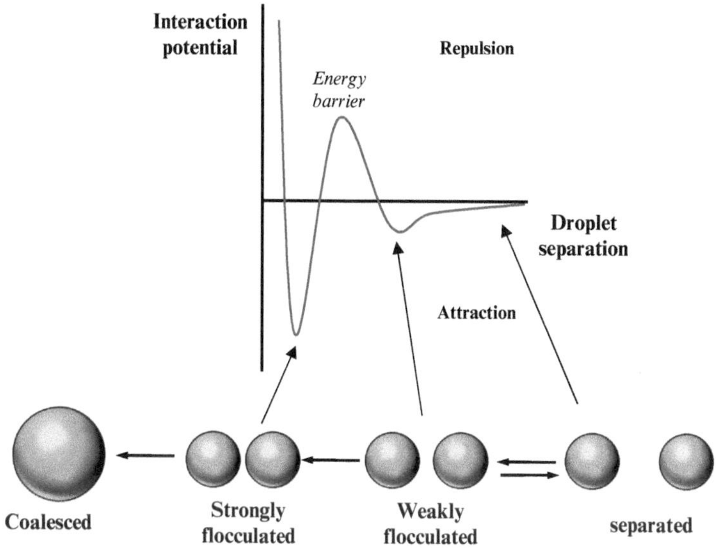

FIG. 2.5 The aggregation of the droplets in nanoemulsions depends on the balance of attractive and repulsive forces between them, which can be described in terms of the interaction potential.

states depending on the nature of the interaction potential profile (McClements, 2011): (i) *separated*—when the attraction is relatively weak and there is a large repulsive-energy barrier; (ii) *weakly flocculated*—when the attraction is moderately strong but there is a large repulsive-energy barrier; (iii) *strongly flocculated*—when the attraction is very strong, the repulsive-energy barrier is small, and the short-range repulsion is very strong; and (iv) *coalescence*—when the attraction is strong, the repulsive-energy barrier is small, and there is no short-range repulsion. It is therefore important to be aware of the different types of colloidal interactions operating within a specific nanoemulsion and of the major factors that impact them. This knowledge can then be used to formulate nanoemulsions with improved physicochemical properties and stability.

2.3.3 Ostwald Ripening

In oil-in-water nanoemulsions, Ostwald ripening is caused by the diffusion of oil molecules from the smaller droplets to the larger droplets through the surrounding liquid (Kabalnov, 2001; Taylor, 1998). This process leads to the growth of the larger droplets at the expense of the smaller ones (Fig. 2.6). The thermodynamic driving force for Ostwald ripening is the fact that the water solubility of oil molecules dispersed within a spherical droplet increases as the droplet radius decreases, which can be estimated using the following equation:

$$S(r) = S(\infty)\exp\left(\frac{2\gamma V_{\mathrm{m}}}{RTr}\right) = S(\infty)\exp\left(\frac{\alpha}{r}\right) \quad (2.3)$$

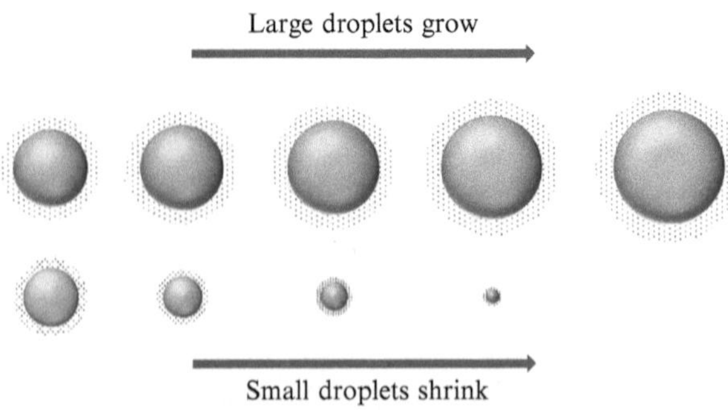

FIG. 2.6 Ostwald ripening involves the growth of large droplets at the expense of smaller ones due to diffusion of dispersed phase through the continuous phase.

Here, $S(\infty)$ is the water solubility of the oil molecules for a planar oil-water system, $S(r)$ is the water solubility of the oil molecules when they are present within a spherical droplet of radius r, $\alpha = 2\gamma V_m/RT$ is a characteristic length scale, V_m is the molar volume of the oil molecules, and γ is the oil-water interfacial tension. This equation predicts that there will be a higher concentration of oil molecules in water that is in contact with a small oil droplet than in contact with a large oil droplet (Fig. 2.6). The oil molecules in the water therefore move from the surfaces of the smaller droplets to the surfaces of the larger ones due to this concentration gradient. This process causes the smaller droplets to shrink and the larger droplets to grow, which leads to an overall increase in the mean droplet size during storage. After steady-state conditions have been achieved, the rate of Ostwald ripening (ω) can be predicted using the following expressions (Kabalnov and Shchukin, 1992):

$$r^3 = r_0^3 + \omega t \tag{2.4a}$$

$$\omega = \frac{4\alpha S(\infty) D}{9} \tag{2.4b}$$

Here, r is the number-weighted mean droplet radius after time t, r_0 is the initial number-weighted mean droplet radius, and D is the translational diffusion coefficient of the oil molecules through water.

In practice, the main factor determining the stability of a nanoemulsion to Ostwald ripening is usually the water solubility of the oil phase. Nanoemulsions formulated from oil phases that are composed primarily of long-chain triacylglycerols (such as canola, corn, fish, olive, palm, peanut, safflower, or sunflower oil) tend to be stable to Ostwald ripening because of the very low water solubility of these lipids. Conversely, nanoemulsions fabricated from oil phases that have an appreciable water solubility (such as short-chain

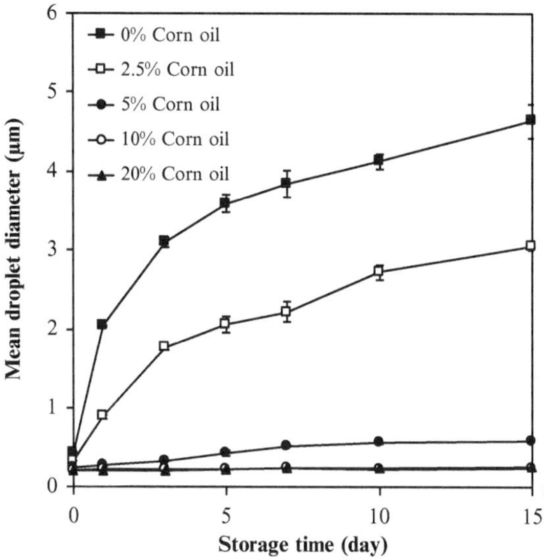

FIG. 2.7 Dependence of mean droplet diameter (d_{43}) on storage time for 5% oil-in-water nanoemulsions stabilized by modified starch (pH 3.5, 20°C). The oil phase contained varying ratios of orange and corn oil. The corn oil acts as a ripening inhibitor.

triacylglycerols, flavor oils, and essential oils) tend to be highly unstable to droplet growth through this process (Li et al., 2009; Rao and McClements, 2012; Chang et al., 2012). An example of this form of instability can be seen in Fig. 2.7, which shows a faster droplet growth rate for nanoemulsions prepared from a short-chain triglyceride (tributyrin) than for nanoemulsions prepared from a long-chain triglyceride (corn oil).

There are methods to improve the stability of nanoemulsions formed from oil phases that have an appreciable water solubility. In this case, a highly water-insoluble oil (known as a ripening agent or inhibitor) is mixed with the more water-soluble oil prior to nanoemulsion formation (Kabalnov and Shchukin, 1992). A variety of different highly hydrophobic molecules can be used as ripening inhibitors, with the most common being long-chain triglycerides. Ripening inhibitors inhibit Ostwald ripening by generating an entropy of mixing effect that counterbalances the curvature effects. For example, the stability of eugenol oil-in-water emulsions can be improved by adding a sufficiently high level of a highly hydrophobic oil (>30% bean oil) prior to homogenization to retard Ostwald ripening (Chen et al., 2016).

2.3.4 Chemical Stability

Nanoemulsions may also breakdown or lose their desirable quality attributes during storage due to various chemical or biochemical reactions that occur within them, e.g., lipid oxidation, color fading, flavor loss, and biopolymer hydrolysis. In this section, we focus on lipid oxidation since this is one of the most common forms of chemical deterioration that occurs in nanoemulsions (Waraho et al., 2011; Berton-Carabin et al., 2014; Walker et al., 2015). Nanoemulsions have large interfacial areas, which can accelerate lipid oxidation because water-soluble prooxidants (such as transition metals) can come into contact with oil-soluble reactants (such as polyunsaturated lipids or hydroperoxides) (McClements and Decker, 2000). Lipid oxidation can be retarded by adding antioxidants or chelating agents, by controlling environmental conditions (such as pH, temperature, oxygen, or light), or by manipulating interfacial properties (such as charge, thickness, and chemical reactivity). It has been indicated that addition of an anionic polysaccharide (alginate) to nanoemulsions decreased the rate and extent of lipid oxidation during storage, which was attributed to the ability of the alginate molecules to chelate metal ions in the aqueous phase (Salvia-Trujillo et al., 2016). In another recent study, it was shown that the stability of nanoemulsions to lipid oxidation could be improved by increasing the lipid droplet concentration, which may have been because there were less transition metal ions available to promote oxidation (Liu et al., 2016).

Food emulsions normally contain a variety of additives to improve their perceived sensory attributes. Many of these additives (e.g., colors, flavors, and preservatives) are highly susceptible to chemical degradation that can be accelerated by elevated temperatures, exposure to light, or high oxygen levels. For example, carotenoids (such as β-carotene, zeaxanthin, astaxanthin, or lycopene) are natural oil-soluble pigments used to give red, pink, orange, or yellow colors to many emulsified food products (Brady, 2013). The presence of ultraviolet or visible light can promote carotenoid oxidation, which results in color fading and loss of desirable visual appearance (Mortensen and Skibsted, 1996). The carotenoids loaded in a transparent nanoemulsion may be more susceptible to this type of degradation than those loaded in opaque conventional emulsions because light can penetrate more easily into the interior of the system. Additional steps may therefore be needed to improve the chemical stability of sensitive ingredients encapsulated in nanoemulsions, e.g., by adding antioxidants or chelating agents (McClements and Decker, 2000; Mao et al., 2009).

2.4 RHEOLOGICAL PROPERTIES

Rheology is the scientific discipline that deals with the deformation and flow of matter (Rao, 2013) and an important attribute for many nanoemulsion-based products. For example, the rheology of a nanoemulsion will determine its ability to flow through a pipe, the efficiency of mixing processes, and its sensory attributes (such as smoothness, creaminess, or mouthfeel). The rheological properties of nanoemulsions may vary considerably depending on their composition, interactions, and structure, ranging from low-viscosity liquids to semisolid materials (Genovese et al., 2007; Berli et al., 2005). The rheological properties of a nanoemulsion are characterized by measuring the relationship between the applied shear stress and the resultant rate of shear strain (liquids or solids). The most common parameters measured on nanoemulsions are usually the apparent shear viscosity (η) or the dynamic shear modulus (G), but other parameters may be more important for certain applications (Mehrnia et al., 2017). Measurements of the rheological properties of nanoemulsions can provide valuable information about the structural organization and interactions of the droplets they contain (Tadros, 2011; Quemada and Berli, 2002). The most important factors determining the rheology of nanoemulsions are the continuous-phase viscosity, droplet concentration, and droplet interactions (Derkach, 2009; Stickel and Powell, 2005).

2.4.1 Dilute Systems

In dilute nanoemulsions, the droplets ($\phi < 0.05$) can be considered to act independently of each other, i.e., the flow profile around one droplet does not impact the flow profile around another droplet.

To a first approximation, the viscosity of a dilute nanoemulsion can be described by the following expression originally derived by Albert Einstein (Hunter, 2001):

$$\eta = \eta_0(1 + 2.5\phi) \tag{2.5}$$

Here, η is the viscosity of the nanoemulsion, η_0 is the viscosity of the continuous phase, and ϕ is the dispersed phase volume fraction. This equation predicts that the viscosity of a dilute nanoemulsion should linearly increase as the droplet concentration increases (Fig. 2.8). A number of assumptions are made in the derivation of this equation, including that the continuous phase is an ideal liquid and that the particles are isolated, rigid spheres. Despite these assumptions, the Einstein equation provides fairly accurate predictions of the viscosity of dilute nanoemulsions containing droplets that do not interact strongly with each other. However, there may be large derivations between this theory and experimental results for nanoemulsions containing droplets that have strong attractive or repulsive interactions between them (McClements, 2015). Typically, flocculated nanoemulsions have a higher viscosity than nonflocculated ones because they trap some of the water between the droplets, which increases the effective dispersed-phase volume fraction. These effects are typically more pronounced in concentrated nanoemulsions than in dilute ones, and therefore, we will discuss them in the next section.

2.4.2 Concentrated Systems

When the droplet concentration in a nanoemulsion exceeds about 5%, the droplets start to interact with each other through hydrodynamic and colloidal interactions that can have a pronounced impact on the shear viscosity

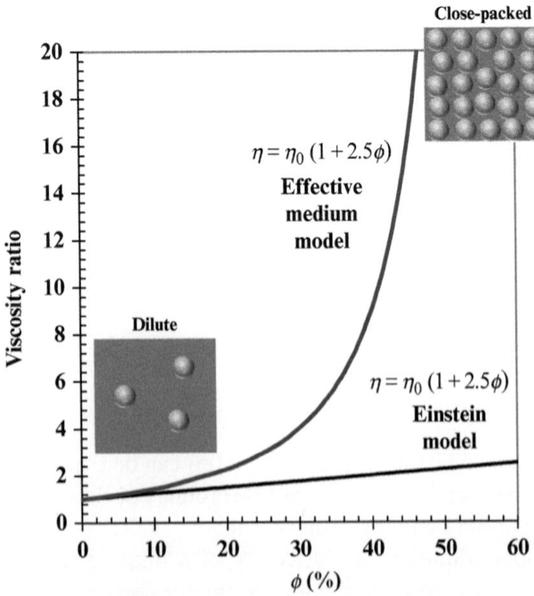

FIG. 2.8 Dependence of viscosity ratio (h/h_0) of colloidal dispersions increases with different disperse-phase volume fraction (the Einstein model showed inset).

(Tadros, 2011). Under these circumstances, the shear viscosity is usually appreciably higher than that predicted by the Einstein equation depending on the nature of the colloidal interactions between the droplets.

2.4.2.1 No Long-Range Colloidal Interactions

In the absence of strong long-range colloidal interactions, the shear viscosity of a nanoemulsion is mainly determined by hydrodynamic interactions that are associated with the fluid flow around the droplets. In concentrated nanoemulsions, the fluid flow around one droplet may be impacted by the presence of the other droplets, which leads to a higher degree of energy dissipation. Hence, the shear viscosity of a concentrated nanoemulsion is larger than that predicted by the Einstein equation. In these systems, the following semiempirical equation can be used to describe the viscosity of a nanoemulsion over a wide droplet concentration range (Derkach, 2009; Pal, 2011):

$$\frac{\eta}{\eta_0} = \left(1 - \frac{\phi}{\phi_c}\right)^{-2} \tag{2.6}$$

Here, ϕ_c is a critical packing parameter related to the volume fraction at which the droplets become close-packed (around 0.65 for noninteracting particles). This equation predicts that there is a large increase in the viscosity of a

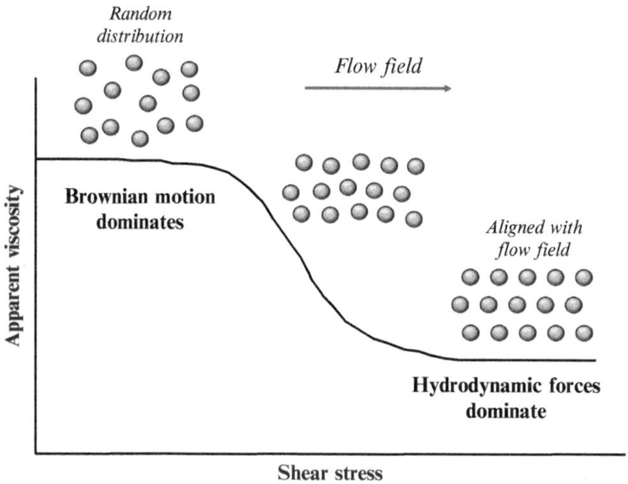

FIG. 2.9 Schematic representation of shear thinning in a concentrated nanoemulsion containing nonflocculated droplets.

nanoemulsion as the droplet concentration increases toward the value where the droplets become close-packed (Fig. 2.8).

The viscosity of concentrated nanoemulsions often exhibits strong shear-thinning behavior because of a change in the relative importance of hydrodynamic effects and Brownian motion (Derkach, 2009). At low shear stresses, the Brownian motion favors a random distribution of the droplets throughout the system, which leads to a nanoemulsion with a relatively high shear viscosity (Larson, 1999). As the shear stress increases, the hydrodynamic forces dominate the Brownian motion effects causing the droplets to become ordered along the fluid-flow lines. As a result, the resistance to fluid flow is reduced, which leads to less energy dissipation and a lower nanoemulsion viscosity (Fig. 2.9).

2.4.2.2 Repulsive Interactions

Nanoemulsions often contain droplets that have a relatively strong electrostatic and/or steric repulsion between them due to the presence of an electrically charged or thick interfacial layer (McClements, 2012; McClements and Rao, 2011). These repulsive interactions prevent the droplets from coming into close proximity when they encounter each other, thereby increasing their effective volume fraction (Tadros, 2011):

$$\phi_{\text{eff}} = \phi\left(1 + \frac{\delta}{r}\right)^3 \tag{2.7}$$

Here, δ is half the distance of closest separation between the two droplets and can be assumed to be an effective "shell thickness." For droplets stabilized by steric repulsion, the value of δ can be taken to be the thickness of the interfacial layer. For droplets stabilized by electrostatic repulsion, the value of δ is related to the Debye screening length (κ^{-1}), which is a measure of how far the electric potential extends into the surrounding aqueous phase. In this case, the shell thickness is given by the following expression at low shear stresses (Macosko and Larson, 1994):

$$\delta = \kappa^{-1} \ln\left(\frac{\alpha}{\ln(\alpha/(\ln\alpha))}\right) \tag{2.8}$$

Here, $\alpha = 4\pi\varepsilon_0\varepsilon_R\Psi_0 r^2\kappa\exp(2r\kappa)/kT$, ε_0 is the dielectric constant of a vacuum, ε_R is the relative dielectric constant of the continuous phase, Ψ_0 is the electric potential at the droplet surface, r is the radius, κ^{-1} is the Debye screening length, k is Boltzmann's constant, and T is the absolute temperature. For charged droplets, the shell thickness decreases with increasing ionic strength.

The impact of repulsive interactions on the rheology of nanoemulsions depends on the relative magnitudes of the shell thickness and the droplet dimensions (Weiss and McClements, 2000). The impact is negligible under conditions where the shell thickness is much smaller than the droplet radius ($\delta << r$), i.e., high ionic strengths or large droplets. Conversely, the impact may be substantial for systems where the shell thickness is comparable with the droplets radius ($\delta \approx r$), i.e., low ionic strengths or small droplets (Tadros, 2011). The importance of repulsive interactions on nanoemulsion rheology is highlighted in Fig. 2.10, which shows the dependence of the shear modulus and phase angle on droplet size (Weiss and McClements, 2000). In this system, the droplets were

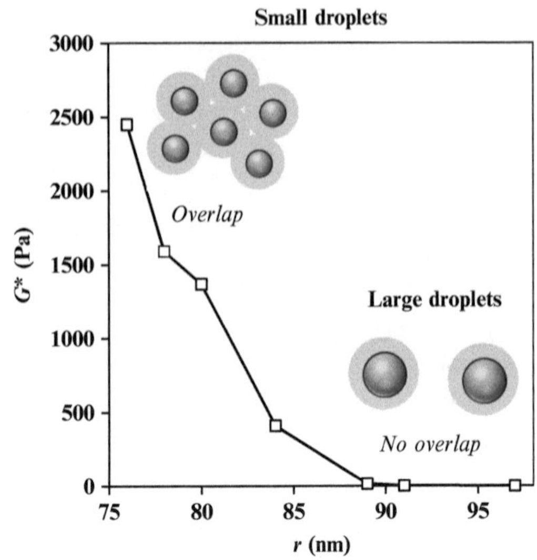

FIG. 2.10 Dependence of complex shear modulus (G^*) and phase angle (δ) on droplet radius for SDS-stabilized n-octadecane oil-in-water nanoemulsions.

stabilized by an anionic surfactant (SDS) that gives them a high negative charge. The shear modulus of the nanoemulsions increased appreciably when the droplet size decreased below a critical value due to overlap of the electrostatic shells surrounding the droplets. This kind of behavior is particularly important in nanoemulsions stabilized by ionic surfactants or amphiphilic polymers that form thick layers (McClements, 2011).

For electrostatically stabilized nanoemulsions, the rheology of the system is strongly dependent on ionic strength under conditions where the shell thickness is fairly similar to the droplet dimensions ($\delta \approx r$). The shell thickness decreases as the ionic strength is increased due to electrostatic screening effects, which can cause a pronounced decrease in the shear modulus or viscosity of a nanoemulsion. Hence, one can go from a gel to a fluid simply by adding salt to the system (Mason et al., 2007).

2.4.2.3 *Attractive Interactions*

In some nanoemulsions, the attractive interactions between the droplets are strong enough to overcome the repulsive interactions, which leads to droplet flocculation (Tadros, 2011). Typically, the viscosity of a flocculated nanoemulsion is higher than that of a nonflocculated one because of the presence of the continuous phase trapped between the aggregated droplets. To a first approximation, the viscosity of flocculated nanoemulsions can be described using the same equations as used for nonflocculated ones by assuming that the particles are flocs, rather than individual oil droplets (Chanamai and McClements, 2000). The apparent viscosity of concentrated flocculated nanoemulsions can thus be described by the following expression (Quemada and Berli, 2002):

$$\frac{\eta}{\eta_0} = \left(1 - \frac{\phi_{\text{eff}}}{\phi_c}\right)^{-2} \tag{2.9}$$

Here, ϕ_{eff} ($=\phi/\phi_{\text{floc}}$) is the *effective* volume fraction of the flocs in the nanoemulsion, where ϕ is the volume fraction of oil droplets in the nanoemulsion and ϕ_{floc} is the volume fraction of oil droplets within the flocs. This equation indicates that at a fixed oil droplet volume fraction (ϕ), the viscosity of a nanoemulsion should increase as the internal packing of the droplets within the flocs decreases (lower ϕ_i) because the effective volume fraction of the flocs then increases.

An example of the influence of droplet flocculation on the rheology of fairly concentrated oil-in-water nanoemulsions is shown in Fig. 2.11. The apparent viscosity of a series of nanoemulsions with the same overall oil content (10%) but with different ratios of cationic and anionic oil droplets was measured (Mao and McClements, 2013). The droplet charge was controlled by using different types of emulsifiers to stabilize them (cationic or anionic). The viscosity was relatively low for nanoemulsions containing only negative or only positive droplets because the strong electrostatic repulsion between them inhibited

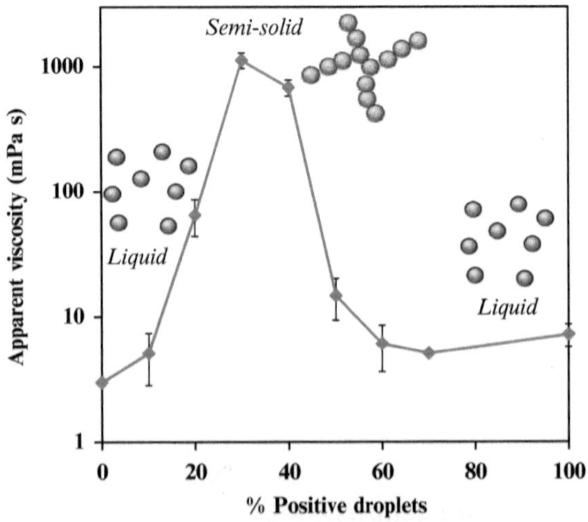

FIG. 2.11 Dependence of apparent viscosity with the particle ratios of mixture emulsion containing positively charged (LF-coated) and negatively charged (β-Lg-coated) droplets (the flocculation of the oppositely charged droplets induces high viscosity of the systems at intermediate particle ratios).

flocculation. Conversely, the viscosity was relatively high in nanoemulsions containing a mixture of cationic and anionic droplets because electrostatic repulsion led to the formation of large flocs.

2.5 APPEARANCE

The optical properties of nanoemulsions are important for many of their applications. Some commercial products should be transparent, and therefore, addition of the nanoemulsions should not cause them to look cloudy or opaque. Conversely, other commercial products are expected to be cloudy or opaque, and therefore, the incorporation of the nanoemulsion should not alter their desirable visual appearance. An appreciation of the relationship between the composition/structure of nanoemulsions and their optical properties is therefore useful for the rational design of commercial products with appropriate appearances. In general, the optical properties of nanoemulsions are determined by their interactions with electromagnetic waves in the visible region (light waves) (McClements, 2002). When light waves encounter the surface of a nanoemulsion, they are both reflected and transmitted, with the relative proportions of these two processes depending on the difference in refractive index between the nanoemulsion and its surroundings. As the light waves travel through the nanoemulsion, they are scattered by the emulsion droplets and absorbed by any chromophores present. The angular scattering pattern of the light waves depends on the size, concentration, and refractive index of the droplets.

The selective absorption of light waves of particular wavelengths by the chromophores in the nanoemulsion determines its color. In this section, the most important factors that determine the overall appearance of nanoemulsions are discussed. Analytic methods for characterizing the optical properties of nanoemulsions are discussed in Chapter 17 of this book and therefore only briefly discussed here.

2.5.1 Measurements of Optical Properties

A variety of analytic instruments are available to measure the optical properties of nanoemulsions (Francis and Clydesdale, 1975; Malacara, 2002). Based on their operating principles, these instruments can be classified into two main groups: spectrophotometric and trichromatic colorimeters.

The optical properties of nanoemulsions can be characterized using UV-visible spectrophotometers that measure the transmission and/or reflection of light waves over the range of wavelengths in the visible region (around 380–780nm). The resulting transmittance or reflectance spectra obtained can be used to characterize the optical properties of a nanoemulsion. Alternatively, the measured spectra can be transformed into tristimulus coordinates such as L^*, a^*, and b^* values, where L^* represents lightness and a^* (red-green) and b^* (yellow-blue) are color coordinates (Chapter 17).

Trichromatic colorimeters are instruments that are specially designed to measure the intensity of the "red," "green," and "blue" components of light that the human eye is sensitive to by using a series of filters. Trichromatic colorimeters are capable of making both transmission and reflectance measurements depending on the nature of the sample being tested. Reflectance measurements are most suitable for determining the optical properties of opaque nanoemulsions, whereas transmission measurements are more suitable for characterizing clear nanoemulsions.

2.5.2 Major Factors Influencing Nanoemulsion Color

The optical properties of nanoemulsions are determined by their interactions with light waves. In general, the transmission or reflection of light waves can be described by light scattering theory, which can be represented simply as

$$T(\lambda, \vartheta) = f(r, \phi, n, \alpha(\lambda)) \tag{2.10a}$$

$$R(\lambda, \vartheta) = f(r, \phi, n, \alpha(\lambda)) \tag{2.10b}$$

Here, λ is the wavelength of light, ϑ is the scattering angle, r is the droplet radius, ϕ is the dispersed phase volume fraction, Δn is the refractive index contrast, and $\alpha(\lambda)$ is the absorption spectrum. The equations for describing these relationships are usually based on the Mie theory and have been discussed in Chapter 16 of this book. The tristimulus color coordinates (L^*, a^*, and b^* values) of nanoemulsions can be calculated from the transmission and reflection spectra using appropriate models (McClements, 2002). These equations highlight that the most important factors affecting the optical properties of nanoemulsions are the droplet size, droplet concentration, refractive index contrast, and absorption spectrum of the components phases.

2.5.2.1 Droplet Size and Concentration

The overall appearance of an oil-in-water nanoemulsion is strongly influenced by the size and concentration of the oil droplets that it contains. The scattering pattern of a nanoemulsion (i.e., light intensity versus scattering angle) depends on the relationship between the droplet size and the wavelength of light (380–780 nm). Typically, relatively strong scattering occurs when the droplet size is fairly similar to the wavelength of light ($r \approx \lambda$), which causes a nanoemulsion to appear turbid or opaque (Fig. 2.12). Conversely, relatively weak scattering occurs when the droplet size is much smaller than the wavelength of light ($r \ll \lambda$), which causes a nanoemulsion to look clear. Experiments have shown that nanoemulsions tend to be transparent when the droplet radius is below about 20–30 nm (Wooster et al., 2008, 2016). Consequently, it may be important to design a nanoemulsion-based product so that it has a droplet size distribution that leads to the required optical properties.

The scattering of light by nanoemulsion typically increases as the droplet concentration increases, which may impact the overall appearance of the final product. For nanoemulsions containing very fine droplets ($r \ll \lambda$), they may remain transparent as the droplet concentration increases. However, for

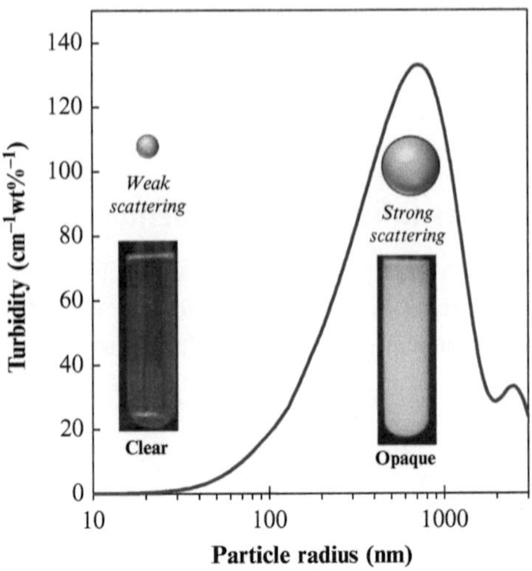

FIG. 2.12 The optical properties of a nanoemulsion depend on the droplet dimensions compared with the wavelength of light waves. Nanoemulsions become transparent when the droplet size is much less than the wavelength of light.

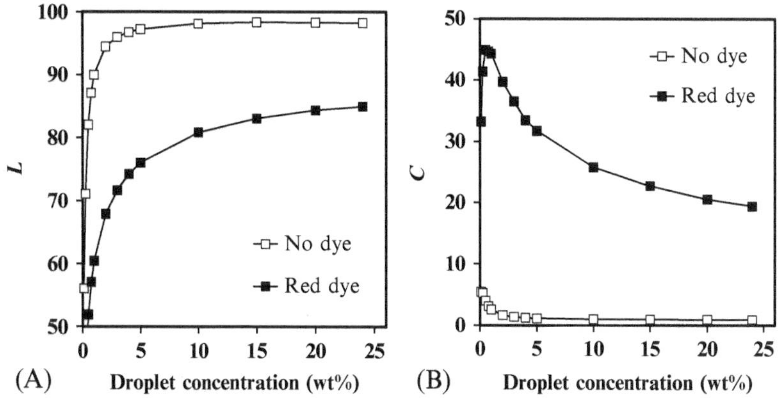

FIG. 2.13 Influence of droplet concentration on the (A) lightness (L^*) and (B) color intensity (C) of n-hexadecane oil-in-water nanoemulsions in the presence and absence of a red dye.

nanoemulsions containing larger droplets, the turbidity or lightness typically increases as the droplet concentration increases. This effect has been demonstrated for oil-in-water nanoemulsions in the absence and presence of a red dye in Fig. 2.13 (Chantrapornchai et al., 1998, 1999b). The lightness (L^*) of the nanoemulsions increases steeply as the oil droplet concentration increases from 0 to 5 wt%, but then, it increases more gradually as the droplet concentration is raised further. The intenseness of the color (C) of a concentrated nanoemulsion containing a dye tends to fade as the droplet concentration is increased because of the increase in multiple light scattering. As a result, the light waves have to travel further through the nanoemulsion and are therefore absorbed more by any chromophores.

2.5.2.2 Refractive Index Contrast

The optical properties of nanoemulsions also depend on the refractive index contrast between the phases, i.e., the difference in refractive index between the dispersed and continuous phases (McClements, 2002). Typically, the intensity of light scattering increases as the refractive index contrast increases, thereby leading to an increase in the turbidity or opacity of nanoemulsions (Helgason et al., 2015). In general, the overall refractive index of a nanoemulsion droplet depends on its structure and composition. The droplets in nanoemulsions can be considered to have a core-shell structure, with the core typically being an oil and the shell being a layer of emulsifier molecules. The overall refractive index of a core-shell particle can be estimated as the weighted average of the refractive indices of the constituents:

$$n_{\text{particle}} = \Phi_S n_S + (1 - \Phi_S) n_C \tag{2.11}$$

Here, n is the refractive index, Φ is the volume fraction, and the subscripts S and C stand for the shell (interfacial layer) and core (oil phase), respectively. The interfacial layer usually consists of emulsifier molecules that have a higher refractive index than oil or water. The overall refractive index of the nanoemulsion droplets therefore tends to increase as the packing of the emulsifier molecules in the interfacial layer increases, interfacial layer thickness increases, or the oil droplets become smaller. For conventional emulsions, the overall refractive index of the disperse phase can be taken to be similar to that of the oil phase, but for nanoemulsions, it may be quite different due to the relatively high volume occupied by the interfacial layer.

The impact of the refractive index contrast on the optical properties of oil-in-water nanoemulsions is shown in Fig. 2.14 (Chantrapornchai et al., 2001). In this case, different amounts of glycerol were added to the aqueous phase to increase its refractive index. At low glycerol levels, the refractive index of the aqueous phase was considerably less than that of the droplets, which led to relatively strong light scattering and an opaque nanoemulsion. As the glycerol level was increased, the refractive index of the aqueous phase becomes

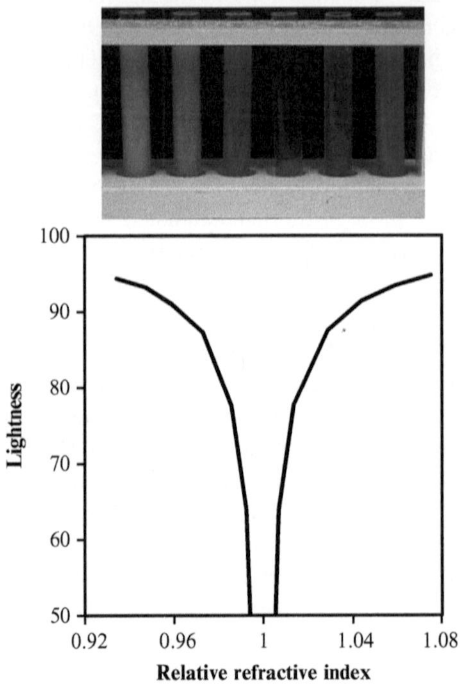

FIG. 2.14 The optical properties of nanoemulsions depend on the refractive index contrast. When the refractive index of the dispersed phase is similar to that of the continuous phase, light scattering is weak, which leads to a clear system.

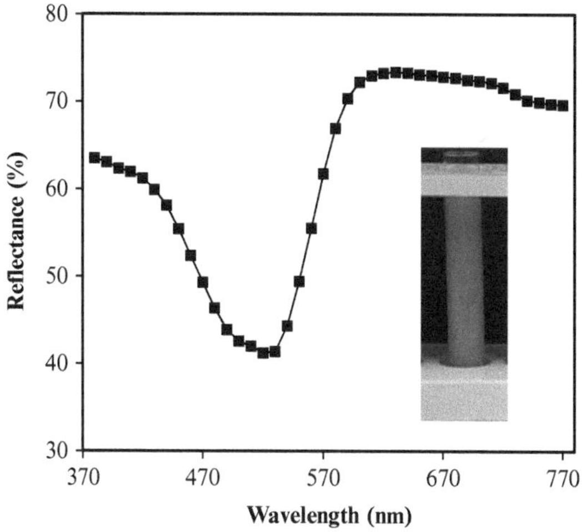

FIG. 2.15 The optical properties of nanoemulsions depend on the presence of any chromophores that selectively absorb light. The reflectance spectra of a nanoemulsion containing a red dye are shown.

closer to that of the droplets, which leads to weak light scattering and an optically transparent system.

2.5.2.3 Absorption Spectrum

Another major factor that determines the overall appearance of nanoemulsions is the presence of any substances (chromophores) that selectively absorb light waves. Selective absorption of light in certain regions of the electromagnetic spectrum will give nanoemulsions a characteristic color, such as red, blue, yellow, or green. The impact of chromophore type and level on the optical properties of oil-in-water nanoemulsions has been investigated by Chantrapornchai et al., 1999b. The reflectance spectrum of oil-in-water nanoemulsions containing a red dye is shown in Fig. 2.15. The minimum observed in the reflectance spectrum of the nanoemulsions (around 520nm) corresponds to a maximum observed in the absorption spectra of red-dye solutions. This phenomenon occurs as light waves passing through the nanoemulsions are selectively absorbed by chromophores in the red dyes at these wavelengths, and therefore, less light is reflected from their surfaces. In general, the incorporation of a dye into a nanoemulsion increases the color intensity because more light is selectively absorbed but decreases the lightness because less light is reflected (Chantrapornchai et al., 1998, 1999a,b).

2.6 CONCLUSIONS

The physicochemical properties of nanoemulsions are important because they impact many of their functional and sensory attributes. Nanoemulsions usually have better stability to gravitational separation and aggregation than conventional emulsions because of their smaller droplet size, which may be advantageous for commercial applications where extended shelf life is required. On the other hand, nanoemulsions may have to be specially designed to reduce Ostwald ripening (e.g., by adding ripening inhibitors) or chemical instability (e.g., by adding antioxidants) for certain types of oil phases, because these instability mechanisms often increase with decreasing droplet size. The composition and structure of nanoemulsions can be controlled to create a variety of different rheological properties, ranging from low-viscosity fluids to semisolid materials. In nanoemulsions where there is a relatively strong attraction or repulsion between the oil droplets, it is often possible to form highly viscous or semisolid materials at much lower droplet concentrations than possible with conventional emulsions. This phenomenon may be useful for creating commercial products with desirable textural attributes, such as foods, cosmetics, or personal care products. The composition and structure of nanoemulsions can also be manipulated to create materials with different optical properties, ranging from transparent to opaque. In particular, transparent nanoemulsions can be produced when the droplet radius is less than about 20 or 30 nm, whereas turbid or opaque nanoemulsions can be produced when the droplet size is larger. Nanoemulsions may therefore have some advantages over conventional emulsions in certain commercial applications.

REFERENCES

Assadpour, E., Jafari, S.-M., Maghsoudlou, Y., 2017. Evaluation of folic acid release from spray dried powder particles of pectin-whey protein nano-capsules. Int. J. Biol. Macromol. 95, 238–247.

Berli, C., Deiber, J., Quemada, D., 2005. On the viscosity of concentrated suspensions of charged colloids. Lat. Am. Appl. Res. 35, 15–22.

Berton-Carabin, C.C., Ropers, M.H., Genot, C., 2014. Lipid oxidation in oil-in-water emulsions: involvement of the interfacial layer. Compr. Rev. Food Sci. Food Saf. 13, 945–977.

Brady, J.W., 2013. Introductory Food Chemistry. Comstock Pub. Associates, Ithaca, NY.

Caivano, J.L., del Pilar, M., 2012. Color in Food: Technological and Psychophysical Aspects. CRC Press, Boca Raton, FL.

Chanamai, R., McClements, D.J., 2000. Dependence of creaming and rheology of monodisperse oil-in-water emulsions on droplet size and concentration. Colloids Surf. A Physicochem. Eng. Asp. 172, 79–86.

Chang, Y., Mclandsborough, L., McClements, D.J., 2012. Physical properties and antimicrobial efficacy of thyme oil nanoemulsions: influence of ripening inhibitors. J. Agric. Food Chem. 60, 12056–12063.

Chantrapornchai, W., Clydesdale, F., McClements, D.J., 1998. Influence of droplet size and concentration on the color of oil-in-water emulsions. J. Agric. Food Chem. 46, 2914–2920.

Chantrapornchai, W., Clydesdale, F., McClements, D.J., 1999a. Influence of droplet characteristics on the optical properties of colored oil-in-water emulsions. Colloids Surf. A Physicochem. Eng. Asp. 155, 373–382.

Chantrapornchai, W., Clydesdale, F., McClements, D.J., 1999b. Theoretical and experimental study of spectral reflectance and color of concentrated oil-in-water emulsions. J. Colloid Interface Sci. 218, 324–330.

Chantrapornchai, W., Clydesdale, F., McClements, D., 2001. Influence of relative refractive index on optical properties of emulsions. Food Res. Int. 34, 827–835.

Chen, J., Stokes, J.R., 2012. Rheology and tribology: two distinctive regimes of food texture sensation. Trends Food Sci. Technol. 25, 4–12.

Chen, H., Khemtong, C., Yang, X., Chang, X., Gao, J., 2011. Nanonization strategies for poorly water-soluble drugs. Drug Discov. Today 16, 354–360.

Chen, H., Jin, X., Li, Y., Tian, J., 2016. Investigation into the physical stability of a eugenol nanoemulsion in the presence of a high content of triglyceride. RSC Adv. 6, 91060–91067.

Date, A.A., Hanes, J., Ensign, L.M., 2016. Nanoparticles for oral delivery: design, evaluation and state-of-the-art. J. Control. Release 240, 504–526.

Derkach, S.R., 2009. Rheology of emulsions. Adv. Colloid Interf. Sci. 151, 1–23.

Francis, F.J., Clydesdale, F.M., 1975. Food Colorimetry: Theory and Applications. AVI Publishing Co. Inc, Roslyn, NY.

Genovese, D., Lozano, J., Rao, M., 2007. The rheology of colloidal and noncolloidal food dispersions. J. Food Sci. 72, R11–R20.

Guzey, D., McClements, D.J., 2006. Formation, stability and properties of multilayer emulsions for application in the food industry. Adv. Colloid Interf. Sci. 128, 227–248.

Güzey, D., McClements, D.J., 2006. Influence of environmental stresses on O/W emulsions stabilized by β-lactoglobulin–pectin and β-lactoglobulin–pectin–chitosan membranes produced by the electrostatic layer-by-layer deposition technique. Food Biophys. 1, 30–40.

Helgason, T., Salminen, H., Kristbergsson, K., McClements, D.J., Weiss, J., 2015. Formation of transparent solid lipid nanoparticles by microfluidization: influence of lipid physical state on appearance. J. Colloid Interface Sci. 448, 114–122.

Helgeson, M.E., 2016. Colloidal behavior of nanoemulsions: interactions, structure, and rheology. Curr. Opin. Colloid Interface Sci. 25, 39–50.

Huang, Q., Yu, H., Ru, Q., 2010. Bioavailability and delivery of nutraceuticals using nanotechnology. J. Food Sci. 75, R50–R57.

Hunter, R.J., 2001. Foundations of Colloid Science. Oxford University Press, New York, NY.

Jafari, S.M., Assadpoor, E., He, Y., Bhandari, B., 2008. Re-coalescence of emulsion droplets during high-energy emulsification. Food Hydrocoll 22 (7), 1191–1202.

Kabalnov, A., 2001. Ostwald ripening and related phenomena. J. Dispers. Sci. Technol. 22, 1–12.

Kabalnov, A.S., Shchukin, E.D., 1992. Ostwald ripening theory: applications to fluorocarbon emulsion stability. Adv. Colloid Interf. Sci. 38, 69–97.

Larson, R.G., 1999. The Structure and Rheology of Complex Fluids. Oxford University Press, New York.

Lesmes, U., McClements, D.J., 2009. Structure–function relationships to guide rational design and fabrication of particulate food delivery systems. Trends Food Sci. Technol. 20, 448–457.

Li, Y., Le Maux, S., Xiao, H., McClements, D.J., 2009. Emulsion-based delivery systems for tributyrin, a potential colon cancer preventative agent. J. Agric. Food Chem. 57, 9243–9249.

Liu, F., Zhu, Z., Ma, C., Luo, X., Bai, L., Decker, E.A., Gao, Y., McClements, D.J., 2016. Fabrication of concentrated fish oil emulsions using dual-channel microfluidization: impact of droplet concentration on physical properties and lipid oxidation. J. Agric. Food Chem. 64, 9532–9541.

Macosko, C.W., Larson, R.G., 1994. Rheology: Principles, Measurements, and Applications. Wiley-VCH, Weinheim.

Malacara, D., 2002. Color Vision and Colorimetry: Theory and Applications. SPIE Press, Washington, DC.

Mao, Y., McClements, D.J., 2013. Modulation of food texture using controlled heteroaggregation of lipid droplets: principles and applications. J. Appl. Polym. Sci. 130, 3833–3841.

Mao, L., Xu, D., Yang, J., Yuan, F., Gao, Y., Zhao, J., 2009. Effects of small and large molecule emulsifiers on the characteristics of b-carotene nanoemulsions prepared by high pressure homogenization. Food Technol. Biotechnol. 47, 336–342.

Mason, T., Wilking, J., Meleson, K., Chang, C., Graves, S., 2007. Nanoemulsions: formation, structure, and physical properties. J. Phys. Condens. Matter. 19.

McClements, D.J., 2002. Theoretical prediction of emulsion color. Adv. Colloid Interf. Sci. 97, 63–89.

McClements, D.J., 2010. Emulsion design to improve the delivery of functional lipophilic components. Annu. Rev. Food Sci. Technol. 1, 241–269.

McClements, D.J., 2011. Edible nanoemulsions: fabrication, properties, and functional performance. Soft Matter 7, 2297–2316.

McClements, D.J., 2012. Nanoemulsions versus microemulsions: terminology, differences, and similarities. Soft Matter 8, 1719–1729.

McClements, D.J., 2015. Food Emulsions: Principles, Practices, and Techniques. CRC Press, Boca Raton, FL.

McClements, D., Decker, E., 2000. Lipid oxidation in oil-in-water emulsions: impact of molecular environment on chemical reactions in heterogeneous food systems. J. Food Sci. 65, 1270–1282.

McClements, D.J., Rao, J., 2011. Food-grade nanoemulsions: formulation, fabrication, properties, performance, biological fate, and potential toxicity. Crit. Rev. Food Sci. Nutr. 51, 285–330.

McClements, D.J., Xiao, H., 2012. Potential biological fate of ingested nanoemulsions: influence of particle characteristics. Food Funct. 3, 202–220.

McClements, D.J., Saliva-Trujillo, L., Zhang, R., Zhang, Z., Zou, L., Yao, M., Xiao, H., 2016. Boosting the bioavailability of hydrophobic nutrients, vitamins, and nutraceuticals in natural products using excipient emulsions. Food Res. Int. 88, 140–152.

Mehrnia, M.-A., Jafari, S.-M., Makhmal-Zadeh, B.S., Maghsoudlou, Y., 2017. Rheological and release properties of double nano-emulsions containing crocin prepared with Angum gum, Arabic gum and whey protein. Food Hydrocoll. 66, 259–267.

Mortensen, A., Skibsted, L.H., 1996. Kinetics of photobleaching of β-carotene in chloroform and formation of transient carotenoid species absorbing in the near infrared. Free Radic. Res. 25, 355–368.

Muller, R.H., Keck, C.M., 2004. Challenges and solutions for the delivery of biotech drugs–a review of drug nanocrystal technology and lipid nanoparticles. J. Biotechnol. 113, 151–170.

Müller, R.H., Radtke, M., Wissing, S.A., 2002. Solid lipid nanoparticles (SLN) and nanostructured lipid carriers (NLC) in cosmetic and dermatological preparations. Adv. Drug Deliv. Rev. 54, S131–S155.

Pal, R., 2011. Rheology of simple and multiple emulsions. Curr. Opin. Colloid Interface Sci. 16, 41–60.

Quemada, D., Berli, C., 2002. Energy of interaction in colloids and its implications in rheological modeling. Adv. Colloid Interf. Sci. 98, 51–85.

Rao, A., 2013. Rheology of Fluid, Semisolid, and Solid Foods: Principles and Applications. Springer Science & Business Media, New York, NY.

Rao, J., McClements, D.J., 2012. Impact of lemon oil composition on formation and stability of model food and beverage emulsions. Food Chem. 134, 749–757.

Saguy, I.S., Singh, R.P., Johnson, T., Fryer, P.J., Sastry, S.K., 2013. Challenges facing food engineering. J. Food Eng. 119, 332–342.

Salvia-Trujillo, L., Decker, E.A., McClements, D.J., 2016. Influence of an anionic polysaccharide on the physical and oxidative stability of omega-3 nanoemulsions: antioxidant effects of alginate. Food Hydrocoll. 52, 690–698.

Selway, N., Stokes, J.R., 2014. Soft materials deformation, flow, and lubrication between compliant substrates: impact on flow behavior, mouthfeel, stability, and flavor. Annu. Rev. Food Sci. Technol. 5, 373–393.

Soukoulis, C., Tsevdou, M., Andre, C.M., Cambier, S., Yonekura, L., Taoukis, P.S., Hoffmann, L., 2017. Modulation of chemical stability and in vitro bioaccessibility of beta-carotene loaded in kappa-carrageenan oil-in-gel emulsions. Food Chem. 220, 208–218.

Stickel, J.J., Powell, R.L., 2005. Fluid mechanics and rheology of dense suspensions. Annu. Rev. Fluid Mech. 37, 129–149.

Tadros, T.F., 2011. Rheology of Dispersions: Principles and Applications. John Wiley & Sons, Hoboken, NJ.

Tadros, T., Izquierdo, P., Esquena, J., Solans, C., 2004. Formation and stability of nano-emulsions. Adv. Colloid Interf. Sci. 108, 303–318.

Taylor, P., 1998. Ostwald ripening in emulsions. Adv. Colloid Interf. Sci. 75, 107–163.

van Vliet, T., 2013. Rheology and Fracture Mechanics of Foods. CRC Press, Boca Raton, FL.

Walker, R., Decker, E.A., McClements, D.J., 2015. Development of food-grade nanoemulsions and emulsions for delivery of omega-3 fatty acids: opportunities and obstacles in the food industry. Food Funct. 6, 42–55.

Walstra, P., 2002. Physical Chemistry of Foods. CRC Press, Boca Raton, FL.

Waraho, T., McClements, D.J., Decker, E.A., 2011. Mechanisms of lipid oxidation in food dispersions. Trends Food Sci. Technol. 22, 3–13.

Weiss, J., McClements, D.J., 2000. Influence of ostwald ripening on rheology of oil-in-water emulsions containing electrostatically stabilized droplets. Langmuir 16, 2145–2150.

Weiss, J., Decker, E.A., McClements, D.J., Kristbergsson, K., Helgason, T., Awad, T., 2008. Solid lipid nanoparticles as delivery systems for bioactive food components. Food Biophys. 3, 146–154.

Wissing, S., Kayser, O., Müller, R., 2004. Solid lipid nanoparticles for parenteral drug delivery. Adv. Drug Deliv. Rev. 56, 1257–1272.

Wooster, T.J., Golding, M., Sanguansri, P., 2008. Impact of oil type on nanoemulsion formation and ostwald ripening stability. Langmuir 24, 12758–12765.

Wooster, T.J., Labbett, D., Sanguansri, P., Andrews, H., 2016. Impact of microemulsion inspired approaches on the formation and destabilisation mechanisms of triglyceride nanoemulsions. Soft Matter 12, 1425–1435.

Yang, Y., McClements, D.J., 2013. Vitamin E bioaccessibility: influence of carrier oil type on digestion and release of emulsified α-tocopherol acetate. Food Chem. 141, 473–481.

Zhang, R., McClements, D.J., 2016. Enhancing nutraceutical bioavailability by controlling the composition and structure of gastrointestinal contents: emulsion-based delivery and excipient systems. Food Struct. 10, 21–36.

Part II

Preparation of Nanoemulsions by Low-Energy Methods

Chapter 3

Catastrophic Phase Inversion Techniques for Nanoemulsification

Antonio Perazzo* and Valentina Preziosi†
*Princeton University, Princeton, NJ, United States, †University of Naples Federico II, Naples, Italy

Chapter Outline

3.1 INTRODUCTION

Nanoemulsions can be produced using a catastrophic phase inversion (CPI) method that exploits either surfactants (Perazzo et al., 2015) or solid particles (Binks, 2002; Binks and Lumsdon, 2000a; Kumar et al., 2015) to stabilize the droplets formed. This emulsification method, which is also known as the "phase inversion composition" or "emulsion inversion point" method, involves titrating one immiscible liquid into another immiscible liquid, which eventually leads to phase inversion and nanoemulsion formation (Perazzo et al., 2015; Roger, 2016; Fryd and Mason, 2012; Solans and Solé, 2012; Ostertag et al., 2012; Gupta et al., 2017) (Fig. 3.1). The two immiscible liquids used are usually oil and water. For example, water may be titrated into a stirred surfactant-oil mixture, which causes the system to invert from a water-in-oil (w/o) emulsion initially to an oil-in-water (o/w) nanoemulsion at the end (Jafari et al., 2017). This process is usually carried out using low-energy flow conditions (<1000 revs/min) (Kotta et al., 2015) and is therefore considered to be a

Nanoemulsions. https://doi.org/10.1016/B978-0-12-811838-2.00003-5

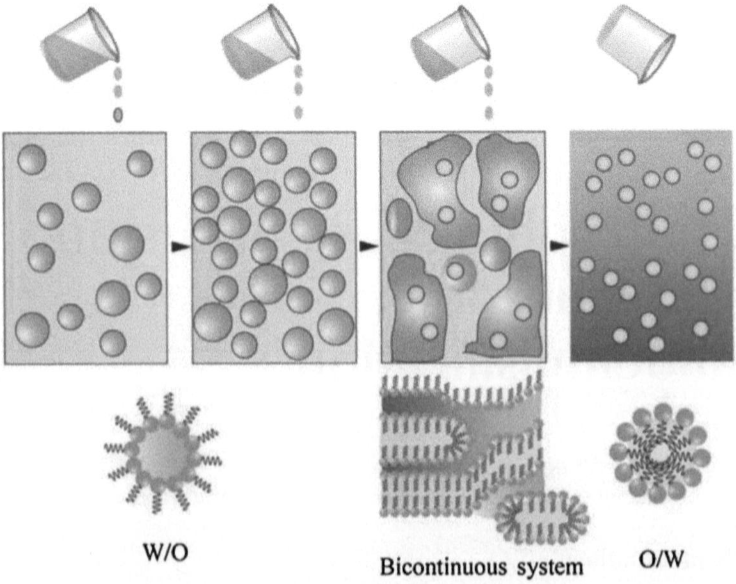

FIG. 3.1 Emulsion CPI pathway. *(Image taken from McClements, D.J., 2011. Edible nanoemulsions: fabrication, properties, and functional performance. Soft Matter 7 (6), 2297–2316.)*

low-energy nanoemulsion formation method. Nanoemulsions containing small droplets can be produced by using blends of surfactants rather than a single one and by carefully controlling the amount of water added to the system at different stages of the emulsification process (Posocco et al., 2016; Heunemann et al., 2011; Jakobs et al., 1999; Rao and McClements, 2010). The underlying mechanisms of the CPI method have been reviewed in the literature (Perazzo et al., 2015; Roger, 2016; Ostertag et al., 2012; Mercuri et al., 2011) and have been shown to have many similarities with the transitional phase inversion (TPI) method, although they are not equivalent (Anton and Vandamme, 2009). In general, CPI methods typically create larger droplets than TPI methods, although they are usually simpler to perform. Knowledge of the underlying mechanism of phase inversion is necessary to understand how this method is capable of generating nanosized droplets. One of the most important factors governing this process is the change in the phase behavior of the system when the surfactant-oil-water ratio is altered (Solans and Solé, 2012). Surfactants in water tend to self-assemble into specific structures, with dimensions ranging from nanometers to micrometers, depending on surfactant type, system composition, and temperature (Rao and McClements, 2010). The inversion of the phases is often preceded by the appearance of a bicontinuous intermediate phase or a lamellar structure that is broken down by dilution with water during the emulsification process, thus giving rise to extremely fine droplets. The effects of temperature are particularly important for nonionic ethoxylated surfactants, which become more hydrophobic when heated. Such a feature can be exploited in combination with the amount of water added to tune the phase behavior of the system and thereby produce nanoemulsions.

Interestingly, it has been reported that multiple emulsions are often formed at intermediate stages along the emulsification pathway (Liu et al., 2012; Jahanzad et al., 2009). Moreover, micellar, hexagonal, lamellar, sponge, reverse sponge, or gel phases may also be formed at intermediate stages, and these may lead to noticeable changes in the rheological properties of the overall system (Liu et al., 2012).

Recently, Roger (2016) showed that it is not necessary to achieve lamellar phases ($L\alpha$) or L_3 sponge phases (both of them have flat/zero curvature) to get nanosized droplets using the CPI method. The smallest droplets are actually obtained when breaking a bicontinuous structure of the L_4 type. These structures have surfactant monolayers that are curved toward the oil phase (i.e., negative curvature) and are obtained with much more water than lamellar phases.

The name "CPI" was originally coined because it was thought that this type of emulsion formation could be described using catastrophe theory (Dickinson, 1982; Dickinson, 1981). However, in reality, this type of inversion process cannot be accurately described using this theory because of the complexity of the hydrodynamic and physicochemical events occurring (Bouchama et al., 2003). The inversion process depends on surfactant, oil and water type and concentration, temperature, agitation rate, flow type, vessel geometry, surface wetting, and electric charge, and so, it is difficult to develop a theory that accurately describes all these factors. On the one hand, from the hydrodynamic point of view, predictive theories have been developed that take into account many of these variables. These hydrodynamic models are based on the concept that two competing phenomena, droplet breakup and droplet coalescence, determine the final droplet size (Perazzo et al., 2015; Kumar et al., 2015). Although predictive, they become unreliable when the system involves high surfactant concentrations or surfactant self-assembly plays an appreciable role. Moreover, they are typically only applicable to processes that generate emulsions rather than nanoemulsions. On the other hand, despite the lack of a comprehensive analytic model that can combine hydrodynamic and physicochemical variables, the mechanistic and thermodynamic understanding of the process has recently improved, thus providing reliable recipes to prepare nanoemulsions. This type of emulsification protocol has been successfully used to encapsulate various pharmaceuticals and nutraceuticals in nanoemulsions (Mayer et al., 2013). Moreover, the bicontinuous phase typically encountered as an intermediate phase in this process has been used as a template to fabricate nanosized particles (Negro et al., 2014).

Here, an overview of the current fundamental understanding of the CPI process, ranging from theory to experiments, is highlighted with an attempt to merge insights from both hydrodynamics and physical chemistry.

3.2 THE ROLE OF SELF-ASSEMBLY AND INTERFACIAL PROPERTIES IN CPI

Surfactants are amphiphilic molecules containing both a nonpolar hydrocarbon tail and a polar head group. The amphiphilic character of such molecules allows for a strongly cooperative association into micelles, whereby the hydrophobic tail forms the micelle core and the polar head forms the micelle-water interface. This cooperative aggregation behavior results from a competition between hydrophobic and hydrophilic forces (Wennerström and Lindman, 1979), which are fundamental for a thermodynamic description of such systems. Hydrophobic interactions (Chandler, 2002, 2005) are the main driving force for surfactant self-assembly, which is related to the tendency for nonpolar molecules to avoid contact with water. Repulsion between the polar head groups of the surfactant also plays a key role in determining the final molecular organization of the micellar systems. Overall, the self-assembly process is driven by the tendency to reach a stable free energy condition, obtained by the means of a structural rearrangement of the surfactant that minimizes the exposure of the hydrophobic tail groups to water. In aqueous solutions and above a certain critical concentration value (CMC), these molecules aggregate to form supermolecular structures, like micelles of various shape and size (Wennerström and Lindman, 1979). In particular, anisotropic micellar solutions, also called L_1 phase (Vitiello et al., 2014), if made of reverse micelles, are named L_2 phases (Garti et al., 2012). For higher surfactant concentrations and depending on surfactant type and temperature (Friberg et al., 2011), self-assembly leads to lyotropic liquid crystalline phases that are large equilibrium structures with a high degree of organization such as hexagonal (filaments of surfactants forming pillars with hexagonal cross section and named H_I for micelles and H_{II} for reverse micelles), bicontinuous cubic (spherical interconnected aggregates packed with a cubic symmetry), Q, or V. Lamellar phases, $L\alpha$, are typically identified by sheets of amphiphiles separated by water, whereas when these amphiphile sheets become interconnected, they form interpenetrating structures, such as the spongelike phase, L_3. A change in system composition may cause the micelles to swell or surfactant-water aggregates to form, thereby increasing the tendency to form structures such as the bicontinuous microemulsion (D) phase, sometimes also called the sponge phase (L_3) when oil swollen or the reverse sponge phase (L_4) when water swollen (Roger et al., 2011).

Surfactant aggregation can be rationalized based on the critical packing parameter (CPP) (V/la) (Israelachvili, 2011), where V is the volume of the surfactant hydrophobic group, l is the length of the hydrocarbon chain, and a is the cross-sectional area of the surfactant hydrophilic core. When $CPP = 1$, the surfactant film has no tendency to curve, and surfactant bilayers are favored. These planar aggregates tend to be formed in the range $1/2 < CPP \leq 1$. For $CPP \ll 1$, micelles having hydrocarbon chains toward the internal part are favored, showing cylindrical aggregates for $1/3 < CPP < 1/2$ and spherical ones for $CPP < 1/3$. Reverse micelles, that is, with water inside, are favored for $CPP > 1$ (Fig. 3.2).

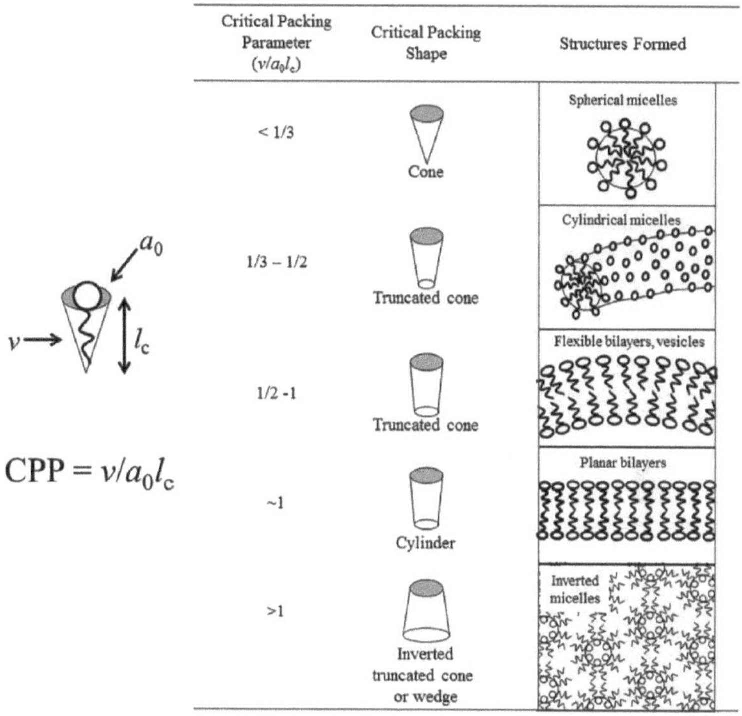

FIG. 3.2 Critical packing parameter (CPP) of surfactants and relative structures formed (Israelachvili, 2011; Salim et al., 2014).

A classification scheme for systems showing microemulsion phases in the case of surfactant-oil-water system has been given by Winsor (1948) (and reviewed by Davies (1957)), who stated the existence of four cases:

Type 1: The surfactants are soluble in water and an o/w microemulsion is formed (Winsor I). This water phase coexists with an oil phase that contains only surfactant monomers.

Type 2: This is the opposite case of the Type 1 system. The surfactant is oil soluble, and a w/o microemulsion is formed. This oil phase is in equilibrium with a water phase that only contains surfactant monomers (Winsor II).

Type 3: A three-phase system is formed where a surfactant-rich middle phase is in equilibrium with two excess phases (oil and water). The emulsion is stratified according to the densities of the three phases (Winsor III or middle-phase microemulsion).

Type 4: A single-phase isotropic micellar solution is formed by using a sufficient quantity of surfactant.

Knowledge of these different structures is useful because as highlighted by Solans and coworkers (Forgiarini et al., 2001; Gutiérrez et al., 2008), Fernandez et al. (2004)), Perazzo et al. (2015), and Roger (2016), the self-assembly of the surfactant structure can be strictly related with the emulsion morphology eventually obtained using the CPI method. A useful way to represent the different aggregation state of a ternary system of surfactant, oil, and water as a function of their relative concentrations is to use phase diagrams.

3.3 DESCRIBING CPI USING PHASE DIAGRAMS AND EMULSIFICATION MAPS

3.3.1 Phase Behavior and Its Role in Phase Inversion

The phase behavior of surfactant-oil-water systems during phase inversion has been reviewed in detail elsewhere (Perazzo et al., 2015; Solans and Solé, 2012). Nonionic ethoxylated surfactants (either abbreviated as C_mE_n, C_iE_j, or C_mEO_n) are affected by temperature-induced modification of the hydrophilic head, making them more hydrophobic and oil soluble with increasing temperature, thus inducing a change in the preferred curvature of the surfactant film. Shinoda reported a flat curvature of the surfactant film near the temperature-induced inversion zone (Shinoda and Saito, 1968). When adding water to an oil-surfactant mixture, the interfacial film formed by nonionic surfactants can change its curvature toward the different phases even when temperature is held constant. This concept is at the base of the CPI method. Wang et al. (2008) and Roger et al. (2010a) stated that the hydration-driven change in surfactant film curvature upon addition of water to a surfactant-oil mixture reduces oil solubilization, thereby leading to oil supersaturation and droplet nucleation. Thus, a mechanism analogous (but not equal) to the temperature-induced curvature change in TPI has been proposed to explain CPI as well. Similarly, other authors have indicated that the addition of water leads to the formation of water droplets that are easily elongated by mixing due to the extremely low interfacial tension (Fernandez et al., 2004; De Gennes and Taupin, 1982). As a result, these droplets merge together and form lamellar structures (Solans and Solé, 2012), which eventually decompose into fine oil droplets.

Similarly, Mercuri et al. (2011) added an oil-surfactant mixture to water in a one-step process and observed a swelling of the organic phase, which gives rise to regions of w/o microemulsions and then liquid crystalline phases at the droplet boundaries. Part of the lyotropic liquid crystalline phases breaks off from the droplet boundary and migrates into the aqueous phase, thus dissociating into fine oil droplets.

CPI is often carried out at a constant temperature using a surfactant-oil mixture with a specific mass ratio and a specific amount of water (see later). The temperature selected is typically between room temperature and phase inversion temperature (PIT). This temperature is often selected to be about 20°C below the PIT of the system so as to facilitate droplet formation but inhibit droplet coalescence. The surfactant-oil mixture is then slowly diluted with water until a bicontinuous/lamellar structure is obtained, and then, a fast dilution is performed using a large amount of water to break down these structures, which gives rise to the spontaneous formation of nanosized droplets (Solans and Solé, 2012; McClements, 2011). The importance of the dilution rate has been highlighted in some studies. For example, increasing the water addition rate fivefold was reported to increase the droplet size from 90 to 130nm (Solans and Solé, 2012; Solè et al., 2010).

Roger et al. (Roger, 2016; Roger et al., 2010b) demonstrated that to obtain the finest droplets, it is more efficient to fast dilute with additional water when the system is beyond its phase inversion composition point (i.e., at higher water content than the one typically performed in CPI). As he stated, a typical CPI lies in the lamellar area of the phase diagram (Roger, 2016) (see Fig. 3.3 as a reference), whereas smaller nanoemulsions are obtained when creating reverse bicontinuous microemulsion/sponge structures (L_4) and then breaking them (under flow) by addition of water. In the latter case, the homogeneous bicontinuous network preferred curvature is turned toward the oil; thus, when adding water, the oil phase separates in the shape of nanosized droplets (the size of which is equal to the curvature of the L_4 sponge phase) and some surfactant is wasted as small micelles in water.

Roger et al. (Roger, 2016; Roger et al., 2010b) define the clearing boundary (CB, Fig. 3.4) as a location in the phase diagram for which, for a certain temperature, a given oil/surfactant ratio and water/surfactant ratio (this one is varied

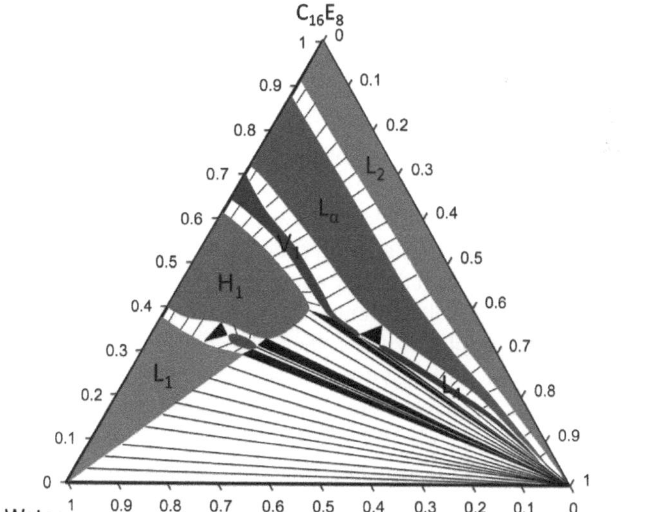

FIG. 3.3 Ternary phase diagram of the $C_{16}E_8$/hexadecane/water system at 50°C. L_1, oil/water micellar solution; L_2, water/oil micellar solution; L_4, water-swollen (reverse) sponge phase; Lα, lamellar phase; V_1, bicontinuous cubic phase; H_1, hexagonal phase. *(From Roger, K., Cabane, B., Olsson, U., 2010. Emulsification through surfactant hydration: the PIC process revisited. Langmuir 27 (2), 604–611.)*

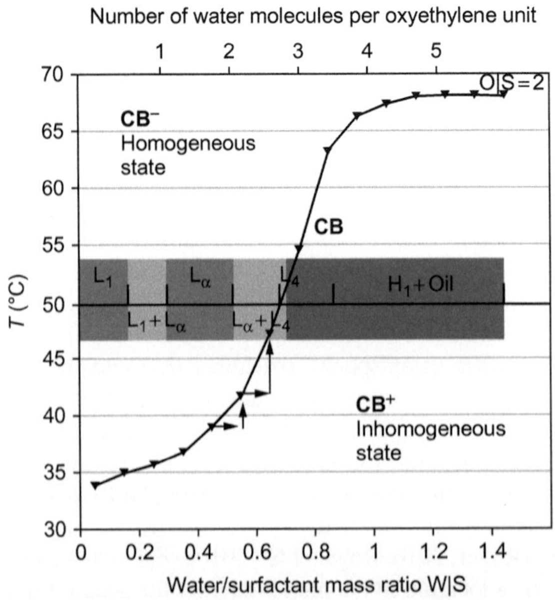

FIG. 3.4 Clearing boundary for oil-surfactant ratio (O/S) = 2 and equilibrium structures at $T = 50°C$. The CB line separates homogeneous states CB^- from inhomogeneous states CB^+. *(From Roger, K., Cabane, B., Olsson, U., 2010. Emulsification through surfactant hydration: the PIC process revisited. Langmuir 27 (2), 604–611.)*

over CPI), a homogenous bicontinuous phase, with no oil or water excess, is formed (i.e., the L_4 phase in Fig. 3.3 or the black sigmoidal line in Fig. 3.4).

This bicontinuous phase is made of a reverse, water swollen, connected bilayer forming a dividing surface between two separated 3-D-continuous oil labyrinths (Fig. 3.5). As well known for such structures, the average mean curvature of the bilayer is zero (Fig. 3.5). But the individual monolayer surfaces have a negative (hence nonzero) mean curvature toward the oil, so the preferred curvature of this water-swollen sponge phase is already turned toward the oil phase (Fig. 3.4) as in the final nanoemulsion.

The final nanosized droplets will assume the size of such preferred curvature once the structure is broken by fast dilution. Roger et al. reported that once fast dilution is performed at this stage, nanosized droplets are obtained regardless of the stirring rate. Knowing the oil-surfactant ratio, this surfactant interfacial film curvature (hence, indirectly, also the droplet size) can be calculated/predicted (Roger et al., 2010b). Nevertheless, it never exactly matches the size of the nanoemulsion droplets obtained. This has been attributed to a certain amount of surfactant that is not present at the interface of the L_4 sponge structure, but instead lies in the bulk phase of the structure and is eventually found at the end of the emulsification process in the form of micelles in the continuous phase surrounding the nanoemulsion droplets. This explains why Roger et al.

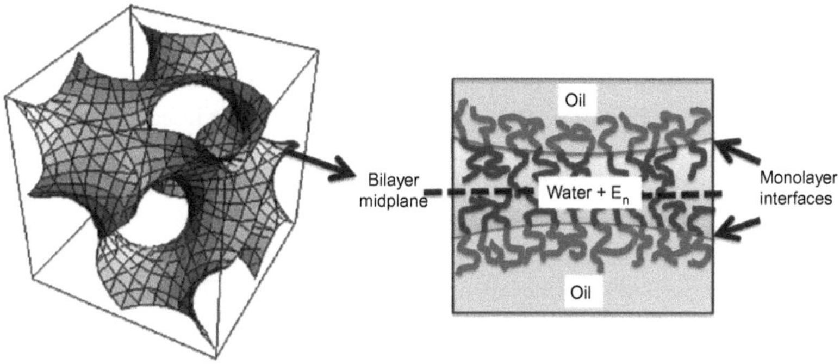

FIG. 3.5 Structure of an oil-rich sponge phase (its acronym being L_4). A reverse, water-swollen, multiply connected bilayer forms a dividing surface between two separate 3-D-continuous oil labyrinths of equal volume. By symmetry, the average mean curvature of the bilayer midplane is zero. The two monolayers, on the other hand, have an average mean curvature toward the oil. *(From Roger, K., Cabane, B., Olsson, U., 2010. Emulsification through surfactant hydration: the PIC process revisited. Langmuir 27 (2), 604–611.)*

refer to the latter as a quantity of "wasted" surfactant, that is, an amount of surfactant that is not used to form nanoemulsion droplets.

The sponge structure that gives rise to the formation of nanoemulsion droplets usually contains more water than is typically used in CPI, where, indeed, lamellar phases that are richer in surfactant content are usually exploited (Roger, 2016). This explains the definition given by Roger et al. of superior-PIC or sup-PIC emulsification, i.e., a PIC performed by breaking an intermediate structure with superior water content. (PIC is the acronym of phase inversion composition that is a synonym of CPI.)

A considerable amount of surfactant is needed for the formation of such oil-rich sponge structures (this puts a limit to the maximum oil-surfactant ratio to be used, <5). Although high surfactant content can be an impairing factor for obtaining nanoemulsions, micelles or microemulsions are formed instead. Thus, an oil-surfactant ratio between 1 and 5 is commonly used, but the precise value depends on the nature of the system used (Roger et al., 2010b).

Another key variable is the initial water content present in the oil-surfactant mixture before starting the CPI. If this amount of water is less than the one necessary to create an oil-rich bicontinuous phase, then one always obtains nanoemulsions (provided that the other composition requirements are met). Conversely, if the initial water content is larger than that necessary to create the sponge phase, then coarse emulsions are obtained rather than nanoemulsions.

In conclusion, at an appropriate temperature, oil-surfactant ratio (roughly between 1 and 5), and initial water content (one that is not already above the value required for homogenous L_4 microemulsion formation), there will be a specific water-surfactant ratio that gives access to the formation of a homogenous oil-rich sponge phase with preferred curvature of the surfactant monolayer toward the oil that by fast water dilution produces a nanoemulsion. This access state is the clearing boundary zone (CB).

Roger et al. (2010b) extended the validity of their results to many pure-grade surfactants of the C_iE_j family and many oil types within the linear alkane family (with comparable viscosity) and confirmed the generality of their results. What it may change in those cases is the area of the different phases within the phase diagram when changing either surfactant or oil.

For nonionic surfactants, the interactions between the surfactant head and water can be modified by either using temperature (in TPI) or using a nonionic surfactant coupled with salt (the latter has the same effect as temperature changes) so that the preferred curvature of surfactant film exactly matches the geometric curvature of the sponge phase. Applying flow to the system, also recently observed by Preziosi et al., the equilibrium bicontinuous network coexisting with excess oil and water breaks into a dispersion of oil-superswollen micelles, named superswollen microemulsions. All the surfactant is used at the interfaces, and no surfactant is present in the bulk. The effect of flow on such complex microstructures has been also recently investigated by Preziosi et al. (2017a).

Regarding systems exploiting a mixture of ionic and nonionic surfactants, the intermediate phase has been found to be more likely a high internal phase emulsion or a structure resembling that of a polyhedral foam rather than a bicontinuous microemulsion (Wolf et al., 2010). Moreover, apart from varying temperature, a change in the interfacial tension could be obtained by adding a cosurfactant to the solution or using surfactant blends (Preziosi et al., 2013; Lv et al., 2014; Posocco et al., 2016; Preziosi et al., 2017b), thus exploiting the so-called synergism effect (Strey, 1996). Here, the use of a couple of surfactants lowers the interfacial bending energy, which in turn decreases the interfacial tension with respect to a system having only one surfactant at the same concentration. This effect can be achieved by simply using a bulky surfactant with a large head group coupled with a smaller one. Recently, nonionic polymeric surfactants, also referred to as microemulsion boosters (Jakobs et al., 1999), have been created to have a large head group so that, coupled with another smaller surfactant, the bending energy at the interface is relatively low, thus providing microemulsion phases with a much lower surfactant content than typically needed. This is an important consideration, since the CPI method is typically limited by the large amounts of surfactant required to form nanoemulsions (Fig. 3.6).

3.3.2 Emulsification Maps Representing CPI

A schematic representation for emulsion morphology and phase behavior as a function of formulation variables (variables depending on the nature of the components) and composition ones (the compound weight or volume fractions) is

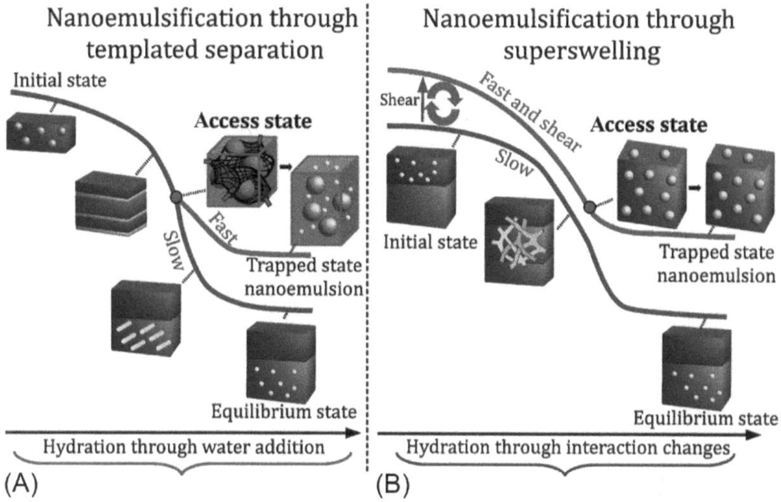

FIG. 3.6 Mechanisms of nanoemulsification through the disruption of bicontinuous structures; in the sketch (A), CPI pathway (or as defined by Roger a sub-PIC) is represented while in (B) mechanism of transitional phase inversion or CPI performed with nonionic surfactant and salt (Roger, 2016).

called an "emulsification map," and its utility and construction have been recently reviewed (Perazzo et al., 2015).

Typical names encountered when dealing with an emulsification map are *normal emulsions* (Bancroft, 1913), that is, emulsions that follow the Bancroft rule, "the phase in which an emulsifier is more soluble constitutes the continuous phase," and *abnormal emulsions* that do not follow this rule (Sajjadi, 2006). A possible criterion for choosing the emulsion optimal formulation is the hydrophile-lipophile balance or HLB number (Griffin, 1946). The HLB number is an index of the affinity of a surfactant toward the oil ($HLB < 10$) or toward the water ($HLB > 10$) and can be used to select the most appropriate surfactant for forming different types of emulsions (Fig. 3.7).

However, this classification scheme is usually too simplistic to be adopted as a general guide for nanoemulsion formation. To broaden the HLB concept, a description of the CPI method from a thermodynamic point of view has been developed (Sajjadi, 2006; Griffin, 1946; Salager et al., 2000, 2004). This approach was first based on a generalized variable called the surfactant affinity difference (SAD) introduced by Salager and then substituted by adopting the so-called hydrophilic-lipophilic deviation (HLD). SAD is defined as

$$\mathrm{SAD} = \mu_{\mathrm{W}}^{*} - \mu_{\mathrm{O}}^{*} = \Delta G_{\mathrm{oil \to water}} = -RT \ln K_{\mathrm{p}} \tag{3.1}$$

where K_p is the partition coefficient of the surfactant between water and oil at the corresponding temperature. The SAD equation can be also expressed as a function of the "formulation variables." (Salager et al., 2000).

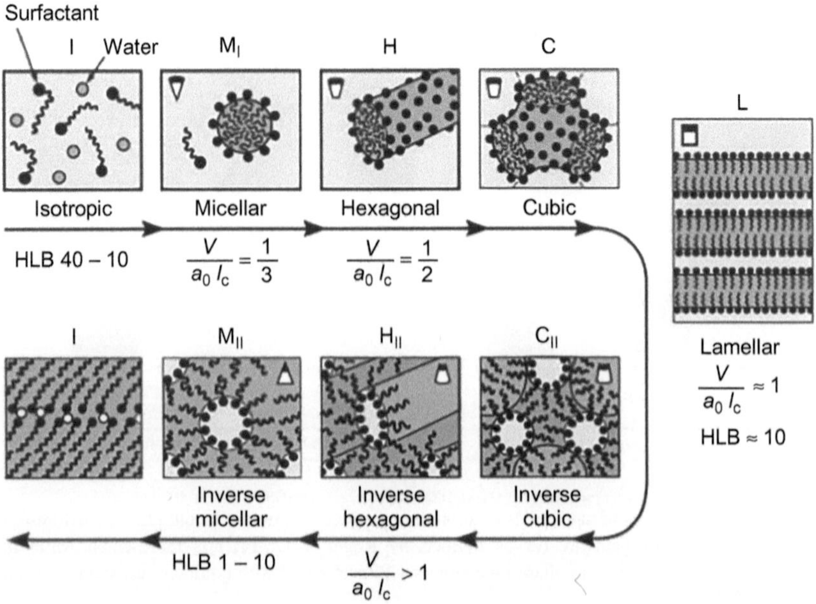

FIG. 3.7 Schematic representation of the different phase structures formed at different critical packing parameters and HLB numbers (Israelachvili, 2011).

HLD is then defined as $\text{HLD} = (\text{SAD} - \text{SAD}_{\text{ref}})/RT$ where SAD_{ref} refers to the value of SAD calculated at the optimum formulation. This formulation has an intermediate bicontinuous structure or lamellar structure with vanishing interfacial tension and zero curvature. This reformulation of SAD was needed to take into account that for nonionic surfactants, SAD is not zero at the optimum formulation, while HLD is indeed equal to zero. Because of its formulation, HLD turns to be a dimensionless parameter that characterizes the behavior of a surfactant within a given surfactant-oil-water mixture and depends on surfactant type, oil type, and aqueous phase properties such as pH, ionic strength, and cosolvent. For $\text{HLD} < 0$, the surfactant has more affinity for water and forms micelles that can stabilize o/w emulsion. The opposite occurs for $\text{HLD} > 0$. When $\text{HLD} = 0$, there is no preferred curvature toward the phases, and a bicontinuous microemulsion or a lamellar phase can be formed. Expressions for the HLD numbers of ionic and nonionic surfactants, respectively, read as given below:

$$\text{HLD} = \ln S - K \cdot \text{ACN} + \sigma - a_{\text{T}}(T - 25) \tag{3.2}$$

$$\text{HLD} = k \cdot \beta + b \cdot S - k \cdot \text{ACN} + t(T - 25) + a \cdot A \tag{3.3}$$

where $\beta = (\alpha - \text{EON})/k$, α, k, and t are surfactant parameters, EON is the average number of ethylene oxide groups per molecule of nonionic surfactants, k and a_{T}

are empirical constants dependent on the nature of the system, *b* and *a* are constant characteristics of each type of salt and alcohol, and *S* and *A* are the salt and alcohol concentration. ACN is the alkane carbon number (or the equivalent alkane carbon number EACN if the oil is not an alkane (Queste et al., 2007)), and σ and α are the characteristic parameters of the surfactant structure, where α depends on the lipophilic group of the surfactant.

In Fig. 3.8, a bidimensional formulation-composition map at constant surfactant concentration is shown, where B indicates an oil-rich phase, C is a water-rich phase, A is an intermediate phase, and the "+" and "−" terms refer to the HLD sign. Following the inversion line, which appears as a stair on the diagram, CPI is achieved when varying the water-to-oil ratio between B− and A− or between A+ and C+ to cross the inversion line. Phase inversion can be delayed (emulsion inversion does not take place immediately) due to multiple emulsion formation. It follows that vertical lines are replaced by an "uncertainty area" (triangularly shaped) on the diagram. In this case, it is not possible to exactly identify the vertical branches of the inversion frontier, but it is known that they fall in the triangular shaded region of the map. The existence of these areas explains the hysteresis (the morphology of the emulsion is not reversible when switching back the water-oil ratio) typically associated with CPI. The horizontal branch of the inversion frontier can get shortened by lowering the surfactant concentration and by increasing the stirring energy or the viscosity of one of the phases.

Mira et al. (Tyrode et al., 2003; Mira et al., 2003; Zambrano et al., 2003) reported a series of studies showing an onward shifted phase inversion point by increasing agitation speed both with and without surfactants.

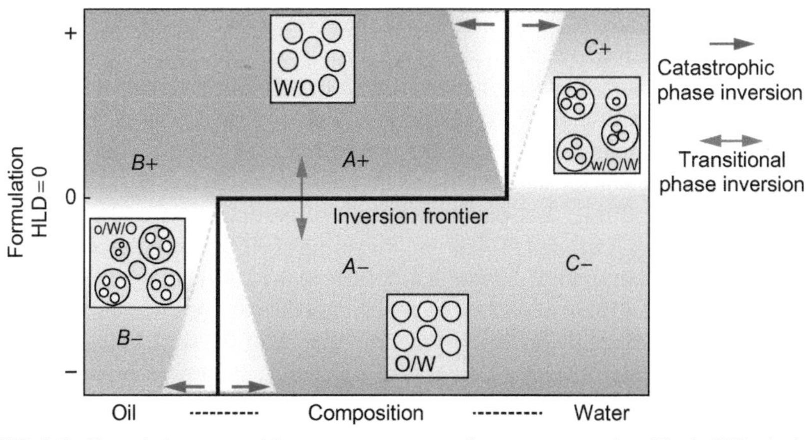

FIG. 3.8 Formulation-composition map at constant surfactant concentration. The *bold line* is the phase inversion frontier. Water-oil ratio is varied moving horizontally in the diagram. *(From Galindo-Alvarez, J., et al., 2011. Viscous oil emulsification by catastrophic phase inversion: influence of oil viscosity and process conditions. Ind. Eng. Chem. Res. 50 (9), 5575–5583.)*

These researchers reported phase inversion to be favored for both low-energy stirring and high-energy stirring, whereas it appears to be inhibited by intermediate-energy stirring. The explanation they gave is based on the low-curvature interface configuration in the low-energy case due to the presence of big droplets, which could favor multiple emulsion creation and hence phase inversion. In the case of high-energy stirring, big droplets and low-curvature interfaces are not favored anymore, but the high agitation energy speeds up the surfactant adsorption from the bulk to the interface, thus favoring the formation of droplets inside droplets (a mechanism different from the one for low stirring energy). In the intermediate-energy stirring regime, droplets are not so big, and surfactant adsorption is not fast enough, thus inhibiting, for both reasons, phase inversion. However, a major role is also played by the HLD. In fact, hydrophilic systems are largely insensitive to rate of addition and to stirring energy, because phase inversion is already fast due to faster surfactant adsorption for hydrophilic surfactants. Conversely, for a hydrophobic surfactant, stirring energy and water rate of addition have a major role. Similar results have been reported also by Tyrode et al. (2003) and Galindo-Alvarez et al. (2011).

This method to describe CPI relies on the existence of an intermediate structure with no preferred curvature (e.g., lamellar phases or bicontinuous phases where the surfactant monolayers have zero curvature and typically zero interfacial tension). However, as mentioned in Section 3.1, Roger et al. (Roger, 2016; Roger et al., 2010b) reported an intermediate reverse bicontinuous structure (L_4) with preferred curvature toward the oil to be the most suitable to obtain CPI (or, alternatively said, sup-PIC).

3.4 CPI USING SOLID PARTICLES

The classical picture of an emulsion is a thermodynamically unstable system made of oil, water, and surfactants, where droplets of one phase are dispersed in the other, and surfactants are needed to stabilize the dispersion by decreasing the interfacial tension between continuous and dispersed phases. However, there are many studies focused on the use of (nano or micro) solid particles in place of surfactants as stabilizers in emulsions (Binks, 2002; Garbin, 2013; Lee et al., 2011). The pioneering work of particle-stabilized emulsions was conducted by Ramsden (1903) and Pickering (1907), hence the term Pickering emulsions (an example in Fig. 3.10), early in the 20th century. Some years later, Finkle et al. (1923) explained more in detail the presence of solid particles on emulsion interface by underlying the importance of the wettability of particles at the oil-water interface (Binks and Lumsdon, 2000a). In particular, he stated that in a liquid-liquid system, in the presence of particles, one liquid wets the solid more than the other liquid, with the more poorly wetting liquid becoming the dispersed phase. The wettability, which allows that particles adsorb irreversibly to the interface, represents one of the main differences with surfactants, which can instead adsorb and desorb at the interface on a relatively fast

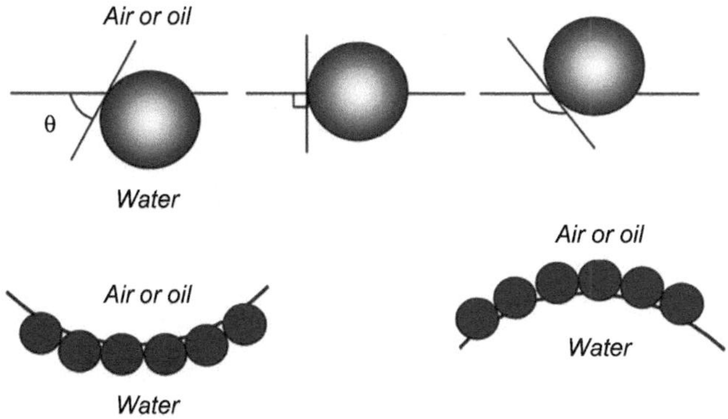

FIG. 3.9 (Top panel) Solid particle position at a planar fluid-water interface. The contact angle ϕ, measured through the water phase, is less than 90 degrees (left)' equal to 90 degrees (center), and greater than 90 degrees (right). (Bottom panel) Possible particle positioning at a curved fluid-water interface. For $\phi > 90$ degrees, solid-stabilized aqueous foams or o/w emulsions may form (left). For $\phi < 90$ degrees, solid-stabilized aerosols or w/o emulsions may form (right) (Binks, 2002).

timescale and whose water- or oil-liking tendency is quantified by the HLB previously described. The wettability can be expressed in terms of the contact angle ϕ that the particles make with the interface. In particular, for hydrophilic particles, the angle φ measured through the water phase is normally < 90 degrees in such a way that a larger part of the particle surface remains in the liquid it wets preferentially, thus giving rise to o/w emulsion. On the other hand, for hydrophobic particles ($\varphi > 90$ degrees), w/o emulsions tend to be stabilized with the particle more exposed to the oil than to the polar phase (Fig. 3.9). Finally, if the droplets are completely wetted in one or other phase, they remain dispersed, and no stable emulsions can be formed.

It's necessary to underline that particle energy adsorption on the oil-water surface is related not only to the contact angle between the droplet and the water phase but also to the tension of the interface γ_{ow} (Binks and Lumsdon, 2000b).

The mechanisms at the basis of emulsion stabilization due to solid particles are the following: according to the first one, particles accumulate at the droplet surface forming a dense film as a mono- or multilayer serving as a mechanical barrier to avoid coalescence, while the second mechanism predicts the formation of a monolayer of particles bridging the surfaces of colliding emulsion droplets (Fig. 3.10). The bridging stabilization mechanism can explain the existence of very stable emulsions even when the droplets are sparsely covered with particles. Moreover, stabilization of emulsions due to the formation of a 3-D network of particles in the continuous phase impeding coalescence has also been reported. However, a thorough comprehension of the stabilization mechanism is still debated (Miller et al., 2006). Improvements of stability are found when there are favorable particle-particle interactions or using particles with high aspect ratio that presumably provide shape-induced capillary forces and

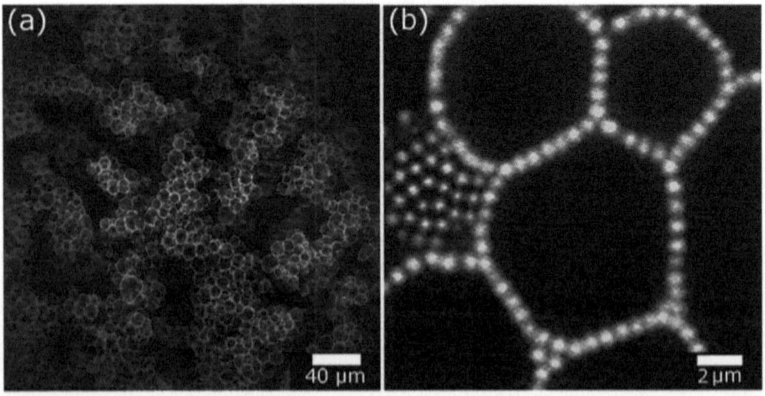

FIG. 3.10 Confocal microscopy images of Pickering emulsion gels. (A) Low-magnification image showing tortuous clusters of droplets in the continuous fluid phase. (B) High-magnification image showing a monolayer of nearly touching particles bridging a number of faceted droplets in the gel interior. *(Image taken from Lee, M.N., Chan, H.K., Mohraz, A., 2011. Characteristics of pickering emulsion gels formed by droplet bridging. Langmuir 28 (6), 3085–3091.)*

packing effects (Madivala et al., 2009). Limitations are shown when particle volumetric fraction is high. In this case, functionalized particles such as Janus particles (Kumar et al., 2013, 2015; Walther and Müller, 2008, 2013; Glaser et al., 2006) (particles that are hydrophobic on one side and hydrophilic on the other one) are used, due to their peculiar configuration, to get a better interfacial efficacy or a faster migration toward the interface. A theoretical study comparing Janus particles with spheres of uniform wettability was performed by Binks and Fletcher (2001), who studied how the particle amphiphilicity can affect the strength of particle adsorption.

The case of surfactant molecules as emulsifiers during the phase inversion process has been deeply investigated, and recently, there is an upsurge of interest in using solid particles for emulsion stabilization due to some advantages as compared with surfactants (Binks, 2002; Binks and Whitby, 2004). The inversion in particle-stabilized emulsions can be achieved either by varying particle hydrophobicity (transitional) or by increasing the volume fraction of the dispersed phase (catastrophic) (Binks and Lumsdon, 2000a). Further difference in emulsions stabilized by particles instead of surfactants is the effect of the oil type on emulsion formation because oil type seems to not affect nanoemulsion morphology when stabilized by particles. Binks and Lumsdon (2000a) showed how emulsion phase inversion methods could be exploited to get tailored particle-stabilized nanoemulsion either tuning particle wettability or varying the volumetric fraction of the immiscible liquid phases. Reasoning on the interfacial bending moment, Kralchevsky et al. (2005) provided a theoretical prediction for the CPI in such systems in agreement with Binks and Lumsdon experiments.

CPI has been used also to prepare particle-stabilized air-in-water foams and water-in-air powders (Binks and Murakami, 2006). In this case, the layer of particles adsorbed on the air or water surfaces may provide a way to control the release of gaseous or liquid components.

Recently, the effect of particle shape and interparticle interactions on the microstructure and rheology of oil-in-water emulsions, stabilized by fumed and spherical particles, has been investigated. In particular, Katepalli et al. (2017) varied interparticle interactions by changing salt concentration in the continuous (aqueous) phase, and then, the effects on the morphology and stability of the emulsions were studied.

There is also a new interest in studying the effect of particles on the stability of oil-in-oil (o/o) emulsions (Fernandez-Rodriguez et al., 2017). Binks and Tyowua (2016) reported an extensive collection of experimental results regarding the behavior of nonaqueous liquid-liquid interfaces with adsorbed particles. They identified three main oil phases that could be used as water substitutes: (i) polar solvents with relatively high dielectric constants, (ii) oils with relatively low dielectric constants, and (iii) liquid polymers. Moreover, Binks also investigated the use of CPI in such systems and demonstrated that the emulsions investigated, made of sunflower oil and PDMS stabilized by SiOH fluorosilica particles, were stable for more than 1 year.

3.5 THE EFFECT OF HYDRODYNAMIC PROCESSING AND PHYSICOCHEMICAL VARIABLES

The impacts of hydrodynamic conditions, wetting characteristics, phase viscosities, and colloidal interactions are some of the most common variables that have been examined on the formation of emulsions and nanoemulsions using the phase inversion method (Perazzo et al., 2015; Kumar et al., 2015; Yeo et al., 2000). Most of the literature in this area is related to emulsion production rather than nanoemulsion production; hence, we will only briefly summarize the most useful information to take into account for nanoemulsion formation. Phase inversion has been herein interpreted and modeled in view of a dynamical phenomenon between droplet coalescence and breakup. In the light of this consideration, phase inversion has been studied in agitated vessels and tube flows to predict the average droplet diameter of the emulsion obtained. Typically, the surface-weighted mean diameter (d_{32}) is used because it relates the total surface area (A) of the disperse phase to its volume fraction (ϕ) in the system: $A = 6\phi/d_{32}$. Phase inversion is believed to be reached when the coalescence rate overwhelms the breakup rate over the majority of the vessel volume. Variables typically considered are temperature, flow regime, wetting, interfacial tension, viscosity, density, and vessel and impeller geometry and material. The concentration range in which two immiscible liquids may both be the continuous phase is known as the "ambivalent range," the range of which depends on how the dispersion is produced, on the volume fractions of the dispersed phases, and on the initial energy level of the system. In an agitated vessel, the ambivalence region initially decreases monotonically with increasing agitation speed, but after a certain point, it becomes independent of the agitation speed.

The input power of the impeller (related to the cube of the rotation speed) transmits mechanical energy to the liquid in the form of turbulent motion, causing the breakage of a phase into small droplets, thereby increasing the interfacial area between the two phases. Higher agitation rates are associated with phase inversion occurring at a smaller amount of dispersed phase. This has been attributed to an increased tendency to form multiple emulsions that are factors favoring catastrophic inversion (Jahanzad et al., 2009).

The propensity of the vessel/impeller material to be preferentially wetted by one of the phases is another variable that has been shown to impact the phase inversion mechanism. Wetting is affected by surface roughness/heterogeneity, temperature, and impurities. It is known that wetting influences the frequency of collision of the droplets and thereby the coalescence rate, thus affecting the inversion of the two phases (Perazzo et al., 2015; Dunstan et al., 2011). Phase inversion from an oil continuous to an aqueous continuous system occurs at much lower oil concentrations when a water-wetting stainless steel vessel/impeller is used rather than an oil-wetting plexiglass vessel/impeller.

Usually, if the dispersed phase is more viscous than the matrix, breakup is slowed down, and high viscosity ratios between the oil and water phases promote the formation of multiple emulsions (Jafari et al., 2008). The tendency of a phase to disperse decreases with the increase of its viscosity. The increase in the viscosity of the two liquids increases the drainage time of the interfacial film between approaching droplets, corresponding to a lower coalescence rate that gives a wider ambivalence zone, thus inhibiting phase inversion.

The viscosity during emulsification assumes a maximum value in the vicinity of the point of inversion, which can be signaled also by an abrupt increase in electric conductivity (Preziosi et al., 2017b). The charge on the surface of the droplets also affects the coalescence process. Hydrophobic interfaces are known to be negatively charged, regardless of surfactant presence. The presence of ions allows to explain why the phase inversion of oil in water occurs at lower values compared with the inversion of water in oil. This interfacial charge is given by the great difference between the dielectric constants of the two immiscible phases. Droplets of oil in water show a repulsion due to an overlap of electric double layers, thus inhibiting coalescence. Conversely, droplets of water in oil do not exhibit this effect, and their higher coalescence efficiency can generate multiple emulsions more readily, thus favoring phase inversion.

Some other basic insights for a proper emulsification, especially regarding the oil type to be used, are available. These hints are not restricted to CPI but are valid in general when mixing oil, water, and surfactants to get stable droplets.

Oil, as reported by Komaiko and McClements (2014, 2016), has a huge effect on the particle size of nanoemulsion produced. For example, in the preparation of food-grade nanoemulsions by spontaneous emulsification, much smaller droplet sizes were obtained when using medium-chain triglycerides as compared with mineral oils.

A common mechanism of nanoemulsion destabilization is Ostwald ripening (Delmas et al., 2011), and Wooster et al. (2008) reported how this process is highly affected by the solubility of the oil phase in the water phase, a feature that is directly related with the oil molar volume. Hydrocarbon oils, such as dodecane, tetradecane, hexadecane, and octadecane, have an increasing molar volume with increasing hydrocarbon number. This means that dodecane (small molar volume and higher water solubility) is more prone to Ostwald ripening than octadecane (large molar volume and lower water solubility). This result was confirmed by Hoffmann et al. (2016) who reported that oil exchange can be slowed down by increasing the chain length of the alkane. The best oils for inhibiting Ostwald ripening were reported to be triglycerides, being highly insoluble in water. However, one difficulty in the use of triglycerides is their high viscosity that could be an impairing factor for proper emulsification. In Komaiko and McClements work, a synergistic effect between an ionic surfactant (SDS) and polyethylene glycol (PEG) has been used to achieve better nanoemulsion stability and smaller droplet sizes. The specific oil-surfactant couple establishes also the PIT of the system that is a reference point useful even in CPI.

Bremond et al. (2011) and Kumar et al. (2016) provided experimental evidence of CPI induced by droplet coalescence. These experiments were performed in well-controlled flow fields such as the low Reynolds number flow of microfluidic channels (Kumar et al., 2015, 2016; Bremond and Bibette, 2012). For such experiments, surfactant concentration was relatively low; hence typically, no nanosized droplets were obtained or shown. Nevertheless, microfluidic flow visualization represents a valuable tool toward further understanding of the CPI in the presence of high surfactant concentrations typical of the intermediate bicontinuous/sponge structures that, once broken down, form nanoemulsions. In general, the effect of flow on the nature of the bicontinuous structures formed is an aspect that has not thoroughly been explored in the literature, but is likely to be important. Some of the most relevant experiments have been conducted on polymeric microemulsions therein highlighting flow-triggered complex phase transitions (Krishnan et al., 2001; Hickey et al., 2016; López-Barrón and Macosko, 2010). No equivalent experiments and results are available for surfactant-oil-water systems. An additional complication, relevant for processing, is that bicontinuous microemulsions are typically shear-thinning fluids, that is, their viscosity decreases as a function of the applied shear rate (Gradzielski and Hoffmann, 1999; Anklam et al., 1995), though the mechanism underlying such rheological behavior is still not thoroughly understood.

3.6 CONCLUSIONS

Nanoemulsions with a high volumetric fraction of nanosized droplets can easily be obtained using phase inversion methods that involved low energy of mixing. For example, in agitated vessels, the typical stirring rate is less than 1000 revs/min. Nanoemulsion droplets can be obtained by CPI by exploiting either surfactants or solid particles. The dilution-induced breakage of lamellar structures or bicontinuous microemulsions structure is already capable to provide nanoemulsion, but a fast dilution step capable of disrupting the reverse sponge-phase structures formed at certain surfactant-oil-water compositions is key to achieving the smallest nanosized droplets. These droplets have dimensions that are comparable in size with the surfactant monolayer curvature within the intermediate reverse sponge phase. Some compositional requirements, such as the oil-surfactant ratio and the water-surfactant ratio, have to be met to generate the required sponge phases. Nonionic surfactant type, ionic/nonionic mixtures, mixtures of nonionic, and nonionic surfactant/salt coupling can be exploited to obtain nanoemulsions using the CPI method. A potential advantage of using surfactant blends is to achieve nanoemulsion formation at relatively small surfactant concentrations. Emulsion stability maps and surfactant-oil-water phase diagrams help to identify system compositions and preparation routes where nanoemulsions can be successfully formed using the CPI method. Hydrodynamic variables involved in the CPI method, such as viscosity, capillary number, and viscosity ratio of the phases, are explored in the regime of high surfactant concentration but modeled only in the regime of low surfactant concentration. Characterization of changes in the morphology of surfactant-oil-water systems during the CPI process under flow is still rare but may provide some valuable mechanistic insights. The mechanistic approach recently described by Roger et al. provides some valuable insights into optimizing system composition and preparation methods required for the fabrication of nanoemulsions containing small droplets.

REFERENCES

Anklam, M.R., Prud'Homme, R.K., Warr, G.G., 1995. Shear thinning in ternary bicontinuous and water-in-oil microemulsions. AIChE J. 41 (3), 677–682.

Anton, N., Vandamme, T.F., 2009. The universality of low-energy nano-emulsification. Int. J. Pharm. 377 (1-2), 142–147.

Bancroft, W.D., 1913. The theory of emulsification V. J. Phys. Chem. 17 (6), 501–519.

Binks, B., Fletcher, P., 2001. Particles adsorbed at the oil-water interface: a theoretical comparison between spheres of uniform wettability and "Janus" particles. Langmuir 17 (16), 4708–4710.

Binks, B., Lumsdon, S., 2000b. Influence of particle wettability on the type and stability of surfactant-free emulsions. Langmuir 16 (23), 8622–8631.

Binks, B.P., 2002. Particles as surfactants—similarities and differences. Curr. Opin. Colloid Interface Sci. 7 (1), 21–41.

Binks, B.P., Lumsdon, S.O., 2000a. Catastrophic phase inversion of water-in-oil emulsions stabilized by hydrophobic silica. Langmuir 16 (6), 2539–2547.

Binks, B.P., Murakami, R., 2006. Phase inversion of particle-stabilized materials from foams to dry water. Nat. Mater. 5 (11), 865–869.

Binks, B.P., Tyowua, A.T., 2016. Oil-in-oil emulsions stabilised solely by solid particles. Soft Matter 12 (3), 876–887.

Binks, B.P., Whitby, C.P., 2004. Silica particle-stabilized emulsions of silicone oil and water: aspects of emulsification. Langmuir 20 (4), 1130–1137.

Bouchama, F., et al., 2003. On the mechanism of catastrophic phase inversion in emulsions. Colloids Surf. A Physicochem. Eng. Asp. 231 (1–3), 11–17.

Bremond, N., Bibette, J., 2012. Exploring emulsion science with microfluidics. Soft Matter 8 (41), 10549–10559.

Bremond, N., Doméjean, H., Bibette, J., 2011. Propagation of drop coalescence in a two-dimensional emulsion: a route towards phase inversion. Phys. Rev. Lett. 106 (21), 214502.

Chandler, D., 2002. Hydrophobicity: two faces of water. Nature 417 (6888), 491.

Chandler, D., 2005. Interfaces and the driving force of hydrophobic assembly. Nature 437 (7059), 640–647.

Davies, J.T., 1957. A Quantitative Kinetic Theory of Emulsion Type, I. Physical Chemistry of the Emulsifying Agent. Butterworths, London.

De Gennes, P.G., Taupin, C., 1982. Microemulsions and the flexibility of oil/water interfaces. J. Phys. Chem. 86 (13), 2294–2304.

Delmas, T., et al., 2011. How to prepare and stabilize very small nanoemulsions. Langmuir 27 (5), 1683–1692.

Dickinson, E., 1981. Interpretation of emulsion phase inversion as a cusp catastrophe. J. Colloid Interface Sci. 84 (1), 284–287.

Dickinson, E., 1982. Thermodynamic aspects of emulsion phase inversion. J. Colloid Interface Sci. 87 (2), 416–423.

Dunstan, T.S., Fletcher, P.D.I., Mashinchi, S., 2011. High internal phase emulsions: catastrophic phase inversion, stability, and triggered destabilization. Langmuir 28 (1), 339–349.

Fernandez, P., et al., 2004. Nano-emulsion formation by emulsion phase inversion. Colloids Surf. A Physicochem. Eng. Asp. 251 (1), 53–58.

Fernandez-Rodriguez, M.A., et al., 2017. Particles adsorbed at various non-aqueous liquid-liquid interfaces. Adv. Colloid Interf. Sci. 247, 208–222.

Finkle, P., Draper, H.D., Hildebrand, J.H., 1923. The theory of emulsification. J. Am. Chem. Soc. 45 (12), 2780–2788.

Forgiarini, A., et al., 2001. Formation of nano-emulsions by low-energy emulsification methods at constant temperature. Langmuir 17 (7), 2076–2083.

Friberg, S.E., Corkery, R.W., Blute, I.A., 2011. Phase inversion temperature (PIT) emulsification process. J. Chem. Eng. Data 56 (12), 4282–4290.

Fryd, M.M., Mason, T.G., 2012. Advanced nanoemulsions. Annu. Rev. Phys. Chem. 63, 493–518.

Galindo-Alvarez, J., et al., 2011. Viscous oil emulsification by catastrophic phase inversion: influence of oil viscosity and process conditions. Ind. Eng. Chem. Res. 50 (9), 5575–5583.

Garbin, V., 2013. Colloidal particles: surfactants with a difference. Phys. Today 66 (10), 68.

Garti, N., Libster, D., Aserin, A., 2012. Lipid polymorphism in lyotropic liquid crystals for triggered release of bioactives. Food Funct. 3 (7), 700–713.

Glaser, N., et al., 2006. Janus particles at liquid-liquid interfaces. Langmuir 22 (12), 5227–5229.

Gradzielski, M., Hoffmann, H., 1999. Rheological Properties of Microemulsions. Marcel Dekker, New York.

Griffin, W.C., 1946. Classification of surface-active agents by "HLB." J. Soc. Cosmet. Chem. 1, 311–326.

Gupta, A., et al., 2017. A general route for nanoemulsion synthesis using low energy methods at constant temperature. Langmuir 33, 7118–7123.

Gutiérrez, J., et al., 2008. Nano-emulsions: new applications and optimization of their preparation. Curr. Opin. Colloid Interface Sci. 13 (4), 245–251.

Heunemann, P., et al., 2011. Formation and structure of slightly anionically charged nanoemulsions obtained by the phase inversion concentration (PIC) method. Soft Matter 7 (12), 5697–5710.

Hickey, R.J., et al., 2016. Structure, viscoelasticity, and interfacial dynamics of a model polymeric bicontinuous microemulsion. Soft Matter 12 (1), 53–66.

Hoffmann, I., et al., 2016. Kinetics of oil exchange in nanoemulsions prepared with the phase inversion concentration method. Langmuir 32 (46), 12084–12090.

Israelachvili, J., 2011. Intermolecular and Surface Forces, Revised third ed. Academic Press, Boston, p. 704.

Jafari, S.M., Assadpoor, E., He, Y., Bhandari, B., 2008. Re-coalescence of emulsion droplets during high-energy emulsification. Food Hydrocoll. 22 (7), 1191–1202.

Jafari, S.M., Paximada, P., Mandala, I., Assadpour, E., Mehrnia, M.A., 2017. 2—Encapsulation by nanoemulsions. Nanoencapsulation Technologies for the Food and Nutraceutical Industries. Academic Press, New York, pp. 36–73.

Jahanzad, F., et al., 2009. Catastrophic phase inversion via formation of multiple emulsions: a prerequisite for formation of fine emulsions. Chem. Eng. Res. Des. 87 (4), 492–498.

Jakobs, B., et al., 1999. Amphiphilic block copolymers as efficiency boosters for microemulsions. Langmuir 15 (20), 6707–6711.

Katepalli, H., et al., 2017. Microstructure and rheology of particle stabilized emulsions: effects of particle shape and inter-particle interactions. J. Colloid Interface Sci. 485, 11–17.

Komaiko, J., McClements, D.J., 2014. Optimization of isothermal low-energy nanoemulsion formation: hydrocarbon oil, non-ionic surfactant, and water systems. J. Colloid Interface Sci. 425, 59–66.

Komaiko, J.S., McClements, D.J., 2016. Formation of food-grade nanoemulsions using low-energy preparation methods: a review of available methods. Compr. Rev. Food Sci. Food Saf. 15 (2), 331–352.

Kotta, S., et al., 2015. Formulation of nanoemulsion: a comparison between phase inversion composition method and high-pressure homogenization method. Drug Deliv. 22 (4), 455–466.

Kralchevsky, P.A., et al., 2005. On the thermodynamics of particle-stabilized emulsions: curvature effects and catastrophic phase inversion. Langmuir 21 (1), 50–63.

Krishnan, K., et al., 2001. Shear-induced nano-macro structural transition in a polymeric bicontinuous microemulsion. Phys. Rev. Lett. 87(9).

Kumar, A., et al., 2013. Amphiphilic Janus particles at fluid interfaces. Soft Matter 9 (29), 6604–6617.

Kumar, A., et al., 2015. Recent developments in phase inversion emulsification. Ind. Eng. Chem. Res. 54 (34), 8375–8396.

Kumar, A., et al., 2016. Flow-induced phase inversion of emulsions in tapered microchannels. Lab Chip 16 (21), 4173–4180.

Lee, M.N., Chan, H.K., Mohraz, A., 2011. Characteristics of pickering emulsion gels formed by droplet bridging. Langmuir 28 (6), 3085–3091.

Liu, Y., et al., 2012. An investigation into the relationship between catastrophic inversion and emulsion phase behaviors. Colloids Surf. A Physicochem. Eng. Asp. 399, 25–34.

López-Barrón, C.R., Macosko, C.W., 2010. Direct measurement of interface anisotropy of bicontinuous structures via 3D image analysis. Langmuir 26 (17), 14284–14293.

Lv, G., et al., 2014. Characterization of the emulsions formed by catastrophic phase inversion. Colloids Surf. A Physicochem. Eng. Asp. 450, 141–147.

Madivala, B., et al., 2009. Exploiting particle shape in solid stabilized emulsions. Soft Matter 5 (8), 1717–1727.

Mayer, S., Weiss, J., McClements, D.J., 2013. Vitamin E-enriched nanoemulsions formed by emulsion phase inversion: factors influencing droplet size and stability. J. Colloid Interface Sci. 402, 122–130.

McClements, D.J., 2011. Edible nanoemulsions: fabrication, properties, and functional performance. Soft Matter 7 (6), 2297–2316.

Mercuri, A., et al., 2011. The effect of composition and gastric conditions on the self-emulsification process of ibuprofen-loaded self-emulsifying drug delivery systems: a microscopic and dynamic gastric model study. Pharm. Res. 28 (7), 1540–1551.

Miller, R., et al., 2006. Composite interfacial layers containing micro-size and nano-size particles. Adv. Colloid Interf. Sci. 128, 17–26.

Mira, I., et al., 2003. Emulsion catastrophic inversion from abnormal to normal morphology. 2. Effect of the stirring intensity on the dynamic inversion frontier. Ind. Eng. Chem. Res. 42 (1), 57–61.

Negro, E., Latsuzbaia, R., Koper, G.J., 2014. Bicontinuous microemulsions for high yield wet synthesis of ultrafine platinum nanoparticles: effect of precursors and kinetics. Langmuir 30 (28), 8300–8307.

Ostertag, F., Weiss, J., McClements, D.J., 2012. Low-energy formation of edible nanoemulsions: factors influencing droplet size produced by emulsion phase inversion. J. Colloid Interface Sci. 388 (1), 95–102.

Perazzo, A., Preziosi, V., Guido, S., 2015. Phase inversion emulsification: current understanding and applications. Adv. Colloid Interf. Sci. 222, 581–599.

Pickering, S.U., 1907. CXCVI.-Emulsions. J. Chem. Soc. Trans. 91 (0), 2001–2021.

Posocco, P., et al., 2016. Interfacial tension of oil/water emulsions with mixed non-ionic surfactants: comparison between experiments and molecular simulations. RSC Adv. 6 (6), 4723–4729.

Preziosi, V., et al., 2013. Phase inversion emulsification. Chem. Eng. Trans. 32, 5.

Preziosi, V., et al., 2017a. Flow-induced nanostructuring of gelled emulsions. Soft Matter 13, 5696–5703.

Preziosi, V., et al., 2017b. Monitoring emulsion microstructure by organic electrochemical transistors. J. Mater. Chem. C 5, 2056–2065.

Queste, S., et al., 2007. The EACN scale for oil classification revisited thanks to fish diagrams. J. Colloid Interface Sci. 312 (1), 98–107.

Ramsden, W., 1903. Separation of solids in the surface-layers of solutions and suspensions(observations on surface-membranes, bubbles, emulsions, and mechanical coagulation)—preliminary account. Proc. R. Soc. Lond. 72, 156–164.

Rao, J., McClements, D.J., 2010. Stabilization of phase inversion temperature nanoemulsions by surfactant displacement. J. Agric. Food Chem. 58 (11), 7059–7066.

Roger, K., 2016. Nanoemulsification in the vicinity of phase inversion: disruption of bicontinuous structures in oil/surfactant/water systems. Curr. Opin. Colloid Interface Sci. 25, 120–128.

Roger, K., Cabane, B., Olsson, U., 2010a. Formation of 10–100 nm size-controlled emulsions through a sub-PIT cycle. Langmuir 26 (6), 3860–3867.

Roger, K., Cabane, B., Olsson, U., 2010b. Emulsification through surfactant hydration: the PIC process revisited. Langmuir 27 (2), 604–611.

Roger, K., et al., 2011. Superswollen microemulsions stabilized by shear and trapped by a temperature quench. Langmuir 27 (17), 10447–10454.

Sajjadi, S., 2006. Nanoemulsion formation by phase inversion emulsification: on the nature of inversion. Langmuir 22 (13), 5597–5603.

Salager, J.L., et al., 2004. Using emulsion inversion in industrial processes. Adv. Colloid Interf. Sci. 108–109, 259–272.

Salager, J.-L., et al., 2000. Current phenomenological know-how and modeling of emulsion inversion. Ind. Eng. Chem. Res. 39 (8), 2665–2676.

Salim, M., et al., 2014. Amphiphilic designer nano-carriers for controlled release: from drug delivery to diagnostics. Med. Chem. Commun. 5 (11), 1602–1618.

Shinoda, K., Saito, H., 1968. The effect of temperature on the phase equilibria and the types of dispersions of the ternary system composed of water, cyclohexane, and nonionic surfactant. J. Colloid Interface Sci. 26 (1), 70–74.

Solans, C., Solé, I., 2012. Nano-emulsions: formation by low-energy methods. Curr. Opin. Colloid Interface Sci. 17 (5), 246–254.

Solè, I., et al., 2010. Nano-emulsions prepared by the phase inversion composition method: preparation variables and scale up. J. Colloid Interface Sci. 344 (2), 417–423.

Strey, R., 1996. Phase behavior and interfacial curvature in water-oil-surfactant systems. Curr. Opin. Colloid Interface Sci. 1 (3), 402–410.

Tyrode, E., et al., 2003. Emulsion catastrophic inversion from abnormal to normal morphology. 3. Conditions for triggering the dynamic inversion and application to industrial processes. Ind. Eng. Chem. Res. 42 (19), 4311–4318.

Vitiello, G., et al., 2014. Phase behavior of the ternary aqueous mixtures of two polydisperse ethoxylated nonionic surfactants. Colloids Surf. A Physicochem. Eng. Asp. 442, 16–24.

Walther, A., Müller, A.H., 2008. Janus particles. Soft Matter 4 (4), 663–668.

Walther, A., Müller, A.H., 2013. Janus particles: synthesis, self-assembly, physical properties, and applications. Chem. Rev. 113 (7), 5194–5261.

Wang, L., et al., 2008. Nanoemulsions prepared by a two-step low-energy process. Langmuir 24 (12), 6092–6099.

Wennerström, H., Lindman, B., 1979. Micelles. Physical chemistry of surfactant association. Phys. Rep. 52 (1), 1–86.

Winsor, P.A., 1948. Hydrotropy, solubilisation and related emulsification processes. Trans. Faraday Soc. 44 (0), 376–398.

Wolf, L., et al., 2010. Cryo-TEM imaging of a novel microemulsion system of silicone oil with an anionic/nonionic surfactant mixture. Soft Matter 6 (21), 5367–5374.

Wooster, T.J., Golding, M., Sanguansri, P., 2008. Impact of oil type on nanoemulsion formation and Ostwald ripening stability. Langmuir 24 (22), 12758–12765.

Yeo, L.Y., et al., 2000. Phase inversion and associated phenomena. Multiph. Sci. Technol. 12 (1), 66.

Zambrano, N., et al., 2003. Emulsion catastrophic inversion from abnormal to normal morphology. 1. Effect of the water-to-oil ratio rate of change on the dynamic inversion frontier. Ind. Eng. Chem. Res. 42 (1), 50–56.

Chapter 4

Transitional Nanoemulsification Methods

Nicolas Anton, Salman Akram and Thierry F. Vandamme
University of Strasbourg, Strasbourg, France

Chapter Outline

4.1 INTRODUCTION

Nanoemulsions have emerged in the last decade as an important tool in the formulation of pharmaceuticals, for applications as nanomedicines and/or as probes for biomedical imaging. Nanoemulsions offer new possibilities for dispersing lipophilic therapeutics or biomedical imaging agents in aqueous phases, such as isotonic aqueous buffer solutions compatible with parenteral administration (Anton et al., 2008a; Attia et al., 2014; Kilin et al., 2014; Li et al., 2013). These lipophilic active pharmaceutical ingredients (API) can be solubilized at high concentrations in the oily core of the nanoemulsion droplets, thereby leading to a uniform distribution throughout the whole system. Nanoemulsions are generally considered as nontoxic nanocarriers with the same potential as liposomes or polymeric systems, with, of course, different encapsulation properties. The composition and structure of nanoemulsions can be optimized to obtain

Nanoemulsions. https://doi.org/10.1016/B978-0-12-811838-2.00004-7

good encapsulation properties, high physicochemical stability, and targeting to specific sites in vivo. Moreover, another fundamental advantage of nanoemulsions lies in the simplicity and robustness of the formulation methods. This chapter focuses on low-energy transitional emulsification methods that are particularly *simple* to perform and that are *robust* in the sense that a wide range of oil compositions can be utilized, for example, including drugs solubilized, even at high concentration, without impacting significantly on the process as will be discussed below.

Let us focus now on the formulation processes of emulsions. In general, the terminology *emulsion* refers to *macroemulsions*, typically with droplet diameters ranging from a micrometer to several tens of micrometers (Leal-Calderon et al., 2007a). The formulation of macroemulsions is relatively simple, requiring widely used and relatively inexpensive mechanical methods such as rotor-stator devices. However, macroemulsions are usually highly unstable due to physical destabilization mechanisms such as gravitational separation, flocculation, and coalescence (Leal-Calderon et al., 2007b). For this reason, macroemulsions are usually homogenized again using another mechanical method that aims at decreasing the droplet size and polydispersity. The resulting nanoemulsions contain droplets with diameters in the nanometer range, that is, typically from about 30 to 300 nm.

The thermodynamic instability of emulsions and nanoemulsions is due to the fact that the free energy ΔG_f associated with their formation is greater than zero. This free energy change is mainly due to the increase in contact between the two immiscible liquids: $\Delta G_f = \Delta A\gamma$, where ΔA is the increase in interfacial area and γ is the interfacial tension. Thus, the interfacial area has a tendency to spontaneously decrease, thereby favoring droplet flocculation and coalescence. On the other hand, once formed, nanoemulsions tend to have very good kinetic stability because their small droplet size reduces the rate of gravitational separation and droplet aggregation. Nevertheless, nanoemulsions are still prone to droplet growth through a phenomenon known as Ostwald ripening, which is the growth of large droplets at the expense of small droplets due to diffusion of oil molecules through the intervening aqueous phase (Leal-Calderon et al., 2007b; Tadros et al., 2004; Anton et al., 2008b). Nanoemulsions must therefore be carefully formulated to avoid Ostwald ripening, for example, by adding ripening inhibitors or ensuring a narrow particle size distribution (Tadros et al., 2004).

This brief overview on the formulation of emulsions and nanoemulsions highlights the potential advantages of emulsification methods that produce small droplets with narrow distributions. In addition, emulsification efficiency depends on the composition, chemical nature of the phases and stabilizing agents, viscosity ratios of the phases, temperature and time of processing, shear rate, cooling time, type of emulsification apparatus, and energy supplied. Rotor/stator apparatuses provide high shearing along with an efficient recirculation of the liquid, thereby giving rise to a premix emulsion between 10 and 100 μm. They induce a strong shearing that favors the breaking up of the drops, but do not permit decreasing the droplet size below about 1 μm, due to the massive dissipation of the mechanical energy in the form of heat. Specially designed mechanical devices capable of supplying huge amounts of energy are necessary to increase ΔA and decrease the droplet size despite heat dissipation. These methods, so-called high-energy methods, are performed by the homogenization of the premix mainly not only with high-pressure homogenization but also with ultrasound-based methods.

These high-energy methods have a number of advantages that have led them to be widely used in industrial processes, such as their versatility and high throughput. However, they also have a number of limitations of high-energy methods: (i) The process typically involves several steps, including dispersion, premixing, and homogenization; (ii) there are high-energy needs; (iii) there is a potential risk of degradation or denaturation of fragile molecules in the formulation; and (iv) there are high costs associated with purchasing, maintaining, and running the equipment. Consequently, there has been interest in alternative emulsification methods, such as low-energy methods, that allow the formulation of very small and monodisperse nanoemulsions without the need for any specialized equipment. These low-energy methods have emerged as highly robust and adaptive emulsification methods that have many potential applications. For example, they can be used to produce nanomedicines that are nontoxic and stable carriers for different sorts of lipophilic active ingredients at the same time.

The purpose of the present chapter is to present an overview of these low-energy methods, mainly related to the concept of *transitional methods*. The term "*transitional*" comes from the *transitional* emulsion phase inversion, a phenomenon traditionally induced by a change in temperature. These systems were first reported in 1969 by Shinoda and Saito (1969). In Section 4.2, a discussion of the underlying principles of the transitional emulsification method is given, with an emphasis on the thermodynamic properties of nonionic surfactants, which play a key role in these processes. In Section 4.3, we focus on the consequences of the transitional phenomena applied to low-energy emulsification. Different low-energy emulsification methods are reviewed, and the impact of nonionic surfactant phase behavior on the process is stressed, including a discussion of the role of ternary phase diagrams of the water/nonionic surfactant/oil system. In addition, differences between transitional nanoemulsions and microemulsions are highlighted, from the formation and formulation processes to the thermodynamic and physicochemical properties of these dispersions. Section 4.4 then deals with the robustness of the transitional nanoemulsification processes, due to the modification of the composition related to the encapsulation of active ingredients in the droplet's core.

4.2 THE ROLE OF PEGYLATED NONIONIC SURFACTANTS ON TRANSITIONAL EMULSIFICATION METHODS

Low-energy emulsification methods are almost exclusively based on the use of surfactants for which the polar head is a poly(ethylene glycol) chain (PEG and

PEGylated surfactants), which are a particular form of nonionic surfactant. Representative examples of nonionic surfactants typically used in the formation of nanoemulsions by low-energy emulsification are presented in Fig. 4.1: (A) PEG-4 dodecyl ether (Rao and McClements, 2010; Astaraki, 2016; Izquierdo et al., 2004), with a total number of ethylene glycols of 4 and a molecular weight of the PEG moiety of 176 g/mol (Brij 30); (B) PEG-20 sorbitan monooleate (Mei et al., 2011; Rao and McClements, 2011), with 20 ethylene glycol units and a molecular weight of the PEG moiety of 880 g/mol (Tween 80); (C) PEG-15 stearate (Klassen et al., 2014), with 15 ethylene glycol units and a molecular weight of the PEG moiety of 660 g/mol (Kolliphor HS15); and (D) PEG-35 ricinoleate (Attia et al., 2014; Li et al., 2013; Anton and Vandamme, 2009), with 35 ethylene glycol units and a molecular weight of the PEG moiety of 1540 g/mol (Kolliphor ELP).

These surfactants all have a relatively high total number of ethylene glycol units, or PEG chain length, with values in a comparable size range (15–35 per lipid chain). Moreover, these surfactants all have hydrophilic-lipophilic balance (HLB) values around 13–15, which is related to the relative sizes of the polar

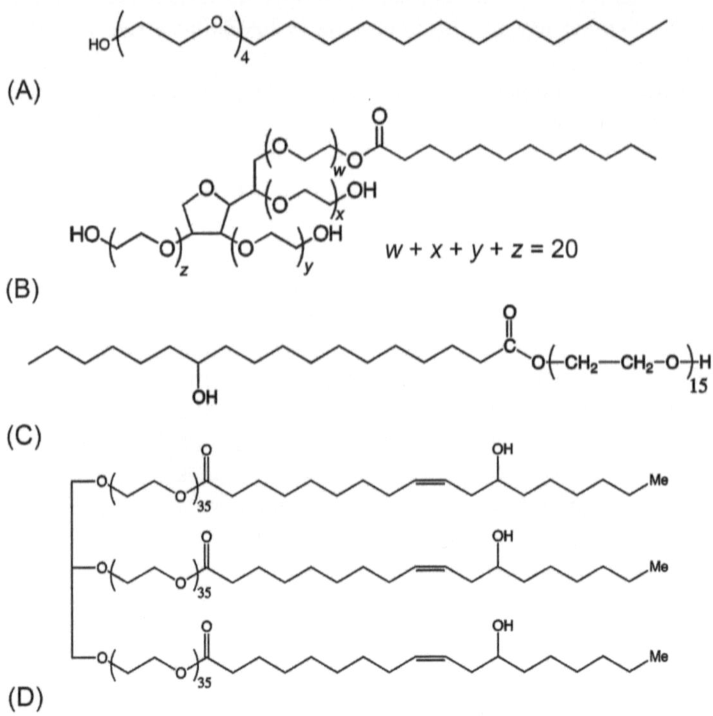

FIG. 4.1 Some representative nonionic surfactant used in transitional emulsification: (A) Brij 30 (PEG-4 dodecyl ether), (B) Tween 80 (PEG-20 sorbitan monooleate), (C) Kolliphor HS15 (PEG-660 stearate), and (D) Kolliphor ELP (PEG-35 ricinoleate).

head and nonpolar chain. Shorter or much longer PEG chains will induce a loss in water solubility or surface activity, respectively. In general, the solubility of PEGylated surfactants is induced not only by interactions between the PEG chain and the bulk water through the formation of hydrogen bounds (Goldstein, 1984; Wartewig et al., 1990) but also by the conformation of the PEG chain that can create a localized dipole moment (in the cis conformation), which enhances the dipole-dipole interactions with the bulk water (Karlstrom, 1985) and thus solubility. Another phenomenon that tends to reduce PEG solubility comes from the structuration of water molecules associated by hydrogen bonds into flickering clusters (so-called linked water) that wraps the PEGylated polar heads. The presence of flickering clusters has been shown to exhibit a buffer or shielding effect that increases with water structuration, that is, with cluster size. In fact, the hydration state of the PEG chains is an important parameter that can be considered as an intrinsic property of the nonionic surfactant, which is strongly related to the thermodynamic environment and composition of the aqueous phase in which it is solubilized. Indeed, temperature, surfactant concentration (Schottand and Han, 1976a), and electrolyte concentration in water induce salting-in or salting-out effects of PEGylated surfactants (Schottand and Han, 1976a; Schick, 1962a; Maclay, 1956; Schott, 1973, 1997, 2001; Schott and Han, 1975; Schottand and Royce, 1984; Schott et al., 1984), thereby shifting the equilibrium $n\mathrm{H_2O} \rightleftarrows (\mathrm{H_2O})_n$ toward the left or the right, respectively, in function of the type of ion. Consequently, the size of the flickering clusters affects the phase diagrams and macroscopic properties of nonionic surfactants, such as the critical micelle concentration (CMC) (Hsiao et al., 1956; Schick, 1962b, 1964; Schottand and Han, 1976b; Rayand and Nemethy, 1971; Malikand and Jhamb, 1970; Malikand and Saleem, 1968; Arai, 1967) and cloud point (cp) (Schick, 1962b; Arai, 1967; Dorenand and Goldfarb, 1970; Shinodaand and Takeda, 1970; Gu et al., 1989).

To summarize, the solubility of nonionic surfactants is primarily determined by interactions between the PEG groups and water and is modified by the size of the flickering clusters. Thus, water solubility is highly dependent on temperature, strongly decreasing with heating because the thermal excitation of the water molecules reduces their interactions with the ethylene oxide groups. This phenomenon is at the origin of the cloud point and nonionic surfactant demixing phenomena (giving rise to a surfactant-rich phase in equilibrium with an aqueous one with surfactant concentration close to the CMC) and at the center of the emulsion phase inversion phenomenon.

A large variety of PEGylated surfactants have been used in phase inversion temperature (PIT), spontaneous emulsification, and self-emulsifying drug delivery system methods, but their global behavior and their role in these emulsification processes follow a universal mechanism. However, if the use of nonionic surfactants is usually necessary to obtain low-energy emulsification, the choice of oil phase is also very important for the success of the process. Indeed, as discussed later, the transitional emulsification methods used to produce nanoemulsions are based on changes in the relative interactions of PEGylated surfactants with the oil and water phases and particularly on the possibility of modifying these interactions using thermodynamic variables. To illustrate these considerations, it is interesting to consider the general phase diagram, reported in Fig. 4.2, of nonionic surfactants with respect to water and oil phases.

At constant composition, it appears that the nonionic surfactants are not fully soluble in the aqueous and oil phases and that this solubility behavior is linked to temperature. Electrolyte concentration influences the location of the phase frontiers (see earlier), while the temperature has a direct incidence on the solubility of the surfactants. The phase diagram shows two demixing *two-phase* regions: (i) for the binary system surfactant/water for which a key parameter is the cloud point in water (cp_β) and (ii) in oil for the binary system surfactant/oil for which a key parameter is the cloud point in oil (cp_α). The phase behavior of nonionic surfactant in water has a crucial role and is principally at the origin of low-energy emulsification processes. More precisely, emulsification comes from the sudden change of the nonionic surfactant solubility, rapidly crossing the boundary of the phase diagram, from the two-phase region to the one-phase region. Let us consider the physicochemical behavior adopted by the nonionic surfactants during the low-energy emulsification or transitional processes, through the binary phase diagram surfactant/water in Fig. 4.2. As an example, point A corresponds to the thermodynamic conditions and composition in which the surfactants are before emulsification, and points B, C, and D correspond to possible points compatible with the emulsification process. Indeed, the process is driven by the transition from a state where the nonionic surfactants are nonsoluble in water to a state where they are fully soluble. Emulsification is therefore possibly performed by a number of routes: (i) *pathway A-B*, which involves dilution with water at constant temperature;

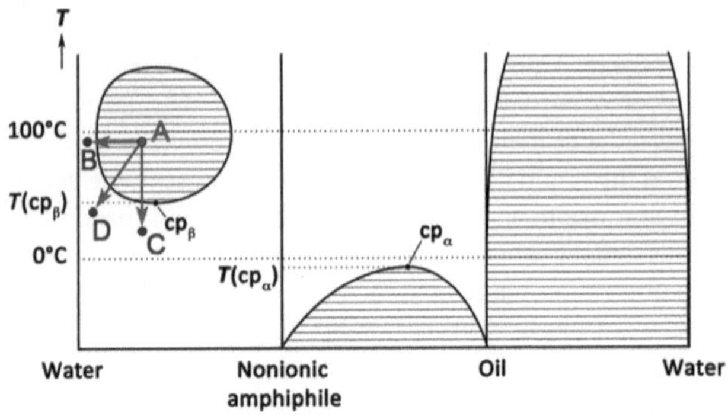

FIG. 4.2 General binary phase diagrams between water, nonionic surfactant, and oil. The cloud points are represented in water cp_β and in oil cp_α. The signification of the pathways A-B, A-C, and A-D is discussed in the text. *(Reproduced with permission from Kahlweit, M., Strey, R., Firman, P., Haase, D., Jen, J., Schomaecker, R., 1988. General patterns of the phase behavior of mixtures of water, nonpolar solvents, amphiphiles, and electrolytes. Langmuir 4, 499–511; Anton, N., Vandamme, T.F., 2011. Nanoemulsions and micro-emulsions: clarifications of the critical differences. Pharm. Res. 28, 978–985.)*

(ii) *pathway A-C*, which corresponds to sudden cooling of the system at constant composition; and (iii) *pathway A-D*, which combines both dilution and cooling (e.g., adding cold water). Pathway A-C occurs at constant composition and may therefore be reversible process, while the other two pathways involve changes in composition and are therefore irreversible. Traditionally, the PIT method was considered to be the only transitional method for producing nanoemulsions. However, the term *transitional* more generally refers to the change in solubility of the nonionic surfactants whatever the pathway (e.g., A-B, A-C, or A-D) and can therefore be used more broadly.

In addition, Fig. 4.2 shows that the nonionic surfactant seems freely soluble in the oil phase whatever the temperature, but below cp_β, the surfactant shows good affinity for both the water and oil phases. This temperature-dependent change in relative solubility of surfactants around cp_β plays a major role in the emulsification process.

4.3 TRANSITIONAL EMULSIFICATION METHODS, EMULSION PHASE INVERSION, SPONTANEOUS EMULSIFICATION, AND UNIVERSALITY OF THE PROCESS

Let us focus now on the fabrication of nanoemulsions using low-energy emulsification methods based on these transitional phenomena. The formation of nanoemulsions by these methods is a dynamic and irreversible process, thereby establishing a direct relationship between ternary phase diagrams (at equilibrium) and emulsification processes (dynamic) that is only justified to emphasize the physicochemical interactions (and their modification with temperature and composition) between the nonionic surfactant and the water and oil phases. Traditionally, the two most important transitional emulsification methods are the PIT and *spontaneous emulsification* methods. In the following sections, the main principles driving these processes are presented, and it is shown that they are based on a universal mechanism related to the behavior of nonionic surfactants.

4.3.1 PIT Method

The particular feature of the PIT method is the presence of all the components (oil, water, and surfactant) from the start of the process. In the PIT method, the ternary system is constantly and mechanically homogenized (e.g., under gentle magnetic stirring), and the temperature is monitored and gradually changed to force the nonionic surfactant to cross the boundary shown in Fig. 4.2. Under these dynamic conditions, the mechanical energy supplied constantly creates new droplets and new interfaces that counterbalance any droplet coalescence that occurs. The droplets formed are stabilized by the nonionic surfactant, and for an equilibrated water-to-oil ratio of around 50 wt%, the type of the emulsion formed (O/W or W/O) is driven by the affinity of the nonionic surfactant for the water or oil phases, respectively. The general principle of the PIT method is summarized in Fig. 4.3, which shows the link between the binary

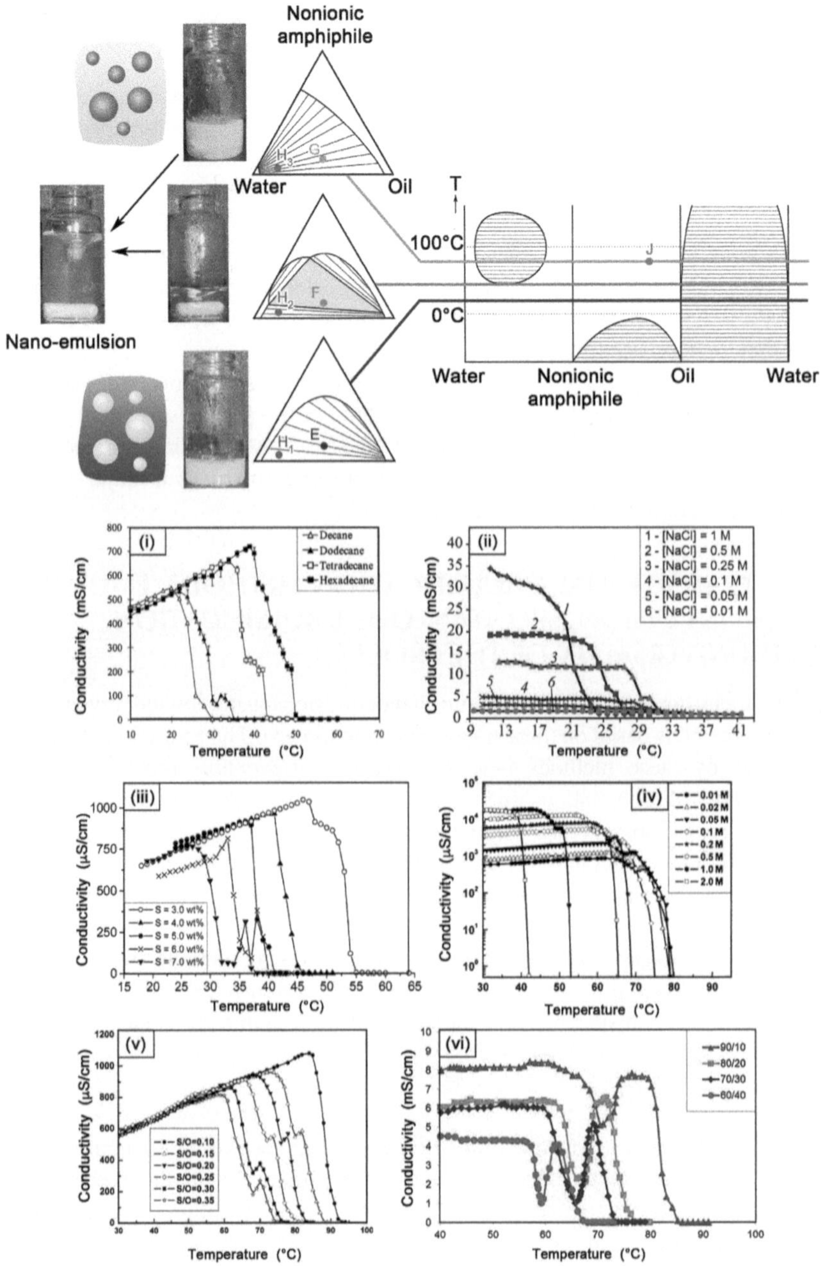

FIG. 4.3 (Top) Schematic presentation of the phase behavior of ternary systems (water/nonionic surfactant/oil) as a function of temperature, below (*lower line*, point E), equal (*middle line*, point F), and above (*upper line*, point G) the phase inversion temperature. The pictures show the different states of the corresponding emulsion during phase inversion (made with MilliQ water/Kolliphor HS15/Labrafac WL) and the resulting nanoemulsion (on the left, point $H_{1,\ 2,\ or\ 3}$). The graphs from (i) to (vi) show emulsion phase inverting with Brij 30 as surfactant in (i) (Rao and McClements, 2010), (ii) (Astaraki, 2016), and (iii) (Izquierdo et al., 2004); Tween 80 in (iv) and (v) (Mei et al., 2011); and Kolliphor HS15/Igepal CO-210 in (iv) (Klassen et al., 2014). The different examples show the influence of (i) the nature of the oil, (ii) and (iv) electrolyte concentration, (iii) and (v) surfactant concentration, and (vi) different nonionic surfactant composition (see details in the text).

diagrams discussed earlier and the ternary diagrams corresponding to the three-component systems studied. Three different characteristic temperatures are emphasized, below (lower curve), equal to (middle curve), and above (upper curve) the cloud point cp_β. A model ternary system (45 wt% water, 45 wt% oil, and 10 wt% nonionic surfactant) is represented as a function of temperature with E, F, and G, along with the pictures showing the nature of the different dispersions.

Ternary phase diagrams clearly show the change of affinities of the surfactants for the water and oil phases through the orientation of the phase lines, as a function of temperature. At *point E*, the ternary diagram corresponds to a Winsor I system, the surfactant shows a better solubility toward the aqueous phase, and the emulsion formed is the oil-in-water (O/W) one; at *point G*, a Winsor II is formed, in which the surfactant is mostly soluble in the oil phase, and the emulsion formed is a water-in-oil (W/O) one; finally, *point F* corresponds to the intermediate configuration, Winsor III, where the nonionic surfactants present similar affinities for the two phases forming a translucent microemulsion (see the picture). The temperature marking the frontier between O/W and W/O emulsions is called PIT, which is intrinsically close to the cloud point (cp_β) but possibly shifted depending on the nature of the oil (for which the surfactant solubility in oil can vary).

Beyond the surfactant behavior and phase diagram, let us consider the emulsification method based on the phase inversion process, for which the determination of the PIT location is crucial. The characterization of the PIT is performed by monitoring the electric conductivity χ of the emulsion, for which an electrolyte is added in water (keeping in mind that the electrolyte can also influence the PIT). This allows the characterization of the emulsion phase inversion and characterization of the parameters that can influence its value. This method is based on the fact that the electric conductivity of the system is much higher when water is the continuous phase (χ of the order of millisiemens per centimeter) than when oil is the continuous phase (χ close to zero). With increasing temperature, the conductivity decreases continuously in the phase inversion region, with conductivity peaks sometimes being observed (attributed to mesostructures and liquid crystals (1)). Overall, the PIT is taken to be located at the middle of the transitional region. Representative examples are reported in Fig. 4.3 (i)–(vi).

The graphs (i), (ii), and (iii) show the emulsion phase inversion obtained with Brij 30 (Fig. 4.1A) and the potential impact of the formulation and composition parameters on the PIT location. In graph (i), the nature of oil is changed by varying the length of the saturated aliphatic chains (C_nH_{2n+2} with $n = 10$, 12, 14, or 16). The PIT clearly increases as the oil chain length increases, showing a loss of Brij 30/oil affinity when the oil phase becomes more lipophilic. In graph (ii), the electrolyte concentration in water is decreased from 1 to 0.01 M, and as a result, the water solubility of the surfactant is increased (due to the decrease in the size of the flickering cluster of water molecules around the polar head group; see Section 4.2) increasing the PIT. Also for Brij 30, in graph (iii), the total surfactant concentration is increased, which decreases their global solubility (as the water volume is constant) and thereby decreases the PIT. This comparison is interesting to see the potential influence of these formulation parameters on the PIT value, but the PIT shifts remain relatively limited, with the location for Brij 30 remaining around 30–50°C. However, for different surfactants with longer PEG chains (e.g., 20 units for Tween 80), the PIT can be up to 80°C, and the effects of NaCl concentration in the water phase (graph (iv)) or of the total surfactant concentration (graph (v)) are much more important. The last example with Kolliphor HS15/Igepal CO-210 (graph (vi)) highlights this observation; the figure shows emulsion phase inversion with a mixture of different surfactants used at different ratios. Kolliphor HS15 is largely hydrophilic, and Igepal CO-210 is largely lipophilic with 15 and 2 ethylene glycol units, respectively. This difference has a major impact on the PIT location: the higher the Kolliphor HS15 proportion, the higher the PIT value. It is noteworthy that other approaches can be used to monitor the emulsion phase inversion process, such as turbidity measurements that are based on a change in the optical properties resulting from alterations in the dimensions of the particles in O/W emulsions, microemulsions, and W/O emulsions (Rao and McClements, 2011; McClements, 2011).

The characterization of the phase diagram and of the phase inversion process is important for the success and optimization of the emulsification using the transitional approach. However, the formation of nanoemulsions is an irreversible process that involves a specific mechanism (Anton and Vandamme, 2009). The principle is summarized in Fig. 4.4A in which the emulsion and location of the nonionic surfactants are schematically represented during the emulsion

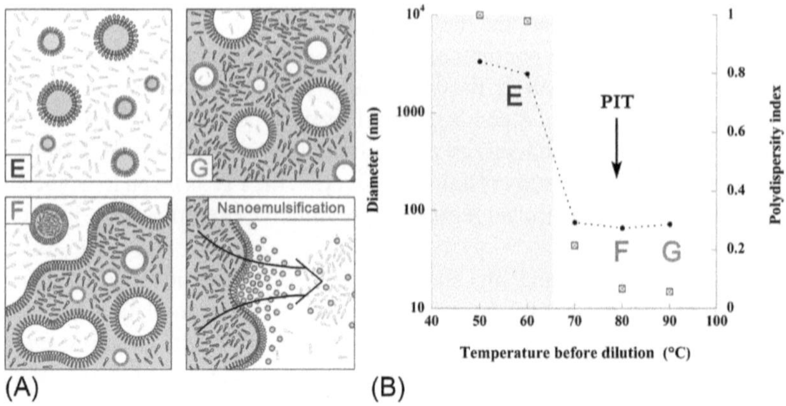

FIG. 4.4 (A) Schematic illustration of the formation of nanoemulsions by the PIT method. The different points E, F, and G correspond to the compositions shown in Fig. 4.3, and the nanoemulsification to the process is described in the text after dilution or rapid cooling of the system. (B) Diameter of the nanoemulsion formed after dilution of the ternary system at the temperatures indicated in the abscissa. *(Reproduced with permission from Anton, N., Vandamme, T.F., 2009. The universality of low-energy nano-emulsification. Int. J. Pharm. 377, 142–147.)*

phase inversion process and linked to the compositions indicated in the relevant phase diagrams (Fig. 4.3). At point E, the majority of nonionic surfactant is solubilized in the aqueous phase, thereby stabilizing O/W emulsions; however, at point G, the surfactants are primarily solubilized in the oil, thereby stabilizing W/O emulsions. Within the PIT zone, at point F, there is a partition of surfactant between both phases, leading to the formation of a nanoscale microemulsion, leading to a transparent appearance of the overall system. If we draw a parallel with the composition pathways presented in Fig. 4.2 (A-B, A-C, or A-D) passing from a lipophilic to a hydrophilic state, herein (see Fig. 4.3), the correspondence is G-H_3, G-E, and G-H_1, respectively. The general pathways gather the dilution and/or cooling (Rao and McClements, 2011; Anton and Vandamme, 2009) and thus can be extended to G-H_3, G-H_2, G-H_1, G-E, F-H_2, and F-H_1. The irreversible emulsification process comes from the rapid change of the water solubility of the nonionic surfactants from a state where they are fully (point G) or partially (point F) located in the oil to a state where they are primarily solubilized in water (points E and $H_{1,2,3}$). As a result, the water phase penetrates very fast into the oil/surfactant mixture, breaking up the oil into the form of nanosized particles, following a spinodal decomposition. The nanodroplets formed are then immediately stabilized by the surfactants present in the water phase, thereby forming a stable dispersion of nanoemulsions (see Fig. 4.4A).

The impact of initial temperature before the nanoemulsification step is shown in Fig. 4.4B (Anton and Vandamme, 2009). The system selected is Kolliphor HS15/Labrafil M1944CS, with a surfactant-to-oil weight ratio (SOR) fixed at 30% and a water-to-oil weight ratio (WOR) fixed at 70%, with a PIT around 80°C. The nanoemulsification process is performed with water dilution at room temperature (a different experiment for each point). The results show that the nanoemulsion droplet size and polydispersity drastically decrease when the system temperature approaches the PIT and then stabilizes above the PIT. Therefore, this corroborates the mechanism proposed and emphasizes the crucial importance of this parameter in the transitional nanoemulsification process.

The last remark on the PIT nanoemulsification method will concern the size of the droplets formed in the nanoemulsion. As seen in Fig. 4.4B, at $T \geq$ PIT, the droplet diameter does not depend strongly on temperature, and so, this parameter does not strongly impact the final size of the droplets in the nanoemulsions formed. However, one fundamental parameter is the concentration of nonionic surfactants used, as shown in Fig. 4.5, for increasing concentrations of Tween 80 (increasing surfactant-to-oil ratio) that the mean size of droplets in the nanoemulsions gradually decreases. Fig. 4.5A shows the size distribution giving an image of the polydispersity, and Fig. 4.5B only reports the mean size with the surfactant concentration. It appears that the value of the mean size is strongly impacted by the surfactant concentration used, from around 95 nm to around 45 nm, and from the shape of the distribution, which becomes more monodisperse. This was expected since the nanoemulsification mechanism is directly

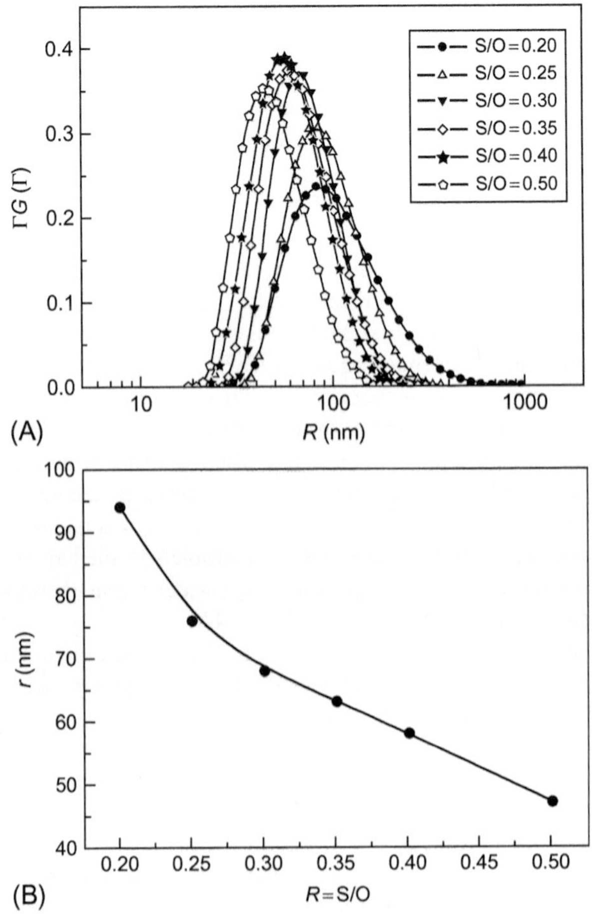

FIG. 4.5 With the ternary system (paraffin oil/water/nonionic surfactant (a mixture of Tween 80 and Span 80)). (A) Size distribution and (B) mean radius, as a function of the surfactant-to-oil ratio (S/O) after nanoemulsification by the PIT method. *(Reproduced with permission from Mei, Z., Xu, J., Sun, D., 2011. O/W nano-emulsions with tunable PIT induced by inorganic salts. Colloids Surf. A Physicochem. Eng. Asp. 375, 102–108.)*

linked to the behavior and displacement of the nonionic surfactants: the higher the nonionic surfactant concentration, the higher the disorder induced in the oil phase during emulsification and the smaller the droplets produced.

4.3.2 Spontaneous Emulsification and the Universality of Transitional Emulsification

The term spontaneous emulsification comes from the simplicity of the emulsification process: the nanoemulsions are generated only by gently mixing two liquid phases. This method is in general considered and classified as a different method than the PIT emulsification method, but in fact, they are very close and based on the same mechanism. As a consequence, compared with the PIT method, the spontaneous emulsification could appear for the formulator as a preferred methodology for forming emulsions by giving the same result with a much higher simplicity. On the other hand, to understand the similarities between these two approaches, it is important to understand the mechanisms driving the PIT method and the transitional processes described earlier.

As illustrated in Fig. 4.6A and B point J (and in Fig. 4.3) and largely supported by the literature (Anton and Vandamme, 2009, 2011; McClements, 2012; Lee et al., 2014), the starting point of the process is a simple oil/surfactant system, set at a temperature for which the surfactant has the best affinity for the oil phase, that is, above the cloud point or PIT. Then, the mixture J is suddenly mixed with the aqueous phase, at temperature lower than PIT (toward H_1) for the best efficiency, but in some cases, the nanoemulsification process is more efficient at higher temperatures (H_2 or H_3) depending on the system, for example, with the oil phase sensitive to temperature (solid lipid, wax, or very viscous oils), and in that case, the water dilution is sufficient to modify the surfactant solubility and induce nanoemulsification. Once the oil/surfactant mixture is in contact with water, changing suddenly the thermodynamic condition to make surfactants more soluble in the aqueous phase than the oily one, Fig. 6A(i) (system composition is, e.g., H_1, H_2, or H_3), the water phase rapidly penetrates the oil/surfactant mixture (Fig. 6A(ii)), breaking up the oil in the form of nanodroplets (Fig. 6A(iii)). In addition, in function of the surfactant-to-oil initial ratio, that is, location of J on the surfactant/oil axis (see Fig. 6B), the size of the resulting nanoemulsion droplets is strongly modified, like in the example reported in Fig. 6C for a system prepared from Kolliphor HS15/Labrafil M1944CS, where the size follows a monotonous decrease according to a power law. Similar results were obtained with a mixture of Tween 80/Span 80, showing similar trends and also emphasizing the strong impact of the temperature of the system before dilution (Tong et al., 2016).

This mechanism is very close to the one described above for the PIT method (Fig. 4.4). It is not surprising because the systems are quite similar, but this is very interesting since it can simplify the nanoemulsification processes. In the literature, the PIT method is often considered quite distinct from the spontaneous emulsification method. However, the above discussion highlights that both methods are actually based on the same physicochemical phenomena and can therefore be considered to be very similar. Therefore, these considerations define the concept of *universality* for the formation of nanoemulsions using low-energy emulsification, which originates from the very specific behavior of nonionic PEGylated surfactants. One experimental observation that supports the fact that the same behavior of nonionic surfactants

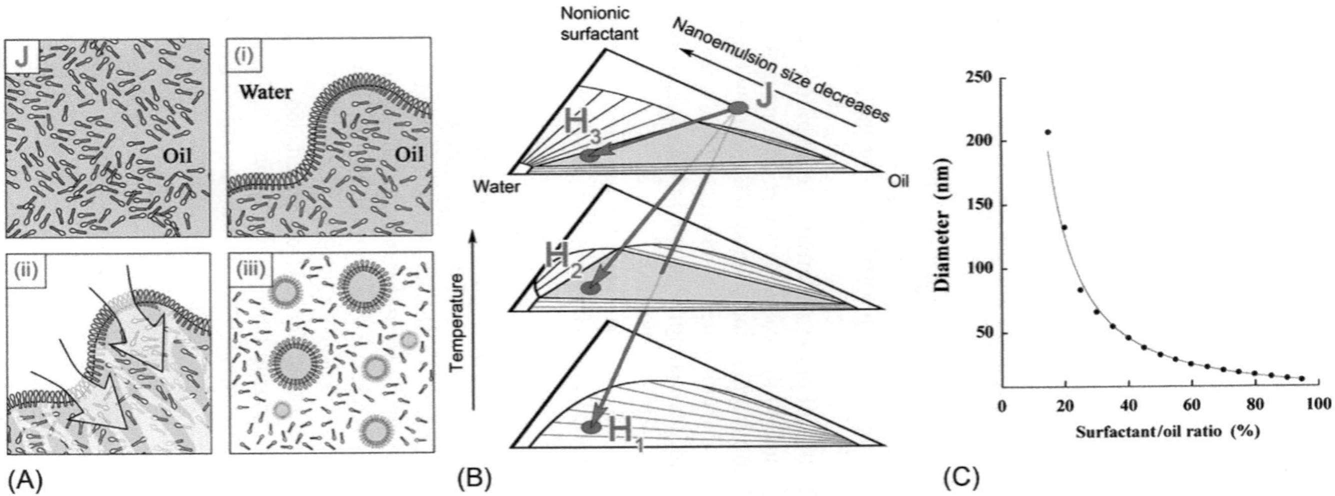

FIG. 4.6 (A) Schematic illustration of nanoemulsion formation using the spontaneous emulsification process. Point J corresponds to the compositions shown in Fig. 4.3, and the schemes (i), (ii), and (iii) show the nanoemulsification mechanism in chronological order after dilution of the oil/surfactant mixture with water. (B) Shown is the thermodynamic configuration of the system, before (J) and after the spontaneous emulsification process, dilution at the same temperature $T >$ PIT (H_3), at $T =$ PIT (H_2), and at $T <$ PIT (H_1). (C) Nanoemulsion diameters are shown as a function of the surfactant content, after nanoemulsification by the spontaneous emulsification method (surfactant = Kolliphor HS15, oil = Labrafil M1944CS (oleoyl macrogolglycerides), and aqueous phase = MilliQ water; curve fit is a power law, to guide the eye). *(Reproduced with permission from Anton, N., Vandamme, T.F., 2009. The universality of low-energy nano-emulsification. Int. J. Pharm. 377, 142–147.)*

drives the nanoemulsification mechanism and thus the concept of universality is the effect of the surfactant-to-oil ratio on the nanoemulsion droplet size, which follows the same trend whatever the method. The results shown in Fig. 4.5 illustrate this fact for the PIT method, and the ones shown in Fig. 4.6C for the spontaneous emulsification method. To further support this idea of universality, the PIT and the spontaneous emulsification methods are compared now with the same system (Kolliphor HS15/Labrafil M1944CS), on the one hand following the spontaneous emulsification procedure for different SOR and, on the other hand, following the PIT method over the same SOR range and also with different WOR.

The results reported in Fig. 4.7 show the direct correspondence between the two methods. In both cases, the droplet size of the nanoemulsions depends neither on the method (at least for SOR $\geq$30%) nor on the WOR value, but only on the surfactant concentration (SOR value), corroborating the predominant role of the nonionic surfactant in the process and its governing role in the universal nanoemulsification process.

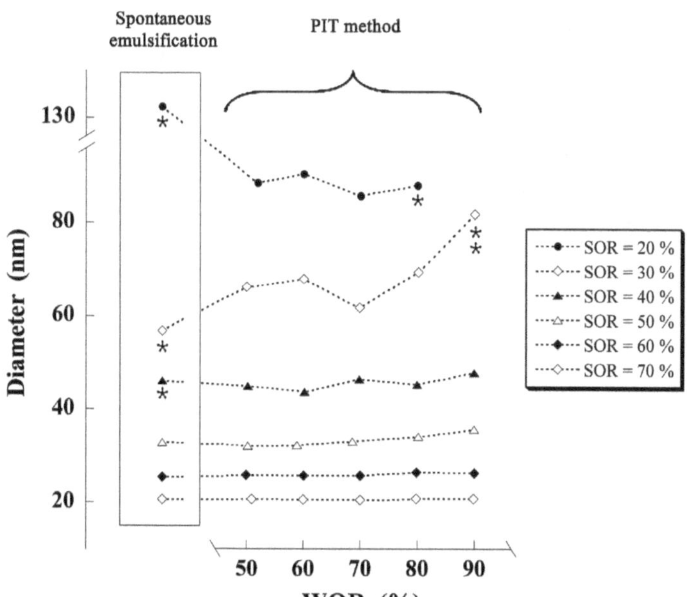

FIG. 4.7 Size of nanoemulsions obtained from the spontaneous emulsification and PIT methods, for different surfactant (SOR) and water (WOR) contents. Surfactant = Kolliphor HS15, oil = Labrafil M1944CS, and aqueous phase, MilliQ water. Two stars = PDI > c0.2; one star = 0.1 < PDI < 0.2, and no star = PDI < 0.1. *(Reproduced with permission from Anton, N., Vandamme, T.F., 2009. The universality of low-energy nano-emulsification. Int. J. Pharm. 377, 142–147.)*

4.3.3 Critical Difference Between Spontaneous Nanoemulsions and Microemulsions

In a chapter on low-energy nanoemulsification, it appears important to emphasize a confusion that is often made between nanoemulsions and microemulsions (Anton and Vandamme, 2011; McClements, 2012). In fact, in some cases, the structure and morphology of microemulsions can be very close to the one of nanoemulsions, that is, in the form of spherical *swollen micelles*. On the other hand, as we have seen in the previous section, the formation of nanoemulsions by spontaneous emulsification can be very close to the methodology used to fabricate microemulsions. These two factors have led to some of the confusion between nanoemulsions and microemulsions, with undeniable impact on their potential applications.

The first comment concerns the stability of nanoemulsions. Like almost all emulsified systems, they exist in a metastable thermodynamic state, rather than in the state with the lowest free energy. However, due to the small droplet size (which reduces gravitational separation and droplet aggregation), the main process inducing their destabilization is Ostwald ripening. This factor results in stability of nanoemulsions for months and even higher if specific additives (ripening inhibitors) slow down interdroplet oil transfer. On the other hand, microemulsions are stable systems from a thermodynamic point of view. They are formed as a result of the equilibrium between oil, water, and surfactants, namely, all the phases are mixed homogeneously, whatever the order of introduction, and the mixture is sealed and set at constant temperature until equilibrium is reached. However, the structure of microemulsions can change considerably when the composition, temperature, and other parameters (such as electrolyte, cosurfactant, or cosolvent concentration) are varied. Let us consider a phase diagram (Fig. 4.8), showing some particular structures of microemulsions formed at a temperature below the cloud point ($T < T(\text{cp}_\beta)$). The point K_1 is the extreme case of a microemulsion (i.e., without oil), which occurs when the surfactants are at a level above the CMC. Under these conditions, the surfactants spontaneously form aggregates (typically containing tens or hundreds of molecules) that have diameters below about and are in equilibrium with water (which contains surfactant monomers at a concentration close to the CMC). Now, if we add to this system a small amount of oil but keep the system in the one-phase region (point K_2), oil molecules are naturally entrapped inside the micelles, forming *swollen micelles*. These swollen micelles show spherical nanodomains dispersed in water and potentially have diameters between about 30 and 100 nm, which actually present a structure very close to the one of nanoemulsions with the fundamental difference in their stability and formulation method. If further oils are added to the systems, making it crossing the phase boundary to a WOR = 50% (point K_3, two-phase region), the micelles cannot "absorb" such a quantity of oil, and the excess oil appears at the top of the flask. If the content of surfactant is further increased, the system will cross

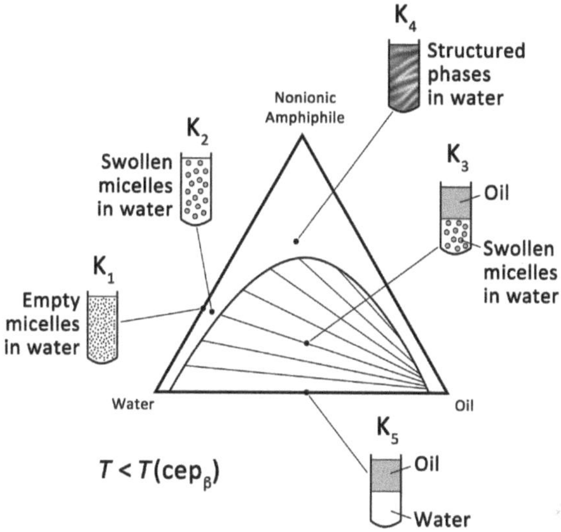

FIG. 4.8 Ternary diagram showing schematically the location of microemulsions and their structures (see details in the text). *(Reproduced with permission from Anton, N., Vandamme, T.F., 2011. Nano-emulsions and micro-emulsions: clarifications of the critical differences. Pharm. Res. 28, 978–985.)*

again the phase boundary (point K_5), and a one-phase bicontinuous microemulsion is generated. Point K_5 simply shows the corresponding system with no surfactant, oil, and water at equilibrium.

This schematic representation lets us understand that, for a certain composition corresponding to K_2, the particles in a microemulsion are morphologically very close to the ones in a nanoemulsion. In addition, the method of forming nanoemulsions by spontaneous emulsification appears very close to the method used to form microemulsions (describe above), with the main difference being that the order of mixing and introduction of the different compounds matters a lot. It is important to clarify this point, because the formulation methodologies for nanoemulsions and microemulsions could appear very similar, but the behaviors of these two colloidal dispersions are quite different. However, even if the formulation of microemulsions seems very simple and more convenient, the intrinsic properties of microemulsions are less compatible with the applications as drug delivery systems. The first reason is the limited concentration of droplets; the feasibility domain of spherical microemulsions shown in Fig. 4.8 is quite small, meaning that the possible WOR values are only high; thus, oil concentration is low, as opposed to nanoemulsions where oil amount can easily increase higher than 20% of the nanoemulsion (Anton and Vandamme, 2009). The second reason is the stability: even if microemulsions are thermodynamically stable, unlike the nanoemulsions, they are not stable against the modification of thermodynamic conditions like temperature or dilution that could correspond to a parenteral administration.

For instance, the dilution of a microemulsion with water makes the system move toward the water corner, which can potentially change the structure and size of the particles and potentially cause precipitation of an encapsulated drug.

4.4 APPLICATIONS OF TRANSITIONAL NANOEMULSIONS FOR ENCAPSULATION OF ACTIVE PRINCIPLE INGREDIENTS

This last section will detail some potential applications of spontaneous emulsification for forming nanoemulsion-based systems. These potential applications include the homogeneous and stable dispersion in water of lipophilic molecules, drugs, contrast agents, imaging probes, and polymers. Therefore, the question we address here deals with the potential influence of the encapsulation of such API on the emulsification process itself. The examples presented in this work support the idea that the chemical nature of the compounds used matters a lot. Due to differences in the specific affinities of nonionic surfactants for different oil and water phases, it may be possible to form a nanoemulsion with one type of oil, but not with another type of oil, using the same procedures. This signifies the fact that a modification of the composition of the oil, for the same surfactant and the same proportions, can have significant consequences on the nanoemulsion droplet size and polydispersity. This effect may also be important when lipophilic API are solubilized in the oil phase and depend on the chemical characteristics and concentrations used. Some examples are reported in Fig. 4.9, showing the impact of the nature of the oil, solubilization of drug, and chemical modification of the oil, on the nanoemulsification process. Fig. 4.9A shows typical size-SOR graphs for nanoemulsions generated by spontaneous emulsification (Anton and Vandamme, 2009) performed with the same surfactant (Kolliphor HS15) and the same protocol (temperature of dilution at 90°C) and different oils: Labrafac WL (medium-chain triglycerides) and Labrafil M1944CS (oleoyl macrogolglycerides). These results clearly illustrate the discussion above; the chemical nature of the oil phase is a key parameter driving the efficiency of spontaneous emulsification, revealed by a difference in the droplet sizes obtained, which can be, for example, around 100 nm for SOR $=40\%$. Interestingly, the two curves meet each other for the highest surfactant concentrations, in accordance with the fact that nonionic surfactants override the effects of oil, for example, at SOR $>80\%$.

The two other examples presented in Fig. 4.9(B_1) and (B_2) show the nanoemulsification processes (Vandamme and Anton, 2010) still performed using the same conditions, for Kolliphor ELP/Labrafil M1944CS and Kolliphor HS15/Labrafac WL, respectively. These two sets of results compare the nanoemulsions encapsulating a lipophilic drug (diclofenac at 1 wt% in oil, *diclofenac-loaded nanoemulsions*) with the empty nanodroplets (*drug-free nanoemulsions*), and the results give very close curves, whatever the surfactant

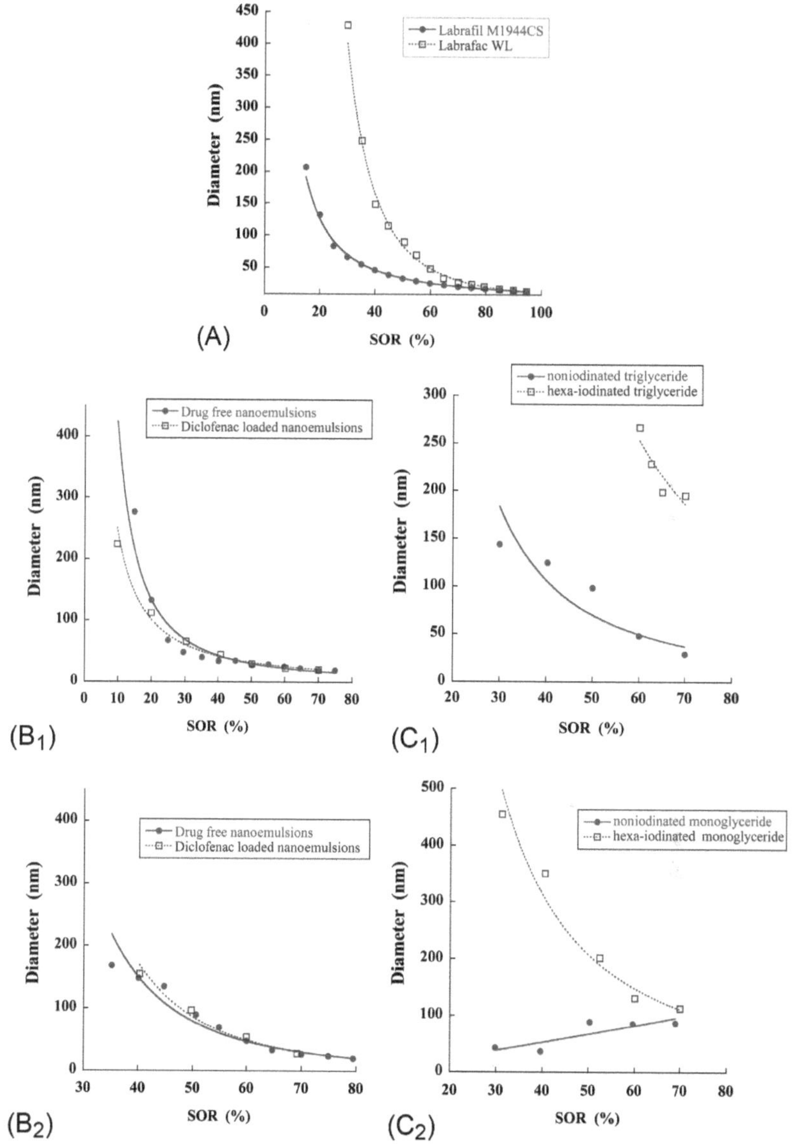

FIG. 4.9 Nanoemulsions formed by spontaneous emulsification, presenting the effect of the surfactant content (SOR) on the mean droplet size of the dispersion: (A) comparing the nature of oil; (B) comparing the presence and absence of lipophilic drugs at 1 wt% in the oil core; and (C) comparing the chemical modification of oils (see the text for details on the compositions for each case). *(Reproduced with permission from (A) Anton, N., Vandamme, T.F., 2009. The universality of low-energy nano-emulsification. Int. J. Pharm. 377, 142–147; (B) Vandamme, T.F., Anton, N., 2010. Low-energy nano-emulsification to design veterinary controlled drug delivery devices. Int. J. Nanomedicine 5, 867–873); (C) Attia, M., Anton, N., Chiper, M., Akasov, R., Anton, H., Messaddeq, N., Fournel, S., Klymchenko, A., Mély, Y., Vandamme, T.F., 2014. Biodistribution of X-ray iodinated contrast agent in nano-emulsions is controlled by the chemical nature of the oily core. ACS Nano 8, 10537–10550.)*

oil system. Even if the curves are not exactly superimposed, in that case, the presence of drug solubilized at 1 wt% does not affect this robust process. On the other hand, if the oil is chemically modified to graft iodinated compounds (two triiodobenzene function per oil molecules) for imaging applications, the process is seriously affected. This is the case shown in Fig. 4.9C_1 and C_2 (Attia et al., 2014), for triglycerides and monoglycerides, respectively, where the noniodinated molecules are presented with the filled circles (plain lines) and the iodinated compounds with empty squares (dashed lines). The difference between these two curves is important, and in fact, the molecule iodination results in a decrease of the efficiency of the process, either significantly increase the droplet size or simply not allow the emulsification to occur as in Fig. 4.9C_1 for SOR values below 60%. This is likely due to an increase of the lipophilicity of the oily molecules after iodination that probably reduces the solubility of nonionic surfactants. This result supports the first observation shown in Fig. 4.9A that the nature of the oil importantly matters. To conclude, this section reports representative examples of nanoemulsification produced by spontaneous emulsification, comparing different oils with the same nonionic surfactant, or encapsulating a lipophilic drug. It appears that the process is robust enough for keeping the same efficiency with including an API solubilized at 1 wt%. Other examples in the literature (Kilin et al., 2014) show that the efficiency of the process is conserved up to 8% (of lipophilic near-infrared dyes in oil). However, it is drastically influenced by the nature of the oil and notably by a change of the lipophilic properties of oils.

As a last remark, it should be noted that spontaneous nanoemulsification has also been adapted to the encapsulation of hydrophilic pharmaceutical ingredients using various formulation strategies. The first strategy was to incorporate hydrophilic species directly into the oil phase using reverse micelles (Vrignaud et al., 2011; Anton et al., 2010), following exactly the same formulation procedure as described above (PIT method) with oils containing model hydrophilic dyes like fluorescein sodium salt or doxorubicin hydrochloride. The second strategy involved adapting the spontaneous emulsification method for the formulation or reverse nanoemulsions, that is, W/O nanoemulsions. This procedure has been described by several research groups (Vrignaud et al., 2013; Anton et al., 2009; Assadpour et al., 2016a,b; Assadpourand and Jafari, 2017; Mohammadi et al., 2016a,b; Esfanjani et al., 2015, 2017; Mehrnia et al., 2016, 2017) and allowed the fabrication of W/O/W nanoemulsions by carrying out a second emulsification of the primary W/O nanoemulsion. These studies allowed the encapsulation of model hydrophilic dyes (methylene blue); proteins (bovine serum albumin) (Vrignaud et al., 2013; Anton et al., 2009); and other bioactive compounds like folic acid, crocin, olive leaf phenolics, and saffron extract (Assadpour et al., 2016a,b; Assadpourand and Jafari, 2017; Mohammadi et al., 2016a,b; Esfanjani et al., 2015, 2017; Mehrnia et al., 2016, 2017).

4.5 CONCLUSION

In this chapter, the principles of nanoemulsion formation by low-energy transitional methods were presented. At first, we presented the main phenomenon driving the low-energy emulsification methods, which is based on the physicochemical behavior of nonionic surfactants. We discussed not only this behavior in terms of the interactions between the PEG moiety and water phase and the role of temperature causing the cloud point but also other factors like the type and level of electrolytes in water, impacting directly on the location of the phase boundary. Then, the link between the physicochemical behavior of nonionic surfactants and nanoemulsification was developed, through a discussion of the PIT and spontaneous emulsification methods. The different ways to perform these nanoemulsification methods, the impact of composition and temperature, and their limitations were presented and critically assessed. A link between phase diagrams (reflecting surfactant behavior) and the nanoemulsification process (due to temperature and/or composition changes) was highlighted. Through the nanoemulsion properties, mainly size and polydispersity, the PIT and spontaneous emulsification methods were compared. We emphasized a universality of the emulsification mechanism based on the behavior of nonionic surfactants. The critical difference between microemulsions and nanoemulsions was highlighted. Finally, the impact of including active ingredients or changing the nature of the oil phase on the formation of nanoemulsions by the transitional emulsification method was presented, because this has important implications for the practical application of this technology.

REFERENCES

Anton, N., Vandamme, T.F., 2009. The universality of low-energy nano-emulsification. Int. J. Pharm. 377, 142–147.

Anton, N., Vandamme, T.F., 2011. Nano-emulsions and micro-emulsions: clarifications of the critical differences. Pharm. Res. 28, 978–985.

Anton, N., Benoit, J.P., Saulnier, P., 2008a. Particular conductive behaviors of emulsions phase inverting. J. Drug Delivery Sci. Tech. 18, 95–99.

Anton, N., Benoit, J.P., Saulnier, P., 2008b. Design and production of nanoparticles formulated from nano-emulsion templates—a review. J. Control. Release 128, 185–199.

Anton, N., Saulnier, P., Gaillard, C., Porcher, E., Vrignaud, S., Benoit, J.P., 2009. Aqueous-core lipid nanocapsules for encapsulating fragile hydrophilic and/or lipophilic molecules. Langmuir 25, 11413–11419.

Anton, N., Mojzisova, H., Porcher, E., Benoit, J.P., Saulnier, P., 2010. Reverse micelles-loaded lipid nano-emulsions: a new technology for the nano-encapsulation of hydrophilic materials. Int. J. Pharm. 398, 204–209.

Arai, H., 1967. Relation between the cloud points and the properties of micelles of nonionic detergents. J. Colloid Interface Sci. 23, 348–351.

Assadpour, E., Maghsoudlou, Y., Jafari, S.M., Ghorbani, M., Aalami, M., 2016a. Optimization of folic acid nano-emulsification and encapsulation by maltodextrin-whey protein double emulsions. Int. J. Biol. Macromol. 86, 197–207.

Assadpour, E., Jafari, S.M., Maghsoudlou, Y., 2016b. Evaluation of folic acid release from spray dried powder particles of pectin-whey protein nano-capsules. Int. J. Biol. Macromol. 95, 238–247.

Assadpour, E., Jafari, S.M., 2017. Spray drying of folic acid within nano-emulsions: optimization by Taguchi approach. Drying Technol. 35 (9), 1152–1160.

Astaraki, A.M., 2016. The effect of concentration of surfactant and electrolyte on the PIT and droplet sizes nanoemulsions of n-dodecane in water. Russ. J. Appl. Chem. 89, 84–89.

Attia, M., Anton, N., Chiper, M., Akasov, R., Anton, H., Messaddeq, N., Fournel, S., Klymchenko, A., Mély, Y., Vandamme, T.F., 2014. Biodistribution of X-ray iodinated contrast agent in nanoemulsions is controlled by the chemical nature of the oily core. ACS Nano 8, 10537–10550.

Dorenand, A., Goldfarb, J., 1970. Electrolyte effects on micellar solutions of nonionic detergents. J. Colloid Interface Sci. 32, 67–72.

Esfanjani, A.F., Jafari, S.M., Assadpour, E., Mohammadi, A., 2015. Nano-encapsulation of saffron extract through double-layered multiple emulsions of pectin and whey protein concentrate. J. Food Eng. 165, 149–155.

Esfanjani, A.F., Jafari, S.M., Assadpour, E., 2017. Preparation of a multiple emulsion based on pectin-whey protein complex for encapsulation of saffron extract nanodroplets. Food Chem. 221, 1962–1969.

Goldstein, R.E., 1984. On the theory of lower critical solution points in hydrogen-bonded mixtures. J. Phys. Chem. 80, 5340–5341.

Gu, T., Qin, S., Ma, C., 1989. The effect of electrolytes on the cloud point of mixed solutions of ionic and nonionic surfactants. J. Colloid Interface Sci. 127, 586–588.

Hsiao, L., Dunning, H.N., Lorenz, P.B., 1956. Critical micelle concentrations of polyoxyethylated non-ionic detergents. J. Phys. Chem. 60, 657–660.

Izquierdo, P., Esquena, J., Tadros, T.F., Dederen, J.C., Feng, J., Garcia-Celma, M.J., Azemar, N., Solans, C., 2004. Phase behavior and nano-emulsion formation by the phase inversion temperature method. Langmuir 20, 6594–6598.

Karlstrom, G.J., 1985. A new model for upper and lower critical solution temperatures in poly(ethylene oxide) solutions. J. Phys. Chem. 89, 4862–4964.

Kilin, V., Anton, H., Anton, N., Steed, E., Vermot, J., Vandamme, T.F., Mely, Y., Klymchenko, A.S., 2014. Counterion-enhanced cyanine dye loading into lipid nano-droplets for single particle tracking in zebrafish. Biomaterials 35, 4950–4957.

Klassen, P.L., George, Z., Warwick, J., Georgiadou, S., 2014. PIT tuning effects of hydrophobic co-surfactants and drugs. Colloids Surf. A Physicochem. Eng. Asp. 455, 1–10.

Leal-Calderon, F., Scfmitt, V., Bibette, J., 2007a. Emulsification. In: Emulsion Science, Basic principles. Springer, New York, pp. 5–40 (Chapter 1).

Leal-Calderon, F., Scfmitt, V., Bibette, J., 2007b. Stability of concentrated emulsions. In: Emulsion Sciences, Basic Principles. Springer, New York, pp. 143–168 (Chapter 5).

Lee, H.S., Morrison, E.D., Frethem, C.D., Zasadzinski, J.A., McCormick, A.V., 2014. Cryogenic electron microscopy study of nanoemulsion formation from microemulsions. Langmuir 30, 10826–10833.

Li, X., Anton, N., Zuber, G., Zhao, M., Messaddeq, N., Hallouard, F., Fessi, H., Vandamme, T.F., 2013. Iodinated alpha-tocopherol nano-emulsions as non-toxic contrast agents for preclinical X-ray imaging. Biomaterials 34, 481–491.

Maclay, W.N., 1956. Factors affecting the solubility of nonionic emulsifiers. J. Colloid Sci. 11, 172–185.

Malikand, W.U., Jhamb, O.P., 1970. Critical micelle concentration of polyoxyethylated nonionic surfactants and the effect of additives. Kolloid Z. Z. Polym. 242, 1209–1211.

Malikand, W.U., Saleem, S.M., 1968. Effect of additives on the critical micelle concentration of some polyethoxylated nonionic detergents. J. Am. Oil Chem. Soc. 45, 670–672.

McClements, D.J., 2011. Edible nanoemulsions: fabrication, properties, and functional performance. Soft Matter 7, 2297–2316.

McClements, D.J., 2012. Nanoemulsions versus microemulsions: terminology, differences, and similarities. Soft Matter 8, 1719–1729.

Mehrnia, M.A., Jafari, S.M., Makhmal-Zadeh, B.S., Maghsoudlou, Y., 2016. Crocin loaded nanoemulsions: factors affecting emulsion properties in spontaneous emulsification. Int. J. Biol. Macromol. 84, 261–267.

Mehrnia, M.A., Jafari, S.M., Makhmal-Zadeh, B.S., Maghsoudlou, Y., 2017. Rheological and release properties of double nano-emulsions containing crocin prepared with Angum gum, Arabic gum and whey protein. Food Hydrocoll. 66, 259–267.

Mei, Z., Xu, J., Sun, D., 2011. O/W nano-emulsions with tunable PIT induced by inorganic salts. Colloids Surf. A Physicochem. Eng. Asp. 375, 102–108.

Mohammadi, A., Jafari, S.M., Assadpour, E., Esfanjani, A.F., 2016a. Nano-encapsulation of olive leaf phenolic compounds through WPC-pectin complexes and evaluating their release rate. Int. J. Biol. Macromol. 82, 816–822.

Mohammadi, A., Jafari, S.M., Esfanjani, A.F., Akhavan, S., 2016b. Application of nano-encapsulated olive leaf extract in controlling the oxidative stability of soybean oil. Food Chem. 190, 513–519.

Rao, J., McClements, D.J., 2010. Stabilization of phase inversion temperature nanoemulsions by surfactant displacement. J. Agric. Food Chem. 58, 7059–7066.

Rao, J., McClements, D.J., 2011. Formation of flavor oil microemulsions, nanoemulsions and emulsions: influence of composition and preparation method. J. Agric. Food Chem. 59, 5026–5035.

Rayand, A., Nemethy, G., 1971. Effects of ionic protein denaturants on micelle formation by nonionic detergents. J. Am. Chem. Soc. 93, 6787–6793.

Schick, M.J., 1962a. Surface films of nonionic detergents. I. Surface tension study. J. Colloid Interface Sci. 17, 801–813.

Schick, M.J., 1962b. Surface films of nonionic detergents. I. Surface tension study. J. Colloid Sci. 17, 801–813.

Schick, M.J., 1964. Effect of electrolyte and urea on micelle formation. J. Phys. Chem. 68, 3585–3592.

Schott, H., 1973. Salting in of nonionic surfactants by complexation with inorganic salts. J. Colloid Interface Sci. 43, 150–155.

Schott, H., 1997. Effect of inorganic additives on solutions of nonionic surfactants. XIV. Effect of chaotropic anions on the cloud point of octoxynol 9 (Triton X-100). J. Colloid Interface Sci. 189, 117–122.

Schott, H., 2001. Effect of inorganic additives on solutions of nonionic surfactants—XVI. Limiting cloud points of highly polyoxyethylated surfactants. Colloids Surf. A Physicochem. Eng. Asp. 186, 129–136.

Schott, H., Han, S.K., 1975. Effect of inorganic additives on solutions of nonionic surfactants. II. J. Pharm. Sci. 64, 658–664.

Schott, H., Royce, A.E., Han, S.K., 1984. Effect of inorganic additives on solutions of nonionic surfactants. VII. Cloud point shift values of individual ions. J. Colloid Interface Sci. 98, 196–201.

Schottand, H., Han, S.K., 1976a. Effect of inorganic additives on solutions of nonionic surfactants. IV: Krafft points. J. Pharm. Sci. 65, 979–981.

Schottand, H., Han, S.K., 1976b. Effect of inorganic additives on solutions of nonionic surfactants. III: CMC's and surface properties. J. Pharm. Sci. 65, 975–978.

Schottand, H., Royce, A.E., 1984. Effect of inorganic additives on solutions of nonionic surfactants. VI: further cloud point relations. J. Pharm. Sci. 73, 793–799.
Shinoda, K., Saito, H., 1969. The stability of O/W type emulsions as functions of temperature and the HLB of emulsifiers: the emulsification by PIT-method. J. Colloid Interface Sci. 30, 258–263.
Shinodaand, K., Takeda, H., 1970. Effect of added salts in water on the hydrophile-lipophile balance of nonionic surfactants: effect of added salts on the phase inversion temperature of emulsions. J. Colloid Interface Sci. 32, 642–646.
Tadros, T.F., Izquierdo, P., Esquena, J., Solans, C., 2004. Formation and stability of nano-emulsions. Adv. Colloid Interf. Sci. 108-109, 303–318.
Tong, K., Zhao, C., Suna, D., 2016. Formation of nanoemulsion with long chain oil by W/O microemulsion dilution method. Colloids Surf. A Physicochem. Eng. Asp. 497, 101–108.
Vandamme, T.F., Anton, N., 2010. Low-energy nano-emulsification to design veterinary controlled drug delivery devices. Int. J. Nanomedicine 5, 867–873.
Vrignaud, S., Anton, N., Gayet, P., Benoit, J.P., Saulnier, P., 2011. Reverse micelle-loaded lipid nanocarriers: a novel drug delivery system for the sustained release of doxorubicin hydrochloride. Eur. J. Pharm. Biopharm. 79, 197–204.
Vrignaud, S., Anton, N., Passirani, C., Benoit, J.P., Saulnier, P., 2013. Aqueous core nanocapsules: a new solution for encapsulating doxorubicin hydrochloride. Drug Dev. Ind. Pharm. 39, 1706–1711.
Wartewig, S., Alig, I., Hergeth, W.D., Lange, J., Lochmann, I., Scherzed, T., 1990. Spectroscopic investigations on aqueous solution of poly(oxyethylene)-poly(oxypropylene)-poly(oxyethylene) triblockcopolymers. J. Mol. Struct. 219, 365–370.

FURTHER READING

Kahlweit, M., Strey, R., Firman, P., Haase, D., Jen, J., Schomaecker, R., 1988. General patterns of the phase behavior of mixtures of water, nonpolar solvents, amphiphiles, and electrolytes. Langmuir 4, 499–511.

Part III

Production of Nanoemulsions by Mechanical Methods

Chapter 5

General Principles of Nanoemulsion Formation by High-Energy Mechanical Methods

Andreas Håkansson* and Marilyn Rayner†
*Kristianstad University, Kristianstad, Sweden, †Lund University, Lund, Sweden

Chapter Outline

Nanoemulsions. https://doi.org/10.1016/B978-0-12-811838-2.00005-9

5.1 INTRODUCTION

Creating a nanoemulsion requires four components: two immiscible liquids, energy, and emulsifier. Energy is needed for dispersing the two fluids into each other and for reducing the drop size of a preexisting emulsion. Turning a coarse "macroemulsion" (an emulsion with drop sizes in the micrometer range) into a nanoemulsion requires a substantial increase in the liquid-liquid interfacial area. This increase represents an increase in the Gibbs free energy of the system, and the process will therefore not occur spontaneously without supplying external energy.

Emulsifiers are needed for decreasing the interfacial energy of the system and for adsorbing at the newly created drop interface. They are also used for stabilizing the drop and protecting it from fusing with nearby drops, a process referred to as (re)coalescence.

Emulsions are, by definition, thermodynamically unstable; given sufficient time, they will eventually phase separate. Obtaining an emulsion with a set of desired properties requires choosing an appropriate emulsifier (or combination thereof) and an appropriate emulsification method. This makes emulsification a difficult and energy-demanding process. Nanoemulsion formation requires that drop sizes are reduced even further and therefore offers a substantial additional challenge.

Methods for creating nanoemulsions are often divided into high-energy techniques (e.g., ultrasonication and high-pressure valve homogenization) and low-energy techniques (e.g., phase inversion techniques) (Gupta et al., 2016b; McClements and Rao, 2011; Tadros et al., 2004). This traditional nomenclature is somewhat misleading, as illustrated by the case of membrane emulsification. Membrane or microchannel techniques generally require higher energy input than phase inversion techniques but substantially less energy than the other high-energy methods. Here, we have chosen to keep the traditional nomenclature but treat membrane emulsification as a high-energy method. This choice is motivated by the similarities in the underlying mechanism of emulsification between the traditional high-energy methods and membrane emulsification.

The low-energy methods are treated in Part II of this book. This chapter introduces the high-energy methods, focusing on the fundamental theoretical principles of emulsification, with special emphasis on what distinguishes nanoemulsion formation from traditional macroemulsion formation. A more in-depth analysis of the most commonly used methods together with examples of applications for nanoemulsion formation is given in the following chapters of this book.

A large number of books, chapters, and review papers on the basic mechanisms of high-energy emulsification methods are available in the literature (McClements, 2016; Rayner, 2015; Schubert et al., 2003; Walstra, 2005; Walstra and Smulders, 1998). Due to the increased interest in nanoemulsions, a number of more specific reviews are also available, some of which include an overview of nanoemulsion formation (Gupta et al., 2016b; McClements and Rao, 2011; Tadros et al., 2004). However, these reviews generally have a very broad scope and do not discuss emulsion formation from high-energy methods in great detail. The objective of this contribution is to provide an overview of the previous work with a focus on the basic principles of emulsion formation with high-energy techniques and the specific challenges arising for nanoemulsion formation.

The chapter is organized as follows. After a brief discussion of the thermodynamics of emulsion formation, Section 5.2 provides an overview of the stress-balance approach for understanding drop fragmentation and for predicting drop diameters. Section 5.3 extends this discussion by adding the dynamic aspects of emulsification, and Section 5.4 introduces the most widely used methods for nanoemulsion formation. The chapter is concluded by a summary, pointing to the main differences between nanoemulsion formation and emulsification in general and providing an outlook toward future research.

5.1.1 The Thermodynamics of Nanoemulsion Formation

Nanoemulsion formation often takes place in two steps. The first step is to create a coarse macroemulsion with large drops, for example, using a low-intensity rotor-stator system. However, the critical part of nanoemulsion formation is the second step, where the drop size is further reduced, for example, from a surface-averaged diameter of $D_{32,1} \approx 10\,\mu\text{m}$ down to the nanometer range, $D_{32,2} \approx 100\,\text{nm}$. The thermodynamics of this process can be described in terms of how it influences the Gibbs free energy of the system, which is a combination of an enthalpy (energy) term and an entropy term. If we let $\Delta_{1\to 2}$ denote the difference operator comparing the nanoemulsion with the preemulsion, the change in Gibbs free energy is

$$\Delta_{1\to 2}G = \Delta_{1\to 2}H - T\Delta_{1\to 2}S \tag{5.1}$$

The enthalpic contribution ($\Delta_{1\to 2}H$) is given by the increase in interfacial energy resulting from the increase in total drop interface area when breaking up larger drops into several smaller drops. The energy increase (per unit volume) can be expressed using the interfacial tension (γ) as

$$\Delta_{1\to 2}H = \gamma\Delta_{1\to 2}A = 6\phi_D\gamma\left(\frac{1}{D_{32,2}} - \frac{1}{D_{32,1}}\right) \tag{5.2}$$

where ϕ_D denotes the volume fraction of the disperse phase. The last step in Eq. (5.2) is obtained by combining the expressions for the average interfacial area and the total number of emulsion drops for the two cases.

To some extent, the increase in interfacial energy following nanoemulsion formation is offset by an increase in the entropy of mixing; fragmenting the drops increases the number of possible ways of rearranging them, and consequently, the entropy increases. This contribution can be quantified, at least for ideal cases. For a monodisperse dispersion of spherical particles of diameter D, the per unit volume of mixing entropy is (Overbeek et al., 1987)

$$S_{mix} = \frac{6 \cdot \phi_D}{\pi \cdot D^3}\left(1 - \ln\phi - \phi\frac{4-3\phi}{(1-\phi)^2} + 3\ln\frac{D}{D_w}\right) \quad (5.3)$$

where D_w denotes the diameter of a single molecule of the continuous phase. The entropic contribution to the Gibbs free energy difference is calculated by subtracting the mixing entropy for the resulting nanoemulsion ($D = D_{32,2}$) with the mixing entropy of the preemulsion ($D = D_{32,1}$).

Fig. 5.1 displays the total resulting Gibbs free energy required to form an O/W emulsion with a low-volume fraction of the disperse phase ($\phi_D = 1\%$) when starting from an emulsion with a monodisperse drop size of 10 μm. The right axis displays the Gibbs free energy, and the left axis shows the relative contribution from the entropic term. As seen in the figure, the entropic part is negligible (<1%), and the change in Gibbs free energy is approximately equal to the change in interfacial energy.

Three points should be noted from Fig. 5.1. First, note that Gibbs free energy is positive. This implies that nanoemulsion formation does not proceed spontaneously. This is an important difference between nanoemulsions and the

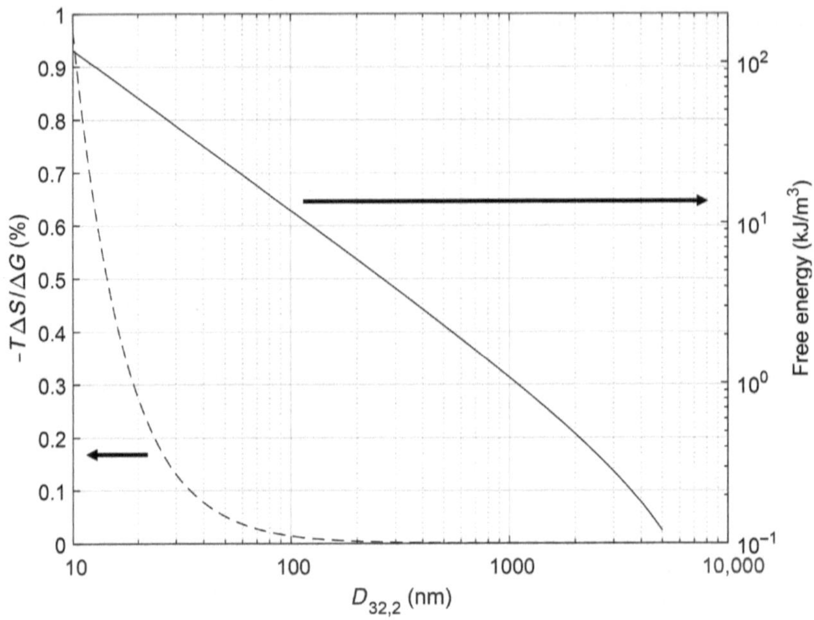

FIG. 5.1 Illustration of the thermodynamics of nanoemulsion formation. Gibbs free energy (ΔG) for transforming an emulsion with $D_{32,1} = 10$ μm into a nanoemulsion with varying surface-averaged diameters ($D_{32,2}$) (right axis). The left axis shows the relative entropic contribution to Gibbs free energy ($D_w = 5$ Å, $T = 298$ K, $\gamma = 20$ mN/m, and $\phi_D = 1\%$).

thermodynamically stable microemulsions. This also implies that the reversed process—destabilization into larger drops or into two separate phases—is spontaneous and will eventually occur for all nanoemulsions. However, if carefully designed, they can achieve a kinetic stability during long time periods (up to several years).

The second point to note from Fig. 5.1 is that the increase in Gibbs free energy is relatively small. High-pressure valve homogenization, a widely used method for forming nanoemulsions, offers an interesting comparison. It often requires an energy density of 10^4–10^6 kJ/m^3 in order to form a nanoemulsion with a drop size of ~100 nm. The applied energy is up to 10^4 times higher than the theoretical limit. This exceptionally low energy efficiency is typical for the high-energy methods discussed in this chapter (with the possible exception of membrane emulsification). However, mechanical methods are still the predominantly used industrial method for nanoemulsion formation, despite their poor energy efficiency, since they have other important advantages; for example, they are often suitable for large-scale continuous production.

The third point to note in relation to Fig. 5.1 is that the energy barrier for drop fragmentation depends strongly on the interfacial tension: the higher the interfacial tension, the more energy must be supplied to obtain small emulsion drops. This is of special importance for nanoemulsion formation since the required surface energy increases rapidly with decreasing drop size. Surfactants and other emulsifiers lower the interfacial tension and, consequently, the energy required for emulsification. This is one reason why the choice of emulsifier is a vital part of nanoemulsion design.

5.2 MECHANICAL BASIS FOR MAKING AND BREAKING DROPLETS

5.2.1 Drop Breakup and the Stress Balance

As seen from the Gibbs free energy discussion above, external energy must be supplied in order to form a nanoemulsion. For the emulsification technologies discussed in this chapter, this energy is supplied in the form of mechanical work acting on the drop interface. With the exception of membrane emulsification, this work comes from hydrodynamic interactions between the drop and the fluid in which it is dispersed. For membrane emulsification (and microchannel techniques), the energy is supplied via a combination of fluid dynamics and physical interactions with the channel.

Any amount of applied disrupting stress, σ, will not be sufficient to break the drop. Breakup will only occur if the destabilizing stress is sufficiently strong to overcome the inherent stability of the drop.

A single drop suspended in a quiescent fluid will spontaneously assume a spherical shape. This follows from thermodynamic reasoning and is illustrated by the Gibbs free energy equation in Eqs. (5.1), (5.2); a spherical shape minimizes the total surface area and thus also minimizes the total free energy of the system. The interfacial energy and the curvature of the drop interface also give rise to a pressure difference between the inside of the drop and the outside, referred to as the Laplace pressure, σ_{La}. For a spherical drop with diameter D, it equals to

$$\sigma_{La} = \frac{4\gamma}{D} \tag{5.4}$$

This pressure will act as a resistance to any stress that is applied in trying to fragment the drop. From this reasoning, it is reasonable to expect that drops will break at approximately the point when the applied stress (σ) equals or exceeds the Laplace pressure (i.e., $\sigma > \sigma_{La}$). This basic insight is often expressed in the form of a dimensionless Weber number, defined as the ratio between the externally applied fragmenting stress and the stabilizing Laplace pressure:

$$We = \frac{\sigma}{\sigma_{La}} = \frac{\sigma D}{4\gamma} \tag{5.5}$$

From this simple stress comparison, it would be expected that all drops are stable toward an external stress as long as $We < 1$ and that all drops will break when the stress is sufficiently strong to allow for $We = 1$. As will become apparent in the following discussion, the situation is, however, somewhat more complex. Nonetheless, the basic stress balance in Eq. (5.5) is still a good starting point for understanding emulsification and for predicting drop sizes resulting from emulsion formation using high-energy techniques. More specifically, experiments reveal that if a drop-size distribution (DSD) is subjected to a stress σ for a sufficient length of time, the size of the largest surviving drops (D_{max}) can be predicted by the stress-balance approach:

$$We = We_{Cr} \Rightarrow D_{max} = \frac{4\gamma We_{Cr}}{\sigma} \tag{5.6}$$

where We_{Cr}, referred to as the critical Weber number, is a constant of order magnitude 1 (Hinze, 1955). The same general scaling (but with a lower proportionality constant) applies for the surface- and volume-weighted average drop diameters (Calabrese et al., 1986).

Traditionally, there has been a difference in nomenclature between studies discussing drop breakup in different flow regimes. The ratio between disrupting and stabilizing stress has been referred to as the Weber number (We) when discussing turbulent drop breakup, whereas the same ratio has often been referred to as a capillary number (Ca) in discussions on laminar breakup. Conceptually, however, they both denote the same ratio. This discrepancy in nomenclature is most likely of historical origin; turbulent (Kolmogorov, 1949; Hinze, 1955) and laminar (Taylor, 1934) drop breakups have often been considered separately. In this chapter, the stress ratio will be referred to as the Weber number, regardless of flow regime, in order to highlight the similarities between the breakup mechanisms.

It is well known that the Weber number is not sufficient for describing the condition for hydrodynamic drop breakup. In particular, the drop viscosity plays a substantial role that is not captured by the Weber number. In many applications of nanoemulsion formation for food or pharmaceutical processing, the dispersed phase viscosity is several times larger than the continuous phase. This gives rise to an additional stabilizing effect. The details of how drop viscosity modulates the simple stress balance in Eq. (5.6) differ between different flow regimes and will be elaborated in the discussion below.

The volume fraction of the disperse phase is another factor known to influence the resulting drop size (Rueger and Calabrese, 2013; Tcholakova et al., 2011) that is not included in the stress analysis above. Increasing the volume fraction of the disperse phase influences both the rate of fragmentation (e.g., by modulating the flow field) and the coalescence rate (by increasing the rate of drop-drop collisions).

5.2.2 Flow Regimes: Laminar and Turbulent Flow

Although fluids can behave in many different ways, the distinction between laminar and turbulent flow is arguably one of the most important, especially for understanding hydrodynamic emulsification. A laminar flow is ordered; flow moves in sheets or lamellae. Intermixing between these is slow and only due to molecular diffusion. This implies that the streamlines emanating from two different points in space will never intersect (Fig. 5.2, left).

Turbulent flows behave very differently. Although there might still be an average flow direction, that is, a net flow from left to right as in Fig. 5.2 (right), individual fluid elements follow seemingly random trajectories, and they will cross repeatedly, as illustrated in Fig. 5.2 (right). Consequently, a turbulent flow is more diffusive than a laminar flow.

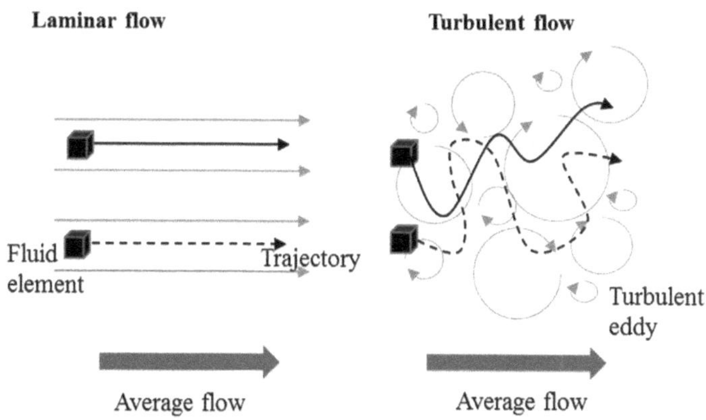

FIG. 5.2 Schematic illustration of the difference between laminar and turbulent flow.

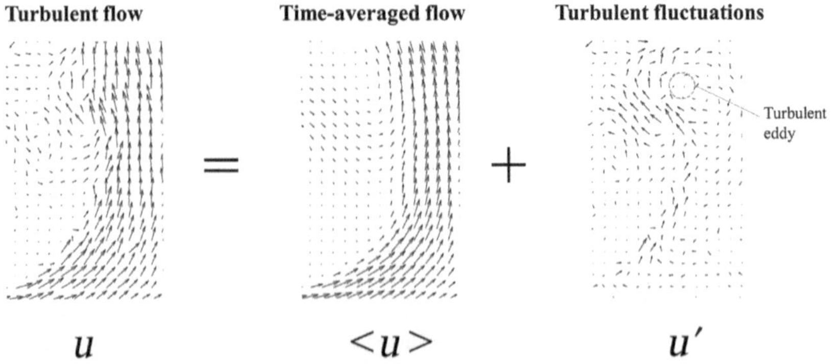

FIG. 5.3 The Reynolds decomposition, illustrating how an instantaneous turbulent velocity (u) can be seen as a superposition of the time-averaged steady velocity ($<u>$) and a field of turbulent fluctuation (u'). *(Based on experimental velocity field data in a rotor-stator mixer from Mortensen, H.H., Calabrese, R.V., Innings, F., Rosendahl, L., 2011. Characteristics of a batch rotor-stator mixer performance elucidated by shaft torque and angle resolved PIV measurements. Can. J. Chem. Eng. 89, 1076–1095.)*

A turbulent flow field can be seen as consisting of two superimposed flow fields (Fig. 5.3). First, the time-averaged flow field that looks similar to a laminar flow field and, second, a chaotic random component are illustrated in Fig. 5.3. The fluid velocity in a single direction, u, is divided in an average component $<u>$ and a fluctuating component u'. This is often referred to as the Reynolds decomposition of the flow field:

$$u = <u> + u' \tag{5.7}$$

These two components of the turbulent flow field are interconnected by the transfer of energy, and this energy cascade is of vital importance for understanding emulsion formation under turbulent conditions. Energy is fed to a turbulent flow in the form of average fluid motion. A pump or a rotating rotor blade will accelerate the fluid leading to an increase in the kinetic energy of the average flow ($<u>$). However, it will also lead to the formation of velocity gradients, for example, close to solid walls or in the shear layers between accelerated and still fluid. Gradients produce random velocity fluctuations (u') in these positions.

When analyzing the random fluctuations, it has been found that they form short-lived but coherent structures often referred to as *turbulent eddies*. One such coherent structure has been marked in Fig. 5.3. A turbulent eddy is characterized by a length scale, l_e. The first eddies that are formed from this transfer of energy from the average flow have length scales of the same order of magnitude as the physical dimensions of the geometry producing the flow. However, these large-scale eddies will almost instantly transfer their energy into smaller eddies, which will, in turn, accelerate even smaller eddies.

This cascade reaction will transfer the kinetic energy to smaller and smaller turbulent eddies, and it will continue until they have become sufficiently small to be dampened out by the viscous forces. Once energy has reached these length scales, it will dissipate as heat.

The total kinetic energy of the turbulent eddies is referred to as the turbulent kinetic energy, denoted k. The rate at which k is dissipated as heat defines the dissipation rate of turbulent kinetic energy, often denoted ε. The smallest eddies that exist before being damped out by turbulence are characterized by the Kolmogorov (micro) length scale of turbulence:

$$\eta = \left(\frac{\mu^3}{\rho^3 \varepsilon}\right)^{1/4} \tag{5.8}$$

where ρ and μ are the fluid density and viscosity, respectively.

Production, transport, and dissipation will eventually reach a dynamic equilibrium, ensuring that there is a distribution of turbulent eddies with different length scales at all times. This is the situation illustrated schematically in Fig. 5.2 (right). The total amount of turbulent kinetic energy available in eddies of different length scales can be described by a turbulent power spectrum, as seen Fig. 5.4. Note that the horizontal axis shows the eddy wave number ($\kappa = 2\pi/l_e$), which is proportional to the inverse of the eddy length scale. (This means that the largest eddies are found to the far left in the figure.) Most of the energy is contained in the *energy-containing subrange* where the average kinetic energy is converted into turbulent kinetic energy. Energy is then transported through the *inertial subrange* and dissipates in the *dissipation range* (Pope, 2000).

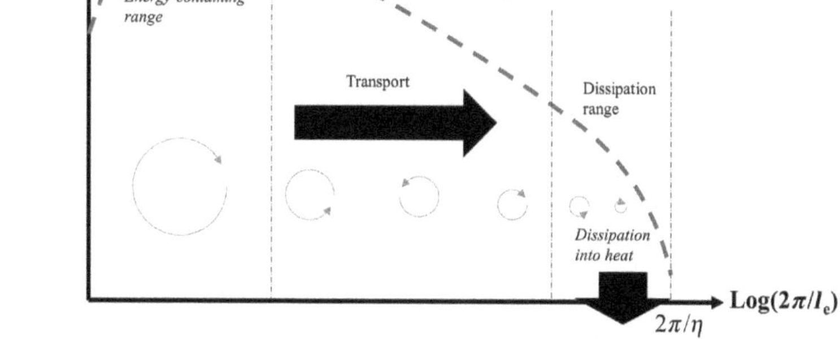

FIG. 5.4 The turbulent kinetic energy cascade and the ideal power spectrum.

In 1883, Reynolds showed that the demarcation between laminar and turbulent flow is determined by the ratio of inertial to viscous forces:

$$Re = \frac{\rho L V}{\mu} \tag{5.9}$$

where L is a characteristic length scale of the average flow and V is a characteristic velocity scale. A fluid flow turns turbulent when Re exceeds the critical Reynolds number, Re_{Cr}.

There have been attempts to propose a general critical Reynolds number describing the transition between laminar and turbulent conditions across all different emulsification techniques (cf. Walstra and Smulders, 1998). However, this should be avoided. The value of the critical Reynolds number depends on how the length and velocity scales are defined. The most commonly stated value, $Re_{Cr} \approx 2300$, refers to pipe flow when the length scale is taken to be equal to the pipe diameter, and the velocity scale is taken to be equal to the average fluid velocity. This specific value should not be taken as a general guideline for other flows. In fact, even for pipes, the critical Reynolds number varies considerably with the surface roughness (White, 1998).

As a further example, the impeller Reynolds number for a rotor-stator mixer is often defined with velocity scaling from the rotor tip speed and with length scale equal to the rotor diameter. Experiments reveal that a transition to fully turbulent condition does not appear until approximately $Re = 10{,}000$ for this commonly used definition (Atiemo-Obeng and Calabrese, 2004). This illustrates the dangers of extrapolating the critical value for pipe flow to other situations. The critical Reynolds number must be determined experimentally in each flow situation, for example, by measuring some characteristic feature that is independent of Re in the turbulent regime only, for example, pressure drop or drag coefficient.

Fragmentation mechanisms are fundamentally different in laminar and turbulent flow. Laminar drop breakup is considered in Section 5.2.3, and turbulent fragmentation is treated in Sections 5.2.4–5.2.6. There is also a third hydrodynamic process causing breakup-cavitation that can be present in both high- and low-Reynolds-number flows; this mechanism is treated in Section 5.2.7.

5.2.3 Laminar Drop Breakup—The Laminar Viscous Mechanism

Consider an emulsion drop with a diameter of ~10 μm that enters a laminar flow field. If the flow is uniform (i.e., if there are no velocity gradients; see Fig. 5.5, upper row), the emulsion drop will experience a drag force in the direction of the flow. This will cause the drop to accelerate in the direction of the flow until it reaches the same velocity as the uniformly flowing fluid. However, the drop will experience no deformation, and it will not break.

Now, consider a slightly different case. The drop is subjected to a simple shear flow: a flow that is uniform in all directions except one where there is a constant velocity gradient, G. This case is illustrated in the middle row of

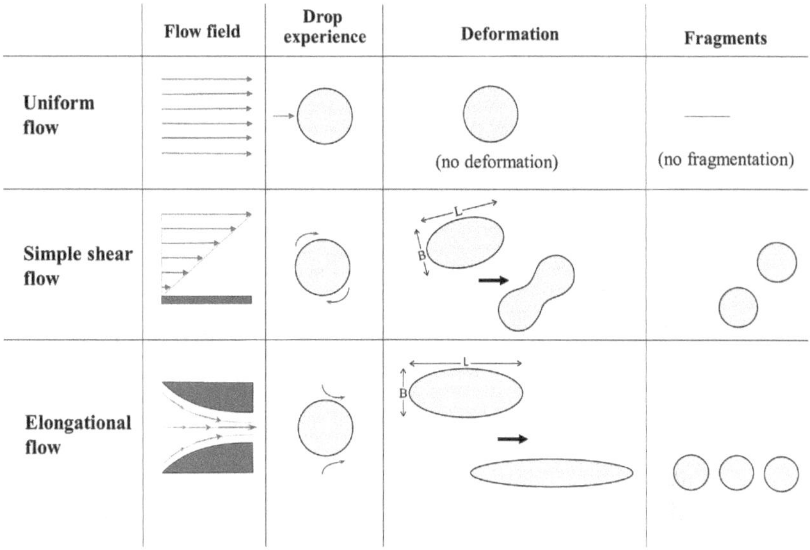

FIG. 5.5 Laminar viscous (LV) breakup, a schematic illustration of three laminar flows, and their effect on an emulsion drop.

Fig. 5.5. The drop will experience different forces on the upper and lower part, causing it to rotate and, subsequently, to deform. As the fluid inside of the drop starts to rotate, a neck is formed (see Fig. 5.5). The neck increases the interfacial area and the interfacial energy until a point where it is large enough to exceed that of two individual drops, at which point the neck snaps, resulting in the creation of two or more fragments.

The amount of deformation, expressed as the ratio between the long major axis and the short major axis (*L*/*B* in Fig. 5.5), reaches ~3 if the viscosity ratio is low (i.e., if $\mu_D/\mu_C = 0.1–1$). At higher ratios, the amount of deformation before breakup is higher (Taylor, 1934); this will be the case for O/W emulsions with triglycerides as the disperse phase, an important special case in food, pharmaceutical, and cosmetic processing.

The disrupting stress applied to the emulsion drop is determined by the fluid viscosity and the magnitude of the velocity gradient (*G*):

$$\sigma = \mu \cdot G \tag{5.10}$$

The relationship between the flow field and the largest drop surviving a laminar simple shear flow can be obtained by combining Eqs. (5.6), (5.10):

$$We = We_{Cr} \Rightarrow D_{max} = \frac{4\gamma We_{Cr}}{\mu G} \tag{5.11}$$

Experiments have also revealed that the critical Weber number depends on the dispersed-to-continuous-phase viscosity ratio with the smallest drops being created when μ_D/μ_C is approximately equal to 0.5. The critical Weber number as

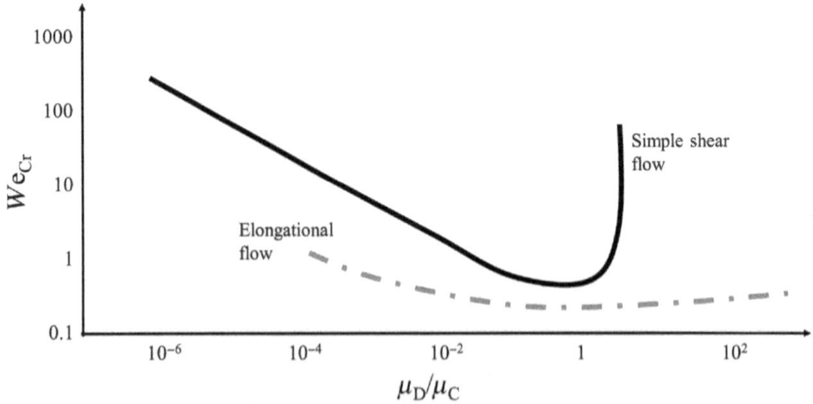

FIG. 5.6 Critical Weber number for simple shear flow and elongational flow as a function of disperse-to-continuous phase viscosity. *(Illustration based on data from Grace, H.P., 1982. Dispersion phenomena in high viscosity immiscible fluid systems and application of static mixers as dispersion device in such systems. Chem. Eng. Commun. 14(3–6), 225–277.)*

a function of the viscosity ratio is illustrated in Fig. 5.6 (Grace, 1982). It should be noted that if the viscosity ratio is above 4, no amount of simple shear will be able to break the drop. Theoretically, this has been described as a consequence of deformation dynamics; stress is not enough for deforming a drop; it must also be given sufficient time. Walstra (2005) has argued that the characteristic timescale for drop deformation is always longer than the characteristic timescale of the flow if the viscosity ratio is larger than 4, at least for simple shear flows. Walstra's argument is discussed in greater detail in Chapter 7 of this book.

Now, consider a preemulsion drop subjected to a third type of laminar flow. The lower row in Fig. 5.5 displays a flow where the streamlines are compressed in one dimension (the vertical dimension in the illustration) and elongated in the other. This is referred to as an elongational flow. A drop subjected to this flow will experience a symmetrical "squeezing" effect, as seen in the illustration of Fig. 5.5, and as a consequence, the drop will be elongated in the streamwise direction. As opposed to the simple shear flow, the symmetry of the applied force ensures that the liquid inside the drop does not start to rotate. Due to this difference, the elongational flow is referred to as irrotational, whereas the simple shear flow is classified as a rotational flow.

The elongated drop will break when it is sufficiently deformed, just as in the simple shear flow case. However, experiments reveal that the velocity gradient necessary to break the drop is substantially lower, as seen in Fig. 5.6 comparing the two cases (Grace, 1982). Moreover, note that breakup is possible for all viscosity ratios in an elongational flow, as opposed to the simple shear case. This comparison shows that knowing which type of laminar flow is obtained in an emulsification process (rotational or irrotational) is at least as interesting as knowing the magnitude of the shear stresses. The situation is further complicated since real emulsification process fluid flows are rarely ideal simple shear flows or ideal elongational flows. The critical Weber number for these nonideal cases is in between those in Fig. 5.6. The ideal cases in Fig. 5.6 can be used as a first approximation to estimate the resulting drop size. However, more detailed investigations require that We_{Cr} is determined experimentally for the specific flow of interest.

5.2.4 Turbulent Drop Breakup—The Turbulent Viscous Mechanism

We now turn to the case where a drop is subjected to a turbulent flow. The motion of an emulsion drop through a turbulent field will depend on the drop inertia (determined by its mass) compared with the inertial forces acting on the drop by the flow (e.g., on velocity and turbulence intensity). A large or heavy drop will have sufficient inertia to resist being accelerated by the random fluctuations and will follow the average flow. A lighter or smaller drop will follow the fluctuations more closely. Note that the turbulent eddies are short-lived; old eddies are lost as energy is transported through the turbulent spectrum, and new ones are formed. The small or light emulsion drop will follow a random motion through the turbulent flow field (as illustrated in Fig. 5.2, right).

The instantaneous velocity field in a turbulent flow will never be completely uniform. Even if the time-averaged mean flow would be free of velocity gradients, the random velocity fluctuations will still be there. As the emulsion drops travels through the turbulent flow field, it will interact with the turbulent eddies. The nature of this interaction will depend on the diameter of the drop compared with the length scales of the eddies.

If the diameter of the emulsion drop is small compared with the length scale of the eddy, the drop will experience a situation that is similar to the laminar case discussed above. The eddy creates a velocity gradient G imposing a viscous shear on the drop interface, elongating it (see Fig. 5.7, left). If the velocity gradient is sufficiently strong and if the interaction is sufficiently long-lived for the drop to have time to deform, it will eventually break the drop (Kolmogorov, 1949; Vankova et al., 2007a; Walstra, 2005). This type of drop breakup is referred to as turbulent viscous (TV) breakup. Most often, a single eddy interacts with the drop, but breakup can also be due to a combination of several eddies (Andersson and Helmi, 2014).

The relationship between which drop diameters are broken up and the applied shear can be expressed using the same theoretical reasoning as for laminar viscous (LV) shear (Eq. 5.11). However, for this expression to be useful, we need to know which shear rates, G, the drops experience in a turbulent flow. In order to accurately determine G or at least a time-averaged representation of it, we would need to know the spectrum of turbulent kinetic energy (cf. Fig. 5.4), which is experimentally difficult to obtain and beyond the scope of the most widely used computer modeling approaches to determine (at least for

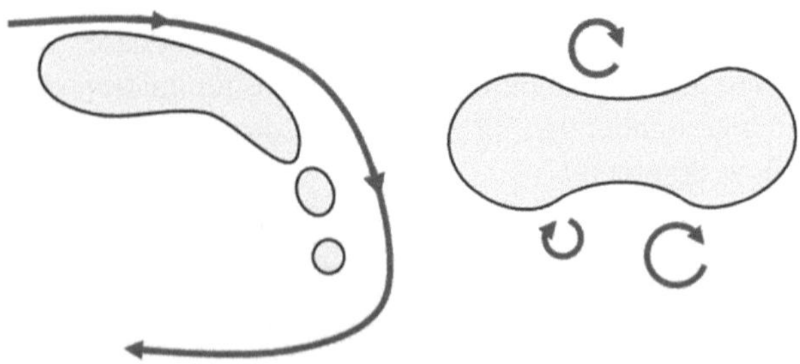

FIG. 5.7 Schematic illustration of the difference between turbulent viscous (TV) breakup (the drop is smaller than the turbulent eddy) and turbulent inertial (TI) breakup (the drop is larger than the turbulent eddy).

commercial emulsification equipment). However, under a set of simplifying assumptions regarding the turbulent flow (isotropy, that there are no velocity gradients in the time-averaged mean flow, and homogeneity, that all properties of the turbulent field are the same in all directions), the entire spectrum can be described from the fluid properties and the dissipation rate of turbulent kinetic energy, ε. A dimensional analysis results in (Kolmogorov, 1949; Hinze, 1955):

$$G = \sqrt{\varepsilon \mu_C \rho_C} \tag{5.12}$$

Although the turbulent flow in emulsification processing equipment is rarely isotropic or homogenous, Eq. (5.12) is often used since experimental data is rarely available. Although it is often a relevant estimation, the underlying assumptions should be kept in mind when applying it.

5.2.5 Turbulent Drop Breakup—The Turbulent Inertial Mechanism

The shorter length scale a turbulent eddy has, the lower is its kinetic energy. However, the smaller eddies are still able to participate in, or even dominate, drop breakup. This is referred to as the turbulent inertial (TI) mechanism of drop breakup.

A drop in a flow with small turbulent eddies will experience pressure fluctuations at the interface due to the fluctuating turbulent field (Fig. 5.7, right). The magnitude of these pressure fluctuations determines the disrupting stress under TI breakup (Hinze, 1955):

$$\sigma = \frac{\rho < u'(l_d)^2 >}{2} \tag{5.13}$$

The interesting part of Eq. (5.13) is $< u'(l_d)^2 >$, the average of the squared velocity fluctuation of eddies with length scales smaller than or equal to the critical length scale, l_d. Before turning to an interpretation of this term, we start with the related quantity, $<u'^2>$, the total averaged squared velocity fluctuations, often referred to as the Reynolds stress. For a homogeneous turbulent flow, this quantity ($<u'^2>$) is proportional to the total turbulent kinetic energy:

$$k = \frac{3}{2} < u'^2 > \tag{5.14}$$

The term in Eq. (5.13) ($<u'(l_d)^2>$) can, therefore, be interpreted as the fraction of the total turbulent kinetic energy arising from turbulent eddies that have length scales between the smallest ones (defined by the Kolmogorov length scale, η) and the critical length scale, l_d.

Once the disruptive stress, σ, is known, it can be used in the stress balance to calculate the diameter of the largest drop surviving a given turbulent field (cf. Eq. 5.6). However, $<u'(l_d)^2>$ is difficult to determine experimentally, just as with the turbulent eddy shear magnitude for the TV mechanism. Under the same assumption as used for the TV, estimation (isotropic and homogenous turbulence) and assuming that l_d is within the inertial subrange (cf. Fig. 5.4), the stress can be estimated from the ideal model spectra:

$$< u'(l_d)^2 > = \int_{2\pi/\eta}^{2\pi/l_d} E(\kappa) d\kappa \approx C\varepsilon^{2/3} l_d^{2/3} \tag{5.15}$$

where C is a proportionality constant ($C \approx 2$, Hinze 1955). Traditionally, it has been assumed that only turbulent eddy length scales smaller than the drop diameter participate in TI breakup (i.e., $l_d = d$). Under this assumption, the disruptive stress can be calculated from (Eqs. 5.13, 5.15):

$$\sigma = \frac{C\rho\varepsilon^{2/3} d^{2/3}}{2} \tag{5.16}$$

More recent experimental investigations suggest that eddy length scales of 1–3 drop diameter also make significant contributions to the disruptive stress (Andersson and Andersson, 2006a), which would result in the same general scaling as in Eq. (5.16) but with a proportionality constant that is a factor 1.6–2.0 times larger.

TI and TV have often been referred to as two different "regimes" of turbulent breakup (Kolmogorov, 1949; Walstra and Smulders, 1998). This is a relevant distinction, both theoretically and experimentally. However, it should be noted that all turbulent flows consist of a range of turbulent eddies. An emulsion drop subjected to a turbulent field will experience turbulent disrupting stresses, both from eddies with smaller length scales (TI stress) and from eddies that are larger than the drop (TV stress). However, there is a clear demarcation line between the cases where the different mechanisms dominate. The stress will be predominantly due to viscous stresses (TV) if the emulsion drop undergoing breakup is smaller than the smallest length scales of the flow. Consequently, cases where $D<\eta$ are referred to as the TV regime. If the drop, on the other hand, is larger than the Kolmogorov length scale, the TI stress will be substantially larger than the TV stress (Boxall et al., 2012). This is referred to as the TI regime of drop breakup.

Exceedingly high dissipation rates of turbulent kinetic energy are needed to lower the Kolmogorov length scale below a few micrometers, as seen from Eq. (5.8). Consequently, turbulent nanoemulsion formation occurs by a TV mechanism; at the very least, this is the case at the last and critical part of the emulsification process, when drops are reduced in size down to the final size.

5.2.6 The Influence of Viscosity on Turbulent Drop Breakup

The turbulent stress-balance approach discussed above does not take drop viscosity into account. However, according to experiments, just as with laminar breakup, dispersed phase viscosity has a profound effect on the resulting emulsion drop diameter (Calabrese et al. 1986).

On dimensional grounds, Hinze (1955) suggested that the effect of drop viscosity can be described in terms of a dimensionless group, often referred to as the Ohnesorge number:

$$Oh = \frac{\mu_D}{\sqrt{\rho_D D \sigma}} \tag{5.17}$$

Moreover, since the basic $We = We_{Cr}$ should still follow if Oh is very small (i.e., if the drop viscosity is low), Hinze suggested that the critical stress depends on Oh as

$$We_{Cr} = C_1(1 + f(Oh)) \tag{5.18}$$

where C_1 is a constant (equal to the critical Weber number of a low dispersed phase viscosity emulsion) and $f(\cdot)$ is an unknown function that tends toward zero when $Oh \to 0$. Decades later, Calabrese et al. (1986) illustrated that Eq. (5.18) can be derived theoretically from an energy balance. Calabrese et al. (1986) assumed that for the largest drops surviving the turbulent flow, the disruptive energy (energy resulting from the external stress) is equal to the total cohesive energy, consisting of both stabilizing surface energy and viscous stabilizing energy. This results in a highly similar relationship to the proposition of Hinze with a linear function f:

$$We_{Cr} = C_2(1 + C_3 \cdot Oh) \tag{5.19}$$

This expression has gained considerable experimental support (e.g., Calabrese et al., 1986; Vankova et al., 2007a) and is the most widely used approached for taking the effect of drop viscosity into account.

More recently, Gupta et al. (2016a,b) have questioned to what extent Eqs. (5.18), (5.19) can be applied to nanoemulsion formation. Gupta points out that the Ohnesorge number is generally higher for nanoemulsions than for traditional macroemulsions (since the smaller drop diameters require higher stresses) and argues that $f(\cdot)$ is not linear at high Oh. The argument is supported by a theoretical discussion based on a stress analysis in the filament formed after the drop has been deformed. It suggests that in the limit of high drop viscosities (such as for the case of triglyceride O/W emulsions) the critical Weber number is:

$$We_{Cr} = C_4 \cdot Oh^{0.4} \tag{5.20}$$

Eq. (5.20) has received experimental support from nanoemulsion formation, both using high-pressure valve homogenization and ultrasonication data (Gupta et al., 2016a).

5.2.7 Drop Break-Up Due to Cavitation

Ideally, an emulsion is a two-phase dispersion with one liquid dispersed in the other. However, in practice, they are often three-phase systems with a small amount of gas (e.g., air) dispersed in the form of small bubbles. If the static pressure of the dispersion would be decreased, this would influence the gas bubbles, since gases in general, and air in particular, are compressible. Consequently, the air bubbles will expand as the pressure decreases. If the pressure is increased again, the drop will be recompressed. If the static pressure is varied periodically, the gas bubble will cycle through a series of expansions and compressions. If the frequency and the amplitude of these fluctuations are sufficiently large, the compression step will eventually lead to a catastrophic collapse, causing the bubble to implode; see Fig. 5.8 (left). The implosion will be fast and forceful. During a short time period (picoseconds), the local temperature increases to more than 10,000°C (Brennen, 1995; Qin et al., 2007). The pressure shock wave from the implosion can hit nearby emulsion drops directly or induce high local turbulence that breaks the drops (Abbas et al., 2013).

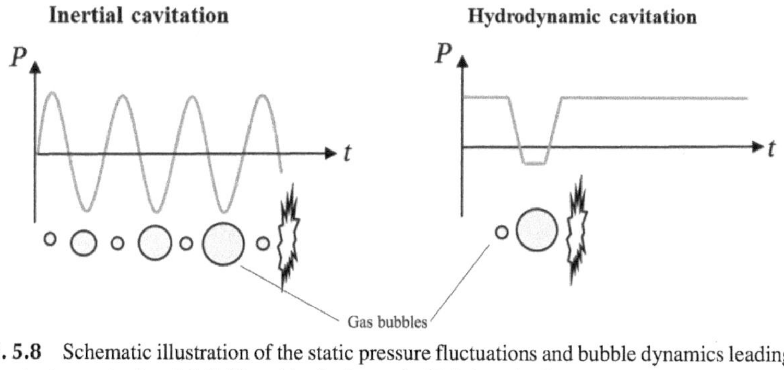

FIG. 5.8 Schematic illustration of the static pressure fluctuations and bubble dynamics leading up to implosion under inertial (left) and hydrodynamic (right) cavitation.

Emulsion drop breakup due to shock waves sent out by imploding gas bubbles is referred to as cavitation-induced emulsification. When the static pressure fluctuation causing the collapse is applied externally (as in the example above), the flow is referred to as a noninertial cavitating flow. Noninertial cavitation can be induced by an ultrasonication probe, where the pressure fluctuations are applied through high-frequency vibrations (see Section 5.4.4).

Noninertial cavitation (also referred to as hydrodynamic cavitation) is based on the same general principle—the energy required for drop breakup is supplied via shock waves from imploding gas bubbles. However, the bubbles are formed in a different process. Consider a flow entering a narrow contraction (see Fig. 5.9); as the cross-sectional area of the channel decreases, the average fluid velocity increases. This corresponds to an increase in the kinetic energy of the fluid. According to Bernoulli's principle, the total energy per unit volume must be conserved between a position before and after the contraction:

$$P+\frac{\rho u^2}{2}=\text{constant} \tag{5.21}$$

Consequently, the increased fluid velocity implies a decrease in the static pressure as the fluid enters the contraction. If the reduction in cross-sectional area is sufficiently large, the local static pressure will decrease below the vapor pressure of the liquid. This results in a flow-induced "boiling"; vapor bubbles nucleate and expand in the flow. These vapor bubbles will be transported by the fluid until reaching a position with a higher pressure, such as the area expansion at the end of the contraction (see Fig. 5.9). These bubbles will implode following a catastrophic collapse, sending out shock waves that can fragment the drops.

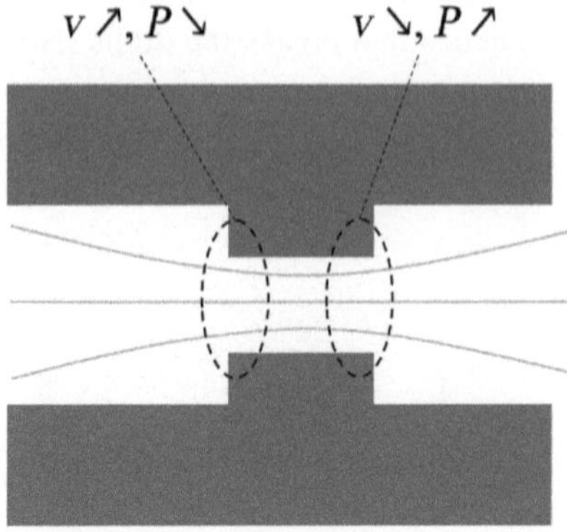

FIG. 5.9 A cause of hydrodynamic cavitation. Static pressure decreases as velocity increases entering the contraction. Static pressure increases again when the fluid exits the contraction (or as it passes the vena contracta).

5.3 DYNAMICS OF DROPLET FORMATION AND STABILIZATION

The stress-balance expressions in Section 5.2 offer a starting point for understanding the underlying principle of emulsification. They have proved to be useful for explaining empirical findings such as the scaling between resulting drop size, emulsion characteristics, and operating conditions. However, there are a number of aspects of emulsification that cannot be understood from the static stress analysis alone. First, emulsification results in a DSD, whereas stress analysis suggests that all drops will fragment down to a critical drop size (D_{max}). Secondly, experiments show that emulsification is dynamic; drops do not break instantly to the steady-state drop size predicted by static stress analysis. Instead, drops are gradually fragmented into smaller drops with processing time or number or passages. Third, stress analysis only describes drop breakup, whereas it has been empirically proved that a substantial extent of coalescence occurs during emulsification. These points suggest that scaling laws must be supplemented by another theoretical perspective in order to provide us with a fundamental understanding of mechanical emulsification.

Many commercial emulsification processes occur by turbulence, cavitation, or combination of the two (see Section 5.4.6). Both turbulence and cavitation are stochastic phenomena, and it is reasonable to postulate that any true understanding of the emulsification process must be based on a stochastic description. Laminar flow is, in principle, deterministic, but laminar breakup is also often described in stochastic terms (Maindarkar et al., 2014; Wieringa et al., 1996).

5.3.1 From Possible to Probable—Population Balance Modelling

Fragmentation and coalescence generally depend on the size of the drops. Larger drops are more readily fragmented and coalesce faster. Population balance modeling (PBM) has become the standard tool for describing and studying the stochastic and drop diameter dependent nature of emulsification processes (Janssen and Hoogland, 2014). The PBM is a framework—based on a differential mass balance approach—used to describe how the dispersed phase is redistributed between drops of different sizes due to the simultaneous fragmentation and coalescence taking place during emulsification. The PBM describes the DSD evolution over time, provided that the rate of fragmentation and coalescence are known for each drop size.

Consider an emulsion where there are $n(D,t)dD$ drops (per unit volume of emulsion) with diameter larger than D but smaller than $D+dD$. If the emulsion is subjected to fragmentation, the number of drops in this size class will decrease when drops with diameters in this interval are fragmented (A in Fig. 5.10) and increase whenever a larger drop breaks and forms at least one fragment with diameter between D and $D+dD$ (B in Fig. 5.10). Similarly,

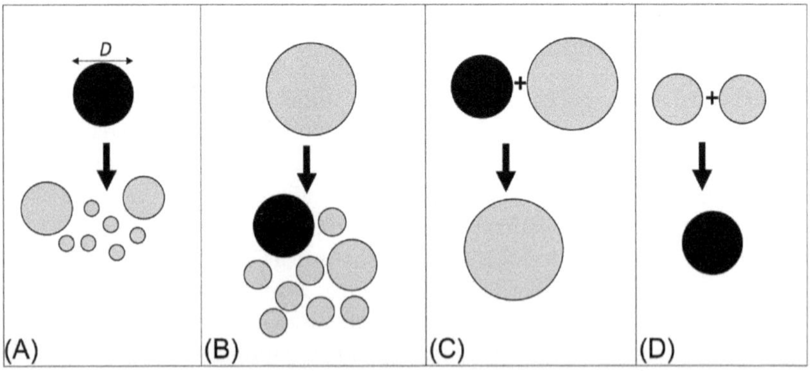

FIG. 5.10 Illustration of the four PBM source terms contributing to the change in the number of drops belonging to the size class representing drops of diameter D. (A) Loss due to fragmentation, (B) Gain due to fragmentation, (C) Loss due to coalescence, and (D) Gain due to coalescence.

if the emulsion is subjected to coalescence, the number of drops will decrease when a drop of this size successfully participates in a coalescence event (C in Fig. 5.10) and increase when a coalescence occurs that results in a new drop with diameters between D and $D+dD$ (D in Fig. 5.10). Under a set of generally permissible assumptions (i.e., binary coalescence only and uncorrelated drop-size concentrations), the DSD evolves according to (Ramkrishna, 2000)

$$\frac{\partial n(D,t)}{\partial t} = \overbrace{-g(D)n(D,t)}^{A} + \overbrace{\int m(D')f(D,D')g(D')n(D',t)dD'}^{B}$$
$$\underbrace{-n(D,t)\int n(D',t)\beta(D',D)dD'}_{C} + \underbrace{\frac{1}{2}\int \beta(D',D-D')n(D',t)n(D-D',t)dD'}_{D} \qquad (5.22)$$

Eq. (5.22) contains three model functions, often referred to as *kernels*, required to close the equations and fully describe the emulsification process:

- $g(D)$, the fragmentation rate of a drop of diameter D
- $\beta(D,D')$, the coalescence rate between a drop of diameter D and D'
- $m(D)$, the number of fragments formed upon fragmenting a drop of diameter D
- $f(D,D')$, the probability distribution of obtaining a fragment of size D from the breakup of a drop of diameter D'

Eq. (5.22) is an integrodifferential equation. Solving it requires special numerical schemes. A comprehensive introduction to PBM and solution method is given by Ramkrishna (2000).

5.3.2 The Rate of Fragmentation

When deriving an expression for the largest drop diameter that can survive a turbulent field (D_{max}) in Section 5.2, we assumed that the turbulence can be characterized by an average turbulent stress, which is, in turn, approximated from the dissipation rate of turbulent kinetic energy, ε. This assumption is part of what is often referred to as the Kolmogorov-Hinze approach to describe emulsification. However, as discussed by Kolmogorov (1949) himself, this is an oversimplification.

Each turbulent eddy will only survive for a short period of time. Looking at any given point in space, the turbulent stress varies stochastically with a mean value $<\sigma>$. Fig. 5.11 displays an experimentally obtained probability distribution of the instantaneous stress, measured in a rotor-stator mixer (Håkansson et al., 2017). At each instant, there is a substantial probability of a stress exceeding the average value, as seen in the figure.

Emulsion drops are often deformed and broken up due to interaction with individual eddies. The size of each drop will therefore depend on the energy content of the eddies that they happen to encounter. This explains why drop breakup is stochastic and why a range of different drop sizes are created under turbulent emulsification.

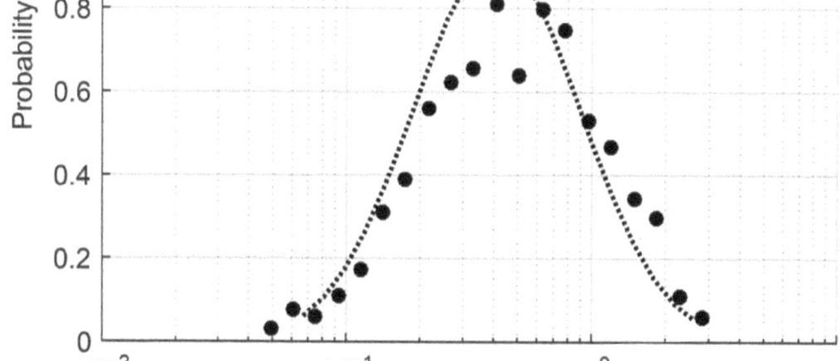

FIG. 5.11 Experimentally measured probability density distribution for the instantaneous stress σ, in relation to the average stress $<\sigma>$ (*markers*), compared with the best-fit lognormal probability distribution (*line*). *(Data obtained in the efficient region of emulsification in a rotor-stator mixer using particle-image velocimetry; see Håkansson, A., Andersson, R., Mortensen, H.H., Innings, F., 2017. Experimental investigations of turbulent fragmenting stresses in a rotor-stator mixer. Part 2. Probability distributions of instantaneous stresses. Chem. Eng. Sci. (in press).)*

Several theoretical fragmentation rate models have been proposed, often starting from models of the stochastic properties of turbulence similar to the distributions in Fig. 5.11. Different models have been proposed using somewhat different approaches. There is, as yet, no generally accepted fragmentation rate kernel for describing turbulent emulsification (see the review by Liao and Lucas, 2009, for an introduction to the field).

The dynamic fragmentation rate models and the more static stress-balance expressions from Section 5.2 can be seen as two different approaches of describing drop breakup. These two approaches are compared in Fig. 5.12. The figure displays the parameter free fragmentation rate of Andersson and Andersson (2006b), showing the fragmentation rate as a function of drop diameter for two different dissipation rates of turbulent kinetic energy (ε). The fragmentation rate increases rapidly with drop size (note the logarithmic scale). The limiting maximum drop size according to the stress-balance approach in Section 5.2 has also been inserted in the figure. Note that the stress balance and the dynamic theory are in close compliance. Below the D_{max} limit, the fragmentation rate is very low, indicating that breakup of these small drops is very unlikely. However, there are also differences between the approaches. The fragmentation rate is small but nonzero, even for drops substantially smaller than D_{max}, indicating that breakup of these drops is rare but possible if the emulsion is subjected to it for a sufficient length of time.

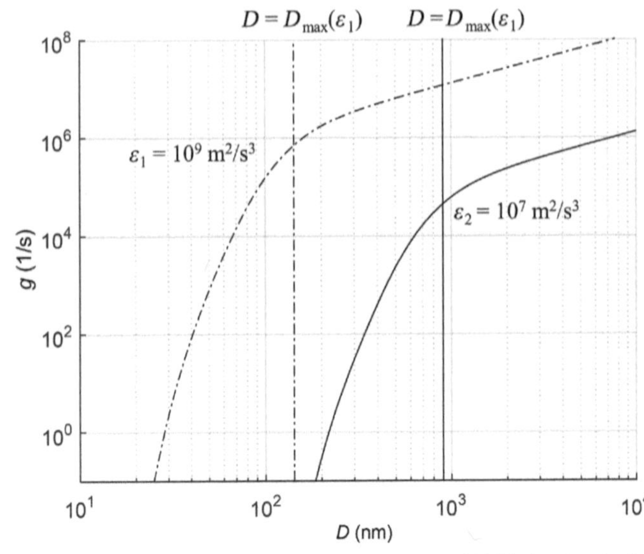

FIG. 5.12 Comparing a rate model (Andersson and Andersson, 2006b) showing the fragmentation rate as a function of drop diameter (*curve*) with the steady-state drop diameter according to the stress-balance approach, Eqs. (5.6), (5.15) (*vertical lines*).

The majority of the literature on emulsification fragmentation rates is theoretical. Studies often propose rate expressions—especially $g(D)$—and compare solutions with the PBM with specific experiments. Independent and direct measurements of fragmentation rates are rare. Some attempts have been made at measuring fragmentation rates directly, for example, by drop breakup visualization (Andersson and Andersson, 2006a) or by back-calculating the rate from an inverse PBM (Vankova et al., 2007b; Håkansson and Hounslow, 2013).

It should also be noted that the fragmentation rate needs to be supplemented by the number of fragments formed and the fragment size probability function to fully describe the breakup process. There is still a considerable amount of disagreement on the shape of the fragment size probability distribution (Liao and Lucas, 2009). Moreover, the number of fragments formed per breakup, $m(D)$, is also controversial. It has long been assumed that breakup is predominantly binary (Liao and Lucas, 2009). However, several independent experimental investigations now indicate that multiple breakup is more often the case (Andersson and Andersson, 2006a; Innings et al., 2011; Tcholakova et al., 2007).

5.3.3 The Importance of Coalescence

Emulsification processes are designed to give a substantial net fragmentation and thus a net decrease in drop sizes. However, it is important to note that coalescence occurs at substantial rates during emulsification (Ganley and van Duijneveldt, 2016; Jafari et al., 2008). Moreover, understanding the conditions under which coalescence takes place is important for emulsification processing design, especially in the case of nanoemulsion formation where exceptionally small drops are needed, and thus, coalescence must be kept at a very low level.

Here, it should also be noted that there is a fundamental difference between the coalescence taking place during emulsification and the coalescence that takes place afterward. There is reason to believe that nanoemulsions are less subjected to coalescence after emulsification compared with micrometer range emulsions. Nanoemulsion drops have a larger ratio of emulsifier adsorption layer thickness to drop radius, which makes them more stable toward coalescence, at least when stabilized by steric-entropic effects (Tadros et al., 2004). However, this factor does not obviously come into effect when considering coalescence during emulsification. Time is needed for emulsifiers to absorb to the newly created interface (Tcholakova et al., 2008), and the coalescence during emulsification can occur before the interface has been stabilized.

The scientific study of coalescence during emulsification is complicated by the fact that it occurs simultaneously with fragmentation. Fragmentation can be studied without coalescence by reducing the volume fraction of the disperse phase, thus making the collision rate very low (e.g., Vankova et al., 2007b). It is, however, significantly more challenging to find conditions where coalescence can be decoupled from fragmentation.

Several mathematical models have been suggested for describing how the coalescence rate depends on factors such as operating conditions and emulsion characteristics; see the review by Liao and Lucas (2010). Unfortunately, there is still no generally accepted expression, and researchers interested in quantifying coalescence for a specific product or range of operating conditions must, therefore, rely on experimental determination.

Due to the importance of coalescence, several decupling techniques for quantifying coalescence during emulsification can be found in literature; see Håkansson (2016) for a critical review. The methods found in literature can, based on their general approach, be divided in two broad classes. The first class applies conditions where coalescence dominates fragmentation for a short time, while the second measures the drop intermixing due to coalescence.

An example of a study in the first class is that of Mohan and Narsimhan (1997), where an emulsion is circulated across a high-pressure valve homogenizer at a high pressure until a dynamic equilibrium drop size is obtained. The homogenizing pressure (and, consequently, turbulent stress) is then drastically reduced. Provided that the step is sufficiently large, so that even the largest drops have a negligible fragmentation rate at the new conditions, a brief period occurs where the drop dynamics is controlled by coalescence alone. The increase rate of drop diameter with time after the step can thus be used to quantify the coalescence rate (Mohan and Narsimhan, 1997).

The second class of coalescence rate quantification methods studies coalescence by investigating how a tracer molecule transfers between emulsion drops. Fig. 5.13 illustrates the general principle. Assume that a fraction of the emulsion drops are marked by a tracer that does not influence the fragmentation or coalescence, but can be distinguished experimentally. As an example, Lobo et al. (2002) marked a small fraction of the drops with a fluorescent probe. After undergoing breakup, the fragments will have the same tracer concentration as the mother drop. Consequently, the variation in tracer concentration seen across a large number of drops is unaffected by fragmentation (Fig. 5.13, left).

However, when two drops with different tracer concentration coalesce, their contents will mix, and the resulting drop will have an intermediate tracer concentration (Fig. 5.13, right). If coalescence is allowed to continue for a sufficient length of time, all drops will reach the same uniform tracer concentration, and the tracer variation across a large number of drops is zero. In the intermediary cases, the variation of the tracer concentration describes the extent of coalescence and can be used to describe the rate of coalescence. This approach has been applied for studying emulsification by several investigators, using different types of tracers and methods for quantifying the tracer variation (Lobo et al., 2002; Taisne et al., 1996; Henry et al., 2009).

Three consistent conclusions on the coalescence rate and how it depends on the emulsification conditions can be drawn from the experimental coalescence rate investigations:

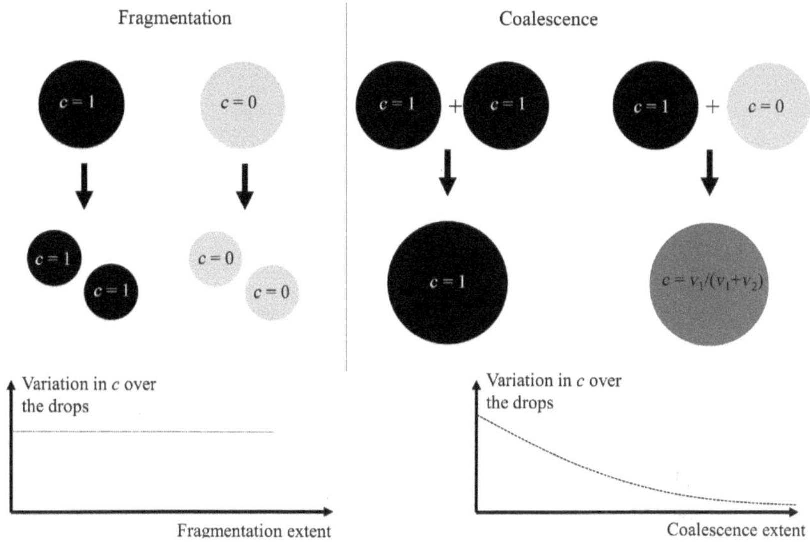

FIG. 5.13 Illustration of the principle of drop-intermixing techniques for measuring the extent of coalescence. *(Modified from Håkansson, A., 2016. Experimental methods for measuring coalescence during emulsification—a critical review. J. Food Eng. 178, 47–59, with permission from the publisher.)*

- Coalescence frequency increases with increasing emulsification intensity, that is, impeller speed or homogenizer pressure (Håkansson and Hounslow, 2013; Lobo et al., 2002; Mohan and Narsimhan, 1997; Taisne et al., 1996). This is expected since the coalescence rate generally increases with hydrodynamic intensity due to the increase in drop-drop collision frequency and the increase in collision energy. The net decrease in drop size obtained when increasing the hydrodynamic intensity is due to the fact that the fragmentation frequency increases faster with intensity than the coalescence frequency (Håkansson and Hounslow, 2013).
- Increasing the volume fraction of oil generally increases the rate of coalescence (Lobo et al., 2002; Mohan and Narsimhan, 1997). This can be attributed to the increased collision frequency due to the larger number density of drops.
- The coalescence rate decreases as a function of the ratio between emulsifier concentration and dispersed phase concentration, and levels off at a very low rate when the emulsifier to dispersed phase ratio is sufficiently high (Henry et al., 2009; Taisne et al., 1996).

The three trends are illustrated schematically in Fig. 5.14.

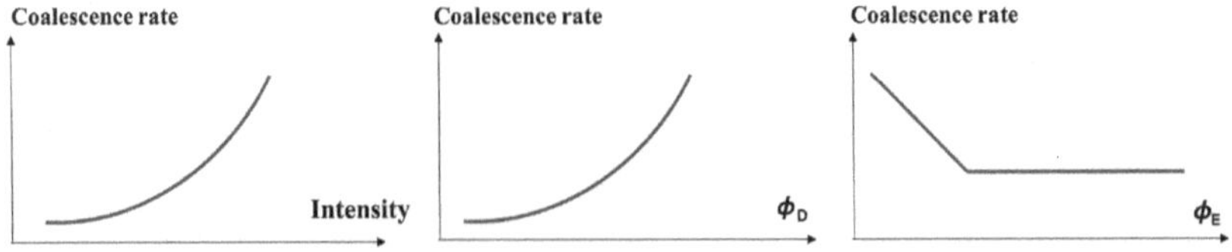

FIG. 5.14 Schematic illustration of how coalescence rate is influenced by turbulence (or laminar shear) intensity, volume fraction of disperse phase (ϕ_D), and emulsifier to dispersed phase concentration (ϕ_E).

5.3.4 Some Additional Complications Related to Hydrodynamics

The hydrodynamic stress plays an important role in describing emulsification, regardless of whether the classical static stress analysis or the PBM framework is used. It is important to note that emulsification often takes place in inhomogeneous flows—where the disrupting stress acting on a drop varies substantially in space. This implies that the local maximum disruptive stress is often orders of magnitude larger than the overall average value. Therefore, a drop entering a high-pressure valve homogenizer, rotor-stator mixer, or other high-energy nanoemulsion formation processes will pass through most of the volume unaffected before entering a narrow region of high disruptive stress. It is in this narrow region where most of the breakup takes place. The average turbulence intensity or shear, averaged over the tank or emulsification system, is therefore a poor description of the hydrodynamic driving force of emulsification. Understanding an emulsification process requires that we understand not only the average energy density but also the maximum local stress (Zhou and Kresta, 1998).

5.4 INTRODUCING THE HIGH ENERGY METHODS

Several high-energy methods are used for emulsion formation in general and nanoemulsion formation in particular. This section introduces the methods treated in more detail in the following chapters.

5.4.1 Rotor-Stator Emulsification

Rotor-stator mixers are used extensively for emulsion formation in different fields of process industry, such as in the food, pharmaceutical, and cosmetic industries (Atiemo-Obeng and Calabrese, 2004). They are often considered to be the standard method for emulsions with intermediate-to-high viscosity and/or dispersed phase volume fraction (Schultz et al., 2004; Jafari et al., 2008; Jafari et al., 2007a). A schematic illustration is shown in Fig. 5.15 (left). The rotor-stator mixer head consists of a rotor, mounted on a rotor shaft, and a perforated stator screen just outside it. The rotor works as the impeller in a centrifugal pump; fluid enters the rotor axially along the rotor shaft. The rotor accelerates the fluid, first tangentially, and then, the blades redirect the fluid,

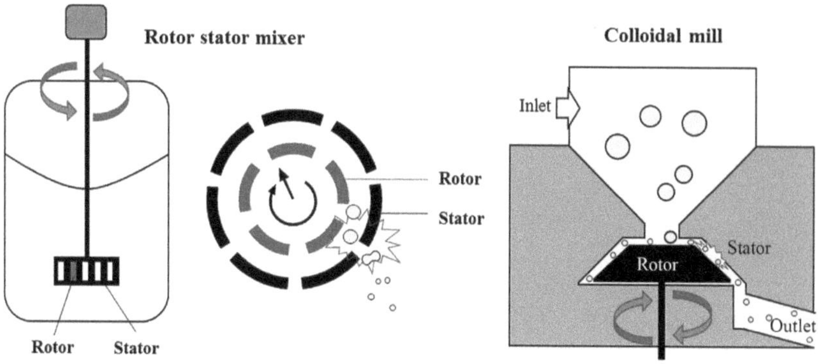

FIG. 5.15 Rotor-stator emulsification devices (highly schematic): a rotor-stator mixer (left) and a colloidal mill (right).

feeding it through the slots in the stator screen. There is still some uncertainty as to where breakup takes place and by which mechanism. However, experimental hydrodynamic investigations show intense hydrodynamic stresses, both in the form of velocity gradients and in the form of intense turbulence inside and downstream of the slot exit (Mortensen et al., 2011).

The hydrodynamic intensity and thus the resulting emulsion drop size depend on the rotor speed, often described in terms of the rotor tip speed, which generally ranges from 10 to 30 m/s in industrial applications (Atiemo-Obeng and Calabrese, 2004). The reduction in mean drop diameter from passing through the rotor-stator region is relatively slow, and each fluid element must generally make a significant number of passages through it in order to reach the steady-state drop diameter, especially to form the small drop sizes required for a nanoemulsion.

Rotor-stator mixers can be operated in batch mode by mounting the rotor stator in a closed tank, in continuous mode by enclosing the rotor stator inside a narrow centrifugal-pump-like casing for in-line operation or in a semicontinuous mode by recirculating the emulsion across an in-line system (Atiemo-Obeng and Calabrese, 2004).

The colloidal mill is another subtype of the rotor-stator emulsification family where the emulsion is squeezed through the narrow gap of an inclined rotor and a stator block. Drops entering the rotor-stator clearance are elongated and breakup is due to LV stresses, see illustration in Fig. 5.15 (right).

Nanoemulsion formation by rotor-stator methods is discussed in greater detail in Chapter 6 of this book.

5.4.2 High Pressure Valve Homogenization

High-pressure valve homogenization is one of the most commonly used methods in industrial nanoemulsion formation (Schubert et al., 2003). It is used

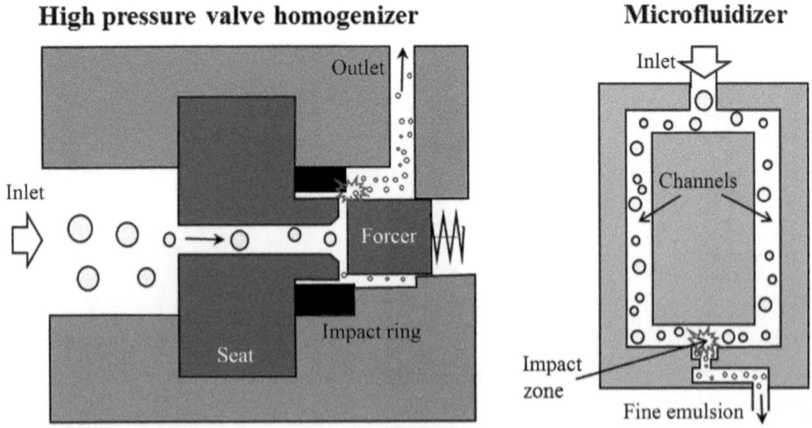

FIG. 5.16 Schematic illustrations of the principle of high-pressure valve homogenization (left) and microfluidization (right).

for reducing the drop size of a coarse preemulsion (often formed using rotor-stator mixing) into a narrow drop distribution with smaller drops. The high-pressure valve homogenizer consists of a pump forcing the preemulsion through a valve forming a narrow gap (10–100 μm) under high pressure; typically, between 50 and 200 MPa is applied for nanoemulsion formation.

The homogenization valve is illustrated in Fig. 5.16 (left). As the fluid enters the valve, it is accelerated to velocities of order magnitude 100 m/s, giving rise to an elongational flow in the valve entrance. The high local velocity also reduces the local pressure below the vapor point forming cavitation bubbles. As these bubbles travel further downstream in the valve, they experience higher local pressures and burst, sending out powerful shock waves.

Flow within the narrow gap is typically laminar or transitional. However, when the fluid exits the narrow gap and enters the much larger outlet chamber, it forms into a powerful turbulent jet (Innings and Trägårdh, 2007). It is here in the turbulent jet formed downstream of the gap that drop breakup has been observed in visualization studies (Bisten and Schuchmann, 2016).

High-pressure valve homogenizers are available in a broad range of scales allowing for continuous production from small laboratory scale (10 L/h) to large production scale (10,000 L/h) (Phipps, 1985).

Nanoemulsion formation by high-pressure valve homogenization is discussed in greater detail in Chapter 7 of this book.

5.4.3 Microfluidization

The microfluidizer is similar in design to the high-pressure valve homogenizer. It consists of a pump forcing a coarse preemulsion into a narrow flow channel often referred to as the reaction chamber (Jafari et al. 2006, 2007b,c). Just as in the high-pressure valve homogenizer, the fluid is accelerated as it enters into the annular narrow channel. However, instead of letting the fluid form a turbulent jet emerging into a stationary outlet chamber at the channel exit, it is angled so as to direct the annular jets toward each other, creating a high-intensity impact zone downstream of the narrow channel; see Fig. 5.16 (right). This design, coupled with high inlet pressures (100–300 MPa), results in the formation of small drop sizes that makes microfluidization a highly interesting method for nanoemulsion formation.

Microfluidizers for continuous production are available in a broad range of production volumes, from batches of 10 mL up to continuous production in the 10,000 L/h range.

Nanoemulsion formation and microfluidization are discussed in greater detail in Chapter 8 of this book.

5.4.4 Ultrasonication

Rotor-stator mixers, high-pressure valve homogenizers, and microfluidizers all supply the energy needed for emulsification by accelerating the preemulsion, either by using a rotor blade or by using a pump to increase pressure in order to achieve acceleration in a flow constriction. Ultrasonication is fundamentally different and achieves breakup of emulsion drops using ultrasonic sound waves, typically applied in the frequency range between 20 and 100 kHz (Abbas et al., 2013). The sound waves give rise to cavitation that fragments the drops.

The ultrasonic probe (Fig. 5.17, left) uses inertial cavitation to provide the necessary energy for drop breakup. Sound waves are generated using a piezotransmitter, which converts electric voltage into mechanical vibrations (Jafari et al., 2006, 2007c). These vibrations are amplified and directed using a probe put into contact with the dispersion. The ultrasonic sound waves give rise to sinusoidal pressure fluctuations in the fluid. Cavitation bubbles are

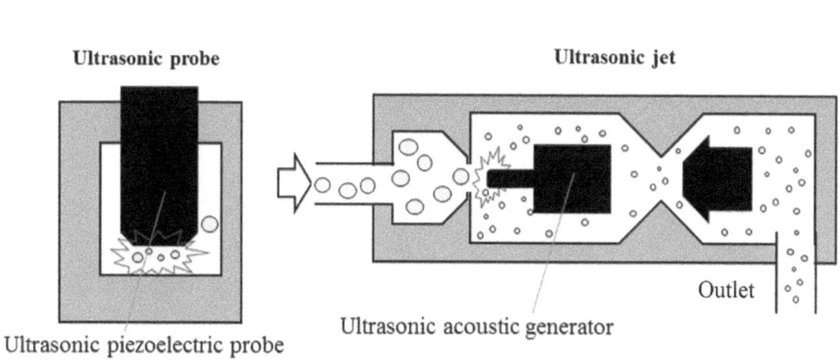

FIG. 5.17 Two methods for emulsion formation using ultrasonication (schematic illustrations): an ultrasonic batch probe (left) and an ultrasonic jet device (right).

formed, and as the pressure fluctuates, the bubbles go through a series of contractions and expansions. When the bubbles implode, they induce high-shear conditions, and emulsion droplets break into smaller droplets (Abbas et al., 2013; Shamsara et al., 2015). The piezoelectric probe can be used for batch mode ultrasonication by mounting it in a beaker or batch reactor or for continuous operation by pumping a fluid stream passed one or several piezotransmitting probes.

Liquid jet generators are an alternative method for generating the ultrasonic pressure fluctuations in continuous mode. A coarse emulsion is pumped through an orifice to impinge on a sharp-edged blade. The flow causes the blade to vibrate rapidly, thus generating the ultrasonic field and cavitation responsible for breaking the drops (McClements, 2016).

Nanoemulsion formation using ultrasonication is discussed in greater detail in Chapter 9 of this book.

5.4.5 Membrane Emulsification

Fig. 5.18 illustrates the principle of membrane emulsification. A porous membrane separates the inner continuous phase and the outer disperse phase. By applying a pressure across the membrane, the dispersed phase is forced through the narrow channels. Simultaneously, a continuous flow of continuous-phase liquid is pumped transversal to the membrane. The disperse phase fills up the membrane pores. As the disperse phase reaches the end of the pore, a neck is formed. The transverse flow applies a laminar shear force on the drop filament and eventually detaches it from the membrane, forming a drop that is swept away by the flow and quickly relaxes into spherical shape.

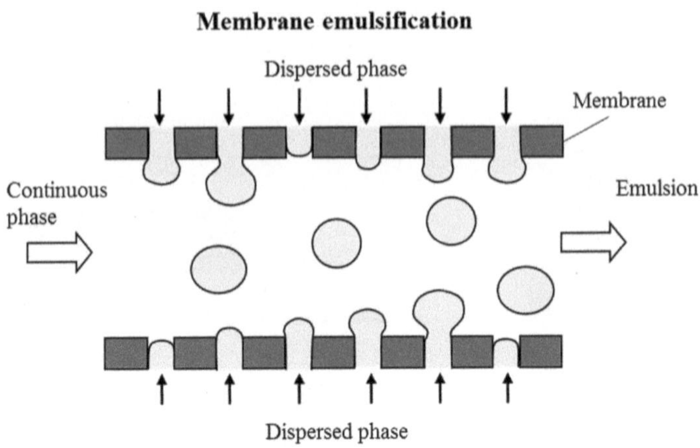

FIG. 5.18 Schematic illustration of drop formation in membrane emulsification.

Membrane emulsification is a relatively novel method for emulsion formation; the technology was invented in the late 1980s by Nakajima and Shimizu (Piacentini et al., 2014). It requires low-energy inputs compared with traditional techniques such as rotor-stator emulsification, high-pressure valve homogenization, or microfluidification. Another advantage with the technique is that it results in relatively uniform drop distributions.

Drop breakup is governed by the membrane properties, geometry, and hydrophobicity. However, it should also be noted that, compared with the other techniques, membrane emulsification has not been as extensively applied for large-scale production.

Nanoemulsion formation using membrane emulsification is discussed in greater detail in Chapter 10 of this book.

5.4.6 Comparing the High-Energy Methods

As was seen in the introduction of this chapter, emulsion formation is thermodynamically unfavored due to the increase in interfacial energy when breaking the large drops into smaller ones. The theoretical amount of energy that needs to be supplied in forming a nanoemulsion is of order magnitude 10–10^2 kJ/m^3, depending on the volume fraction of the disperse phase (see Fig. 5.1). For most of the abovementioned methods, the applied energy is several orders of magnitude greater. Table 5.1 provides some estimates of the energy densities that are applied under standard operation of the different techniques. With the exception of membrane emulsification, the energy density is generally in the range of 10^3–10^5 kJ/m^3 with relatively small variations between different techniques. This can be attributed to the fact that emulsion formation is indirect. Rotor-stator and high-pressure devices rely on accelerating the whole fluid, and only a small fraction of this energy will be converted to the high-intensity turbulence required to break drops. A similar case can be made against ultrasonication; much of the energy is used for vibrating the fluid, and apparently, not all the shock wave energy is transferred directly to drop interfaces.

Membrane emulsification is an exception and requires a substantially lower energy density because energy is applied more directly to the disperse phase during deformation. Nonetheless, membrane emulsification is relatively rare in industrial applications of nanoemulsion formation, partly because of the challenges involved in making the resulting drops sufficiently small and partly due to challenges in scale-up (Piacentini et al., 2014).

The different high-energy techniques differ substantially in terms of how well they handle dispersions with high viscosity. Table 5.1 provides a comparison between the abovementioned techniques. High-pressure valve homogenizers and microfluidizers are most suitable for low viscosity products, ultrasonication, and membrane emulsification that are used up to intermediary viscosities, whereas high viscosity products are most suitably processed using rotor-stator devices.

TABLE 5.1 A Comparison of Some Emulsification Processes

Type	Source of Energy	Breakup Mechanism	Energy Density (kJ/m^3)	Viscosity Range (m Pas)
Rotor-stator mixer	Rotor	TI, TV, LV	10^3–10^5	20–500
Colloidal mill	Rotor	LV	10^3–10^5	20–500
High-pressure valve homogenization	Pump	TI, TV, CI	10^4–10^6	1–200
Microfluidization	Pump	TI, TV, CI	10^4–10^6	1–200
Ultrasonication	Ultrasound	CI	10^3–10^5	Low to medium
Membrane emulsification	Transmembrane pressure	LV[a]	10 –10^3	Low to medium

[a]*The drop formation under membrane emulsification is similar but not identical to the laminar viscous mechanism, since the membrane geometry plays an important role; see Sugiura et al. (2002).*
CI, cavitation; *LV*, laminar viscous; *TI*, turbulent inertial; *TV*, turbulent viscous.
Based on data reported by Abbas, A., Hayat, K., Karangwa, E., Bashari, M., Zhang, X., 2013. An overview of ultrasound-assisted food-grade nanoemulsions. Food Eng. Rev. 5, 139–157; Davies, J.T., 1985. Drop sizes of emulsions related to turbulent energy dissipation rates. Chem. Eng. Sci. 40(5), 839–842; Jafari, S.M., Assadpoor, E., He, Y., Bhandari, B., 2008. Re-coalescence of emulsion droplets during high-energy emulsification. Food Hydrocoll. 22(7), 1191–1202; Rayner, M., 2015. Scales and forces in emulsification. In: Rayner, M., Dejmek, P. (Eds.), Engineering Aspects of Food Emulsification and Homogenization, CRC Press, Bocca Raton, FL pp. 3–32; Schultz, S., Wagner, G., Urban, K., Ulrich J., 2004. High-pressure homogenization as a process for emulsification. Chem. Eng. Technol. 27(4), 361–368; Walstra, P., 2005. Emulsions. In: Lyklema, J. (Ed.), Fundamentals of Interface and Colloid Science. Elsevier, Amsterdam, pp. 8.1–8.94.

5.5 SUMMARY AND NOTES ON THE PARTICULARITIES OF NANOEMULSION FORMATION

We have reason to believe that nanoemulsion formation has much in common with the formation of emulsions in general. The most obvious difference is that the resulting drop diameters are smaller than for emulsions in the traditional micrometer range. In order to highlight similarities and differences in the macroemulsion case, we will conclude this chapter by considering the journey of an emulsion drop, from the point where emulsification start until it has been reduced to a nanoemulsion drop. The first step is often to form a coarse preemulsion. Assume that this first step has resulted in a macroemulsion with a DSD characterized by a surface-weighted average diameter of 10 μm. We will follow this drop as it is transformed to a nanoemulsion.

As discussed, the drop is stabilized by the Laplace pressure and a viscous stress. Feeding this preemulsion through an emulsification process will subject it to external disrupting hydrodynamic stresses. Most frequently, the drops only experience these high hydrodynamic stresses during short time periods or in a small fluid volume, for example, a high-turbulence zone close to the stator slot in rotor-stator mixing or downstream of the narrow gap in a high-pressure valve homogenizer. The type and magnitude of the stress will depend on the specific process to which the drop is subjected.

The stresses applied will deform and break the drop, creating two or more fragments. Note that these fragments are not sufficiently small to form a nanoemulsion. The resulting emulsion will need to reenter the high-intensity zone several times in order to refragment the fragments. Moreover, drop breakup is rarely deterministic in the sense that all drops subjected to the stress field will break at once. This further increases the requirement for long processing times (in the case of batch processing) or several recirculations over the emulsification process (for continuous mode of operation). It should also be noted that nanoemulsions, due to their small drop sizes, will generally be smaller than the Kolmogorov length scale, at least during the final steps of the emulsification process. Thus, regardless of whether the breakup is due to turbulent or laminar stresses, it will be viscous in the last and critical step. Viscosity will, therefore, play an important role in most cases of nanoemulsion formation, which is not always the case for micrometer range emulsions (e.g., emulsification by the TI mechanism, which is not uncommon for macroemulsion formation; see Hinze, 1955).

As soon as new drop fragments are formed, emulsifiers are transported toward the interface, either via Brownian motion (for small surfactants) or though convective transport (for macromolecular emulsifiers) (Tcholakova et al., 2008). The same stresses that act in breaking a drop can also push drops into contact. If this transport of emulsifier to the interface is not sufficiently fast or if the emulsifier is unable to provide sufficient drop-drop repulsion, the drop fragments might recoalesce before exiting the high-intensity zone. This must be avoided in order to ensure the exceedingly small drop sizes required for nanoemulsion formation.

It must also be remembered that the fundamental knowledge on the mechanism of emulsion formation in commercial emulsification equipment is still somewhat limited. Most emulsification processes have been developed without extensive knowledge of how design, geometry, and operation influence the hydrodynamic stresses acting on emulsion drops. Commercial designs are still more frequently based on trial and error and on the manufacturer's practical experience. During the last few decades, there has, however, been an increase in the fundamental understanding of emulsification processes. This has resulted in new suggestions on how to improve design and operating conditions (examples thereof can be seen in the following chapters). Improved process design based on an increased fundamental knowledge will be of special importance for the future of nanoemulsion formation because of the high-performance requirements when forming nanometer-scale drops.

It should also be noted that most of the fundamental work on emulsification is based on micrometer range emulsions, as illustrated by the recent debate of how to extend the effect of drop viscosity on turbulent fragmentation to nanoemulsion formation (Gupta et al., 2016a,b) and the shift in the relative importance of the relative strength of destabilizing mechanisms when comparing macroemulsions and nanoemulsions (Tadros et al., 2004); more studies on the particularities of nanoemulsion formation are needed to improve efficiency and extend applicability.

REFERENCES

Abbas, A., Hayat, K., Karangwa, E., Bashari, M., Zhang, X., 2013. An overview of ultrasound-assisted food-grade nanoemulsions. Food Eng. Rev. 5, 139–157.

Andersson, R., Andersson, B., 2006a. On the breakup of fluid particles in turbulent flows. AICHE J. 52 (6), 2020–2030.

Andersson, R., Andersson, B., 2006b. Modeling the breakup of fluid particles in turbulent flows. AICHE J. 52 (6), 2031–2038.

Andersson, R., Helmi, A., 2014. Computational fluid dynamics simulation of fluid particle fragmentation in turbulent flows. Appl. Math. Model. 38 (17–18), 4186–4196.

Atiemo-Obeng, V.A., Calabrese, R., 2004. Rotor-stator mixing devices. In: Paul, E.L., Atiemo-Obeng, V.A., Kresta, S.M. (Eds.), Handbook of Industrial Mixing. Wiley, Hoboken, NJ, pp. 479–505.

Bisten, A., Schuchmann, H.P., 2016. Optical measuring methods for the investigation of high-pressure homogenisation. Processes 4 (4), 41.

Boxall, J.A., Koh, C.A., Sloan, E.D., Sum, A.K., Wu, D.T., 2012. Droplet size scaling of water-in-oil emulsions under turbulent flow. Langmuir 28 (1), 104–110.

Brennen, C.E., 1995. Cavitation and Bubble Dynamics. Oxford University Press, Oxford.

Calabrese, R.V., Chang, T.P.K., Dang, P.T., 1986. Drop breakup in turbulent stirred-tank contractors. AICHE J. 32 (4), 657–666.

Ganley, W.J., van Duijneveldt, J.S., 2016. Steady-state droplet size in montmorillonite stabilised emulsions. Soft Matter 12 (30), 6481–6489.

Grace, H.P., 1982. Dispersion phenomena in high viscosity immiscible fluid systems and application of static mixers as dispersion device in such systems. Chem. Eng. Commun. 14 (3–6), 225–277.

Gupta, A., Eral, H.B., Hatton, T.A., Doyle, P.S., 2016a. Controlling and predicting droplet size of nanoemulsions: scaling relations with experimental validation. Soft Matter 12 (5), 1452–1458.

Gupta, A., Eral, H.B., Hatton, T.A., Doyle, P.S., 2016b. Nanoemulsions: formation, properties and applications. Soft Matter 12 (5), 2826–2841.

Håkansson, A., 2016. Experimental methods for measuring coalescence during emulsification—a critical review. J. Food Eng. 178, 47–59.

Håkansson, A., Hounslow, M.J., 2013. Simultaneous determination of fragmentation and coalescence rates during pilot-scale high-pressure homogenization. J. Food Eng. 116, 7–13.

Håkansson, A., Andersson, R., Mortensen, H.H., Innings, F., 2017. Experimental investigations of turbulent fragmenting stresses in a rotor-stator mixer. Part 2. Probability distributions of instantaneous stresses. Chem. Eng. Sci. 171 (Suppl. C), 638–649. https://doi.org/10.1016/j.ces.2017.06.038.

Henry, J.V.L., Fryer, P.J., Frith, W.J., Norton, I.T., 2009. Emulsification mechanism and storage instabilities of hydrocarbon in-water sub-micron emulsions stabilised with Tweens (20 and 80), Brij 96v and sucrose monoesters. J. Colloid Interface Sci. 338, 201–206.

Hinze, J., 1955. Fundamentals of the hydrodynamic mechanism of splitting in dispersion processes. AICHE J. 1 (3), 289–295.

Innings, F., Trägårdh, C., 2007. Analysis of the flow field in a high-pressure homogenizer. Exp. Thermal Fluid Sci. 32 (2), 345–354.

Innings, F., Fuchs, L., Trägårdh, C., 2011. Theoretical and experimental analyses of drop deformation and break-up in a scale model of a high-pressure homogenizer. J. Food Eng. 103 (1), 21–28.

Jafari, S.M., He, Y., Bhandari, B., 2006. Nano-emulsion production by sonication and microfluidization—a comparison. Int. J. Food Prop. 9 (3), 475–485.

Jafari, S.M., He, Y., Bhandari, B., 2007a. Effectiveness of encapsulating biopolymers to produce sub-micron emulsions by high energy emulsification techniques. Food Res. Int. 40 (7), 862–873.

Jafari, S.M., He, Y., Bhandari, B., 2007b. Optimization of nano-emulsions production by microfluidization. Eur. Food Res. Technol. 225 (5–6), 733–741.

Jafari, S.M., He, Y., Bhandari, B., 2007c. Production of sub-micron emulsions by ultrasound and microfluidization techniques. J. Food Eng. 82 (4), 478–488.

Jafari, S.M., Assadpoor, E., He, Y., Bhandari, B., 2008. Re-coalescence of emulsion droplets during high-energy emulsification. Food Hydrocoll. 22 (7), 1191–1202.

Janssen, J.J.M., Hoogland, H., 2014. Modelling strategies for emulsification in industrial practice. Can. J. Chem. Eng. 92, 198–202.

Kolmogorov, A.N., 1949. On the breakage of drops in a turbulent flow. Dokl. Akad. Nauk. SSSR 66, 825–828(Originally in Russian. Reprinted and translated in Selected Works of Kolmogorov, A.N., 1991. Volume 1: Mathematics and Mechanics. In: Tikhomirov, V.M. (Ed.), pp. 339–343).

Liao, Y., Lucas, D., 2009. A literature review of theoretical models for drop and bubble breakup in turbulent dispersions. Chem. Eng. Sci. 64, 3389–3406.

Liao, Y., Lucas, D., 2010. A literature review on mechanisms and models for the coalescence process. Chem. Eng. Sci. 65, 2851–2864.

Lobo, L., Svereika, A., Nair, M., 2002. Coalescence during emulsification. 1. Method development. J. Colloid Interface Sci. 253, 409–418.

Maindarkar, S., Dubbelboer, A., Meuldijk, J., Hoogland, H., Henson, M., 2014. Prediction of emulsion drop size distributions in colloid mills. Chem. Eng. Sci. 118, 114–125.

McClements, D.J., 2016. Food Emulsions: Principles, Practices, and Techniques. CRC Press, Boca Raton, FL.

McClements, D.J., Rao, J., 2011. Food-grade nanoemulsions: formation, fabrication, properties performance, biological fate, and potential toxicity. Crit. Rev. Food Sci. Nutr. 51 (4), 285–330.

Mohan, S., Narsimhan, G., 1997. Coalescence of protein-stabilized emulsions in a high-pressure homogenizer. J. Colloid Interface Sci. 192, 1–15.

Mortensen, H.H., Calabrese, R.V., Innings, F., Rosendahl, L., 2011. Characteristics of a batch rotor-stator mixer performance elucidated by shaft torque and angle resolved PIV measurements. Can. J. Chem. Eng. 89, 1076–1095.

Overbeek, J.T.G., Verhoeckx, G.J., de Bruyn, P.L., Lekkerkerker, H.N.W., 1987. On understanding microemulsions. II. Thermodynamics of droplet-type microemulsions. J. Colloid Interface Sci. 119 (2), 422–441.

Phipps, L.W., 1985. The High Pressure Dairy Homogenizer. The National Institute for Research in Dairying, Reading, UK.

Piacentini, E., Drioli, E., Giorno, L., 2014. Membrane emulsification technology: twenty-five years of innovation and research through patent survey. J. Membr. Sci. 468, 410–422.

Pope, S.B., 2000. Turbulent Flows. Cambridge University Press, Cambridge.

Qin, Z., Bremhorst, K., Alehossein, H., Meyer, T., 2007. Simulation of cavitation bubbles in a convergent-divergent nozzle water jet. J. Fluid Mech. 573, 1–25.

Ramkrishna, D., 2000. Population Balances. Academic Press, San Diego, CA.

Rayner, M., 2015. Scales and forces in emulsification. In: Rayner, M., Dejmek, P. (Eds.), Engineering Aspects of Food Emulsification and Homogenization. CRC Press, Bocca Raton, FL, pp. 3–32.

Rueger, P.E., Calabrese, R.V., 2013. Dispersion of water into oil in rotor-stator mixer. Part 2: effect of phase fraction. Chem. Eng. Res. Des. 91 (11), 2134–2141.

Schubert, H., Ax, K., Behrend, O., 2003. Product engineering of dispersed systems. Trends Food Sci. Technol. 14, 9–16.

Schultz, S., Wagner, G., Urban, K., Ulrich, J., 2004. High-pressure homogenization as a process for emulsification. Chem. Eng. Technol. 27 (4), 361–368.

Shamsara, O., Muhidinov, Z.K., Jafari, S.M., Bobokalonov, J., Jonmurodov, A., Taghvaei, M., Kumpugdee-Vollrath, M., 2015. Effect of ultrasonication, pH and heating on stability of apricot gum-lactoglobuline two layer nanoemulsions. Int. J. Biol. Macromol. 81, 1019–1025.

Sugiura, S., Nakajima, M., Kumazawa, N., Iwamoto, S., Seki, M., 2002. Characterization of spontaneous transformation-based droplet formation during microchannel emulsification. J. Phys. Chem. B 106 (36), 9405–9409.

Tadros, T., Izquierdo, P., Esquena, J., Solans, C., 2004. Formation and stability of nano-emulsions. Adv. Colloid Interf. Sci. 108–109, 303–318.

Taisne, L., Walstra, P., Cabane, B., 1996. Transfer of oil between emulsion droplets. J. Colloid Interface Sci. 184 (2), 378–390.

Taylor, G.I., 1934. The formation of emulsions in definable fields of flow. Proc. R. Soc. Lond. A 146, 501–523.

Tcholakova, S., Vanova, N., Denkov, N.D., Danner, T., 2007. Emulsification in turbulent flow: 3. Daughter drop-size distribution. J. Colloid Interface Sci. 310, 570–589.

Tcholakova, S., Denkov, N.D., Lips, A., 2008. Comparison of solid particles, globular proteins and surfactants as emulsifiers. Phys. Chem. Chem. Phys. 10 (12), 1068–1627.

Tcholakova, S., Lesov, I., Golemanov, K., Denkov, N.D., Judat, S., Engel, R., Danner, T., 2011. Efficient emulsification of viscous oils at high drop volume fraction. Langmuir 27, 14783–14796.

Vankova, N., Tcholakova, S., Denkov, N.D., Ivanov, I., Vulchev, V.D., Danner, T., 2007a. Emulsification in turbulent flow 1. Mean and maximum drop diameters in inertial and viscous regimes. J. Colloid Interface Sci. 312, 363–380.

Vankova, N., Tcholakova, S., Denkov, N.D., Vulchev, V.D., Danner, T., 2007b. Emulsification in turbulent flow 2. Breakage rate constants. J. Colloid Interface Sci. 313, 612–629.

Walstra, P., 2005. Emulsions. In: Lyklema, J. (Ed.), Fundamentals of Interface and Colloid Science. Elsevier, Amsterdam, pp. 8.1–8.94.

Walstra, P., Smulders, P.E.A., 1998. Emulsion formation. In: Binks, B.P. (Ed.), Modern Aspects of Emulsion Science. Royal Society of Chemistry, Cambridge, pp. 56–99.

White, F., 1998. Fluid Mechanics, fourth ed. McGraw-Hill, Boston, MA.

Wieringa, J.A., van Dieren, F., Janssen, J.J.M., Agterof, W.G., 1996. Droplet breakup mechanisms during emulsification in colloid mills at high dispersed phase volume fraction. Chem. Eng. Res. Des. 74 (5), 554–562.

Zhou, G., Kresta, S.M., 1998. Correlation of mean drop size and minimum drop size with the turbulence energy dissipation and the flow in an agitated tank. Chem. Eng. Sci. 53 (11), 2063–2079.

FURTHER READING

Davies, J.T., 1985. Drop sizes of emulsions related to turbulent energy dissipation rates. Chem. Eng. Sci. 40 (5), 839–842.

Chapter 6

Fabrication of Nanoemulsions by Rotor-Stator Emulsification

Ulrike S. van der Schaaf and Heike P. Karbstein (formerly Schuchmann)
Karlsruhe Institute of Technology, Karlsruhe, Germany

Chapter Outline

6.1 INTRODUCTION

Rotor-stator devices are widely used in the industry not only for emulsification purposes but also for general dispersion and comminution tasks. They are relatively easy to install into existing vessels and tanks and require comparably low costs of investment. For these reasons, rotor-stator processes are often the emulsification process of choice in many industrial sectors.

In processes aimed at the production of nanoemulsions, rotor-stator devices are often used to prepare a coarse emulsion and before additional comminution steps by, e.g., high-pressure homogenization. The reason is that rotor-stator emulsification is probably the least favorable method for the one-step production of nanoemulsions. It is very difficult to achieve droplet sizes

Nanoemulsions. https://doi.org/10.1016/B978-0-12-811838-2.00006-0

below 1 μm with this technique (Jafari et al., 2006, 2007a,b). As a top-down approach of emulsification, a large volume of bulk disperse phase is slowly comminuted into very small single droplets. High external stresses are necessary to create a huge new interfacial area. Achieving these high stresses is a challenge in rotor-stator devices. Nevertheless, several studies prove that it is nonetheless possible to achieve droplets in the nanorange by rotor-stator processes alone. This requires a careful selection of process and formulation parameters though. Therefore, the aim of this chapter is to create a general understanding of how emulsion droplet sizes can be influenced in rotor-stator systems. At the end, readers that will start to produce nanoemulsions will hopefully have gained enough knowledge to make an informed decision about which equipment to choose and how to design their emulsification process. Readers with prior knowledge in emulsification should have received the necessary tools to improve existing emulsification processes using already existing equipment.

For this purpose, the chapter is structured as follows: at the beginning, different rotor-stator devices commonly found in laboratories and in industrial plants are presented. Furthermore, different process designs, in which rotor-stator devices can be used, are outlined shortly. Next, basic concepts are explained that help to manipulate the emulsion droplet size generated in rotor-stator devices. Finally, some examples from literature are listed that show which combination of process and formulation design can lead to stable nanoemulsions. The overall aim is to highlight the interacting parameters in rotor-stator processes and how to control them.

6.2 CLASSIFICATION OF ROTOR-STATOR EMULSIFICATION DEVICES

Rotor-stator emulsification devices are commercially available in a wide variety of constructive designs by a large number of suppliers. All of these designs have in common that one part of the device is moving while the other one is stationary. Designs in which the two parts of the apparatus are moving at different speeds relative to each other are available as well. These devices are also known as rotor-rotor emulsification devices.

There are different ways to classify rotor-stator or rotor-rotor devices. One common way is to differentiate between batch and in-line devices. The latter ones are suitable for the continuous production of emulsions. The most common constructive designs of rotor-stator devices are explained in the following subchapters. This overview is by no means complete as a lot of company-specific designs that are often patented can be found as well.

Many companies all over the world market rotor-stator devices: Chemineer, Ekato, Gea, IKA Werke, Kinematica, Proxes, Ross, Rayneri, Silverson, Symex, and Ystral are examples, but this list is by no means complete. Rotor-stator device suppliers are also common guests at chemical engineering fairs where equipment can be inspected.

6.2.1 Batch Devices

6.2.1.1 *High-Shear Mixers*

High-shear mixers, or radial discharge mixers, are long dispersing elements inserted into a vessel holding the liquids to be emulsified. These elements typically consist of two parts: (1) a hollow stator with—sometimes exchangeable—screens or an axial slotted rim at the tip and (2) a coaxially arranged rotating shaft inserted into the stator. Different rotor designs are in use such as blades or gear-rim units (see Fig. 6.1A and B). The gap between the rotor and stator rim is typically between 0.5 mm and a few millimeters (Pacek et al., 2013).

High-shear mixers come in various sizes, and their constructive design can exhibit various complexities. The diameter of the dispersing element can be as small as 1 cm suitable for the emulsification of only a few milliliters of liquid volume. In this scale, high-shear mixers are standard laboratory equipment for screening experiments or for the preparation of small batches of emulsions containing extremely valuable ingredients. For emulsification, the dispersing element, which can be handheld or mounted to a stand rod, is lowered into a beaker and switched on. Different rotational speeds of typically up to 25,000 rpm can be set (Urban et al., 2006a).

For large-scale industrial applications, suppliers also offer dispersing elements with diameters up to half a meter. Several tons of products can be processed at once with these devices. Here, the operating mode is comparable to lab-scale equipment, but the dispersing element is fixed installed into the reactor. Production-scale high-shear mixers are usually capable of rotational speeds between 1000 and 5000 rpm resulting in peripheral velocities of up to 50 m/s (Urban et al., 2006a). Top-mounted high-shear mixers are easy to install and maintain, but, due to the long shaft inserted into the vessel, strong vibrations can occur at high rotational speeds. This issue can be avoided by bottom-mounting high-shear mixers as this reduces the necessary shaft length significantly. This type of installation requires more effort to ensure a proper sealing of the installation opening and of the mixer itself (Pacek et al., 2013).

(A) (B)

FIG. 6.1 (A and B) Examples of high-shear mixers for batch processes.

Because of the fixed installation of the dispersing element, thorough circulation of the fluid within the vessel can be a challenge particularly at high fluid viscosity. Therefore, special attention needs to be put on a proper scaling of the high-shear mixer. Suppliers can usually help with finding the optimal ratio between rotor-stator diameter and vessel volume. Furthermore, to ensure proper circulation, high-shear mixers can be installed eccentrically, and they are often used in combination with anchor or spiral agitators.

Droplet breakup in high-shear mixers usually takes place in inertial-turbulent flow conditions. Due to the typically large vessel volumes, high-shear mixing processes are characterized by an inhomogeneous energy input, which often leads to emulsions with wide droplet size distributions (Urban et al., 2006a).

6.2.1.2 Disperser Discs

Disc systems, or disperser or dissolver discs, are a specific type of agitators for vessels. A flat disc with toothed rim is attached to a rotating shaft, which is top-mounted to a vessel (see Fig. 6.2). The rim of the disc can be designed in various ways to increase turbulence and shear forces. Typically, the rim is radially or axially toothed. Disc systems are often used in combination with other agitators or with baffles attached to the wall of the vessel to improve the mixing effect (Urban et al., 2006b).

Mostly, disperser discs are used to disperse solids in a liquid. They are very suitable for high viscous products, which is why they are very common in the dye and lacquer industry. They are also able to produce fine droplets in high viscous emulsions of up to 10 Pa s, and they are very robust against solid

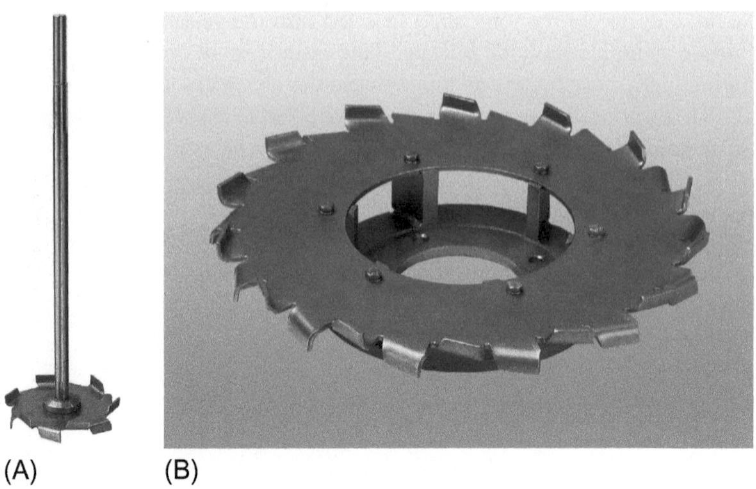

FIG. 6.2 (A) Dissolver disc and rotating shaft. (B) Close-up of a dissolver disc. *(From Vollrath GmbH, with permission.)*

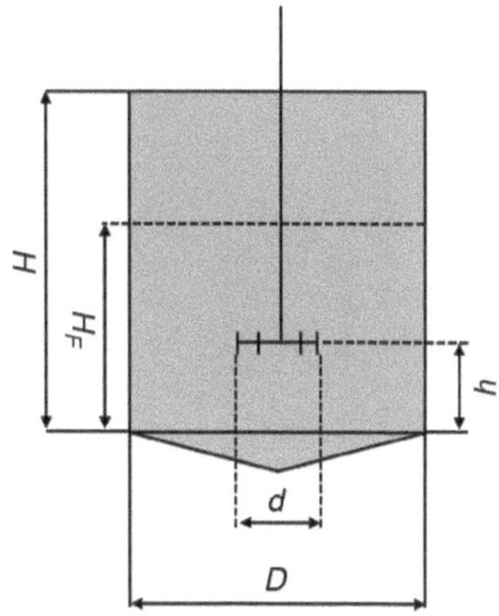

Diameter of disperser disc	d
Height of disperser disc over vessel bottom	$h = 0.5 - 1\ d$
Diameter of vessel	$D = 2 - 3\ d$
Height of filling	$H_F = 2\ d$
Height of vessel	$H > 2\ d$

FIG. 6.3 Dimensions and installation conditions for disperser discs optimized to disperse solid particles in a highly viscous fluid. *(Graph adapted from Goldschmidt, A., Streitberger, H.-J., 2002. BASF-Handbuch Lackiertechnik. Vincentz Verlag, Hannover; Urban, K., Wagner, G., Schaffner, D., Röglin, D., Ulrich, J., 2006a. Rotor-stator and disc systems for emulsification processes. Chem. Eng. Technol. 29, 24–31.)*

particles or fibers in the emulsion (Urban et al., 2006b). The optimal operating mode for dispersing solid particles in a high viscous fluid is shown in Fig. 6.3 (Goldschmidt and Streitberger, 2002). Depending on the disc diameter, peripheral velocities of up to 25 m/s can be reached. Furthermore, disc systems usually generate a vortex within the vessel that needs to be accounted for when dimensioning the vessel. This vortex can be used to easily incorporate the disperse phase into a high viscous continuous phase (Urban et al., 2006a).

6.2.2 Continuous Devices

6.2.2.1 Gear-Rim Dispersing Units

Gear-rim dispersing units are rotor-stator devices that consist of at least two coaxially meshed rings equipped with radial holes or slots of different size (see Fig. 6.4A and B). The diameter of these gear rims varies between few centimeters for lab-scale devices and several tenths of centimeters for industrial equipment resulting in peripheral velocities of the rotor of up to 40 m/s (Karbstein and Schubert, 1995). The gap between the two rims is typically in the order of a few millimeters. Typically, one of the gear rims is static; however, devices in which both gear rims are rotating at different relative velocities are available as well. On the one hand, these so-called cotwisters can enhance emulsion throughput by rotating in the same direction at different speed. On the other hand, they can enhance energy input into the system by increasing the emulsion residence time in the gap when rotating in opposite directions.

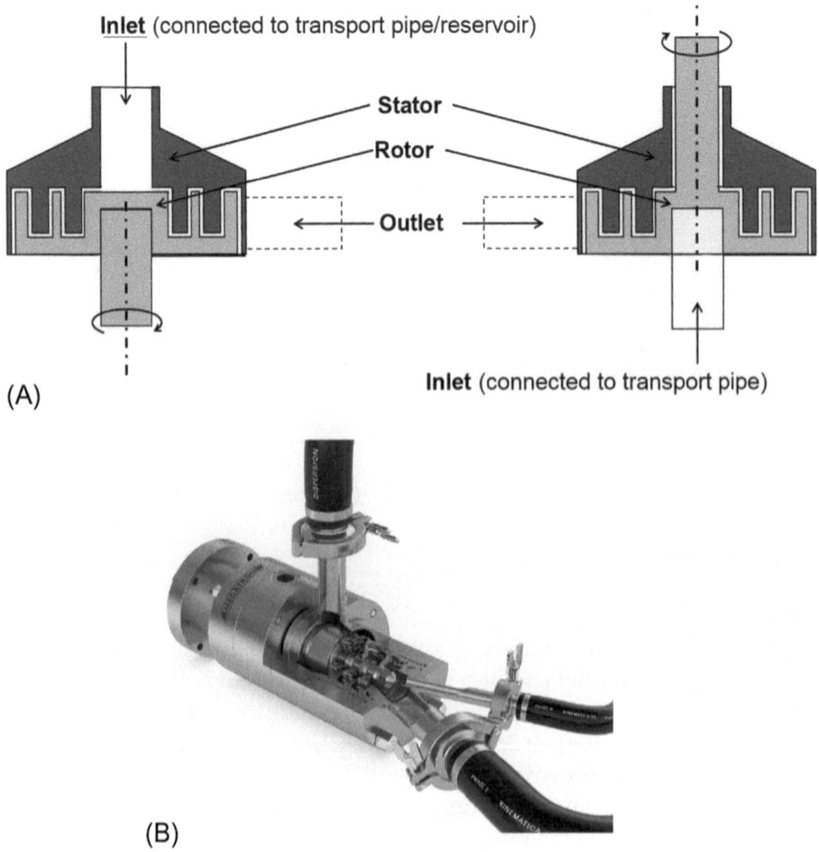

FIG. 6.4 (A) Schematics of two different in-line gear-rim setups with labeling of the most important elements. (B) Example of an in-line gear-rim device with injection of the disperse phase directly into the mixing chamber.

During emulsification, the liquid is sucked into the device in axial direction and is accelerated due to centrifugal forces caused by the moving rotor (see Fig. 6.5). The liquid is then redirected and flows through the holes or slots where it is slowed down and accelerated in both radial and tangential direction (Schubert and Armbruster, 1989). At the end, the liquid often leaves the gear-rim unit in radial direction. Droplet breakup in gear-rim dispersing units typically occurs in turbulent flow conditions. Moreover, gear-rim devices are self-feeding making additional external pumps only necessary in case of very high liquid viscosities or ensuring a defined throughput independent of gear-rim speed especially when very high throughputs are required. In order to intensify turbulence and thus to improve droplet breakup, multiple gear rims—or rotor-stator units—can be connected in series (Karbstein, 1994). Furthermore,

FIG. 6.5 Schematics of droplet comminution in in-line gear-rim devices.

in certain devices, it is possible to inject the disperse phase directly into the high-shear zone, which ensures a faster mixing of components.

6.2.2.2 *Colloid Mills*

Colloid mills consist of a cone-shaped rotor and stator, which are coaxially aligned. Liquid flows in axial direction through the narrow gap between rotor and stator (see Fig. 6.6A and B). The surface of both elements can be either smooth to promote laminar flow or grooved to promote vortex formation and an earlier transition to turbulent flow (Karbstein, 1994). The surface structure of both rotor and stator can vary in the number and depth of the grooves and in the angle in which the grooves are arranged. Rotor cone diameters are between few centimeters in bench-top devices and tenths of centimeters for large-scale devices, which result in peripheral velocities typically between 5 and 40 m/s. The gap between rotor and stator is very narrow in relation to the cone diameter and can be varied by adjusting the cone in axial direction. By this, gap widths between approx. 100 μm in lab-scale devices and a few millimeters in industrial-scale devices can be realized (Karbstein and Schubert, 1995). By reducing the gap width, higher shear forces can be achieved, and the throughput can be reduced due to the smaller cross-sectional area. The latter results in longer residence times in the shearing zone (Karbstein, 1994). These features mean that on the one hand, a higher energy input can be achieved, which can promote droplet breakup. On the other hand, the pumping capacity of the colloid mill is reduced, which can make external pumps necessary. Colloid mills require a minimum product viscosity of about 20 mPa s (Schubert and Armbruster, 1989). At lower viscosities, product is prone to fast draining by gravity even when the gap width between rotor and stator is very small.

6.3 MODES OF OPERATION OF ROTOR-STATOR DEVICES

Rotor-stator devices are available for batch and for continuous processes (see schematics in Fig. 6.7). Stirrers in agitated vessels and high-shear mixers can

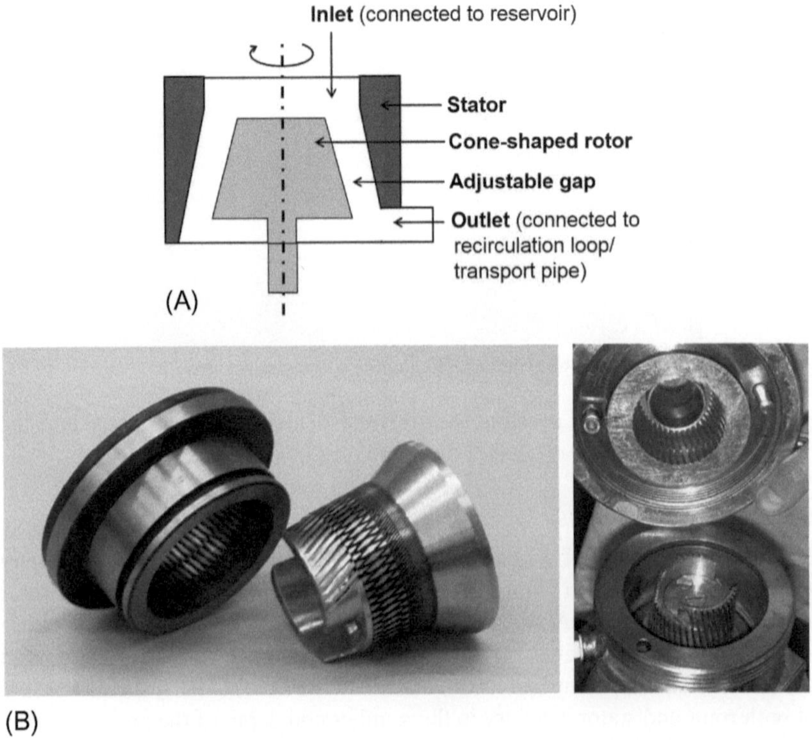

FIG. 6.6 (A) Schematics of a colloid mill with labeling of the most important elements. (B) Left: stator and rotor cone of a colloid mill. Right: view into a pilot-scale colloid mill. The static upper cone is detached. In the bottom, the rotating cone can be seen.

only be used in batch processes (see Fig. 6.8). They are useful for the production of small volumes and for products where individual batches need to be traced, e.g., pharmaceutical products. They also allow for combining several unit operations in one apparatus only, such as emulsifying, pasteurizing, cooling, and mixing. The latter is often the reason for producers to work with this type of equipment, e.g., in the specialty food sector. The disadvantage of agitated vessels and high-shear mixers is their inhomogeneous power input, which results in wide emulsion droplet size distributions (Schuchmann and Köhler, 2012). For narrower droplet size distributions, longer process times are necessary in order to ensure that all volume elements of the product pass the zones of highest shear within the vessel (Jasińska et al., 2014). Furthermore, the power necessary to achieve a certain homogenization result depends on the vessel volume and filling degree. If very large product volumes need to be homogenized by stirrers or high-shear mixers, extremely large motors are necessary to generate the required power. If the product viscosity is high, the required power is even higher as well (Pacek et al., 2013). Then, the limit of batch devices is quickly

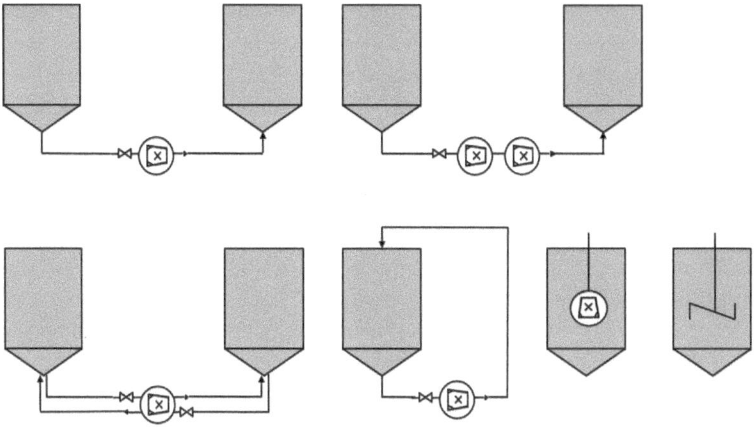

FIG. 6.7 Flow chart of possible modes of operations of rotor-stator devices. Top left: continuous process with single in-line device. Top right: continuous process with serial in-line devices (cascade mode). Bottom left: batch process with in-line device using interchanging storage tanks (pendulum mode). Bottom center: batch process with in-line device using a recirculation loop. Bottom right: batch processes with high-shear mixer or stirrer.

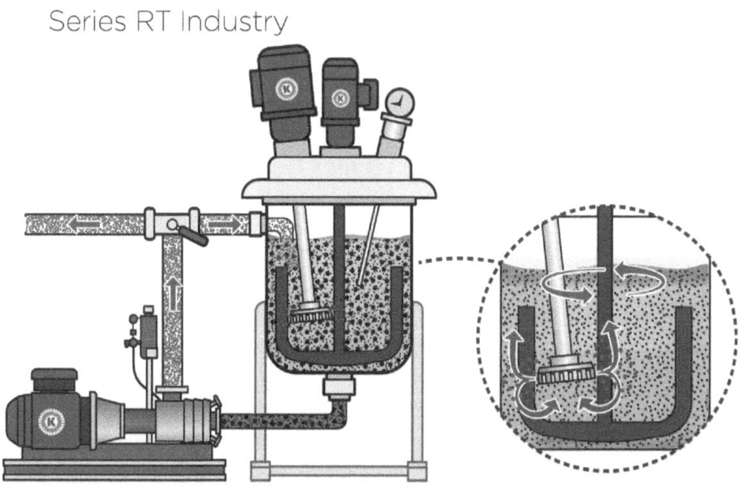

FIG. 6.8 The vessel is equipped with a high-shear mixer. An anchor stirrer ensures thorough mixing within the vessel. Additionally, the product is fed through an in-line device that is operated in recirculation mode.

reached, and it becomes unavoidable to switch the process to one that uses in-line devices. Due to their high pumping capacity, smaller units are necessary to achieve the same power input.

In-line devices are mostly used in continuous processes; see Fig. 6.7. This type of process is often desired by the producing industry because of its higher process efficiency: higher product output can be achieved in a shorter production time. Moreover, storage tanks as required by batch processes are not necessary, which can reduce investment costs. In a continuous process using external in-line devices and separate storage tanks for premix and finished product, it is guaranteed that all the liquid passes the emulsification device so that narrow emulsion droplet size distributions can be achieved in a short process time. If there is a risk of too large droplets when the product passes the homogenizer only once, several in-line units can be installed serially (cascade mode), or the storage tanks can be used interchangeably (pendulum mode). This generates a continuous multipass process.

However, in-line devices can also be used in batch processes. One possible process design is to use two storage tanks that are connected by a pipe in which the in-line device is installed. The product is now pumped from one storage tank to the other passing the in-line homogenizer (Schuchmann and Köhler, 2012). When the first tank is entirely emptied, the process is reversed, and the product is pumped from the second storage tank back into the first one (pendulum mode). The other possible process design for a semibatch process is to use a recirculation loop connecting the outlet of the in-line device with the storage tank so that the product is pumped back into the storage tank (Fig. 6.8). In this case, the advantage of an in-line process (defined narrow droplet size distributions within short process times) is partly gone: by refeeding processed product into the storage tank, it mixes with the unprocessed medium so that longer process times are necessary.

6.4 ENGINEERING DESCRIPTION OF ROTOR-STATOR EMULSIFICATION

The mechanical emulsification process comprises the steps of droplet breakup and immediate droplet stabilization (Karbstein, 1994). Emulsion droplets are broken up when they are deformed long enough by external stresses and when these deforming stresses surpass a critical value. The stresses themselves are a result of various flow patterns generated by the emulsification device (Schuchmann and Köhler, 2012; Walstra, 1993). Several process and formulation parameters influence the type of flow within the emulsification device (see Chapter 5) and therefore affect the generated stresses and the droplet breakup mechanism. Droplet breakup in rotor-stator devices occurs mostly in turbulent flow, but for colloid mills, laminar and transitional flow conditions are of particular importance as well (Karbstein, 1994).

When droplets are broken up, new interface is created, which needs to be stabilized by the adsorption of emulsifier molecules. Stabilization kinetics competes with kinetics of collision of unstabilized droplets (Karbstein and Schubert, 1995; Karbstein, 1994). If emulsifier molecules do not adsorb fast enough, droplets recoalesce leading to a coarsening of the emulsion (Håkansson et al., 2016; Jafari et al., 2008). The stabilization step is controlled by the interplay of the adsorption kinetics of the emulsifier and the process conditions providing the emulsifier molecules with enough time to actually adsorb (Stang et al., 1994).

Mechanical emulsification processes are complex processes characterized by the occurrence of spatial and temporal inhomogeneities of the generated stresses. In order to improve the understanding of rotor-stator processes and therefore to gain better control of the resulting emulsion morphology, different attempts have been made. Experimental approaches involved the investigation of optically accessible rotor-stator devices in order to visualize flow patterns within the device (Armbruster, 1990; Karbstein, 1994). Furthermore, simulation studies were conducted to gain an understanding of local stress distributions within rotor-stator geometries (Jasińska et al., 2014; Pacek et al., 2013). Such elaborate practices are often not feasible in an industrial setting. Several characteristic numbers were defined to describe emulsification processes and predict mean emulsion droplet sizes. Two relevant concepts—one for batch and one for continuous processes—are presented in the following chapters.

6.4.1 The Power Density Concept as a Tool to Scale Batch Processes

In batch processes, the magnitude of the stresses that can be used for droplet breakup is determined by the power density P_V:

$$P_V = \frac{P}{V}$$

where P is the power and V is the liquid volume that the power is applied to (Schuchmann and Köhler, 2012). P_V is proportional to the energy dissipation rate ε that is applied by several authors (Hall et al., 2011b; Pacek et al., 2013; Walstra, 1993). Droplet breakup in batch processes, e.g., in agitated vessels, mainly occurs under turbulent flow conditions (Knoch, 2002). The power required by a stirrer to achieve turbulent flow depends on various parameters but is also a characteristic feature of the individual stirrer type. For a detailed description of the dimensioning of agitated vessels, the reader is referred to Zlokarnik (1999, 2000). An exemplary description for disperser discs will shortly be discussed below.

Flow in agitated vessels is considered to be turbulent at Reynolds numbers $Re > 100$ in baffled vessels and $Re > 50{,}000$ in unbaffled vessels (Zlokarnik, 2000). Here, the Reynolds number of stirrers is defined as

$$Re = \frac{n \cdot d^2 \cdot \rho_m}{\eta_e}$$

with n as the stirrer speed, d as the stirrer diameter, ρ_m as the mean density of the emulsion, and η_e as the viscosity of the emulsion. Using this equation, the stirrer speed necessary for turbulent flow can be calculated.

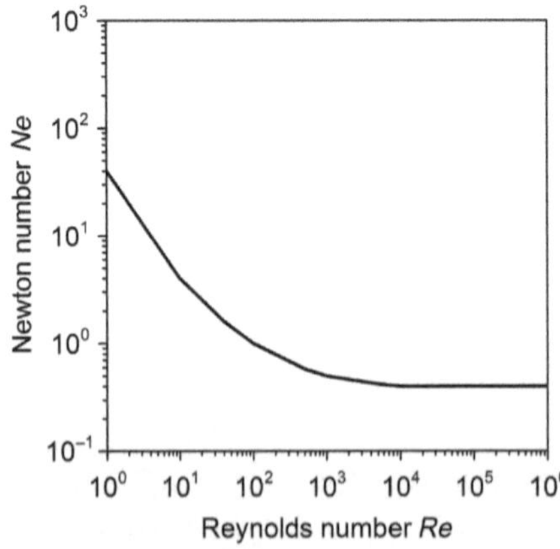

FIG. 6.9 Power characteristics of a propeller stirrer in a baffled vessel. *(Characteristic curve adapted from Zlokarnik, M., 2000. Stirring, Ullmann's Encyclopedia of Industrial Chemistry. Wiley-VCH Verlag GmbH & Co. KGaA, Weinheim.)*

For each stirrer type, there exists a Newton number Ne characterizing the power requirements at a given Reynolds number Re so that $Ne = f(Re)$. Ne is defined as follows:

$$Ne = \frac{P}{\rho_m \cdot n^3 \cdot d^5}$$

The relationship between Ne and Re is called the power characteristics of a stirrer and graphs, and tables exist that visualize this relationship and from which the relevant Newton numbers can be read for each stirrer type (Zlokarnik, 2000). As an example (see Fig. 6.9), in case of fully turbulent flow caused by a propeller in a baffled vessel ($Re > 50{,}000$), the corresponding Newton number is $Ne = 0.4$. The required power can then be calculated from the above equation for the set stirrer speed.

Following that, the Sauter mean diameter $d_{3,2}$ as the mean emulsion droplet size that can be expected from that power input is proportional not only to the power density P_V but also to the emulsification time t_e, i.e., the time period over which the power is applied (see Section 6.5.1.4), and to the disperse phase viscosity η_d as a formulation parameter:

$$d_{3,2} \sim P_V^{-b} \cdot f(t_e) \cdot \eta_d^c$$

The exponents b and c modify this relationship to account for different droplet breakup mechanisms (Schuchmann and Köhler, 2012).

6.4.2 The Energy Density Concept as a Tool to Compare Continuous Processes

In continuous emulsification processes, the magnitude of deforming stresses does depend not only on the power density but also on the residence time t_{res}

in the disruption zone of the in-line rotor-stator device (Koglin et al., 1981). The local stress distribution within a continuous rotor-stator device is quite inhomogeneous. The longer the residence time, the higher the probability that a droplet passes the areas of high stresses (Jasińska et al., 2014). Consequently, the characteristic item to describe continuous emulsification processes comprises both the power density P_V and the residence time t_{res}. It is summarized in the energy density E_V that is defined as (Karbstein, 1994)

$$E_V = P_V \cdot t_{res}$$

Similar to batch processes, in turbulent flow, the Sauter mean diameter $d_{3,2}$ of emulsions is proportional to the energy density E_V and to the disperse phase viscosity η_d:

$$d_{3,2} \sim E_V^{-b} \cdot \eta_d^c$$

In steady laminar shear flow, however, not the disperse phase viscosity but the viscosity ratio between disperse and surrounding phase is relevant for the mean droplet size:

$$d_{3,2} \sim E_V^{-b} \cdot f(\eta_d / \eta_e)$$

with η_e as the viscosity of the phase surrounding the droplets (see also subchapter on viscosity ratio) (Schuchmann and Köhler, 2012). Depending on the rotor-stator device geometry, laminar elongational flow sometimes superimposes laminar shear flow (Feigl et al., 2007; Windhab et al., 2007). In this case, $f(\eta_d/\eta_e)$ changes having less effect on the mean droplet size, which is in agreement with the theory on drop breakup in linear flow by Bentley and Leal (1986).

The mean droplet size of emulsions produced under various conditions can be plotted over the energy density applied to prepare these emulsions. The slope of the resulting curve can be used to analyze emulsification processes concerning their efficacy and to compare different processes (Karbstein and Schubert, 1995; Karbstein, 1994). This is shown schematically in Fig. 6.10: a model emulsion was produced in three different emulsification devices. Only the energy density applied during emulsification was varied. Devices 1 and 2 produce smaller mean droplet sizes than device 3 at any given energy density. Therefore, the emulsification efficacy of devices 1 and 2 is higher. One possible reason for this observation might be a different droplet breakup mechanism. No difference in the emulsification efficacy between devices 1 and 2 is found because all data points fall onto the same curve. The only difference is that device 2 is capable of generating higher energy densities than device 1 and thus smaller droplets.

Once such curves have been created for a formulation and an emulsification device, they can also be used for the scale-up of emulsification processes: in order to produce an emulsion with the desired droplet size, the corresponding energy density is simply read from the diagram. The process parameters necessary to achieve this energy density, however, can freely be chosen (Karbstein, 1994). The next chapter will explain which process (and formulation) parameters are relevant for rotor-stator devices and how they impact the energy density.

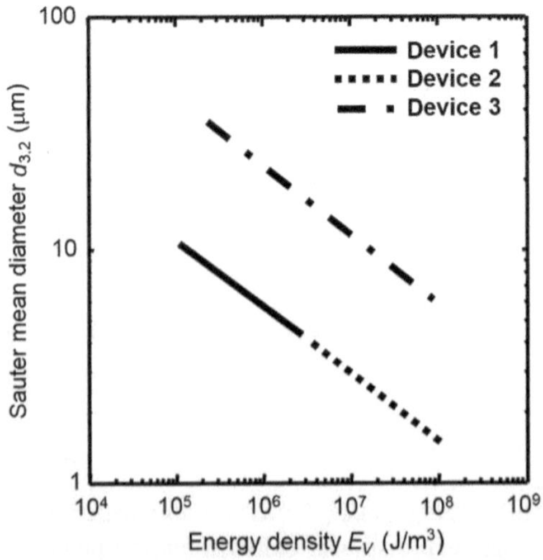

FIG. 6.10 Schematic of the graphic representation of the energy density concept. Sauter mean diameters of emulsions prepared by different emulsification devices are plotted over the energy density E_V applied to prepare these emulsions. The used emulsification devices differ in their emulsification efficiency (e.g., device 1 vs device 3) and in the generable energy density (device 1 vs device 2).

6.5 STRATEGIES TO MINIMIZE EMULSION DROPLET SIZES

6.5.1 Influence of Process Parameters

When recoalescence is suppressed, higher energy density leads to smaller emulsion droplet sizes. Various process parameters are available, which allow for intensifying the energy density in rotor-stator devices. These are foremost the speed, size, and design of the rotor and the rotor-stator size ratio. All of these parameters and how they can be used to obtain nanoemulsions are described in the next chapters.

6.5.1.1 Rotational Speed

Increasing the rotational speed is perhaps the most intuitive way to reduce the droplet size in rotor-stator emulsification processes. Fig. 6.11 (left) shows how the mean droplet size of an emulsion produced by an in-line gear-rim device in a continuous single-pass process reduces when the speed of the rotor is increased. In fact, by increasing the number of revolutions per minute, the energy input into the emulsion rises. Therefore, the same results can also be plotted versus the energy density that is shown in Fig. 6.11 (right) (Karbstein, 1994). It can be seen that a threefold increase in energy density halves the Sauter mean diameter of the emulsion.

The described relationship between droplet size and rotational speed is universally true for all rotor-stator systems provided that all other process

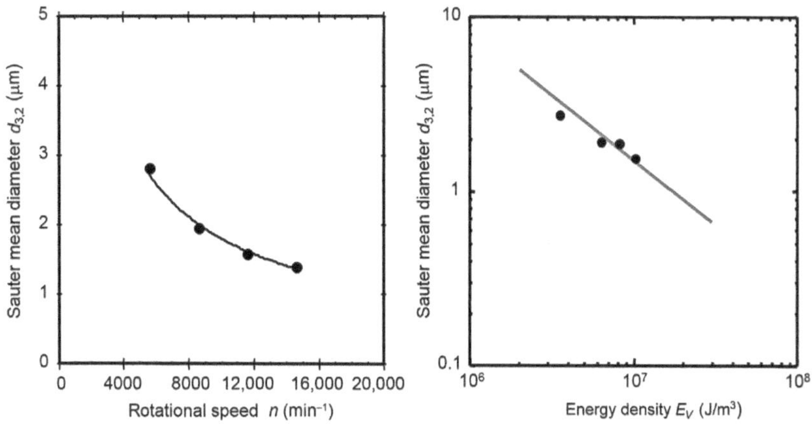

FIG. 6.11 Sauter mean diameter of oil-in-water emulsions produced by a continuous gear-rim device in single pass. The emulsions consisted of 30% canola oil in water and polyoxyethylene-(10)-lauryl ether as a typical emulsifier used in cosmetic products. The rotor comprised two concentric gear rims with a diameter of 48 mm. Left: dependency on the rotational speed. Right: the same results are plotted versus the energy density obtained by varying the rotational speed. *(Graphs adapted from Karbstein, H.P., 1994. Untersuchungen zum Herstellen und Stabilisieren von Öl-in-Wasser-Emulsionen. Universität Karlsruhe.)*

and formulation parameters are kept constant. The results for in-line gear-rim devices were confirmed by Hall et al. in a more recent publication (Hall et al., 2011a). Wolf et al. produced food-grade water-in-oil emulsions in a colloid mill and were able to reduce the droplet size from around 2 μm to 200 nm by increasing the rotational speed from 3000 to 10,000 rpm (Wolf et al., 2013). Schuch et al. used the same device to produce water-in-oil-in-water emulsions from the aforementioned water-in-oil emulsion. In this case, a reduction of the outer droplet size from 130 to 40 μm was feasible by again increasing the rotational speed from 3000 to 10,000 rpm (Schuch et al., 2013).

Several studies report reduced droplet sizes at higher rotational speeds when emulsions were prepared by high-shear mixers (El-Jaby et al., 2007; Pérez-Mosqueda et al., 2015; Rueger and Calabrese, 2013a). El-Jaby et al. (2007) obtained droplets as small as 300 nm in an emulsion system intended for miniemulsion polymerization. Pérez-Mosqueda et al. (2015) were able to reduce the Sauter mean diameter of d-limonene-based emulsions from 2500 to 800 nm by increasing the rotational speed of a lab-scale device from 6000 to 17,500 rpm.

Increasing the rotational speed of the impeller reduces the droplet size in agitated vessels as well. Knoch (2002) could show an inversely proportional relationship of the droplet size evolution upon increased rotational speed of a pitched blade impeller. Furthermore, the rotational speed was found to be more important for achieving small droplets than other process parameters in vessel agitated by disperser discs (Catte et al., 2002).

6.5.1.2 Rotor Size and Size Ratio

Rotors with larger radius R_R achieve higher peripheral velocities v_u when operated at the same rotational speed n:

$$v_u = 2 \cdot \pi \cdot n \cdot R_R$$

In in-line rotor-stator devices, this leads to smaller droplets because of higher resulting energy densities. Karbstein investigated gear-rim dispersing units of different rotor diameters (see Fig. 6.12). It can clearly be seen that by using a larger rotor, higher energy densities can be achieved, which result in smaller Sauter mean diameters of the investigated emulsion formulation.

In colloid mills, adjusting the gap width between rotor and stator is a further option to increase energy densities. Frank et al. (2011) emulsified a water-in-oil emulsion in a second water phase using a colloid mill to produce a double emulsion. Different hydrophilic emulsifiers were applied (a protein, a polysaccharide, and a synthetic emulsifier). For each emulsifier, the Sauter mean diameter of the emulsion was halved by reducing the gap width from 0.24 to 0.16 mm.

However, a possible second effect needs to be kept in mind: reducing the gap width also reduces the cross-sectional area causing the liquid to flow faster through the device when a constant volume throughput is set. Production lines typically run at constant volume throughput so that the product experiences

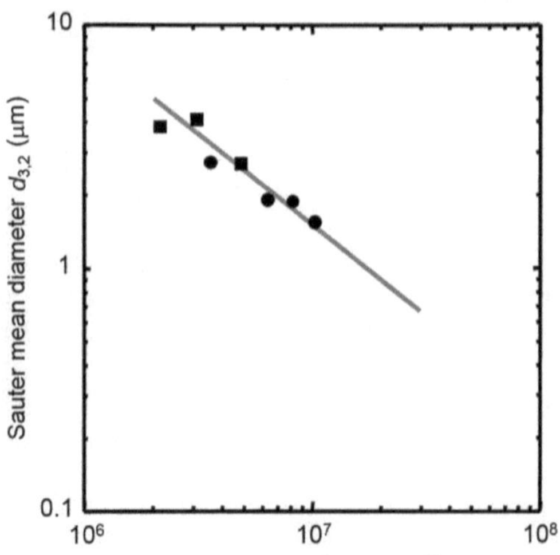

FIG. 6.12 Sauter mean diameter of oil-in-water emulsions produced by a continuous gear-rim device. The rotor comprised two concentric gear rims with a diameter of 36 mm *(squares)* and of 48 mm *(circles)*, respectively. The emulsions consisted of 30% canola oil in water and polyoxyethylene-(10)-lauryl ether as a typical emulsifier used in cosmetic products. *(Graph adapted from Karbstein, H.P., 1994. Untersuchungen zum Herstellen und Stabilisieren von Öl-in-Wasser-Emulsionen. Universität Karlsruhe.)*

shorter residence times within the disruption zone. Therefore, both higher shear forces and reduced residence times are a consequence of smaller gap widths. Both features influence the energy density; however, they counteract each other. Several parameters such as product viscosity determine which of the two effects will dominate droplet breakup in the end.

Emulsions produced by batch devices show a comparable dependency of the droplet size on the rotor diameter as in-line gear-rim devices. Maa and Hsu (1996) showed that by using a high-shear mixer tip with 2 cm diameter, much smaller droplets can be obtained at a given rotational speed than by using a tip with 1 cm diameter. Furthermore, the maximum stable droplet size could be reached much faster using the larger dispersing unit.

Corresponding results have been reported for disperser discs of different size. Discs with diameters between 50 and 110 mm were used to emulsify vegetable oil in a concentrated solution of modified starch (Urban et al., 2006a,b). While smaller disc diameters allowed to obtain mean droplet sizes as small as 1500 nm, the larger discs made droplets as small as 140 nm possible. However, in case of disperser discs, the stirrer diameter cannot be enlarged endlessly for a further reduction of the droplet size. In vessels agitated by spinning discs, droplet breakup occurs in turbulent flow in the vicinity of the disc. In order for turbulent flow to be fully developed, the ratio between rotor diameter and vessel diameter must be optimized according to the dimensions laid out in Fig. 6.3. If the rotor-vessel size ratio is much larger than 0.3, droplet breakup is hindered, and only coarse emulsions can be produced (Catte et al., 2002).

6.5.1.3 Rotor Design

In order to achieve a higher energy density and therefore to produce finer emulsions, the design of the rotor or of the rotor-stator combination can be modified. In gear-rim devices, it is possible to vary the slot design of each gear-rim and to increase the number of concentrically arranged gear rims within each rotor-stator unit. Higher numbers of gear rims increase the shear intensity and reduce the volume stream of the emulsion, resulting in a higher energy input into the emulsion (Karbstein, 1994). For a given rotor radius, this leads to smaller mean droplet sizes. This relationship is visualized in Fig. 6.13, which is Fig. 6.12 (black symbols) extended by data points (gray symbols) representing the Sauter mean diameters of emulsions produced with the same rotor diameters as the emulsions in Fig. 6.12 except that more gear rims were used. The gray symbols are located in the bottom right indicating that a larger number of gear rims result in smaller droplets at the same rotor speed. This is visualized exemplarily for a rotor speed of 8000 min^{-1} in Fig. 6.13. It can also be seen that all data points fall onto the same curve, which indicates that changing the number of gear rims does not alter the droplet breakup mechanism itself.

The influence of the slot size of the gear rims on the mean droplet size cannot be predicted that easily from theory. With decreasing slot size, droplets

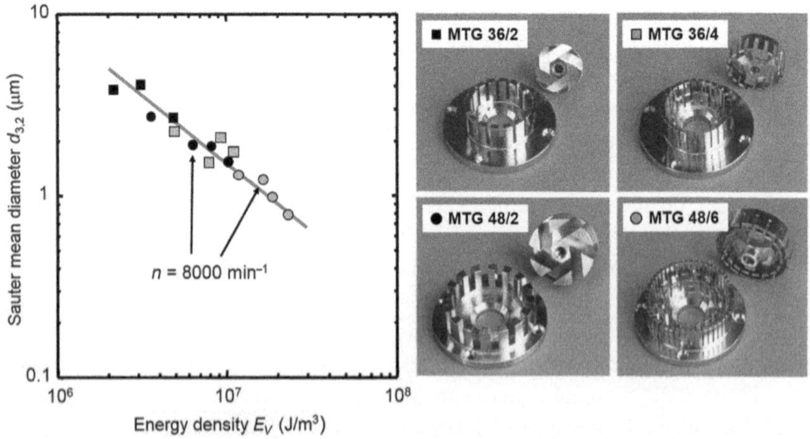

FIG. 6.13 Sauter mean diameter of oil-in-water emulsions produced by a continuous gear-rim device. Two rotor diameters (*squares*, 36mm; *circles*, 48mm) with a different number of gear rims are compared. *Black symbols*, two gear rims; *gray squares*, four gear rims; *gray circles*, six gear rims. The emulsions consisted of 30% canola oil in water and polyoxyethylene-(10)-lauryl ether as a typical emulsifier used in cosmetic products. *(Graph adapted from Karbstein, H.P., 1994. Untersuchungen zum Herstellen und Stabilisieren von Öl-in-Wasser-Emulsionen. Universität Karlsruhe.)*

experience higher stresses, but the residence time in the high-shear zone decreases simultaneously, and both effects counteract each other (see the definition of energy density in Section 4.2). As an example, Scholz and Keck (2015) investigated an in-line gear-rim device run in recirculation mode. The slot sizes of the gear rim were 0.5 and 1 mm. They found that both rotor designs were equally suited to produce emulsions with droplets of around 150nm. However, the rotor design with the smaller slot size required less homogenization time to obtain the mentioned droplet size.

Colloid mills often come with an interchangeable set of rotating cones that have grooves varying in shape and geometry. Karbstein, 1994 investigated the effect of these grooves on droplet breakup and on the resulting emulsion morphology. Similar to the slot size of gear-rim devices, only the emulsion throughput in self-feeding modus but not the mean droplet size of emulsions was affected by the groove design (Fig. 6.14). However, for an effective droplet breakup, it is necessary that there are grooves at all. A comparison with a colloid mill with both a smooth rotor and stator of the same size showed that much larger droplet sizes are obtained when there are no grooves (Karbstein, 1994). It could be shown that droplet breakup in smooth colloid mills occurs in dominantly laminar shear flow. Grooves on the cone promote a much faster transition to turbulent flow. At a given energy density, droplet breakup in turbulent flow conditions is much more effective than in laminar flow so that smaller droplet sizes can be obtained with grooved cones.

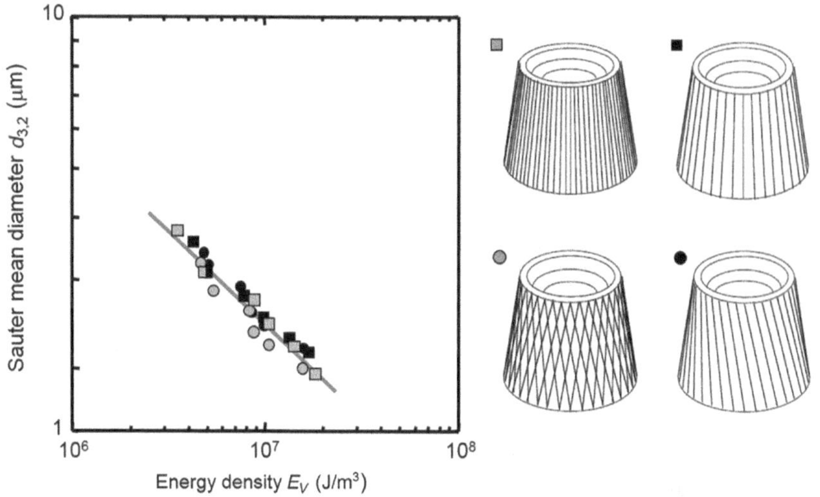

FIG. 6.14 Sauter mean diameter of oil-in-water emulsions produced by a continuous colloid mill. Different cone designs are compared: *black squares*, square-shaped grooves, 0° angle; *gray squares*, sawtooth-shaped grooves; *black circles*, square-shaped grooves, 30° angle; *gray circles*, square-shaped grooves, crossed at 30° angle. *(Graph adapted from Karbstein, H.P., 1994. Untersuchungen zum Herstellen und Stabilisieren von Öl-in-Wasser-Emulsionen. Universität Karlsruhe.)*

In an agitated vessel, it is recommended to use propeller stirrers or pitched blade impellers in order to obtain smaller droplets at equal power input (Knoch, 2002). Such impeller types cause flow primarily in axial direction that has proved advantageous for droplet breakup (Henzler and Biedermann, 1996). When using disperser discs, axially slotted discs should be preferred over radially toothed ones. Urban et al. (2006b) showed that the effective equilibrium droplet size of emulsions produced by axially and radially slotted discs was almost identical. However, the axially slotted disc required only around 10 min homogenization time to create emulsion droplets of around 130 nm, while the radially slotted disc required more than double the time. Furthermore, axially slotted discs with a higher number of slots produce emulsions with smaller mean droplet size (Urban et al., 2006a).

6.5.1.4 Emulsification Time in Batch Devices

Despite the common approach to average power and energy input into the emulsion, it could be shown, e.g., by numerical simulations (Pacek et al., 2013), that flow in rotor-stator devices is not constant leading to a locally quite inhomogeneous power density distribution (Tesch, 2002). As a result, not all emulsion droplets pass the zones of highest shear when being transported through the rotor-stator device (Jasińska et al., 2014). The obtained mean droplet sizes of the emulsion are larger than what might be achieved theoretically. In order to reduce the droplet sizes further and to obtain on average finer emulsions, the emulsification process can be run as a multipass homogenization process. In in-line devices, this can be achieved by feeding the processed emulsion back to the rotor-stator inlet either manually or by a recirculation loop to create a semi-batch process. Tesch and later on Jasinska et al. showed that feeding an emulsion through an in-line gear-rim device multiple times leads to a gradual reduction of the average droplet size with every pass until an equilibrium mean droplet size is reached (Jasińska et al., 2014; Tesch, 2002). At the same time, droplet size distributions become narrower with each pass. In in-line devices with recirculation setup, multipass homogenization corresponds to a longer processing time. In these devices, similar results, i.e., on average smaller droplets and narrower size distributions with longer processing time, could be found (Scholz and Keck, 2015). It has to be mentioned that the concept of reducing average droplet size by increasing processing times (or production passes) has limitations: once every droplet has passed the zone of highest stresses, no more breakup will result, and droplet sizes do not decrease further.

In true batch devices, the power density distribution is even more inhomogeneous making longer processing times necessary to obtain fine emulsions. In lab-scale high-shear mixers, typical processing times are between 2 and 5 min to obtain the minimum equilibrium droplet size for a given formulation (Maa and Hsu, 1996). However, much longer homogenization times of up to 4 h have been reported as well (El-Jaby et al., 2007, 2009). For vessels agitated by spinning discs, a comparable dependence of the mean droplet size on the processing time was found (El Kinawy et al., 2012; Urban et al., 2006b). Due to the typically larger vessel volumes, longer processing times between 30 and 60 min are necessary to obtain the smallest possible droplet size.

6.5.2 Influence of Formulation Parameters

The droplet size in emulsions produced by rotor-stator devices is not only influenced by process parameters. Various formulation parameters can impact the emulsion morphology as well. As droplets are broken up by stresses transmitted onto one liquid phase by the surrounding second liquid phase, formulation parameters affecting the viscosity of the system and thus the magnitude of the transmitted stresses are of importance. This can be the oil type, the disperse phase concentration, or the use of stabilizers. These parameters mainly affect droplet breakup. Other formulation parameters such as the type of emulsifier and its concentration impact the stabilization of the newly formed droplets. By reducing immediate recoalescence, these parameters also influence the droplet size of emulsions.

6.5.2.1 Viscosity of the Continuous Phase

In rotor-stator devices, droplets are broken up by stresses transmitted onto them by the surrounding liquid phase. Therefore, the viscosity of the continuous

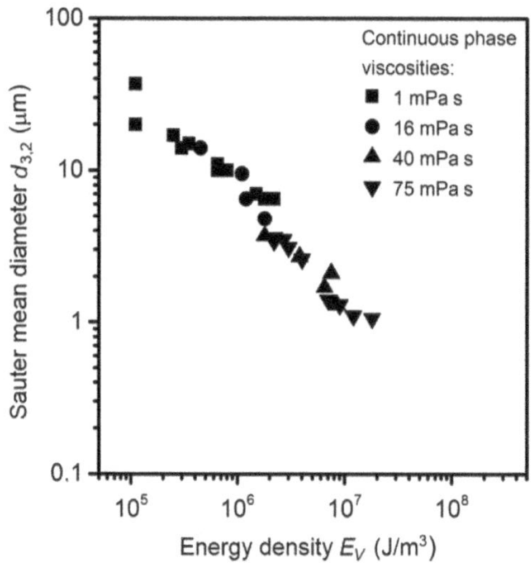

FIG. 6.15 Sauter mean diameter of oil-in-water produced by a continuous gear-rim device. The viscosity of the continuous phase was adjusted by adding the polymer polyethylene glycol. All emulsions consisted of 30% disperse phase with a viscosity of 60 mPa s. The hydrophilic emulsifier was polyoxyethylene-(10)-lauryl ether. *(Graph adapted from Karbstein, H.P., 1994. Untersuchungen zum Herstellen und Stabilisieren von Öl-in-Wasser-Emulsionen. Universität Karlsruhe.)*

phase is one of the key parameters to influence the droplet breakup and the emulsion droplet size. Most of the time, higher viscosity of the continuous phase leads to smaller droplets as shown in Fig. 6.15 (Karbstein, 1994). Here, the Sauter mean diameter of several oil-in-water emulsions produced in an in-line gear-rim device is plotted over the energy density. In these emulsions, the viscosity of the continuous phase was adjusted by adding the polymer polyethylene glycol (PEG 12,000 or 20,000). It can be seen that the droplet size decreases sharply with increasing energy density. As all data points fall onto the same curve, the droplet breakup mechanism is independent of the continuous phase viscosity. However, with increasing viscosity of the continuous phase, data points acquired at the same rotor speed range are shifted to the right part, i.e., to smaller droplets due to higher energy densities. Furthermore, for high viscosities of the continuous phase, the overall viscosity of the emulsion increases, which reduces the fluid transport within the in-line rotor-stator device. Consequently, the residence time of the emulsion in the disruption zone increases, and even more energy can be introduced into the system and is available for droplet disruption.

Similar results have been reported for in-line gear-rim systems, for high-shear mixers, for disperser discs (El Kinawy et al., 2012; Hall et al., 2011a; Maa and Hsu, 1996), and for W/O-type emulsions (Tesch, 2002). The droplet sizes of these emulsions were in the same size range as those shown in Fig. 6.15. For the production of nanoemulsions, it is therefore favorable to choose a formulation with high continuous phase viscosity that is typically achieved by adding polymeric stabilizers to the continuous phase (El Kinawy et al., 2012; Karbstein, 1994).

However, the emulsion droplet size cannot be reduced infinitively by increasing the viscosity of the continuous phase. When the viscosity becomes too high, the Sauter mean diameter can also increase as the type of flow within the rotor-stator device may be affected as well. Higher emulsion viscosities restrict eddy formation and turbulence and instead promote laminar flow reducing droplet breakup efficacy.

6.5.2.2 *Viscosity of Disperse Phase*

In contrast to the effect of the continuous phase viscosity, a high viscosity of the disperse phase usually leads to coarser emulsions and larger droplets. Fig. 6.16 shows the Sauter mean diameter of emulsions prepared in a colloid mill. The disperse phase viscosity was varied by using mineral oils of different viscosities, while the formulation of the continuous phase was kept constant. It can be seen that the mean droplet size still decreases with increasing energy density. However, the curves are shifted toward larger droplet sizes when the disperse viscosity is increased. The reason for this behavior is the fact that droplets behave more and more like solid particles when their viscosity increases. As a result, they are more difficult to deform and to break up. The illustrated relationship is also valid for inverse emulsions (water-in-oil emulsions) (Tesch, 2002) and for

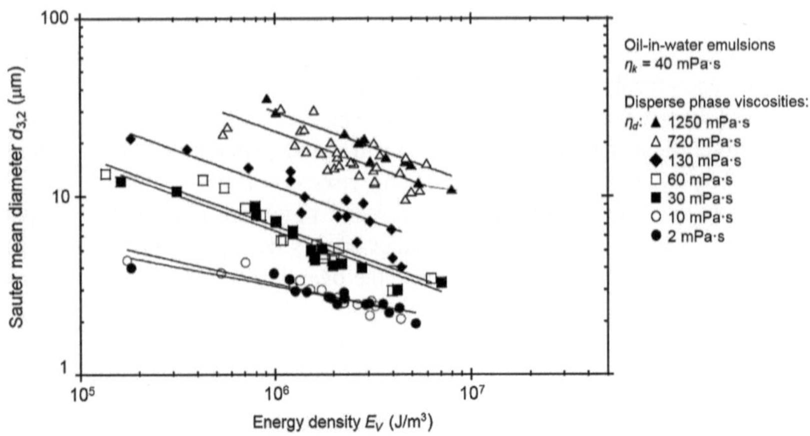

FIG. 6.16 Sauter mean diameter of oil-in-water emulsions produced by a colloid mill. The viscosity of the disperse phase was adjusted by using mineral oils with different viscosities. The disperse phase concentration of the emulsions was 30%, and polyoxyethylene-(10)-lauryl ether was used as emulsifier. *(Graph adapted from Karbstein, H.P., 1994. Untersuchungen zum Herstellen und Stabilisieren von Öl-in-Wasser-Emulsionen. Universität Karlsruhe.)*

different emulsification devices as comparable results have also been reported for in-line gear-rim devices (Hall et al., 2011a,b; Karbstein, 1994).

Altering the disperse phase viscosity as a formulation parameter is often not feasible in practical applications: Specific disperse phases are required, e.g., by formula; exchanging the disperse phase often drastically alters the product properties of the entire emulsion. However, when using synthetic oils (mineral or silicone oil), variations are possible as these oil types typically come in a large variety of viscosities. Viscosity also decreases with increasing temperature. However, heating up usually also involves an increased temperature of the continuous phase. Consequently, droplet breakup might be worse due to a reduced continuous phase viscosity.

6.5.2.3 *Viscosity Ratio*

When rotor-stator devices and process conditions are chosen in which droplet breakup occurs mainly in steady laminar shear flow, not only the viscosities of the individual phases but also the ratio between them can have an influence on droplet breakup. Well-known studies (Bentley and Leal, 1986; Grace, 1982) show the influence of the viscosity ratio between disperse and continuous phase on droplet breakup in steady laminar shear flow. It was found that droplet breakup of single droplets is easiest when the viscosity ratio is around 1, i.e., when the viscosity of the disperse and continuous phase is equal. This finding was confirmed for emulsion systems ($\phi > 0.05$) produced in colloid mills by Armbruster (1990); however, with one modification, the overall emulsion viscosity was used instead of the continuous phase viscosity in order to calculate the viscosity ratio. The presence of droplets usually makes the emulsion highly viscous than the pure continuous phase. The so-called effective medium approach assumes that in concentrated emulsions ($\phi > 0.05$), stresses are transmitted onto droplets not only by the continuous phase but also by the entire emulsion surrounding each emulsion droplet. These results were later on confirmed by Jansen et al. (2001), and corresponding results have also been reported for in-line gear-rim devices. Fig. 6.17 shows the Sauter mean diameter of water-in-oil emulsions plotted over the viscosity ratio. It can be seen that a minimum droplet size can be obtained at viscosity ratios close to 1 independent of the energy input applied during emulsification. At lower viscosity ratios, i.e., when the viscosity of the continuous phase is higher, a slight increase in droplet size can be seen. Comparable results were reported for high-shear mixer devices by Calabrese et al. (Rueger and Calabrese, 2013b). At viscosity ratios >1, i.e., when the viscosity of the disperse phase is higher, the Sauter mean diameter increases steeply. This is in agreement with the results described in the previous chapter. Consequently, droplet phase viscosities higher than the continuous phase viscosity or emulsion viscosity should strongly be avoided when small droplets are to be produced. If fine emulsions need to be obtained anyway, it is recommended to increase the viscosity of the phase surrounding the droplets.

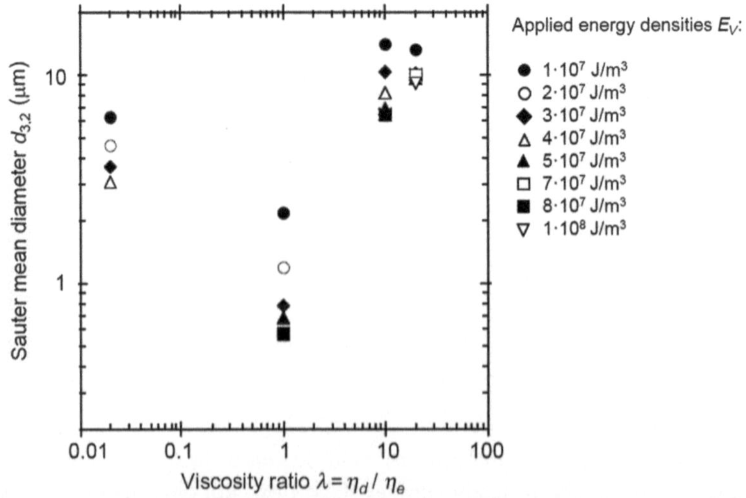

FIG. 6.17 Sauter mean diameter of water-in-oil emulsions produced by a continuous gear-rim device at different energy densities. Polyglycerol polyricinoleate was used as emulsifier and $\varphi = 0.3$. Energy densities varied between 107 and 108 J/m^3 with higher energy densities resulting in smaller Sauter mean diameters. *(Graph adapted from Tesch, S., 2002. Charakterisieren mechanischer Emulgierverfahren: Herstellen und Stabilisieren von Tropfen als Teilschritte beim Formulieren von Emulsionen. Universität Karlsruhe.)*

This can be achieved by adding stabilizers or viscosity enhancers to the continuous phase or by producing emulsions at higher disperse phase contents as will be described in the following subchapter.

The influence of the viscosity ratio on the emulsification result is only important for devices and process conditions in which droplet breakup occurs in dominantly laminar flow such as colloid mills or certain gear-rim devices. For devices and process conditions in which mainly turbulent flow occurs (many stirrers and gear-rim device setups), the impact of viscosity ratio was found to be negligible (Urban et al., 2006b). This is why these devices prove particularly useful for the production of very fine emulsions when the disperse phase has a much higher viscosity than the continuous phase. Applying elongational flow regimes prior to breakup is also reported to be useful in reducing droplet sizes in rotor-stator devices (Windhab et al., 2007).

6.5.2.4 *Disperse Phase Ratio*

Droplets are broken up by stresses transmitted onto them by the surrounding liquid phase. It was previously explained that finer emulsions can be achieved when the continuous phase viscosity is higher. The "effective medium approach" extends this concept by laying out that it is not so much the continuous phase viscosity but rather the emulsion viscosity that impacts droplet breakup. The emulsion viscosity, however, does not only depend on the continuous phase

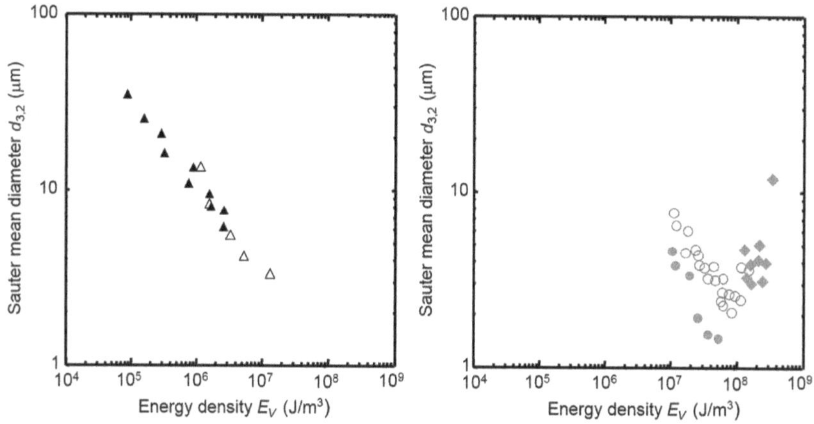

FIG. 6.18 Sauter mean diameter of emulsions produced by a continuous gear-rim device. Left: recoalescence does not occur. Oil-in-water emulsions with polyoxyethylene-(10)-lauryl ether as emulsifier. *Filled triangles*, $0.1 < \varphi < 0.5$; *empty triangles*, $0.6 < \varphi < 0.8$. Right: significant amount of recoalescence. Water-in-oil emulsions with polyglycerol polyricinoleate as emulsifier. *Filled circles*, $\varphi = 0.01$; *empty circles*, $0.1 < \varphi < 0.5$; *diamonds*, $0.6 < \varphi < 0.7$. Every set of data points were acquired at comparable process conditions (rotor speed). *(Graphs adapted from Karbstein, H.P., 1994. Untersuchungen zum Herstellen und Stabilisieren von Öl-in-Wasser-Emulsionen. Universität Karlsruhe; Tesch, S., 2002. Charakterisieren mechanischer Emulgierverfahren: Herstellen und Stabilisieren von Tropfen als Teilschritte beim Formulieren von Emulsionen. Universität Karlsruhe.)*

viscosity but also on the disperse phase ratio. At higher disperse phase ratios, the emulsion viscosity is higher, and more energy can be introduced into the emulsion system (Karbstein, 1994). As a result, smaller droplets can be produced provided that recoalescence of droplets is minimized. In Fig. 6.18, the Sauter mean diameter of several O/W emulsions prepared in an in-line gear-rim device is plotted over the energy density. The formulation of these emulsions was always the same except for the disperse phase ratio $\phi = V_d/V_e$ that varied from 0.1 to 0.8 (with V_d being the volume of the disperse phase and V_e being the volume of the emulsion). It can be seen that with increasing ϕ, the energy density resulting from comparable process parameters increases and the average droplet sizes decrease. Similar to a higher continuous phase viscosity, the higher emulsion viscosity caused by the high disperse phase ratio reduces the fluid transport within the device that further increases the mechanical energy input. The same relationship between disperse ratio and mean droplet size was reported for colloid mills (Karbstein, 1994) and batch gear-rim devices (Maa and Hsu, 1996; Pérez-Mosqueda et al., 2015).

In terms of process and formulation design, these results mean that emulsions should preferably be produced at high disperse phase ratios to achieve small droplets. The emulsions can then be diluted to obtain an emulsion of the desired disperse phase ratio. However, this strategy only works out when recoalescence in the system is suppressed by, e.g., a proper choice of emulsifier (see following chapter).

If the final droplet size is influenced significantly by coalescence, increasing the disperse phase ratio can even lead to much bigger droplets. This is also visualized in Fig. 5.8 but this time for water-in-oil emulsions. Again, higher energy densities can be achieved in water-in-oil emulsions due to the higher emulsion viscosity. Furthermore, with higher disperse phase ratio, a gradual increase in the mean droplet size can be seen. This is due to an increased droplet collision frequency at high disperse phase ratios, which is even enhanced by the higher energy densities (Tesch, 2002).

Studies exist in which the chosen process and formulation parameters lead to an emulsion system in which droplet breakup and coalescence are more or less balanced. In these studies, the mean droplet size is reported to be independent of the disperse phase ratio. For example, Tesch (2002) found that the mean droplet size is independent of the disperse phase ratio when the same water-in-oil emulsions shown in Fig. 5.8 are prepared in a colloid mill instead of a gear-rim device. Furthermore, Hall et al. (2011a) found the droplet size produced in a continuous gear-rim device to be nearly independent of the disperse phase ratio in a range of $0.01 < \phi < 0.2$. Rueger and Calabrese (2013a) investigated the emulsification process in high-shear mixers and report the mean droplet size to be independent of the disperse phase ratio only at $\phi > 0.1$. Below, a significant increase in droplet size with increasing disperse phase ratio was found.

6.5.2.5 Emulsifier Concentration and Adsorption Kinetics

When droplets are broken up, they need to be stabilized immediately against coalescence. For this purpose, emulsifiers that adsorb to the newly created interface are used. However, adsorption kinetics of emulsifiers differ significantly, depending on their molecular structure (Dukhin et al., 1995; Fainerman et al., 1998; Karbstein, 1994; Miller, 2011; Schuchmann and Köhler, 2012). At any disperse phase ratio, the interfacial area of a fine emulsion is by far larger than that of a coarse emulsion so that higher emulsifier concentrations are necessary when small droplets and fine emulsions need to be stabilized. This is no special property of emulsions produced by rotor-stator devices, but it applies to all mechanical emulsification processes. However, smaller droplets cannot be achieved sometimes, even if the emulsifier concentration is adjusted. This is the case when recoalescence occurs directly after droplet breakup because the emulsifier is not able to stabilize the interface fast enough (Karbstein, 1994; Stang et al., 1994). Recoalescence is mostly an issue in emulsification processes with short residence times and high volume flow rates, which is why emulsions produced in high-pressure homogenization processes are particularly susceptible (Karbstein, 1994; Schuchmann and Köhler, 2012).

Fig. 6.19 visualizes the advantages of rotor-stator devices when using an emulsifier with slow adsorption kinetics (Karbstein, 1994). The depicted

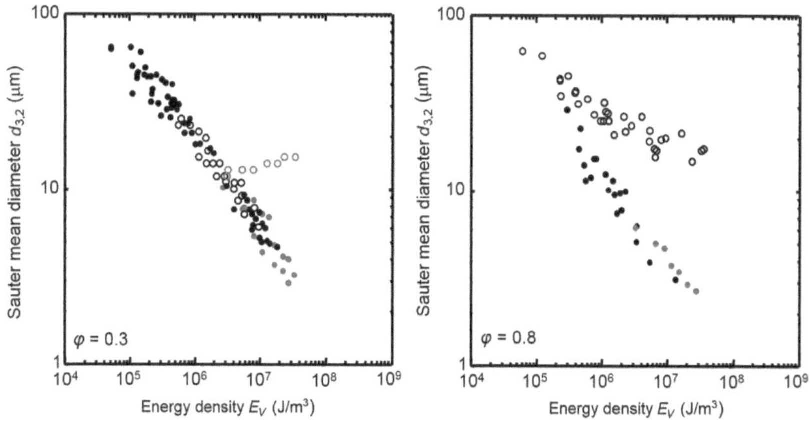

FIG. 6.19 Sauter mean diameter of oil-in-water emulsions with different disperse phase ratio φ produced by a colloid *(black symbols)* and by high-pressure homogenization *(gray symbols). Filled symbols*: the fast-adsorbing emulsifier polyoxyethylene-(10)-lauryl ether was used. *Open symbols*: the slow-adsorbing emulsifier egg yolk was used. In the right graph, there are no *gray open symbols* shown because it was not possible to produce highly concentrated emulsions stabilized with egg yolk by high-pressure homogenization. *(Graph adapted from Karbstein, H.P., 1994. Untersuchungen zum Herstellen und Stabilisieren von Öl-in-Wasser-Emulsionen. Universität Karlsruhe.)*

emulsions differ in formulation and in the emulsification process. A slow-adsorbing emulsifier (egg yolk) is compared with a fast-adsorbing one (polyoxyethylene-(10)-lauryl ether) in emulsions produced at lower (left image) and at very high disperse phase ratio (right image). Furthermore, the effect of rotor-stator emulsification is compared with high-pressure homogenization. For the emulsifier with fast adsorption kinetics, the mean droplet size decreases with increasing energy density E_V regardless of the emulsification device or of the disperse phase ratio. For the slow-adsorbing emulsifier, however, big differences can be seen. At low disperse phase ratios, the mean droplet size decreases only when the emulsion is produced in the rotor-stator device. Emulsions produced by high-pressure homogenization show a gradual increase in the mean droplet size due to an increased collision frequency at higher energy densities. At high disperse phase ratio, this effect is even more pronounced. When using the emulsifier with slow adsorption kinetics, stable emulsions can *only* be produced by the rotor-stator device. Again, the mean droplet size decreases with increasing energy density E_V, however with a much flatter slope. This is an indication that under the given experimental conditions, recoalescence also occurs in the rotor-stator process to a significant extent. In contrast to the rotor-stator process, high-pressure homogenization led to immediate phase separation, so that there are no data points available for these emulsions.

The long residence times in rotor-stator devices combined with reduced stresses are one of their big advantages as they allow for the production of fine emulsions even with slow-adsorbing emulsifiers. However, depending on the exact process conditions and formulation parameters, recoalescence can be a challenge in rotor-stator devices, too, as was also seen in Fig. 6.19 for water-in-oil emulsions.

Therefore, if the emulsifier can be chosen as a formulation parameter, emulsifiers with fast adsorption kinetics should be preferred. If the formulation cannot be changed, it can be necessary to give the emulsifier more time for adsorption, which can be achieved by two means:

(1) Increase the residence time in the disruption zone of the rotor-stator device by reducing the flow rate. Reduced flow rates can be created by, e.g., emulsifying at high viscosities.
(2) Emulsify at enhanced turbulence in the emulsification device. This reduces the droplet coalescence rate as the contact time between two droplets becomes too short for droplets to actually coalesce (Stang et al., 2001; Tcholakova et al., 2004).

6.6 EXAMPLES OF THE SUCCESSFUL PRODUCTION OF NANOEMULSIONS IN ROTOR-STATOR PROCESSES

Rotor-stator processes are usually not the emulsification method of choice for the production of nanoemulsions. Still, several studies can be found, which document the successful preparation of emulsions with very small droplet sizes in various rotor-stator devices (see Table 6.1). Many of these studies rely on fast-adsorbing emulsifiers to prevent recoalescence as much as possible. For example, Scholz and Keck (2015) used an in-line gear-rim device equipped with a recirculation loop to prepare emulsions in a batch setup. The emulsions had a disperse phase ratio of $\phi=0.05$ and consisted of medium-chain triglyceride oil dispersed in water with 5% Tween 80 and Span 80 as emulsifier. It was possible to achieve mean droplet sizes between 150 and 270 nm by varying the rotor speed and emulsification time. The produced emulsions had smaller droplet sizes than emulsions of the same formulation that were prepared by high-pressure homogenization.

A comparable formulation was used to prepare cosmeceutical emulsions (Han et al., 2012): the dispersed oil phase contained antioxidants as active ingredients and was used at a ratio of $\phi=0.3$. The emulsifier system and its overall concentration were the same as in the above study. Furthermore, the water phase was thickened by 0.8% xanthan. Emulsions with a mean droplet size of 126 nm were obtained by high-shear mixing at 6000 rpm for 5 min.

True nanoemulsions, i.e., emulsions containing nanoobjects according to ISO/TS 27687:2008, were prepared by Karthik and Anandharamakrishnan (2016). They produced nanoemulsions from docosahexaenoic acid as oil phase at a ratio of $\phi=0.1$ with 2.8% of Tween 40 as emulsifier. Emulsification by

TABLE 6.1 Overview of Examples for Successful Nanoemulsion Production Using Rotor-Stator Devices

Emulsion Type	Formulation	Rotor-Stator System	Strategy to Reduce Droplet Size	Achieved Mean Droplet Size	Reference
O/W	5% Miglyol 812 2.5% Tween 80 2.5% Span 80 Water	In-line gear rim with recirculation loop	Fast-adsorbing emulsifier, low disperse phase ratio, high rotational speed	150–270 nm	Scholz and Keck (2015)
O/W	30.5% oil phase +lipophilic active ingredients 4% Tween 80 1% Span 80 0.8% xanthan in water	High-shear mixer	Fast-adsorbing emulsifier, increased viscosity of continuous phase	126 nm	Han et al. (2012)
O/W	10% algae oil 2.8% Tween 40 Water	High-shear mixer	Fast-adsorbing emulsifier, low disperse phase ratio, high rotational speed	87 nm	Karthik and Anandharamakrishnan (2016)
O/W	38%–55% methyl methacrylate and butyl acrylate (1:1) <1% reaction initiators 0.4%–1.2% sodium dodecylbenzenesulfonate 4% or 10% octadecyl acrylate Water	High-shear mixer	High rotational speed, long homogenization time	311 nm 180–200 nm	El-Jaby et al. (2007) El-Jaby et al. (2009)

Continued

TABLE 6.1 Overview of Examples for Successful Nanoemulsion Production Using Rotor-Stator Devices—cont'd

Emulsion Type	Formulation	Rotor-Stator System	Strategy to Reduce Droplet Size	Achieved Mean Droplet Size	Reference
W/O	30% water 1% NaCl 3% polyglycerol polyricinoleate Vegetable oil	Colloid mill	Narrow gap width, high rotational speed	300 nm	Wolf et al. (2013)
W/O	60% water 0.5% NaCl 5% polyglycerol polyricinoleate Canola oil	Colloid mill with recirculation loop	High emulsion viscosity, high rotational speed	300 nm	Schuch et al. (2013)
O/W	10% or 15% corn oil 45% or 40% starch Water	Disperser discs	High continuous phase viscosity, large disc diameter, long emulsification time	120 nm	Urban et al. (2006b)

The type of emulsion (oil-in-water O/W or water-in-oil W/O) is given as well as the used type of rotor-stator device and the strategy employed to reduce the droplet size. The formulation of the respective emulsions is summarized by first stating the ingredients of the disperse phase, then the emulsifier system and finally the continuous phase ingredients with the main continuous phase substance adding up to the full volume or weight of the emulsion.

high-shear mixing at 24,000 rpm for 25 min resulted in emulsions with narrow droplet size distributions and mean droplet sizes <100 nm.

For miniemulsion polymerization, emulsions with mean droplet sizes <500 nm are a prerequisite. El-Jaby et al. (2007, 2009) prepared such emulsions by high-shear mixing at up to 3000 rpm and up to 90 min emulsification time. Their emulsions consisted of an organic phase comprising the two monomers methyl methacrylate and butyl acrylate at a disperse phase ratio of $\phi = 0.4$ with 1% sodium dodecylbenzenesulfonate as emulsifier. The obtained emulsions had mean droplet sizes well below 500 nm. For the longest emulsification times, it was even possible to achieve droplet sizes below 200 nm that resulted in the formation of nanoparticles of around 160 nm.

Although the use of rotor-stator systems for nanoemulsion production is rarely documented, these systems do have advantages compared with, e.g., high-pressure homogenizers. The strength of rotor-stator devices lies where high-pressure homogenizers cannot achieve satisfying emulsification results anymore, that is, emulsion of high viscosity stabilized by slow-adsorbing emulsifiers. Schuch et al. (2013) and Wolf et al. (2013) produced water-in-oil emulsions for the preparation of double emulsions in continuous rotor-stator devices. Such food-grade water-in-oil emulsions are typically very highly viscous because of the high viscosity of the continuous oil phase. In both studies, the water phase consisted of a sodium chloride solution that was dispersed in canola oil. Polyglycerol polyricinoleate (PGPR) was used as emulsifier. Wolf et al. prepared emulsions with 3% PGPR and a disperse phase ratio of $\phi = 0.3$ in a colloid mill with 0.15 mm gap width at 10,000 rpm. The resulting emulsions had a mean droplet size of around 300 nm. Droplets of the same size were achieved by Schuch et al. in an in-line gear-rim device by using not only higher amounts of PGPR (5%) but also a higher disperse phase ratio ($\phi = 0.6$) that increased the viscosity even further.

Finally, disperser discs are reportedly well suited for extremely high viscous emulsions. Urban et al. prepared oil-in-water emulsions in which up to 15% corn oil was dispersed into a starch solution (Urban et al., 2006b). Here, modified starch acted as both a viscosity enhancer and a hydrocolloid emulsifier with slow adsorption kinetics. By using 45% starch in the continuous phase, a viscosity of the continuous phase of almost 1 Pa s was achieved. The resulting emulsions showed mean droplet sizes below 300 nm after only 5 min of emulsification time, which could be reduced down to around 120 nm by prolonging the emulsification time to 30 min.

From these examples, it can be seen that rotor-stator emulsification is particularly suitable for the emulsification of highly viscous liquids. If an emulsion does not meet this criterion, the formulation can be adapted by relatively easy means in order to improve the processing conditions in rotor-stator devices. Stabilizers can be added to the continuous phase in order to increase the continuous phase viscosity as seen in Han et al. (2012) and Urban et al. (2006b). Otherwise, the emulsion can be prepared at a higher disperse phase ratio that increases the overall emulsion viscosity. After the emulsification step, the emulsion can be diluted with pure continuous phase to achieve the desired disperse phase ratio without altering the droplet size as described by Schuch et al. (2013).

6.7 CONCLUSION

In order to produce nanoemulsions in rotor-stator processes, very high energy densities must be generated by the device, and all emulsion droplets within the product must be equally stressed. For this reason, rotor-rotor devices are particularly useful as they can generate higher stresses than rotor-stator devices. In both rotor-rotor and rotor-stator devices, high energy densities can be achieved (a) by high peripheral velocities (i.e., large rotor diameter and high rotational speed), (b) by high product viscosities, (c) by a longer emulsification time, and (d) by higher disperse phase ratios. It is, for example, possible to prepare an emulsion at a higher disperse phase ratio so that the overall emulsion viscosity is increased. After the emulsification step, the emulsion can be diluted with pure continuous phase to achieve the desired disperse phase ratio without altering the droplet size as described by Schuch et al. (2013). However, this requires the use of fast-adsorbing emulsifiers. In case of emulsifiers with slow adsorption kinetics, rotor-stator devices perform much better than high-pressure homogenizers. Due to the longer residence times, very fine droplets can be stabilized even by slow-adsorbing emulsifiers in rotor-stator processes. This makes these processes an important alternative for emulsion producers in various industrial sectors.

REFERENCES

Armbruster, H., 1990. Untersuchungen zum kontinuierlichen Emulgierprozess in Kolloidmühlen unter Berücksichtigung spezifischer Emulgatoreigenschaften und der Strömungsverhältnisse im Dispergierspalt. Universität Karlsruhe (TH), Karlsruhe.

Bentley, B.J., Leal, L.G., 1986. An experimental investigation of drop deformation and breakup in steady two-dimensional linear flows. J. Fluid Mech. 167, 241–283.

Catte, M., Poprawski, J., Aubry, J.M., Van Hecke, E., 2002. Analysis of an emulsification process using fractional experimental design. J. Dispers. Sci. Technol. 23, 323–331.

Dukhin, S.S., Kretzschmar, G., Miller, R., 1995. Dynamics of Adsorption at Liquid Interfaces: Theory, Experiment, Application. Elsevier B.V., Amsterdam

El Kinawy, O.S., Petersen, S., Helmdach, L., Ulrich, J., 2012. Parameter selection of emulsification processes: conditions for nano- and macroemulsions. Chem. Eng. Technol. 35, 1604–1608.

El-Jaby, U., Cunningham, M., McKenna, T.F.L., 2009. Comparison of emulsification devices for the production of miniemulsions. Ind. Eng. Chem. Res. 48, 10147–10151.

El-Jaby, U., McKenna, T.F.L., F Cunningham, M., 2007. Miniemulsification: an analysis of the use of rotor stators as emulsification devices. Macromol. Symp. 259, 1–9.

Fainerman, V.B., Lucassen-Reynders, E.H., Miller, R., 1998. Adsorption of surfactants and proteins at fluid interfaces. Colloids Surf. A Physicochem. Eng. Asp. 143, 141–165.

Feigl, K., Megias-Alguacil, D., Fischer, P., Windhab, E.J., 2007. Simulation and experiments of droplet deformation and orientation in simple shear flow with surfactants. Chem. Eng. Sci. 62, 3242–3258.

Frank, K., Köhler, K., Schuchmann, H.P., 2011. Formulation of labile hydrophilic ingredients in multiple emulsions: influence of the formulation's composition on the emulsion's stability and on the stability of entrapped bioactives. J. Dispers. Sci. Technol. 32, 1753–1758.

Goldschmidt, A., Streitberger, H.-J., 2002. BASF-Handbuch Lackiertechnik. Vincentz Verlag, Hannover.

Grace, H.P., 1982. Dispersion phenomena in high viscosity immiscible fluid systems and application of static mixers as dispersion devices in such systems. Chem. Eng. Commun. 14, 225–277.

Håkansson, A., Askaner, M., Innings, F., 2016. Extent and mechanism of coalescence in rotor-stator mixer food-emulsion emulsification. J. Food Eng. 175, 127–135.

Hall, S., Cooke, M., El-Hamouz, A., Kowalski, A.J., 2011a. Droplet break-up by in-line Silverson rotor–stator mixer. Chem. Eng. Sci. 66, 2068–2079.

Hall, S., Cooke, M., Pacek, A.W., Kowalski, A.J., Rothman, D., 2011b. Scaling up of silverson rotor-stator mixers. Can. J. Chem. Eng. 89, 1040–1050.

Han, N.S., Basri, M., Abd Rahman, M.B., Abd Rahman, R.N.Z.R., Salleh, A.B., Ismail, Z., 2012. Preparation of emulsions by rotor-stator homogenizer and ultrasonic cavitation for the cosmeceutical industry. J. Cosmet. Sci. 63, 333–344.

Henzler, H.-J., Biedermann, A., 1996. Modelluntersuchungen zur Partikelbeanspruchung in Reaktoren. Chem. Ing. Tech. 68, 1546–1561.

Jafari, S.M., He, Y., Bhandari, B., 2006. Nano-emulsion production by sonication and microfluidization—a comparison. Int. J. Food Prop. 9 (3), 475–485.

Jafari, S.M., He, Y., Bhandari, B., 2007a. Optimization of nano-emulsions production by microfluidization. Eur. Food Res. Technol. 225 (5–6), 733–741.

Jafari, S.M., He, Y., Bhandari, B., 2007b. Production of sub-micron emulsions by ultrasound and microfluidization techniques. J. Food Eng. 82 (4), 478–488.

Jafari, S.M., Assadpoor, E., He, Y., Bhandari, B., 2008. Re-coalescence of emulsion droplets during high-energy emulsification. Food Hydrocoll., Volume 22, Issue 7, October 2008, Pages 1191-1202.

Jansen, K.M.B., Agterof, W.G.M., Mellema, J., 2001. Droplet breakup in concentrated emulsions. J. Rheol. (N. Y. N. Y) 45, 227.

Jasińska, M., Bałdyga, J., Hall, S., Pacek, A.W., 2014. Dispersion of oil droplets in rotor-stator mixers: experimental investigations and modeling. Chem. Eng. Process. Process Intensif. 84, 45–53.

Karbstein, H.P., 1994. Untersuchungen zum Herstellen und Stabilisieren von Öl-in-Wasser-Emulsionen. Universität Karlsruhe, Karlsruhe.

Karbstein, H.P., Schubert, H., 1995. Developments in the continuous mechanical production of oil-in-water macro-emulsions. Chem. Eng. Process. Process Intensif. 34, 205–211.

Karthik, P., Anandharamakrishnan, C., 2016. Fabrication of a nutrient delivery system of docosahexaenoic acid nanoemulsions via high energy techniques. RSC Adv. 6, 3501–3513.

Knoch, A., 2002. Emulgieren im Rührapparat. Chem. Ing. Tech. 74, 1369–1379.

Koglin, B., Pawlowski, J., Schnöring, H., 1981. Kontinuierliches Emulgieren mit Rotor/Stator-Maschinen: Einfluß der volumenbezogenen Dispergierleistung und der Verweilzeit auf die Emulsionsfeinheit. Chem. Ing. Tech. 53, 641–647.

Maa, Y.-F., Hsu, C., 1996. Liquid-liquid emulsification by rotor/stator homogenization. J. Control. Release 38, 219–228.

Miller, R., 2011. Drop and Bubble Interfaces. Koninklijke Brill NV, Leiden.

Pacek, A.W., Hall, S., Cooke, M., Kowalski, A.J., 2013. Emulsification in rotor-stator mixers. In: Emulsion Formation and Stability. Wiley-VCH Verlag GmbH & Co. KGaA, Weinheim, pp. 127–167.

Pérez-Mosqueda, L.M., Trujillo-Cayado, L.A., Carrillo, F., Ramírez, P., Muñoz, J., 2015. Formulation and optimization by experimental design of eco-friendly emulsions based on d-limonene. Colloids Surf. B: Biointerfaces 128, 127–131.

Rueger, P.E., Calabrese, R.V., 2013a. Dispersion of water into oil in a rotor–stator mixer. Part 2: effect of phase fraction. Chem. Eng. Res. Des. 91, 2134–2141.

Rueger, P.E., Calabrese, R.V., 2013b. Dispersion of water into oil in a rotor–stator mixer. Part 1: drop breakup in dilute systems. Chem. Eng. Res. Des. 91, 2122–2133.

Scholz, P., Keck, C.M., 2015. Nanoemulsions produced by rotor–stator high speed stirring. Int. J. Pharm. 482, 110–117.

Schubert, H., Armbruster, H., 1989. Prinzipien der Herstellung und Stabilitat von Emulsionen. Chem. Ing. Tech. 61, 701–711.

Schuch, A., Deiters, P., Henne, J., Köhler, K., Schuchmann, H.P., 2013. Production of W/O/W (water-in-oil-in-water) multiple emulsions: droplet breakup and release of water. J. Colloid Interface Sci. 402, 157–164.

Schuchmann, H.P., Köhler, K. (Eds.), 2012. Emulgiertechnik: Grundlagen, Verfahren und Anwendungen. third ed. Behr's Verlag, Hamburg.

Stang, M., Karbstein, H.P., Schubert, H., 1994. Adsorption kinetics of emulsifiers at oil-water interfaces and their effect on mechanical emulsification. Chem. Eng. Process. Process Intensif. 33, 307–311.

Stang, M., Schuchmann, H., Schubert, H., 2001. Emulsification in high-pressure homogenizers. Eng. Life Sci. 1, 151.

Tcholakova, S., Denkov, N.D., Banner, T., 2004. Role of surfactant type and concentration for the mean drop size during emulsification in turbulent flow. Langmuir 20, 7444–7458.

Tesch, S., 2002. Charakterisieren mechanischer Emulgierverfahren: Herstellen und Stabilisieren von Tropfen als Teilschritte beim Formulieren von Emulsionen. Universität Karlsruhe, Karlsruhe.

Urban, K., Wagner, G., Schaffner, D., Röglin, D., Ulrich, J., 2006a. Rotor-stator and disc systems for emulsification processes. Chem. Eng. Technol. 29, 24–31.

Urban, K., Wagner, G., Schaffner, D., Ulrich, J., 2006b. Dispergierscheiben, die bessere Alternative zum Hochdruck beim Emulgieren? Chem. Ing. Tech. 78, 1069–1077.

Walstra, P., 1993. Principles of emulsion formation. Chem. Eng. Sci. 48, 333–349.

Windhab, E.J., Eisner, V., Troxler, B., Dürig, A.K., Grohmann, F.-R., 2007. Europäische Patentschrift, EP1781402 B1.

Wolf, F., Koehler, K., Schuchmann, H.P., 2013. Stabilization of water droplets in oil with PGPR for use in oral and dermal applications. J. Food Process Eng. 36, 276–283.

Zlokarnik, M., 1999. Rührtechnik. Springer Berlin Heidelberg, Berlin, Heidelberg.

Zlokarnik, M., 2000. Stirring, Ullmann's Encyclopedia of Industrial Chemistry. Wiley-VCH Verlag GmbH & Co. KGaA, Weinheim.

Chapter 7

Fabrication of Nanoemulsions by High-Pressure Valve Homogenization

Andreas Håkansson
Kristianstad University, Kristianstad, Sweden

Chapter Outline

Nanoemulsions. https://doi.org/10.1016/B978-0-12-811838-2.00007-2

7.1 INTRODUCTION

High-pressure valve homogenization is used for emulsification in a wide range of processing industries. The basic technology was developed in the early 1900s by August Gaulin for reducing the fat globule size in milk as a means of increasing stability and shelf life (Gaulin, 1904). The mechanism by which fat globules are disrupted in the high-pressure valve homogenizer (HPH) has long been unknown and the source of many speculations; see discussion in Håkansson (2015). The contemporary view is that the valve design combined with the high pressures applied gives rise to an intense and narrow region of high-intensity hydrodynamic stress that effectively breaks the drops into smaller fragments.

The HPH has two main advantages compared with other high-energy methods of emulsification. Firstly, it produces very high local stresses. Secondly, it is a well-tested mature technology for continuous production. Because of these advantages, the HPH is still used extensively in dairy production. However, it has also found many other uses in food, pharmaceutical, cosmetics, and chemical processing. HPH is often considered the standard technology for emulsification of fluids with low to intermediate viscosities (i.e., 1–200 mPa s), regardless of application (Jafari et al., 2008; Schultz et al., 2004). HPH has also found uses outside of the emulsification field, mainly in cell harvesting where the high local stresses are used to rupture cells and release intracellular components (Siddiqi et al., 1997).

The two general advantages of the HPH technology have also made it especially interesting for nanoemulsion formation. The highly localized and intense hydrodynamic stresses are useful in creating the small drops required for nanoemulsions to form. And the long experience of large-scale continuous production has made HPH interesting for academics and industries developing production-scale nanoemulsification.

This chapter summarizes the current understanding of HPH with special attention to nanoemulsion formation. In order to discuss strengths and limitations of the technology and for interpreting empirical investigations of HPH, a sound understanding of the general operating principles is first needed. The design, geometry, and operation of HPH are discussed in Section 7.2. As discussed in Chapter 5 of this book, the ability of high-energy emulsification methods to create nanoemulsions depends on their ability to create small drops, i.e., give high fragmentation rates and low coalescence rates. These underlying processes and how they are influenced by operating parameters and product properties are discussed in Section 7.3.

As previously mentioned, the availability of large-scale processing equipment, able to process up to 50,000 L/h, is a marked advantage of HPH. However, the academic research on nanoemulsion formation in HPHs is carried out on small-scale benchtop models. The scalability of HPH is therefore a vital aspect for validating the efficiency of the method. The scale-up is discussed in Section 7.4.

An often noted challenge with nanoemulsion formation in HPH is the high energy input giving rise to substantial product temperature increase. This might have detrimental effects on the emulsion or any carrier molecules, e.g., nutrients or pharmaceuticals. Temperature effects are discussed in Section 7.5.

A large number of studies have been conducted on nanoemulsion formation in HPH. Section 7.6 reviews the available information and compares it with the principles developed in earlier chapters. Special attention is turned to three specific challenges in using HPH in nanoemulsion formation: the required homogenization pressure, the risk of overprocessing, and the need for using multiple passages. Finally, Section 7.7 concludes the chapter.

It should also be mentioned that several previous reviews of HPH, emulsification, and nanoemulsion formation are available in literature. These have informed the present chapter; however, they have somewhat different approaches then in this chapter. Phipps (1985) discussed HPH design and suitability for milk processing. Walstra published a series of reviews of emulsification in general, often with special attention to HPH (Walstra, 1983; Walstra and Smulders, 1998; Walstra, 2005). More recent reviews of HPH design and drop breakup can be found in Innings (2015) and Håkansson (2015), respectively. However, none of these treat the special issues related to nanoemulsions. Previous reviews on nanoemulsion formation discuss HPH but on a general level (Gupta et al., 2016; McClements and Rao, 2011; Tadros et al., 2004). This chapter extends these previous reviews by combining the fundamental approach with recent experimental advances in light of the special conditions required for nanoemulsion formation.

7.2 DESIGN AND PRINCIPLES OF OPERATION

7.2.1 HPH Valve Design

With regard to weight and volume, the largest part of the HPH consists of motors and pumps used to pressurize the flow. However, from the perspective of emulsion formation, the most interesting part of the HPH is the homogenizing valve, the only part that the emulsion passes through. A schematic illustration of an HPH valve with two stages can be seen in Fig. 7.1. And a magnification of the first stage can be seen in Fig. 7.2. Each stage consists of three parts: an annular seat, a cylindrical forcer (sometimes referred to as the plunger), and an annular impact ring. Stainless steel is sometimes used; however, modern valves are most often designed in ceramic materials in order to better withstand the intense hydrodynamic forces (i.e., cavitation).

Emulsion formation in HPH requires that the emulsion is first pre-emulsified to create a coarse pre-emulsion. The pre-emulsion is fed into the HPH valve using a set of piston pumps—often between three and five serially connected pumps for production-scale HPHs, fewer for laboratory-scale machines. When the pre-emulsion enters the HPH valve and approaches the forcer, it is redirected

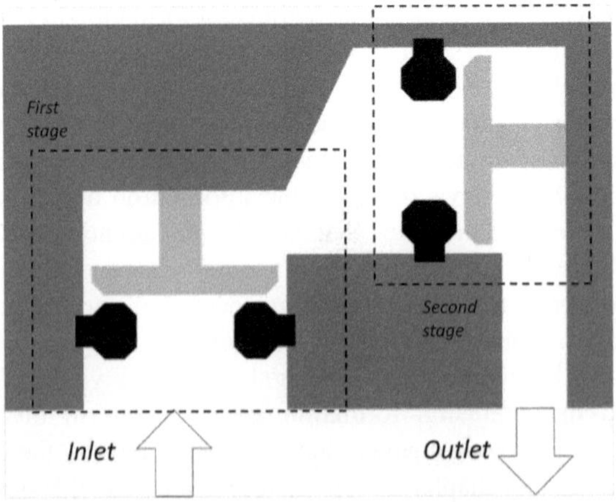

FIG. 7.1 Schematic illustration of an HPH valve with two stages.

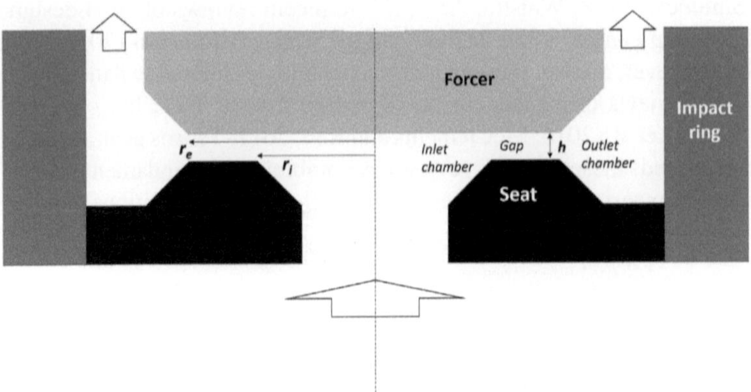

FIG. 7.2 Magnification of the first stage of the HPH valve in Fig. 7.1.

and forced radially through the gap formed between seat and forcer. A narrow gap height is maintained by applying a high hydraulic back pressure on the forcer. After passing through the HPH valve, drops have broken up into smaller fragments forming a homogenous emulsion.

A common misconception among earlier investigators was that drops are broken up by physical interaction, e.g., squeezing, in the narrow gap (Gaulin, 1904). However, later research has shown that the homogenizing gap is at least an order magnitude larger than the drop and, consequently, drop breakup is not due to direct physical interaction between the valve and the drops.

7.2.2 Geometry, Flowrate and Homogenizing Pressure

The degree of homogenization and consequently the size of the resulting drops are controlled by the homogenizing pressure, Δp, defined as the pressure difference across the valve. Traditional applications such as milk homogenization are often carried out at homogenizing pressures ranging from 20 to 40 MPa. This reduces the Sauter mean diameter (D_{32}) from approximately 5 to 1 μm. Substantially higher pressures are often needed for nanoemulsion formation (~100 MPa).

The high pressures give rise to high linear velocities in the gap. If fed with a volumetric flowrate, Q, the radial velocity at the gap exit is (McKillop et al., 1955)

$$U_g = \frac{Q}{2\pi r_e h}, \tag{7.1}$$

where h is the gap height and r_e is the gap exit radius (see Fig. 7.2). The full velocity profile as a function of radial position is illustrated in Fig. 7.3.

The stresses experienced by an emulsion drop passing through the HPH valve increase monotonically with the gap velocity. However, the gap height and consequently the gap velocity and the hydrodynamic stresses cannot be measured during operation. The gap height is set implicitly by adjusting the hydraulic pressure on the forcer. The only information accessible during an HPH experiment is the applied homogenizing pressure. This has made it difficult to understand the hydrodynamics inside the HPH.

An empirical correlation for estimating the gap height as a function of valve geometry can be obtained by assuming that the total pressure drop across the valve can be divided in three parts, (1) an entrance loss when entering the gap, (2) a friction loss for the annular channel flow in the gap, and (3) an exit loss term when exiting the gap (Phipps, 1975):

$$\Delta p = \frac{\rho}{4}\left(\frac{Q}{2\pi \cdot r_i \cdot h}\right)^2 + \frac{5\rho \upsilon^{3/5}}{h^3}\left(\frac{Q}{2\pi}\right)^{7/5}\left(\frac{1}{r_i^{2/5}} - \frac{1}{r_e^{2/5}}\right) + \frac{\rho}{2}\left(\frac{Q}{2\pi \cdot r_e \cdot h}\right)^2. \tag{7.2}$$

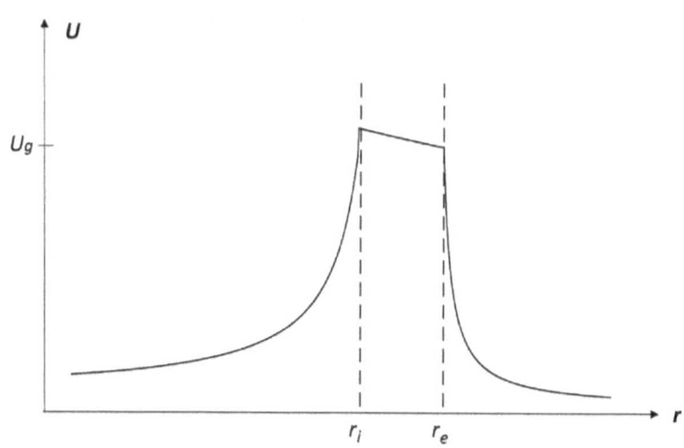

FIG. 7.3 Schematic radial velocity profile in the HPH valve.

In Eq. 7.2, r_i denotes the gap inlet radius, ρ fluid density, and ν kinematic viscosity.

Note that the volumetric flowrate, Q, is set independently of the homogenizing pressure. In commercial applications, the flowrate is often given by an external feed pump. The internal piston pumps are only used for increasing the static pressure.

The specific details of the valve designs, such as gap length, seat and forcer inclines, seat shape, and impact ring distance, differ between manufacturers. Some manufacturers of laboratory-scale (benchtop) HPH designs use a spherical ball forcer in order to ensure a short gap.

Due to its special relevance for nanoemulsion formation, the design used by Stansted also deserves a special mention. The valve is similar to the one shown in Fig. 7.2. However, the flow is applied in an opposite direction; fluid converges inward through the valve. See Floury et al. (2004a) for an in-depth analysis of the geometry and flow field of this design.

7.2.3 Thermodynamic Efficiency

The thermodynamic efficiency of the HPH is low, despite its widespread use and long history. Theoretically, reducing the Sauter mean drop diameter from D_0 to D requires an energy input corresponding to the increase in surface energy, which is proportional to the total increase in drop surface area (Mohr, 1987b):

$$P_{surface} = 6\gamma\phi_D Q\left(\frac{1}{D} - \frac{1}{D_0}\right), \tag{7.3}$$

where ϕ_D is the disperse phase volume fraction. The actual power input of HPH is determined by the flowrate and homogenizing pressure:

$$P_{input} = Q\Delta p. \tag{7.4}$$

The thermodynamic efficiency of HPH emulsification is obtained as the ratio between required surface energy increase and actual energy input (Mohr, 1987b):

$$\eta_{HPH} = \frac{P_{surface}}{P_{input}} = \frac{6\gamma\phi_D}{\Delta P}\left(\frac{1}{D} - \frac{1}{D_0}\right). \tag{7.5}$$

Producing a 10% (v/v) emulsion with a diameter of 2 μm at a surface tension of 5 mN/m using a homogenizing pressure of 25 MPa results in a thermodynamic efficiency of approximately 0.05%. The thermodynamic efficiency is lower still for nanoemulsions. Experimental investigations suggest a power-law dependence between resulting drop diameter and homogenizing pressure,

$$D \propto \Delta p^{-q}, \tag{7.6}$$

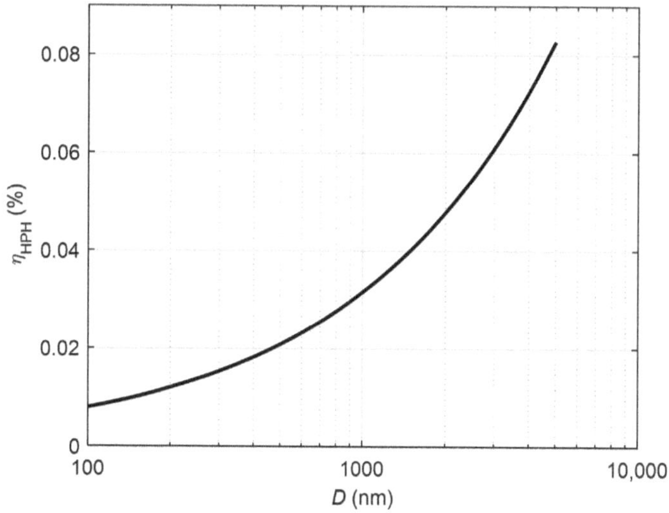

FIG. 7.4 Thermodynamic efficiency of emulsification in creating an emulsion with Sauter mean diameter D (Eqs. 7.5–7.6), assuming $q=0.6$, $\phi_D=10\%$, $\gamma=5\,\text{mN/m}$, and $D_0=5\,\mu\text{m}$.

with a constant q in the range of 0.6–0.9 (Phipps, 1985; Walstra and Smulders, 1998). Since the pressure exponent is less than one, the net effect when inserted in Eq. (7.5) is a decrease in efficiency with decreasing drop diameters; see Fig. 7.4. The thermodynamic efficiency of creating a nanoemulsion with an average drop diameter of 200 nm is approximately 0.01%, a fifth of the value for a conventional micrometer range emulsion.

As seen from Fig. 7.4, only a small fraction of the energy supplied to the HPH is actually transformed into surface energy. Hence, it is not the efficiency in converting mechanical energy into surface energy that has made HPH the standard method for low-viscosity emulsification. The advantage of the HPH lies in the favorable dynamics; it gives rise to high breakup rates and relatively low coalescence rates, as discussed in Section 7.3.

7.2.4 One-Stage or Two-Stage Design

High-pressure homogenizers are often designed with two valves mounted in series (see Fig. 7.1). This is referred to as a two-stage design. The homogenizing pressure in a two-stage design is defined as the total pressure difference between the inlet of the first valve and the outlet of the second one. The popularity of two-stage design has arisen purely from empirical findings; it has been shown to give rise to somewhat higher thermodynamic efficiencies (Mohr, 1987b). The fraction of the total pressure drop in the first and second stage is described in terms of the Thoma number, defined as the first-stage exit pressure, p_2, divided by the homogenizing pressure:

$$Th = \frac{p_2}{\Delta p}. \tag{7.7}$$

The finding of a maximum efficiency (often reported close to $Th = 0.2$, Kurzhals, 1977; Loo and Carleton, 1953) is somewhat surprising given that the advantage of the HPH is in delivering a high local stress and high local breakup rates. Dividing the energy input in two stages necessarily decreases the maximum stress and would, thus, be expected to decrease the efficiency.

Two main hypotheses have been suggested to explain why two-stage devices give increased efficiency. First, as emulsion drops pass the high-intensity region, fragmentation is the dominating mechanism. However, due to the intense hydrodynamic forces, collisions between drops or fragments occur giving rise to flocculation. It has been suggested that the increased efficiency of the second stages comes from breakup of these flocs (Phipps, 1985).

Secondly, the Thoma number influences the back pressure to the first HPH valve that in turn influences the static pressure distribution in the valve and the risk of cavitation. The *Th*-optimum has therefore been suggested as representing the optimum amount of cavitation; see discussion in Section 7.3.5.

7.3 DROP FRAGMENTATION AND COALESCENCE MECHANISMS

As discussed in Chapter 5, emulsification is a dynamic process. The drop-size reduction obtained upon passing an emulsion through an HPH valve depends on the initial drop-size distribution, the rate of breakup, and the rate of coalescence. These processes are, in turn, influenced by the fluid-flow conditions in the HPH valve and the interfacial phenomena. Hence, in order to understand nanoemulsion formation in high-pressure homogenization, two factors must be studied: the hydrodynamics in the valve and the interfacial chemistry resulting from drop-emulsifier interactions.

With a few exceptions, we have reason to believe that the interfacial chemistry (e.g., adsorption, film rupture, and surface rheology) follows the same general trend as discussed in Chapter 5. The exceptions are discussed in Section 7.3.8.

The hydrodynamics, however, is highly specific to the HPH valve. HPH hydrodynamics was, for a long time, a contested field with a lack of experimental data to corroborate theoretical suggestions. Different investigators made different suggestions on the role of laminar shear, turbulence, and cavitation; on the position of drop breakup; on the role of coalescence; and on the effect of disperse phase volume fraction. However, recent advances in experimental methodologies have formed the foundation of a consensus on where and how emulsification is linked to the geometry and flow field of the HPH.

7.3.1 Three Approaches for Studying HPH Emulsification

Three main approaches for studying HPH hydrodynamics and its influence on drop breakup can be identified in literature. The first and oldest is based on a combination of emulsification experiments and simplistic theoretical scaling arguments. In an influential and representative study, Walstra (1969) argued that emulsification proceeds by turbulent inertial fragmentation in HPHs by showing that the homogenization pressure coefficient (i.e., q in Eq. 7.6) corresponds to the value that would be expected under turbulent inertial conditions. These comparisons between scaling and theoretical suggestions could prove a valuable tool (see, e.g., Lee and Norton, 2013; Lee et al., 2014) but are limited to testing specific hypotheses and generally assume that HPH valve hydrodynamics is dominated by one type of flow and that the stresses experienced in the valve display no spatial variations.

The second approach is to model the flow, the dispersion process, or a combination of the two using mathematical models, i.e., computational fluid dynamics (CFD) for the flow fields and population balance modeling (PBM) for the dispersion process. CFD studies have proved useful for visualizing the flow field (Floury et al., 2004a; Håkansson et al., 2012; Kleinig and Middelberg, 1996; Steiner et al., 2006; Stevenson and Chen, 1997) and provided the first insights into the spatial variation of flow field, static pressure, and turbulence. Moreover, with increasing computational power, it becomes easier to use CFD for valve design. However, it should be kept in mind that the choice of turbulence model in the mathematical formulation is vital for obtaining a representative flow field and that the standard turbulence models often perform poorly as compared to experiments (Håkansson et al., 2012).

With regard to dispersion modeling, several PBM studies have been reported for emulsion formation in HPH (Håkansson et al., 2009, 2013a; Maindarkar et al., 2015; Raikar et al., 2009, 2011). However, there is no general conclusion on the form of the breakage, coalescence, and adsorption rate constants; on how to handle emulsifier transport; or on how to couple it to CFD (cf. Håkansson et al., 2013a; Maindarkar et al., 2015). Moreover, setting up and solving PBM is a time-consuming task. Comparisons between simulated and experimental drop-size distributions show that PBM frameworks are valuable for confirming or rejecting suggested mechanisms. However, they are less suitable for formulating new hypotheses.

The third and newest approach is based on optical measurements of the flow, either to measure the flow fields or to visualize cavitation or drop breakup. Bisten and Schuchmann (2016) have provided a comprehensive review of these optical investigations and on what could be concluded on HPH emulsification based on them. The optical methods have many advantages over the earlier approaches. First, they allow for direct measurement of basic phenomena (such as the velocity fields and drop breakup), whereas the previous two approaches rely on experimental data on drop diameter only. Therefore, the optical methods can (as opposed to the previous approaches) be used to generate new suggestions of mechanisms. Secondly, optical methods allow for understanding of local effects.

The main disadvantage of the optical methods is that they cannot be applied to HPH valves directly due to the asymmetrical geometry (see Fig. 7.2). The optical experiments have therefore been carried out in geometrically modified or scaled HPH valves. The different ways of modifying or scaling the valves by different investigators can explain some of the variations seen among different studies (Bisten and Schuchmann, 2016). Still the optical methods are the most promising way for increased understanding. The proceeding sections summarize the current understanding, primarily based on these optical methods.

7.3.2 Laminar Shear and the Inlet Chamber

As can be seen from the velocity profile in Fig. 7.3, the fluid is accelerated radially as it enters the inlet chamber. This gives rise to an elongational stress, G, on drops as they pass through the inlet chamber. Drop visualization experiments show that this deforms the drops, from spherical to ellipsoidal shape (Innings and Trägårdh, 2005). From a static-stress-balance perspective, breakup is expected once the deformation becomes sufficiently large, i.e., when the capillary number,

$$Ca = \frac{G\mu_C D}{2\gamma}, \tag{7.8}$$

exceeds the critical limit, Ca_{Cr}, of order magnitude 0.1–1 for an O/W emulsion (Grace, 1982). The maximum elongation rate scales with the gap velocity and gap height (Floury et al., 2004b; Håkansson et al., 2011; Innings and Trägårdh, 2007):

$$G \propto \frac{U_g}{h}. \tag{7.9}$$

None of the drop visualization studies found any drop breakup taking place in the inlet chamber despite high elongational rates (Innings and Trägårdh, 2005; Innings et al., 2011a; Kelemen et al., 2015b). However, these experiments were conducted at lower homogenizing pressures than used for nanoemulsion formation. Hence, from a static-stress balance, it is not obvious why elongation could not become sufficiently strong to break the drops at the elevated pressures used for nanoemulsion formation.

However, as originally argued by Walstra (1983) and elaborated by later investigators (Floury et al., 2004b; Kelemen et al., 2015b), dynamic effects can be used to explain why elongation is unable to break the drop, regardless of the homogenizing pressure. A drop subjected to an elongation rate G needs a time of order magnitude (Walstra and Smulders, 1998),

$$\tau_{def} = \frac{\mu_D}{\mu_C G}, \tag{7.10}$$

in order to deform. However, the time spent in the elongational flow, τ, scales with the linear velocity and is inversely proportional to G,

$$\tau \propto \frac{1}{G}, \tag{7.11}$$

implying that the drop will only have time sufficient to deform ($\tau > \tau_{def}$) if the viscosity ratio is sufficiently low:

$$\frac{\mu_D}{\mu_C} < C. \tag{7.12}$$

Walstra (1983) suggests a value $C=4$. According to Eq. 7.12, laminar breakup is unfeasible, regardless of shear, for emulsions with relative high disperse phase viscosity, such as many O/W emulsions. Hence, inlet chamber elongation is often unable to break drops regardless of homogenizing pressure.

7.3.3 Shear and Turbulence in the Gap

The Reynolds number at the gap exit is (McKillop et al., 1955)

$$Re = \frac{Q}{\pi r_e \nu}. \tag{7.13}$$

At least for large HPHs with high production volumes, the Reynolds number is sufficiently high to indicate turbulent flow in the gap ($Re \sim 100{,}000$, Håkansson, 2017). This has led earlier investigators to suggest that turbulent eddies in the gap are responsible for drop fragmentation in the HPH valve. However, this hypothesis has been disproved in several drop breakup visualizations (Innings and Trägårdh, 2005; Innings et al., 2011a; Kelemen et al., 2015b). Drops pass through the gap largely unaffected; in some studies, they even relaxed back to a less deformed state (Innings and Trägårdh, 2005).

An indication as to why no breakup takes place in the gap despite the high Reynolds numbers comes from velocity field measurements (Håkansson et al., 2011; Kelemen et al., 2015a), showing that high levels of shear and turbulent kinetic energy are predominantly found in the narrow region next to the seat and forcer where radial velocities are low. Since the flow of drops through these high-stress regions is low, breakup probability is consequently also low.

It should be noted that these observations are based on relatively short gaps. Since the boundary layers and the high-shear region grow as a function of distance from the gap inlet, gap fragmentation becomes more likely in designs with relatively long gaps, such as laboratory-scale HPHs; see discussion in Section 7.4.

7.3.4 Turbulence in the Outlet Chamber

As the fluid exits the gap, it forms a jet (Gothsch et al., 2015; Innings and Trägårdh, 2005; Kelemen et al. 2015a). At high Reynolds numbers ($Re > 14{,}000$, according to Kelemen et al., 2014), a turbulent jet is formed, and the average kinetic energy is converted into turbulence. There is a broad agreement between breakup visualizations in different geometries and scales and with different drop viscosities; on that, the breakup takes place in the jet created downstream of the gap (Innings and Trägårdh, 2005; Innings et al., 2011a; Kelemen et al., 2015b).

Under turbulent conditions, the hydrodynamic intensity is often described in terms of the dissipation rate of TKE, ε, in the high-intensity region (cf. Zhou and Kresta, 1998), which for the HPH has been estimated as

$$\varepsilon = \frac{CU_g^3}{h} \tag{7.14}$$

with $C_1 = 1/20$ (Mohr, 1987a) or 1/80 (Innings and Trägårdh, 2007). Note that in order to use Eq. (7.14) to estimate the dissipation rate of TKE for a given HPH valve design and operation, the gap velocity (U_g) and gap height (h) must first be calculated from Eqs. (7.1), (7.2).

Neither of the breakup visualizations show breakup taking place directly at the gap exit, or even where the turbulent kinetic energy is at its maximum, but rather at a position further downstream (Innings et al., 2011a; Kelemen et al., 2015b). This has been attributed to the turbulent energy cascade. Turbulent energy is produced in the form of large turbulent structures. These large eddies are broken up in consecutively smaller eddies until they are sufficiently small to be damped out by the fluid viscosity, and the energy is dissipated as heat. Eddies of length scales of the same size as the drops are more efficient in breaking the drops. Moreover, time is required for turbulent energy to be transported from the large length scales where it is produced to the intermediary length scales that are more efficient in breaking up the drops. The position of drop breakup has been showed to correspond to the position where the turbulent cascade has transported energy down to the appropriate length scales (Håkansson et al., 2011).

In the laminar case, with lower Reynolds numbers, the jet is laminar, and the drop is further elongated as it travels downstream of the gap exit until it breaks into multiple fragments (Kelemen et al., 2015b).

7.3.5 Cavitation

As the fluid approaches and enters the valve, the velocity magnitude and hence the kinetic energy of the fluid increase by several orders of magnitude (see Fig. 7.3). As a consequence of energy conservation, the static pressure decreases with a corresponding amount. Thus, although the static pressure at the valve inlet is high (~10–100 MPa), it decreases down to the vapor pressure of the continuous phase, causing cavitation (i.e., flow-induced vapor formation) inside or downstream of the valve.

The presence of cavitation in the HPH valve has been verified using a number of different experimental techniques. Indirect measurements have been based on the ultrasonic acoustic emissions associated with cavitation (Håkansson et al., 2010; Kurzhals, 1977; Schlender et al., 2015) and the formation of free radicals leading to the decomposition of potassium iodide (Floury et al., 2004b; Shirgaonkar et al., 1998). In addition, more direct observation of cavitation in specially designed valves with optical access has been performed in several studies since the 1950s (Håkansson et al., 2010; McKillop et al., 1955; Phipps, 1974; Schlender et al., 2015).

Cavitation is a substantial problem for HPH design. As the cavitation bubbles continue downstream of the point of maximum radial velocity (located at the gap entrance; see Fig. 7.3), the static pressure increases, and the bubbles implode giving rise to intense shock waves. When this occurs in close proximity to the valve walls, it leads to material wear that over time can severely deform the HPH valve and cause decreased emulsification efficiency (Innings et al., 2011b; Phipps, 1985).

Cavitation intensity in the HPH valve increases with the homogenizing pressure (Floury et al., 2004b; Håkansson et al., 2010; Kurzhals, 1977; Shirgaonkar et al., 1998) and is reduced by applying a sufficiently high back pressure on the valve outlet (Gothsch et al., 2015; Schlender et al., 2015). Some studies also report that cavitation intensity passes through a maximum level at an intermediary back pressure; however, other studies report a monotonically decreasing cavitation with increasing back pressure ratios (see discussion in Håkansson, 2015).

Given the well-known presence of cavitation in the HPH valve and its strength—as apparent from its ability to cause wear—it has often been suggested that cavitation either contributes or even dominates drop breakup in HPH valves (McKillop et al., 1955; Kurzhals, 1977; Loo et al., 1950; Schlender et al., 2015). However, the question as to what extent cavitation influences drop breakup under emulsion formation is still somewhat controversial since experimental investigations are not in agreement.

Several experimental observations support the cavitation hypothesis. As mentioned in Section 7.2.4, the well-established optimum thermodynamic efficiency as a function of Thoma number is often interpreted as an indication of cavitation-dominated emulsification. Moreover, studies comparing the Th giving rise to maximal cavitation with the Th giving rise to minimal drop sizes have found good agreement between the two (Kurzhals, 1977; Schlender et al., 2015).

However, there are also contradictory studies. Gothsch et al. (2015, 2016) investigate cavitation over a wide range of back pressures, showing a monotonic suppression of cavitation with increasing back pressure. However, when comparing with emulsification experiments run at the different back pressures, they observed decreasing drop sizes with increasing back pressures and concluded that cavitation does not cause breakup but it suppresses it. As discussed by Bisten and Schuchmann (2016), the different investigators use different measurement techniques and use different HPH valve models that could explain part of the observed differences.

In summary, there is ample experimental support for cavitation taking place in HPH. Moreover, cavitation is a major challenge in HPH valve design due to wear. However, the effect of cavitation on drop breakup is very much an open question.

7.3.6 Coalescence During Emulsification

As discussed in Chapter 5, the intense hydrodynamic conditions in high-energy emulsification give rise to substantial (re)coalescence between drops. The overall extent of coalescence is generally lower than the overall extent of fragmentation, which is the reason why drop sizes do decrease when being subjected to high-pressure homogenization. However, it is important to note that coalescence is generally present during emulsification and influences the resulting drop-size distribution. The coalescence rate increases with the hydrodynamic intensity, partly because it gives rise to an increased drop-drop collision frequency. Because of the comparatively high hydrodynamic intensity in HPHs, it can be argued that coalescence during high-pressure valve homogenization is even more influential than in many other emulsification processes (Håkansson and Hounslow, 2013; Mohan and Narsimhan, 1997; Taisne et al., 1996).

The same general difficulties and solutions to quantify coalescence rates apply for HPHs as for the general case of high-energy method; see the general discussion in Section 5.3. In particular, it should be noted that coalescence rates increase with increasing homogenizing pressure (Floury et al., 2004b; Håkansson and Hounslow, 2013; Mohan and Narsimhan, 1997; Jafari, et al., 2008).

7.3.7 The Role of Disperse Phase Volume Fraction

As discussed in Chapter 5, increasing the volume fraction of disperse phase increases the coalescence rate. However, volume fraction also influences the fragmentation processes. Dispersing drops in a flow increases the viscosity, which in turn increases the viscous stresses. Moreover, it also influences the turbulent energy cascade. This can result in either enhancement or attenuation of the turbulent stresses depending on the size of the drops and the turbulent flow (Crowe, 2000; Hetstroni, 1989; Poelma and Ooms, 2006). Several semi-empirical correlations have been suggested to predict turbulence modulation by disperse phase volume fraction (e.g., Crowe, 2000; Hetstroni, 1989). However, it is still unknown to what extent these apply to HPH flows.

Note that standard emulsification experiments cannot be used to investigate what effect disperse phase volume fraction has on the rate of drop fragmentation. Two experimental conditions with different volume fraction will differ both in terms of turbulence modulation and in coalescence rate, making it difficult to distinguish the two. Instead, insight must be based on direct measurement of HPH flow fields and turbulence characteristics for varying volume fraction. As previously mentioned, studies of one-phase HPH flow have relied on optical methods such as particle imaging velocimetry (PIV). However, conducting PIV experiments with disperse phase presents is experimentally challenging since the PIV technique requires a transparent fluid. Two-phase studies must therefore use refractive index matching of dispersed and continuous phase. Two such studies on HPH similar systems have been reported. Håkansson et al. (2013b) used refractive index matching and studied low volume fractions, 0%–3.3% (v/v). They concluded that disperse phase volume fraction has limited effects on the flow and turbulence in a scale-up model of an HPH valve. The turbulent gap exit-jet width increases with increasing volume fraction of oils, and the turbulent kinetic energy was somewhat increased, specifically in terms of an increased amount of energy of large-scale turbulent eddies.

More recently, Kollhoff et al. (2015) studied higher volume fractions, up to 30% (v/v), using μPIV in a plane HPH valve similar geometry. It was concluded that there are no measurable differences to the flow field in the inlet chamber or in the gap when altering the volume fraction of disperse phase.

In summary, the effect of volume fraction of dispersed phase on the flow fields and fragmentation rate in HPH is still a field in need of further investigations. However, as seen from a number of recent investigations (Kelemen et al., 2015a; Kollhoff et al., 2015), the advances in μPIV methodology are progressing rapidly, and a deeper understanding will most likely be available in the near future.

7.3.8 The Role of Surfactants and Emulsifiers

Surfactants and emulsifiers play the same role as during other types of mechanical emulsification processes. They decrease the interfacial tension of the drops, thus enabling drop fragmentation, and decrease the rate of coalescence by decreasing the rate of film thinning or giving rise to repulsive drop-drop interactions (see Chapter 5 for a comprehensive treatment).

However, when applied to HPH, some special cases should be mentioned. First, the passage time through the high-intensity region is very short. In order for the emulsifier to stabilize the newly formed fragments against coalescence and in order for them to allow secondary fragmentation of large drops, the adsorption process must be very fast. The HPH valve gap is often approximately

1 mm long (somewhat shorter for small laboratory-scale machines but rarely less than 0.5 mm), whereas the radial velocities are of order magnitude 100 m/s. Hence, it takes a drop approximately 10 μs to pass the gap. As seen in visualization studies, drop breakup takes place in the outlet chamber, where velocities are somewhat lower; still the time for adsorption is limited. This can be problematic, especially for particle stabilizers or macromolecular emulsifiers such as proteins or hydrophobically modified starches where mass transport is relatively slow (Jafari et al., 2007a). Also note that the passage time through the high-intensity region decreases with increasing homogenizing pressure, and thus, this effect is expected to be more pronounced for nanoemulsion formation than for microrange emulsions.

The second difference between the more general case of emulsification and HPH is related to the very high pressures and intense hydrodynamics in the valve. Several studies have suggested that these conditions might decrease the emulsifying efficiency of macromolecular emulsifiers, either by causing conformational changes or by breaking covalent bonds (Floury et al., 2002; Paquin, 1999; Rojas et al., 2008). This effect is, again, more pronounced for nanoemulsion formation and might limit the applicability of these emulsifiers. The effects are discussed in greater detail in relation to the empirical findings of overprocessing in Section 7.6.3.

7.4 SCALE-UP AND SCALE-DOWN

HPHs are available in different sizes, from the smaller laboratory or benchtop models with a production volume of ~10 L/h or less, via the pilot-scale machines with ~100 L/h, to the largest production-scale HPHs at well over 10,0000 L/h. This is often seen as an advantage of the technique as opposed to many other nanoemulsion production methods, especially so for food applications where large production volumes are often required.

The wide range of HPH scales also suggests that scale-up from laboratory-scale investigations and academic investigations to large-scale industrial production would be relatively straightforward and that feasibility studies in terms of, e.g., stability of encapsulated nutrients or pharmaceuticals could be directly translated from one scale to another. However, this assumption might be mistaken. HPHs of different scale differ substantially, and it is not obvious so that the laboratory- and production-scale machines are identical in mechanism of operation or consequently in terms of nanoemulsion formation. The interesting question is therefore as follows: how similar are HPHs of different scales with regard to nanoemulsion formation really?

7.4.1 Experimental Insights on the Effect of HPH Scale

Few nanoemulsion formation studies investigate scalability of the process. An exception is the study of Baspinar et al. (2010) comparing two laboratory-scale

HPHs allowing for sample volume of either 7 or 40 mL. They conclude that both scales result in the same drop size. However, both of the utilized machines are of laboratory scale and fairly similar in production volume when comparing with the large span of HPH scales available. Similarities between two laboratory-scale HPHs do not necessarily imply similarity between laboratory scale and production scale.

Even in the more general HPH emulsification literature, relatively few empirical investigations comparing HPHs of different scales are available. An often cited first finding is that the homogenizing pressure exponent (q in Eq. (7.6)) is higher for smaller HPHs ($q=0.9$ or 1.0) than for production-scale HPHs ($q=0.6$). This has often been interpreted as a sign of different fragmentation mechanisms in HPHs of different scales, e.g., turbulent fragmentation in production-scale HPHs and laminar shear-driven fragmentation in smaller machines (Walstra and Smulders, 1998). However, due to the lack of production-scale experiments, there is relatively little experimental support for a difference in q-values, and some high-quality studies see no such effect at all (Phipps, 1982a,b).

Some further information on scalability can be obtained when extending the review to other applications. HPHs have also seen extensive use in cell disruption, as a step in harvesting intracellular components from fermentation. An interesting work on scalability can be found from this field. Siddiqi et al. (1997) compared the efficiency in terms of rate of protein release from HPHs ranging from laboratory scale to semiproduction scale and concluded that the three investigated machines were highly similar. A similar finding of scale independence was reported by Donsi et al. (2009), showing that the efficiency of HPHs in microbial inactivation did not show any pronounced effect of the HPH scale when comparing a laboratory-scale with a pilot-scale machines. However, cell disruption and microbial inactivation differ from nanoemulsion formation, and it is unclear what relevance these findings have on emulsification.

7.4.2 Scaling, Fluid Velocity and Pressure Distribution

Since we are lacking clear experimental evidence on the effect of HPH scale on nanoemulsion formation, a theoretical analysis can give some insight into the question. Such an analysis has been presented by Håkansson (2017), and this section is based on that discussion.

As seen from Eq. (7.1), the characteristic velocity in the HPH valve depends on three factors, the flowrate, the gap exit radius, and the gap height. As previously mentioned, the flowrates between laboratory-scale and production-scale HPH differ by a factor 1000. The gap height depends mainly on the homogenizing pressure. It is often smaller for laboratory-scale machines, but not by more than a factor 10. Finally, the gap dimension and hence exit radius are somewhat larger for production-scale HPHs but again only by a factor 10.

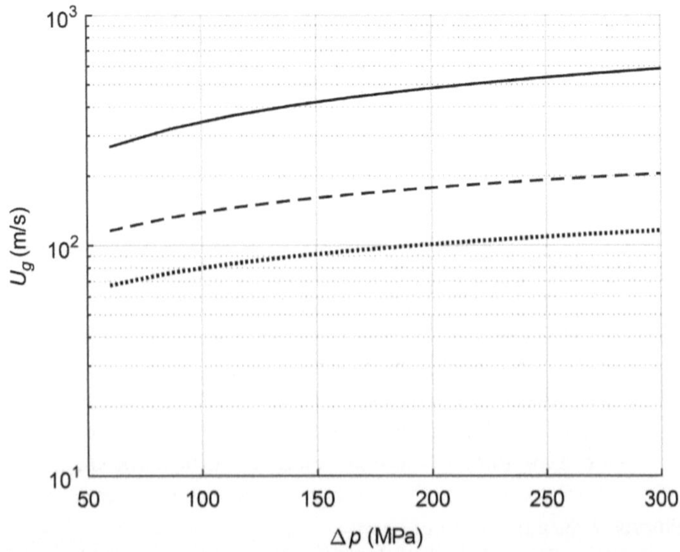

FIG. 7.5 Gap velocity as a function of homogenizing pressure for 10,000 L/h production-scale HPH *(solid)*, 100 L/h pilot-scale HPH *(dashed)*, and 10 L/h laboratory-scale HPH *(dotted)*. See Håkansson (2017) for geometric dimensions.

The effect of these three differences is a substantially higher fluid velocity in production-scale HPHs than in pilot- or laboratory-scale HPHs. Fig. 7.5 displays the gap velocity as a function of homogenizing pressure for three representable scales.

Another principal difference is in terms of the pressure profile in the gap and consequently in how much of the homogenizing pressure that is lost in the inlet chamber, the gap, and the outlet chamber of the HPH valve. Fig. 7.6 displays the proportion of pressure spent in the three regions for three HPH scales. For the production-scale HPH, most of the pressure is lost in the outlet chamber. For the smaller scales, the gap friction becomes increasingly important—due to the relatively large gap length-to-height ratios in laboratory-scale designs. For the laboratory-scale HPH in Fig. 7.6, more than 90% of the pressure loss occurs in the gap. This is an important difference since the pressure loss is the source of energy available for emulsification. Thus, energy is available in different parts of the valves for HPHs of different scales. Furthermore, since the different parts of the valve are characterized by different hydrodynamics, these differences could result in differences in both position and mechanism of breakup.

7.4.3 Fragmentation Mechanisms and Scale

Due to the smaller gap heights in laboratory-scale HPH, the elongational strain is larger compared with production-scale HPH. However, as discussed in

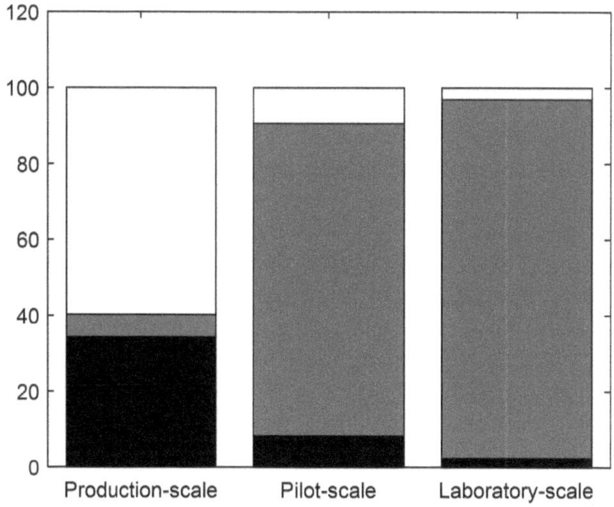

FIG. 7.6 Percentage of pressure loss in the inlet chamber *(black)*, gap *(gray)*, and outlet chamber *(white)* for a production-scale HPH (10,000 L/h), a pilot-scale HPH (100 L/h), and a laboratory-scale HPH (10 L/h). See Håkansson (2017) for geometric dimensions.

Section 7.3.2, the dynamic time scales limit elongational disruption, and thus, the mechanism is not expected to cause considerable breakup regardless of HPH scale.

As for turbulent breakup, the Reynolds number is highly scale-dependent. As seen in Eq. 7.13, the HPH valve Reynolds number depends on flowrate and the gap exit radius. Since flowrate differs substantially more than gap dimensions, the production-scale HPH has a Reynolds number that is several orders of magnitude larger. By comparing with the full turbulence transition limit suggested by Kelemen et al. (2014, 2015a), Håkansson (2017) concluded that only production-scale HPHs operate in the fully turbulent regime. Laboratory regime machines range from laminar to transitional, whereas pilot-scale HPHs operate in the transitional regime. This suggests that turbulence fragmentation is more pronounced in large volume HPH and is lower or even absent in the laboratory-scale HPH. This could help explain some inconsistencies between different experimental studies. For example, turbulent shear is often suggested as the main disruptive force during HPH. However, Lee and Norton (2013) and Lee et al. (2014) note that viscosity ratio scaling observed during nanoemulsion formation is not consistent with turbulent shear. A possible explanation for this discrepancy is that their study was conducted on small laboratory-scale HPHs that might differ in terms of dominant mechanism. Laboratory-scale HPHs often have relatively large gap length-to-height ratios suggesting that gap breakup would be more likely in these designs.

A final difference between HPHs of different scale is in terms of the cavitation number,

$$N_{cav} = \frac{\Delta p}{\rho U_g^2/2}, \tag{7.15}$$

often used to describe the propensity of a flow to induce cavitation (Brennen, 1995). Due to the high gap velocities in production-scale HPHs, the cavitation number is substantially lower in production-scale HPH, suggesting more cavitation in larger-scale HPHs. Cavitation is generally expected for $N_{cav} < 1$, and theoretical comparisons show that only the production-scale HPH reaches this value (Håkansson, 2017). This suggests that cavitation is more influential in production scale than laboratory scale. This is of high relevance for the scale-up of nanoemulsion formation processes. As previously mentioned (Section 7.3.5), cavitation influences breakup. Moreover, cavitation gives rise to free-radical formation that might influence chemical stability of encapsulated products, such as nutrients or pharmaceuticals.

7.4.4 Implications for Scale-up of Nanoemulsion Formation

In summary, HPHs are available in widely different scales. This is an advantage since it allows for large-scale production, which is often difficult with the competing techniques. However, the fact that these different emulsification devices are all called high-pressure valve homogenizers and share the same general valve design does not necessarily imply that emulsification proceeds according to the same mechanisms or that they have the same influence on product quality. Theoretical analyses of Reynolds numbers, pressure losses, and cavitation numbers suggest considerable differences. It has even been suggested that these theoretical differences are so large that HPHs of different scales should be considered different emulsification processes and that a production-scale HPH might be more similar to an inline rotor-stator mixer than to a laboratory-scale HPH, at least when it comes to the mechanism of drop fragmentation (Håkansson, 2017, p. 37).

On the other hand, these theoretical suggestions have not been experimentally tested yet. And some experiments point to relatively small differences between HPHs of different scales (Donsi et al., 2009; Phipps, 1982a,b; Siddiqi et al., 1997).

The conclusion is that more experimental investigations of the scalability of HPH are in great need and studies evaluating the suitability of HPH as a method of nanoemulsion formation should optimally either be based on experiments carried out in the HPH scale relevant for production or contain scale-up experiments including both the effect of obtained drop sizes and chemical stability.

7.5 HEAT GENERATION AND TEMPERATURE RISE

The small drop sizes required for nanoemulsion formation require high hydrodynamic stresses, which in the case of HPH corresponds to high homogenizing pressures. Combined with the exceedingly low thermodynamic efficiency of HPH (Section 7.2.3), this gives rise to substantial temperature increase during nanoemulsion formation in HPHs.

7.5.1 Local Increase in Temperature

Experimental investigations show that the product temperature increases linearly with homogenizing pressure, with between 19 and 23°C/100 MPa (Benzaria et al., 2014; Mao et al., 2010; Zamora and Guamis, 2015). This temperature increase can be predicted for a given homogenizing pressure and product using an energy balance across the HPH valve. The power draw of an HPH is the product of flowrate and homogenizing pressure (Eq. 7.4). Some of this energy is converted to surface energy when the Sauter drop diameter decreases. However, this portion is very small (<0.1%; see Fig. 7.4) and can be neglected in the energy balance. The temperature increase of the product can thus be estimated by assuming that all the energy input is converted to thermal energy:

$$\rho Q C_P T_{in} + Q\Delta p = \rho Q C_P T_{out} \Leftrightarrow \Delta T = T_{out} - T_{in} = \frac{\Delta p}{\rho C_P}, \tag{7.16}$$

where T_{in} and T_{out} are the temperatures in and out of the valve, respectively, and C_P is the specific heat of the emulsion. Fig. 7.7 displays the temperature increase as a function of homogenizing pressure for three volume fractions of disperse phase ($\phi_D = 0.1\%$, 10%, and 20%).

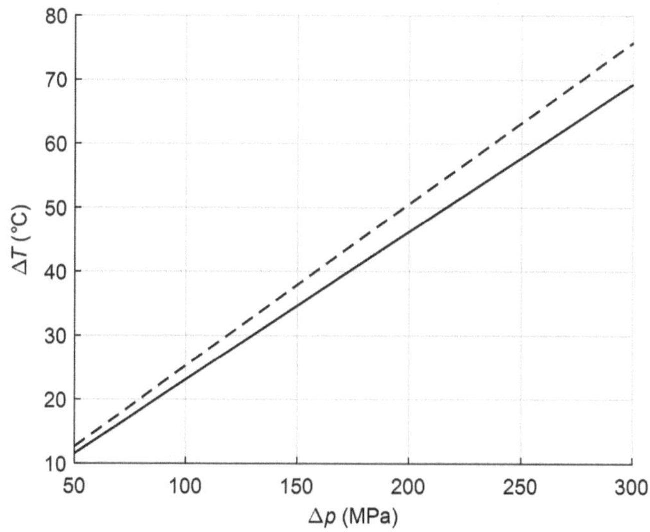

FIG. 7.7 Temperature increase for one passage through the HPH valve at different homogenizing pressures for a low volume fraction emulsion (0.1 % (v/v), *solid line*) and an intermediate volume fraction O/W emulsion (20% (v/v), *dashed line*).

7.5.2 Product Quality and HPH Temperature Increase

As seen in Fig. 7.7, a pre-emulsion fed to the HPH at room temperature can reach temperatures above 90°C upon exiting the valve. This has led to concern since nanoemulsions are often used for delivering nutrients, pharmaceuticals, or other heat-sensitive molecules. However, it should be remembered that the passage time through the HPH valve is short due to the small geometric dimensions and the high velocities. Provided that cooling is applied directly after the emulsification stage, the temperature increase need not influence the stability of the product (Benzaria et al., 2014).

7.6 SUITABILITY FOR NANOEMULSION FORMATION

The preceding sections have focused on the general principles of the high-pressure valve homogenization. And most commercial applications of the technique are on microsized emulsions. This section discusses how suitable the HPH technique is for the specific case of nanoemulsion formation.

7.6.1 Applications and Required Homogenizing Pressure

HPH has been used extensively for nanoemulsion formation; a few representative recent examples of academic studies are summarized in Table 7.1. The studies generally fall in two classes: studies attempting to form nanoemulsions for delivering lipophilic nutrients (labeled *Food* in Table 7.1) and studies forming nanoemulsions as delivery systems for pharmaceuticals (labeled *Pharma* in Table 7.1). The general conclusion from the studies in Table 7.1 is that nanoemulsions (at least when defined broadly as emulsions with drops <500 nm) can be formulated with HPH, that the emulsions are stable during storage and that they retain the desired properties. This shows that HPH can be a useful method for nanoemulsion formation.

As seen in Table 7.1, the homogenizing pressure required to form stable nanoemulsions differs widely between experiments; the different studies report pressures between 40 and 200 MPa for creating drops of approximately the same size ($D \sim 200$ mm). No general suggestion on the homogenizing pressure required to form nanoemulsion with HPH can therefore be given. Each new application needs extensive empirical testing in order to arrive at required set of operating conditions.

Moreover, the previous investigations on nanoemulsion formation in HPHs point to two important challenges that should be kept in mind when designing and interpreting this type of experiments: first, the need for several HPH passages to obtain a nanoemulsion, as discussed further in Section 7.6.2, and, secondly, the risk of overprocessing when applying too high pressure or too many passages, as discussed in Section 7.6.3.

TABLE 7.1 Summary of Example Studies Using HPH for Nanoemulsion Formation

Δp (MPa)	Passages	Emulsifier	ϕ_D (%)	D (nm)	Emulsion Type	Field	Reference
200	1	LME	37	260	O/W	Pharma	Benzaria et al. (2014)
40	3	LME	10	220	O/W	Food	Mao et al. (2010)
80	3	LME	10	190	O/W	Food	Mao et al. (2010)
120	3	LME	10	160	O/W	Food	Mao et al. (2010)
80	5	LME	5	200–400	O/W	Pharma	Tubesha et al. (2013)
50	1	LME	1	200	W/O	Food	El Kinawy et al. (2012)
100	4	PRE	5	200	W/O	Pharma	Ali et al. (2016)
70–140	>10	LME	5–15	170–340	O/W	Food	Silva et al. (2015)
50	3	LME	20	150–240	O/W	Pharma	Mitri et al. (2011)
75	16	LME	5	120	O/W	Pharma	Klang et al. (2011)

All with vegetable oil or triglycerides as the oil phase. *LME*, low molecular emulsifier (e.g., polysorbates); *PE*, polymeric emulsifier (e.g., polyglycerol polyricinoleate); *PR*, protein emulsifier (e.g., beta-lactoglobulin or whey protein isolate).

7.6.2 HPH Passages

One of the advantages of HPH, as opposed to many alternative methods for nanoemulsion formation, is that it allows for continuous production, which is generally more cost-effective and a prerequisite for efficient large-scale production. However, it is well known that the steady-state drop size is not obtained after a single HPH passage; the drop continues to decrease when recirculating the emulsion over the HPH. This can be attributed to the dynamic breakup and coalescence process; the time spent in the high-stress region is not sufficiently large to allow for all drops to break down to their final steady-state size, or there is not sufficient time for the larger fragments that are formed in a primary breakup event to undergo a second breakup resulting in fragments exiting the valve with diameters above the theoretical limit as obtained from a steady-state stress analysis.

As seen in Table 7.1, several studies require multiple HPH passages before reaching the desired nanoemulsion size range. For academic purposes, this is often unproblematic and can be achieved by recirculating the emulsion over the HPH. However, in an industrial setting, it is more problematic. Two or three passages could be obtained by connecting several HPHs in series; however, this greatly increases the production cost, since the capital cost is a substantial part of the total cost of high-pressure valve homogenization (Innings, 2015). This limits the relevance of some of the case studies presented in academic literature, at least if continuous production is a desired outcome.

7.6.3 Overprocessing

Although both the rate of fragmentation and the rate of coalescence increase with increasing homogenizing pressure, the fragmentation rate increases faster. Consequently, one expects monotonically smaller drop sizes as a function of increasing homogenization pressure. A similar argument can be made concerning the number of HPH passages; since the drop-size distribution exiting an HPH must be larger or equal to the steady-state drop sizes, recirculating an emulsion several times over the HPH is expected to give rise to continuously decreasing drop sizes until it reaches the steady-state drop distribution after which drop sizes are expected to remain constant.

Some nanoemulsion emulsification studies do observe this monotonic decrease with increasing homogenizing pressure and number of HPH passages (El Kinawy et al., 2012). However, investigators who have systematically varied the homogenizing pressure and number of HPH passages find a point beyond which the drops increase substantially in size (Ali et al., 2016; Desrumaux and Marcand, 2002; Lee et al., 2014). This observation of decreasing emulsification efficiency with increasing energy input has been previously observed in other systems, such as ultrasonic homogenizers (Tornberg, 1980) and microfluidizers (Jafari et al., 2006, 2007b,c), and is referred to as "overprocessing" or recoalescence (Jafari et al., 2008). This observation has been reported for emulsion in the micrometer range as well (Tornberg, 1980), but it is observed more commonly for nanoemulsions due to the higher emulsification pressures and a large number of HPH passages compared with other applications.

Different suggestions have been offered to explain what causes this effect, all of which starts from the notion that an increase in drop size implies an increase in coalescence rate in relation to fragmentation rate, thus resulting in a net increase of the coalescence. However, it is not always clear what causes this relative change in rates. For example, Lee et al. (2014) observed that the drop size increases when subjecting the emulsion to more than five passages and suggested that it is due to increasing coalescence rates and attribute this increase to an increase in temperature (Lee et al., 2014, p. 35). However, it is not obvious why the coalescence rate would increase with increasing temperature. A temperature increase mainly decreases the fluid viscosity, and there is no general agreement on increasing coalescence rates with decreasing fluid viscosity (see review in Liao and Lucas, 2010).

Ali et al. (2016) also observed that the drop size starts to increase after a certain number of passages and discussed two possible mechanisms as to why coalescence rates have increased: (i) that the emulsifiers are denatured or structurally modified by passing through the HPH valve and (ii) that it is an effect of limited adsorption time. Starting with the first suggestion (i), it is well known that HPH can cause conformational changes or even degradation of macromolecules and possibly also of high-molecular emulsifiers such as proteins or hydrophobically modified starches. Paquin (1999) summarizes experimental evidence of how both proteins and polysaccharides undergo conformational changes when subjected to relatively low homogenizing pressures (5–30 MPa). Several studies have confirmed and extended these findings, e.g., Floury et al. (2002) report the degradation of methyl cellulose during HPH, and Rojas et al. (2008) report the degradation and conformational changes of barley wax at low to intermediate homogenizing pressures (5–20 MPa).

Studies have attempted to clarify if degradation or conformational changes of emulsifiers could explain the observed optima in energy intensity. Desrumaux and Marcand (2002) investigated a protein emulsifier (whey) before and after passing through the HPH at the homogenizing pressure where the optima were observed, but could not see any indication of protein degradation. Ali et al. (2016) investigated if the macromolecular emulsifier (β-lacto globulin) has experienced conformational damage by investigating the interfacial properties and surface dilatational rheology before and after high-pressure homogenization, but could not detect any significant loss of emulsifying capabilities. Thus, although it is well known that macromolecules could be structurally impaired or even broken up within the HPH, it has not been clarified to what extent this is the cause of the observed overprocessing.

The second suggestion (ii) as to why coalescence should increase faster than fragmentation at high pressures was originally suggested by Tornberg (1980) and is related to the adsorption rate of emulsifier to the newly created drop interfaces. When drops pass through the high-intensity region in the HPH valve, they start to break. However, in the intense and random flow, the two newly formed drops have a high probability of colliding with either each other or other newly formed fragments. In order to avoid recoalescence, emulsifier must be transported to and adsorb on the newly created interface. Tornberg (1980) suggested that with a very high fragmentation rate, the new surface area becomes very large in relation to the available emulsifier, which leads to a relatively low adsorption rate. When the emulsifier lacks time to adsorb, the coalescence frequency increases giving rise to larger drops.

However, this adsorption hypothesis of overprocessing could also be questioned, at least as an explanation of increasing drop sizes as a function of passage number. For the drops that are not undergoing breakup, no extra surface is created, and thus, newly created surface cannot be used to explain why they experience a net rate of coalescence. For the drops that are broken up, if the formed fragments coalesced with each other, the drop would not become larger than it was before this passage. If the breakup results in one large and several satellite drops, it could be the case that two of the large fragments coalesce, thus creating a larger drop. However, provided that the initial drop is covered by emulsifier (as is the case if it is stabile), the larger a fragment is, the less devoid of surfactant it would be, and the less likely it would be of it to coalesce. More investigations into adsorption and coalescence process and how they interact under high intensities of hydrodynamic stresses are needed to finally conclude on to what extent it can explain observations of increasing drop sizes with increasing homogenization pressure or number of HPH passages.

In summary, overprocessing is often observed when creating nanoemulsions using HPH, i.e., there is an optimum homogenization pressure and an optimum number of passages beyond which drops do not get smaller. However, as seen in Table 7.1, the minimum is often found at sufficiently small drop sizes that HPH can still be used to form stabile nanoemulsions. Although several mechanistic explanations of overprocessing have been put forward, there is of yet no generally accepted explanation to the observed phenomena.

7.6.4 Future Perspectives on of HPH Nanoemulsion Research and Development

As previously mentioned, HPH has been applied extensively to nanoemulsion formation in numerous studies. However, several important questions remain to be answered. Three such questions—or classes of question—are as follows:

- Principle understanding and HPH hydrodynamics
- Scalability and scale-up effects
- Overprocessing, explanation, and effects

The first point relates to the principle understanding of HPH emulsification, both in terms of the flow field obtained with a certain valve design and method of operation and in terms of which mechanisms control drop breakup. This is a rapidly growing field, but still many questions remain unanswered, not at least in terms of the effect of coalescence and cavitation and on how disperse phase volume fraction influences the flow. These questions are not specific to nanoemulsion formation but often highly relevant since the small desired drop sizes set high demands on increased efficiency.

The second point relates to scalability. Few, if any, studies on HPH and nanoemulsion investigate if the findings from laboratory-scale investigations apply for the HPH scale intended in the production stage. Theoretical comparisons indicate that substantial differences between laboratory and production stage HPHs might exist.

Third, many experimental studies find that drop sizes increase as a function of homogenizing pressure or number of HPH passages above a critical limit. This overprocessing effect is of high relevance for nanoemulsion formation since it sets an effective limit on applicability. Continued investigations on the mechanisms leading to overprocessing and how it could be avoided to create smaller drops are therefore needed.

In addition to these points, it should also be noted that several advances in improving the general HPH efficiency have been presented. This is of special interest for nanoemulsion where efficient fragmentation is needed to create the very small drop sizes. Several suggestions on improved HPH devices have been proposed by redesigning the basic valve geometry, e.g., by Finke et al. (2012, 2014) and Gall et al. (2016). Other recent examples include the study by Calligaris et al. (2016) showing that nanoemulsion formation can be made more efficient by combining HPH with another emulsification process (i.e., by combining it with ultrasonication). As seen from these examples, although HPH is a mature technology, research continues on improving both understanding and efficiency.

7.7 CONCLUSIONS AND FINAL REMARKS

High-pressure homogenization is a standard method for nanoemulsion formation, especially suitable for applications with low product viscosities requiring high production volumes and continuous operation. HPH is a mature technology, it has been commercially applied for more than a century, and a substantial amount of engineering experience and scientifically based knowledge is available. This is an advantage compared with many alternatives but newer techniques. Furthermore, HPHs are commercially available in a wide range of scales and can therefore be used for anything from small-scale specialty products to large-scale production. However, the technology also has some drawbacks that should be kept in mind. The thermodynamic efficiency of traditional microrange emulsion formation HPH is low, and the efficiency is even lower for nanoemulsion formation. This implies relatively high-energy expenditures per unit mass of emulsion. Furthermore, the energy losses give rise to heat that substantially increases the product temperature (by 20–60°C, depending on homogenizing pressure). Cooling of the HPH valve is generally insufficient to entirely stop this temperature increase. However, if cooling is applied directly at the valve outlet, the high-temperature time can be limited, and experimental investigations reveal that heat-sensitive chemicals can thus be protected from degradation. It should also be noted that most HPH valves were originally designed for drops larger than those in nanoemulsions. Several experimental studies reveal that more than one passage through the HPH is needed in order to obtain the desired size. When more than one passage is needed, the HPH loses much of its advantage as a continuous process. Finally, it should be noted that our basic understanding of HPH and emulsification has greatly expanded during the last decade. Combined with the many new suggestions on improved design and operation and the growing number of studies on HPH nanoemulsion formation, both efficiency and applicability can be expected to increase in the near future.

REFERENCES

Ali, A., Mekhloufi, G., Huang, N., Agnely, F., 2016. β-Lactoglobulin stabilized nanoemulsions—formation and process factors affecting droplet size and nanoemulsion stability. Int. J. Pharm. 500, 291–304.

Baspinar, Y., Keck, C.M., Borchert, H.-H., 2010. Development of a positively charged prednicarbate nanoemulsion. Int. J. Pharm. 383 (1-2), 201–208.

Benzaria, A., Gràcia-Julià, A., Picart-Palmade, L., Hue, P., Chevalier-Lucia, D., Marti-Mestres, G., Hodor, N., Dumay, E., 2014. UHPH-processed O/W submicron emulsions stabilised with a lipid-based surfactant: physicochemical characteristics and behaviour on in vitro TC7-cell monolayer and ex vivo pig's ear. Colloids Surf. B: Biointerfaces 116, 237–246.

Bisten, A., Schuchmann, H.P., 2016. Optical measuring methods for the investigation of high-pressure homogenisation. Processes 4 (4), 41.

Brennen, C.E., 1995. Cavitation and Bubble Dynamics. Oxford University Press, New York.

Calligaris, S., Plazzotta, S., Bot, F., Grasselli, S., Malchiodi, A., Anese, M., 2016. Nanoemulsion preparation by combining high pressure homogenization and high power ultrasound at low energy densities. Food Res. Int. 83, 25–30.

Crowe, C.T., 2000. On models for turbulence modulation in fluid-particle flows. Int. J. Multiphase Flow 26 (5), 719–727.

Desrumaux, A., Marcand, J., 2002. Formation of sunflower oil emulsions stabilized by whey proteins with high-pressure homogenization (up to 350 MPa): effect of pressure on emulsion characteristics. Int. J. Food Sci. Technol. 37 (3), 263–269.

Donsi, F., Ferrari, G., Lenza, E., Maresca, P., 2009. Main factors regulating microbial inactivation by high-pressure homogenization: operating parameters and scale of operation. Chem. Eng. Sci. 64 (3), 520–532.

El Kinawy, O.S., Petersen, S., Ulrich, J., 2012. Technological aspects of nanoemulsion formation of low-fat foods enriched with vitamin E by high-pressure homogenization. Chem. Eng. Technol. 35 (5), 937–940.

Finke, J.H., Schur, J., Richter, C., Gothsch, T., Kwade, A., Büttgenbach, S., Müller-Goymann, C.C., 2012. The influence of customized geometries and process parameters on nanoemulsion and solid lipid nanoparticle production in microsystems. Chem. Eng. J. 209, 126–137.

Finke, J.H., Niemann, S., Richter, C., Gothsch, T., Kwade, A., Büttgenbach, S., Müller-Goymann, C.C., 2014. Multiple orifices in customized microsystem high-pressure emulsification: the impact of design and counter pressure on homogenization efficiency. Chem. Eng. J. 248, 107–121.

Floury, J., Desrumaux, A., Axelos, M.A.V., Legrand, J., 2002. Degradation of methylcellulose during ultra-high pressure homogenisation. Food Hydrocoll. 16 (1), 47–53.

Floury, J., Bellettre, J., Legrand, J., Desrumaux, A., 2004a. Analysis of a new type of high pressure homogeniser. A study of the flow pattern. Chem. Eng. Sci. 59 (4), 843–853.

Floury, J., Legrand, J., Desrumaux, A., 2004b. Analysis of a new type of high pressure homogeniser. Part B. Study of droplet break-up and recoalescence phenomena. Chem. Eng. Sci. 59 (6), 1285–1294.

Gall, V., Runde, M., Schuchmann, H.P., 2016. Extending applications of high-pressure homogenization by using simultaneous emulsification and mixing (SEM)—an overview. Processes 4 (4), 46.

Gaulin, A., 1904. Process of treating milk or similar liquids. US Patent No. 753,792.

Gothsch, T., Schilcher, C., Richter, C., Beinert, S., Dietzel, A., Büttgenbach, S., Kwade, A., 2015. High-pressure microfluidic systems (HPMS): flow and cavitation measurements in supported silicon microsystems. Microfluid. Nanofluid. 18 (1), 121–130.

Gothsch, T., Richter, C., Beinert, S., Schilcher, C., Schilde, C., Büttgenbach, S., Kwade, A., 2016. Effect of cavitation on dispersion and emulsification process in high-pressure microsystems (HPMS). Chem. Eng. Sci. 144, 239–248.

Grace, H.P., 1982. Dispersion phenomena in high viscosity immiscible fluid systems and application of static mixers as dispersion device in such systems. Chem. Eng. Commun. 14 (3–6), 225–277.

Gupta, A., Eral, H.B., Hatton, T.A., Doyle, P.S., 2016. Nanoemulsions: formation, properties and applications. Soft Matter 12, 2826–2841.

Håkansson, A., 2015. Droplet breakup in high-pressure homogenizers. In: Rayner, M., Dejmek, P. (Eds.), Engineering Aspects of Food Emulsification and Homogenization. CRC Press, Boca Raton, FL, pp. 125–148.

Håkansson, A., 2017. Scale-down failed—dissimilarities between high-pressure homogenizers of different scales due to failed mechanistic matching. J. Food Eng. 195, 31–39.

Håkansson, A., Hounslow, M.J., 2013. Simultaneous determination of fragmentation and coalescence rates during pilot-scale high-pressure homogenization. J. Food Eng. 116 (1), 7–13.

Håkansson, A., Trägårdh, C., Bergenståhl, B., 2009. Dynamic simulation of emulsion formation in a high pressure homogenizer. Chem. Eng. Sci. 64 (12), 2915–2925.

Håkansson, A., Fuchs, L., Innings, F., Revstedt, J., Bergenståhl, B., Trägårdh, C., 2010. Visual observations and acoustic measurements of cavitation in an experimental model of a high-pressure homogenizer. J. Food Eng. 100 (3), 504–513.

Håkansson, A., Fuchs, L., Innings, F., Revstedt, J., Trägårdh, C., Bergenståhl, B., 2011. High resolution experimental measurement of turbulent flow field in a high pressure homogenizer model and its implications on turbulent drop fragmentation. Chem. Eng. Sci. 66 (8), 1790–1801.

Håkansson, A., Fuchs, L., Innings, F., Revstedt, J., Trägårdh, C., Bergenståhl, B., 2012. Experimental validation of k-ε RANS-CFD on a high-pressure homogenizer valve. Chem. Eng. Sci. 71, 264–273.

Håkansson, A., Innings, F., Trägårdh, C., Bergenståhl, B., 2013a. A high-pressure homogenization emulsification model—improved emulsifier transport and hydrodynamic coupling. Chem. Eng. Sci. 91, 44–53.

Håkansson, A., Fuchs, L., Innings, F., Revstedt, J., Trägårdh, C., Bergenståhl, B., 2013b. Velocity measurements of turbulent two-phase flow in a high-pressure homogenizer model. Chem. Eng. Commun. 200 (1), 93–114.

Hetstroni, G., 1989. Particle-turbulence interactions. Int. J. Multiphase Flow 15 (5), 735–746.

Innings, F., 2015. High-pressure homogenizer design. In: Rayner, M., Dejmek, P. (Eds.), Engineering Aspects of Food Emulsification and Homogenization. CRC Press, Boca Raton, FL, pp. 149–168.

Innings, F., Trägårdh, C., 2005. Visualization of the drop deformation and break-up process in a high pressure homogenizer. Chem. Eng. Technol. 28 (8), 882–891.

Innings, F., Trägårdh, C., 2007. Analysis of the flow field in a high-pressure homogenizer. Exp. Thermal Fluid Sci. 32 (2), 345–354.

Innings, F., Fuchs, L., Trägårdh, C., 2011a. Theoretical and experimental analyses of drop deformation and break-up in a scale model of a high-pressure homogenizer. J. Food Eng. 103 (1), 21–28.

Innings, F., Hultman, E., Forsberg, F., Prakash, B., 2011b. Understanding and analysis of wear in homogenizers for processing liquid food. Wear 271 (9-10), 2588–2598.

Jafari, S.M., He, Y., Bhandari, B., 2006. Nano-emulsion production by sonication and microfluidization – a comparison. Int. J. Food Prop. 9 (3), 475–485.

Jafari, S.M., He, Y., Bhandari, B., 2007a. Effectiveness of encapsulating biopolymers to produce sub-micron emulsions by high energy emulsification techniques. Food Res. Int. 40 (7), 862–873.

Jafari, S.M., He, Y., Bhandari, B., 2007b. Optimization of nano-emulsion production by microfluidization. Eur. Food Res. Technol. 225 (5), 733–741.

Jafari, S.M., He, Y., Bhandari, B., 2007c. Production of sub-micron emulsions by ultrasound and microfluidization techniques. J. Food Eng. 82 (4), 478–488.

Jafari, S.M., Assadpoor, E., He, Y., Bhandari, B., 2008. Re-coalescence of emulsion droplets during high-energy emulsification. Food Hydrocoll. 22 (7), 1191–1202.

Kelemen, K., Schuch, A.C., Schuchmann, H.P., 2014. Influence of flow conditions in high-pressure orifices on droplet disruption of oil-in-water emulsions. Chem. Eng. Technol. 37 (7), 1227–1234.

Kelemen, K., Crowther, F.E., Cierpka, C., Hecht, L.L., Kähler, C.J., Schuchmann, H.P., 2015a. Investigations on the characterization of laminar and transitional flow conditions after high pressure homogenization orifices. Microfluid. Nanofluid. 18 (4), 599–612.

Kelemen, K., Gepperth, S., Koch, R., Bauer, H.-J., Schuchmann, H.P., 2015b. On the visualization of droplet deformation and breakup during high-pressure homogenization. Microfluid. Nanofluid. 19 (5), 1139–1158.

Klang, V., Schwarz, J.C., Matsko, N., Rezvani, E., El-Hagin, N., Wirth, M., Valenta, C., 2011. Semisolid sucrose stearate-based emulsions as dermal drug delivery systems. Pharmaceutics 3 (2), 275–306.

Kleinig, A.R., Middelberg, A.P.J., 1996. The correlation of cell disruption with homogenizer valve pressure gradient determined by computation fluid dynamics. Chem. Eng. Sci. 51 (23), 5103–5110.

Kollhoff, R.T., Kelemen, K., Schuchmann, H.P., 2015. Local multiphase flow characterization with micro particle image velocimetry using refractive index matching. Chem. Eng. Technol. 38 (10), 1774–1782.

Kurzhals, H.-A., 1977. Untersuchungen über die physikalisch-technischen Vorgänge beim Homogenisieren von Milch in Hochdruck-Homogenisiermaschinen. (Doctoral Thesis). University of Hannover, Hannover.

Lee, L., Norton, I.T., 2013. Comparing droplet breakup for a high-pressure valve homogenizer and a microfluidizer for the potential production of food-grade nanoemulsions. J. Food Eng. 114 (2), 158–163.

Lee, L., Hancoks, R., Noble, I., Norton, I.T., 2014. Production of water-in-oil nanoemulsions using high pressure homogenisation: a study on droplet break-up. J. Food Eng. 131, 33–37.

Liao, Y., Lucas, D., 2010. A literature review on mechanisms and models for the coalescence process. Chem. Eng. Sci. 65 (10), 2851–2864.

Loo, C.C., Carleton, W.M., 1953. Further studies of cavitation in the homogenization of milk products. J. Dairy Sci. 36 (1), 64–75.

Loo, C.C., Slatter, W.L., Powell, R.W., 1950. A study of the cavitation effect in the homogenization of dairy products. J. Dairy Sci. 33 (10), 692–702.

Maindarkar, S.N., Hoogland, H., Henson, M.A., 2015. Predicting the combined effects of oil and surfactant concentrations on the drop size distributions of homogenized emulsions. Colloids Surf. A Physiochem. Eng. Aspects 467, 18–30.

Mao, L., Yang, J., Xu, D., Yuan, F., Gao, Y., 2010. Effects of homogenization models and emulsifiers on the physicochemical properties of β-carotene nanoemulsions. J. Dispers. Sci. Technol. 31 (7), 986–993.

McClements, D.J., Rao, J., 2011. Food-grade nanoemulsions: Formulation, fabrication, properties, performance, biological fate, and potential toxicity. Crit. Rev. Food Sci. Nutr. 51 (4), 285–330.

McKillop, A.A., Dunkley, W.L., Brockmeyer, R.L., Perry, R.L., 1955. The cavitation theory of homogenization. J. Dairy Sci. 38 (3), 273–283.

Mitri, K., Shegokar, R., Gohla, S., Anselmi, C., Müller, R.H., 2011. Lipid nanocarriers for dermal delivery of lutein: preparation, characterization, stability and performance. Int. J. Pharm. 414 (1-2), 267–275.

Mohan, S., Narsimhan, G., 1997. Coalescence of protein-stabilized emulsions in a high-pressure homogenizer. J. Colloid Interface Sci. 192 (1), 1–15.

Mohr, K.-H., 1987a. High-pressure homogenization. Part I. Liquid-liquid dispersion in turbulence fields of high energy density. J. Food Eng. 6 (3), 177–186.

Mohr, K.-H., 1987b. High-pressure homogenization. Part II. The influence of cavitation on liquid-liquid dispersion in turbulence fields of high energy density. J. Food Eng. 6 (4), 311–324.

Paquin, P., 1999. Technological properties of high pressure homogenizers: the effect of fat globules, milk proteins, and polysaccharides. Int. Dairy J. 9 (3-6), 329–335.

Phipps, L.W., 1974. Cavitation and separated flow in a simple homogenizing valve and their influence on the break-up of fat globules in milk. J. Dairy Res. 41 (1), 1–8.

Phipps, L.W., 1975. The fragmentation of oil drops in emulsion by a high-pressure homogenizer. J. Phys. D. Appl. Phys. 8, 448–462.

Phipps, L.W., 1982a. Homogenizing valve design and its influence on milk fat globule dispersion. I. Low rate of flow (100 l/h, Re $\leq$ 3000). J. Dairy Res. 49 (2), 309–315.

Phipps, L.W., 1982b. Homogenizing valve design and its influence on milk fat globule dispersion. II. High rate of flow (1000 l/h, 6000 $<$ Re $<$ 15 000). J. Dairy Res. 49 (2), 309–315.

Phipps, L.W., 1985. The High Pressure Dairy Homogenizer. The National Institute for Research in Dairying, Reading, MA.

Poelma, C., Ooms, G., 2006. Particle-turbulence interaction in a homogeneous, isotropic turbulent suspension. Appl. Mech. Rev. 59 (2), 78–90.

Raikar, N.B., Bhatia, S.R., Malone, M.F., Henson, M.A., 2009. Experimental studies and population balance equation models for breakage prediction of emulsion drop size distributions. Chem. Eng. Sci. 64 (10), 2433–2447.

Raikar, N.B., Bhatia, S.R., Malone, M.F., McClements, D.J., Henson, M.A., 2011. Predicting the effect of the homogenization pressure on emulsion drop-size distributions. Ind. Eng. Chem. Res. 50 (10), 6089–6100.

Rojas, C.C., Wahlund, K.-G., Bergenståhl, B., Nilsson, L., 2008. Macromolecular geometries determined with field-flow fractionation and their impact on the overlap concentration. Biomacromolecules 9 (6), 1684–1690.

Schlender, M., Spengler, A., Schuchmann, H.P., 2015. High-pressure emulsion formation in cylindrical coaxial orifices: influence of cavitation induced pattern on oil drop size. Int. J. Multiphase Flow 74, 84–95.

Schultz, S., Wagner, G., Urban, K., Ulrich, J., 2004. High-pressure homogenization as a process for emulsification. Chem. Eng. Technol. 27 (4), 361–368.

Shirgaonkar, I.Z., Lothe, R.R., Pandit, A.B., 1998. Comments on the mechanism of microbial cell disruption in high-pressure and high-speed devices. Biotechnol. Prog. 14 (4), 657–660.

Siddiqi, S.F., Titchener-Hooker, N.J., Shamlou, P.A., 1997. High pressure disruption of yeast cells: the use of scale down operation for the prediction of protein release and cell debris size distribution. Biotechnol. Bioeng. 55 (4), 642–649.

Silva, H.D., Cerqueira, M.A., Vicente, A.A., 2015. Influence of surfactant and processing conditions in the stability of oil-in-water nanoemulsions. J. Food Eng. 167 (Part B), 89–98.

Steiner, H., Teppner, R., Brenn, G., Vankova, N., Tcholakova, S., Denkov, N., 2006. Numerical simulation and experimental study of emulsification in a narrow-gap homogenizer. Chem. Eng. Sci. 61 (17), 5841–5855.

Stevenson, M.J., Chen, X.D., 1997. Visualization of the flow patterns in a high pressure homogenizing valve using a CFD package. J. Food Eng. 33 (1-2), 151–165.

Tadros, T., Izquierdo, P., Esquena, J., Solans, C., 2004. Formation and stability of nano-emulsions. Adv. Colloid Interf. Sci. 108-109, 303–318.

Taisne, L., Walstra, P., Cabane, B., 1996. Transfer of oil between emulsion droplets. J. Colloid Interface Sci. 184 (2), 378–390.

Tornberg, E., 1980. Functional characteristics of protein stabilized emulsions: emulsifying behavior of proteins in a sonifier. J. Food Sci. 45 (6), 1662–1668.

Tubesha, Z., Abu Bakar, Z., Ismail, M., 2013. Characterization and stability evaluation of thymoquinone nanoemulsions prepared by high-pressure homogenization. J. Nanomater. 2013, 1–6.

Walstra, P., 1969. Preliminary note on the mechanism of homogenization. Neth. Milk Dairy J. 23, 290–292.

Walstra, P., 1983. Formation of emulsions. In: Becher, P. (Ed.), Encyclopedia of Emulsion Technology Volume I: Basic Theory. Marcel Dekker Inc., New York, pp. 57–127.

Walstra, P., 2005. Emulsions. In: Lyklema, J. (Ed.), Fundamentals of Interface and Colloid Science. Elsevier, Amsterdam, pp. 8.1–8.94.

Walstra, P., Smulders, P.E.A., 1998. Emulsion formation. In: Binks, B.P. (Ed.), Modern Aspects of Emulsion Science. Royal Society of Chemistry, Cambridge, pp. 56–99.

Zamora, A., Guamis, B., 2015. Opportunities for ultra-high-pressure homogenization (UHPH) for the food industry. Food Eng. Rev. 7 (2), 130–142.

Zhou, G., Kresta, S.M., 1998. Correlation of mean drop size and minimum drop size with the turbulence energy dissipation and the flow in an agitated tank. Chem. Eng. Sci. 53 (11), 2063–2079.

Chapter 8

Fabrication of Nanoemulsions by Microfluidization

Fidel Villalobos-Castillejos*, Virginia G. Granillo-Guerrero*, Diana E. Leyva-Daniel*, Liliana Alamilla-Beltrán*, Gustavo F. Gutiérrez-López*, Amor Monroy-Villagrana* and Seid Mahdi Jafari†

*National Polytechnic Institute (Instituto Politécnico Nacional), Mexico City, Mexico, †Gorgan University of Agricultural Sciences and Natural Resources, Gorgan, Iran

Chapter Outline

8.1 INTRODUCTION

Nowadays, many techniques and devices specifically designed to produce nanoemulsions have been implemented, such as the microfluidization, a high-energy homogenization method that is applicable in industry operations

Nanoemulsions. https://doi.org/10.1016/B978-0-12-811838-2.00008-4

due to the flexible control of the emulsion droplet size (EDS) it provides, and its ability to produce fine emulsions from a wide diversity of materials such as vitamins, bioactive lipids, drugs, antioxidants, and flavors (McClements, 2011; McClements et al., 2007; Weiss et al., 2009).

Nanoemulsions are less susceptible to gravitational separation (creaming and sedimentation), since the Brownian motion prevails over the gravitational forces (Jafari et al., 2017). On the other hand, a small EDS improves the nanoemulsion's stability against flocculation or coalescence because the range of attractive forces among particles decreases with particle-size reduction (McClements, 2005; Monroy-Villagrana et al., 2015; Tadros et al., 2004).

The EDS in nanoemulsions influences the appearance, bioavailability, bioaccessibility, texture and stability on the final product, and parameters such as opacity, lightness, and color, influencing the perception of consumers in food, pharmaceutical, and cosmetic products, whose manufacture must specify the optimum EDS (Acosta, 2009; Buranasuksombat et al., 2011; McClements et al., 2007; Qian et al., 2012; Qian and McClements, 2011; Salvia-Trujillo et al., 2013b; Tang and Liu, 2013; Chung and McClements 2014; Piorkowski and McClements, 2014).

The microfluidizer has been used to produce nanoemulsions, based on the action of simultaneous forces of deformation, collision, and cavitation (Jafari et al., 2007a,b, 2008a; Quintanilla-Carvajal et al., 2014; Salvia-Trujillo et al., 2013a; Yang et al., 2012). This will produce a homogeneous EDS distribution (Paquin, 1999), which will improve the shelf life of the products and decrease the creaming rate. The minimum size will depend on the design of interaction chamber (IXC), operating conditions, residence time of emulsion in the IXC, emulsion composition, interfacial tension, and rheological properties.

8.2 MICROFLUIDIZER ELEMENTS

The microfluidizer, a high-pressure homogenizer, is considered a static and mechanical mixer by Maa and Hsu (1999) and Sembira et al. (1988). A high-pressure homogenizer consists of a high-pressure generator (pump) assembled to a valve or nozzle (Fig. 8.1) specially designed to withstand the application of high pressures (> 200 MPa). In any type of homogenizing valve, the processed fluid passes through a converging section called the valve space and then expands (Pereda et al., 2008; Stang, 2001). This principle of operation is applicable to the microfluidizers, there being significant differences between these and the conventional high-pressure homogenizer in terms of the pump type, the pressures that it reaches, and the nozzle, which is replaced by an interaction chamber (IXC) (Stang, 2001).

As for the type of pump, the microfluidizer uses a pneumatic pump that oscillates in two strokes or movements (one of suction and the other of compression), being able to pressurize the air from about 150–650 kPa to 275 MPa (40,000 Psi) (Jafari et al., 2007a). On the other hand, the IXC is considered a continuous microreactor that uses turbulent mixing, energy dissipation, and a

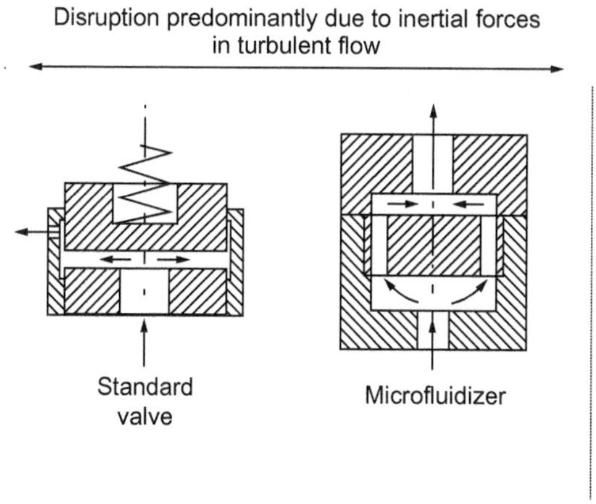

FIG. 8.1 Homogenizing nozzle and interaction chamber (IXC) classified by mechanism of droplet disruption (Jafari et al., 2008a,b).

fixed geometry to create a uniform pressure profile, thus obtaining a precise size and distribution of particles (Microfluidics, 2008).

Fig. 8.2 shows a general scheme of an air-driven microfluidizer (Model M-110 L, Microfluidics, the United States) with different types that are commercially available. This equipment is composed of an inlet/outlet reservoir, a high-pressure pump, a pressure gauge or transducer, an interaction chamber (IXC), and an auxiliary processing mode (APM), which is another IXC.

Usually, during the microfluidization, a coarse emulsion enters the system, via the inlet reservoir, and is forced into the IXC by the high-pressure pneumatic pump, and the fluid is accelerated to a high velocity and finally impinged simultaneously with each other, from two opposite channels of the IXC (Horiba, 2010; McClements, 2005; Microfluidics, 2008, 2016; Stang et al., 2001), causing a maximum shear rate of $10^7\ s^{-1}$ (Horiba, 2010). If more cycles are necessary, the emulsion can be passed through the microfluidizer processor again (Marangoni and Pink, 2014). The sample exits via the heat exchanger or cooling coil outlet to remove the excess heat. This process results in small EDS with a uniform distribution (Microfluidics, 2008, 2016). When the emulsion is generated through a coarse emulsion, the process is called single-channel method (Bai and McClements, 2016a).

Through this method, oil-in-water (O/W) and water-in-oil (W/O) nanoemulsions (100–1000 nm) can be produced (Jafari et al., 2007a; McClements, 2011; Windhab et al., 2005).

8.3 EDS REDUCTION BY MICROFLUIDIZATION

The formation of nanoemulsion occurs in the IXC under two principal types of flow (laminar and turbulent) and under cavitation stress (McClements, 2005; Monroy-Villagrana et al., 2015; Stang, 2001; Walstra, 2003). The possibility

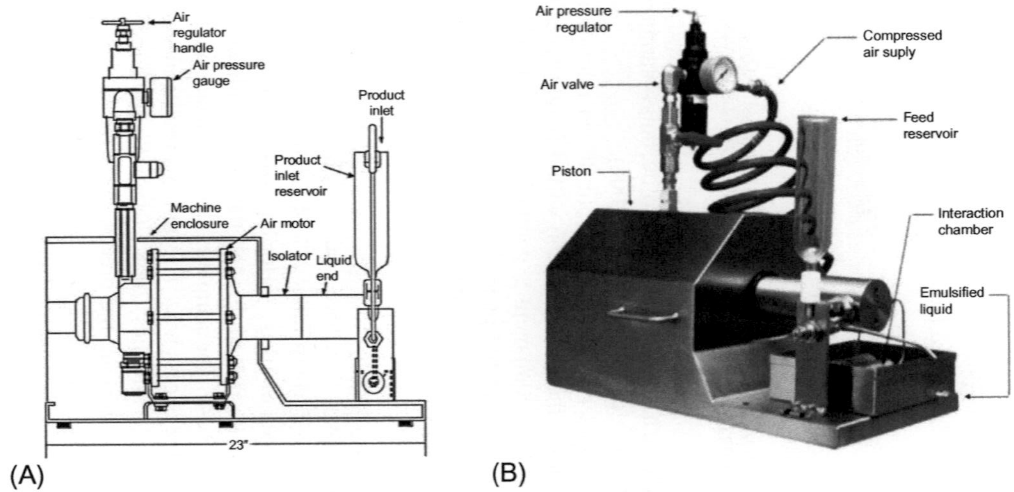

FIG. 8.2 Schematic (A) and real representation of a microfluidizer (B). *(Taken from Microfluidics, Model M-110 Processor User Guide, 2006, Microfluidics, Newton MA, 1–2.)*

of laminar or turbulent flow depends on the frictional and inertial force balance acting on the fluid, characterized by the Reynolds number (McClements, 2005):

$$\mathrm{Re} = \frac{L\bar{v}\rho_c}{\mu_c} = \frac{\text{Inertial forces}}{\text{Viscous forces}} \tag{8.1}$$

where L is the length, $\bar{v}$ is the velocity, ρ_c is the density, and μ_c is the viscosity of the emulsion. These parameters are given by the feeding pressure, length of the pipes and IXC of the microfluidizer, and density and viscosity of the sample. Then, three phenomena could happen:

1. Laminar flow. The laminar flow is predominant at low Reynolds numbers ($Re < 8000$, low flow rates), and the fluid follows a well-defined pattern. The droplet disruption in laminar flow conditions occurs by a combination of viscous and elongational contributions. The droplets are broken by viscous forces; therefore, the formation of nanoemulsions is difficult under these conditions. In the microfluidizer, the laminar flow mainly occurs in the pipes, outside the microchannels (McClements, 2005; Walstra and Smulder, 1998).
2. Turbulent flow. It is characterized by the presence of eddies (vortices and whirls). Due to the turbulence, the surface of the droplets is subjected to random velocity and pressure fluctuations, resulting in a deformation of the drop surface. The shear and pressure gradients are associated with an increase of the number of eddies, forming bigger, intermediate, and very small ones, where intermediate size eddies are assumed to be responsible for size reduction. Eddies, in general, cause droplet disruption under turbulent flow. Besides, the turbulent flow is characteristic in the interaction chamber and is responsible for nanoemulsion formation by microfluidization (Hasenhuettl and Hartel, 2008; McClements, 2005; Walstra, 1993, 2003). According to a study conducted in 2014, an alpha-tocopherol emulsion with 20% solids and a viscosity of 0.017 (Pa s) reaches this type of flow at a feeding pressure above 69 MPa/10,000 psi, with *Re* numbers between 10,110 and 13,589 (Monroy-Villagrana et al., 2015).
3. Cavitation. This phenomenon occurs in a fluid when the pressure variations are highly fluctuating (as in the microfluidizer) and when the formation of cavities implodes and generates shock waves. During the compression, the cavity collapses generating an intense shock wave that propagates into the fluid and causes a deformation and disruption in the nearby drops. When the pressure, acting on a fluid, increases, it produces a compression in this fluid. On the other hand, when the pressure decreases, the difference in pressures induce the formation of cavities which induce the system to reach a critical value of pressure within the channel so promoting cavitation. In cavitation, the phase change takes place at constant temperature, and it happens due to a reduction in pressure generated by the flow itself. In the microfluidizer, the internal diameter of the pipes is higher than the microchannels, thus

promoting the compression and decompression of the emulsions, triggering the cavitation process (d'Agostino and Salvetti, 2008; Lickiss and McGrath, 1996; McClements, 2005).

8.4 FACTORS INFLUENCING THE PROPERTIES OF NANOEMULSIONS PRODUCED BY MICROFLUIDIZATION

Considering the high levels of turbulence that must be attained to carry out the generation of nanoemulsions within the microfluidizer interaction chamber, a suitable choice of such a chamber and its arrangements within the equipment must be done considering the required minimum size of the particle, the rheological properties of the dispersed and continuous phases, the residence time in the IXC, and the type and the concentrations of oil and emulsifier (Brösel and Schubert, 1999; Lee and Norton, 2013; Monroy-Villagrana et al., 2015; Qian and McClements, 2011).

8.4.1 Type of Interaction Chamber

There are two main types of IXC, "Y type" and "Z type," and each one can be used for high and low shears and from laboratory to production scales (Microfluidics, 2008, 2014). Fig. 8.3 shows a schematic representation of these chambers, where the high-shear and high-impact zones are pointed up. Most of the chambers are made of stainless steel (exterior) and either polycrystalline diamond or aluminum oxide ceramic (interior).

Different types of Y-type and Z-type chambers and their minimum internal dimensions are shown in Table 8.1. In the Y-type and Z-type chambers, the

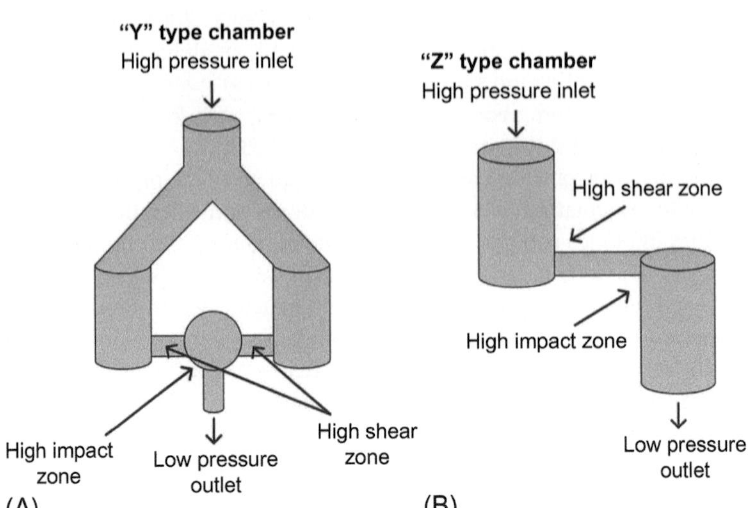

FIG. 8.3 (A) Y-type interaction chamber and (B) Z-type interaction chamber.

TABLE 8.1 Different Kinds of Y-Type and Z-Type Chambers and Their Minimum Internal Dimensions (Microfluidics, 2008)

Y-Type Chamber		Z-Type Chamber	
Style of Chamber	*Microchannel Dimensions (μm)*	*Style of Chamber*	*Microchannel Dimensions (μm)*
F12Y	75	G10z	87
F20Y	75	H10z	100
J20Y	125	L10z	150
J30Y	125	H210z	200
		H30z	200
		L210Z	250
		L30Z	300

minimum diameter of the microchannels can be 75 μm, and these very small orifices create microvolumes that experience a constant pressure profile up to 275 MPa/40,000 psi (Microfluidics, 2014).

The IXC is a continuous microreactor that uses intense disruptive forces [inertial forces in turbulent flow and cavitation (Stang, 2001)], impact, energy dissipation, and fixed geometry to create a uniform pressure profile (Microfluidics, 2008). When the fluids are intermingled, the droplets are disrupted, resulting in small particle sizes with a narrow-size distribution (McClements, 2005; Microfluidics, 2008, 2016).

For nanoemulsion formation, a Y-type chamber with the narrowest size of microchannels (75 μm) (Microfluidics, 2008, 2014; Bai and McClements, 2016a,b) is preferred to afford the maximum deformation level at different pressures of homogenization. Due to the high levels of turbulence achieved by the equipment during the formation of nanoemulsions, the Y-type chamber is usually used in conjunction with an APM, which is always a Z-type chamber.

The APM can be located either downstream or upstream the Y-type chamber:

- *Downstream*: to add backpressure, increasing the lifetime of the Y-type chamber, stabilizing the flow rate, and enhancing the effectiveness of homogenization.
- *Upstream*: to help in the predispersion of formulations that can contain solids or crystals, prior to the Y-type chamber usage (Marangoni and Pink, 2014; Microfluidics, 2008, 2014).

In the first one, the minimum volume used is around 10 mL, and on the industrial scale, it could reach up from 12,000 to 13,000 lh^{-1} (McClements, 2005; Paquin, 1999). For scaling and accordingly to the volume of production, the interaction chambers can be single or multislotted. The first one has a single microchannel and is used for lab-scale processing or small batches. The second one comprises multiple microchannels in parallel, which can be used at industrial scales, by increasing the volumetric flow rate through the IXC, and then, larger volumes of product can be continuously processed in an efficient time (Microfluidics, 2014). The microfluidization can be used in laboratory and on industrial scales. The selection of the most suitable chamber must be based on the initial and final EDS, application and level of deformation, and impact required.

8.4.2 Single-Channel to Dual-Channel Microfluidization Method

A recent development to produce nanoemulsions is the use of a dual-channel microfluidization where the dispersed and continuous phases are forced through an air-driven high-pressure microfluidizer. For the single-channel method (the conventional nanoemulsion elaboration process), a preemulsion (coarse emulsion) is made by mixing the lipid phase with the aqueous phase using a high-speed mixer, which is fed to the microfluidizer through the reservoir. Additionally, through the dual-channel method, the lipid and aqueous phases are fed into two different glass reservoirs; thereafter, the formation of emulsions occurs simultaneously by forcing the lipid and aqueous phases to pass through the microfluidizer (Bai and McClements, 2016a).

Dual-channel microfluidization generates smaller droplets compared with the single-channel method under similar conditions. The main advantage of dual-channel microfluidization is the production of nanoemulsions with high oil-phase levels (up to 50%) in a single step, reducing cost and time of production (Bai and McClements, 2016b). In a dual-channel microfluidization, natural emulsifiers such as whey protein isolate (WPI) and Q-Naturale (natural emulsifier extracted from *Quillaja* trees) were more effective to produce nanoemulsions at low emulsifier concentrations. Also, the nanoemulsions had a good stability toward droplet aggregation and gravitational separation (Bai et al., 2016).

8.4.3 Rheological Properties of Microfluidized Nanoemulsions

The main factors that determine the rheology of the emulsion are the dispersed-phase volume fraction, components of different phases, EDS, colloidal interactions, and particle charge (McClements, 1999; McClements and Rao, 2011). Emphasis has been placed on dispersed-phase volume fraction (Φ) (Dimitrova and Leal-Calderon, 2004; Tang and Liu, 2013) and viscosity of the dispersed phase (Jafari et al., 2008b).

In order to control the final EDS of nanoemulsions, which is related to quality and stability of the final products, the rheological properties of the feed coarse emulsion is an important characteristic to evaluate in microfluidizer processing (Wang et al., 2015). According to Schultz et al. (2004), the correlation between mean droplet diameter (d) and viscosity of dispersed phase (η_d) is

$$d = C\eta_{\mathrm{d}}^{b} \tag{8.2}$$

where the parameter C is defined after the experiments for each emulsification system. The exponent b depends on the homogenization used. For microfluidizers, $b = 0.4–0.9$.

Some studies have reported the effect of these factors on EDS, and different strategies have been used to modify these properties and minimize the droplet size during microfluidization:

- *Changes in the viscosities of the continuous and dispersed phase.* Jafari et al. (2008b) reported that when the viscosity of the dispersed phase increases, droplet disruption and breakup would be more difficult, and the droplet size will increase. In relation to this, Wooster et al. (2008) disclosed that smaller droplets of emulsions could be produced in O/W emulsions when low-viscosity alkane oils were used as the oil phase rather than high-viscosity triglyceride oils. These results were verified by Qian and McClements (2011), who also concluded that the surfactant type has a lot to do with the minimum particle size obtained. This is related to the role of inertial to viscous forces in droplet breakup in turbulent flow in the equipment and the fast stabilization of new interfaces by surfactant (Monroy-Villagrana et al., 2015; Levenspiel, 1999).
- *Increasing the number of microfluidization cycles and pressure.* Jafari et al. (2007a) described a variation of apparent viscosity of emulsions, where the increment of shear stress decreased the viscosity of emulsions (due to microfluidization, 20–100 MPa, two cycles) of D-limonene, maltodextrin (MD), and Hi-Cap. However, they concluded that it was not possible to obtain EDS below 500 nm under the conditions evaluated, mainly because increasing the microfluidization energy input beyond moderate pressures (40–60 MPa) and cycles (1–2) leads to "overprocessing" of emulsion droplets due to recoalescence. In this case, these changes were attributed to the effect of microfluidization on the modified-starch biopolymer used in the study (Hi-Cap). It has been reported that the EDS tends to decrease with increasing number of cycles (Qian and McClements, 2011), but for certain emulsifiers (typically biopolymers), this may lead to "overprocessing," resulting in an increase in particle size (Jafari et al., 2008b). In this study, it was found that the adsorption of the emulsifiers was more likely to be associated with higher concentrations of collision and coalescence. In this sense, Qian and McClements (2011) studied the influence of a greater number of microfluidization cycles (up to eight cycles) and an increase in

pressure (400–1400 MPa) on the particle size in O/W emulsions of corn oil (5%) using different surfactants (SDS, Tween 20, β-lactoglobulin, and sodium caseinate). In the case of β-lactoglobulin (2 wt%), the difference in particle size was mainly between the coarse emulsion (1130 nm) and the emulsion obtained during the first microfluidization cycle (165 nm) at 14 kbar (1400 MPa). After this cycle, the particle size continued to decrease until reaching 146 nm in cycle 8, not decreasing significantly from cycle 6. Also, an emulsion elaborated for one cycle of microfluidization under the same conditions but with varying pressure (400, 600, 800, 1000, 12,000, and 1400 MPa) showed a reduction of the particle size by increasing the latter parameter, obtaining particle sizes of 231, 214, 180, 177, 168, and 165 nm, respectively. This is due to the increase in disruptive forces generated within the interaction chamber. From this, it can be deduced that beyond the number of cycles and the pressure employed, the type of stabilizer and its concentration are decisive factors for the reduction of particle size.

8.4.4 Type of Surfactant or Emulsifier

As mentioned above, the emulsifiers and stabilizers can influence the results of the emulsification process in different ways: it is possible (1) that the surfactants facilitate the deformation and the rupture of the droplets because they decrease the interfacial tension or (2) that the surfactants act to stabilize the interface by preventing the coalescence (Brösel & Schubert, 1999) or by modifying the viscosity of the continuous phase, causing a reduction in the gravitational separation (Klinkesorn et al., 2004).

According to Qian and McClements (2011) and McClements (2004), during microfluidization, there are different mechanisms that influence the changes in continuous phase when small-molecule surfactants such as sodium dodecyl sulfate (SDS) are used: (i) increment of droplet fragmentation due to increased disrupted shear stresses, (ii) reduction of droplet collision frequency due to a decreased droplet recoalescence, and (iii) reduction in emulsifier adsorption rate due to an increased droplet recoalescence, which is also consistent with the Jafari et al. (2007a) report. Qian and McClements (2011) reported that the smallest droplets ($d=57$ nm) can be produced after six cycles at 1400 MPa, using SDS as emulsifier, containing 50 wt% glycerol in the aqueous phase and 100% octadecane in oil phase. In the same way, Wulff-Pérez et al. (2013) and Kim et al. (2013) studied the effects of rheological properties of nanoemulsions by using polymeric surfactants (steric surfactants, Pluronics and Myrjs), low-molecular weight PEG, and ionic surfactants. These authors found that the structure of the surfactants substantially affects the rheological properties of the emulsions. When steric surfactants (Pluronics and Myrjs) are used, the behavior of the emulsions is more elastic than viscous, and in all cases, the storage moduli (G') is dominant over loss moduli (G'') within the linear viscoelastic range. They reported a hydrodynamic diameter, measured by dynamic light scattering, between 322 and 280 nm. They also reported that the hydrophobic part of Pluronics caused an efficient structural barrier at the interface rather than affecting the interfacial rheological directly. In the case of Kim et al. (2013), the nanoemulsions exhibited a thermos-reversible viscoelasticity at low temperatures, changing from a low-viscosity liquid to a viscoelastic material with decreasing temperature.

In the case of emulsions processed by using biopolymers with emulsifying and stabilizing properties, for example, for encapsulation purposes, it has been found that these biopolymers produce more stable emulsions as compared with those integrated by low-molecular weight emulsifiers (Jafari et al., 2007a; Monroy-Villagrana et al., 2014; Assadpour et al., 2016a,b). The size of the biopolymers and their interfacial properties directly influence the final size of the emulsions (Bouyer et al., 2013; Chaparro-Mercado et al., 2012). A comparative information table about the influence of the surfactant/emulsifier on the EDS in nanoemulsions is shown in Table 8.2.

8.4.5 Recoalescence of Emulsion Droplets During Microfluidization

As explained earlier, the distribution of EDS produced by a microfluidizer appears to be narrower and smaller than those produced by traditional homogenization (Jafari et al., 2006). The dependence of EDS on the emulsification pressure and cycles in the microfluidization has been studied; there is a decrease of EDS when the homogenization pressure and number of cycles increase. However, higher pressures and longer emulsification times may lead to "overprocessing" or "recoalescence," resulting in an increase in the particle size due to the mechanical damage produced by the high shear, turbulence, and cavitation presented in the device (Bai and McClements, 2016a,b; Jafari et al., 2006, 2007b, 2008b; Piorkowski and McClements, 2014).

We must bear in mind that the microfluidization process includes the deformation and disruption of the droplets, increasing the specific surface area of the emulsion and stabilization of this fresh interface by using the emulsifier (Jafari et al., 2008b) The increase of the input energy during microfluidization promotes the competition between these two opposite processes, drop breakage and drop-drop coalesce (Jafari et al., 2006). A fresh interface is created whenever a droplet is formed from breakage of an original droplet. As from its formation and its subsequent encounter with other droplets, some surfactant will adsorb onto this fresh interface. If the timescale of collision is shorter than the timescale of adsorption, the fresh interface of the newly formed droplets will depend on the relative rates of surfactant adsorption and drop collision.

According to Jafari et al. (2006, 2008b), there are two ways of avoiding recoalescence immediately after droplet disruption:

TABLE 8.2 Emulsifiers Used in Nanoemulsion Production by Microfluidization

Emulsifier	EDS (nm)	Ratio (Emulsifier-Oil)	Microfluidizer Features	Cycles	Pressure (MPa)	References
Arabic gum/maltodextrin	384	1:2.5	M-110Y Y-75 μm	5	76	Monroy-Villagrana et al. (2014)
β-Lactoglobulin	165	1:2.5	Air driven	1	1400	Qian and McClements (2011)
High methoxyl pectin/Tween 80	<50	3:1	M-110P	5	150	Guerra-Rosas et al. (2016)
Sodium alginate/Tween 80	6	2:1	M-110P	10	150	Salvia-Trujillo et al. (2013a,b)
Tween 20, 40, 60, and 80/Brij 35/Span 80	<100	1:1	M-110P F20Y-75 μm	10	172	Goh et al. (2015)
Tween 20	120	1.5:4	M-101	3	27	Salvia-Trujillo et al. (2013a, b)
Tween 80	121	1:10	Y-Z IXC	1	130	Bai and McClements (2016b)
Tween 80 and 85	<100	<1:1	M110L F20Y-75 μm	3	68.95	Yang et al. (2012)
Rhamnolipid	<150	1:10	Y-Z IXC	3	62–130	Bai and McClements (2016a,b)

(1) Fast stabilization of new interfaces by sufficient surfactant molecules.
(2) Stabilization through hydrodynamic effects or increasing the collision time by turbulence flow field after dispersion zone.

The occurrence of recoalescence depends on different reasons like (1) residence time of the emulsification zone, (2) energy input (pressure and cycles), (3) design of the system, (4) conditions of the emulsion, (5) concentration of the emulsifier, (6) adsorption rate of the emulsifier, (7) adsorption efficiency, and (8) interaction of the emulsifier with other ingredients (Jafari et al., 2008b).

8.4.6 Residence Time Distributions and Energy Density

A way to characterize the process inside the IXC during nanoemulsion formation is through the evaluation of the regimes under the microfluidization and emulsification process. Monroy-Villagrana et al. (2015) discussed that despite the high surface-volume ratios in microchannels that positively influence the mass and heat transfer, the small cross section is a major hurdle for achieving a proper mixing. For this purpose, a hydrodynamic characterization on the formation of α-tocopherol emulsions using MD and gum arabic (GA) as emulsifying agents (1:1, respectively) was performed. These emulsions were produced by changing the oil-phase concentrations from 17% to 28% and by comparing the hydrodynamics of the emulsion colored with a red-dye tracer in water (0.05% w/v red 40 dye). The processing conditions used were 69, 86, and 103 MPa of the operating pressure (working pressures for nanoemulsion formulation) and one cycle of microfluidization, evaluating also the rheological indicators (viscosity coefficient and density) of the samples. Some findings of this study were the following:

1. The pressure drop from the inlet reservoir to the IXC is negligible for the pressures used on the study.
2. The flow in the pipes of microfluidizer before reaching the IXC was laminar (emulsion $Re \approx 219$–349).
3. The flow in IXC under these conditions was found to be turbulent (emulsion $Re \approx 8$ and 700–13,500). Besides, emulsification process on microfluidization is primarily caused by the inertial forces in a turbulent regime.
4. The pressure and viscosity results are the most important parameters on turbulence and particle-size reduction, with an influence of 40% on each one, in an inversely proportional manner.
5. When surface-active biopolymers for stabilization purposes are used, small particle sizes can be obtained by increasing the oil proportion and increasing the residence time in the IXC using the lowest pressure (69 MPa).

8.5 APPLICATIONS AND RECENT DEVELOPMENTS IN NANOEMULSIONS PRODUCED BY MICROFLUIDIZATION

The application of nanoemulsions in food, pharmaceutical, and cosmetic industries is increasing due to its capacity to greatly augment the bioavailability of highly lipophilic substances and due to the high stability toward particle aggregation and gravitational separation (McClements, 2011).

8.5.1 Pharmaceuticals

In the pharmaceutical area, nanoemulsions are widely used in delivery of drugs and as biological or diagnostic agents. In the present time, nanoemulsions are used for targeted drug delivery of various anticancer drugs, photosensitizers, or therapeutic agents. Additionally, nanoemulsions can provide prolonged action of medicaments (Jaiswal et al., 2015). The preparation of nanoemulsions with uniform droplet size and the selection of its preparation method are of prime importance in the pharmaceutical area. This is due to the nanoemulsion properties such as stability, color, appearance, texture, and rheology that are greatly affected by the droplet size of emulsion (McClements, 2004).

Comparative studies of the operating efficiency of microfluidizer as compared with ultrasound in order to produce pharmaceutical-grade nanoemulsions have been made, using aspirin as a model drug. An important aspect in the microfluidization process is the influence of prehomogenization and drug loading on the resultant mean droplet diameter (146–1.53 nm) and the independence from parameters such as the pressure employed (200–600 bar) and the number of passes (up to 10) on the abovementioned droplet diameter (Tang et al., 2013).

Additionally, there are some patents related to nanoemulsions produced by microfluidization in the pharmaceutical industry as a way to protect intellectual property. One example is the patent developed by Nanobio Corporation, in which a nanoemulsion ($\leq$200 nm) with antiinflammatory activity is produced by using a microfluidizer. Okumu et al. (2016) presented an invention that includes O/W nanoemulsion (200 nm) of celecoxib for parenteral administration. This invention includes treating patients in pain.

More examples of the use of nanoemulsions produced by microfluidization in the pharmaceutical area are shown in Table 8.3.

8.5.2 Cosmetics

The cosmetic industry has shown an increased interest in nanoemulsions due to its active release efficiency on the skin, thus reducing the probability of skin irritation; excellent sensorial and aesthetic aspects; low viscosity; need for less surfactant; lower production costs; and higher stability, compared with those of microemulsions (Bouchemal et al., 2004; Maruno and Rocha-Filho, 2009). The O/W nanoemulsion type has been studied more than W/O, and the high-energy method was the most recently reported in the literature (Koroleva and Yurtov, 2012).

One of the most important cosmetic products prepared with nanoemulsion are moisturizers. Its application includes the preventive action against xerosis, delaying premature aging, and dermatologic therapies in a wide variety of skin

TABLE 8.3 Pharmaceutical Applications of Nanoemulsions Elaborated by Microfluidization

Dispersed Phase	Emulsifier	Pharmaceutical Application	References
Ritonavir	(Hydroxypropyl) methyl cellulose, polyvinylpyrrolidone K30	Preparation of ritonavir nanosuspensions, an antiHIV protease inhibitor, to solve its poor water solubility	Karakucuk et al. (2017)
Capmul PG-8 NF and olive oil	Pluronic surfactants	Colloidal system with potential application for medical and synthetic applications	Mountain et al. (2014)
Chlorambucil, soybean oil, cholesterol, Lipoid-80	Glycerol	Improvement of delivery efficiency of chlorambucil in solid tumors	Ganta et al. (2008)
Soybean oil, avocado oil, squalene oil, etc. Organic phosphate solvents and/or alcohols Antiinflammatory agents, steroidal and nonsteroidal	Polysorbate detergents, phenoxypolyethoxylethanols	Nanoemulsion with low toxicity is an effective treatment of bacterial and viral infections	Nanobio Corporation (2007)
Polyethoxylated castor oil or polyethoxylated hydrogenated castor oil Cyclosporine	Unnecessary	Ophthalmic nanoemulsion that contain immunosuppressant and does not cause irritation to eyes	Huons Co., Ltd. (2015)

disorders (Ribeiro et al., 2015). Presently, the market trends look for products of vegetal material with high moisturizing capacity, without toxicity and irritation on the skin surface (Ribeiro et al., 2015; Pereira et al., 2016).

The patent US 8318181 B2 published in 2012, proposed botulinum nanoemulsions prepared by microfluidization. The particle-size distribution was 10 and 300 nm. The application of nanoemulsions can be applied to cosmetic and medical treatment of muscular contracture states (see Table 8.4).

8.5.3 Food

In the food industry, the nanoemulsion production is focusing on the use of natural biopolymers to obtain "clean-label" products (Jafari et al., 2015; Mohammadi et al., 2016a,b; Esfanjani et al., 2017; Assadpour et al., 2017). Different biopolymers such as GA, WPI, modified GA, modified starch, and soy lecithin have been studied to fabricate emulsion-based delivery systems for

TABLE 8.4 Cosmetic Applications of Nanoemulsions Elaborated by Microfluidization

Dispersed Phase	Emulsifier	Cosmetic Application	References
Soybean oil	2-Methacryloyloxyethyl phosphorylcholine *n*-butyl methacrylate copolymer (PMB)	Lotion used to treat skin diseases and can be applied to the scalp	Ishikawa et al. (2016)
Botulinum toxin, saturated and unsaturated almond, soybean, sunflower	Combination of one or more emulsifiers and surfactants such as phosphoglycerides, Tween 20, lecithin, and fatty acid amides	Nanoemulsions comprising botulinum toxins for the cosmetic and medical treatment of muscular contracture states	Edelson and Nicolosi (2016)
Peptides, almond, apricot kernel, avocado oil	Combination of one or more emulsifiers and surfactants such as phosphoglycerides, phosphatidylcholines, Tween 20, and acetyl palmitate	The invention provides a formula to incorporate short peptides (2–30 amino acids long) with beneficial cosmetic and therapeutic effects on the skin	Edelson and Kotyla (2016)

TABLE 8.5 Food Applications of Nanoemulsions Elaborated by Microfluidization

Dispersed Phase	Emulsifier	Food Application	References
Fish oil	Mesquite gum (*Prosopis laevigata*)	Stable nanoemulsions of polyunsaturated fatty acids	García-Márquez et al. (2016)
β-Carotene, corn oil	Whey protein isolate, Quillaja saponin	β-Carotene nanoemulsion-based delivery system	Luo et al. (2017)
High-oleic palm oil	Whey powder, Tween 20	Increasing of oleic palm oil bioavailability, solubility, and targeted delivery	Ricaurte et al. (2016)
N,N-Dimethyl decanamide, D-limonene	Polyoxyethylene glycerol fatty acid ester	Producing of nanoemulsions for eco-friendly solvents and a green surfactant	Santos et al. (2016)
Lemongrass essential oil	Sodium alginate, Tween 80	Evaluation of nanoemulsions bactericidal action against *E. coli*	Salvia-Trujillo et al. (2014)
Soy oil	Soy protein isolate	Elaboration of a kind of cold, gel-like emulsions	Tang and Liu (2013)

vitamin E, orange oil terpenes, fish oil, and essential oils (Amaral and Bhargava, 2015; Bai and McClements, 2016a; Ozturk et al., 2015) (Table 8.5).

8.6 A CASE STUDY ON PRODUCTION OF β-CAROTENE NANOEMULSIONS BY MICROFLUIDIZATION FOR ENCAPSULATION PURPOSES

Microencapsulation by spray drying of nanoemulsions has been widely used in the food and pharmaceutical areas in order to improve the solubility and control the release of active agents, protection against degradation, and water removal (Esfanjani et al., 2015; Jafari et al., 2008a; McClements et al., 2007; Weiss et al., 2009; McClements, 2011). The selection of biopolymers used for microencapsulation is an important factor due to the characteristics that these substances give to microcapsules. There are several biopolymers used in spray drying; the most common are (1) GA, a compound that has emulsifying properties, good solubility, and low viscosity at high concentrations, and (2) MD, a compound that despite not possessing emulsifying properties provides support to the capsules obtained by spray drying.

Since spray drying is carried out immediately after emulsification, only the properties of the emulsion at the time of manufacture have been considered. This has restricted the knowledge of the stability of nanoemulsions over time and the possible use of biopolymers in spray drying as emulsifiers, despite being considered slow emulsifiers due to the time required by their surface adsorption, which can be shortened with the use of the microfluidizer (Becher, 2001; Jafari et al., 2007a; McClements, 2005; Paquin, 1999). On the other hand, the increasing demand for clean-label products must be taken into account, so it is important to consider the natural biopolymers according to their properties as biodegradable, digestible, and biocompatible, instead of considering them as synthetic surfactants in the production of nanoemulsions (Bouyer et al., 2013; Miao et al., 2014).

In order to observe the influence of the homogenization method and biopolymer used, nanoemulsions of β-carotene were prepared by triplicate using the techniques of microfluidization (69 MPa and two cycles) and rotor/stator (10,000 rpm, 3 min), a conventional homogenization method, according to the mixture design D-optimal (Table 8.6). GA and MD were used as wall materials and/or emulsifiers. The stability of the nanoemulsions, EDS, and lightness were evaluated by using Turbiscan Lab Expert, Formulaction, France; a particle-size analyzer (Malvern Instrument Ltd. System 2600, England); and a colorimeter (Color reader CR-10, Konica Minolta, Japan), respectively.

TABLE 8.6 Experimental Design D-Optimal Mixture

Run	AA (%)	MD (%)	GA (%)
1	5	0	95
2	2	0	98
3	2	98	0
4	2	98	0
5	3.5	48.25	48.25
6	2.75	24.13	73.13
7	5	0	95
8	5	95	0
9	3	0	98
10	4.25	71.62	24.13

AA, active agent; *GA*, gum arabic; *MD*, maltodextrin

TABLE 8.7 EDS, Stability Index, and Lightness for Emulsion Obtained by Two Different Homogenizers ($n = 3 \pm SD$)

Run	Emulsion Droplet Size (nm)		Stability Index (ΔBS/h)		Lightness (L*)	
	Rotor/Stator	*Microfluidization*	*Rotor/Stator*	*Microfluidization*	*Rotor/Stator*	*Microfluidization*
1	1143 ± 77.8^{e}	$417.3 \pm 1.06^{2,3}$	2.03 ± 0.49^{b}	$0.76 \pm 0.18^{2,3,4}$	49.74 ± 0.78^{a}	$64.22 \pm 1.19^{1,2}$
2	2210.5 ± 65.8^{a}	$410.9 \pm 9.05^{2,3}$	7.54 ± 0.53^{a}	0.3 ± 0.06^{4}	44.04 ± 0.57^{a}	$56.98 \pm 0.25^{1,2,3}$
3	$2007 \pm 11.3^{a,b}$	$509 \pm 81.95^{1,2}$	$4.30 \pm 0.18^{a,b}$	$1.75 \pm 0.24^{2,3}$	44.16 ± 0.79^{a}	$52.95 \pm 4.82^{2,3,4}$
4	$1650 \pm 22.6^{b,c}$	$678.2 \pm 163.13^{1,2}$	$4.45 \pm 0.06^{a,b}$	1.9 ± 0.74^{2}	42.98 ± 0.91^{a}	48.38 ± 6.02^{4}
5	$1639.5 \pm 375.4^{b,c}$	$431.8 \pm 9.62^{2,3}$	$6.31 \pm 0.27^{a,b}$	0.33 ± 0.06^{4}	45.76 ± 0.82^{a}	64.56 ± 0.45^{1}
6	$1589 \pm 101.8^{c,d}$	$414.1 \pm 10.68^{2,3}$	$6.10 \pm 0.49^{a,b}$	0.27 ± 0.04^{4}	43.92 ± 0.64^{a}	$60.4 \pm 1.19^{1,2,3}$
7	$1213.5 \pm 34.6^{d,e}$	$402.6 \pm 5.8^{2,3}$	3.40 ± 0.78^{b}	$0.74 \pm 0.25^{2,3,4}$	48.42 ± 0.99^{a}	63.94 ± 1.81^{1}
8	$1228.5 \pm 19.1^{d,e}$	$721 \pm 4.17^{1,2}$	3.28 ± 1.45^{b}	3.28 ± 0.45^{1}	47 ± 0.82^{a}	$52.64 \pm 5.43^{3,4}$
9	981.9 ± 23.7^{e}	362.4 ± 29.2^{3}	$5.18 \pm 0.57^{a,b}$	$0.49 \pm 0.12^{2,3,4}$	45.22 ± 0.37^{a}	$63.24 \pm 3.59^{1,2}$
10	$1328.5 \pm 277.9^{c,d,e}$	$420.3 \pm 95.67^{2,3}$	$5.79 \pm 1.07^{a,b}$	$0.45 \pm 0.21^{2,3,4}$	49.9 ± 0.07^{a}	65.32 ± 1.36^{1}

Values in the same columns that do not present significant differences were marked with the equal letter or number ($P < .05$).

The microfluidization has a significant effect on the reduction of EDS (362–721 nm), thus producing more stable (0.27–3.28 ΔBS (backscattering)/h) and lighter (L*, 48–65) emulsions in comparison with those elaborated by rotor/stator (981–2210 nm and 2.03–7.54 ΔBS/h, 44–49) (Table 8.7). To describe the reduction phenomena on the EDS, it is necessary to exclude the pressure drop from the inlet reservoir of the IXC, the laminar flow before the IXC, and the turbulent flow on IXC in nanoemulsions elaborated with the same wall materials (MD and GA) and α-tocopherol as bioactive agent, according to the results obtained by Monroy-Villagrana et al. (2015). The authors concluded that the emulsification process on microfluidization is primarily caused by the inertial forces in a turbulent regime. In this work, pressure was maintained in all the runs, denoting the influence of the amount of wall material over the nanoemulsion properties.

The GA is the wall material that generates the most stable emulsions due to its emulsifying property. When surface-active biopolymers (GA) for stabilization purposes are used, small EDS can be obtained by increasing the oil and GA proportion. There is an inverse proportional relationship between EDS and lightness values. The method for emulsion formation is a key factor on the properties of the emulsions obtained. Although microfluidization and rotor/stator are considered as "high-energy methods," they are substantiated on a different basis, and therefore, the type of forces applied varies. In the high-velocity homogenization, the reduction of the EDS occurs because of the shearing forces. On the other hand, in the high-pressure homogenization, the shear, turbulence, and cavitation forces are presented.

8.7 CONCLUSIONS

This chapter showed the suitability of microfluidization on nanoemulsion production from a wide variety of materials. The principles, features, and conditions under the emulsification process occur within the equipment. Small-molecule surfactants formed smaller droplets than biopolymers, which was attributed to their ability to rapidly adsorb onto the interface of oil surfaces during microfluidization. We should also consider the influence of rheology and viscosity of dispersed and continuous phases. The minimum EDS can be obtained at low-viscosity ratios, by using a stabilizer. This information will facilitate the design of nanoemulsions for different applications by microfluidization.

REFERENCES

Acosta, E., 2009. Bioavailability of nanoparticles in nutrient and nutraceutical delivery. Curr. Opin. Colloid Interface Sci. 14 (1), 3–15.

Assadpour, E., Maghsoudlou, Y., Jafari, S.-M., Ghorbani, M., Aalami, M., 2016a. Optimization of folic acid nano-emulsification and encapsulation by maltodextrin-whey protein double emulsions. Int. J. Biol. Macromol. 86, 197–207.

Assadpour, E., Maghsoudlou, Y., Jafari, S.-M., Ghorbani, M., Aalami, M., 2016b. Evaluation of folic acid nano-encapsulation by double emulsions. Food Bioprocess Technol. 9 (12), 2024–2032.

Assadpour, E., Jafari, S.-M., Maghsoudlou, Y., 2017. Evaluation of folic acid release from spray dried powder particles of pectin-whey protein nano-capsules. Int. J. Biol. Macromol. 95, 238–247.

Amaral, D.M.F., Bhargava, K., 2015. Essential oil nanoemulsions and food applications. Adv. Food Technol. Nutr. Sci. Open J. 1 (4), 84–87.

Bai, L., McClements, D.J., 2016a. Formation and stabilization of nanoemulsions using biosurfactants: rhamnolipids. J. Colloid Interface Sci. 479, 71–79.

Bai, L., McClements, D.J., 2016b. Development of microfluidization methods for efficient production of concentrated nanoemulsions: comparison of single-and dual-channel microfluidizers. J. Colloid Interface Sci. 466, 206–212.

Bai, L., Huan, S., Gu, J., McClements, D.J., 2016. Fabrication of oil-in-water nanoemulsions by dual-channel microfluidization using natural emulsifiers: Saponins, phospholipids, proteins, and polysaccharides. Food Hydrocoll. 61, 703–711.

Becher, P., 2001. Emulsions: Theory and Practice, third ed. Oxford University Press, Oxford.

Bouchemal, K., Briançon, S., Perrier, E., Fessi, H., 2004. Nano-emulsion formulation using spontaneous emulsification: solvent, oil and surfactant optimisation. Int. J. Pharm. 280, 241–251.

Bouyer, E., Mekhloufi, G., Huang, N., Rosilio, V., Agnely, F., 2013. β-Lactoglobulin, gum arabic, and xanthan gum for emulsifying sweet almond oil: formulation and stabilization mechanisms of pharmaceutical emulsions. Colloids Surf. A Physicochem. Eng. Asp. 433, 77–87.

Brösel, S., Schubert, H., 1999. Investigations on the role of surfactants in mechanical emulsification using a high-pressure homogenizer with an orifice valve. Chem. Eng. Process. Process Intensif. 38 (4), 533–540.

Buranasuksombat, U., Kwon, Y.J., Turner, M., Bhandari, B., 2011. Influence of emulsion droplet size on antimicrobial properties. Food Sci. Biotechnol. 20 (3), 793–800.

Chaparro-Mercado, M.C., García-Ochoa, F., Hernández-Sánchez, H., Alamilla-Beltrán, L., Quintanilla-Carvajal, M.X., Cornejo-Mazón, M., Pedroza-Islas, R., Gutierrez-López, G.F., 2012. Diseño de una estructura intersticial para una emulsión de aceite de semilla de uva por diseño de experimentos y superficie de respuesta. Rev. Mex. Ing. Quim. 11 (1), 11–21.

Chung, C., McClements, D.J., 2014. Structure-function relationships in food emulsions: improving food quality and sensory perception. Food Struct. 1, 106–126.

d'Agostino, L., Salvetti, M.V. (Eds.), 2008. Fluid Dynamics of Cavitation and Cavitating Turbopumps. Springer Science & Business Media, New York, NY.

Dimitrova, T.D., Leal-Calderon, F., 2004. Rheological properties of highly concentrated protein-stabilized emulsions. Adv. Colloid Interf. Sci. 108, 49–61.

Edelson, J., Kotyla, T., 2016. U.S. Patent No. 9,486,409. U.S. Patent and Trademark Office, Washington, DC.

Edelson, J., Nicolosi, R., 2016. U.S. Patent No. 9,486,408. U.S. Patent and Trademark Office, Washington, DC.

Esfanjani, A.F., Jafari, S.M., Assadpoor, E., Mohammadi, A., 2015. Nanoencapsulation of saffron extract through double-layered multiple emulsions of pectin and whey protein concentrate. J. Food Eng. 165, 149–155.

Esfanjani, F.A., Jafari, S.M., Assadpour, E., 2017. Preparation of a multiple emulsion based on pectin-whey protein complex for encapsulation of saffron extract nanodroplets. Food Chem. 221, 1962–1969.

Ganta, S., Devalapally, H., Shahiwala, A., Amiji, M., 2008. A review of stimuli-responsive nanocarriers for drug and gene delivery. J. Control. Release 126 (3), 187–204.

García-Márquez, E., Higuera-Ciapara, I., Espinosa-Andrews, H., 2016. Design of fish oil-in-water nanoemulsion by microfluidization. Innovative Food Sci. Emerg. Technol. 40, 87–91.

Goh, P.S., Ng, M.H., Choo, Y.M., Amru, N.B., Chuah, C.H., 2015. Production of nanoemulsions from palm-based tocotrienol rich fraction by microfluidization. Molecules 20 (11), 19936–19946.

Guerra-Rosas, M.I., Morales-Castro, J., Ochoa-Martínez, L.A., Salvia-Trujillo, L., Martín-Belloso, O., 2016. Long-term stability of food-grade nanoemulsions from high methoxyl pectin containing essential oils. Food Hydrocoll. 52, 438–446.

Hasenhuettl, G.L., Hartel, R.W. (Eds.), 2008. Food Emulsifiers and Their Applications. In: vol. 19. Springer, New York.

Horiba, I., 2010. Applications Note: Monitoring Particle Size Reduction of Pharmaceutical Products by Microfluidizers Using Laser Diffraction. L.A. Series, pp. 1–3.

Huons Co., Ltd, 2015. Cyclosporine-Containing, Non-Irritative Nanoemulsion Ophthalmic Composition. WO 2013/165074.

Ishikawa, A., Fujii, M., Morimoto, K., Yamada, T., Koizumi, N., Kondoh, M., Watanabe, Y., 2016. Oil-in-water emulsion lotion providing controlled release using 2-methacryloyloxyethyl phosphorylcholine n-butyl methacrylate copolymer as emulsifier. Results Pharma Sci. 2, 16–22.

Jafari, S.M., He, Y., Bhandari, B., 2006. Nano-emulsion production by sonication and microfluidization—a comparison. Int. J. Food Prop. 9, 475–485.

Jafari, S.M., He, Y., Bhandari, B., 2007a. Optimization of nano-emulsions production by microfluidization. Eur. Food Res. Technol. 225 (5–6), 733–741.

Jafari, S.M., He, Y., Bhandari, B., 2007b. Production of sub-micron emulsions by ultrasound and microfluidization techniques. J. Food Eng. 82 (4), 478–488.

Jafari, S.M., Assadpoor, E., Bhandari, B., He, Y., 2008a. Nano-particle encapsulation of fish oil by spray drying. Food Res. Int. 41 (2), 172–183.

Jafari, S.M., Assadpoor, E., He, Y., Bhandari, B., 2008b. Re-coalescence of emulsion droplets during high-energy emulsification. Food Hydrocoll. 22 (7), 1191–1202.

Jafari, S.M., Fathi, M., Mandala, I., 2015. Emerging product formation. In: Food Waste Recovery: Processing Technologies and Industrial Techniques. Elsevier Inc., New York, pp. 293–317

Jafari, S.M., Paximada, P., Mandala, I., Assadpour, E., Mehrnia, M.A., 2017. 2—Encapsulation by nanoemulsions. In: Nanoencapsulation Technologies for the Food and Nutraceutical Industries. Academic Press, New York, pp. 36–73.

Jaiswal, M., Dudhe, R., Sharma, P.K., 2015. Nanoemulsion: an advanced mode of drug delivery system. 3 Biotech. 5 (2), 123–127.

Karakucuk, A., Celebi, N., Teksin, Z.S., 2017. Preparation of ritonavir nanosuspensions by microfluidization using polymeric stabilizers: I. A design of experiment approach. Eur. J. Pharm. Sci. 95, 111–121.

Kim, J., Gao, Y., Hebebrand, C., Peirtsegaele, E., Helgeson, M.E., 2013. Polymer–surfactant complexation as a generic route to responsive viscoelastic nanoemulsions. Soft Matter 9 (29), 6897–6910.

Klinkesorn, U., Sophanodora, P., Chinachoti, P., McClements, D.J., 2004. Stability and rheology of corn oil-in-water emulsions containing maltodextrin. Food Res. Int. 37 (9), 851–859.

Koroleva, M.Y., Yurtov, E.V., 2012. Nanoemulsions: the properties, methods of preparation and promising applications. Russ. Chem. Rev. 81, 21–43.

Lee, L., Norton, I.T., 2013. Comparing droplet breakup for a high-pressure valve homogeniser and a microfluidizer for the potential production of food-grade nanoemulsions. J. Food Eng. 114 (2), 158–163.

Levenspiel, O., 1999. Chemical reaction engineering. Ind. Eng. Chem. Res. 38 (11), 4140–4143.

Lickiss, P.D., McGrath, V.E., 1996. Breaking the sound barrier. Chem. Br. 32 (3), 47–50.

Luo, X., Zhou, Y., Bai, L., Liu, F., Deng, Y., McClements, D.J., 2017. Fabrication of β-carotene nanoemulsion-based delivery systems using dual-channel microfluidization: physical and chemical stability. J. Colloid Interface Sci. 490, 328–335.

Maa, Y.F., Hsu, C.C., 1999. Performance of sonication and microfluidization for liquid–liquid emulsification. Pharm. Dev. Technol. 4 (2), 233–240.

Marangoni, A.G. and Pink, D. (eds.) (2014). Edible Nanostructures: A Bottom-Up Approach. London: Royal Society of Chemistry. Available from: https://books.google.com.mx/books?id=igUWBQAAQBAJ&pg=PA1&lpg=PA1&dq=marangoni,+Edible+Nanostructures&source=bl&ots=5RV7jgPYwx&sig=5ciad8D_iNdpbOERiWx8VncRAqw&hl=es&sa=X&ved=0ahUKEwiC9e_JvM7QAhXo4IMKHdFrCLEQ6AEIOjAD#v=onepage&q=marangoni%2C%20Edible%20Nanostructures&f=false (accessed 08 October 2016).

Maruno, M., Rocha-Filho, P.A.D., 2009. O/W nanoemulsion after 15 years of preparation: a suitable vehicle for pharmaceutical and cosmetic applications. J. Dispers. Sci. Technol. 31 (1), 17–22.

McClements, D.J., 1999. Food Emulsions: Principles, Practices, and Techniques. CRC Press, Boca Raton, FL.

McClements, D.J., 2004. Food Emulsions: Principles, Practices, and Techniques, second ed. CRC Press, Boca Raton, FL.

McClements, D.J., 2005. Food emulsions in practice. In: Food Emulsions: Principles, Practices and Techniques. CRC Press, Boca Raton, FL, pp. 515–544.

McClements, D.J., 2011. Edible nanoemulsions: fabrication, properties, and functional performance. Soft Matter 7 (6), 2297–2316.

McClements, D.J., Rao, J., 2011. Food-grade nanoemulsions: formulation, fabrication, properties, performance, biological fate, and potential toxicity. Crit. Rev. Food Sci. Nutr. 51 (4), 285–330.

McClements, D.J., Decker, E.A., Weiss, J., 2007. Emulsion-based delivery systems for lipophilic bioactive components. J. Food Sci. 72 (8), R109–R124.

Miao, M., Li, R., Jiang, B., Cui, S.W., Zhang, T., Jin, Z., 2014. Structure and physicochemical properties of octenyl succinic esters of sugary maize soluble starch and waxy maize starch. Food Chem. 151, 154–160.

Microfluidics, 2008. Microfluidizer Processor User Guide. Microfluidics, Newton MA, pp. 1–11.

Microfluidics, 2014. Microfluidizer Processor User Guide. Available from: http://www.alfatest.it/keyportal/uploads/chamber-user-guide-jan-2015.pdf. Accessed 26 September 2016.

Microfluidics, 2016. Particle Size Reduction Solutions. Available from: http://www.microfluidicscorp.com/applications/particle-size-reduction/. Accessed 21 September 2016.

Mohammadi, A., Jafari, S.M., Esfanjani, A.F., Akhavan, S., 2016a. Application of nano-encapsulated olive leaf extract in controlling the oxidative stability of soybean oil. Food Chem. 190, 513–519.

Mohammadi, A., Jafari, S.M., Assadpour, E., Faridi Esfanjani, A., 2016b. Nano-encapsulation of olive leaf phenolic compounds through WPC-pectin complexes and evaluating their release rate. Int. J. Biol. Macromol. 82, 816–822.

Monroy-Villagrana, A., Cano-Sarmiento, C., Alamilla-Beltrán, L., Hernández-Sánchez, H., Gutiérrez-López, G.F., 2014. Coupled Taguchi-RSM optimization of the conditions to emulsify α-tocopherol in an Arabic gum-maltodextrin matrix by microfluidization. Rev. Mex. Ing. Quim. 13 (3), 679–688.

Monroy-Villagrana, A., Alamilla-Beltrán, L., Hernández-Sánchez, H., Gutiérrez-López, G.F., 2015. Hydrodynamic characterization of the formation of alpha-tocopherol nanoemulsions in a microfluidizer. In: Food Nanoscience and Nanotechnology. Springer Science & Business Media, New York, NY, pp. 163–175.

Mountain, G.A., Jelier, B.J., Bagia, C., Friesen, C.M., Janjic, J.M., 2014. Design and formulation of nanoemulsions using 2-(poly (hexafluoropropylene oxide)) perfluoropropyl benzene in combination with linear perfluoro (polyethylene glycol dimethyl ether). J. Fluor. Chem. 162, 38–44.

Nanobio Corporation, 2007. Nanoemulsion Compositions Having Anti-Inflammatory Activity. US 20070036831 A1.

Okumu, F., Chu, J.J., Webb, J.A., Sabino, R.A., Chen, A.X., 2016. U.S. Patent No. 9,446,056. U.S. Patent and Trademark Office, Washington, DC.

Ozturk, B., Argin, S., Ozilgen, M., McClements, D.J., 2015. Formation and stabilization of nanoemulsion-based vitamin E delivery systems using natural biopolymers: whey protein isolate and gum arabic. Food Chem. 188, 256–263.

Paquin, P., 1999. Technological properties of high pressure homogenizers: the effect of fat globules, milk proteins and polysaccharides. Int. Dairy J. 9, 329–335.

Pereda, J., Jaramillo, D.P., Quevedo, J.M., Ferragut, V., Guamis, B., Trujillo, A.J., 2008. Characterization of volatile compounds in ultra-high-pressure homogenized milk. Int. Dairy J. 18, 826–834.

Pereira, T.A., Guerreiro, C.M., Maruno, M., Ferrari, M., Rocha-Filho, P.A., 2016. Exotic vegetable oils for cosmetic O/W nanoemulsions: in vivo evaluation. Molecules 21 (3), 248.

Piorkowski, D.T., McClements, D.J., 2014. Beverage emulsions: recent developments in formulation, production, and applications. Food Hydrocoll. 42, 5–41.

Qian, C., McClements, D.J., 2011. Formation of nanoemulsions stabilized by model food-grade emulsifiers using high-pressure homogenization: factors affecting particle size. Food Hydrocoll. 25 (5), 1000–1008.

Qian, C., Decker, E.A., Xiao, H., McClements, D.J., 2012. Nanoemulsion delivery systems: influence of carrier oil on β-carotene bioaccessibility. Food Chem. 135 (3), 1440–1447.

Quintanilla-Carvajal, M.X., Hernandez-Sanchez, H., Alamilla-Beltran, L., Zepeda-Vallejo, G., Jaramillo-Flores, M.E., de Jesús Perea-Flores, M., Jimenez-Aparicio, A., Gutiérrez-López, G.F., 2014. Effects of microfluidisation process on the amounts and distribution of encapsulated and non-encapsulated α-tocopherol microcapsules obtained by spray drying. Food Res. Int. 63, 2–8.

Ribeiro, R.C.D.A., Barreto, S.M.A.G., Ostrosky, E.A., Rocha-Filho, P.A.D., Veríssimo, L.M., Ferrari, M., 2015. Production and characterization of cosmetic nanoemulsions containing Opuntia ficus-indica (L.) mill extract as moisturizing agent. Molecules 20 (2), 2492–2509.

Ricaurte, L., Perea-Flores, M.J., Martinez, A., Quintanilla-Carvajal, M.X., 2016. Production of high-oleic palm oil nanoemulsions by high-shear homogenization (microfluidization). Innovative Food Sci. Emerg. Technol. 35, 75–85.

Salvia-Trujillo, L., Rojas-Graü, M.A., Soliva-Fortuny, R., Martín-Belloso, O., 2013a. Effect of processing parameters on physicochemical characteristics of microfluidized lemongrass essential oil-alginate nanoemulsions. Food Hydrocoll. 30 (1), 401–407.

Salvia-Trujillo, L., Qian, C., Martín-Belloso, O., McClements, D.J., 2013b. Influence of particle size on lipid digestion and β-carotene bioaccessibility in emulsions and nanoemulsions. Food Chem. 141 (2), 1472–1480.

Salvia-Trujillo, L., Rojas-Graü, M.A., Soliva-Fortuny, R., Martín-Belloso, O., 2014. Impact of microfluidization or ultrasound processing on the antimicrobial activity against Escherichia coli of lemongrass oil-loaded nanoemulsions. Food Control 37, 292–297.

Santos, J., Trujillo-Cayado, L.A., Calero, N., Alfaro, M.C., Muñoz, J., 2016. Development of eco-friendly emulsions produced by microfluidization technique. J. Ind. Eng. Chem. 36, 90–95.

Schultz, S., Wagner, G., Urban, K., Ulrich, J., 2004. High-pressure homogenization as a process for emulsion formation. Chem. Eng. Technol. 27 (4), 361–368.

Sembira, A.N., Merchuck, J.C., Wolf, D., 1988. Characteristics of a motionless mixer for dispersion of immiscible fluids—III. Dynamic behavior of the average drop size and dispersed phase hold-up. Chem. Eng. Sci. 43, 373–377.

Stang, M., 2001. Zerkleinern und Stabilisieren von Tropfen beim mechanischen Emulgieren. VDI Res. Rep., Ser. 3, No. 527. 1998, VDI Verlag, Düsseldorf, (also Ph. D. Thesis, University of Karlsruhe). In: Stang, M., Schuchmann, H., Schubert, H. (Eds.), Emulsification in high-pressure homogenizers. Eng. Life Sci. 1(4), 151.

Stang, M., Schuchmann, H., Schubert, H., 2001. Emulsification in high-pressure homogenizers. Eng. Life Sci. 1 (4), 151–157.

Tadros, T., Izquierdo, P., Esquena, J., Solans, C., 2004. Formation and stability of nano-emulsions. Adv. Colloid Interf. Sci. 108, 303–318.

Tang, C.H., Liu, F., 2013. Cold, gel-like soy protein emulsions by microfluidization: emulsion characteristics, rheological and microstructural properties, and gelling mechanism. Food Hydrocoll. 30 (1), 61–72.

Tang, S.Y., Shridharan, P., Sivakumar, M., 2013. Impact of process parameters in the generation of novel aspirin nanoemulsions—comparative studies between ultrasound cavitation and microfluidizer. Ultrason. Sonochem. 20, 485–497.

Walstra, P., 1993. Principles of emulsion formation. Chem. Eng. Sci. 48 (2), 333–349.

Walstra, P., 2003. Formation of emulsions and foams. In: Physical Chemistry of Foods. Marcel Dekker, New York.

Walstra, P., Smulder, P.E.A., 1998. Emulsion formation (in McClements, D.J., 2004. Food Emulsions: Principles, Practices, and Techniques, second ed. CRC Press, Boca Raton, FL.)

Wang, J., Jing, H., Wang, Y., 2015. Possible effects of complex internal structures on the apparent viscosity of multiple emulsions. Chem. Eng. Sci. 135, 381–392.

Weiss, J., Gaysinsky, S., Davidson, M., Mcclements, J., 2009. Nanostructured encapsulation systems: food antimicrobials. In: Barbosa-Cánovas, G.V., Mortimer, A., Lineback, D., Spiess, W., Buckle, K., Colonna, P. (Eds.), Global Issues in Food Science and Technology. Academic Press, New York, pp. 425–479.

Windhab, E.J., Dressler, M., Feigl, K., Fischer, P., Megias-Alguacil, D., 2005. Emulsion processing—from single-drop deformation to design of complex processes and products. Chem. Eng. Sci. 60 (8), 2101–2113.

Wooster, T.J., Golding, M., Sanguansri, P., 2008. Impact of oil type on nanoemulsion formation and Ostwald ripening stability. Langmuir 24 (22), 12758–12765.

Wulff-Pérez, M., Martín-Rodríguez, A., Gálvez-Ruiz, M.J., de Vicente, J., 2013. The effect of polymeric surfactants on the rheological properties of nanoemulsions. Colloid Polym. Sci. 291 (3), 709–716.

Yang, Y., Marshall-Breton, C., Leser, M.E., Sher, A.A., McClements, D.J., 2012. Fabrication of ultrafine edible emulsions: comparison of high-energy and low-energy homogenization methods. Food Hydrocoll. 29 (2), 398–406.

FURTHER READING

Adeosun, J.T., Lawal, A., 2009. Numerical and experimental studies of mixing characteristics in a T-junction microchannel using residence-time distribution. Chem. Eng. Sci. 64 (10), 2422–2432.

Adeosun, J.T., Lawal, A., 2010. Residence-time distribution as a measure of mixing in T-junction and multilaminated/elongational flow micromixers. Chem. Eng. Sci. 65 (5), 1865–1874.

Fryd, M.M., Mason, T.G., 2012. Advanced nanoemulsions. Annu. Rev. Phys. Chem. 63, 493–518.

Gharsallaoui, M., Benincasa, C., Ayadi, M., Perri, E., Khlif, M., Gabsi, S., 2011. Study on the impact of wastewater irrigation on the quality of oils obtained from olives harvested by hand and from the ground and extracted at different times after the harvesting. Sci. Hortic. 128 (1), 23–29.

Gregory, A., Mountain, G., Jelier, B.J., Bagia, C., Friesen, C.M., Janjic, J.M., 2014. Design and formulation of nanoemulsions using 2-(poly (hexafluoropropylene oxide)) perfluoropropyl benzene in combination with linear perfluoro (polyethylene glycol dimethyl ether). J. Fluor. Chem. 162, 38–44.

Hessel, V., Löwe, H., Schönfeld, F., 2005. Micromixers—a review on passive and active mixing principles. Chem. Eng. Sci. 60 (8), 2479–2501.

Kabri, T.H., Arab-Tehrany, E., Belhaj, N., Linder, M., 2011. Physico-chemical characterization of nano-emulsions in cosmetic matrix enriched on omega-3. J. Nanobiotechnol. 9, 1–8.

Komaiko, J., Sastrosubroto, A., McClements, D.J., 2016. Encapsulation of ω-3 fatty acids in nanoemulsion-based delivery systems fabricated from natural emulsifiers: Sunflower phospholipids. Food Chem. 203, 331–339.

Liu, F., Zhu, Z., Ma, C., Luo, X., Bai, L., Decker, A.D., Gao, Y., McClements, D.J., 2016. Fabrication of concentrated fish oil emulsions using dual-channel microfluidization: impact of droplet concentration on physical properties and lipid oxidation. J. Agric. Food Chem. 64, 9532–9541.

Lobo, L., Svereika, A., 2003. Coalescence during emulsification: 2. Role of small molecule surfactants. J. Colloid Interface Sci. 261 (2), 498–507.

Microfluidics, 2012. A Coaxial Microreactor System. Available from: http://www.prager-elektronik.at/datenblaetter/microfluidics/PureNanoCR5.pdf. Accessed 25 January 2017.

Nguyen, N.T., Wu, Z., 2004. Micromixers—a review. J. Micromech. Microeng. 15 (2), R1.

Pipe, C.J., McKinley, G.H., 2009. Microfluidic rheometry. Mech. Res. Commun. 36 (1), 110–120.

Sonneville-Aubrun, O., Simonnet, J.T., L'alloret, F., 2004. Nanoemulsions: a new vehicle for skincare products. Adv. Colloid Interf. Sci. 108, 145–149.

Chapter 9

Fabrication of Nanoemulsions by Ultrasonication

Seyed M.T. Gharibzahedi* and Seid Mahdi Jafari†
*Islamic Azad University, Lahijan, Iran, †Gorgan University of Agricultural Sciences and Natural Resources, Gorgan, Iran

Chapter Outline

9.1 INTRODUCTION

Selection of an appropriate homogenization technology to consistently produce nanoemulsions (NEs) with well-defined droplet characteristics is one of the most important steps in many commercial applications (Jafari, et al., 2008; McClements, 2011; Mehrnia et al., 2016, 2017). Control of the droplet characteristics is important because this impacts the optical, rheology, textural, and release characteristics of NEs (Ghasemi et al., 2017, 2018) which has been explained in Chapter 2. As discussed elsewhere in this book, various high-energy (e.g., high-pressure homogenization, ultrasound-assisted emulsification (UAE), and microfluidization) or low-energy (e.g., spontaneous emulsification,

Nanoemulsions. https://doi.org/10.1016/B978-0-12-811838-2.00009-6

emulsion inversion point, phase inversion composition, and phase inversion temperature) methods are available to produce NEs (Chapters 3–5). Each of these methods has certain advantages and disadvantages for producing NEs with particular properties from specific ingredients (Jafari et al., 2015). High-energy homogenization procedures are all characterized by the fact that they utilize mechanical devices that generate intense mechanical forces that break up the immiscible liquids (usually oil and water) to form fine droplets (Kentish et al., 2008).

In the past two decades, UAE has attracted considerable attention for the fabrication of NEs because of its ability to produce small droplets from a variety of starting materials at good energy efficiency and relatively low cost (Gaikwad and Pandit, 2008; Periasamy et al., 2016). The propagation of high-intensity ultrasonic waves through liquids leads to the production of rapidly oscillating microbubbles that violently collapse and generate intense cavitation forces in their immediate vicinity. These intense disruptive forces efficiently break down larger droplets into smaller ones (< 100 nm) inside the homogenizer (Jadhav et al., 2015). The NEs produced using this method can have very fine droplets and narrow particle-size distributions, which makes them particularly suitable for many commercial applications (Carpenter and Saharan, 2017). For instance, the food and beverage industry often wishes to fortify transparent and/or semi-transparent products with hydrophobic bioactive ingredients (such as oil-soluble vitamins, nutraceuticals, flavors, or colors), and so, the NEs produced by UAE are particularly suitable for this purpose (Tadros et al., 2004; Gharibzahedi et al., 2013).

The design of NEs with desirable physicochemical and functional attributes depends on understanding the key parameters that impact the UAE process. Therefore, this chapter provides a comprehensive overview of our current understanding of the design and operation of sonication equipment for producing NEs. In addition, recent studies of the utilization of NEs produced by sonication for the encapsulation and delivery of bioactive components are reviewed.

9.2 A HISTORICAL PROSPECTIVE OF UAE

The first report of the use of high-intensity ultrasound to produce emulsions was published by Wood and Loomis (1927). Two years later, more detailed studies of the utilization of this technology to create emulsions were carried out by Richards and William (Richards, 1929; William and Richards, 1929). In 1944, the first patent on ultrasonic emulsification was recorded (Swiss Patent; no. 394.390). Between 1930 and 1970, a series of studies on the application of high-intensity ultrasound for producing emulsions suitable for use in the food and pharmaceutical industries were presented, with an emphasis on understanding the relationship between operating parameters and droplet size (Higgins and Skauen, 1972; Mason et al., 1996; Tadros et al., 2004). In the late 1970s, two mechanisms for UAE operation were proposed: (i) induction of an interfacial instability

"Rayleigh-Taylor instability" at the boundary of the oil and water phases by the acoustic waves and (ii) generation of localized shock waves of high pressure (~100 MPa) and temperature (4000 K) as a result of the collapse of microbubbles formed by cavitation-induced turbulence (Li and Fogler, 1978a, b).

In the late 1990s and early 2000s, a number of studies compared the efficiency of UAE with other more conventional homogenization techniques, such as rotor-stator high-shear mixers and high-pressure valve homogenizers. These studies showed that UAE is a good alternative homogenization method for many applications because it can produce small droplets with good energy efficiency (Abismail et al., 1999; Ma and Hsu, 1999). Around this time, a review article concluded that the cavitation phenomenon is the principal reason for droplet size reduction using high-intensity ultrasound (Canselier et al., 2002). Then, Tadros et al. (2004) introduced ultrasonication as an efficient procedure with many technological and economic benefits to prepare NEs. In the same year, Landfester et al. (2004) ultrasonically prepared polymerizable NEs and demonstrated that an increase in sonication time (ST) at various amplitudes can effectively improve the emulsification process. One of the first practical efforts to produce NEs using ultrasound was conducted by Freitas et al. (2006). They designed a new continuous and aseptic ultrasonic emulsification system to develop submicron-sized droplets/particles that could encapsulate pharmaceutical components. Tiwari and Amiji (2006) also developed paclitaxel NE formulations using UAE to improve its oral delivery. Some of the first studies to produce food-grade oil-in-water NEs using ultrasound were performed by Jafari et al. (2006, 2007). They fabricated D-limonene NEs by ultrasonication and subsequently converted them into a powdered form using a spray dryer. The achievements of these early researchers were an important step in the development of a wide number of applications of ultrasonic homogenization in the food industry.

9.3 ADVANTAGES AND DISADVANTAGES OF ULTRASOUND EMULSIFICATION

Table 9.1 illustrates some advantages and disadvantages of UAE compared with other high-energy emulsification processes. The UAE method has been compared with microfluidization, where the former method was shown to be more cost-effective (Abismail et al., 1999; Delmas et al., 2011), simpler to implement, and faster (Jafari et al., 2006; Leong et al., 2009; Periasamy et al., 2016). Other advantages of this approach are lower equipment contamination, more aseptic processing conditions (Abismail et al., 1999; Delmas et al., 2011), and less energy consumption with more balanced local power densities (Canselier et al., 2002; Tadros et al., 2004; Gaikwad and Pandit, 2008; Tang et al., 2013). The formation of NEs using UAE has also been compared with other high-energy homogenization methods, such as rotor-stator mixing, colloid

TABLE 9.1 A List of *Pros* and *Cons* of UAE Compared With Other High-Energy Emulsification Methods

Emulsion Type	Advantages	Disadvantages	Compared With	References
Food-grade NE	Similar local power densities, easy cleaning	–	Microfluidization	Leong et al. (2009)
Food-grade NE	More practicable regarding production cost, contamination of equipment, and aseptic processing conditions	Less efficiency	Microfluidization	Abismail et al. (1999) and Delmas et al. (2011)
Food-grade NE[a]	Reduction of the surfactant use, lower energy consumption, higher emulsion stability with less PDI	–	Mechanical agitation	Tadros et al. (2004), Canselier et al. (2002), and Gaikwad and Pandit (2008)
Food-grade NE Citronella oil NE	Little requirement to the surfactant, emulsion production with smaller droplet sizes within a narrow size distribution	–	Conventional procedures	Mason et al. (1996) and Agrawal et al. (2017)
Ketchup, mayonnaise	Less time, emulsion stability	–	Mechanical treatments	Chemat et al. (2011)
D-Limonene NE	–	Longer time, larger droplets with a wider size distribution	Microfluidization	Jafari et al. (2007)
D-Limonene NE Black cumin EO NE	The NE production by incorporating higher amounts of D-limonene; easy, cost-effective, clean, fast, and aseptic	–	Microfluidization	Jafari et al. (2006) and Periasamy et al. (2016)

Lemongrass EO NE	–	A detrimental effect on the EO antimicrobial activity	Microfluidization	Salvia-Trujillo et al. (2014)
Sunflower oil NE	Comparable or smaller size in emulsion droplets	–	High-pressure homogenization	Calligaris et al. (2016)
Aspirin NE	An identical result with less energy input	–	Microfluidization	Tang et al. (2013)
Pharmaceutical NE Food-grade NE	A satisfactory control over the particle-size distribution, high dispersion stability, more energy efficiency	–	Inline rotor-stator mixers, colloid mills, and high-pressure homogenizers	Sivakumar et al. (2014) and Ghosh et al. (2013a)
Food-grade NE	–	Scaling-up, irregular power distribution, possible deterioration (protein denaturation, polysaccharide depolymerization, or lipid oxidation) due to high local intensities, cross-contamination	–	McClements and Rao (2011) and Abbas et al. (2013)
Diesel emulsion	More stability	–	Mechanical homogenization	Lin and Chen (2008)
Diesel emulsion	Smaller droplets with more uniform distribution		Mechanical treatments	Wilhelm et al. (1999)

[a]*Developed by ultrasound at low frequency.*

milling, and high-pressure homogenization. These studies showed that the NEs formed by UAE were more stable (Canselier et al., 2002; Gaikwad and Pandit, 2008; Chemat et al., 2011), required less surfactant (Mason et al., 1996; Tadros et al., 2004; Lin and Chen, 2008; Ghosh et al., 2013a), and could be used to form more concentrated systems (Jafari et al., 2006). The UAE method has been utilized to produce NEs containing small droplets and narrow particle-size distributions that are suitable for application in the food (Mason et al., 1996; Canselier et al., 2002; Tadros et al., 2004; Gaikwad and Pandit, 2008; Calligaris et al., 2016), pharmaceutical (Sivakumar et al., 2014), and petroleum (Wilhelm et al., 1999) industries.

For certain systems, the UAE method has been reported to be less efficient than other types of high-energy homogenization. For instance, Jafari et al. (2007) reported that D-limonene oil-in-water NEs produced using microfluidization contained smaller droplets than those produced by sonication. Similarly, Salvia-Trujillo et al. (2014) reported that the antimicrobial activity of lemongrass oil-loaded NEs (against *Escherichia coli*) fabricated by sonication was less effective than those produced by microfluidization. This fact was attributed to differences in the ability of the two methods at controlling the temperature of the NEs during homogenization. The microfluidizer was able to maintain the NEs at a lower temperature throughout the homogenization process than the sonicator; consequently, there was less damage to the antimicrobial constituents in the essential oils (Salvia-Trujillo et al., 2014). Indeed, this is a potential drawback of the sonication method for certain applications because the high local temperatures and pressures generated by cavitation can dissociate water molecules into free-radical species that degrade labile substances (Riesz et al., 1985). Another mechanism presented by Jiang et al. (2002) for the ultrasound degradation of substances at high temperatures is direct pyrolysis process inside and at the interface of the collapsing hot bubbles. Uncontrolled temperature rises during sonication can also lead to chemical deterioration of heat-labile substances (McClements and Rao, 2011). Difficulties in scaling up the UAE system and no adequate implementation of a regular power distribution during the UAE process are some other potential disadvantages for UAE (McClements and Rao, 2011; Abbas et al., 2013).

9.4 PRINCIPLES OF ULTRASONIC HOMOGENIZATION

NEs are usually fabricated ultrasonically using an apparatus that is composed of a probe, a metal horn, a generator (provider of electric waves with a standard frequency of 20–24 kHz), and a piezoelectric transducer (Fig. 9.1; Abbas et al., 2013). The piezoelectric transducer is a device that converts electric oscillations into mechanical ones of a similar frequency. Quartz and certain ceramics (e.g., lithium sulphate, barium titanate, zirconate titanate, and lead titanate) are commonly used to fabricate piezoelectric transducers. The mechanical vibrations generated by the piezoelectric transducer are amplified and directed by

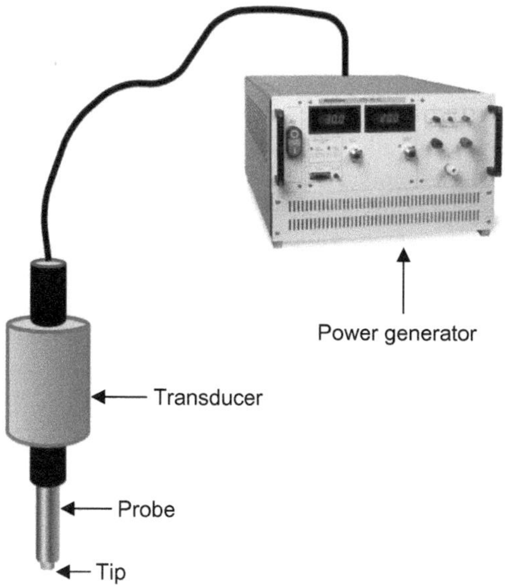

FIG. 9.1 An image of ultrasonic probe apparatus and its elements.

the metal horn and then converted into acoustic waves at its tip. By submerging the tip into the sample, the ultrasonic waves are transmitted into the liquid where they induce an oscillating acoustic pressure (P_a) (Feng and Yang, 2011). According to Muthukumaran et al. (2006), the amplitude of the acoustic pressure wave is determined by three main parameters: time (t), frequency (f), and maximum pressure amplitude ($P_{a,max}$):

$$P_a = P_{a,max} \sin(2\pi ft) \tag{9.1}$$

The high-intensity pressure waves generated by ultrasonic transducers lead to the formation of microbubbles near the tip of the probe that oscillate violently and then implode, thereby generating intense shear forces (Leong et al., 2017). This form of acoustic cavitation is the most important factor that leads to the generation of small droplets during homogenization (Thompson and Doraiswamy, 1999). Eq. (9.1) suggests that $P_{a,max}$ is directly related to the electric energy (power) applied to the ultrasonic transducer. Therefore, it can be concluded that acoustic streaming and cavitation are the major mechanisms at low and high amplitudes, respectively.

9.5 RECENT ADVANCES IN ULTRASOUND EQUIPMENT DESIGN FOR NANOEMULSIFICATION

Generally, the formation of NEs using UAE is performed using either batch or continuous systems, with the batch processes mainly being used in the laboratory and the continuous processes mainly being used in industrial settings. The batch ultrasonic systems used in many research laboratories allow NEs to be generated simply and cheaply on a small scale. These systems usually involve dipping an ultrasonic probe into a vessel that contains the oil and water phases and then generating high-intensity ultrasonic waves to disrupt the two phases and form small droplets (Kumar et al., 2006; Abbas et al., 2013). Typically, $<2\%$ of the liquid volume in a batch experiment is affected by the sonication energy, and therefore, this approach is not highly efficient, unless the samples can be simultaneously stirred to ensure that the fluids all pass through the disruption zone. Another potential disadvantage of the batch-type system is the relatively long processing times required, with the droplet size decreasing with increasing processing time. The emulsion volume and surface area of the ultrasonic probe are two of the main parameters determining this processing time.

The limitations of the batch process have led to interest in the development of continuous-flow ultrasonic homogenization processes. In this case, a coarse emulsion is made to flow through a disruption zone generated by an ultrasonic probe. Typically, the homogenization time required in a continuous-flow system is shorter due to the smaller ratio of the chamber volume to the surface area of the probe tip. As a result, the emulsion droplets required a shorter residence time in the interior of the acoustic field to obtain adequate interfacial coverage with surfactant molecules (Kumaresan et al., 2006; O'Sullivan et al., 2015). However, it should be noted that the growth of emulsion droplets as a result of recoalescence can occur, especially in systems containing low surfactant levels (Jafari et al., 2008). Applying the continuous-flow ultrasonic system in an industrial scale can thus be a successful achievement in the implementation of UAE processes. The efficiency of these devices depends on factors such as the ST, ultrasonic applied power (UAP), ultrasonic amplitude (UA), ultrasonic frequency (UF), and ultrasonic temperature (UT).

Table 9.2 shows some recent advances in ultrasound equipment design for cavitation-assisted emulsification and compares their benefits with conventional technologies. For instance, Freitas et al. (2006) developed a continuous system with a glass tube containing pressurized water, which was located within a steel jacket to transmit the sonicated liquid through vibrating the jacket with a piezoelectric transducer (Fig. 9.2). This sanitary processing system has considerable potential for the formation of pharmaceutical NEs (Table 9.2). Fig. 9.3 shows the preemulsions with an increased coalescence tendency and NEs fabricated by UAE at two frequencies of 25 and 32 W. There were no microscopically visible droplets at the higher UAP, while droplets with a size between 5 and 10 μm were found at a lower UAP.

Peshkovsky et al. (2013) and Peshkovsky and Bystryak (2014) applied Barbell horn ultrasonic technology (BHUT) to emulsify oil and water. This technology indicates that a decrease in output end diameter of an acoustic horn is not essential to amplify UAs. These special horns with large output diameters

TABLE 9.2 Recent Advances in Ultrasound Equipment Design for Emulsification

Ultrasonic Setup/ Configuration	Produced NEs	Structural Changes (Action Mechanism)	Advantages	References
Continuous-flow ultrasonic	Pharmaceutical NEs	The ultrasonic probe was welded to the steel jacket, existence of pressurized water in the space between the jacket and the glass tube as an acoustic conductor	Avoiding the direct contact of sonotrode with the processing medium, eliminating the contamination from ultrasonic pitting	Freitas et al. (2006)
High-power ultrasonic (BHUT[a])	Translucent O/W NEs	Formation of high-intensity cavitation in both below and above the output segments, the maximum diameter of output tip is ~75 mm for ultrasonic amplitude of 100 μm peak to peak	Scale up the process using BHUT by a factor of 10 without reducing the ultrasonic amplitude or compromising the product quality	Peshkovsky et al. (2013)
Continuous-flow ultrasonic	MF59-analog pharmaceutical NE	BHUT supports the horn designs where amplification and output surface are autonomous variables. It can implement large-scale commercial processors at high amplitudes	BHUT ability to scale up horn dimensions without having to decrease the amplitudes (necessary for the industrial implementation)	Peshkovsky and Bystryak (2014)
High-frequency ultrasound using piezoelectric transducer	Emulsions containing different vegetable oils	The piezoelectric transducers were equipped to a piezoelectric ceramic with 20 mm diameter vibrating under electric field influence. The reactor volume, ultrasonic frequency, operation voltage, and electric current intensity were 6 L, 1700 kHz, 24 V, and 0.82–0.90 A, respectively	Possible production of emulsifier-free emulsions with high physical stability and without any degradation of fatty acid profile of used oil	Kaci et al. (2014)

Continued

TABLE 9.2 Recent Advances in Ultrasound Equipment Design for Emulsification—cont'd

Ultrasonic Setup/ Configuration	Produced NEs	Structural Changes (Action Mechanism)	Advantages	References
Continuous-type ultrasonic (lab-scale)[b]	Rapeseed oil NE (emulsifiers, Tween-80 or milk protein isolate)	The ultrasonic probe was placed at an orthogonal position to the preemulsion flow path using a brass tee junction covered by ice to alleviate the heat increase	Shorter emulsification timescale in comparison with batch one, a more efficient production of smaller emulsion droplets size due to the further effective use of acoustic energy	O'Sullivan et al. (2015)
Continuous-type ultrasonic (pilot-scale)[c]	Rapeseed oil NE (emulsifiers, Tween-80 or milk protein isolate)	The ultrasonic probe was placed 20 mm from the inlet of the coarse emulsion, and the outlet was positioned perpendicular to the sonotrode	A considerable improvement in emulsification efficiency compared with the lab-scale continuous-type ultrasonic	O'Sullivan et al. (2015)
Different sonochemical reactor designs[d]	Coconut oil/ water NE (surfactants, Tween 80 and Span 80)	The best configuration was an ultrasonic flow cell with longitudinal horn with a 1000 W rated power and a 2 L capacity having a provision of inlet and outlet for the continuous emulsifying operation	High-power ultrasonic processor could be efficiently applied for the improved nanoemulsification process	Ramisetty et al. (2015)
Hydrodynamic cavitation-assisted emulsification (HCAE)	Citronella oil NE	Converting the pressure energy into kinetic one by passing the liquid via the venture throat leads to a decrease in local pressure below the liquid vapor pressure that produces bubbles like sonication. The bubble collapse provides hot spots and pressure shock waves to disrupt oil droplets	Better scale-up potential in comparison with ultrasonic cavitation, the emulsions production with lower PDI compared with sonication, higher cavitation yield (3–4 times) than sonication method, a comparable physical stability with sonication process	Agrawal et al. (2017)

Cavitation intensifying bags (CIBs)	Hexadecane emulsions	The CIB containing the two unmixed liquids was placed inside an ultrasonic bath [frequencies (37 kHz (80°C, 0.5 h) and 80 kHz (20°C, 2 h)) and amplitude (100%)], straight above one of the transducers of the ultrasonic bath using a custom-made hanging rod. 1/3 tap water and 2/3 demi water at 21°C were used to fill the bath	A scaled-up microreactor with enhanced energy efficiencies, the effectiveness of CIBs was 60 times more than conventional bags in breaking up droplets, the CIBs has a similar score to the membrane emulsification without depending on the wettability changes of the membrane, energy norm in the nonoptimized tests was equal to HPH	van Zwieten et al. (2017)

[a] *BHUT, Barbell horn ultrasonic technology.*
[b] *Volumetric flow rate (using a peristaltic pump) and ultrasonic amplitude were 25–250 mL/min and 20%–40%, respectively.*
[c] *Volumetric flow rate (using a centrifugation pump) and ultrasonic amplitude were 2700–5700 mL/min and 50%–100%, respectively.*
[d] *With diverse processing capacities (50 mL–2 L) and power ratings (165–240 W).*

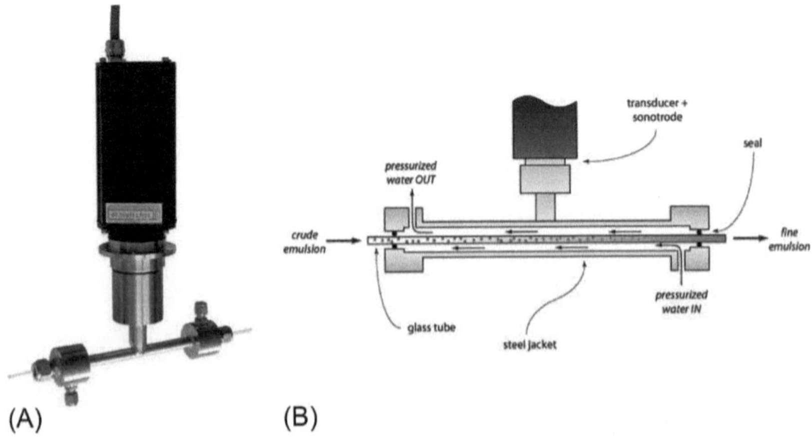

FIG. 9.2 The ultrasonic flow-through cell: (A) real and (B) schematic illustration.

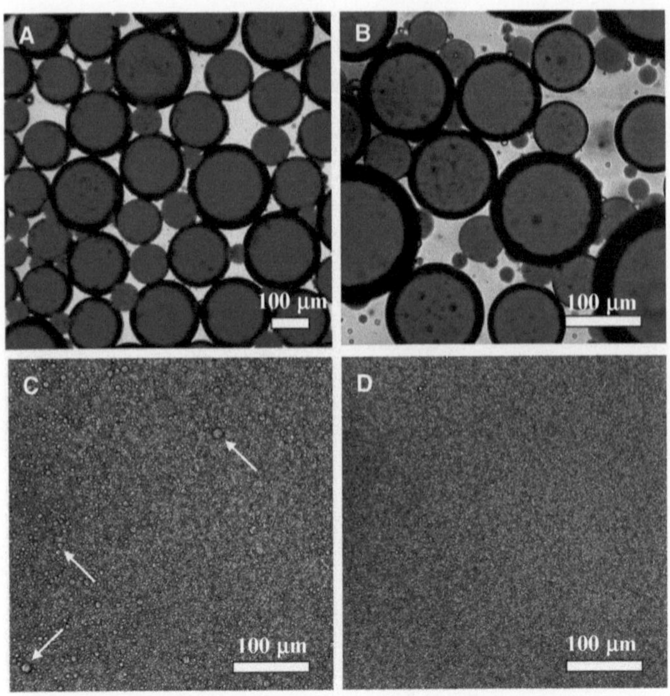

FIG. 9.3 Microstructural images of 20% olive oil-in-water coarse emulsions (A and B) and their submicron NEs prepared using UAE method at (C) 25 W and (D) 32 W UAP.

[e.g., half-wave Barbell horn (HBH)] like conventional horns have a high capability to produce similar UAs. However, HBH can support two high-amplitude radiating regions with a meaningfully greater cumulative output surface area. Besides, there is an extra output surface above the output portion in the HBH with relatively the same ability to generate high UAs as the one below it (Fig. 9.4). These researchers finally reported the successful production of translucent and high-quality oil-in-water NEs at high UAs using BHUT in a commercial scale.

Kaci et al. (2014) showed that a relatively high-frequency sonicator (1.7 MHz) could produce emulsifier-free colloidal systems with an acceptable stability. Large oil droplets were initially formed through an interfacial destabilization mechanism, and then smaller droplets were produced by cavitation. O'Sullivan et al. (2015) used both laboratory-scale (batch/continuous) and pilot-scale (continuous) sonicators with relatively high UAs to produce NEs in low processing volumes in relatively short times. Therefore, a remarkable improvement in emulsification efficiency was achieved by focusing the ultrasonic energy into a smaller volume (O'Sullivan et al., 2015). Ramisetty et al. (2015) utilized another novel sonicator (Fig. 9.5) operated at different

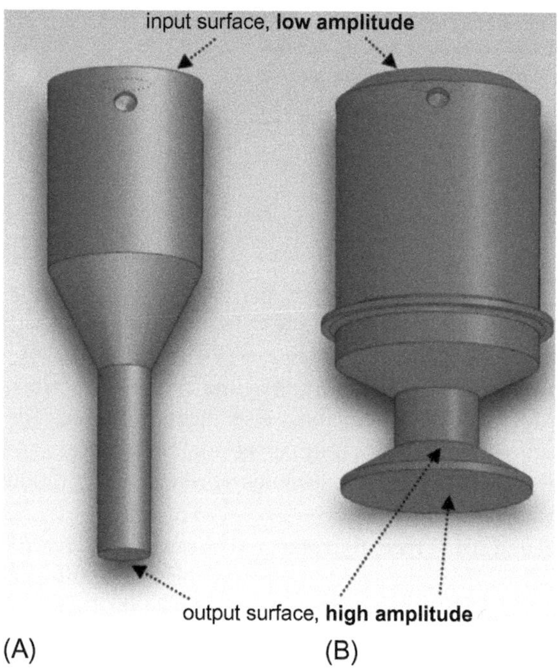

FIG. 9.4 High gain ultrasonic horns: (A) conventional horn and (B) HBH. *(Reproduced with permission from Peshkovsky, A.S., Bystryak, S., 2014. Continuous-flow production of a pharmaceutical nanoemulsion by high-amplitude ultrasound: process scale-up. Chem. Eng. Process. 82, 132–136.)*

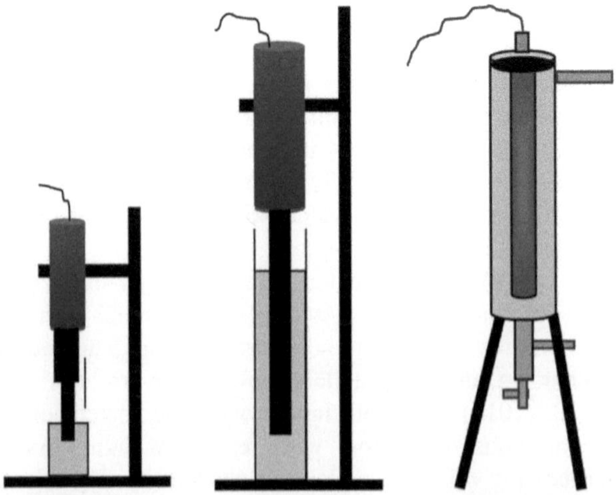

FIG. 9.5 Equipment (1): ultrasonic processor (VCX 750 model, Sonics & Materials, Inc., the United States) at operation conditions of 20 kHz UF, 750 W UAP, and 2.12×10^{-3} m^2 surface area; equipment (2): ultrasonic processor (VC 1500 model, Sonics & Materials, Inc., the United States) applied at 20 kHz UF, 1500 W UAP, and 3.20×10^{-2} m^2 surface area; equipment (3): Roop Telsonic Ultrasonix utilized at 20 kHz UF, 1000 W UAP, and 7.40×10^{-2} m^2 surface area. *(Reproduced with permission from Ramisetty, K.A., Pandit, A.B., Gogate, P.R., 2015. Ultrasound assisted preparation of emulsion of coconut oil in water: understanding the effect of operating parameters and comparison of reactor designs. Chem. Eng. Process. 88, 70–77.)*

power dissipation rates (750, 1000, and 1500 W) and surface areas (2.12×10^{-3}, 3.20×10^{-2}, and 7.40×10^{-2} m^2) and at a constant frequency (20 kHz) to produce coconut oil-in-water NEs with high stability. This study showed that the emulsification process could be improved by using higher ultrasonic power levels to decrease the time required to reach a minimum droplet size. A similar effect has also been reported for both batch (Cucheval and Chow, 2008; Delmas et al., 2011; Kaltsa et al., 2013; O'Sullivan et al., 2015) and continuous (Freitas et al., 2006; O'Sullivan et al., 2015) sonication methods.

Agrawal et al. (2017) recently compared ultrasonic and hydrodynamic cavitation methods with synthesize citronella oil NEs. The ultrasonic cavitation method was reported to be less energy-efficient than the hydrodynamic cavitation method, but when used under optimal conditions, it produced NEs containing smaller droplets. Recently, van Zwieten et al. (2017) developed a novel microreactor, which consisted of cavitation intensifying bags combined with an ultrasound bath (Fig. 9.6), to efficiently generate hexadecane oil-in-water emulsions (Table 9.2).

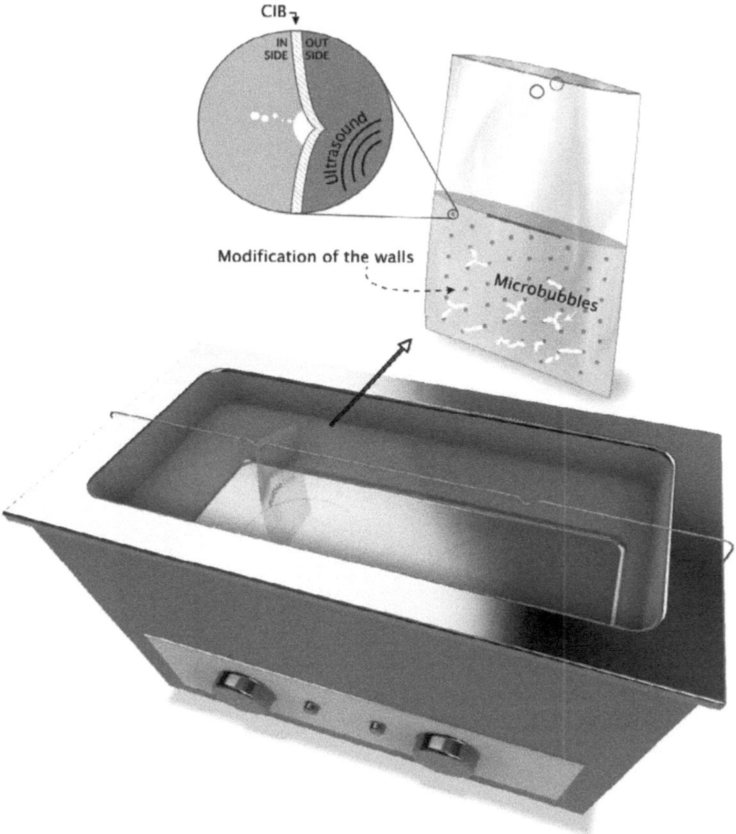

FIG. 9.6 The cavitation intensifying bag (CIB, upper part) located within an ultrasonic bath (lower part) in overhead one of its transducers. *(Reproduced with permission from van Zwieten, R., Verhaagen, B., Schroën, K., Rivas, D.F., 2017. Emulsification in novel ultrasonic cavitation intensifying bag reactors. Ultrason. Sonochem. 36, 446–453.)*

9.6 FACTORS AFFECTING THE EFFICIENCY OF UAE PROCESS

The factors that impact the fabrication of NEs produced by sonication can be divided into two major groups: (i) formulation parameters and (i) processing conditions.

9.6.1 Effect of Formulation Parameters

The reliable production of NEs with predictable and controlled characteristics is a key challenge for any industrial implementation of UAE processes. The nature of the oil and water phases, the oil-to-water ratio, the type and concentration of surfactants and other stabilizers used, and the surfactant-to-oil ratio (SOR) are usually the most important factors influencing NE production. Table 9.3 summarizes some investigations on the impact of these formulation parameters on the fabrication of NEs using sonication methods.

TABLE 9.3 Formulation and Emulsification Conditions of Different NEs Manufactured by Ultrasonication

Oil Phase	Formulation Parameters				Ultrasonic Process Parameters[a]					Droplet Size (nm)	Stability (%)	References
	Φ_{Oil} (%)	*Surfactant/ Cosurfactant*	*SOR*	*Stabilizer*	*ST (min)*	*UAP (W)*	*UA (%, μm)*	*UF (kHz)*	*UT (°C)*			
Black cumin EO[b]	3	3%–9% Polysorbate 80	1–3	–	60	750[c]	–	20	25	37–64	86.1–93	Periasamy et al. (2016)
Orange peel EO	1	2% Tween 80	2	PG+GT[b] (0.25%)	2.3	–	94%	20	37	12.68	>90	Mirmajidi Hashtjin and Abbasi (2015a)
Orange peel EO	1	2% Tween 80	2	–	2.3	–	94%	20	37	18.16	~ 94	Mirmajidi Hashtjin and Abbasi (2015b)
Orange peel EO, EAE[b]	3–20	Tween 20 (0.3%–24%) Ethanol (0.3%–16%)	–	–	30	–	70%	20	20<	72–110	>95	Hernández-Jaimes et al. (2013)
Lemongrass EO	1	1% Tween 80	1	1% Na alginate	0–3	400	100 μm	24	20–47	4.31	–	Salvia-Trujillo et al. (2013)
Lemongrass EO	1	1% Tween 80	1	1% Na alginate	3	400	100 μm	24	47	4–35	–	Salvia-Trujillo et al. (2014)
Citronella oil	1	Tween 80, Span 80	1	–	5	750[c]	50%	22	–	<60	–	Agrawal et al. (2017)
Eucalyptus oil	16.66	16.66% Tween 80	1	–	30	750[c]	40%	20	–	3.8	–	Sugumar et al. (2014)
Mustard oil	10	Tween 80, Span 80 (8%)	0.8	–	30	750[c]	40%	20	≤5	87.38	~ 100	Carpenter and Saharan (2017)

Nettle/canola oils	5–20	0.57% Tween 80	–	–	3.5	220	–	20	<20	134.1	~90	Gharibzahedi et al. (2015a)
Nettle/canola oils	10	20% Tween 40	2	12% jujube gum	20	175	–	24	25	63–240.1	>90	Gharibzahedi and Mohammadnabi (2017)
Nettle/canola oils	10	20% Tween 40	2	1.5% jujube gum	23.64	165.87	–	24	20<	~ 87	>95	Gharibzahedi (2017)
Basil oil	6	6%–24% Tween 80	1–4	–	15	750[c]	–	20	<10	29.3	–	Ghosh et al. (2013a)
Lemon oil	10	0.5%–23% Tween 80: PGMC[b] (2:1)	1–2	–	1	–	70%	–	~23	<200	–	Rao and McClements (2011)
Lemon oil	7	Tween 80	1.33	–	3	120	20%	–	25	25–35	–	Kaur et al. (2017b)
D-Limonene oil	10	Span 85, Brij 98/ ethylene glycol	1.06–3.84	–	1.16–2.84	7.9–28.1	–	–	20<	<100	–	Li and Chiang (2012)
D-Limonene oil	–	Labrasol, plurol oleique, propylene glycol, transcutol, oleic acid, isostearyl isostearate	–	–	3	–	–	–	25<	18–20	85.7–97.34	Sandig et al. (2013)
D-Limonene oil	10	Span 85, polyoxyethylene (20) oleyl ether/ ethylene glycol	–	–	2	18	–	20	20 <	23	–	Lu et al. (2014)
D-Limonene oil	10	Hi-Cap100 (waxy corn starch modified, 10%)	1	MD[b](DE = 16–20, 30%)	1.6	100%[d]	100 μm	24	~ 45	~ 500	–	Jafari et al. (2007)

Continued

TABLE 9.3 Formulation and Emulsification Conditions of Different NEs Manufactured by Ultrasonication—cont'd

Oil Phase	Formulation Parameters				Ultrasonic Process Parameters					Droplet Size (nm)	Stability (%)	References
	Φ_{Oil} (%)	*Surfactant/ Cosurfactant*	*SOR*	*Stabilizer*	*ST (min)*	*UAP (W)*	*UA (%, μm)*	*UF (kHz)*	*UT (°C)*			
D-Limonene oil	20, 25	Hi-Cap 100 or WPC[b](10%)	0.4–0.5	MD (DE = 16–20, 30%)	0.3–1	100%[d]	100 μm	24	32–39	243/298	–	Jafari et al. (2006)
Trans-cinnamaldehyde oil	5,10,15	Tween 80 (optimal) or Tween 20 (5%, 10%)	0.33, 0.5, 1, 2	–	5, 10,15 (optimal)	400[c]	–	20.5	–	27.76	–	Moghimi et al. (2017)
Cinnamon oil	6	6, 12, 18% Tween 80	1, 2, 3	–	10–30	750	–	20	–	65–254	–	Ghosh et al. (2013b)
Clove oil	2.5–15	Tween 80, Span 80 (2.5%–7.5%)	0.5–1	–	2.5–7.5	70	–	20	<30	<50	>95	Shahavi et al. (2016)
Coconut oil	0.1–0.4	Tween 80, Span 80	–	–	5–60	750, 1000, 1500	40%–100%	20	–	<200	–	Ramisetty et al. (2015)
Pine nut oil	20	Egg lecithin, stearylamine, deoxycholic acid	–	–	10	–	21%	–	–	90–120	–	Tiwari and Amiji (2006)
Castor oil	~ 4	~ 10% Transcutol, 0.075% Pluronic F68, ~ 10% Lauroglycol 90, 5% Span 80	–	1% gelatin	1–1.5	–	20%, 40%	–	–	~400	–	Tang et al. (2012b)

Olive oil	20	Lacprodan 80 (WPC, 3%)	–	XG, LBG, GG[b](0.1%–0.5%)	4	150[c]	119/173 μm	20–24	~30–33	615	–	Kaltsa et al. (2013)
Virgin olive oil	5,10,20	WPI	–	2% FG	12	–	40%	–	30	~ 200		Kaltsa et al. (2016)
Palm kernel oil	~ 15	Tween 80: Span 80 (4:1, 7.68%)	–	0.9% XG, 1 %Beeswax	1.5	400	50%	24	25<	153.6	–	Loong et al. (2014)
Palm oil	15	Kolliphor RH40/ Labrafil M1944-CS (70:30)	0.33	β-D-glucan	5	700	–	20	30	263	–	Alzorqi et al. (2016)
Sunflower oil	17.05	13.67% Span 80, 3.04% Tween 80	0.3	–	5	–	100%	24	20	112	–	Calligaris et al. (2016)
Sunflower oil	0.22	HPMC[b] (E5LV, 2%)	9	–	20	–	20%	20	25 <	300	–	Camino and Pilosof (2011)
Sunflower oil	8.5	12% Tween 80/ Span 20	–	–	3.5	220	–	20	20 <	105.5	~ 90	Gharibzahedi et al. (2015a,b)
Sunflower oil	–	0.2% LgC[b]/ apricot gum (8:1)	–	–	10	130	25%	20	–	< 200	–	Shamsara et al. (2015)
Soybean oil	–	PC[b]-SFAEs	–	–	60	500	–	–	55	58.2	–	Takegami et al. (2008)
Soybean oil	–	Lipoid E80, Pluronic F68, Tween 80, Sodium oleate	–	–	60	40	–	–	4	43.5	–	Li et al. (2008)
Soybean oil	–	Tween 80, Labrasol, Lipoid E80	3.66	–	1.66	–	–	–	–	153	–	Gao et al. (2011)
Soybean oil	5	1% Tween 80	0.2	–	1–15	–	–	20	20	700	92	Cucheval and Chow (2008)

Continued

TABLE 9.3 Formulation and Emulsification Conditions of Different NEs Manufactured by Ultrasonication—cont'd

Oil Phase	Formulation Parameters				Ultrasonic Process Parameters					Droplet Size (nm)	Stability (%)	References
	Φ_{Oil} (%)	*Surfactant/ Cosurfactant*	*SOR*	*Stabilizer*	*ST (min)*	*UAP (W)*	*UA (%, μm)*	*UF (kHz)*	*UT (°C)*			
Soybean oil	10	8% Tween 80, 2% Span 80	1	–	–	133	90 μm	–	65	79	> 95	Peshkovsky et al. (2013)
Rapeseed oil	10	Tween 80, milk protein isolate (0.1%–3%)	–	–	Up to 5 min	8.5–32	20%–40%	–	–	~ 200	–	O'Sullivan et al. (2015)
Linseed oil	–	3% Tween 40	–	–	0.42	17, 25, 32	60%, 80%, 100%	24	–	~ 500	–	Freitas et al. (2006)
Flaxseed oil	15	5.6% Tween 40	0.37	–	5	400	–	20–24	50–60	120	–	Kentish et al. (2008)
Flaxseed/ safflower oil	–	Lipoid E80, deoxycholic acid	–	–	10	–	21	–	–	218	–	Vyas et al. (2008)
Flaxseed oil	–	(120 mg) egg lecithin, ethanol	–	–	5	500	21%	20	–	132	–	Ganta and Amiji (2009)
Sunflower oil, canola oil	15	5.3% Tween 80, 4.4% Span 80, 5.6% SDS[b], 13.6% PEG[b]	–	–	< 5	400	30 μm	–	21–23	< 40	–	Leong et al. (2009)
Canola oil	0.25	3% SPC/SPI	–	–	5	0.7 W/mL[e]	–	–	< 50	< 100	–	Lee et al. (2016)
Olive, sesame, or soybean oil	25	Pluronic F68 (Poloxamer 188)	3	–	1	450	–	20	–	< 380	–	Wulff-Perez et al. (2009)

Corn oil	5,10,20	PEG[b] 300, PEG 1000 (15%, 20%)	–	15 % BSA[b]	–	200/400	100 μm	20	< 10	75	–	Tabibiazar et al. (2015)
Corn oil, lemon oil	4	1% Saponin, 5% Tween 80	1.5	–	7	40	20%	–	–	52	> 90	Kaur et al. (2016)
Wheat bran oil	1	Optimal: Span 80 (37.4%) + Tween 80 (62.6%)[f]	7.3		0.83 (5 s–pulses)	500	20%	20	–	39.4	–	Rebolleda et al. (2015)
Thistle oil	10	Polysorbate 80, decyl glucoside, polyglyceryl-4 ester blend (2%, 4%, 6%)	0.2, 0.4, 0.6	–	5	–	20%, 40%, or 60%	20	–	<200	–	Miastkowska et al. (2017)
Omega-3 oil (*Echium plantagineum* L. seed)	5, 10	0.7% EGP[b]	–	1–20 % WPI	–	750	20%–40%	20	30–45	225.83	> 90	Chalothorn and Warisnoicharoen (2012)
Fish oil	5	1% Tween 20/Span 80	–	–	30	200	–	–	27	76–94.2	~ 100	Dey et al. (2012)
Fish oil	5	Tween 80, Span 80	0.5–1.5	–	10		80%	–	25	82	–	Nejadmansouri et al. (2016)
MCT	10	3.3% Tween 80/1.88% Span 80, 2% Lipoid-S75 (Lecithin), 1.68% Pluronic F68	~ 0.37–~ 0.57	–	2	–	–	–	–	~ 200 PDI = 0.16	–	de Araújo et al. (2007)
MCT	–	Tween 80, 1%–2.9% ethanol	0.05, 19	–	–	–	–	–	–	20–60	–	Amani et al. (2010)

Continued

TABLE 9.3 Formulation and Emulsification Conditions of Different NEs Manufactured by Ultrasonication—cont'd

Oil Phase	Formulation Parameters				Ultrasonic Process Parameters					Droplet Size (nm)	Stability (%)	References
	Φ_{Oil} *(%)*	*Surfactant/ Cosurfactant*	*SOR*	*Stabilizer*	*ST (min)*	*UAP (W)*	*UA (%, μm)*	*UF (kHz)*	*UT (°C)*			
MCT	5	OSA[b]-modified starches (three types: Purity Gum 2000, Hi-Cap 100, and Purity Gum Ultra)	–	–	1–13	1200[c]	10%–70%	20	40–45	< 200	–	Abbas et al. (2014)
MCT, safflower oil	6	Lecithin: Tween 80 (1%: 2%)	–	–	15	400[c]	60%	24	60	43.21	–	Tan et al. (2016)
MCT or olive oil	5	10% ω-3-enriched phosphatidyl choline	2	–	12	–	30%	20	–	91	–	Cavazos-Garduño et al. (2015)
Compritol1 ATO 888	5	5% Span 80, 4% SLS/Labrasol: Plurol oleique, 53.47:13.27%	1.8	–	20	240	40%	–	–	<275	–	Clares et al. (2014)
Isopropyl myristate, Lauroglycol 90	5	1% Lipoid E80, 1% Pluronic F68, 1% hydroxypropyl-β-cyclodextrin/ PVP, 1% Tween 80	–	0.2 % HPMC	15	–	21%	–	< 25	~ 50	–	Doh et al. (2013)
Capryol 90	5–16	Cremophor RH 40/Transcutol HP (2:1, [illegible]	–	1 % oleic acid	23.42	156	10–40%	25	<25	10–30	–	Shi et al. (2015)

Liquid/solid lipids (3:7)	5	Tween 80, lecithin	0.55	–	4	100	50%	–	78	<100	>78	Tamjidi et al. (2014)
Paraffin wax	20	SLS (0%–10%)	–	–	15	750[c]	40%	20	–	160.9	–	Jadhav et al. (2015)
α-Tocopherol-based oil	1.5	SSL, Tween 80/ethanol	3		3	60	20%	–	–	<60	–	Kaur et al. (2017a)

[a]*ST, UAP, UA, UF, and UT are sonication time, ultrasonic applied power, ultrasonic amplitude, ultrasonic frequency, and ultrasonic temperature, respectively.*
[b]BSA, *bovine serum albumin;* EAE, Equisetum arvense *extract;* EGP, *ethylene glycol monophenyl ether;* EO, *essential oil;* FG, *fenugreek gum;* GG, *guar gum;* GT, *gum tragacanth;* HPMC, *hydroxypropylmethylcellulose;* LBG, *locust bean gum;* LgC, *whey lactoglobulin concentrate;* MCT, *medium-chain triglyceride;* MD, *maltodextrin;* OSA, *octenyl succinic anhydride;* PC, *phosphatidylcholine;* PEG, *polyethylene glycol;* PG, *Persian gum;* PGMC, *propylene glycol monocaprylate;* PVP, *poly(vinylpyrrolidone);* SDS, *sodium dodecyl sulfate; SFAEs are RYOTO sugar ester including sucrose laurate (SL) L-1695, sucrose myristate (SM) M-1695, sucrose palmitate (SP) P-1670, and sucrose stearate (SS) S-1670;* SLS, *sodium lauryl sulphate;* SPC, *soy protein concentrate;* SPI, *soy protein isolates;* SSL, *sodium stearoyl lactate; WPC, whey protein concentrate;* WPI, *whey protein isolate;* XG, *xanthan gum.*
[c]*As maximum power output.*
[d]*Full power.*
[e]APD, *acoustic power density.*
[f]*Tween 20, span 20, and DATEM (diacetyl tartaric acid ester of mono- and diglycerides) in this study were used.*

9.6.1.1 Type of the Dispersed Phase (Oil)

The type of oil used in preparing oil-in-water NEs can substantially affect many of their properties including their physicochemical properties (e.g., ingredient solubility, viscosity, aggregation stability, and oxidation), nutritional attributes (e.g., omega-3 content), and biological characteristics (e.g., bioavailability/bioaccessibility). Different kinds of lipids have been used as the dispersed phase in oil-in-water NEs fabricated by ultrasonication such as vegetable oils (from canola, corn, sunflower, safflower, rapeseed, linseed, flaxseed, sesame, thistle, soybean, wheat bran, castor, and basil seed), essential oils (from black cumin, orange peel, nettle, lemongrass, citronella, eucalyptus, D-limonene/lemon, *trans*-cinnamaldehyde/cinnamon, and clove), fruit oils [from coconut, (virgin) olive, and palm (kernel)], nuts (pine nut), omega-based oils (from fish and *Echium plantagineum* L. seed), and medium-chain triglycerides (MCTs) as shown in Table 9.3. It is important to carefully choose the oil type used in an NE formulation because it influences the size of the droplets produced at a particular applied ultrasonic energy.

NEs containing small droplets can be produced from essential oils because of their low viscosity and low interfacial tension; however, they are often highly unstable to droplet growth after their production due to Ostwald ripening (McClements and Rao, 2011). Ostwald ripening is the process whereby oil molecules move from the small droplets to the large droplets through the intervening aqueous phase. For this reason, a highly nonpolar oil with a very low water solubility, which is referred to as a ripening inhibitor, is typically mixed with the essential oils prior to NE formation so as to inhibit this process (Wooster et al., 2008). The ability of ripening inhibitors to retard Ostwald ripening is a result of an entropy of mixing effect (McClements, 2011).

Another potential challenge associated with the use of NEs produced by the ultrasonic method is their high susceptibility to oxidation due to the relatively high surface area of the oil phase and their exposure to high temperatures and free radicals during sonication (Chemat et al., 2004). These problems can be overcome by selecting oil phases that are stable to oxidation, adding antioxidants, avoiding long STs, and minimizing temperature rises during sonication (Abbas et al., 2013).

9.6.1.2 Volume Fraction of the Dispersed Phase

NEs have been fabricated using a wide range of dispersed phase concentrations using sonication methods, varying from around 0.25% to 25% (Table 9.3). In general, the droplet size tends to increase with increasing disperse phase concentration under fixed homogenization conditions, which can be attributed to the increased viscosity, greater surfactant depletion, and enhanced coalescence rate at higher droplet levels (Jafari et al., 2008). The greater viscosity of concentrated NEs leads to a reduction in the shear forces generated and a rise in the cavitation threshold, thereby reducing the efficiency of droplet disruption during sonication (Wooster et al., 2008). This fact has been reflected in the results obtained by Tabibiazar et al. (2015) for corn oil-in-water NEs stabilized with albumin (5%–20% oil); Gharibzahedi et al. (2015b) for nettle/canola oil-in-water NEs (5%–20% oil); Ramisetty et al. (2015) for coconut oil-in-water NEs (0.1%–0.4% oil); Kaltsa et al. (2016) for virgin olive oil-in-water NEs (5%–20% oil); and Moghimi et al. (2017) for *trans*-cinnamaldehyde oil-in-water NEs (5%–15% oil). In summary, the rate of droplet coalescence during sonication is reduced when the oil concentration is reduced and/or the surfactant concentration is increased, thereby leading to finer droplets being produced.

9.6.1.3 Type and Concentration of Surfactants and Other Stabilizers

Table 9.3 shows the types of surfactants that have been utilized to fabricate NEs using the sonication method. Small-molecule surfactants, such as Pluronic F68, Tween (20, 40, and 80), Span (20 and 80), and egg lecithin, are some of the most widely used surfactants in ultrasound-mediated NE formulations. The surfactant-to-oil ratios typically used are in the range from 0.2 to 19, with most of them being between 1 and 3 (Table 9.3). A significant decrease in droplets size has been observed by increasing the surfactant concentration, for instance, Tween 60 (1–10 g/L) in kerosene-in-water NEs (Abismail et al. 1999), Tween 80 (7%–13%) in lemon oil-in-water NEs (Rao and McClements, 2011), whey protein isolate (1%–20%) in *E. plantagineum* L. seed oil-in-water NEs (Chalothorn and Warisnoicharoen, 2012), Tween 80 (6%–18%) in cinnamon oil-in-water NEs (Ghosh et al., 2013b), Tween 40 and 80 (2.5%–20%) in nettle/canola oil-in-water NEs (Gharibzahedi and Mohammadnabi, 2016), and Tween 80 (0.33%–2%) in *trans*-cinnamaldehyde oil-in-water NEs (Moghimi et al., 2017). Smaller droplets can be produced at higher surfactant levels because they can cover more surface area (McClements, 1994). However, if the surfactant level used is too high, then there may be a high concentration of surfactant micelles formed in the aqueous phase that can transport oil molecules between oil droplets through Ostwald ripening or that can promote droplet flocculation through a depletion mechanism.

Polysaccharide and protein stabilizers have also been used in the formulation of ultrasound-induced NEs. For instance, tragacanth, locust bean, xanthan, guar, fenugreek, jujube and Persian gums, hydroxypropylmethylcellulose (HPMC), β-D-glucan, maltodextrin, beeswax, sodium alginate, gelatin, bovine serum albumin, and whey protein have been used to stabilize NE formulations by various mechanisms (Table 9.3). Some of these stabilizers act as emulsifiers that adsorb to the droplet surfaces, whereas others act as thickening or gelling agents that inhibit droplet movement. It should be noted that some protein emulsifiers may become denatured during sonication due to the high local temperatures and pressures generated, which can lead to the loss of functionality.

9.6.2 Effect of Operating Parameters

The formation of NEs by sonication is usually a two-step procedure: (i) A coarse emulsion is first produced using a suitable stirring or mixing device; (ii) a fine NE is then produced using a sonicator. The operating parameters of both of these mechanical devices impact the size of the droplets produced in the final NEs and must therefore be carefully controlled to optimize NE formation (Tables 9.3 and 9.4).

9.6.2.1 Preparation Method of Coarse Emulsions

Prior to sonication, coarse emulsions are usually produced using either magnetic stirring or high-shear mixing (Shamsara et al., 2017). Magnetic stirring has been conducted using different agitation rates (100–1200 rpm) and times (2–30 min) to produce coarse emulsions with different characteristics (Table 9.4). High-shear mixing is usually carried out using either a rotor-stator system or a colloid mill operated at different agitation rates (3400–25,000 rpm) and times (1–30 min). A few studies have focused on the use of high-shear homogenizing to fabricate emulsion premixes for preparing NEs based on flaxseed, palm kernel, sunflower, and D-limonene oils (Table 9.4).

It has been demonstrated that the production of good coarse emulsions prior to sonication can considerably decrease the subsequent energy density required to generate NEs. For instance, Tang et al. (2013) reported that NEs with smaller droplets could be produced by sonication when a coarse emulsion was first produced using a high-shear mixer. Tal-Figiel (2007) reported that the generation of coarse emulsions using a magnetic stirrer at 1000 rpm made it possible to produce very stable cosmetic oil-in-water NEs (<100 nm). Kentish et al. (2008) reported that the use of a high-speed mixer is necessary to maintain a continuous fluid flow during sonication, but the formation of a coarse emulsion did not have a considerable impact on the polydispersity of the final NEs produced in a batch-flow system. Also, the application of an ultrasonic bath to prepare a coarse emulsion prior to sonication was reported to produce NEs with narrower particle-size distributions (Sandig et al., 2013; Tamjidi et al., 2014; Kaur et al., 2016).

9.6.2.2 Sonication Time

The STs used to produce NEs by the ultrasonic method have varied widely, ranging from around 0.3 to 60 min (Table 9.3). The optimization of the ST is important to produce small droplets with the minimum amount of energy expenditure. Typically, the mean droplet diameter and polydispersity decrease with increasing ST, since then more of the droplets have a change to pass through

TABLE 9.4 A Summary on Functionalities, Preparation Procedures of Coarse Emulsions, and Storage Stability Studies for Bioactive-Loaded NEs Fabricated With the Sonication Probe System

NE Type	Functionality	Preparation Method for Coarse Emulsion	Sample Amount (mL/g)	Probe Type/ Diameter	Storage Stability Survey	References
Black cumin EO	Therapy effect on the cancer cells	n.r[a]	n.r	Ultrasonic tip/13 mm	A significant increase in the *Z*-average from during storage for 6 days at room temperature was observed	Periasamy et al. (2016)
Orange peel EO	Delivery vehicle	Magnetic stirrer (700 rpm, 15 min)	100 g	Cylindrical titanium sonotrode/ 19.1 mm	The effect of storage time and temperature on the *Z*-average value was very significant. The PDI had a descending trend and then was almost steady during the first 8 weeks of storage (at 25°C and 45°C)	Mirmajidi Hashtjin and Abbasi (2015a)
Orange peel EO	Delivery vehicle	Magnetic stirrer (700 rpm, 15 min)	100 g	High gain sonotrode/ 19.1 mm	*Z*-average and PDI values of the NEs stored at 5°C and 25°C for 4 weeks had a decreasing trend; then, they showed an increasing and almost steady trend, respectively	Mirmajidi Hashtjin and Abbasi (2015b)

Continued

TABLE 9.4 A Summary on Functionalities, Preparation Procedures of Coarse Emulsions, and Storage Stability Studies for Bioactive-Loaded NEs Fabricated With the Sonication Probe System—cont'd

NE Type	Functionality	Preparation Method for Coarse Emulsion	Sample Amount (mL/g)	Probe Type/ Diameter	Storage Stability Survey	References
Orange peel EO, *Equisetum arvense* extract	Delivery of antioxidant flavonoids of *E. Arvense* extract	Ultra-Turrax T10 basic (8000 rpm, 3 min)	n.r	Tip horn/ 6 mm	NEs at 4°C did not suffer significant changes, but those at 25°C showed a significant increase in droplet size after 8 weeks of storage	Hernández-Jaimes et al. (2013)
Lemongrass EO	Delivery vehicle	T25 digital Ultra-Turrax mixer (3400 rpm, 2 min)	100 mL/min[b]	Sonotrode/ 22 mm	n.r	Salvia-Trujillo et al. (2013)
Lemongrass EO	Delivery vehicle, antibacterial agent	T25 digital Ultra-Turrax mixer (3400 rpm, 2 min)	100 mL/min[b]	Sonotrode/ 22 mm	n.r	Salvia-Trujillo et al. (2014)
Citronella oil	Transportation of immobile, insoluble bioactive compounds	Mixing using a pitched blade turbine (700 rpm)	100 mL	Sonotrode containing a piezoelectric crystal/n.r	No phase separation in mechanical agitation, thermal and freeze-thaw tests was observed after 2–3 weeks	Agrawal et al. (2017)

Eucalyptus oil	Antibacterial agent, wound-healing activity, drug delivery vehicle	Magnetic stirrer (250 rpm, 10 min)	n.r	Titanium sonotrode (horn)/n.r	NEs after 25 min-emulsification time were highly stable for 90 days without any phase separation or creaming	Sugumar et al. (2014)
Mustard oil	Delivery vehicle in food and cosmetics industry	The oil dropwise adding into the aqueous mixture and mixing	100 mL	Sonotrode probe/n.r	NEs had a long-term stability after 90 days of storage at room temperature without any visual evidence of creaming and phase separation and also remained kinetically stable	Carpenter and Saharan (2017)
Nettle EO, canola oil	Delivery vehicle, antioxidant	High-speed blender (Ultra-Turrax, 2.5 min)	100 mL	Titanium sonotrode probe/13 mm	An insignificant change in Ostwald ripening and the growth rate of droplets for 30 days were obtained	Gharibzahedi et al. (2015a,b)
Nettle EO, canola oil (8.75% + 1.25%)	Production of bioactive edible coatings for sturgeon fillets/antioxidant agent	T25 digital Ultra-Turrax mixer (12,500 rpm, 2.5 min)	150 mL	Titanium sonotrode/ 13 mm	The *Z*-average and PDI of 5.0% nettle oil NEs notably increased than 2.0% and 3.5% nettle oil NEs during 7-day storage at 4°C	Gharibzahedi and Mohammadnabi (2017)

Continued

TABLE 9.4 A Summary on Functionalities, Preparation Procedures of Coarse Emulsions, and Storage Stability Studies for Bioactive-Loaded NEs Fabricated With the Sonication Probe System—cont'd

NE Type	Functionality	Preparation Method for Coarse Emulsion	Sample Amount (mL/g)	Probe Type/ Diameter	Storage Stability Survey	References
Nettle EO, canola oil (8.75% + 1.25%)	Delivery vehicle, antioxidant, antimicrobial	Rotor-stator homogenizer (T25 digital Ultra-Turrax mixer 12,500 rpm, 2.5 min)	150 mL	Cylindrical titanium sonotrode probe/13 mm	Although there was no substantial difference in the droplets growth between 4°C and 25°C for NEs stored for 45 days, a lower oxidative stability rate (24.1%) was monitored at 4°C	Gharibzahedi (2017)
Basil oil	Antioxidant, antimicrobial	Magnetic stirrer	n.r	Sonotrode containing a piezoelectric crystal/ 13 mm	There was no alteration in droplet size and phase characteristics after 30-day storage	Ghosh et al. (2013a)
Lemon oil	Delivery vehicle/ flavor agent	Magnetic stirrer (30 min)	n.r	n.r	Creation of a separated phase and a cream layer after overnight storage was observed	Rao and McClements (2011)
Lemon oil	Tocopherol/quinine delivery vehicle, antioxidant, cytotoxicity inhibition	Constant stirring (1200 rpm, 10 min)	n.r	S26d2 horn tip/n.r	No significant change was found in the *Z*-average diameter during 60-day storage	Kaur et al. (2017b)

D-Limonene oil	Delivery vehicle/ flavor and antioxidant agents	Rotor/stator system (Polytron, PT-MR 3000)	n.r	Tip horn/ 20 mm	Increasing the droplet size during 14 weeks at 4°C (99.1 nm) was more than 25°C (58.8 nm)	Li and Chiang (2012)
D-Limonene oil	Imipramine/doxepin delivery vehicle, analgesic and antiallodynic activities	Magnetic stirrer/ Ultrasonic bath p-selecta 514	n.r	–/n.r	A good storage stability was reported for 3 months at ambient temperature	Sandig et al. (2013)
D-Limonene oil	Delivery vehicle	Polytron homogenizer	n.r	Tip horn/ 20 mm	n.r	Lu et al. (2014)
D-Limonene oil	Flavor agent	Rotor-stator system	400 mL	Tip horn/ 22 mm	n.r	Jafari et al. (2007)
D-Limonene oil	Flavor agent	High-speed IKA blender	~ 500 mL	Tip horn/ 22 mm	n.r	Jafari et al. (2006)
Trans-cinnamaldehyde oil	Antibacterial, food delivery vehicle	n.r	3 mL	Piezoelectric crystal sonotrode/ 8 mm	A significant stability for 6 months was monitored	Moghimi et al. (2017)
Cinnamon oil	Bactericidal activity	Magnetic stirrer (400 rpm)	n.r	Sonotrode containing a piezoelectric crystal/ 13 mm	n.r	Ghosh et al. (2013b)
Clove oil	Antibacterial agent	Premixing using a vortex mixer for 1 min	n.r	Sonotrode probe made of titanium/ 3 mm	n.r	Shahavi et al. (2016)

Continued

TABLE 9.4 A Summary on Functionalities, Preparation Procedures of Coarse Emulsions, and Storage Stability Studies for Bioactive-Loaded NEs Fabricated With the Sonication Probe System—cont'd

NE Type	Functionality	Preparation Method for Coarse Emulsion	Sample Amount (mL/g)	Probe Type/ Diameter	Storage Stability Survey	References
Castor oil	Antiinflammatory, analgesic activity, aspirin vehicle	Premixing using a vortex mixer for 15 and 5 min during two steps	n.r	Tapered microtip/n.r	The optimal NE formulations contained aspirin had an excellent stability over 30-day storage	Tang et al. (2012b)
Coconut oil	Delivery vehicle	Conventional stirring	Up to 2000 mL	Bench top ultrasonic horn/n.r	n.r	Ramisetty et al. (2015)
Pine nut oil	Paclitaxel delivery vehicle, anticancer	Stirring and gentle heating	n.r	probe-type sonicator/n.r	n.r	Tiwari and Amiji (2006)
Olive oil	Delivery vehicle	Ultra-Turrax T25 device (6500 rpm, 4 min)	100 mL	VS 70T probe/13 mm	NEs formulated with 5 wt% stabilizers were stable after 10 day storage at 5°C	Kaltsa et al. (2013)
Virgin olive oil	Delivery vehicle	Ultra-Turrax T25 device (3500 rpm, 2 min)	10 mL	Titanium probe (MS 73)/3 mm	NEs formulated with the minimum content olive oil (2.5%) had an improved stability during a long-term storage of 6 months	Kaltsa et al. (2016)

Palm kernel oil	Fullerene delivery vehicle	High-shear homogenizer (4500 rpm, 15 min)	n.r	Titanium probe tip/ 14 mm	NE made under optimum conditions was stable at 25°C and 45°C for over 90 days of storage time	Loong et al. (2014)
Palm oil	Delivery vehicle, antioxidant	Continuous stirring	300 mL	Probe tip/ 25 mm	Storage time (90 days) had no significant changes that were observed at 4°C and 25° C	Alzorqi et al. (2016)
Sunflower oil	Delivery vehicle	High-speed blender (8000 rpm, 1 min)	150 mL	Sonicator horn tip/n.r	No changes in the droplet size were found as a function of 15-day storage at 4°C	Calligaris et al. (2016)
Sunflower oil	Delivery vehicle	Ultra-Turrax (25,000 rpm, 1 min): In some experiments was necessary	5 mL	High-grade titanium alloy probe threaded to a 3 mm tapered microtip/ 13 mm	An interesting result was obtained. The droplet size during 20-day storage was nearly constant at room temperature	Camino and Pilosof (2011)
Sunflower oil	Canthaxanthin vehicle	T25 digital Ultra-Turrax mixer (24,000 rpm, 5 min)	100 mL	Sonotrode probe/13 mm	A low droplet growth rate was recorded for 21 days at ambient temperature	Gharibzahedi et al. (2015a,b)

Continued

TABLE 9.4 A Summary on Functionalities, Preparation Procedures of Coarse Emulsions, and Storage Stability Studies for Bioactive-Loaded NEs Fabricated With the Sonication Probe System—cont'd

NE Type	Functionality	Preparation Method for Coarse Emulsion	Sample Amount (mL/g)	Probe Type/ Diameter	Storage Stability Survey	References
Sunflower oil	Delivery vehicle	High-speed homogenizer (12,000 rpm)	5 mL	Tapered titanium sonotrode/ 3 mm	A full stability for NEs during 10 days storage in ambient temperature was recorded	Shamsara et al. (2015)
Soybean oil	Delivery vehicle	Without preparation/ n.r	~ 30 mL	Probe-type sonicator (VC-501)/n.r	There was no remarkable alteration in size and PDI of NE droplets during 1-year storage at ambient temperature	Takegami et al. (2008)
Soybean oil	Delivery vehicle	Magnetic stirrer (10 min)	60 mL	Ultrasonic tip (Branson probe)/n.r	There was no a variation in droplets size of NE stored for 1 day at 5°C	Cucheval and Chow (2008)
Soybean oil	Delivery vehicle	Magnetic stirrer (500 rpm, 2 min)	50 mL (10 mL/ min[b])	Tip horn/ 15.6 mm	High stability at 25°C for nearly 12 months	Peshkovsky et al. (2013)
Soybean oil	Bufadienolides delivery vehicle, antitumor, analgesic activity	Magnetic stirrer, high-shear mixing (10,000 rpm, 10 min)	~ 100 mL	n.r	The NEs stored in a desiccator with silica-gel at 25°C had a good stability for 3 months	Li et al. (2008)

Soybean oil	Fisetin delivery vehicle, antitumor	Mixing under agitation	n.r	n.r	High stability for NEs stored for 30 days at 4°C in the dark conditions	Gao et al. (2011)
Rapeseed oil	Delivery vehicle	High-shear mixing (8000 rpm, 2 min)	3.02–151.13 mL	Ultrasonic probe/55 mm	n.r	O'Sullivan et al. (2015)
Linseed oil	Pharmaceutical delivery	Magnetic stirrer	1.27^{b}	Sonotrode fixed to a piezoelectric transducer	n.r	Freitas et al. (2006)
Flaxseed oil	Delivery vehicle	Ultra-Turrax (13,500 rpm, 2 min)	50 mL	Tip horn/ 19 mm Sonotrode tip/ 22 mm	n.r	Kentish et al. (2008)
Flaxseed/ safflower oil	Saquinavir delivery vehicle for HIV/AIDS therapy	Constant stirring	n.r	Vibra Cell VC 505 probe sonicator/n.r	A high stability was monitored for samples for 2 months stored at 4° C in the dark	Vyas et al. (2008)
Flaxseed oil	Paclitaxel and curcumin delivery	Silverson homogenizer (6000 rpm, 2 min)	5 mL	n.r	Although a minor rise in droplet size was observed after 90-day storage, there was no phase separation	Ganta and Amiji (2009)
Sunflower, canola oil	Delivery vehicle	Ultra-Turrax mixer (low shear rate, 1–2 min)	40 mL	Tip horn/ 12 mm	The sonication process had no significant effect on the oil stability	Leong et al. (2009)

Continued

TABLE 9.4 A Summary on Functionalities, Preparation Procedures of Coarse Emulsions, and Storage Stability Studies for Bioactive-Loaded NEs Fabricated With the Sonication Probe System—cont'd

NE Type	Functionality	Preparation Method for Coarse Emulsion	Sample Amount (mL/g)	Probe Type/ Diameter	Storage Stability Survey	References
Canola oil	Nanoparticle preparation to deliver bioactive compounds (e.g., vitamin D3)	Magnetic stirrer	100 mL	VC-750 probe sonicator/ 12.5 mm	n.r	Lee et al. (2016)
Olive, sesame, or soybean oil	Delivery vehicle	Without preparation/ n.r	7 mL	n.r	NEs for 9 h had a high stability against coalescence, even with a relatively slight quantity of surfactant and with oil volumes more than 25%	Wulff-Perez et al. (2009)
Corn oil	Delivery vehicle for food bioactive components	High-speed Ultra-Turrax blender (22,000 rpm, 5 min)	5 mL	Sonotrode H3 containing a piezoelectric crystal with a titanium probe/3 mm	No remarkable enhancement in the mean droplet diameter of NEs that contained PEG with 5%, 10%, and 20% oil fractions over 40 days was found	Tabibiazar et al. (2015)
Corn oil, lemon oil	Quercitin delivery vehicle,	Mixing/ heating at 70° C for 15 min	n.r	Sonication bath/probe/ n.r	The prepared NEs were stable at room temperature for 45 days	Kaur et al. (2016)

Wheat bran oil	Antioxidant, tyrosinase inhibitory activity, delivery vehicle	High-speed blender during 5 min.	n.r	Titanium alloy microtip probe/3 mm	Although no significant change in droplet size for the optimal NEs after 15 and 60 days of storage at 4°C was found, there was the coalescence of oil droplets for 60 days at 25°C	Rebolleda et al. (2015)
Thistle oil	Antioxidant, moisturizing, and skin-regenerating cosmetic	Rotation of magnetic stirrer (100, 300, or 500 rpm for 5, 10, or 15 min)	n.r	Tip horn/n.r	The average droplet size and PDI were nearly unchanged after 60 days of storage	Miastkowska et al. (2017)
Omega-3 (*Echium plantagineum* L. seed) oil	Strong antioxidant/ delivery vehicle	High-speed stirrer	10 mL	Ultrahigh-intensity sonotrode/ 3 mm	Although there was no instability in 5% oil NEs, an increase in size of NEs stored for 60 days was observed	Chalothorn and Warisnoicharoen (2012)
Fish oil	Strong antioxidant/ delivery vehicle	Magnetic stirrer (700 rpm, 30 min) and T25 digital Ultra-Turrax mixer (8000 rpm, 30 min)	200 mL	n.r	A high 2-day stability against centrifugation experiments and heating-cooling and freeze-thaw cycles	Dey et al. (2012)

Continued

TABLE 9.4 A Summary on Functionalities, Preparation Procedures of Coarse Emulsions, and Storage Stability Studies for Bioactive-Loaded NEs Fabricated With the Sonication Probe System—cont'd

NE Type	Functionality	Preparation Method for Coarse Emulsion	Sample Amount (mL/g)	Probe Type/ Diameter	Storage Stability Survey	References
Fish oil	Delivery vehicle	Magnetic stirrer (700 rpm, 15 min), a dispenser/ homogenizer (20,000 rpm, 5 min)	n.r	n.r	Ostwald ripening rate and oxidative instability of NEs stored at 25°C were less than those determined at 4°C	Nejadmansouri et al. (2016)
MCT	Delivery vehicle	Magnetic stirrer (150 rpm, 30 min)	10 mL	The tapered microtip ultrasound probe 1/8/ 3 mm	The size and zeta potential of the cationic NEs continued constant during a 60-day storage	de Araújo et al. (2007)
MCT	Budesonide vehicle	Without preparation/n.r	20 mL	VCX500 probe sonicator/n.r	Physically stable formulation was obtained for samples stored at 30°C for 1 month	Amani et al. (2010)
MCT	Curcumin delivery vehicle, Antioxidant	High-speed blender (Ultra-Turrax T25, 14,000 rpm, 2 min)	50 mL	Probe-type ultrasonic/ 20 mm	Curcumin content of ≤ 6 mg/mL MCT was successfully encapsulated into the NE without the sediment growth during 30-day storage	Abbas et al. (2014)

MCT, safflower oil (20:80)	Valproic acid delivery vehicle	Magnetic stirrer ($1.34 \times g$)	n.r	Ultrasonic tip/n.r	NEs formulated with 1.5% of valproic acid had a good stability as there are no significant changes in droplet size and PDI	Tan et al. (2016)
MCT or olive oil	Anticancer, delivery vehicle for bioactive compounds	High-speed stirring (Ultra-Turrax T25, 20,000 rpm, 3 min)	n.r	Probe-type ultrasonic/ 13 mm	The NE stability was highly affected by storage temperature, so that the smallest growth in globule size at 5°C was observed	Cavazos-Garduño et al. (2015)
Compritol1 ATO 888	Retinyl palmitate delivery vehicle, treatment of skin diseases	Ultra-Turrax T10 (16,000 rpm, 1 min)	n.r	Probe-type VCX500 ultrasonic/ 13 mm	n.r	Clares et al. (2014)
Isopropyl myristate, Lauroglycol 90	Granisetron delivery vehicle, antiemetic	Constant stirring using a mechanical stirrer	n.r	Vibra Cell VC750 probe-type sonicator/n.r	High stability for NEs stored for 3 months at 4° C in the dark place	Doh et al. (2013)
Capryol 90	Emodin delivery vehicle	Continuous magnetic stirring			Around 80.8% of emodin was found to be released from NE *in vitro* after 5 days	Shi et al. (2015)
Liquid/solid lipids (3:7)	Astaxanthin delivery vehicle, antioxidant	Stirring (2000 rpm, 3 min) + bath sonicator (15 min)	n.r	Probe-type sonicator/n.r	No astaxanthin leakage (as crystallized) was found in the developed astaxanthin-loaded nanostructured lipid carriers after 25 days of storage	Tamjidi et al. (2014)

Continued

TABLE 9.4 A Summary on Functionalities, Preparation Procedures of Coarse Emulsions, and Storage Stability Studies for Bioactive-Loaded NEs Fabricated With the Sonication Probe System—cont'd

NE Type	Functionality	Preparation Method for Coarse Emulsion	Sample Amount (mL/g)	Probe Type/ Diameter	Storage Stability Survey	References
Paraffin wax	Delivery vehicle	Magnetic stirrer (1000 rpm)	n.r	n.r	The NEs prepared ultrasonically were stable for more than 3 months	Jadhav et al. (2015)
α-Tocopherol-based oil	Benzylisothiocyanate and curcumin delivery vehicle, antioxidant	Continuous stirring (10 min)	n.r	S26d2 horn tip/2 mm	NEs had a good loading and release efficiency for the studied antioxidant agents against different environmental stresses such as salt, pH, and dilution. The high storage stability up to 3 months at 25°C, 37°C, and 4°C was monitored	Kaur et al. (2017a)

[a] *Not reported (n.r).*
[b] *Flow rate.*

the disruption zone (Jena and Das, 2006). This trend has been reported by Leong et al. (2009) for sunflower/canola oil NEs (5–20 min); Ghosh et al. (2013a) for basil oil NEs (5–15 min); Salvia-Trujillo et al. (2013) for lemongrass oil NEs (0–3 min); Tang et al. (2013) for aspirin NEs (0.16–1.66 min); and Gharibzahedi (2017) for nettle oil NEs (5–25 min).

It should be noted that an increase in droplet size and polydispersity may occur if the ST is too long (Kentish et al., 2008; Li and Chiang, 2012). This effect may occur due to an increase in temperature or due to damage to the surfactant or other stabilizers during prolonged sonication (Shi et al., 2015). Moreover, increasing the ST can promote oxidation reactions due to the generation of transient hydroxyl (OH·) and hydrogen (H·) free radicals (Gharibzahedi, 2017), which may be detrimental for certain types of bioactive ingredients.

9.6.2.3 *Ultrasonic Applied Power*

A broad range of power levels have been used to form NEs by the sonication method, ranging from around 8 to 1500 W, with the majority being from 400 to 750 W (Table 9.3). NEs have been produced at power levels ≤200 W for various types of oil phase, including α-tocopherol-based oils, fish oils, nettle/canola, lemon/corn, corn, linseed, rapeseed, soybean, sunflower, D-limonene, and clove oils (Table 9.3). Optimization of the power level is necessary to obtain the preferred drop size distribution in the shortest possible time, thereby minimizing costs, which is important for the industrial production of NEs.

Up to a certain value, increasing the power level facilitates droplet disruption and decreases droplet size due to the corresponding increase in acoustic cavitation (Jadhav et al., 2015). However, if the power level is too high, then the droplet size may actually increase, due to the increase in temperature and/or the loss of surfactant functionality. Higgins and Skauen (1972) pointed out that the low effectiveness of emulsification under this condition might be due to the rise of the cavitation nuclei number and the fluid vapor pressure, which can dampen the shock waves generated and attenuate the maximum pressures produced at the microbubble implosion sites. Moreover, the droplet size may also increase at high power levels because of the higher rate of droplet-droplet collisions that occur in strongly turbulent shear flows. Finally, the formation of a denser cloud of cavitation bubbles at high power levels may promote the coalescence of oil droplets (Tang et al., 2012a).

9.6.2.4 *Ultrasonic Amplitude*

The amplitude of the ultrasonic waves applied to a NE also impacts the size of the droplets produced. This value is usually modulated by altering the amplitude of the electric wave applied to the piezoelectric transducer. Typically, the UA is ≤100 μm peak to peak. Since there is a direct relationship between the UA and the cavitation intensity, this parameter can be adjusted to alter the efficiency of droplet disruption during sonication. However, it has been shown that an increase in UA can only improve homogenization over a limited range. The reason why the disruption efficiency is reduced above this range has been attributed to the increased density of microbubbles formed close to the probe surface, which restricts the amount of ultrasonic energy transmitted into the fluid (Raso et al., 1999). Hence, the establishment of an optimal UA is essential for producing small droplets but without damaging the ultrasonic transducer (Santos et al., 2009). Ultrasound emulsifications of lemongrass oil (Salvia-Trujillo et al., 2013, 2014), D-limonene oil (Jafari et al., 2006, 2007), and corn oil (Tabibiazar et al., 2015) were performed at an UA of 100 μm (Table 9.4). Peshkovsky et al. (2013) for soybean oil NEs and Leong et al. (2009) for sunflower/canola oil NEs applied an UA of 90 and 30 μm, respectively (Table 9.3). A higher UA (119 or 173 μm) has been utilized to fabricate olive oil NEs (Kaltsa et al., 2013). Other researchers have reported the UA as a percentage, ranging from around 10% to 100% (Table 9.3). Abbas et al. (2013) recommended the use of higher UAs to emulsify viscous systems.

9.6.2.5 *Ultrasonic Frequency*

UFs between about 20 and 25 kHz are typically used to produce NEs using sonication methods (Table 9.3). There is an inverse correlation between the square of the UF and the power level, and so, the utilization of ultrasonic waves with higher power levels usually requires lower frequencies (Canselier et al., 2002). Although the application of high-power ultrasound can lead to the production of NEs with high physical stability, generation of considerable heat and shear forces can chemically destabilize them due to the development of free radicals of H· and OH· via pyrolysis of the water molecule (Paniwnyk, 2016). However, mechanical disruption and free-radical generation during sonication can be an effective means of creating new molecular structures with novel functionalities (Leong et al., 2017). The decreased magnitude of cavitation at a high UF can be attributed to the reduction in time required for the rapid development and collapse of microbubbles (Abbas et al., 2013).

9.6.2.6 *Ultrasonic Temperature*

The oil and water phases have been held at a range of different temperatures during the sonication process, ranging from about 4°C to 78°C (Table 9.3). However, this parameter is usually maintained at lower than 25°C (Table 9.3). Generally, the temperature of a NE tends to increase with increasing ultrasonic power level and ST. This temperature increase typically leads to a decrease in emulsion viscosity, interfacial tension, and Laplace pressure, which may facilitate droplet disruption, thereby leading to smaller droplets. However, the recoalescence rate of droplets can also increase due to a loss of emulsifier functionality at elevated temperatures, leading to an increase in droplet size (Jafari et al., 2008). Droplet growth at increased temperatures usually occurs in high-energy homogenization devices such as sonication and high-pressure homogenization

(Marie et al., 2002; Jafari et al., 2007). At higher temperatures, heat-labile active components present in NEs can be damaged due to thermal decomposition or generation of free radicals leading to lipid oxidation, protein denaturation, or polysaccharide depolymerization (McClements and Rao 2011). The rise in temperature that occurs during sonication can be minimized by optimizing the power level, reducing the ST, or using a cooling jacket (Jafari et al., 2007).

9.7 STORAGE STABILITY AND FUNCTIONALITY OF ULTRASOUND-MEDIATED NEs

After formation, it is often important that NEs remain physically and chemically stable during storage. NEs may physically break down through a variety of instability mechanisms, including creaming, sedimentation, flocculation, coalescence, and Ostwald ripening (Chapter 2). They may also chemically break down through mechanisms such as oxidation, reduction, or hydrolysis of key components.

9.7.1 Physical Storage Stability

Previous studies on the physical stability of NEs fabricated using sonication under different storage conditions are summarized in Table 9.4. The size and location of the droplets in NEs may change appreciably during storage (Gharibzahedi and Mohammadnabi, 2016). For certain systems, a significant increase in mean droplet size has been reported in both short-term (Rao and McClements, 2011; Periasamy et al., 2016; Gharibzahedi and Mohammadnabi, 2017) and long-term (Mirmajidi Hashtjin and Abbasi, 2015a; Li and Chiang, 2012; Hernández-Jaimes et al., 2013; Rebolleda et al., 2015) storage studies. However, in other systems, little change in droplet size has been reported during storage. These effects can be attributed to differences in the composition and structure of the initial NEs and to differences in storage conditions.

A good stability has been reported for NEs produced by sonication when stored for 10 days (olive and sunflower oils), 20–21 days (sunflower oil), 30 days (MCT, canola/nettle, basil, castor, and soybean oils), 40 days (corn oil), 45 days (nettle and corn/lemon oils), 60 days (MCT, lemon, flaxseed/safflower, wheat bran, thistle, and *E. plantagineum* L. seed oils), 90 days (eucalyptus, D-limonene, palm, palm kernel and soybean oils, and α-tocopherol-based oil), 180 days (*trans*-cinnamaldehyde and olive oils), and 12 months (soybean oil) (Table 9.4). No visual evidence of creaming and phase separation of NEs exposed to mechanical agitation, thermal treatment, and freeze-thaw treatment has been reported (Dey et al., 2012; Agrawal et al., 2017). However, storage at higher temperatures (25–45°C) led to a greater rate of droplet coalescence than storage at lower ones (4°C). Conversely, Li and Chiang (2012) reported that droplet growth due to Ostwald ripening was inversely related to temperature for of D-limonene oil-in-water NEs. This effect was attributed to the changes in diffusion coefficient, solubility, and interfacial tension with temperature. Hence, an increase in instability rate at the low temperature (4°C) was found in comparison with that at the high temperature (25°C).

9.7.2 Chemical Storage Stability

A few efforts have been made to study the oxidative stability of NEs produced by sonication during storage at different temperatures. Lipid oxidation tends to occur at oil droplet surfaces and depends on various factors, including lipid type, interfacial area, interfacial composition, antioxidant/prooxidant levels, and the presence of metal ion chelators (Cavazos-Garduño et al., 2015). Reducing the droplet size during sonication can promote lipid oxidation because of an increased surface area. The rate of lipid oxidation may also increase as the degree of unsaturation of the lipid increases, that is, polyunsaturated > monounsaturated > saturated lipids (Waraho et al., 2011; Gharibzahedi et al., 2012).

Some emulsifiers can sterically inhibit the ability of prooxidants (e.g., metal ions) in the aqueous phase to interact with the lipids inside the droplets, or they may act as antioxidants that can quench free radicals (Gharibzahedi, 2017). Sonication can promote lipid oxidation in NEs by causing an increase in temperature or by generating free radicals, particularly at high power levels or long STs. Consequently, the operating parameters used during sonication should be carefully selected to avoid lipid oxidation while still producing small droplets. Gharibzahedi (2017) reported a lower oxidation rate in nettle oil NEs stored at 4°C than those stored at 25°C for 45 days (Table 9.4). Conversely, Nejadmansouri et al. (2016) reported an opposite result so that fish oil NEs stored at refrigerator temperature showed a faster oxidation rate than those stored at room temperature (Table 9.4). The higher oxidative stability of NEs stored at ambient temperature may be due to a decrease in the water solubility of oxygen at higher temperatures (Coupland and McClements, 1996), changes in the direction (parallel or perpendicular) of the unsaturated lipid molecules at the droplet surfaces (Hiemenz and Rajagopalan, 1997), and changes in the molecular arrangements of fatty acids in the interior of droplets (Nejadmansouri et al., 2016).

9.7.3 Functionality of Ultrasound-Mediated NEs

Table 9.4 summarizes studies of the functional attributes of NEs fabricated using sonication. One of the most interesting functional attributes of these NE-based delivery systems is their antioxidant activities. An increase in antioxidant activity has been observed in a dose-dependent manner, so that the oxidation rate decreased at high contents of the encapsulated bioactive components (Alzorqi et al., 2016; Gharibzahedi, 2017). Reducing the droplet size to the nanoscale using sonication significantly increases the droplet surface area, which can improve their antimicrobial activity by promoting droplet interactions with negatively charged microbial cell membranes, leading to the leakage of intracellular constituents of the microorganisms (Sugumar et al., 2014). This fact has been especially observed by loading essential oils (e.g., lemongrass, nettle, eucalyptus, clove, *trans*-cinnamaldehyde, cinnamon, and basil) into NEs in the presence of phenolic compounds (such as chlorogenic and caffeic acids) or terpenes (such as neophytadiene and phytol) (Table 9.4).

It is also important to bear in mind that the emulsification of lemongrass essential oils using microfluidization compared with sonication could lead to the development of delivery systems with improved antimicrobial activity (Salvia-Trujillo et al., 2014). Overall, incorporating functional constituents into NE formed using sonication can considerably extend their shelf life because of the high antioxidant and antimicrobial contents. A superior inhibition in cancer cell proliferation and enhanced cell apoptosis in a dose- and time-dependent pattern was reported for NE-based delivery systems loaded with paclitaxel (Tiwari and Amiji, 2006), betulinic acid (Cavazos-Garduño et al., 2015), and black cumin essential oil (Periasamy et al., 2016). Easy penetration of the NE droplets toward the membranes of cancer cells with improved solubilization capacity of therapeutic hydrophobic compounds facilitates the rate of apoptosis, necrosis, and cell death (Periasamy et al., 2016). Activities of analgesic (Li et al., 2008; Tang et al., 2012b; Sandig et al., 2013), antiallodynic (Sandig et al., 2013), antiinflammatory (Tang et al., 2012b), antiemetic (Doh et al., 2013), antitumor (Li et al., 2008; Gao et al., 2011), tyrosinase inhibitory (Rebolleda et al., 2015), wound healing (Sugumar et al., 2014), and skin regeneration and moisturizing (Clares et al., 2014; Miastkowska et al., 2017) are some other functionalities of NEs fabricated by UAE (Table 9.4).

9.8 CONCLUSION AND FURTHER REMARKS

The present chapter overviews historical and current knowledge of ultrasonic homogenization, with a focus on the development of NE systems for application in the food, pharmaceutical, cosmetic, and other industrial applications. Advantages and disadvantages present in the ultrasound emulsification method in comparison with the high-energy methods were discussed from the viewpoint of sanitary conditions of processing, production costs (e.g., energy consumption, process time, and quantity of surfactant required), stability, droplet size, and process impacts on some functionalities. Ultrasonic homogenization is an easy, quick, aseptic, and cost-effective method that can lead to NEs with high physical stability. Moreover, NE delivery systems can be used to improve the antioxidant and antibacterial capacities of functional agents. There have been a number of advances in the technologies available to produce NEs using sonication that may lead to greater commercial applications, for example, continuous-flow high-power ultrasonic processors. The optimal control of formulation and process variables involved in ultrasonic homogenization provides manufacturers with an opportunity to develop NE delivery systems with high stability, small droplet sizes, and narrow polydispersity. Careful design of these systems can lead to the fabrication of NEs with high antioxidant and antimicrobial activities.

A number of technological challenges and research gaps still need to be resolved for the optimized production of NE-based delivery systems using sonication. Most scientific investigations have been conducted using laboratory-scale ultrasonic devices, and it is often difficult to scale them up for industrial applications. The utilization of pilot-scale sonicators along with the optimization of process variables can aid in design and scale-up for industrial facilities. Therefore, a fruitful area of future research would be the development and utilization of ultrasonic continuous-flow systems and evaluation of the best models for process design and optimization of operating conditions and formulations. Although encapsulation of various bioactive compounds in NEs produced by sonication has been examined, there is inadequate information on the toxicity, oxidative stability, and bioavailability rate of many water-insoluble functional ingredients encapsulated in these systems. A successful and attractive evolution in this field will be the design of smart emulsifier-free delivery systems containing two or several bioactive constituents with controlled release profiles in the body. Clinical and in vivo studies are thus essential to understand the absorption of lipid nutrients in NEs and their effects on a number of body functions.

REFERENCES

Abbas, S., Hayat, K., Karangwa, E., Bashari, M., Zhang, X., 2013. An overview of ultrasound-assisted food-grade nanoemulsions. Food Eng. Rev. 5, 139–157.

Abbas, S., Bashari, M., Akhtar, W., Li, W.W., Zhang, X., 2014. Process optimization of ultrasound-assisted curcumin nanoemulsions stabilized by OSA-modified starch. Ultrason. Sonochem. 21, 1265–1274.

Abismail, B., Canselier, J., Wilhelm, A., Delmas, H., Gourdon, C., 1999. Emulsification by ultrasound: drop size distribution and stability. Ultrason. Sonochem. 6, 75–83.

Agrawal, N., Maddikeri, G.L., Pandit, A.B., 2017. Sustained release formulations of citronella oil nanoemulsion using cavitational techniques. Ultrason. Sonochem. 36, 367–374.

Alzorqi, I., Ketabchi, M.R., Sudheer, S., Manickam, S., 2016. Optimization of ultrasound induced emulsification on the formulation of palm-olein based nanoemulsions for the incorporation of antioxidant β-D-glucan polysaccharides. Ultrason. Sonochem. 31, 71–84.

Amani, A., York, P., Chrystyn, H., Clark, B., 2010. Factors affecting the stability of nanoemulsions-use of artificial neural networks. Pharm. Res. 27, 37–45.

Calligaris, S., Plazzotta, S., Bot, F., Grasselli, S., Malchiodi, A., Anese, M., 2016. Nanoemulsion preparation by combining high pressure homogenization and high power ultrasound at low energy densities. Food Res. Int. 83, 25–30.

Camino, N.A., Pilosof, A.M.R., 2011. Hydroxypropylmethylcellulose at the oil–water interface. Part II. Submicron-emulsions as affected by pH. Food Hydrocoll. 25, 1051–1062.

Canselier, J.P., Delmas, H., Wilhelm, A.M., Abismail, B., 2002. Ultrasound emulsification-an overview. J. Dispers. Sci. Technol. 23, 333–349.

Carpenter, J., Saharan, V.K., 2017. Ultrasonic assisted formation and stability of mustard oil in water nanoemulsion: effect of process parameters and their optimization. Ultrason. Sonochem. 35, 422–430.

Cavazos-Garduño, A., Ochoa Flores, A.A., Serrano-Niño, J.C., Martínez-Sanchez, C.E., Beristain, C.I., García, H.S., 2015. Preparation of betulinic acid nanoemulsions stabilized by ω-3 enriched phosphatidylcholine. Ultrason. Sonochem. 24, 204–213.

Chalothorn, K., Warisnoicharoen, W., 2012. Ultrasonic emulsification of whey protein isolate-stabilized nanoemulsions containing omega-3 oil from plant seed. Am. J. Food Technol. 7, 532–541.

Chemat, F., Grondin, I., Costes, P., Moutoussamy, L., Sing, A.S.C., Smadja, J., 2004. High power ultrasound effects on lipid oxidation of refined sunflower oil. Ultrason. Sonochem. 11, 281–285.

Chemat, F., Huma, Z., Khan, M.K., 2011. Applications of ultrasound in food technology: processing, preservation and extraction. Ultrason. Sonochem. 18, 813–835.

Clares, B., Calpena, A.C., Parra, A., Abrego, G., Alvarado, H., Fangueiro, J.F., Souto, E.B., 2014. Nanoemulsions (NEs), liposomes (LPs) and solid lipid nanoparticles (SLNs) for retinyl palmitate: effect on skin permeation. Int. J. Pharm. 473, 591–598.

Coupland, J.N., McClements, D.J., 1996. Lipid oxidation in food emulsions. Trends Food Sci. Technol. 7, 83–91.

Cucheval, A., Chow, R.C.Y., 2008. A study on the emulsification of oil by power ultrasound. Ultrason. Sonochem. 15, 916–920.

de Araújo, S.C., de Mattos, A.C., Teixeira, H.F., Coelho, P.M., Nelson, D.L., de Oliveira, M.C., 2007. Improvement of *in vitro* efficacy of a novel schistosomicidal drug by incorporation into nanoemulsions. Int. J. Pharm. 337, 307–315.

Delmas, T., Piraux, H., Couffin, A.C., Texier, I., Vinet, F., Poulin, P., Cates, M.E., Bibette, J., 2011. How to prepare and stabilize very small nanoemulsions. Langmuir 27, 1683–1692.

Dey, T.K., Ghosh, S., Ghosh, M., Koley, H., Dhar, P., 2012. Comparative study of gastrointestinal absorption of EPA & DHA rich fish oil from nano and conventional emulsion formulation in rats. Food Res. Int. 49, 72–79.

Doh, H.J., Jung, Y., Balakrishnan, P., Cho, H.J., Kim, D.D., 2013. A novel lipid nanoemulsion system for improved permeation of granisetron. Colloids Surf. B: Biointerfaces 101, 475–480.

Feng, H., Yang, W., 2011. Ultrasonic processing. In: Zhang, H.Q., Barbosa-Cánovas, G.V., Balasubramaniam, V.M., Dunne, C.P., Farkas, D.F., Yuan, J.T.C. (Eds.), Nonthermal Processing Technologies for Food. Wiley-Blackwell and IFT Press, UK.

Freitas, S., Hielscher, G., Merkle, H.P., Gander, B., 2006. Continuous contact- and contamination-free ultrasonic emulsification—a useful tool for pharmaceutical development and production. Ultrason. Sonochem. 13, 76–85.

Gaikwad, S.G., Pandit, A.B., 2008. Ultrasound emulsification: effect of ultrasonic and physicochemical properties on dispersed phase volume and droplet size. Ultrason. Sonochem. 15, 554–563.

Ganta, S., Amiji, M., 2009. Coadministration of paclitaxel and curcumin in nanoemulsion formulations to overcome multidrug resistance in tumor cells. Mol. Pharm. 6, 928–939.

Gao, F., Zhang, Z., Bu, H., Huang, Y., Gao, Z., Shen, J., Zhao, C., Li, Y., 2011. Nanoemulsion improves the oral absorption of candesartan cilexetil in rats: performance and mechanism. J. Control. Release 149, 168–174.

Gharibzahedi, S.M.T., 2017. Ultrasound-mediated nettle oil nanoemulsions stabilized by purified jujube polysaccharide: process optimization, microbial evaluation and physicochemical storage stability. J. Mol. Liq. 234, 240–248.

Gharibzahedi, S.M.T., Mohammadnabi, S., 2016. Characterizing the novel surfactant-stabilized nanoemulsions of stinging nettle essential oil: thermal behaviour, storage stability, antimicrobial activity and bioaccessibility. J. Mol. Liq. 224, 1332–1340.

Gharibzahedi, S.M.T., Mohammadnabi, S., 2017. Effect of novel bioactive edible coatings based on jujube gum and nettle oil-loaded nanoemulsions on the shelf-life of Beluga sturgeon fillets. Int. J. Biol. Macromol. 95, 769–777.

Gharibzahedi, S.M.T., Mousavi, S.M., Hamedi, M., Khodaiyan, F., Razavi, S.H., 2012. Development of an optimal formulation for oxidative stability of walnut-beverage emulsion based on gum arabic and xanthan gum using response surface methodology. Carbohydr. Polym. 87, 1611–1619.

Gharibzahedi, S.M.T., Razavi, S.H., Mousavi, S.M., 2013. Ultrasound-assisted formation of the canthaxanthin emulsions stabilized by Arabic and xanthan gums. Carbohydr. Polym. 96, 21–30.

Gharibzahedi, S.M.T., Razavi, S.H., Mousavi, S.M., 2015a. Optimal development of a new stable nutraceutical nanoemulsion based on the inclusion complex of 2-hydroxypropyl-β-cyclodextrin with canthaxanthin accumulated by *Dietzia natronolimnaea* HS-1 using ultrasound-assisted emulsification. J. Dispers. Sci. Technol. 36, 614–625.

Gharibzahedi, S.M.T., Rostami, H., Yousefi, S., 2015b. Formulation design and physicochemical stability characterization of nanoemulsions of nettle (*Urtica Dioica*) essential oil using a model-based methodology. J. Food Process. Preserv. 39 (6), 2947–2958.

Ghasemi, S., Jafari, S.M., Assadpour, E., Khomeiri, M., 2017. Production of pectin-whey protein nano-complexes as carriers of orange peel oil. Carbohydr. Polym. 177 (Suppl. C), 369–377.

Ghasemi, S., Jafari, S.M., Assadpour, E., Khomeiri, M., 2018. Nanoencapsulation of d-limonene within nanocarriers produced by pectin-whey protein complexes. Food Hydrocoll. https://doi.org/10.1016/j.foodhyd.2017.09.030.

Ghosh, V., Mukherjee, A., Chandrasekaran, N., 2013a. Ultrasonic emulsification of food-grade nanoemulsion formulation and evaluation of its bactericidal activity. Ultrason. Sonochem. 20, 338–344.

Ghosh, V., Saranya, S., Mukherjee, A., Chandrasekaran, N., 2013b. Cinnamon oil nanoemulsion formulation by ultrasonic emulsification: investigation of its bactericidal activity. J. Nanosci. Nanotechnol. 13, 114–122.

Hernández-Jaimes, C., Fouconnier, B., Pérez-Alonso, C., Munguía-Guillén, J.L., Vernon-Carter, E.J., 2013. Antioxidant activity degradation, formulation optimization, characterization, and stability of *Equisetum arvense* extract nanoemulsion. J. Dispers. Sci. Technol. 34, 64–71.

Hiemenz, P.C., Rajagopalan, R., 1997. Principles of Colloid and Surface Chemistry, thith ed. Marcel Dekker, New York.

Higgins, D.M., Skauen, D.M., 1972. Influence of power on quality of emulsions prepared by ultrasound. J. Pharm. Sci. 61, 1567–1570.

Jadhav, A.J., Holkar, C.R., Karekar, S.E., Pinjari, D.V., Pandit, A.B., 2015. Ultrasound assisted manufacturing of paraffin wax nanoemulsions: process optimization. Ultrason. Sonochem. 23, 201–207.

Jafari, S.M., He, Y., Bhandari, B., 2006. Nano-emulsion production by sonication and microfluidization-a comparison. Int. J. Food Prop. 9, 475–485.

Jafari, S.M., He, Y., Bhandari, B., 2007. Production of sub-micron emulsions by ultrasound and microfluidization techniques. J. Food Eng. 82, 478–488.

Jafari, S.M., Assadpoor, E., He, Y., Bhandari, B., 2008. Re-coalescence of emulsion droplets during high-energy emulsification. Food Hydrocoll. 22, 1191–1202.

Jafari, S.M., Fathi, M., Mandala, I., 2015. Emerging Product Formation Food Waste Recovery: Processing Technologies and Industrial Techniques. Elsevier, New York, pp. 293–317.

Jena, S., Das, H., 2006. Modeling of particle size distribution of sonicated coconut milk emulsion: effect of emulsifiers and sonication time. Food Res. Int. 39, 606–611.

Jiang, Y., Pétrier, C., Waite, T.D., 2002. Kinetics and mechanisms of ultrasonic degradation of volatile chlorinated aromatics in aqueous solutions. Ultrason. Sonochem. 9, 317–323.

Kaci, M., Meziani, S., Arab-Tehrany, E., Gillet, G., Desjardins-Lavisse, I., Desobry, S., 2014. Emulsification by high frequency ultrasound using piezoelectric transducer: formation and stability of emulsifier free emulsion. Ultrason. Sonochem. 21, 1010–1017.

Kaltsa, O., Michon, C., Yanniotis, S., Mandala, I., 2013. Ultrasonic energy input influence on the production of sub-micron o/w emulsions containing whey protein and common stabilizers. Ultrason. Sonochem. 20, 881–891.

Kaltsa, O., Spiliopoulou, N., Yanniotis, S., Mandala, I., 2016. Stability and physical properties of model macro- and nano/submicron emulsions containing fenugreek gum. Food Hydrocoll. 61, 625–632.

Kaur, K., Kumar, R., Mehta, S.K., 2016. Formulation of saponin stabilized nanoemulsion by ultrasonic method and its role to protect the degradation of quercitin from UV light. Ultrason. Sonochem. 31, 29–38.

Kaur, K., Kaur, J., Kumar, R., Mehta, S.K., 2017a. Formulation and physiochemical study of α-tocopherol based oil in water nanoemulsion stabilized with non toxic, biodegradable surfactant: sodium stearoyl lactate. Ultrason. Sonochem. 38, 570–578. https://doi.org/10.1016/j.ultsonch.2016.08.026.

Kaur, K., Kumar, R., Arpita, Goel, S., Uppal, S., Bhatia, A., Mehta, S.K., 2017b. Physiochemical and cytotoxicity study of TPGS stabilized nanoemulsion designed by ultrasonication method. Ultrason. Sonochem. 34, 173–182.

Kentish, S., Wooster, T.J., Ashokkumar, M., Balachandran, S., Mawson, R., Simons, L., 2008. The use of ultrasonics for nanoemulsion preparation. Innovative Food Sci. Emerg. Technol. 9, 170–175.

Kumar, A., Kumaresan, T., Pandit, A.B., Joshi, J.B., 2006. Characterization of flow phenomena induced by ultrasonic horn. Chem. Eng. Sci. 61, 7410–7420.

Kumaresan, T., Kumar, A., Pandit, A.B., Joshi, J.B., 2006. Modeling flow pattern induced by ultrasound: the influence of modeling approach and turbulence models. Ind. Eng. Chem. Res. 46, 2936–2950.

Landfester, K., Eisenblätter, J., Rothe, R., 2004. Preparation of polymerizable miniemulsions by ultrasonication. J. Coat. Technol. Res. 1, 65–68.

Lee, H., Yildiz, G., dos Santos, L.C., Jiang, S., Andrade, J.E., Engeseth, N.J., Feng, H., 2016. Soy protein nano-aggregates with improved functional properties prepared by sequential pH treatment and ultrasonication. Food Hydrocoll. 55, 200–209.

Leong, T.S.H., Wooster, T.J., Kentish, S.E., Ashokkumar, M., 2009. Minimising oil droplet size using ultrasonic emulsification. Ultrason. Sonochem. 16, 721–727.

Leong, T.S.H., Martin, G.J.O., Ashokkumar, M., 2017. Ultrasonic encapsulation—a review. Ultrason. Sonochem. 35, 605–614.

Li, P.H., Chiang, B.H., 2012. Process optimization and stability of d-limonene-in-water nanoemulsions prepared by ultrasonic emulsification using response surface methodology. Ultrason. Sonochem. 19, 192–197.

Li, M.K., Fogler, H.S., 1978a. Acoustic emulsification. Part 1. The instability of oil-water interface to form the initial droplets. J. Fluid Mech. 88, 499–511.

Li, M.K., Fogler, H.S., 1978b. Acoustic emulsification. Part 2: breakup of the large primary oil droplets in a water medium. J. Fluid Mech. 88, 513–528.

Li, F., Wang, T., He, H., Tang, X., 2008. The properties of bufadienolides-loaded nanoemulsion and submicro-emulsion during lyophilization. Int. J. Pharm. 349, 291–299.

Lin, C.Y., Chen, L.W., 2008. Comparison of fuel properties and emission characteristics of two and three-phase emulsions prepared by ultrasonically vibrating and mechanically homogenizing emulsification methods. Fuel 87, 2154–2161.

Loong, N.C., Basri, M., Fang, L.F., Fard Masoumi, H.R., Tripathy, M., Abedi Karjiban, R., Abdul-Malek, E., 2014. Comparison of Box–Behnken and central composite designs in optimization of fullerene loaded palm-based nano-emulsions for cosmeceutical application. Ind. Crop. Prod. 59, 309–317.

Lu, W.C., Chiang, B.H., Huang, D.W., Li, P.H., 2014. Skin permeation of D-limonene-based nanoemulsions as a transdermal carrier prepared by ultrasonic emulsification. Ultrason. Sonochem. 21, 826–832.

Ma, Y.F., Hsu, C.C., 1999. Performance of sonication and microfluidization for liquid-liquid emulsification. Pharm. Dev. Technol. 4, 233–240.

Marie, P., Perrier-Cornet, J.M., Gervais, P., 2002. Influence of major parameters in emulsification mechanisms using a high-pressure jet. J. Food Eng. 53, 43–51.

Mason, T.J., Paniwnyk, L., Lorimer, J.P., 1996. The uses of ultrasound in food technology. Ultrason. Sonochem. 3, 253–260.

McClements, D.J., 1994. Ultrasonic determination of depletion flocculation in oil-in-water emulsions containing a non-ionic surfactant. Colloids Surf. A Physicochem. Eng. Asp. 90, 25–35.

McClements, D.J., 2011. Edible nanoemulsions: fabrication, properties, and functional performance. Soft Matter 7, 2297–2316.

McClements, D.J., Rao, J., 2011. Food-grade nanoemulsions: formulation, fabrication, properties, performance, biological fate, and potential toxicity. Crit. Rev. Food Sci. Nutr. 51, 285–330.

Mehrnia, M.A., Jafari, S.M., Makhmal-Zadeh, B.S., Maghsoudlou, Y., 2016. Crocin loaded nano-emulsions: factors affecting emulsion properties in spontaneous emulsification. Int. J. Biol. Macromol. 84, 261–267.

Mehrnia, M.-A., Jafari, S.-M., Makhmal-Zadeh, B.S., Maghsoudlou, Y., 2017. Rheological and release properties of double nano-emulsions containing crocin prepared with Angum gum, Arabic gum and whey protein. Food Hydrocoll. 66, 259–267.

Miastkowska, M.A., Banach, M., Pulit-Prociak, J., Sikora, E.S., Głogowska, A., Zielina, M., 2017. Statistical analysis of optimal ultrasound emulsification parameters in thistle-oil nanoemulsions. J. Surfactant Deterg. 20, 233–246.

Mirmajidi Hashtjin, A., Abbasi, S., 2015a. Nano-emulsification of orange peel essential oil using sonication and native gums. Food Hydrocoll. 44, 40–48.

Mirmajidi Hashtjin, A., Abbasi, S., 2015b. Optimization of ultrasonic emulsification conditions for the production of orange peel essential oil nanoemulsions. J. Food Sci. Technol. 52, 2679–2689.

Moghimi, R., Aliahmadi, A., Rafati, H., 2017. Ultrasonic nanoemulsification of food grade trans-cinnamaldehyde: 1,8-Cineol and investigation of the mechanism of antibacterial activity. Ultrason. Sonochem. 35, 415–421.

Muthukumaran, S., Kentish, S.E., Stevens, G.W., Ashokkumar, M., 2006. Application of ultrasound in membrane separation processes: a review. Rev. Chem. Eng. 22, 155–194.

Nejadmansouri, M., Hosseini, S.M.H., Niakosari, M., Yousefi, G.H., Golmakani, M.T., 2016. Physicochemical properties and oxidative stability of fish oil nanoemulsions as affected by hydrophilic lipophilic balance, surfactant to oil ratio and storage temperature. Colloids Surf. A Physicochem. Eng. Asp. 506, 821–832.

O'Sullivan, J., Murray, B., Flynn, C., Norton, I., 2015. Comparison of batch and continuous ultrasonic emulsification processes. J. Food Eng. 167, 114–121.

Paniwnyk, L., 2016. Applications of ultrasound in processing of liquid foods (a review). Ultrason. Sonochem. 38, 794–806. https://doi.org/10.1016/j.ultsonch.2016.12.025.

Periasamy, V.S., Athinarayanan, J., Alshatwi, A.A., 2016. Anticancer activity of an ultrasonic nanoemulsion formulation of *Nigella sativa* L. essential oil on human breast cancer cells. Ultrason. Sonochem. 31, 449–455.

Peshkovsky, A.S., Bystryak, S., 2014. Continuous-flow production of a pharmaceutical nanoemulsion by high-amplitude ultrasound: process scale-up. Chem. Eng. Process. 82, 132–136.

Peshkovsky, A.S., Peshkovsky, S.L., Bystryak, S., 2013. Scalable high power ultrasonic technology for the production of translucent nanoemulsions. Chem. Eng. Process. 69, 77–82.

Ramisetty, K.A., Pandit, A.B., Gogate, P.R., 2015. Ultrasound assisted preparation of emulsion of coconut oil in water: understanding the effect of operating parameters and comparison of reactor designs. Chem. Eng. Process. 88, 70–77.

Rao, J., McClements, D.J., 2011. Formation of flavor oil microemulsions, nanoemulsions and emulsions: influence of composition and preparation method. J. Agric. Food Chem. 59, 5026–5035.

Raso, J., Mañas, P., Pagán, R., Sala, F.J., 1999. Influence of different factors on the output power transferred into medium by ultrasound. Ultrason. Sonochem. 5, 157–162.

Rebolleda, S., Sanz, M.T., Benito, J.M., Beltrán, S., Escudero, I., San-José, M.L.G., 2015. Formulation and characterisation of wheat bran oil-in-water nanoemulsions. Food Chem. 167, 16–23.

Richards, W.T., 1929. The chemical effects of high frequency sound waves I. A study of emulsifying action. J. Am. Chem. Soc. 51, 1724–1729.

Riesz, P., Berdahl, D., Christman, C.L., 1985. Free radical generation by ultrasound in aqueous and nonaqueous solutions. Environ. Health Perspect. 64, 233–252.

Salvia-Trujillo, L., Rojas-Graü, A., Soliva-Fortuny, R., Martín-Belloso, O., 2013. Physicochemical characterization of lemongrass essential oil-alginate nanoemulsions: effect of ultrasound processing parameters. Food Bioprocess Technol. 6, 2439–2446.

Salvia-Trujillo, L., Rojas-Graü, M.A., Soliva-Fortuny, R., Martín-Belloso, O., 2014. Impact of microfluidization or ultrasound processing on the antimicrobial activity against *Escherichia coli* of lemongrass oil-loaded nanoemulsions. Food Control 37, 292–297.

Sandig, A.G., Campmany, A.C.C., Campos, F.F., Villena, M.J.M., Naveros, B.C., 2013. Transdermal delivery of imipramine and doxepin from newly oil-in-water nanoemulsions for an analgesic and anti-allodynic activity: development, characterization and *in vivo* evaluation. Colloids Surf. B: Biointerfaces 103, 558–565.

Santos, H.M., Lodeiro, C., Capelo-Martinez, J.L., 2009. The power of ultrasound. In: Ultrasound in Chemistry. Wiley-VCH Verlag GmbH & Co. KGaA, Weinheim, pp. 1–16.

Shahavi, M.H., Hosseinia, M., Jahanshahi, M., Meyer, R.L., Najafpour Darzi, G., 2016. Clove oil nanoemulsion as an effective antibacterial agent: Taguchi optimization method. Desalin. Water Treat. 57, 18379–18390.

Shamsara, O., Jafari, S.M., Muhidinov, Z.K., 2017. Fabrication, characterization and stability of oil in water nano-emulsions produced by apricot gum-pectin complexes. Int. J. Biol. Macromol. 103 (Suppl. C), 1285–1293.

Shamsara, O., Muhidinov, Z.K., Jafari, S.M., Bobokalonov, J., Jonmurodov, A., Taghvaei, M., Kumpugdee-Vollrathd, M., 2015. Effect of ultrasonication, pH and heating on stability of apricotgum–lactoglobuline two layer nanoemulsions. Int. J. Biol. Macromol. 81, 1019–1025.

Shi, Y., Li, H., Li, J., Zhi, D., Zhang, X., Liu, H., Wang, H., Li, H., 2015. Development, optimization and evaluation of emodin loaded nanoemulsion prepared by ultrasonic emulsification. J. Drug Deliv. Sci. Technol. 27, 46–55.

Sivakumar, M., Tang, S.Y., Tan, K.W., 2014. Cavitation technology—a greener processing technique for the generation of pharmaceutical nanoemulsions. Ultrason. Sonochem. 21, 2069–2083.

Sugumar, S., Ghosh, V., Nirmala, M.J., Mukherjee, A., Chandrasekaran, N., 2014. Ultrasonic emulsification of eucalyptus oil nanoemulsion: antibacterial activity against *Staphylococcus aureus* and wound healing activity in Wistar rats. Ultrason. Sonochem. 21, 1044–1049.

Tabibiazar, M., Davaran, S., Hashemi, M., Homayonirad, A., Rasoulzadeh, F., Hamishehkar, H., Mohammadifar, M.A., 2015. Design and fabrication of a food-grade albumin-stabilized nanoemulsion. Food Hydrocoll. 44, 220–228.

Tadros, T., Izquierdo, P., Esquena, J., Solans, C., 2004. Formation and stability of nanoemulsions. Adv. Colloid Interf. Sci. 108–109, 303–318.

Takegami, S., Kitamura, K., Kawada, H., Matsumoto, Y., Kitade, T., Ishida, H., Nagata, C., 2008. Preparation and characterization of a new lipid nano-emulsion containing two cosurfactants, sodium palmitate for droplet size reduction and sucrose palmitate for stability enhancement. Chem. Pharm. Bull. 56, 1097–1102.

Tal-Figiel, B., 2007. The formation of stable w/o, o/w, w/o/w cosmetic emulsions in an ultrasonic field. Chem. Eng. Res. Des. 85, 730–734.

Tamjidi, F., Shahedi, M., Varshosaz, J., Nasirpour, A., 2014. Design and characterization of astaxanthin-loaded nanostructured lipid carriers. Innovative Food Sci. Emerg. Technol. 26, 366–374.

Tan, S.F., Fard Masoumi, H.R., Abedi Karjiban, R., Stanslas, J., Kirby, B.P., Basri, M., Basri, H.B., 2016. Ultrasonic emulsification of parenteral valproic acid-loaded nanoemulsion with response surface methodology and evaluation of its stability. Ultrason. Sonochem. 29, 299–308.

Tang, S.Y., Manickam, S., Wei, T.K., Nashiru, B., 2012a. Formulation development and optimization of a novel Cremophore EL-based nanoemulsion using ultrasound cavitation. Ultrason. Sonochem. 19, 330–345.

Tang, S.Y., Sivakumar, M., Ng, A.M.H., Shridharan, P., 2012b. Anti-inflammatory and analgesic activity of novel oral aspirin-loaded nanoemulsion and nano multiple emulsion formulations generated using ultrasound cavitation. Int. J. Pharm. 430, 299–306.

Tang, S.Y., Shridharan, P., Sivakumar, M., 2013. Impact of process parameters in the generation of novel aspirin nanoemulsions—comparative studies between ultrasound cavitation and microfluidizer. Ultrason. Sonochem. 20, 485–497.

Thompson, L., Doraiswamy, L., 1999. Sonochemistry: science and engineering. Ind. Eng. Chem. Res. 38, 1215–1249.

Tiwari, S.B., Amiji, M.M., 2006. Improved oral delivery of paclitaxel following administration in nanoemulsion formulations. J. Nanosci. Nanotechnol. 6, 3215–3221.

van Zwieten, R., Verhaagen, B., Schroën, K., Rivas, D.F., 2017. Emulsification in novel ultrasonic cavitation intensifying bag reactors. Ultrason. Sonochem. 36, 446–453.

Vyas, T.K., Shahiwala, A., Amiji, M.M., 2008. Improved oral bioavailability and brain transport of Saquinavir upon administration in novel nanoemulsion formulations. Int. J. Pharm. 347, 93–101.

Waraho, T., McClements, D.J., Decker, E.A., 2011. Impact of free fatty acid concentration and structure on lipid oxidation in oil-in-water emulsions. Food Chem. 129, 854–859.

Wilhelm, A.M., Delmas, H., Gourdon, C., 1999. Emulsification by ultrasound: drop size distribution and stability. Ultrason. Sonochem. 6, 75–83.

William, B., Richards, I., 1929. The chemical effects of high frequency sound waves II. A study of emulsifying action. J. Am. Chem. Soc. 51, 1724–1729.

Wood, W., Loomis, A.L., 1927. The physical and biological effects of high frequency sound waves of great intensity. Philos. Mag. J. Sci. 44, 417–436.

Wooster, T.J., Golding, M., Sanguansri, P., 2008. Impact of oil type on nanoemulsion formation and Ostwald ripening stability. Langmuir 24, 12758–12765.

Wulff-Perez, M., Torcello-Gomez, A., Galvez-Ruiz, M.J., Martin-Rodriguez, A., 2009. Stability of emulsions for parenteral feeding: preparation and characterization of o/w nanoemulsions with natural oils and Pluronic F68 as surfactant. Food Hydrocoll. 23, 1096–1102.

Chapter 10

Fabrication of Nanoemulsions by Membrane Emulsification

Goran T. Vladisavljević
Loughborough University, Loughborough, United Kingdom

Chapter Outline

Nanoemulsions. https://doi.org/10.1016/B978-0-12-811838-2.00010-2

ABBREVIATIONS

Chemicals and materials

AOT	docusate sodium salt
APTES	(3-aminopropyl)triethoxysilane
APTMS	(3-aminopropyl)trimethoxysilane
Brij® 35	polyethylene glycol dodecyl ether
CA	cellulose acetate
CATO	glyceryl behenate
CR-310	condensed ricinoleic acid tetraglycerin ester
CR-500	hexaglycerol polyricinoleate
DCM	dichloromethane
DD	dodecane
DOX	doxorubicin
EA	ethyl acetate
GA	glutaraldehyde
Gelucire 44/14	lauroyl polyoxylglycerides
HS-9	HS-11stearic acid hexaglycerol esters
L-7D	lauric acid decaglycerol ester
Labrasol®	lauroyl polyoxylglycerides
Lipoid® E80	egg yolk lecithin manufactured by Lipoid GmbH
Lipoid S75 and Lipoid S100	soybean phospholipids manufactured by Lipoid GmbH
MCE	mixed cellulose ester
MCT	medium-chain triglycerides (Miglyol® 812)
ML-310	tetraglycerol monolaurate
MCA	mixed cellulose acetate
ODS	octadecyldimethylchlorosilane
PAH	poly(allylamine hydrochloride)
PBS	phosphate-buffered saline
PC	polycarbonate
PE	polyester
PES	polyethersulfone
PGPR	polyglycerol polyricinoleate
PG10-L	polyglyceryl-10 laurate
PLA	polylactic acid or polylactide
PLGA	poly(lactic-*co*-glycolic acid)
PLLA	poly(L-lactic acid)
PO-500	hexaglycerin penta ester

Pol	poloxamer 188 (Pluronic® F-68)
PrOH	2-propanol
PS	polysulfone
PTFE	polytetrafluoroethylene
PVA	polyvinyl alcohol
PVDF	polyvinylidene fluoride
Q_{10}	ubidecarenone
SBO	soybean oil
SDS	sodium dodecyl sulfate
SL	sucrose laurate
Span 20	sorbitan monolaurate
Span 80	sorbitan monooleate
SPG	Shirasu porous glass
S-24D	S-28Dstearic acid decaglycerol esters
RIF	rifampicin
TM	trimyristin
TMS	trimethylchlorosilane
Transcutol HP	highly purified diethylene glycol monoethyl ether
Tween 20	polyoxyethylene (20) sorbitan monolaurate
Tween 80	polyoxyethylene (20) sorbitan monooleate
Tyl	tyloxapol

Emulsions/dispersions

O/W	oil in water
O/W/O	oil in water in oil
S/O/W	solid in oil in water
W/O	water in oil
W/O/W	water in oil in water

Other

DGU	droplet generation unit
DME	direct membrane emulsification
ME	membrane emulsification
HLB	hydrophile-lipophile balance
NP	nanoparticle
SEM	scanning electron microscopy
UCST	upper critical solution temperature
XMT	X-ray microtomography
XL	cross-linker

Symbols

A_m	cross-sectional area of membrane, m^2
A	amplitude of membrane oscillation, m
a	ratio of droplet size in premix and pore size, –
b	height of stirrer blade, m

Ca	capillary number, –
D	stirrer diameter, m
D_{cell}	internal diameter of stirred cell, m
D_1	hydraulic diameter of dispersed-phase channel, m
D_2	hydraulic diameter of continuous-phase channel, m
d_{50}	diameter of the drop that 50% of a sample's volume is smaller than, m
d_d	drop diameter, m
d_{pm}	drop diameter in premix, m
d_{mi}	inner diameter of membrane tube, m
d_p	pore diameter, m
E_a	activation energy for diffusion during phase separation, J mol^{-1}
F_{ca}	capillary force, N
F_d	drag force, N
f	frequency of membrane oscillation, Hz
h	height of cross flow channel, m
J	transmembrane flux, $m^3\ m^{-2}\ s^{-1}$
K	constant in Eq. (10.1)
K'	proportionality constant between pore size and drop size, –
K''	proportionality constant in Eq. (10.9)
m_m	total mass of dry membrane, kg
N	total number of pores in membrane, –
n_b	number of stirrer blades, –
Q	volume flow rate, $m^3\ s^{-1}$
P	pressure, Pa
P_{cap}	capillary pressure, Pa
R	universal gas constant (8314 kJ $kmol^{-1}\ K^{-1}$)
Re	rotating Reynolds number of continuous phase, –
R_m	hydraulic resistance of membrane
r_b	inner radius of a cylinder in which the membrane is rotating
r_{mo}	outer radius of membrane tube, m
r_p	pore radius, m
r_{trans}	radial distance from the axis of rotation at which the shear is maximal, m
T	temperature, K
T_m	melting point, K
T_1	initial temperature of mother glass, K
T_2	phase-separation temperature, K
t	time, s
V_p	total volume of all pores in a membrane, m^3
W	width of cross flow channel, m
γ	interfacial tension between water and oil phase, N m^{-1}
δ	membrane thickness, m
ε	membrane porosity, –
η	viscosity, Pa s

ρ_c	density of continuous phase, kg m^{-3}
ξ	pore tortuosity, –
ϕ	dispersed-phase content, –
θ	contact angle, rad
τ_w	shear stress on surface, Pa
ω	angular velocity of membrane or stirrer, rad s^{-1}

Subscripts

c	continuous phase
cr	critical
d	dispersed phase
e	emulsion
o	oil phase
w	water phase
skin	skin layer
sup	support layer
sym	symmetrical
p	pore

10.1 INTRODUCTION

Synthetic microporous membranes are mainly used for separation and purification purposes and to assist in chemical or biochemical conversions (membrane reactors/bioreactors). In a membrane filtration process such as ultrafiltration or microfiltration, the feed stream is divided into two product streams of different compositions, retentate and permeate (Fig. 10.1A), as a result of selective permeation of the feed-stream components through the membrane (Mulder, 1996).

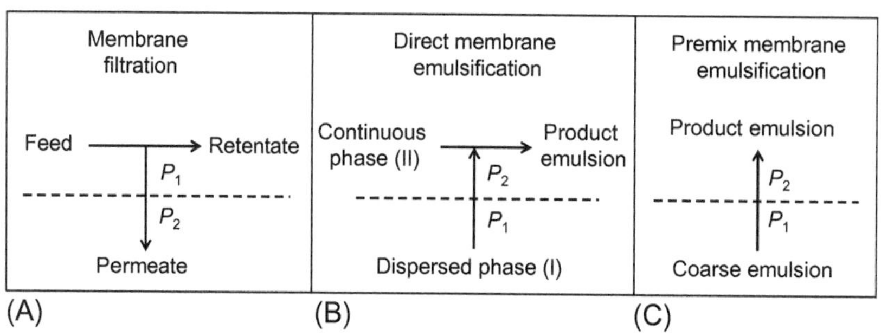

FIG. 10.1 A comparison between membrane filtration (ultrafiltration, microfiltration, and reverse osmosis) and membrane emulsification, where $P_1 > P_2$. In a membrane filtration process (A), feed stream (a dispersion or solution) is split into two product streams of different compositions (Mulder, 1996). In a direct membrane emulsification (B), two immiscible liquids are brought into contact by injecting one liquid (I) through the membrane into another liquid (II) to form an emulsion. Premix membrane emulsification (C) involves passing the coarse emulsion through the membrane, which results in the droplet-size reduction.

A shear force is applied at the retentate/membrane interface to limit concentration polarization and accumulation of the rejected solids on the high-pressure side of the membrane, thereby improving the transmembrane flux. In the last two decades, membrane dispersion processes are gaining in popularity, as a result of rising global interest in microfluidic technologies and significant progress in microfabricated membranes. In a direct membrane dispersion process, liquid phase I is injected through the membrane into liquid phase II for the purpose of (i) mixing two miscible solvents, for example, water and ethanol, which is usually followed by nanoprecipitation of solute(s) initially dissolved in liquid I and insoluble in liquid II (Chen et al., 2004) or (ii) forming droplets or bubbles of phase I into phase II. The creation of an emulsion directly from two separate liquids by passing one liquid (the dispersed or internal phase) through the membrane into another immiscible liquid (the continuous or external phase) is defined as direct membrane emulsification (direct ME), Fig. 10.1B (Nakashima et al., 1991, 2000). The reduction in size of the droplets in an already existing emulsion by passing a coarse emulsion (premix or pre-emulsion) through the membrane is defined as premix membrane emulsification (premix ME), Fig. 10.1C (Suzuki et al., 1996).

10.2 DIRECT ME VS. PREMIX ME

In direct ME, the dispersed phase is injected through the membrane into the continuous phase at controlled rate, which leads to the formation of droplets at the membrane/continuous-phase interface. A shear force is applied at the membrane/continuous-phase interface to facilitate the detachment of the drops from the membrane surface by providing a drag force on the forming droplets that counteracts the capillary force. Hydrophobic membranes are needed to produce water-in-oil (W/O) and oil-in-water-in-oil ($O_1/W/O_2$) emulsions (Cheng et al., 2008; Jing et al., 2006), whereas hydrophilic membranes are required to prepare oil-in-water (O/W) and water-in-oil-in-water ($W_1/O/W_2$) emulsions (Fig. 10.2A). If the driving pressure is lower than the critical (capillary) pressure, no drop generation will occur.

In premix ME, a pre-emulsion prepared by gentle mechanical stirring is forced through the membrane (Suzuki et al., 1996) or a packed bed of uniform particles (van der Zwan et al., 2008; Yasuda et al., 2010; Laouini et al., 2014) to reduce the droplet size as a result of shear stresses with the membrane pores. Hydrophobic and hydrophilic membranes are needed to produce W/O and O/W emulsions, respectively (Fig. 10.2B). If the applied pressure difference is lower than the critical pressure, the membrane will reject the droplets while allowing a pure continuous-phase liquid to pass through, which will lead to separation of the emulsion into a droplet-free continuous phase and concentrated emulsion (Koltuniewicz et al., 1995; Park et al., 2001).

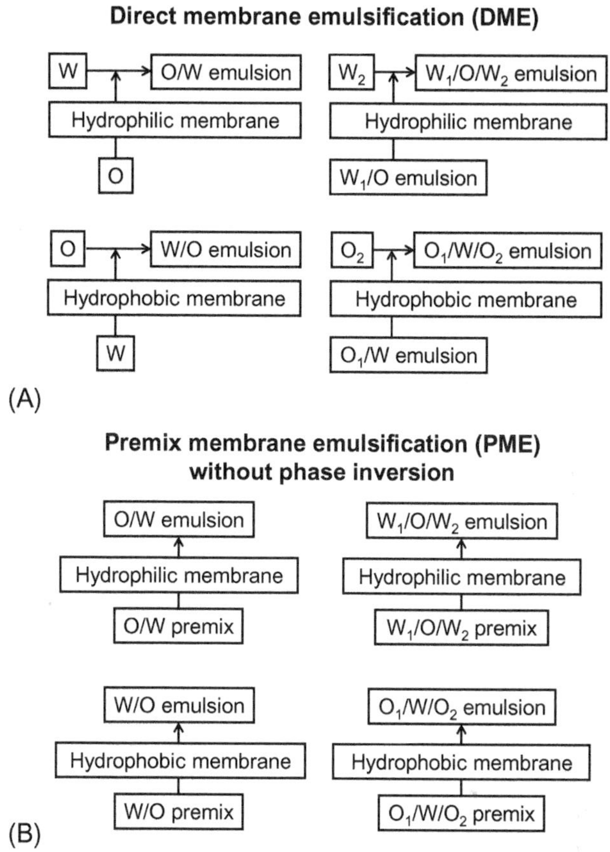

FIG. 10.2 Emulsion types that can be produced using membrane emulsification and the required wettability of the membrane.

If the membrane surface is wet by the dispersed-phase liquid of the pre-emulsion, the rate of droplet coalescence in the membrane pores is faster than the rate of droplet breakup, which will lead to phase inversion in the emulsion passing through the membrane (Suzuki et al., 1999) or its separation into two distinct phases (Kukizaki and Goto, 2008). For the phase separation to occur, the pre-emulsion should contain a blend of low- and high-HLB surfactants, for example, tetraglycerol monolaurate, ML-310 (HLB $\approx$11) dissolved in the aqueous phase, and hexaglycerol polyricinoleate, CR-500 (HLB $<$ 1) dissolved in the oil phase (Suzuki et al., 1999). A W/O emulsion was inverted into a O/W emulsion by passing through a hydrophilic membrane, whereas a phase-inverted W/O emulsion was made by passing an O/W or W/O/W emulsion through a hydrophobic membrane (Suzuki et al., 1999; Kawashima et al., 1991). Premix membrane emulsification with phase separation can be used to prepare emulsions with high disperse-phase volume fractions ($\phi > 0.8$) from pre-emulsions with low disperse-phase volume fractions ($\phi < 0.2$). The highest disperse-phase volume fraction achieved in phase-separated emulsions prepared by membrane-induced phase inversion was 0.90 and 0.84 for O/W and W/O emulsion type, respectively (Suzuki et al., 1999).

The major limitation of direct ME is a low transmembrane flux, which requires recirculation of the product stream for long periods of time to reach the desired dispersed-phase content. The advantages of premix over direct ME are smaller drop sizes and higher transmembrane fluxes that can be achieved with the same membrane. On the other hand, in premix ME, more severe membrane fouling and broader particle-size distributions can be expected.

10.3 COMPARISON BETWEEN MEMBRANE EMULSIFICATION AND MICROFLUIDIC EMULSIFICATION

In microfluidic devices, the drops are usually made by injecting the dispersed phase through a perpendicular channel into a cross flowing stream of the continuous phase flowing through the main channel (T junction, Fig. 10.3B) (Thorsen et al., 2001) or forcing a coaxial jet of two immiscible liquids to flow through a narrow orifice (flow-focusing device, Fig. 10.3C) (Anna et al., 2003). The drops generated in microfluidic systems are highly uniform with a variation

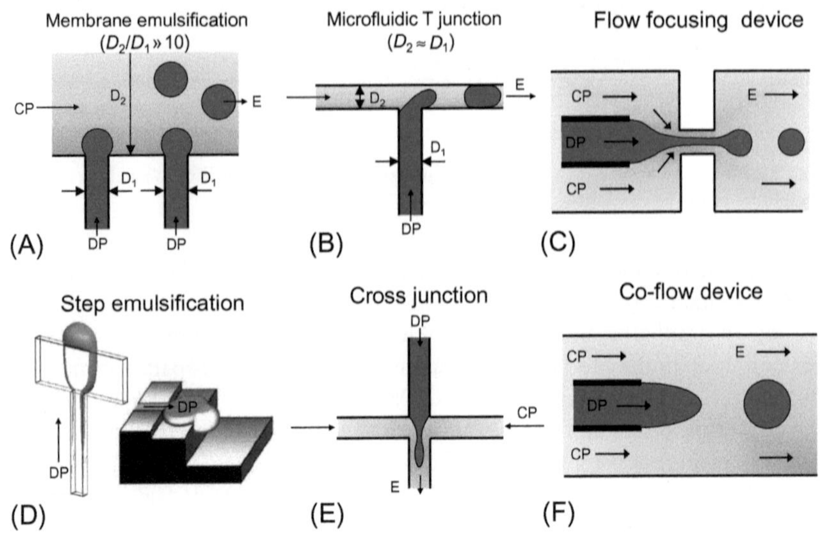

FIG. 10.3 Comparison of droplet generation in direct membrane emulsification (ME) and microfluidic devices. D_1 and D_2 are the hydraulic diameters of the channel for the dispersed phase and continuous phase, respectively. *(Adapted from Vladisavljević, G.T., Khalid, N., Neves, M.A., Kuroiwa, T., Nakajima, M., Uemura, K., Ichikawa, S., Kobayashi, I., 2013. Industrial lab on a chip: design, applications and scale up for drug discovery and delivery. Adv. Drug Deliv. Rev., 65, 1626–1663.)*

of their sizes in dripping regime of less than 3%, and the span values, $(d_{90}-d_{10})/d_{50}$, are less than 0.05, fulfilling the definition of a monodisperse distribution. Although the drop generation frequency can exceed 10,000 Hz (Yobas et al., 2006), the volume flow rate of the dispersed phase is very low, 10^{-2}–10 mL h^{-1}, due to only one droplet-forming channel, thus only one drop generation unit (DGU). On the other hand, depending on the pore size, Shirasu porous glass (SPG) membrane has 10^9–10^{14} m^{-2} pores, which are DGUs in ME. Although the percentage of droplet forming (active) pores in SPG ME is only 1%–10% (Vladisavljević et al., 2004), the total number of active DGUs is still 10^7–10^{13} m^{-2}.

The ratio of the hydraulic diameter of the continuous-phase channel to that of the dispersed-phase channel, D_2/D_1, is much greater than 100 in direct ME and 1–10 in a T-junction geometry (Fig. 10.3A and B). Consequently, in a T junction, the formed drops occupy almost the entire cross section of the main channel at low capillary numbers of the continuous phase ($Ca_c < 0.015$), and droplet breakup is caused by the pressure gradient across the droplet, while it is protruding into the main channel, which is known as the plugging-squeezing mechanism (Seemann et al., 2012; Vladisavljević et al., 2012). In direct ME, the size of the main channel is much greater than the size of the formed drops, and thus, the pressure gradient across the emerging droplet is negligible, preventing the occurrence of the squeezing regime. In addition, at high capillary numbers of the continuous phase ($Ca_c > 0.3$), T-junction operates in the jetting regime characterized by the formation of a long thread of the dispersed phase prior to jet breakup. In direct ME, the main channel is so large that a shear force from the surrounding continuous phase is not sufficiently high to pull the emerging dispersed phase into a long jet. Instead, at high dispersed-phase flow rates, direct ME operates in the continuous outflow regime (Kobayashi et al., 2003) characterized by the formation of big droplets that retain its spherical shape during the entire growth phase without the formation of a long thread, as evidenced by real-time microscopic observations of drop generation on the membrane surface (Vladisavljević et al., 2007, 2011).

The main advantage of ME compared with microfluidic emulsification is in much higher droplet throughputs that can be achieved. The volume flow rate of the dispersed phase in premix ME is above 10^4 mL h^{-1} for SPG membrane with a pore size of 11 μm and a cross-sectional area of 4 cm^2, leading to the generation of oil drops with a diameter of 10 μm (Vladisavljević et al., 2004). Considering that commercial membrane modules have a membrane area of 10–100 m^2, while the scale-up of microfluidic drop generators is challenging due to significant pressure drop in microfluidic channels and difficulties in providing uniform flow distribution in long microchannel networks, it is clear that membrane devices are more suitable for high-throughput applications. On the other hand, the variation of droplet sizes in ME is above 10% with span factor between 0.3 and 0.6. Another advantage of microfluidic over membrane emulsification is in the ability to produce droplets with a complex internal structure (Vladisavljević, 2016) and to manipulate (move, trap, merge, split, and sort) individual drops with high precision after generation. In ME, only one type of dispersed-phase liquid is injected through the membrane, and droplets are collected in a bulk channel. Therefore, the generated droplets cannot be individually manipulated, and multiple emulsions can only be formed via a two-step emulsification process involving primary emulsification in a conventional homogenizer and secondary emulsification in membrane homogenizer. On the other hand, in microfluidic drop generators, different dispersed-phase liquids can be simultaneously supplied through different entry channels, enabling generation of complex drops (double, triple, quadruple, and quintuple) in a single chip and the multiple emulsion drops with controlled number of inner drops, with distinct inner drops, and with two or three distinct surface regions (Janus and ternary drops).

10.4 COMPARISON BETWEEN MEMBRANE AND CONVENTIONAL HOMOGENIZATION

Compared with high-pressure valve homogenization, high-shear rotor-stator mixing, and ultrasonic homogenization, membrane emulsification operates under mild shear conditions, allowing operation at room temperature and production of multiple emulsions with high encapsulation yields (Surh et al., 2007; Vladisavljević and Williams, 2008; Dragosavac et al., 2012). Conventional emulsification techniques are not suitable for preparation of emulsions containing shear- and heat-sensitive ingredients, such as proteins, due to high energy inputs per unit volume and high shear forces used to disrupt large drops, resulting in heating (Karbstein and Schubert, 1995).

Shear rate. In direct ME, the shear rate on the membrane/continuous-phase interface is in the range of (1–50) × 10^3 s^{-1}, but micron-sized drops can be produced at zero shear by exploiting the "push-off force" arising from steric hindrance between droplets on the membrane surface (Kukizaki, 2009a; Kukizaki and Goto, 2009; Kosvintsev et al., 2008) or Laplace pressure gradient as a result of nonspherical pores (Kobayashi et al., 2011). In some membrane emulsification devices, the shear is strategically localized on the membrane surface and negligible at any distance from the membrane surface greater than 0.5 mm (Silva et al., 2015). The shear rate in high-shear in-line mixers and colloid mills is in the range of (1–2) × 10^5 s^{-1} and can reach 10^7 s^{-1} in Microfluidizers®. Furthermore, the shear in conventional homogenizers is not spatially uniform. In rotor-stator devices, shear rates are high in close proximity to a rotor and low in "dead zones," leading to the production of polydisperse emulsions. On the other hand, in the majority of membrane emulsification devices, shear rate is uniformly distributed over the whole membrane surface.

Energy input per unit volume, E/V. In premix ME, the driving pressure for pushing the feed emulsion through SPG membrane with a pore size of

0.1–10 μm is 1–50 bar, and thus, E/V is 10^5–5×10^6 J m^{-3} in each pass. In direct ME, E/V depends on the mean drop size and the volume percent of the dispersed phase, ϕ, and ranges from 10^3 J m^{-3} for a 30 μm drop size and $\phi = 1$ vol% to 10^6 J m^{-3} for a 1 μm drop size and $\phi = 50$ vol% (Gijsbertsen-Abrahamse et al., 2004). In high-pressure valve homogenizers, E/V is from 10^6 J m^{-3} for a 1 μm drop size and $\phi = 30$ vol% to 10^8 J m^{-3} for a 0.4 μm drop size and $\phi = 30$ vol%.

10.5 MICROPOROUS MEMBRANES FOR EMULSIFICATION

Membranes used for emulsification should have the following properties: (i) narrow pore-size distribution with a wide range of available pore sizes to prepare emulsions over a broad range of mean drop sizes, (ii) low hydrodynamic resistance to achieve high throughputs through the membrane at reasonable low pressure differences, (iii) high mechanical strength and high thermal and chemical resistance, (iv) high tolerance to organic solvents used in emulsion formulations, (v) ease of surface modification to easily change surface wettability from hydrophilic to hydrophobic or surface charge from negative to positive or vice versa, and (vi) low fabrication costs per unit membrane area.

Microporous membranes commonly used for emulsification are SPG, polymeric, and microengineered.

10.5.1 SPG Membrane

SPG meets the majority of the abovementioned criteria, and it is the most widely used membrane for emulsification purposes. Advantages of SPG membrane over microengineered membranes are higher porosity, more versatile surface chemistry that can be used to modify the surface charge and wettability, the broader range of available pore sizes, and lower fabrication costs.

10.5.1.1 Fabrication of SPG Membrane

SPG membrane is fabricated from Na_2O-CaO-Al_2O_3-B_2O_3-SiO_2- or Na_2O-CaO-MgO-Al_2O_3-B_2O_3-SiO_2-type mother glass by spinodal decomposition (Nakashima and Kuroki, 1981; Nakashima and Shimizu, 1986; Kukizaki and Nakashima, 2004). The mother glass is prepared by melting Shirasu, limestone, boric acid, and soda ash at 1350°C. Typical amounts of the raw materials are given in Table 10.1. MgO and ZrO_2 can be added to adjust the rate of phase separation and alkaline durability of SPG glass. Shirasu is a volcanic ash deposit from southern Kyushu, which contains 72–77 wt% SiO_2, 10–15 wt% Al_2O_3, and small amounts of other inorganic oxides, such as Na_2O, K_2O, and Fe_2O_3 (Table 10.2). Molten mother glass is shaped into tubes and then cooled down to 650–750°C for phase separation. During phase separation that takes between several hours and several tens of hours, a homogeneous glass melt separates into an acid-insoluble (Al_2O_3-SiO_2-rich) phase and acid-soluble (CaO-B_2O_3-rich)

TABLE 10.1 Typical Percentages of Raw Materials in the Na_2O-CaO-Al_2O_3-B_2O_3-SiO_2 Mother Glass Used for the Fabrication of SPG Membrane (Nakashima, 2002)[a]

	wt%
Shirasu	51
Limestone	23
Boric acid	22
Soda ash	4

[a]*Up to 5 wt% of MgO can also be added.*

TABLE 10.2 Composition of the Primary Glass[a], SPG[a], Porous Vycor Glass, and Pyrex Glass (Nakashima, 2002; Nakashima et al., 1992)

	Primary Glass for SPG, wt%	SPG wt%	Vycor® Glass wt%	Pyrex® Glass wt%
SiO_2	49	69	94–99.5	81
Al_2O_2	10	13	0–0.5	2
CaO	17	2	–	–
B_2O_3	16	7	0.2–6.0	13
Na_2O	5	5	<0.1	4
K_2O	2	4	–	–
Fe_2O_3	1	0.4	–	–

[a]*Based on the relative amounts of the raw materials shown in Table 10.1.*

phase (Fig. 10.4). The phase-separated glass is then treated with a hydrochloric acid solution to dissolve CaO-B_2O_3-rich phase, which results in the formation of porous Al_2O_3-SiO_2-rich skeleton, whose composition is shown in Table 10.2. The porosity of SPG membrane depends on the volume fraction of the acid-soluble phase in the phase-separated glass and ranges between 50% and 60% (Vladisavljević et al., 2005).

Fig. 10.5 depicts the spinodal decomposition process induced by cooling the mother glass from a temperature of $T_1 \sim 1350°C$, at which all components are miscible in all proportions, to the phase separation temperature T_2, which lies

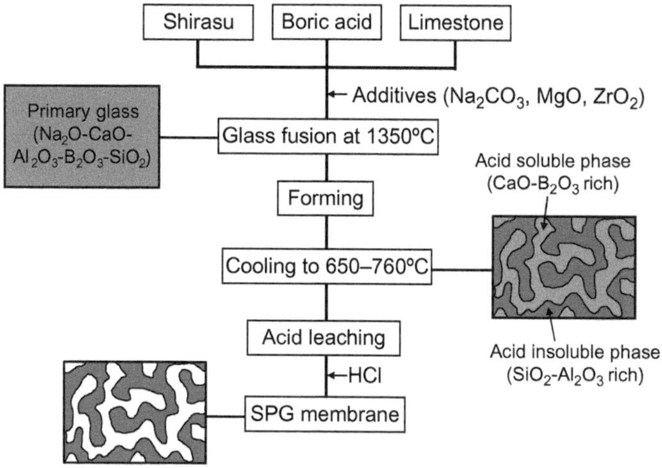

FIG. 10.4 A process flow diagram of different steps involved in the fabrication of SPG membrane. *(Reprinted from Vladisavljević, G.T., Kobayashi, I., Nakajima, M., 2012. Production of uniform droplets using membrane, microchannel and microfluidic emulsification devices. Microfluid. Nanofluid. 13, 151–178, with permission from Springer.)*

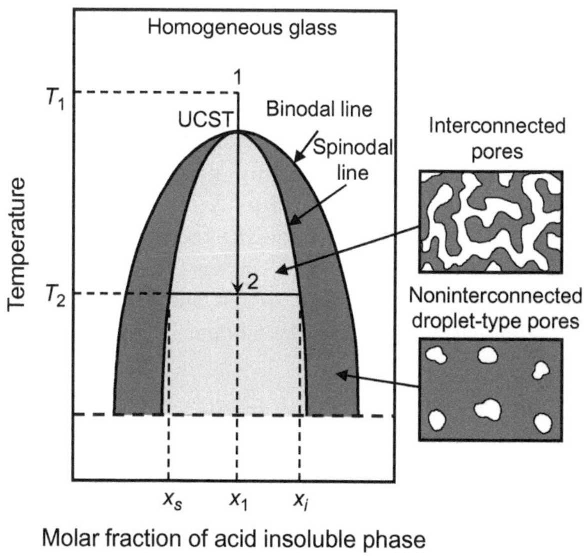

FIG. 10.5 Spinodal decomposition of glass induced by cooling homogeneous mother glass from an initial temperature T_1 to phase separation temperature T_2 lying in the spinodal region (within the spinodal line). To prevent phase separation via nucleation, a transition from stable to spinodal region on the phase diagram must proceed quickly or through the upper critical solution temperature (UCST).

within the spinodal region. A homogeneous glass with the fraction of acid-insoluble phase of x_1 is separated into two immiscible phases with the fractions of x_s and x_i. The ratio of the acid-soluble to acid-insoluble phase (x_s/x_i) in the phase-separated glass is $(x_i - x_1)/(x_1 - x_s)$. If the fraction of acid-insoluble phase is less than x_s or above x_i, the glass separation will take place by the nucleation and growth mechanism. The nucleation and growth mechanism occurs in the metastable region of the phase diagram, between the spinodal and binodal line, and leads to the formation of discrete spherical domains of one phase embedded in a continuous matrix of the other. This morphology is undesirable in the fabrication of SPG membrane since it leads to nonpermeable structure.

The mean pore size, d_p, depends on the time, t, and temperature, T_2, of the phase-separation process (Kukizaki, 2010):

$$d_p = 4K^{1/2}\left(V_p/m_m\right)t^{1/2}\exp\left[-E_a/(2RT_2)\right] \tag{10.1}$$

where K is the constant for each composition of mother glass, $E_a =$ 400–600 kJ mol^{-1} is the activation energy for diffusion during phase separation (Nakashima, 2002; Kukizaki, 2010), $R = 8314$ kJ kmol^{-1} K^{-1} is the universal gas constant, and V_p/m_m is the total pore volume per unit mass of nonporous glass. The mean pore size is proportional to the square root of the heat treatment time and decreases exponentially with $1/T_2$. Thus, small pores can be obtained using high temperatures during phase separation and short heat treatment times.

10.5.1.2 Properties of SPG Membrane

SPG membrane is available from SPG Technology Co., Ltd (Sadowara, Japan) with a mean pore size from 40 nm to 40 μm (Table 10.3). SPG membrane has a uniform internal microstructure with highly interconnected cylindrical pores extending in all directions, as confirmed by X-ray microtomography (XMT) (Vladisavljević et al., 2007). A pore tortuosity factor of $\xi \approx 1.3$ is independent on the mean pore size. On SEM and XMT images, the pores have a noncylindrical shape due to pore junctions (Fig. 10.6A and B). The number of pores per unit cross-sectional area of SPG membrane is given by (Vladisavljević et al., 2005)

$$N/A_m = 0.56/d_p^2 \tag{10.2}$$

where N/A_m and d_p are in m^{-2} and m, respectively. The hydraulic resistance of isotropic SPG membrane is given by (Vladisavljević et al., 2005)

$$R_{m,\mathrm{sym}} = 32\xi^2\delta_m/\left(d_p^2\varepsilon\right) \tag{10.3}$$

where δ_m is the membrane thickness and ε is the membrane porosity. The hydraulic resistance of isotropic SPG membrane with a thickness of 400–1000 μm is 10^8–10^{12} m^{-1} (Table 10.3) but can be reduced by fabricating

TABLE 10.3 Properties of Commercial Isotropic (Symmetrical) SPG Membrane (Nakashima, 2002; Nakashima et al., 1992; Vladisavljević et al., 2005; Kukizaki, 2009b)

Shape	Tubes or Flat Disks
Thickness, δ_m	0.4–1 mm
Compressive strength	200–280 MPa
Pore diameter, d_p	0.04–40 μm
Porosity, ε	50%–60%
True density, ρ	2000–2500 kg m^{-3}
Zeta potential at pH = 3–10 and C_{NaCl} = 1–100 mol m^{-3}	−15 to −45 mV
Pore tortuosity, ξ	1.25–1.4
Number of pores per unit cross-sectional area, N/A_m	10^9–10^{14} m^{-2}
Specific pore volume, V_p/m_m	0.5–0.6 dm^3 kg^{-1}
Hydraulic resistance, $R_{m,sym}$	10^8–10^{12} m^{-1}

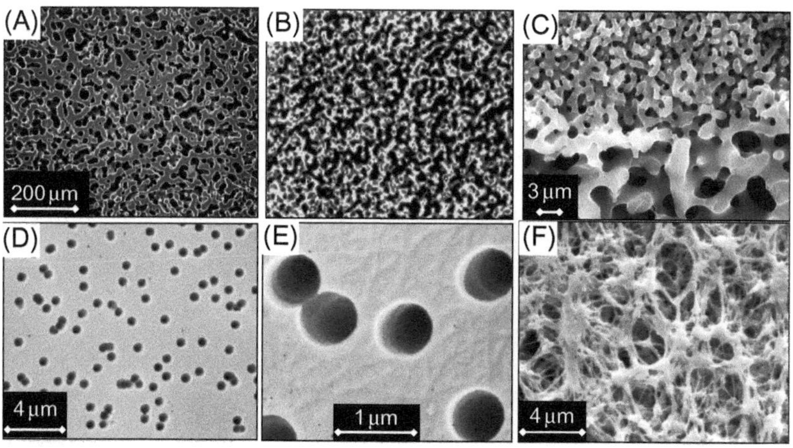

FIG. 10.6 (A) SEM image of the surface of isotropic SPG membrane polished with diamond paste and used for visualization of droplet generation process by metallurgical microscope; (B) XMT image of the internal cross section of isotropic SPG membrane. No difference between the internal structure and surface morphology was found implying that isotropic SPG membrane is fully homogeneous (Vladisavljević et al., 2007); (C) SEM image of the cross section of an anisotropic SPG membrane (Kukizaki and Goto, 2007b); (D–E) SEM image of track-etched polycarbonate (PC) membrane with a pore size of 670 nm manufactured by irradiating nonporous PC film with Cu^{3+} (Chen et al., 2008); (F) SEM image of hydrophilized polypropylene membrane filter with a pore size of 200 nm (He and Ulbricht, 2006).

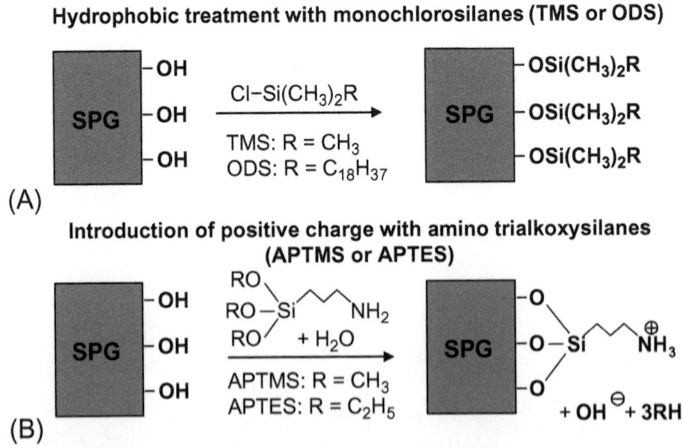

FIG. 10.7 Chemical modification of SPG surface by treatment with organosilane compounds: (A) hydrophobic treatment with monochlorosilanes (Kai et al., 2006); (B) introduction of amino groups by amino trialkoxysilanes to render the surface positively charged (Nakashima et al., 1994).

SPG membrane with anisotropic structure, as shown in Fig. 10.7C (Kukizaki and Goto, 2007a). Since the pore tortuosity ξ and membrane porosity ε are independent on the pore size, the hydraulic resistance of anisotropic SPG membrane is (Kukizaki and Goto, 2007a)

$$R_{m,asym} = 32\xi^2\delta_{skin}/\left(d_{p,skin}^2\varepsilon\right) + 32\xi^2\delta_{sup}/\left(d_{p,\,\mathrm{sup}}^2\varepsilon\right) \tag{10.4}$$

where δ_{skin} and δ_{sup} are the thicknesses of the skin and support layer, respectively, and $d_{p,skin}$ and $d_{p,sup}$ are their mean pore diameters. The thickness of the skin layer is 6% of the overall membrane thickness, and the ratio of the pore diameters in the skin and support layer is about 7 (Kukizaki and Goto, 2007a):

$$R_{m,asym} = 32\left(\xi^2/\varepsilon\right)\left(\delta_{skin}/d_{p,skin}^2 + \delta_{sup}/d_{p,sup}^2\right) = R_{m,sym}\left(0.06/1 + 0.94/7^2\right)$$
$$\approx 0.08R_{m,sym}$$

Therefore, the hydraulic resistance of anisotropic SPG membrane is just 8% of that for an anisotropic membrane of the same total thickness.

A comparison of chemical composition of SPG and common commercial porous glasses is shown in Table 10.2. SPG is more stable in water and alkaline solutions than porous Vycor® glass, because it contains less SiO_2 and more Al_2O_3. The durability of both glasses in highly alkaline solutions is poor, due to cleavage of siloxane (Si—O—Si) bonds:

$$\equiv \text{Si—O—Si} \equiv + \text{OH}^- \rightarrow \equiv \text{Si—O}^- + \equiv \text{Si—OH}$$

Alkaline durability of SPG can be improved by incorporating 3 mol% ZrO_2 into the glass skeleton, which results into the formation of stable Zr—O—Si bonds in the silicate network (Kukizaki, 2010). The compressive strength of SPG of 200–280 MPa is much higher than that of porous alumina or zirconia of the same porosity (Nakashima et al., 1992), because SPG is made up of a continuous glass skeleton with very few defects, while porous-sintered alumina or zirconia is composed of skeletal grains joined together discontinuously via grain boundaries.

10.5.1.3 Surface Modification of SPG Membrane

The surface of SPG membrane can be made hydrophobic by chemical reaction with organosilane compounds (Kukizaki and Wada, 2008) or physical coating with silicone resin (Vladisavljević et al., 2005), such as polymethylsilsesquioxane. Monochlorosilanes such as trimethylchlorosilane (TMS) and octadecyldimethylchlorosilane (ODS) are the most suitable chlorosilanes for hydrophobic treatment due to only one chlorine atom per molecule, which means that no polymerization between silane molecules can occur while they react with a silanol group on the pore surface (Fig. 10.7A) (Kai et al., 2006). The longer the carbon chain in the organosilane molecule, the more hydrophobic the membrane surface becomes (Kukizaki and Wada, 2008).

The surface of untreated SPG has a negative zeta potential between −15 and −45 mV within a pH range of 2–8, due to dissociation of silanol groups ($\equiv Si—OH \leftrightarrow \equiv SiO^- + H^+$) (Kukizaki, 2009b). A positive charge on the membrane surface can be induced by treating the membrane with amino trialkoxysilanes, such as (3-aminopropyl)-trimethoxysilane (APTMS) and (3-aminopropyl)-triethoxysilane (APTES) (Fig. 10.7B). Amino trialkoxysilanes undergo hydrolysis in aqueous solution resulting in the formation of reactive silanol groups, which can be then condensed with a silanol group on the SPG surface to form stable siloxane bond (Si—O—Si).

10.5.2 Polymeric Membranes

Track-etched polymeric membranes made from polycarbonate (PC), polyesters (PEs) such as polyethylene terephthalate (PET), and polyimide (PI) are the most common polymeric membranes used for emulsification. The track-etch method involves bombarding a nonporous polymer sheet with fission fragments or accelerated heavy ion beams from a nuclear reactor to create latent (sensitized) tracks where the chemical bonds in the polymer backbone are broken, followed by chemical etching (typically using concentrated NaOH solutions) to convert the latent tracks into hollow channels (Apel, 2001). Different etching methods have been developed that allow for the formation of cylindrical, conical, funnel-like, cigar-like, and other shapes of track-etched pores. Nuclepore™ track-etched PC membranes are available with a pore size ranging between 0.015 nm and 12 μm and the membrane diameter of 10–300 mm. The pore density is controlled by the residence time in the irradiator, while the pore shape and diameter are controlled by the residence time in the etching bath.

Track-etched membranes are less prone to fouling than SPG membranes because they are more than 10 times thinner and contain noninterconnected, straight-through pores (Fig. 10.6D and E). Both membranes are sterilizable, thus preventing cross contamination during emulsification, but track-etched polymer membranes are cheaper than SPG membranes and can be thus disposed after use (Joseph and Bunjes, 2012). A hydrophilicity of track-etched PC membrane can be enhanced by plasma treatment (Gehrmann and Bunjes, 2017a). The most popular plasmas used for hydrophilic treatment are those obtained from Ar, He, O_2, N_2, CO_2, and air. The thickness of track-etched polymeric membranes is typically less than 30 μm, but the porosity is relatively small (5%–30%), which limits the transmembrane flux.

Disposable syringe filters made from polytetrafluoroethylene (PTFE), polyethersulfone (PES), cellulose acetate (CA), mixed cellulose ester (MCE) (cellulose nitrate with a small content of CA), nylon, polyesters (PEs), and polyvinylidene fluoride (PVDF) are increasingly used in premix ME (Gehrmann and Bunjes, 2016a). Porous structure of syringe filters is typically created by phase-inversion process, whereby a homogeneous polymer solution is transformed from a liquid to a solid state by (Liu et al., 2011) (i) thermally induced phase separation; (ii) controlled evaporation of solvent from a mixture of polymer, solvent, and nonsolvent; (iii) precipitation from the vapor phase; and (iv) immersion precipitation. Membrane filters have a thickness between 20 and 200 μm and highly porous spongelike structure with a porosity of 60%–76% (Fig. 10.6F). Membrane filters are thicker than track-etched membrane and exhibit a broader pore-size distribution (Gehrmann and Bunjes, 2017a).

10.5.3 Microengineered or Microsieve Membranes

Microengineered (microsieve) membranes are microfiltration or ultrafiltration membranes with highly ordered, uniform pores manufactured by semiconductor fabrication methods. Typical microengineered membranes used in ME are electroformed nickel membranes manufactured using UV-LIGA process (Nazir et al., 2011; Schadler and Windhab, 2006; Egidi et al., 2008), silicon nitride Aquamarijn™ microsieves fabricated by reactive-ion etching (RIE) (van Rijn et al., 1997), stainless steel membranes fabricated by pulsed laser drilling (Dowding et al., 2001; Vladisavljević and Williams, 2006), single-crystal silicon microchannel arrays fabricated by deep reactive-ion etching (DRIE) (Kobayashi et al., 2003), and alumina membranes manufactured by anodic oxidation of aluminum in an acidic electrolyte (Yanagishita et al., 2010; Lee and Mattia, 2013). Alumina membranes with ideally ordered pore arrangements are highly preferred in ME. Formation of uniform pore arrays without defects and dislocations can be enhanced by imprinting aluminum substrate using a mold before anodization (Yanagishita et al., 2017). The pore diameter, interpore spacing, and membrane thickness can be controlled by the applied voltage, anodization time, electrolyte concentration, and temperature.

The pore diameter and interpore spacing are proportional to the applied voltage and inversely proportional to the electrolyte concentration (Belwalkar et al., 2008). The membrane thickness increases with increasing acid concentration in the bath.

10.6 EQUIPMENT FOR MEMBRANE EMULSIFICATION

Common batch devices for ME are cross flow tubular membrane modules with a membrane tube of about 10 mm in diameter and 100–500 mm in length, SPG microkits with a length of the membrane tube of 7–20 mm, and membrane extruders. Recently, rotating, oscillating, and flow pulsing ME systems were developed to achieve continuous operation.

10.6.1 Batch Cross-Flow Membrane Emulsification

In a batch cross flow system, a continuous phase/emulsion is circulated from a storage tank through the membrane tube and back to the storage tank (Fig. 10.8).

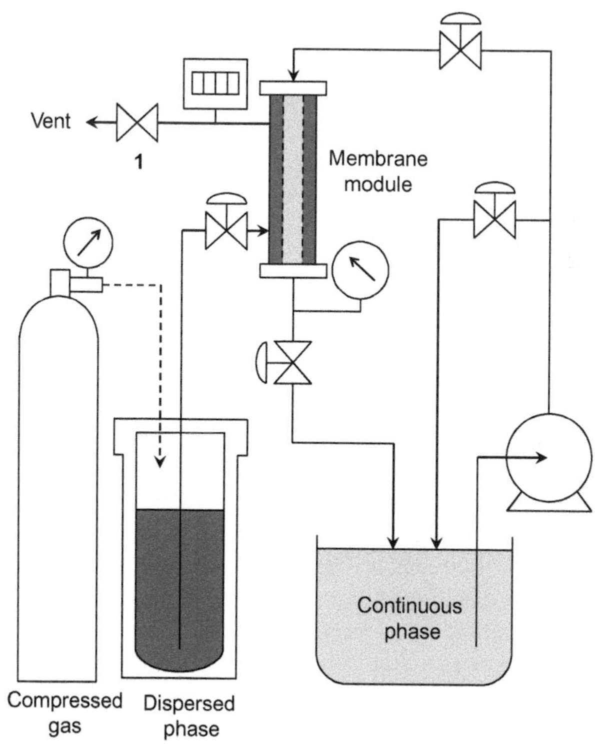

FIG. 10.8 Apparatus for cross flow direct ME using tubular SPG membrane. During start-up, valve **1** is open to remove air from the shell side of the module (Nakashima et al., 1994).

A dispersed phase-forming liquid is introduced to the shell side of the module and slowly penetrates through the membrane in the continuous-phase stream driven by a compressed gas. The driving pressure is 1.1–5 times the capillary pressure (Vladisavljević and Schubert, 2003). The system is operated until the desired dispersed-phase concentration is achieved in the emulsion. Cross flow systems are easy to scale up and offer a constant shear along the membrane surface. However, the dispersed-to-continuous-phase flow rate ratio, Q_d/Q_c, is very low, because the transmembrane flux should be kept at 1–30 L m^{-2} h^{-1} to obtain uniform drops. Therefore, the product stream must be recirculated to achieve reasonably high dispersed-phase contents, which can lead to secondary droplet breakup. Two methods have been used to decrease the flow rate of the product stream in the recirculation loop while keeping the constant shear stress at the membrane surface: insertion of static turbulence promoters in the lumen of the membrane tube (Koris et al., 2011) and introduction of back-and-forward pulsations in the continuous-phase flow (Piacentini et al., 2013; Holdich et al., 2013). The particle-size distribution in the product emulsion containing 30 vol% dispersed phase was compared by running the process for 8.5 h in steady cross flow and pulsed cross flow systems. In a pulsed cross flow system, the mean particle size and size distribution remained constant over 8.5 h, while in a steady cross flow system, the emulsion became polydispersed after 3 h due to high recirculation flow rate (Piacentini et al., 2013).

10.6.2 Batch SPG Micro Kits

Cross flow membrane emulsification systems must be operated with at least several hundred milliliters of the continuous phase to fill in the membrane module, pipes, cross flow pump, and the recirculating tank. SPG microkits shown in Fig. 10.9 can operate with only 50 mL of the continuous phase, and the dead volume is very low, which is useful for expensive samples, such as pharmaceutical W/O/W emulsions for clinical applications (Higashi and Setoguchi, 2000) and formulations with new drugs entering early clinical development because at that stage very limited amounts of drug with unknown toxicology are available (Balbach and Korn, 2004). In the external pressure-type microkit for direct ME (Fig. 10.9A), the continuous phase is stirred with a magnetic bar or mechanical impeller at 300–1000 rpm, and the membrane tube is immersed in the continuous phase. The dispersed phase is injected through the membrane from outside to inside, and the drops are formed at the inner wall of the SPG tube. Membrane tube serves as a draft tube, which results in more efficient mixing of the continuous phase than in the internal pressure SPG kit. In order to emulsify lipids that are solid at the room temperature, a water-jacketed pressure vessel attached to a recirculating water bath is used, and the continuous-phase beaker is placed on a hot plate (Joseph and Bunjes, 2014).

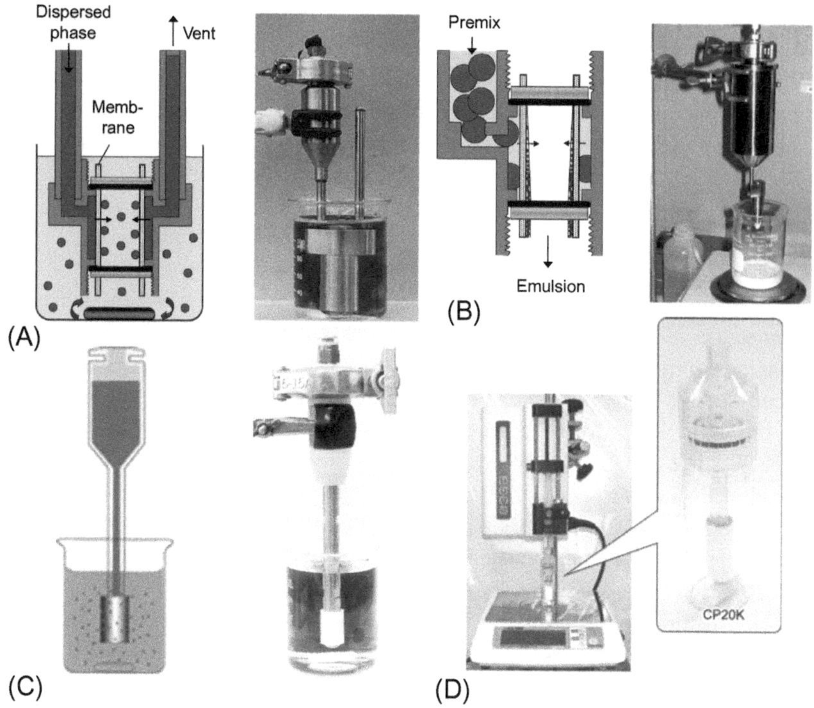

FIG. 10.9 Small-scale laboratory kits for membrane emulsification: (A) external pressure-type microkit for DME; (B) external pressure-type microkit for PME; (C) internal pressure-type microkit for DME; and SPG filter kit. All kits are commercially available from SPG Technology Co., Ltd (Sadowara, Japan). The kits are supplied with SPG membrane tube with an effective length of 10–15 mm.

The external pressure-type microkit for premix ME is shown in Fig. 10.9B. In that case, the membrane tube is placed above the collection beaker. A pressurized premix from the pressure vessel is passed through the membrane tube from outside to inside at the pressure difference that ranges from 1–3 bar (for a 10 μm membrane) to more than 10 bar (for a 1 μm membrane) and up to 50 bar for 0.1 μm membrane. The product emulsion is discharged by gravity from the membrane tube and collected in the beaker. In order to reduce the droplet size additionally and improve droplet-size uniformity, the product emulsion is repeatedly passed through the same membrane without cleaning between successive homogenization cycles (Vladisavljević et al., 2004, 2006,b). Membrane homogenization using repeated cycles was developed by Olson et al. (1979) and was used for homogenizing liposomal suspensions using track-etched polycarbonate filters.

In internal pressure-type microkit shown in Fig. 10.9C, the pressurized dispersed phase permeates through the membrane tube from inside to outside. Since the membrane tube is pressurized from the interior, the maximum pressure difference that can be applied is 3 bar, while the external pressure-type kit can withstand the pressures up to 5 bar. In SPG filter kit (Fig. 10.9D), the dispersed phase is loaded into a stainless steel syringe, which is attached to the syringe dispensing device driven by a controlled speed motor to control the rate of permeation through the membrane. The membrane is dipped in a glass vial filled with 6–10 mL of the continuous phase stirred with a magnetic bar.

10.6.3 Membrane Extruders

The Avestin LiposoFast-Basic extruder consists of two gas-tight syringes mounted on a stainless steel holder that holds a PC membrane sandwiched between two nylon meshes (Joseph and Bunjes, 2012). The extruder produces nanoemulsions by extruding manually premix through the membrane and passing the sample multiple times back and forth between two syringes. A disposable syringe filter with a pore size from 100 to 220 nm can be used instead of PC membrane (Gehrmann and Bunjes, 2016a). The process involves two steps (Fig. 10.10A): (i) mixing the two phases together by passing the liquid repeatedly back and forth between the two syringes without installing a syringe filter in between and (ii) preparation of fine emulsion by pushing the premix back and forth through the syringe filter mounted between the two syringes.

The LiposoFast-Basic extruder can be placed in the LiposoFast-Stabilizer to facilitate the extrusion of highly concentrated samples through membranes with small pore sizes. The LiposoFast-Stabilizer can accommodate 0.5 and 1.0 mL syringes. For the preparation of nanoemulsions from molten lipids, the extruder should be immersed in a water bath to maintain the extrusion temperature above the melting point of the lipid.

The advantages of the LiposoFast-Basic extruder are small sample volumes (0.2–1 mL), virtually no dead volume, and short production time (several minutes), making it convenient for preparation of large numbers of sample for screening purposes and handling of expensive, scarce, or highly potent drugs used in small quantities (Joseph and Bunjes, 2012). The EmulsiFlex-C5 homogenizer can be used to process larger sample volumes (7–1000 mL) at the flow rate of 1–5 L/h, depending on the homogenizing pressure that can reach 207 MPa. It has an air-driven pump and can be equipped with a standard homogenizing valve or filter/extruder unit for premix ME (Gehrmann and Bunjes, 2016a). Unlike in the LiposoFast extruder, the sample passes through the membrane only in one direction. Gehrmann and Bunjes (2016b) developed an instrumented small-scale membrane extruder (Fig. 10.10B) composed of a stainless steel syringe mounted on a high-pressure syringe pump and a stainless steel membrane holder, which is the filter/extruder unit of the EmulsiFlex-C5 homogenizer. The membrane holder was connected to an open collection vial, which in turn was connected to the inlet of the membrane holder via nonreturn valve to allow sample recycling. The syringe pump was computer-controlled to run the process at different constant flow rates and log the extrusion pressure every 0.1 s.

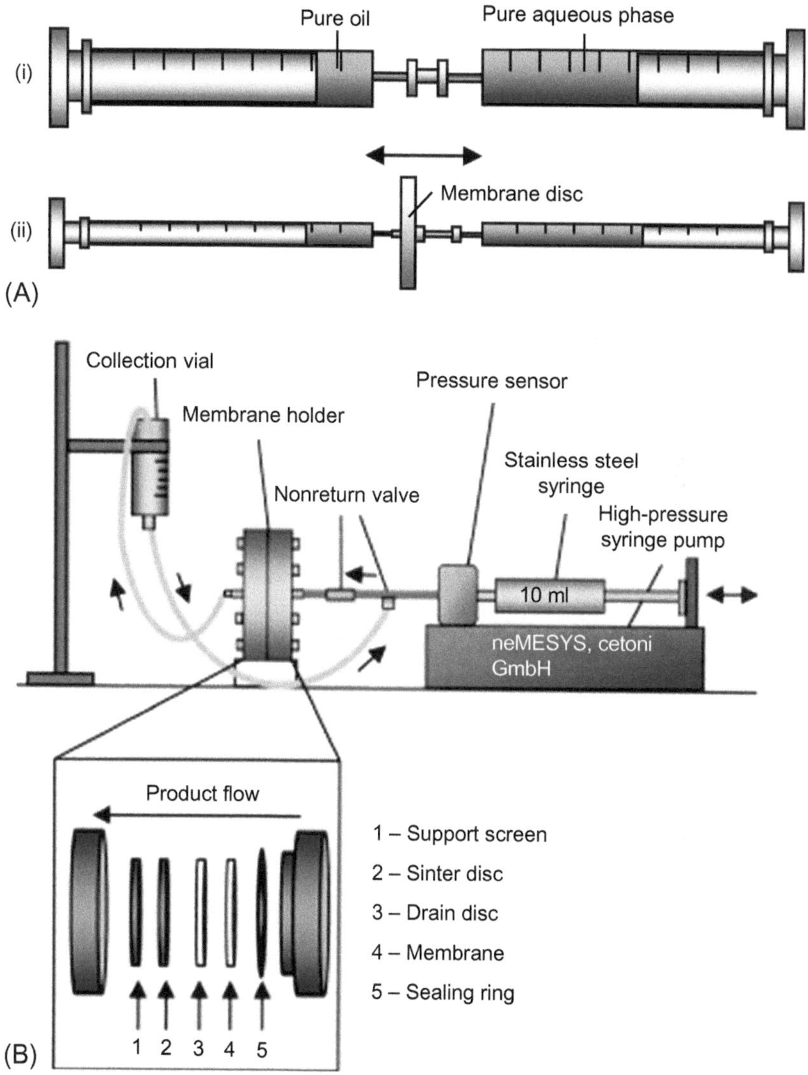

FIG. 10.10 Preparation of nanoemulsions by membrane extrusion: (A) manual extrusion involving (i) mixing together pure oil and aqueous phases by pushing the liquids back and forth multiple times between the two 20 mL syringes and (ii) preparation of nanoemulsion by pushing the premix back and forth through the membrane filter multiple times between the two 1 mL syringes (Gehrmann and Bunjes, 2016a) and (B) instrumented small-scale membrane extruder composed of a computer-controlled high-pressure syringe pump, pressure transducer, membrane holder, collection vial, and nonreturn valves for sample recycling (Gehrmann and Bunjes, 2016b).

10.6.4 Rotating Membrane Emulsification Systems

In static membrane systems, shear on the membrane surface is controlled by providing monodirectional flow or agitation over the membrane surface. In rotating membrane systems, shear on the membrane surface is provided by spinning the membrane tube in the continuous phase at the speeds from 100–1500 rpm in laboratory systems (Yuan et al., 2009; Vladisavljević and Williams, 2006; Aryanti et al., 2006) to 6000–10,000 rpm in commercial MEGATRON® devices (Graber, 2010). Rotating SPG membrane can be operated batchwise (Pawlik and Norton, 2012, 2013) or continuously. In a continuous system, the flow rate of the continuous phase is relatively low, so that shear on the membrane is controlled by the rotational speed of the membrane rather than the product flow rate. Emulsions with a high dispersed-phase concentration can be produced in rotating membrane systems continuously without emulsion recycling, which is useful for shear-sensitive emulsions.

10.6.5 Oscillating Membrane Emulsification Systems

In oscillating membrane emulsification systems, tubular vertical membrane oscillates within a static continuous phase clockwise and counterclockwise (Silva et al., 2015) or upward and downward (Holdich et al., 2010) with the frequencies from 10 to 90 Hz and the amplitudes from 0.1 to 7 mm to generate shear on the membrane surface. Metallic membranes are commonly used in oscillating membrane systems, either stainless steel membranes with laser-drilled pores (Silva et al., 2015), electroformed nickel membranes (Holdich et al., 2010), or stainless steel microscreens (Zeng et al., 2013).

10.7 PREDICTION OF MEAN DROP SIZE IN DIRECT ME

The size distribution of droplets produced in direct ME depends on the membrane properties (surface wettability and charge, porosity, pore-size distribution, pore morphology, and spatial arrangement of the pores), formulation parameters (viscosity of the dispersed and continuous phase, type and concentration of the surfactants, and additives used), and process parameters (shear stress at the membrane surface and transmembrane flux) (Joscelyne and Trägårdh, 2000).

10.7.1 Effects of Transmembrane Pressure and Flux

The minimum transmembrane pressure for driving the dispersed phase through the membrane pores is known as the capillary pressure, P_{cap}, and is given by the Young-Laplace equation:

$$P_{cap} = P_o - P_w = \frac{4\gamma_{wo}\cos\theta}{d_p} \tag{10.5}$$

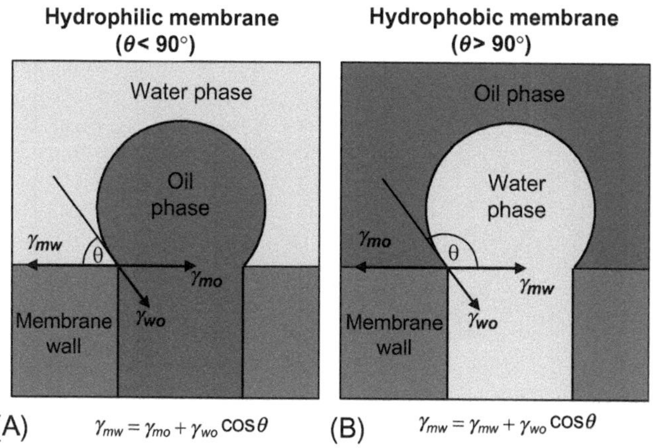

FIG. 10.11 Contact angles through the water phase formed by the dispersed phase at the three-phase boundary between the oil phase, water phase, and membrane surface: (A) production of O/W emulsion ($\theta < 90°$ and $P_o > P_w$) and (B) production of W/O emulsion ($\theta > 90°$ and $P_o < P_w$). The contact angle θ is the angle measured through the water phase, where a liquid-liquid interface meets the membrane surface (γ_{mw} = interfacial tension between the membrane and water phase, γ_{mo} = interfacial tension between the membrane and oil phase, and γ_{wo} = interfacial tension between the water and oil phase).

where γ_{wo} is the equilibrium interfacial tension between the dispersed and continuous phase and θ is the contact angle, that is, angle at the three-phase boundary where the continuous phase, dispersed phase, and membrane intersect (Fig. 10.11). A hydrophilic membrane ($\theta < 90°$) is used for preparation of O/W emulsions, and thus, $P_{cap} > 0$ and $P_o > P_w$. A hydrophobic membrane ($\theta > 90°$) is used for preparation of W/O emulsions, and thus, $P_{cap} < 0$ and $P_o < P_w$, that is, the water-phase pressure should be higher than the oil-phase pressure by at least P_{cap} to drive the water phase through the membrane.

The two principal mechanisms of drop formation in DME are (a) shear-controlled detachment as a result of shear stress on the membrane surface and (b) spontaneous detachment driven by the gradient in Laplace pressure (Sugiura et al., 2002). Shear-controlled detachment takes place at circular rectilinear pores of microengineered membranes fabricated by track etching, laser micromachining, or electroforming (Kosvintsev et al., 2005). Spontaneous detachment occurs at irregular pores of SPG membrane, terraced channels (Sugiura et al., 2002), and slotted channels with high aspect ratio (Kobayashi et al., 2004) as a result of curvature imbalance along the liquid-liquid interface (Fig. 10.3D). In a shear-controlled droplet formation process, larger drops are formed at higher transmembrane fluxes (line 1 in Fig. 10.12) because the drops are not instantaneously pinched off but after a finite time, the necking time, during which a certain amount of the dispersed phase will flow into the forming droplet (van der Graaf et al., 2006). The influx of the dispersed phase into the forming droplet during the necking step increases with increasing the

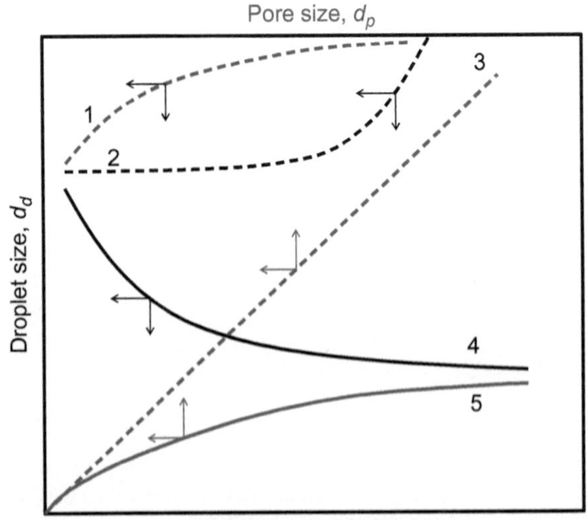

FIG. 10.12 Mean drop size, d_d, in DME *(dashed lines)* and PME *(solid lines)* as a function of mean pore size, d_p, and transmembrane flux, J: (1) d_d vs. J in shear-controlled DME, (2) d_d vs. J in interfacial tension-driven DME, (3) d_d vs. d_p in interfacial tension-driven DME in dripping regime, (4) d_d vs. J in PME, and (5) d_d vs. d_p in PME.

dispersed-phase flow rate causing the droplet size to increase. At higher fluxes, the push-off force as a result of droplet interactions on the membrane surface facilitates the detachment of the drops, causing the line 1 to reach a plateau at high fluxes (Egidi et al., 2008).

Droplet generation regime in a spontaneous membrane emulsification process is defined by the capillary number of the dispersed phase: $Ca_d = U_d\eta_d/\gamma_{wo}$, where U_d is the velocity of the dispersed phase in the pore and μ_d is the viscosity of the dispersed phase. At low Ca_d values, ($Ca_d < Ca_{cr}$, where $Ca_{cr} \approx 0.05$), drops are formed in the dripping regime. In this regime, the interfacial tension force dominates the shear force (Sugiura et al., 2002), and the drop size depends solely on the pore geometry but not on the transmembrane flux and shear stress (line 2 at low flux values in Fig. 10.12), making the process highly robust. At $Ca_d > Ca_{cr}$, large drops with $d_d > 10d_p$ are formed in the continuous outflow regime (Kobayashi et al., 2003) (line 2 at high flux values). In this regime, the viscous force due to flow of the dispersed phase dominates the interfacial tension, and the drop size strongly depends on the flux, but does not depend on the pore size. The velocity of the dispersed phase at which the transition from a dripping to continuous outflow occurs is independent on the pore size (Kobayashi et al., 2011) and increases with decreasing the viscosity of the dispersed phase (Vladisavljević et al., 2011). At the η_d/η_c viscosity ratio of 55, the critical velocity of the dispersed phase in the pores is

2–5 mm/s (Sugiura et al., 2002; Kobayashi et al., 2011). Direct ME must be operated in dripping regime, since emulsions produced in the continuous outflow regime are polydisperse, due to random nature of the droplet formation process. The transition from a dripping to continuous outflow does not occur simultaneously for all the pores, leading to bimodal particle-size distribution of the drops as a result of the large difference in size of the drops produced in the dripping and continuous outflow regime.

10.7.2 Effects of Pore Size and Shear Stress

For spontaneous drop formation, a linear relationship between the mean drop size and the mean pore size exists in dripping regime: $d_d = K'd_p$ (line 3 in Fig. 10.12), where $K' = 2.8$–3.5 for SPG membrane (Kukizaki and Goto, 2007b, 2009; Nakashima et al., 1991; Vladisavljević et al., 2006). A slope of d_d vs. d_p line increases with decreasing the shear stress on the membrane surface, but the effect is very limited at low flux. Even without any shear, $K' = 3.3$ for O/W emulsions stabilized with 1% Tween 80 produced using SPG membrane at low flux (Kukizaki and Goto, 2009).

In a shear-controlled droplet formation process, the mean droplet size depends on the balance between the shear force, F_d, exerted on the forming drop by the continuous phase and the capillary force, F_{ca} (Kosvintsev et al., 2005):

$$F_{ca} = \pi d_p \gamma \quad (10.6)$$

$$F_d = 9\pi\tau_w d_d \sqrt{(d_d/2)^2 - r_p^2} \quad (10.7)$$

where r_p is the pore radius and τ_w is the shear stress on the membrane surface. At $J \to 0$, the effect of transmembrane flux can be neglected, and the droplet detaches from the membrane surface when $F_{ca} = F_d$:

$$d_d = \frac{\sqrt{18\tau_w^2 r_p^2 + 2\sqrt{81\tau_w^4 r_p^4 + 4r_p^2\tau_w^2\gamma^2}}}{3\tau_w} \quad (10.8)$$

According to Eq. (10.8), the mean drop size decreases with increasing shear stress on the membrane surface and reaches $2r_p$ at very high τ_w values. Depending on the membrane system, τ_w in direct ME can be calculated using the equations presented in Table 10.4. The variations of the shear stress with time or position along the membrane surface are shown in Fig. 10.13 (Laouini et al., 2013). In a pulsed cross flow system (Holdich et al., 2013), flow oscillations are superimposed on a low cross flow velocity, but the contribution of the cross flow to the overall shear stress can be neglected (Piacentini et al., 2013).

TABLE 10.4 Shear Stress on the Membrane Surface in Direct ME for Different Membrane Shapes and Shear Generation Techniques

Shear Generation Technique	Membrane Shape	Shear Stress on Membrane Surface
Cross flow	Cylindrical	$\tau_W = \frac{8\eta_c U_c}{d_{mi}}$
Cross flow	Flat	$\tau_W = \frac{3Q_c\eta_c}{2h^2 W}$
Rotating	Cylindrical	$\tau_W = \frac{2\eta_c r_{mo}^2 \omega}{r_b^2 - r_{mo}^2}$
Paddle stirrer	Flat	$\tau_{W,\max} = \frac{0.825\eta_c \omega r_{trans}}{\sqrt{\eta_c/(\rho_c \omega)}}$ [a]
Oscillating	Any	$\tau_{W,\max} = (2\pi Af)^{3/2}(\eta_c\rho_c)^{1/2}$

[a] $r_{trans} = 1.23\frac{D}{2}\left(0.57 + 0.35\frac{D}{D_{cell}}\right)\left(\frac{b}{T}\right)^{0.036} n_b^{0.116}\,\mathrm{Re}/(1000 + 1.43\,\mathrm{Re})$

10.7.3 Effect of Surfactant

The role of surfactant in ME is to rapidly adsorb to newly formed oil-water interface to reduce the interfacial tension and facilitate droplet detachment from the pore, thus reducing the final drop size and stabilizing the formed drops against coalescence. The kinetics of adsorption of surfactants at oil-aqueous interface has a significant impact on the drop size during direct ME (Schröder et al., 1998; van der Graaf et al., 2004; Rayner et al., 2005). As a rule, the faster the surfactant molecules adsorb to the interface, the smaller the size of the resultant drops. Surfactant molecules must not adsorb to the membrane surface, since otherwise the dispersed phase will spread over the membrane surface instead of being confined within the pores, which would lead to uncontrolled drop generation. This means that the functional groups of surfactant molecules must not carry a positive charge to avoid electrostatic deposition onto negatively charged surface of SPG membrane (Nakashima et al., 1993). The use of cationic surfactants, such as cetyltrimethylammonium bromide (CTAB), leads to polydispersed O/W emulsions with $d_d/d_p > 20$ (Nakashima et al., 1993). The use of zwitterionic surfactants must also be avoided, even when they carry a net negative charge. For example, lecithin at pH 3 fouls SPG membrane due to electrostatic interactions between positively charged patches ($—N(CH_3)_3^+$ and $—NH_3^+$) on phospholipid molecules and negatively charged silanol groups on SPG surface; although at pH 3, the net charge on lecithin molecules is negative (Surh et al., 2008). To produce droplets stabilized with cationic surfactants, the SPG membrane must be treated with amino trialkoxysilanes to induce positive charge on the membrane surface (Fig. 10.7B), or the charge of anionic droplets must be altered after ME by surfactant displacement (Vladisavljević and McClements, 2010).

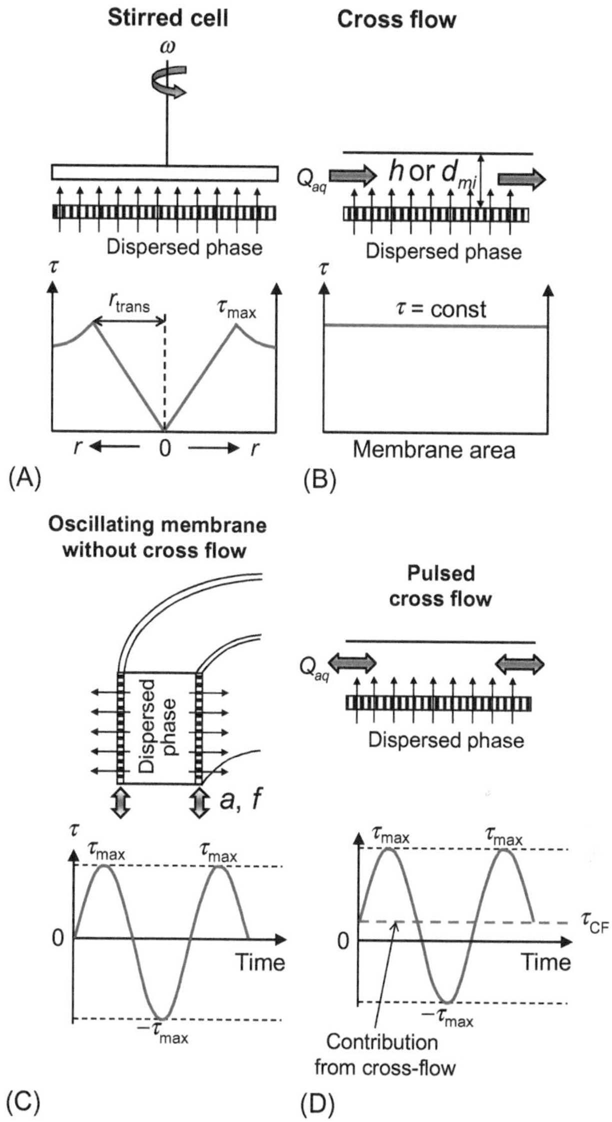

FIG. 10.13 Spatial or temporal variations of shear stress at the membrane surface in direct ME devices. The variation of the shear stress with time for pulsed cross flow (D) is the same as in the oscillation membrane system (C), except that temporal variations are superimposed on constant shear stress τ_{CF} generated due to continuous-phase cross flow.

10.8 FACTORS AFFECTING DROPLET SIZE IN PREMIX ME

In premix ME, large drops of the premix are disrupted using a combination of several effects including snap-off effects due to localized shear forces, breakup processes due to interfacial tension effects (Rayleigh and Laplace instabilities), and steric hindrance effects between droplets (van der Zwan et al., 2006). The final drop size in premix ME depends on the mean pore size, transmembrane pressure, number of extrusion cycles, viscosity of the continuous and dispersed phases, and interfacial tension (Nazir et al., 2010). The mean drop size is a non-linear function of the mean pore size (line 5 in Fig. 10.12):

$$d_d = K''\left(d_p\right)^n \tag{10.9}$$

where $n < 1$. The d_d/d_p ratio decreases with increasing the mean pore size and ranges from 1 to 1.5 for SPG membranes with $d_p = 5$–20 μm at the shear stress on the pore wall of 200 Pa (Vladisavljević et al., 2006). The critical pressure in premix ME is given by (Park et al., 2001)

$$P_{cap} = \frac{\gamma\left[2 + 2a^6/\sqrt{2a^6 - 1} \times \arccos\left(1/a^3\right) - 4a^2\right]}{a + \sqrt{a^2 - 1}} \tag{10.10}$$

where $a = d_{pm}/d_p$ and d_{pm} is the mean drop size in premix. If $a \gg 1$, the capillary pressure is given by Eq. (10.5). In premix ME, the optimum transmembrane pressure is 10–50 times larger than P_{cap} (Vladisavljević et al., 2004). The mean drop size decreases with increasing mean shear stress on the pore walls, which is given by

$$\tau_{w,p} = 8\eta_e J\xi/\left(\varepsilon d_p\right) \tag{10.11}$$

where η_e is the viscosity of emulsion inside the pores. Therefore, premix ME results in smaller droplets, which are generated at higher transmembrane pressures, as shown by line 4 in Fig. 10.12. In repeated premix ME, the drop size is reduced after each extrusion cycle until it reaches a constant minimum value (Vladisavljević et al., 2004; Laouini et al., 2014).

10.9 MICROEMULSIONS VS. NANOEMULSIONS

An oil-in-water microemulsion is a thermodynamically stable colloidal dispersion consisting of small spheroid particles (composed of oil, surfactant, and possibly cosurfactant and lipophilic active ingredients) dispersed within an aqueous medium (McClements, 2012). An oil-in-water nanoemulsion is a thermodynamically unstable but kinetically stable colloidal dispersion consisting of two immiscible liquids, with one of the liquids being dispersed as small spherical droplets in the other liquid. In both dispersions, the mean droplet diameter is below 200 nm (McClements, 2012), but the fabrication process is significantly different. For a microemulsion, the free energy of the dispersed system is lower than the free energy of the separate phases (oil and water). Thus, microemulsions can be formed spontaneously by bringing the oil and water phase together at a particular temperature without supplying any external energy. However, it is often necessary to apply limited amount of external energy (in the form of stirring or heating) to form a microemulsion because there is a kinetic energy

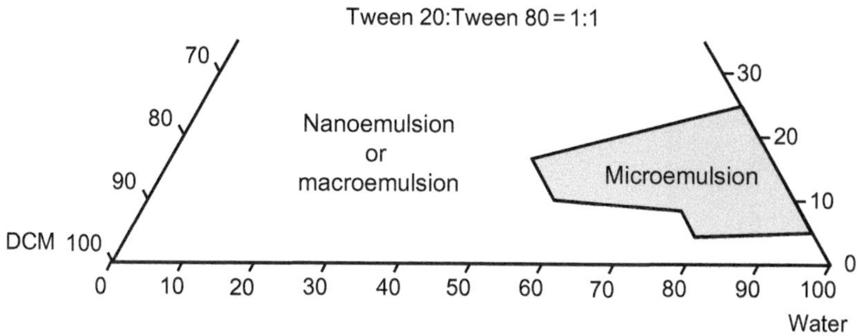

FIG. 10.14 The pseudoternary phase diagram for the dispersed system composed of dichloromethane (DCM), Tween 80/Tween 20 surfactant mixture (in the 1:1 volume ratio), and water (Oh et al., 2011). Within the gray microemulsion region, the drop size in direct ME is much smaller than the membrane pore size and independent on the pore size. In the nanoemulsion or macroemulsion region, the drop size in direct ME is larger than the pore size.

barrier (the activation energy, ΔG^*) that must be overcome. In direct ME, the external energy required for formation of a microemulsion is supplied by extruding the dispersed phase through the membrane pores at certain pressure and providing a gentle agitation of the continuous phase (Oh et al., 2011). The droplet-size distribution depends on the shear conditions in the continuous phase and emulsion formulation and very slightly on the membrane pore size. The role of membrane in the formation of microemulsion is primarily to achieve injection of the dispersed phase into the continuous phase at controlled rate and to provide uniform and controllable shear conditions during emulsification. Microemulsions can only be formed for certain mass ratios of the main system constituents (water, oil, and surfactants). Fig. 10.14 shows the pseudoternary phase diagram for the system comprised of water, DCM and Tween 20/Tween 80 surfactant mixture (1:1 volume ratio). An O/W microemulsion can only be formed within the gray-shaded area, that is, for relatively high surfactant concentrations (low interfacial tensions) and low oil contents.

The minimum concentration of surfactant required for microemulsion formation can be determined by making a series of O/W emulsions with increasing concentration of surfactant and measuring the drop size (Fig. 10.15). For low surfactant (AOT) concentrations in the dispersed phase of 10^{-5}–10^{-2} M, the average drop size in direct ME was in the micron size range (2–5 μm) and at least five times larger than the pore size of 0.4 μm. In addition, the drop size was in good agreement with an existing force-balance model (the solid line in Fig. 10.15), indicating that a macroemulsion was formed. A decrease of the average drop size with an increase in surfactant concentration in this region is a consequence of the decrease in interfacial tension, which facilitates detachment of the growing drop from the membrane surface under agitation. At the AOT concentration above 10^{-2} M, the average droplet size was two to five times smaller than the pore size, pointing to the formation of microemulsions.

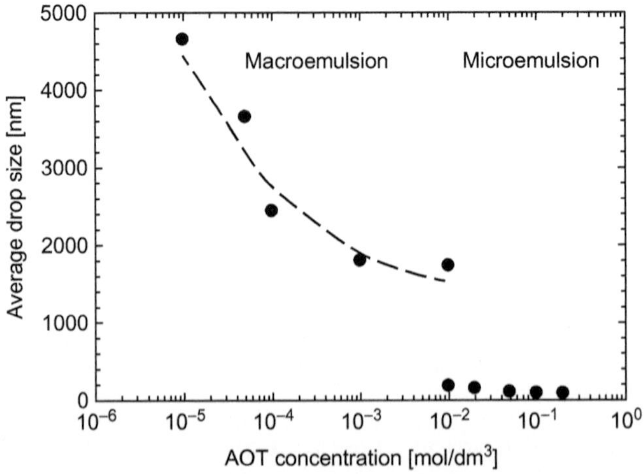

FIG. 10.15 The effect of concentration of surfactant AOT (docusate sodium salt) in the dispersed phase on the average size of emulsion drops formed by direct ME in a stirred cell using ceramic membrane with a pore size of 0.4 μm (Adamczak et al., 2014). In the macroemulsion region, the average drop size was several times larger than the membrane pore size, while in the microemulsion region, the average drop size was two to five times smaller than the membrane pore size. The dispersed phase was AOT solution in squalene, and the continuous phase was 100 ppm PAH in 0.015 M NaCl solution.

A sudden decrease in drop size at the critical AOT concentration of 10^{-2} M indicates the change in drop generation from membrane-assisted drop generation at the higher interfacial tension to self-emulsification of the dispersed phase at very low interfacial tensions. In Tables 10.5 and 10.6, the term nanoemulsions will be used for both micro- and nanoemulsions.

10.10 FACTORS AFFECTING FORMATION OF MICRO/NANOEMULSIONS VIA MEMBRANE EMULSIFICATION

10.10.1 Direct Membrane Emulsification

Temperature. The temperature of the continuous phase is one of the most important parameters in membrane emulsification affecting the viscosity of the dispersed and continuous phases and also the HLB value of the surfactant and its solubility. The change of interfacial tension between the oil and the water phase at higher temperatures could lead to coarsening of the emulsion via Ostwald ripening and/or coalescence. It has been reported that temperature influences the uniformity of emulsions during membrane emulsification (Oh et al., 2011). Choi et al. (2012) prepared O/W microemulsions at 25 and 35°C and found that smaller drop sizes with narrower particle-size distributions were obtained at 25°C, and the same trend was observed by Laouini et al. (2012).

TABLE 10.5 Examples of Nanoemulsions Prepared Using Direct Membrane Emulsification

Process	Product	Formulation	Authors
A. Internal pressure-type microkit, 2.5 μm SPG membrane	O/W nanoemulsions loaded with itraconazole (d_d=410–550 nm)	*DP*: 40 wt% itraconazole in DCM *CP*: 5% (w/v) surfactant mixture (Transcutol HP/ Span 20 at 1:1 mass ratio)+2.5% (w/v) dextran in water	Choi et al. (2012)
B. Internal pressure-type micro kit, 2.5 μm SPG membrane	O/W nanoemulsion loaded with flurbiprofen (d_d=360 nm)	*DP*: 20 wt% flurbiprofen in DCM *CP*: 5% (w/v) surfactant mixture (Tween 20/Tween 80 at 1:1 mass ratio)+1% (w/v) PVA in water	Oh et al. (2011)
C. Internal pressure-type micro kit, 1.1 μm SPG membrane	O/W nanoemulsion loaded with fenofibrate (d_d=157 nm)	*DP*: 0.2 wt% flurbiprofen+ 0.3 wt% Labrafil M 1944 CS+0.2 wt % Capryol PGMC in ethanol *CP*: 0.075 wt% Labrasol in water	Pradhan et al. (2013)
D. Stirred cell, ceramic 0.45 μm ceramic membrane	O/W nanoemulsions (d_d=80–180 nm; and ϕ=0.001)	*DP*: 0.01–0.5 M AOT in squalene *CP*: 100 ppm PAH in 0.015 M NaCl aqueous solution	Adamczak et al. (2014)
E. Cross flow SPG membrane, 0.9 μm pore size	O/W nanoemulsion loaded with vitamin E (d_d=106 nm; span=0.3; and ϕ=0.18)	*DP*: 5 wt% vitamin E in MCT *CP*: 2.7% (w/v) Tween 80+2.7% (w/v) Brij 35 in water	Laouini et al. (2012)
F. External pressure-type micro kit with jacketed pressure	O/W nanoemulsions (ϕ=0.01–0.05	*DP*: MCT, SBO, 0.5 wt% Span 80 in SBO	Joseph and Bunjes (2014)

Continued

TABLE 10.5 Examples of Nanoemulsions Prepared Using Direct Membrane Emulsification—cont'd

Process	Product	Formulation	Authors
vessel, 0.1 μm SPG membrane	and d_d=400–900 nm)	*CP*: 5 wt% SDS or 1–2 wt% Tween 20	
G. Anodic porous alumina membrane with a pore size of 0.13, 0.22, and 0.32 μm.	O/W nanoemulsions (d_d=196, 418, and 744 nm)	*DP*: PAK-02 acrylic monomer/oleic acid (1:1 v/v) *CP*: 0.3 wt% SDS in water	Yanagishita et al. (2010)
H. SPG filter kit, 0.1 μm SPG membrane	$W_1/O/W_2$ nanoemulsions loaded with calcein (d_d=117–788 nm)	W_1: 1.0% (w/v) calcein in 100 mM PBS at pH 7.4 *O*: 28.6 wt% CR-310 in soybean oil W_2: 1% (w/v) HS-11, S-24D, L-7D, Gelucire 44/14, or Labrasol +1% (w/v) PVA +5% (w/v) glucose in water	Koga et al. (2010)
I. Internal pressure-type micro kit, 1.4 μm SPG membrane	$W_1/O/W_2$ nanoemulsion loaded with doxorubicin (d_d=440 nm)	W_1: 0.5% (w/v) DOX in water *O*: 5% (w/v) PGPR in soybean oil W_2: 3% (w/v) Tween 20 in water	Pradhan et al. (2014)

d_d, *average droplet size;* DP, *dispersed phase;* CP, *continuous phase.*

Hydrodynamic conditions. Choi et al. (2012) investigated the effect of agitator speed over a range of 150–700 rpm on the average size of nanoemulsion drops formed in internal pressure-type SPG microkit. They found that the z-average and PDI increased as the agitator speed increased to 700 rpm probably because the higher shear increased the probability of collisions between droplets after formation, thereby causing the coalescence of droplets and increase in droplet size. On the other hand, Laouini et al. (2012) obtained nanoemulsions with smaller drops and narrower particle-size distributions at higher cross flow velocities.

TABLE 10.6 Examples of Nanoemulsions Prepared Using Premix Membrane Emulsification

Process	Product	Formulation	Authors
A. Repeated premix ME, LiposoFast-basic extruder, 50, 100, and 200 nm PC membrane	O/W nanoemulsions (d_d=100–200 nm and ϕ=0.05–0.2)	*DP*: MCT or SBO *CP*: Aqueous solution containing 18.75 wt% SDS or 18.75 wt% Pol or 18.75 wt% SL or 18.75 wt% PG10-L	Joseph and Bunjes (2012)
B. External pressure-type microkit with jacketed pressure vessel, 0.1, 0.2, 0.3, 0.5, and 1.1 μm SPG membrane	O/W nanoemulsions (d_d=100–200 nm and ϕ=0.1)	*DP*: MCT *CP*: Aqueous solution containing 7.5 wt% SDS, 7.5 wt% Pol, or 7.5 wt% Tween 20	Joseph and Bunjes (2014)
C. Syringe filters, 0.2 +0.1 μm PES, MCA, CE, and nylon membrane	O/W nanoemulsions (d_d< 500 nm)	*DP*: MCT *CP*: Aqueous solution containing 2.5 wt% Lipoid S100 or 2.5 wt % Lipoid S75 or 5 wt% Tween 80 or 5 wt% SL	Gehrmann and Bunjes (2016a)
D. Syringe filters, 0.2 μm pore-sized PC, PE, PES, PS, PVDF, nylon and CA membrane	O/W nanoemulsions (d_d=0.08–11 μm and ϕ=0.2)	*DP*: MCT or peanut oil *CP*: 6.25 wt% aqueous solution of SDS, Pol, Tween 80, Tyl, or SL	Gehrmann and Bunjes (2017a)
E. Anodisc® aluminum oxide membrane filter, 0.2 μm pore size	O/W nanoemulsions (d_d=77–250 nm and ϕ=0.2)	*DP*: MCT or peanut oil *CP*: 6.25 wt% aqueous solution of SDS, Pol, Tween 80, Tyl, or SL	Gehrmann and Bunjes (2017a)
F. Instrumented small-scale extruder, 0.2 μm pore-sized PC, PE, PES, PS, and PVDF membrane	O/W nanoemulsions (d_d=100–200 nm and ϕ=0.2)	*DP*: MCT *CP*: 18.75 wt% aqueous solution of SDS	Gehrmann and Bunjes (2016b)

Pressure difference. The pressure difference used in direct ME process depends on the type of emulsion formed. If a microemulsion is formed, the applied pressure difference is often less than 100 kPa, because the membrane pore size is typically above 1 μm. Oh et al. (2011) prepared microemulsions with z-average values between 50 and 100 nm at the pressure difference ranging from 15 to 80 kPa using 2.5 μm pore-sized SPG membrane. They reported a decrease in droplet size with an increase in pressure difference. However, using the same membrane, Choi et al. (2012) observed an increase in the z-average value from 410 to 550 nm when the pressure difference increased from 15 to 40 kPa. The driving pressure below 15 kPa was not enough to force the dispersed phase through the membrane. In most cases, an increase in driving pressure led to higher PDI values (Choi et al., 2012; Laouini et al., 2012). If a nanoemulsion is formed, the applied pressure difference is typically above 100 kPa, because the required membrane pore size is well below 1 μm. Joseph and Bunjes (2014) made O/W nanoemulsions composed of soybean and MCT oil using SPG membrane with a pore size of 0.1 μm at the pressure difference of 800 kPa.

Surfactant concentration. Laouini et al. (2012) investigated the effect of surfactant concentration on the formation of O/W nanoemulsions in cross flow emulsification with SPG membrane. The presence of surfactant in the aqueous phase lowers the interfacial tension between the oil and aqueous phase and facilitates droplet disruption given that the interfacial tension is a drop holding force during the process. In addition, surfactants stabilize the formed droplets against coalescence and aggregation. For the system composed of MCT, water, and Tween 80/Brij 35 1:1 w/w mixture at the mass ratio of 14/80/5/6, two peaks were observed in particle-size distribution curve, the main peak at 125 nm corresponding to oil droplets and another peak at 11 nm corresponding to micelles. When the total surfactant content was reduced from 6 to 2.25 wt%, the peak corresponding to micelles decreased progressively and disappeared at 2.25 wt%, which was accompanied by a decrease in the size distribution span from 0.52 to 0.25. Further decrease in the total surfactant content from 2.25 to 2 wt% led to broadening of the size distribution curve, and the main peak occurred at 290 nm. Therefore, there is an optimum surfactant concentration for the preparation of microemulsions with monomodal size distribution and small average droplet size.

Drug loading. The effect of loading vitamin E in MCT oil in the range of 5–10 wt% on the particle-size distribution in cross flow membrane emulsification was investigated by Laouini et al. (2012). Increasing vitamin E loading led to an increase in the average droplet size and broadening of the particle-size distribution curve, which was explained by the higher dispersed-phase viscosity and possible effect on the interfacial tension. Oh et al. (2011) found that the z-average of O/W microemulsions increased from 98 to 360 nm when DCM drops were loaded with 20 wt% of flurbiprofen (Oh et al., 2011). Joseph and Bunjes (2014) prepared trimyristin nanoemulsions loaded with ubidecarenone using external pressure-type microkit with jacketed pressure vessel. The emulsification temperature was 10°C higher than the melting point of trimyristin. The particle size was independent on the drug loading in the range between 0 and 5 wt%.

Membrane pore size. The ratio between the drop size and the pore size depends on whether a microemulsion or nanoemulsion is formed. The mean drop size of a microemulsion formed by direct ME is significantly smaller than the mean pore size. Oh et al. (2011) prepared an O/W microemulsion with a z-average of 98.5 nm using 2.5 μm SPG membrane in the presence of 5% (w/v) surfactant (Tween 20/Tween 80 mixture) and 1% (w/v) PVA in the aqueous phase. Therefore, the mean droplet size was 25 times smaller than the mean pore size, which means that the drop size was not controlled by the pore size. If a nanoemulsion is formed with the same membrane using the smaller amount of surfactant, the drop size would be about 8 μm. Laouini et al. (2012) prepared O/W nanoemulsions using three different pore sizes of SPG membrane. The average particle size was very similar using 0.4 and 0.9 μm pore-sized membranes, but the injection rate was very low with 0.4 μm membrane. When 10.2 μm membrane was used, the injection rate was very high, but the resultant particle-size distribution was very broad with a large average particle size. It clearly shows that the pore size of SPG membrane for preparation of microemulsions should be above 0.4 μm but well below 10 μm. In most cases, microemulsions were formed using SPG membrane with a pore size between 0.45 and 2.5 μm, and the mean drop size was two to nine times smaller than the pore size (Table 10.5A–E). On the other hand, O/W nanoemulsions can be formed by direct ME if the pore size of SPG membrane is 0.1 or 0.2 μm (Joseph and Bunjes, 2014).

10.10.2 Premix Membrane Emulsification

Contact angle. A low contact angle between the membrane and the surfactant solution is crucial for the generation of droplets in the nanorange. An incomplete wetting of the membrane with the continuous phase resulted in polydisperse emulsions with larger droplets irrespective of the formulation (Gehrmann and Bunjes, 2017a).

Membrane pore-size distribution. Smaller droplet sizes can be achieved using membranes with smaller pore sizes, but the effect of pore size on the drop size is much less pronounced in premix than in direct ME (Joseph and Bunjes, 2012, 2014). In repeated membrane extrusion, a narrow pore-size distribution of the membrane is not critically important for the drop-size uniformity, because the probability that all droplets will be disrupted by the smallest pores increases with every extrusion cycle (Gehrmann and Bunjes, 2017a). That is why polymeric membranes are more often used in premix ME, while SPG membranes are dominant in direct ME. The median size of SDS-stabilized MCT oil drops in the prepared nanoemulsions was smaller or equal to the pore size after 21 cycles through a track-etched PC membrane with a pore size of 200 nm (Gehrmann and Bunjes, 2016b). The ratio of the median drop size to the membrane pore size for a 200 nm track-etched PC membrane ranged from 1:1 at the transmembrane flux of 0.27 $m^3\ m^{-2}\ h^{-1}$ to 0.6:1 at the flux of 3.8 $m^3\ m^{-2}\ h^{-1}$. After one passes through SPG membrane with a mean pore size of 100–300 nm at 900 kPa, the ratio of the median drop size to the membrane pore size was in the range from 0.4:1 to 1.1:1 for nanoemulsions composed of 2.5–10 wt% MCT oil stabilized with SDS. The prepared nanoemulsions were monomodal with a median drop size smaller than 200 nm.

Membrane structure. The use of track-etched PC or PE membranes with straight-through pores led to smaller median drop sizes in the produced emulsions than the use of spongelike PES, PS, and nylon membranes (Gehrmann and Bunjes, 2016b). On the other hand, track-etched PC membranes are less efficient in reducing the drop size in premix ME than SPG membranes due to their nontortuous, rectilinear pores and smaller thickness (in most cases less than 20 nm) as compared with SPG membranes (700–900 nm). For example, one extrusion cycle through SPG membrane with a pore size of 100–300 nm led to smaller drops and narrower-size distributions than 11 or 21 extrusion cycles through tracked-etched PC membranes with the same pore size (Joseph and Bunjes, 2014). Zhou et al. (2009) also observed that an increasing thickness of a branched membrane leads to an increased extent of droplet disruption and therefore to smaller particle sizes.

Number of extrusion cycles. Not surprisingly, with increasing number of extrusion cycles at constant transmembrane flux, the drop size and transmembrane pressure decreased, because the smaller droplets passed through the membrane with less friction with the pore walls. After certain number of cycles (typically about 20), constant median drop size and transmembrane pressure were established. Track-etched membranes with well-defined pore structure required a smaller number of extrusion cycles to reach a steady-state particle size than polymeric membrane filters with highly branched pores (Gehrmann and Bunjes, 2017a). After only two extrusions through the track-etched PE membrane with a pore size of 200 nm, a monomodal particle-size distribution of SDS-stabilized oil droplets with a d_{50} value of 320 nm was obtained (Gehrmann and Bunjes, 2016b). The greatest reduction in drop size and transmembrane pressure were observed during the first three extrusion cycles.

Transmembrane flux and extrusion pressure. At constant transmembrane flux, the extrusion pressure has the highest value during first extrusion cycle and then declines to reach a plateau value after large number of cycles (Gehrmann and Bunjes, 2016b). This trend is caused by the reduction of particle size with increasing cycle numbers leading to reduced resistance to flow. The extrusion pressure observed increased with increasing the flow rate through the membrane due to higher hydrodynamic resistance (Gehrmann and Bunjes, 2016b). Transmembrane flux was found to increase with increasing extrusion pressure. In most cases, smaller droplet sizes and narrower particle-size distributions of nanoemulsions were observed at higher extrusion pressures (fluxes). The extrusion pressure of 3 bar was sufficient to produce SDS-stabilized oil droplets with a median size of just above 200 nm after 21 cycles through a track-etched PC membrane with a pore size of 200 nm (Gehrmann and Bunjes, 2016b). At constant extrusion pressure, an increase in flux was observed with increasing number of extrusion cycles (Vladisavljević et al., 2004).

Quality of premix. The quality of the pre-emulsion had no influence on the quality of the nanoemulsion after five extrusion cycles (Gehrmann and Bunjes, 2017b).

10.11 PREPARATION OF MICRO/NANOEMULSIONS USING DIRECT ME

Colloidal O/W dispersions prepared by ME can be used as delivery systems for oral, parenteral, and transdermal administration of poorly water-soluble drugs such as itraconazole, flurbiprofen, and fenofibrate (Table 10.5A–C). An O/W microemulsion composed of itraconazole, dichloromethane (DCM), water, Transcutol® HP/Span 20 1:1 surfactant mixture, and dextran at the weight ratio of 2/5/100/5/2.5 was prepared using an internal pressure-type microkit with a 2.5 μm SPG membrane at an agitator speed of 150 rpm, a feed pressure of 15 kPa, and a temperature of 25°C (Choi et al., 2012). Transcutol® HP was added as a good solubilizing agent for itraconazole, and Span 20 was used as a cosurfactant to expand the microemulsion region of the formulation (Fig. 10.16). The resultant emulsion was solidified by spray drying to produce solid particles composed of surfactant-coated drug nanoparticles embedded within a hydrophilic dextran matrix. The release rate of the drug from the solid emulsion was significantly higher than that of pure itraconazole powder due to formation of a microemulsion after reconstitution that permitted much faster dissolution rates, leading to higher drug bioavailability. A pharmacokinetic study in rats showed that the total plasma concentration of itraconazole after oral administration was significantly higher than that of pure itraconazole powder with the same drug dose. The z-average diameters of the liquid emulsion drops and reconstituted solid emulsion drops were 410 and 453 nm, respectively. The differential scanning calorimetry (DSC) results revealed that itraconazole was present in the solid emulsion as a molecular dispersion or in drug-rich amorphous domains.

Laouini et al. (2012) prepared O/W nanoemulsions loaded with vitamin E for pulmonary delivery of vitamin E using cross flow SPG membrane (Table 10.5E). The role of cosurfactant in expanding the size of stable microemulsion region is shown in Fig. 10.17 for the ternary system composed of Tween 80/Brij 35 surfactant mixture (1:1 w/w), MCT oil, and water. The system with combined used of two surfactants gave the maximum microemulsion region and the smallest drop size as compared with the systems composed of individual surfactants. Under optimal operating conditions, a nanoemulsion composed of water, MCT, and Tween 80/Brij 35 surfactant mixture at the weight ratio of 80/17.75/2.25 was prepared with an average drop size of 78 nm and a drop distribution span of 0.25.

Joseph and Bunjes (2014) prepared O/W nanoemulsions using external pressure-type SPG microkit with a pore size of 0.1–1.1 μm (Fig. 10.5F). The ratio of the median drop size to the membrane pore size ranged from 2.8:1 to

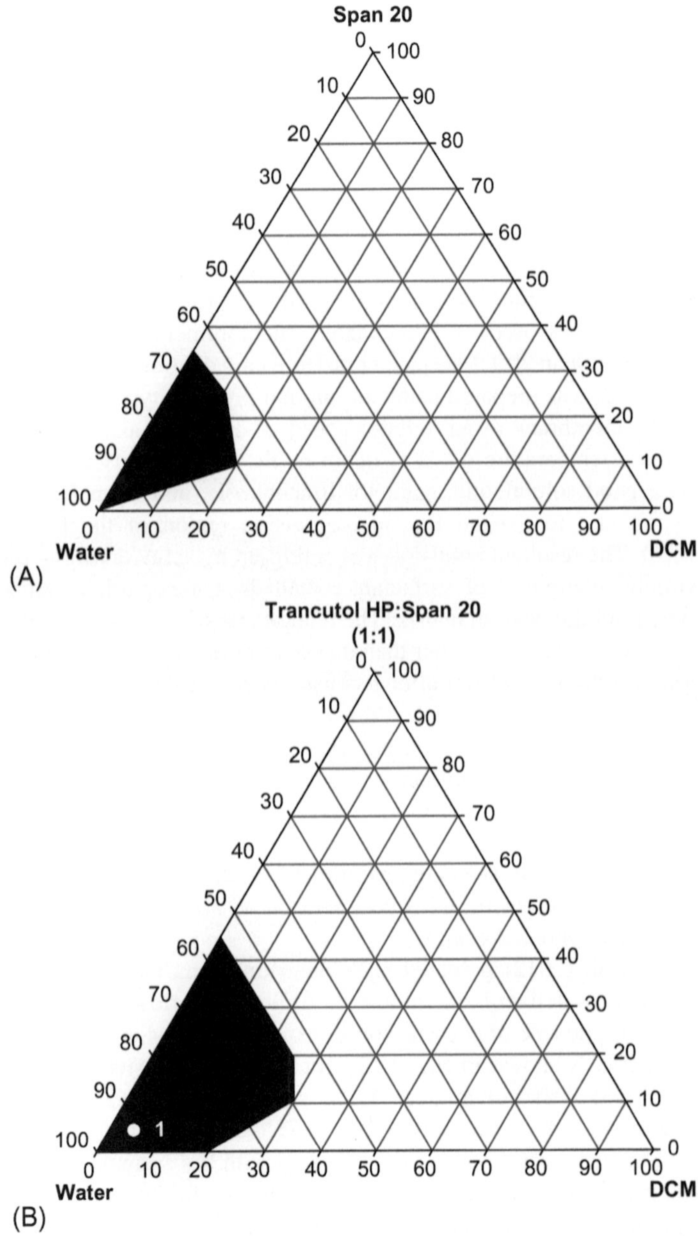

FIG. 10.16 The pseudoternary phase diagram for (A) Span 20, DCM, and water and (B) Span 20: Transcutol HP (1:1 mixture), DCM, and water. *Black* regions are stable microemulsion regions, and point 1 represents the composition of the blend used to prepare a microemulsion by direct ME in SPG microkit, Table 10.5A (Choi et al., 2012).

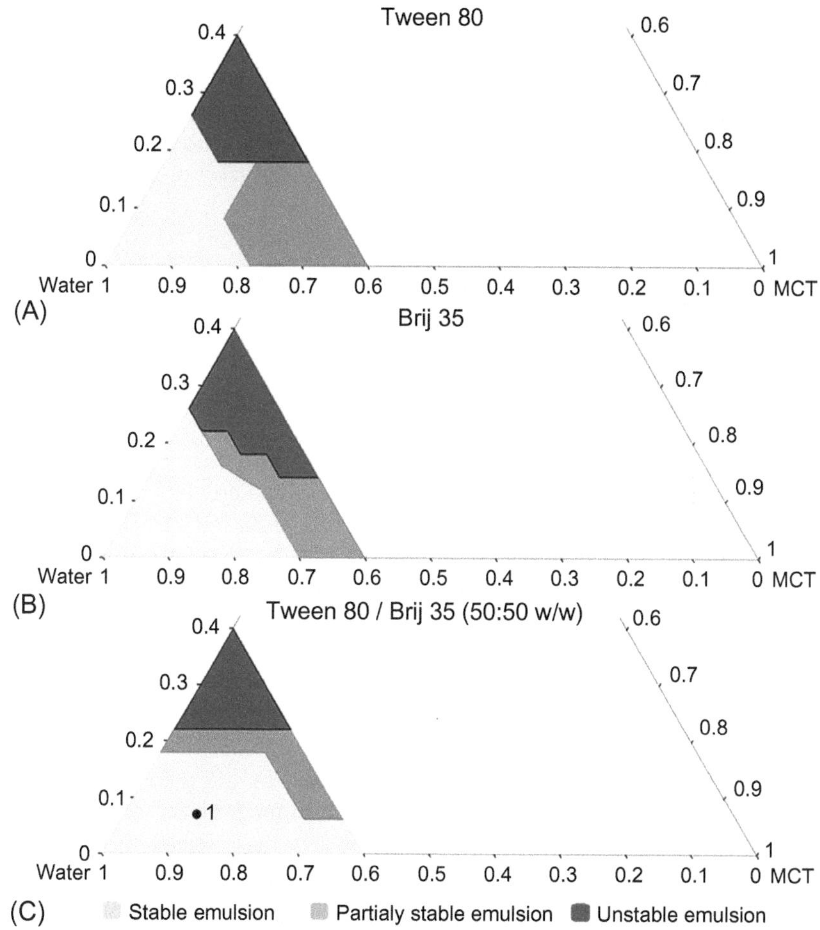

FIG. 10.17 Pseudoternary phase diagram for (A) Tween 80, MCT, and water; (B) Brij 35, MCT, and water; and (C) Tween 80: Brij 35 (1:1 w/w mixture), MCT, and water. *Light gray* regions are stable microemulsion regions, *gray* regions are partially stable, and *dark gray* regions are unstable regions. Point 1 represents the composition of the blend that resulted in the highest drop-size uniformity in cross flow DME using SPG membrane, Table 10.5E (Laouini et al., 2012).

8.2:1 when nonpolar oils such as soybean or MCT oil were dispersed at 200–800 kPa into gently stirred aqueous phase containing 5 wt% SDS or 1–2 wt% Tween 20. Submicron drops were consistently produced only with a 0.1 μm pore-sized SPG membrane. However, injecting more polar Gelucire® 44/14 lipid melt at 55°C and 10 kPa through a 1.1 μm SPG membrane into a heated aqueous solution containing 2 wt% Tween 20 resulted in an emulsion with a median drop size of 70 nm indicating that a microemulsion was formed. Self-emulsifying properties of Gelucire® 44/14 were confirmed by preparing NPs by mixing the molten lipid at 55°C with 2% Tween 20 solution using a magnetic stirrer at 300 rpm without using any membrane. After cooling the mixture to 5°C, the particle size was the same as after membrane emulsification. It shows that emulsification of Gelucire® 44/14 melt was not controlled by the membrane.

Yanagishita et al. (2010) prepared nanodroplets of photocurable acrylic monomer in 0.3 wt% SDS solutions using anodic porous alumina membrane with a pore size ranging from 130 to 320 nm (Table 10.5G). The ratio of the median drop size to the membrane pore size ranged from 1.5:1 to 2.3:1, and a coefficient of variation of drop sizes was less than 10%. The formed drops were polymerized using a UV lamp without any shrinkage during polymerization.

W/O/W emulsions prepared by ME can be used as colloidal carriers for highly water-soluble drugs (Table 10.5H–I). Koga et al. (2010) prepared W_1/O/W_2 nanoemulsions for enhanced intestinal absorption of BCS class III drugs using SPG filter kit. BCS class III drugs are highly hydrophilic drugs with a low permeability across the lipophilic intestinal membrane, which can be encapsulated in a W_1/O/W_2 or S/O/W emulsion to enhance their bioavailability. W_1/O/W_2 nanoemulsions with calcein entrapped in inner water droplets were prepared by adding CR-310, a lipophilic surfactant in the oil phase (soybean oil) and different hydrophilic surfactants with and without PVA in the outer aqueous phase. A primary W_1/O emulsion was prepared using a Polytron® high-speed rotor-stator homogenizer at 16,000 rpm and was extruded at 0.1 mL/h through a 0.1 μm pore-sized SPG membrane into a glass vial containing 9 mL of the stirred outer phase. The mean size of W_1/O drops, calcein entrapment efficiency, and intestinal absorption of calcein by rat intestine were significantly affected by the choice of the hydrophilic surfactant. In the absence of any hydrophilic surfactant, the drop size was 322 nm not affected by the presence of PVA. The smallest drop size of 105 nm was achieved in the presence of 1% HS-11 in the outer aqueous phase. On the other hand, in the presence of HS-9 and S-28D surfactants, the drop size was above 1 μm. After intraduodenal administration of 1.0 mg/kg calcein from the prepared nanoemulsion into rats, the maximum concentration of calcein reached in the blood serum was 118 ±47 ng/mL. After administration of calcein solution with the same calcein dose, the maximum concentration of calcein in the serum was only 13.7 ±5.4 ng/mL. Pradhan et al. (2014) prepared W_1/O/W_2 microemulsions loaded with doxorubicin using SPG membrane with a pore size of 1.4 μm. The oil phase was 5% (w/v) PGPR in soybean oil, and the external aqueous phase was 3% (w/v) Tween 20 solution (Table 10.5I).

10.12 PREPARATION OF NANOEMULSIONS USING PREMIX ME

Joseph and Bunjes (2012) prepared pharmaceutical O/W nanoemulsions by repeated premix ME through PC membranes with a pore size of 50, 100, or 200 nm using LiposoFast-Basic extruder or EmulsiFlex-C5 homogenizer equipped with a high-pressure filter/extruder. The dispersed phase was MCT or soybean oil, and the aqueous solutions of sodium dodecyl sulfate (SDS), poloxamer 188 (Pol), sucrose laurate (SL), or polyglyceryl-10 laurate (PG10-L) were used as the continuous phase. The oil was predispersed in the surfactant solution with an Ultra-Turrax device, and the resulting premix containing 20 wt% oil, 15 wt% surfactant, and 65% water was extruded through the membrane 11 or 21 times. For extrusion through a 50 nm membrane, the premixes were preextruded 11 times using a 200 nm membrane, followed by 11 or 21 extrusion cycles with a 50 nm membrane. The nanodispersions had mean particle sizes between 100 and 200 nm after 21 extrusion cycles, depending on the membrane pore size, the composition of the emulsion, and the extrusion device employed. The smallest particle size was obtained with poloxamer 188 and the largest with PG10-L; the narrowest particle-size distribution was observed with SDS and the broadest with PG10-L. The minimum amount of surfactant in the aqueous phase depended on the membrane pore size. For nanoemulsions composed of 20 wt% MCT oil prepared using membranes with 200 and 100 nm pore sizes, the concentration of SDS could be reduced from 15% to 4% without significant effect on the median drop size. In premix ME processes with smaller pore sizes, SDS concentration of at least 8% was needed to achieve monomodal particle-size distributions after 21 cycles.

Gehrmann and Bunjes (2016a) prepared O/W nanoemulsions stabilized with phospholipids (Lipoid S75 and S100), Tween 80, and SL by extruding premix through disposable syringe filters equipped with polyethersulfone (PES), cellulose acetate (CA), mixed cellulose ester (MCE), or nylon membranes (Table 10.6C). The premix was prepared by pushing the liquids back and forth between the two syringes starting from two separate phases in two different syringes (Fig. 10.10A). The premix with a median particle size between 10 and 24 μm was preextruded through a 1 μm glass-fiber filter to reduce the particle size to 1–6 μm, followed by 21 extrusion cycles through a 200 nm membrane and one subsequent cycle through a 100 nm membrane. In most cases, the last cycle was only possible after a 1:5 dilution of the nanoemulsion with the continuous aqueous phase. The final median drop size achieved with PES membrane was 400 nm or below irrespective of the surfactant. In all cases, the most substantial decrease in drop size was achieved after the first extrusion. When oil drops were stabilized with Lipoid S100 and sucrose laurate (SL), the median drop size was below 1 μm already after first cycle, whereas the drops stabilized with Lipoid S75 and Tween 80 were still in the micrometer range. The most uniform size distributions (span $\approx$0.5) were achieved with Tween 80 and SL, and the largest median particle size after 21 cycles was obtained using CA and nylon membranes.

Gehrmann and Bunjes (2017a) extruded premixed MCT emulsions composed of five different pharmaceutically relevant water-soluble surfactants through seven different hydrophilic polymeric membranes with a pore size of 200 nm. The emulsions obtained after 21 cycles differed strongly in particle-size distribution with a range of median drop sizes between 150 nm and 11 μm. A key factor determining the median drop size was the contact angle between the continuous phase and membrane material, defined by the combination of the surfactant and membrane material used. The microstructure and thickness of the membrane and the type of dispersed phase were less important as similar results were obtained with peanut oil and MCT with widely different viscosities (60 and 30 mPa s, respectively). Based on their ability to create nanoemulsions with a d_{50} value smaller than 500 nm after 21 extrusion cycles at the transmembrane flow rate of 25 mL/s, the hydrophilic polymeric membranes were divided into three groups (Gehrmann and Bunjes, 2017a). The first group (Nylon and CA) is the membranes that can produce emulsions with $d_{50} < 500$ nm and low span values with all pharmaceutically relevant surfactants (SDS, Pol, Tween 80, Tyl, and SL). The second group (PC, PS, and PVDF) yielded $d_{50} < 500$ nm only with SDS. The third group (PES and PE) showed the performance between the two groups. Depending on the type of the surfactant and membrane material, the contact angle between the surfactant solution and the membrane surface varied between 17° and 77° (Fig. 10.18). The contact angle between the surfactant solution and the membrane surface of less than 49° led to nanoemulsions with $d_{50} < 500$ nm in all cases, whereas nanoemulsions with $d_{50} > 3$ μm were obtained for all contact angles higher than 55°. To prove the assumption that membrane hydrophilicity is a key factor in determining the drop size in repeated extrusion through polymeric membranes, PC and PE membranes were treated with nitrogen and oxygen plasma at 25–75 W for 5–10 min to enhance their hydrophilicity. After hydrophilization, smaller

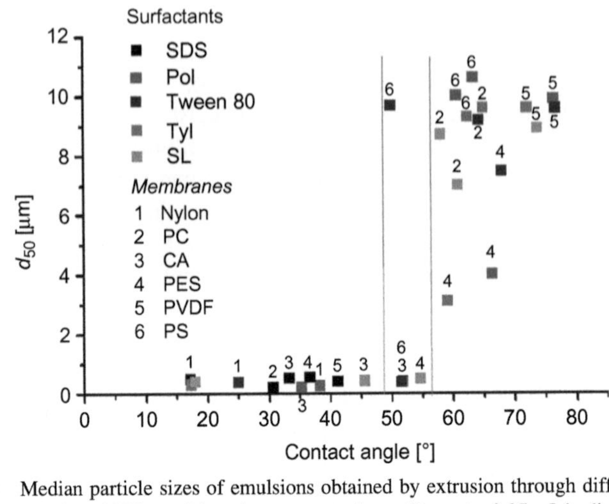

FIG. 10.18 Median particle sizes of emulsions obtained by extrusion through different syringe filters with a pore size of 200 nm (21 cycles; transmembrane flow rate, 0.25 mL/s; dispersed phase, MCT oil; and continuous phase, 6.25 wt% surfactant solution) as a function of contact angle between surfactant solution and membrane surface. Each point is a mean of three repeated experiments.

d_{50} values were achieved with both membranes. Premix ME using 21 extrusion cycles through Anodisc™ membrane filter containing highly hydrophilic aluminum oxide Anopore™ membrane with a pore size of 200 nm led to d_{50} values smaller than 250 nm with all surfactants. Using Anopore™ membrane, the minimum d_{50} value of 77 ±4 nm was achieved with SDS (Gehrmann and Bunjes, 2017a).

10.13 PRODUCTION OF NANOPARTICLES FROM NANOEMULSIONS PREPARED BY ME

Nanodroplets prepared by membrane emulsification can be solidified into nanoparticles (NPs) by implementing various physicochemical or chemical processes within the generated drops. Depending on the formulation, solidification reaction can include ionic cross-linking, solvent evaporation, cooling crystallization, spray drying, and free-radical polymerization.

10.13.1 Hydrogel Nanoparticles

Hydrogel NPs are nanospheres composed of cross-linked hydrophilic polymers, which can absorb and hold large amount of water, due to their hydrophilic character (Hoare and Kohane, 2008; Mokhtari, et al., 2017). Hydrogel NPs were produced by emulsifying an aqueous solution of hydrophilic polymer (chitosan, sodium alginate, etc), in an organic solution of oil-soluble surfactant, using hydrophobic membrane, followed by ionic or covalent cross-linking of the polymer within the formed nanodroplets. Chitosan NPs have been produced by repeated premix ME using a 1.4 μm SPG membrane (Table 10.7A) or direct ME using a 0.5 μm SPG membrane (Table 10.7B). The organic continuous phase was a mixture of paraffin and petroleum ether containing 4 wt% of hexaglycerin penta ester (PO-500). Covalent cross-linking was achieved by dripping toluene saturated with glutaraldehyde in the prepared W/O nanoemulsion (Fig. 10.19A and B).

In ionic gelation, charged functional groups along the polymer chains interact with oppositely charged divalent or polyvalent ions to form ionic bridges that cross-link the polymer. Alginate NPs with an average size of 370 nm were produced by mixing together two W/O emulsions: nanoemulsion 1 containing aqueous droplets of 1 wt% alginate solution and miniemulsion 2 containing aqueous droplets of 5 M $CaCl_2$ solution (Nan et al., 2014). The cross-linking was achieved by merging two droplets (1 and 2) upon mechanical agitation, which triggers a cross-linking reaction between alginate chains and Ca^{2+} ions within fused droplets (Fig. 10.19C). Emulsion 1 was prepared by premix ME at the transmembrane pressure of 10 bar using 1.4 μm SPG membrane, and miniemulsion 2 was prepared by ultrasonication. Hybrid chitosan/alginate nanospheres were prepared by dispersing the produced alginate NPs into 0.7 wt% chitosan solution.

TABLE 10.7 Hydrogel Nanoparticles (NPs) Produced From Nanodroplets Prepared by Membrane Emulsification

Process	Product	Emulsion Formulation	Authors
External gelation			
A. Repeated premix ME, 1.4 μm SPG membrane, five cycles	Chitosan NPs (d_{NP}=370 nm)	*DP*: 0.2 wt% chitosan+1 wt% acetic acid in water *CP*: 4 wt% PO-500 in liquid paraffin/petroleum ether (1:2, v/v) *XL*: Toluene saturated by glutaraldehyde	Lv et al. (2009); Ma et al. (2010)
B. Direct ME, external pressure-type microkit, 0.5 μm SPG membrane	W/O nanoemulsion (d_d=780 nm and ϕ=0.09) Chitosan NPs (d_{NP}=390 nm)	*DP*: 1.5 wt% chitosan+1 wt% acetic acid+0.9 wt% NaCl in water *CP*: 4 wt% PO-500 in liquid paraffin/petroleum ether (7:5, v/v) *XL*: toluene saturated with glutaraldehyde	Wang et al. (2005)
Gelation triggered by droplet merging			
C. Premix ME, 1.4 and 2.8 μm SPG membrane	Alginate NPs (d_{NP}=370 and 700 nm, respectively)	*DP1*: 1 wt% alginate in water *DP2*: 5 mol L^{-1} $CaCl_2$ in water *CP*: 4 wt% PO-500 in liquid paraffin/petroleum ether (1/2, v/v)	Nan et al. (2014)

DP, *dispersed phase;* CP, *continuous phase;* XL, *cross-linking agent.*

10.13.2 Solid Lipid Nanoparticles

Solid lipid NPs are composed of lipids that are solid at room temperature, such as saturated fatty acids (palmitic acid, stearic acid, and behenic acid); triesters of glycerol and fatty acids; and mono-, di-, and triesters of glycerol and polyethylene glycol (Jafari, et al., 2015; Katouzian and Jafari, 2016; Katouzian et al., 2017).

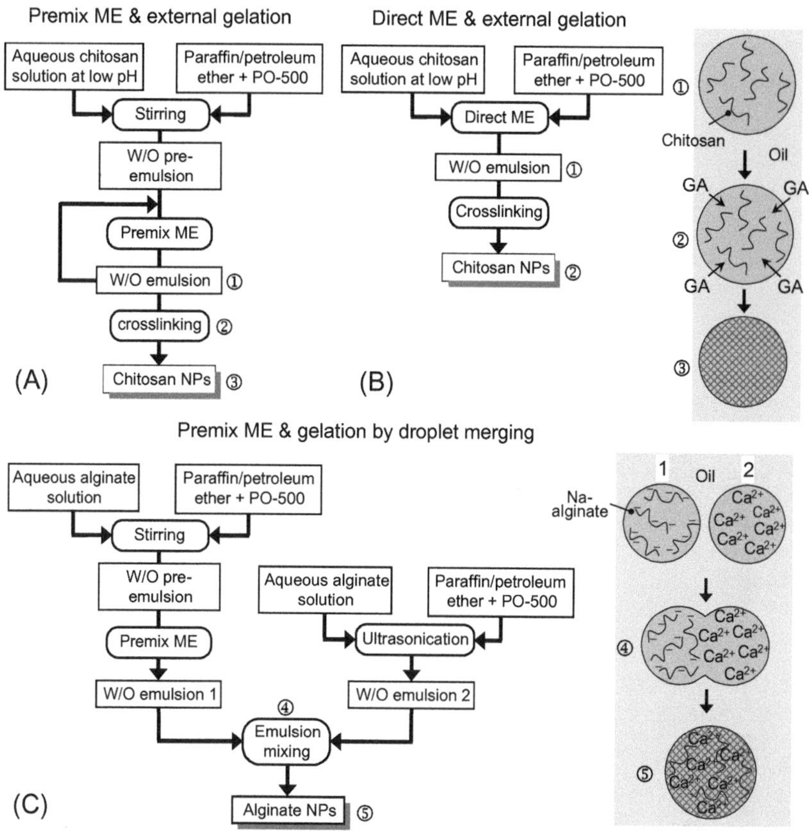

FIG. 10.19 Preparation of hydrogel NPs from W/O nanoemulsions: (A) premix ME and subsequent external cross-linking by glutaraldehyde (GA) (Lv et al., 2009), (B) direct ME and external cross-linking by GA (Wang et al., 2005), and (C) premix ME and internal ionic cross-linking triggered by droplet merging (Nan et al., 2014).

The melting point, T_m, of solid lipids dispersed using membranes was between 44 and 75°C, and ME was carried out at the temperature that was at least 5°C above the melting point of the matrix lipid to prevent pore clogging by fat particles. The resulting nanodroplets were then crystallized into solid NPs by rapid cooling to 5°C in an ice bath. D'oria et al. (2009) produced Compritol 888 and Gelucire 44/14 NPs with a mean size between 50 and 750 nm by cross flow direct ME using SPG membrane with a pore size ranging from 0.2 to 1 μm at the temperature of 80 or 65°C and the flux up to $0.84\,m^3\,m^{-2}\,h^{-1}$ (Table 10.8A). Joseph and Bunjes (2012) prepared trimyristin (TM) NPs by premix ME through a PC track-etched membrane with a pore size of 200 nm using EmulsiFlex-C5 homogenizer equipped with a high-pressure filter/extruder (Table 10.8B). The continuous phase was 18.75 wt% aqueous solution of sodium dodecyl sulfate (SDS), poloxamer 188 (Pol), sucrose laurate (SL), or

TABLE 10.8 Solid Lipid NPs Fabricated by Solidification of Melted Lipid Nanodroplets Prepared by Membrane Emulsification

Process	Product	Emulsion Formulation	Authors
A. Direct ME, cross flow system, 0.2, 0.4, and 1 μm SPG membrane	Gelucire 44/14 NPs (T_m=44°C and d_{NP}=50–130 nm) and Compritol 888 NPs (T_m=69−74°C and d_{NP}=560–760 nm)	*DP*: Gelucire 44/14 melt at 65°C or Compritol 888 melt at 80°C *CP*: Aqueous 0.125 wt% Tween 20 or 1.26 wt% Pluronic F68 solution	D'Oria et al. (2009)
B. Repeated premix ME, EmulsiFlex-C5 homogenizer with filter/extruder unit, 0.2 μm PC membrane	TM NPs (T_m=55–58°C and d_{NP}=100–200 nm)	*DP*: TM melt at 65°C *CP*: Aqueous solution containing 18.75 wt% SDS or 18.75 wt% Pol or 18.75 wt% SL or 18.75 wt% PG10-L	Joseph and Bunjes (2012)
C. Direct ME, external pressure-type microkit with jacketed pressure vessel, 0.1, 0.2 and 0.3 μm SPG membrane	TM, CATO, Gelucire 44/14, and Q_{10}-loaded TM NPs (d_{NP}=70–400 nm)	*DP*: 0–5 wt% Q_{10} in TM melt at 66°C, CATO melt at 80°C or Gelucire 44/14 melt at 54°C *CP*: 5 wt% SDS or 15 wt% SL in water	Joseph and Bunjes (2014)

polyglyceryl-10 laurate (PG10-L). The mean particle size ranged between 100 and 200 nm after 11–21 membrane extrusion cycles at the operating pressure ranging between 25 and 180 bar. Joseph and Bunjes (2014) prepared glyceryl behenate (CATO) NPs by direct ME at 60 kPa using 0.2 μm SPG membrane. CATO NPs stabilized with 5 wt% SDS were much smaller (d_{50}=80 nm) than those stabilized with 5 wt% poloxamer 188 (d_{50}=950 nm). On the other hand, sucrose laurate-stabilized TM NPs were smaller (d_{50}=72 nm) than SDS-stabilized TM NPs (d_{50}=360 nm) when prepared with a 0.1 μm SPG membrane at 800 kPa.

10.13.3 Biodegradable Polymeric Nanoparticles

Biodegradable polymeric NPs were prepared by dispersing an organic polymer solution in an aqueous surfactant solution using membrane emulsification and subsequent evaporation of volatile organic solvent from the formed emulsion droplets. Ito et al. (2011) prepared PLGA NPs by premix ME using SPG membranes with a pore size ranging from 0.5 to 2 μm (Table 10.9 and Fig. 10.20A). The maximum concentration of PLGA in acetone that can be used in the dispersed phase without significant clogging of the membrane pores was 3 wt %. The most uniform NPs with a CV of 17% and a particle diameter of 190 nm were prepared using 2 wt% PLGA solution and SPG membrane with a pore size of 0.48 μm. Doan et al. (2011) prepared PLGA NPs loaded with 18 wt% of rifampicin (RIF) by premix ME using three extrusion cycles through a 5.9 μm SPG membrane at 100 kPa. Zhang et al. (2014b) prepared PLA NPs for vaccine formulations with a diameter of 820 nm by premix ME using five extrusion cycles through a 2.8 μm SPG membrane. The droplets were solidified into NPs by stirring overnight under ambient conditions to completely remove the organic solvent (DCM).

Oil-core/polymer-shell NPs were produced by adding a nonvolatile nonsolvent in the organic phase and inducing internal phase separation during solvent evaporation (Sawalha et al., 2008). The drops are composed of a mixture of PLLA (polymer), DCM (a volatile water-immiscible good solvent for PLLA), and dodecane (DD, a nonvolatile water-immiscible nonsolvent). As DCM evaporates, the drops shrink and become enriched in PLLA and DD (Fig. 10.20B). When the drop composition reaches the binodal boundary, the phase separation occurs, leading to the formation of tiny droplets of polymer-enriched phase (rich in DCM and PLLA) dispersed in a polymer-poor phase (mainly consisting of DD). As PLLA is more hydrophilic than DD, polymer-rich droplets migrate within the drop to the interface with the aqueous phase, where they fuse and spread to engulf the oil droplet (Loxley and Vincent, 1998). Further DCM removal causes PLLA to precipitate in the shell, forming a coherent solid envelope around a DD core. The presence of 25–30 wt% alcohol (methanol, ethanol, or 2-propanol) in the continuous phase increases the rate of DCM removal, as DCM is more soluble in alcohol-water mixtures than in pure water (Sawalha et al., 2008). As a result, the drops solidify more quickly, and smaller and more uniform PLLA particles were obtained. The premix was extruded through a 1 μm glass-fiber syringe filter using 15 cycles. The smallest of 350 nm was achieved using 25 wt% 2-propanol in the continuous phase followed by 30 wt% ethanol (800 nm) and 30 wt% methanol (1.4 μm). At higher alcohol concentrations in the continuous phase (>50% for methanol, > 5% for ethanol, and >35% for 2-propanol), DCM is completely miscible in the continuous phase, which results in direct nanoprecipitation of PLLA without emulsification and formation of very small PLLA NPs.

Biodegradable polymeric NPs loaded with hydrophilic active ingredients were prepared by double emulsion-solvent evaporation method (Table 10.9C). The process involves the preparation of W_1/O emulsion containing a hydrophilic active (i.e., a protein) entrapped in the inner water phase (W_1) by

TABLE 10.9 Biodegradable Polymeric Nanospheres and Nanocapsules Prepared by Membrane Emulsification/Solvent Evaporation

Process	Product	Emulsion Formulation	Authors
A. Nanospheres prepared via O/W route			
Premix ME, external pressure SPG microkit, d_p=0.5–2 μm	PLGA NPs (d_{NP}=150–250 nm and CV=15%–20%)	*DP*: 1–3 wt% PLGA in acetone *CP*: 1 wt% PVA in water	Ito et al. (2011)
Repeated (n=3) premix ME at 100 kPa, external pressure microkit, 5.9 μm SPG membrane	RIF-loaded PLGA NPs (d_{NP}=640 nm)	*DP*: 3 wt% PLGA+ 2 wt% RIF in EA *CP*: 3 wt% PVA+ 8.5 vol% EA in water, pH 4	Doan et al. (2011)
Repeated (n=5) premix ME, 2.8 μm SPG membrane	PLA NPs as vaccine adjuvants (d_{NP}=820 nm)	*DP*: 4 wt% PLA in DCM *CP*: 1.9% (w/v) PVA in water	Zhang et al. (2014b)
B. Core/shell nanocapsules prepared via O/W route			
Repeated (n=15) premix ME, 1 μm glass-fiber syringe membrane	DD core/PLLA shell NPs (d_{NP}=350 nm)	*DP*: 0.6 wt% PLLA+ 9.1 wt% DD in DCM *CP*: 0.3% PVA+ 25 wt% PrOH in water	Sawalha et al. (2008)
C. Matrix nanocapsules prepared via W/O/W route			
Repeated premix ME (n=8), 2.8 μm SPG membrane	PLGA particles loaded with ovalbumin (d_{part}=591 nm and PDI=0.17)	W_1: 10% (w/v) ovalbumin in water *O*: 5% (w/v) PLGA in ethyl acetate W_2: 1.5% (w/v) PVA + 0.9% (w/v) NaCl in water	Zhang et al. (2014a)

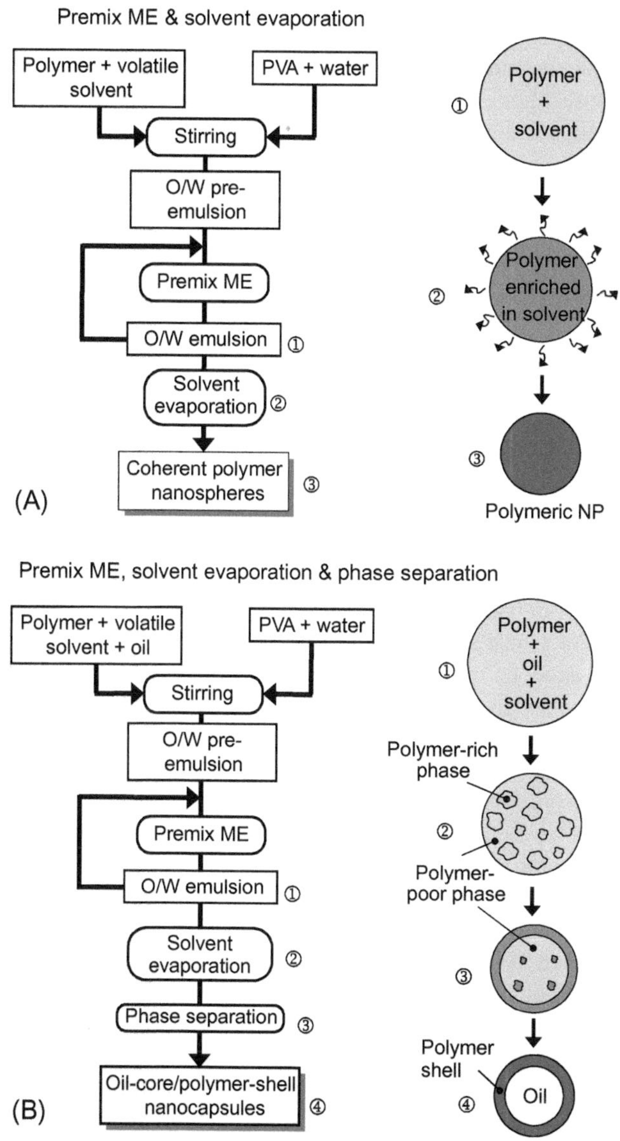

FIG. 10.20 Preparation of biodegradable polymeric NPs by membrane emulsification/solvent evaporation: (A) nanospheres (Ito et al., 2011) and (B) core/shell nanocapsules made by addition of nonvolatile poor solvent ("oil") in the dispersed phase and evaporation of good solvent (Sawalha et al., 2008).

ultrasonication or rotor-stator homogenization. Ultrasonication results in smaller droplets that are more uniformly distributed in a polymer matrix after solvent evaporation (Qi et al., 2014). The second step is the formation of $W_1/O/W_2$ emulsion by extruding the $W_1/O/W_2$ premix through the membrane multiple times (Table 10.9C).

10.14 CONCLUSIONS

Microporous membranes are increasingly used for preparation of micro- and nanoemulsions. Membrane emulsification is a milder process than high-pressure homogenization and enables preparation of emulsions with narrow particle-size distribution and controlled average drop size. The two membrane-assisted processes used are direct membrane emulsification, where a pure dispersed-phase liquid is injected through the membrane at controlled injection rate and shear on the membrane surface, and premix membrane emulsification, where a pre-emulsified mixture of the dispersed and continuous phase is extruded (usually repeatedly) through the membrane. Microemulsions are usually prepared with direct membrane emulsification using membranes with a pore size in the range from 0.45 to 2.5 μm, and the median drop size is two to nine times smaller than the pore size. Nanoemulsions can be prepared with direct or premix membrane emulsification using nanoporous membranes.

The most common membranes used for preparation of micro- and nanoemulsions are SPG, polymeric, and microsieve membranes. Nanoemulsions can be prepared by direct membrane emulsification using SPG membranes with a pore size of 0.1 or 0.2 μm, and the ratio of the median drop size to the pore size ranges from 2.8:1 to 8.2:1 when soybean or MCT oil was dispersed into the gently stirred aqueous solutions of SDS or Tween 20. Due to its higher ability to generate smaller drops, premix membrane emulsification is a more convenient process for preparation of nanoemulsions than direct process. The median drop size achieved in premix membrane emulsification can be reduced by decreasing the pore size and increasing the number of extrusion cycles and transmembrane pressure. Track-etched polycarbonate membranes were found to be less efficient in reducing the drop size in premix membrane emulsification than SPG membranes due to their rectilinear pores and smaller thickness (less than 20 nm) as compared with SPG membranes (700–900 nm). One extrusion of premix through SPG membrane with a pore size of 100–300 nm led to smaller drops than 11 or 21 extrusion cycles through tracked-etched PC membranes with the same pore size. The median drop size in premix membrane emulsification using polymeric membrane primarily depended on the contact angle between the continuous aqueous phase and the membrane surface. The contact angle between the surfactant solution and the membrane surface must be less than 49° to create nanoemulsions with a d_{50} value smaller than 500 nm after 21 extrusion cycles at the flow rate of 25 mL/s through a polymeric membrane filter.

Nanodroplets prepared by membrane emulsification can be solidified into hydrogel, solid lipid, or polymeric nanoparticles by implementing various physicochemical or chemical reactions within the generated drops. Depending on the formulation, solidification reaction can include ionic or covalent cross-linking, solvent evaporation, cooling crystallization, spray drying, or free-radical polymerization.

REFERENCES

Adamczak, M., Kupiec, A., Jarek, E., Szczepanowicz, K., Warszyński, P., 2014. Preparation of the squalene-based capsules by membrane emulsification method and polyelectrolyte multilayer adsorption. Colloids Surf. A Physicochem. Eng. Asp. 462, 147–152.

Anna, S.L., Bontoux, N., Stone, H.A., 2003. Formation of dispersions using 'flow focusing' in microchannels. Appl. Phys. Lett. 82, 364–366.

Apel, P., 2001. Track-etching technique in membrane technology. Radiat. Meas. 34, 559–566.

Aryanti, N., Williams, R.A., Hou, R., Vladisavljević, G.T., 2006. Performance of rotating membrane emulsification for O/W production. Desalination 200, 572–574.

Balbach, S., Korn, C., 2004. Pharmaceutical evaluation of early development candidates 'the 100 mg-approach'. Int. J. Pharm. 275, 1–12.

Belwalkar, A., Grasing, E., Van Geertruyden, B., Huang, Z., Misiolek, W.Z., 2008. Effect of processing parameters on pore structure and thickness of anodic aluminum oxide (AAO) tubular membranes. J. Membr. Sci. 319, 192–198.

Chen, G.G., Luo, G.S., Xu, J.H., Wang, J.D., 2004. Membrane dispersion precipitation method to prepare nanopartials. Powder Technol. 139, 180–185.

Chen, Y., Zhao, Z., Dai, J., Liu, Y., Ma, H., Nie, R., 2008. Etching characteristic for tracks of multicharged ions in polymer. Radiat. Meas. 43, S111–S115.

Cheng, C.J., Chu, L.Y., Xie, R., Wang, X.W., 2008. Preparation of highly monodisperse W/O emulsions with hydrophobically modified SPG membranes. Chem. Eng. Technol. 31, 377–383.

Choi, Y.K., Poudel, B.K., Marasini, N., Yang, K.Y., Kim, J.W., Kim, J.O., Choi, H.G., Yong, C.S., 2012. Enhanced solubility and oral bioavailability of itraconazole by combining membrane emulsification and spray drying technique. Int. J. Pharm. 434, 264–271.

Doan, T.V.P., Couet, W., Olivier, J.C., 2011. Formulation and in vitro characterization of inhalable rifampicin-loaded PLGA microspheres for sustained lung delivery. Int. J. Pharm. 414, 112–117.

D'oria, C., Charcosset, C., Barresi, A.A., Fessi, H., 2009. Preparation of solid lipid particles by membrane emulsification—influence of process parameters. Colloids Surf. A Physicochem. Eng. Asp. 338, 114–118.

Dowding, P.J., Goodwin, J.W., Vincent, B., 2001. Production of porous suspension polymer beads with a narrow size distribution using a cross-flow membrane and a continuous tubular reactor. Colloids Surf. A 180, 301–309.

Dragosavac, M.M., Holdich, R.G., Vladisavljević, G.T., Sovilj, M.N., 2012. Stirred cell membrane emulsification for multiple emulsions containing unrefined pumpkin seed oil with uniform droplet size. J. Membr. Sci. 392–393, 122–129.

Egidi, E., Gasparini, G., Holdich, R.G., Vladisavljević, G.T., Kosvintsev, S.R., 2008. Membrane emulsification using membranes of regular pore spacing: droplet size and uniformity in the presence of surface shear. J. Membr. Sci. 323, 414–420.

Gehrmann, S., Bunjes, H., 2016a. Preparation of lipid nanoemulsions by premix membrane emulsification with disposable materials. Int. J. Pharm. 511, 741–744.

Gehrmann, S., Bunjes, H., 2016b. Instrumented small scale extruder to investigate the influence of process parameters during premix membrane emulsification. Chem. Eng. J. 284, 716–723.

Gehrmann, S., Bunjes, H., 2017a. Influence of membrane material on the production of colloidal emulsions by premix membrane emulsification. Eur. J. Pharm. Biopharm. https://doi.org/10.1016/j.ejpb.2016.11.006.

Gehrmann, S., Bunjes, H., 2017b. Preparation of nanoemulsions by premix membrane emulsification: which parameters have a significant influence on the resulting particle size? J. Pharm. Sci. https://doi.org/10.1016/j.xphs.2017.04.066.

Gijsbertsen-Abrahamse, A.J., van der Padt, A., Boom, R.M., 2004. Status of cross-flow membrane emulsification and outlook for industrial application. J. Membr. Sci. 230, 149–159.

Graber, M., 2010. Transport phenomena in rotating membrane processed W/O/W emulsions (PhD Thesis). Institute of Food, Nutrition, and Health, Laboratory of Food Process Engineering, ETH Zurich, Switzerland.

He, D., Ulbricht, M., 2006. Surface-selective photo-grafting on porous polymer membranes via a synergist immobilization method. J. Mater. Chem. 16, 1860–1868.

Higashi, S., Setoguchi, T., 2000. Hepatic arterial injection chemotherapy for hepatocellular carcinoma with epirubicin aqueous solution as numerous vesicles in iodinated poppy-seed oil microdroplets: clinical application of water-in-oil-in-water emulsion prepared using a membrane emulsification technique. Adv. Drug Deliv. Rev. 45, 57–64.

Hoare, T.R., Kohane, D.S., 2008. Hydrogels in drug delivery: progress and challenges. Polymer 49, 1993–2007.

Holdich, R.G., Dragosavac, M.M., Vladisavljević, G.T., Kosvintsev, S.R., 2010. Membrane emulsification with oscillating and stationary membranes. Ind. Eng. Chem. Res. 49, 3810–3817.

Holdich, R.G., Dragosavac, M., Vladisavljević, G.T., Piacentini, E., 2013. Continuous membrane emulsification with pulsed (oscillatory) flow. Ind. Eng. Chem. Res. 52, 507–515.

Ito, F., Kanakubo, Y., Murakami, Y., 2011. Rapid preparation of monodisperse biodegradable polymer nanospheres using a membrane emulsification technique under low gas pressure. J. Polym. Res. 18, 2077–2085.

Jafari, S.M., Fathi, M., Mandala, I., 2015. Emerging product formation. In: Food Waste Recovery: Processing Technologies and Industrial Techniques. Elsevier Inc., New York, pp. 293–317

Jing, C., Chin, C.Y., Xie, R., 2006. Preparation of highly monodisperse W/O emulsions with hydrophobically modified SPG membranes. J. Colloid Interface Sci. 300, 375–382.

Joscelyne, S.M., Trägårdh, G., 2000. Membrane emulsification—a literature review. J. Membr. Sci. 169, 107–117.

Joseph, S., Bunjes, H., 2012. Preparation of nanoemulsions and solid lipid nanoparticles by premix membrane emulsification. J. Pharm. Sci. 2012, 2479–2489.

Joseph, S., Bunjes, H., 2014. Evaluation of Shirasu Porous Glass (SPG) membrane emulsification for the preparation of colloidal lipid drug carrier dispersions. Eur. J. Pharm. Biopharm. 87, 178–186.

Kai, T., Suma, Y., Ono, S., Yamaguchi, Nakao, S., 2006. Effect of the pore surface modification of an inorganic substrate on the plasma-grafting behavior of pore-filling-type organic/inorganic composite membranes. J. Polym. Sci. Part A-1: Polym. Chem. 44, 846–856.

Karbstein, H., Schubert, H., 1995. Developments in the continuous mechanical production of oil-water macro-emulsions. Chem. Eng. Process. 34, 205–211.

Katouzian, I., Jafari, S.M., 2016. Nano-encapsulation as a promising approach for targeted delivery and controlled release of vitamins. Trends Food Sci. Technol. 53, 34–48.

Katouzian, I., Esfanjani, A.F., Jafari, S.M., Akhavan, S., 2017. Formulation and application of a new generation of lipid nano-carriers for the food bioactive ingredients. Trends Food Sci. Technol. 68 (Suppl. C), 14–25.

Kawashima, Y., Hino, T., Takeuchi, H., Niwa, T., Horibe, K., 1991. Shear-induced phase inversion and size control of water/oil/water emulsion droplets with porous membrane. J. Colloid Interface Sci. 145, 512–523.

Kobayashi, I., Mukataka, S., Nakajima, M., 2004. Effect of slot aspect ratio on droplet formation from silicon straight-through microchannels. J. Colloid Interface Sci. 279, 277–280.

Kobayashi, I., Nakajima, M., Mukataka, S., 2003. Preparation characteristics of oil-in-water emulsions using differently charged surfactants in straight-through microchannel emulsification. Colloids Surf. A 229, 33–41.

Kobayashi, I., Vladisavljević, G.T., Uemura, K., Nakajima, M., 2011. CFD analysis of microchannel emulsification: droplet generation process and size effect of asymmetric straight flow-through microchannels. Chem. Eng. Sci. 66, 5556–5565.

Koga, K., Takarada, N., Takada, K., 2010. Nano-sized water-in-oil-in-water emulsion enhances intestinal absorption of calcein, a high solubility and low permeability compound. Eur. J. Pharm. Biopharm. 74, 223–232.

Koltuniewicz, A.B., Field, R.W., Arnot, T.C., 1995. Cross-flow and dead-end microfiltration of oily-water emulsion. Part I: Experimental study and analysis of flux decline. J. Membr. Sci. 102, 193–207.

Koris, A., Piacentini, E., Vatai, G., Bekassy-Molnar, E., Drioli, E., Giorno, L., 2011. Investigation on the effects of a mechanical shear-stress modification method during cross-flow membrane emulsification. J. Membr. Sci. 371, 28–36.

Kosvintsev, S.R., Gasparini, G., Holdich, R.G., 2008. Membrane emulsification: droplet size and uniformity in the absence of surface shear. J. Membr. Sci. 313, 182–189.

Kosvintsev, S.R., Gasparini, G., Holdich, R.G., Cumming, I.W., Stillwell, M.T., 2005. Liquid-liquid membrane dispersion in a stirred cell with and without controlled shear. Ind. Eng. Chem. Res. 44, 9323–9330.

Kukizaki, M., 2009a. Shirasu porous glass (SPG) membrane emulsification in the absence of shear flow at the membrane surface: influence of surfactant type and concentration, viscosities of dispersed and continuous phases, and transmembrane pressure. J. Membr. Sci. 327, 234–243.

Kukizaki, M., 2009b. Relation between salt rejection and electrokinetic properties on Shirasu porous glass (SPG) membranes with nano-order uniform pores. Sep. Sci. Technol. 69, 87–96.

Kukizaki, M., 2010. Large-scale production of alkali-resistant Shirasu porous glass (SPG) membranes: influence of ZrO_2 addition on crystallization and phase separation in Na_2O-CaO-Al_2O_3-B_2O_3-SiO_2 glasses; and alkali durability and pore morphology of the membranes. J. Membr. Sci. 360, 426–435.

Kukizaki, M., Goto, M., 2007a. Preparation and characterization of a new asymmetric type of Shirasu porous glass (SPG) membrane used for membrane emulsification. J. Membr. Sci. 299, 190–199.

Kukizaki, M., Goto, M., 2007b. Preparation and evaluation of uniformly sized solid lipid microcapsules using membrane emulsification. Colloids Surf. A Physicochem. Eng. Asp. 293, 87–94.

Kukizaki, M., Goto, M., 2008. Demulsification of water-in-oil emulsions by permeation through Shirasu-porous-glass (SPG) membranes. J. Membr. Sci. 322, 196–203.

Kukizaki, M., Goto, M., 2009. A comparative study of SPG membrane emulsification in the presence and absence of continuous-phase flow. J. Chem. Eng. Jpn 42, 520–530.

Kukizaki, M., Nakashima, T., 2004. Acid leaching process in the preparation of porous glass membranes from phase-separated glass in the Na_2O-CaO-MgO-Al_2O_3-B_2O_3-SiO_2 system. Membranes 29, 301–308.

Kukizaki, M., Wada, T., 2008. Effect of the membrane wettability on the size and size distribution of microbubbles formed from Shirasu-porous-glass (SPG) membranes. Colloids Surf. A Physicochem. Eng. Asp. 317, 146–154.

Laouini, A., Charcosset, C., Fessi, H., Holdich, R.G., Vladisavljević, G.T., 2013. Preparation of liposomes: a novel application of microengineered membranes: from laboratory scale to large scale. Colloids Surf. B. Biointerfaces 112, 272–278.

Laouini, A., Charcosset, C., Fessi, H., Schroen, K., 2014. Use of dynamic membranes for the preparation of vitamin E-loaded lipid particles: an alternative to prevent fouling observed in classical cross-flow emulsification. Chem. Eng. J. 236, 498–505.

Laouini, A., Fessi, H., Charcosset, C., 2012. Membrane emulsification: a promising alternative for vitamin E encapsulation within nano-emulsion. J. Membr. Sci. 423–424, 85–96.

Lee, K.P., Mattia, D., 2013. Manufacturing of nanoemulsions using nanoporous anodized alumina membranes: experimental investigation and process modeling. Ind. Eng. Chem. Res. 52, 14866–14874.

Liu, F., Hashim, N.A., Liu, Y., Abed, M.R.M., Li, K., 2011. Progress in the production and modification of PVDF membranes. J. Membr. Sci. 375, 1–27.

Loxley, A., Vincent, B., 1998. Preparation of poly(methylmethacrylate) microcapsule with liquid cores. J. Colloid Interface Sci. 208, 49–62.

Lv, P.P., Wei, W., Gong, F.L., Zhang, Y.L., Zhao, H.Y., Lei, J.D., Wang, L.Y., Ma, G.H., 2009. Preparation of uniformly sized chitosan nanospheres by a premix membrane emulsification technique. Ind. Eng. Chem. Res. 48, 8819–8828.

Ma, G.H., Yang, J., Lv, P.P., Wang, L.Y., Wei, W., Tian, R., Wu, J., Su, Z.G., 2010. Preparation of uniform microspheres and microcapsules by modified emulsification process. Macromol. Symp. 288, 41–48.

McClements, D.J., 2012. Nanoemulsions versus microemulsions: terminology, differences, and similarities. Soft Matter 8, 1719–1729.

Mokhtari, S., Jafari, S.M., Assadpour, E., 2017. Development of a nutraceutical nano-delivery system through emulsification/internal gelation of alginate. Food Chem. 229, 286–295.

Mulder, M., 1996. Basic Principles of Membrane Technology. Kluwer Academic Publishers, Dordrecht.

Nakashima, T., 2002. In: Porous glass material and its recent applications.Paper Presented at 38th International SPG Forum on Membrane and Particle Science and Technology in Food and Medical Care in Sadowara, Japan.

Nakashima, T., Kuroki, Y., 1981. Effect of composition and heat treatment on the phase separation of NaO-B_2O_3-SiO_2-Al_2O_3-CaO glass prepared from volcanic ashes. Nippon Kagaku Kaishi 8, 1231–1238.

Nakashima, T., Shimizu, M., 1986. Porous glass from calcium alumino boro-silicate glass. Ceram. Jpn. 21, 408–412.

Nakashima, T., Shimizu, M., Kukizaki, M., 1991. Membrane emulsification by microporous glass. Key Eng. Mater. 61–62, 513–516.

Nakashima, T., Shimizu, M., Kukizaki, M., 1992. Mechanical strength and thermal resistance of porous glass. J. Ceram. Soc. Jpn. Int. Ed. 100, 1389–1393.

Nakashima, T., Shimizu, M., Kukizaki, M., 1993. Effect of surfactant on production of monodispersed O/W emulsion in membrane emulsification. Kag. Kog. Ronbunshu 19, 991–997.

Nakashima, T., Shimizu, M., Kukizaki, M., 1994. Monodisperse single and double emulsions and method of producing same. US Patent 5,326,484.

Nakashima, T., Shimizu, M., Kukizaki, M., 2000. Particle control of emulsion by membrane emulsification and its applications. Adv. Drug Deliv. Rev. 45, 47–56.

Nan, F., Wu, J., Qi, F., Liu, Y., Ngai, T., Ma, G., 2014. Uniform chitosan-coated alginate particles as emulsifiers for preparation of stable Pickering emulsions with stimulus dependence. Colloids Surf. A Physicochem. Eng. Asp. 456, 246–252.

Nazir, A., Schroën, K., Boom, R., 2010. Premix emulsification: a review. J. Membr. Sci. 362, 1–11.

Nazir, A., Schroën, K., Boom, R., 2011. High-throughput premix membrane emulsification using nickel sieves having straight-through pores. J. Membr. Sci. 383, 116–123.

Oh, D.H., Balakrishnan, P., Oh, Y.K., Kim, D.D., Yong, C.S., Choi, H.G., 2011. Effect of process parameters on nanoemulsion droplet size and distribution in SPG membrane emulsification. Int. J. Pharm. 404, 191–197.

Olson, F., Hunt, C.A., Szoka, F.C., 1979. Preparation of liposomes of defined size distribution by extrusion through polycarbonate membranes. Biochim. Biophys. Acta 557, 9–23.

Park, S.H., Yamaguchi, T., Nakao, S., 2001. Cut-off of dilute O/W emulsions through a microfiltration membrane. J. Membr. Sci. 190, 167–178.

Pawlik, A.K., Norton, I.T., 2012. Encapsulation stability of duplex emulsions prepared with SPG cross-flow membrane, SPG rotating membrane and rotor-stator techniques—a comparison. J. Membr. Sci. 415–416, 459–468.

Pawlik, A.K., Norton, I.T., 2013. SPG rotating membrane technique for production of food grade emulsions. J. Food Eng. 114, 530–537.

Piacentini, E., Drioli, E., Giorno, L., 2013. Pulsed back-and-forward cross-flow batch membrane emulsification with high productivity to obtain highly uniform and concentrate emulsions. J. Membr. Sci. 453, 119–125.

Piacentini, E., Giorno, L., Dragosavac, M.M., Vladisavljević, G.T., Holdich, R.G., 2013. Microencapsulation of oil droplets using cold water fish gelatine/gum Arabic complex coacervation by membrane emulsification. Food Res. Int. 53, 362–372.

Pradhan, R., Kim, Y.I., Jeong, J.H., Choi, H.G., Yong, C.S., Kim, J.O., 2014. Fabrication, characterization and pharmacokinetic evaluation of doxorubicin-loaded water-in-oil-in-water microemulsions using a membrane emulsification technique. Chem. Pharm. Bull. 62, 875–882.

Pradhan, R., Lee, D.W., Choi, H.G., Yong, C.S., Kim, J.O., 2013. Fabrication of a uniformly sized fenofibrate microemulsion by membrane emulsification. J. Microencapsul. 30, 42–48.

Qi, F., Wu, J., Hao, D., Yang, T., Ren, Y., Ma, G., Su, Z., 2014. Comparative studies on the influences of primary emulsion preparation on properties of uniform-sized exenatide-loaded PLGA microspheres. Pharm. Res. 31, 1566–1574.

Rayner, M., Trägårdh, G., Trägårdh, C., 2005. The impact of mass transfer and interfacial expansion rate on droplet size in membrane emulsification processes. Colloids Surf. A Physicochem. Eng. Asp. 266, 1–17.

Sawalha, H., Fan, Y., Schroën, K., Boom, R., 2008. Preparation of hollow polylactide microcapsules through premix membrane emulsification—effects of nonsolvent properties. J. Membr. Sci. 325, 665–671.

Schadler, V., Windhab, E.J., 2006. Continuous membrane emulsification by using a membrane system with controlled pore distance. Desalination 189, 130–135.

Schröder, V., Behrend, O., Schubert, H., 1998. Effect of dynamic interfacial tension on the emulsification process using microporous, ceramic membrane. J. Colloid Interface Sci. 202, 334–340.

Seemann, R., Brinkmann, M., Pfohl, T., Herminghaus, S., 2012. Droplet based microfluidics. Rep. Prog. Phys. 75. https://doi.org/10.1088/0034-4885/75/1/016601.

Silva, P.S., Stillwell, M., Williams, B., Dragosavac, M., Vladisavljević, G.T., Bandulasena, H., Holdich, R.G., 2015. Azimuthally oscillating membrane emulsification for controlled droplet production. AIChE J 61, 3607–3615.

Sugiura, S., Nakajima, M., Kumazawa, N., Iwamoto, S., Seki, M., 2002. Characterization of spontaneous transformation-based droplet formation during microchannel emulsification. J. Phys. Chem. B 106, 9405–9409.

Surh, J., Jeong, Y.G., Vladisavljević, G.T., 2008. On the preparation of lecithin-stabilized oil-in-water emulsions by multi-stage premix membrane emulsification. J. Food Eng. 89, 164–170.

Surh, J., Vladisavljević, G.T., Mun, S., McClements, D.J., 2007. Preparation and characterization of water/oil and water/oil/water emulsions containing biopolymer-gelled water droplets. J. Agric. Food Chem. 55, 175–184.

Suzuki, K., Fujiki, I., Hagura, Y., 1999. Preparation of high concentration of O/W and W/O emulsions by the membrane phase inversion emulsification using PTFE membranes. Food Sci. Technol. Int. Tokyo 5, 234–238.

Suzuki, K., Shuto, I., Hagura, Y., 1996. Characteristics of the membrane emulsification method combined with preliminary emulsification for preparing corn oil-in-water emulsions. Food Sci. Technol. Int. Tokyo 2, 43–47.

Thorsen, T., Roberts, R.W., Arnold, F.H., Quake, S.R., 2001. Dynamic pattern formation in a vesicle-generating microfluidic device. Phys. Rev. Lett. 86, 4163–4166.

van der Graaf, S., Nisisako, T., Schroën, C.G.P.H., van der Sman, R.G.M., Boom, R.M., 2006. Lattice Boltzmann simulations of droplet formation in a T-shaped microchannel. Langmuir 22, 4144–4152.

van der Graaf, S., Schroën, C.G.P.H., Van der Sman, R.G.M., Boom, R.M., 2004. Influence of dynamic interfacial tension on droplet formation during membrane emulsification. J. Colloid Interface Sci. 277, 456–463.

van der Zwan, E.A., Schroën, C.G.P.H., Boom, R.M., 2008. Premix membrane emulsification by using a packed layer of glass beads. AIChE J 54, 2190–2197.

van der Zwan, E., Schroën, K., van Dijke, K., Boom, R., 2006. Visualization of droplet break-up in pre-mix membrane emulsification using microfluidic devices. Colloids Surf. A Physicochem. Eng. Asp. 277, 223–229.

van Rijn, C., van der Wekken, M., Nijdam, W., Elwenspoek, M., 1997. Deflection and maximum load of microfiltration membrane sieves made with silicon micromachining. J. Microelectromech. Syst. 6, 48–54.

Vladisavljević, G.T., 2016. Recent advances in the production of controllable multiple emulsions using microfabricated devices. Particuology 24, 1–17.

Vladisavljević, G.T., Kobayashi, I., Nakajima, M., 2011. Effect of dispersed phase viscosity on maximum droplet generation frequency in microchannel emulsification using asymmetric straight-through channels. Microfluid. Nanofluid. 10, 1199–1209.

Vladisavljević, G.T., Kobayashi, I., Nakajima, M., 2012. Production of uniform droplets using membrane, microchannel and microfluidic emulsification devices. Microfluid. Nanofluid. 13, 151–178.

Vladisavljević, G.T., Kobayashi, I., Nakajima, M., Williams, R.A., Shimizu, M., Nakashima, T., 2007. Shirasu Porous Glass membrane: characterisation of microstructure by high resolution X-ray microtomography and visualisation of droplet formation in real time. J. Membr. Sci. 302, 243–253.

Vladisavljević, G.T., Lambrich, U., Nakajima, M., Schubert, H., 2004. Production of O/W emulsions using SPG membranes, ceramic α-Al_2O_3 membranes, microfluidizer and a microchannel plate: a comparative study. Colloids Surf. A Physicochem. Eng. Asp. 232, 199–207.

Vladisavljević, G.T., McClements, D.J., 2010. Modification of interfacial characteristics of monodisperse droplets produced using membrane emulsification by surfactant displacement and/or polyelectrolyte electrostatic deposition. Colloids Surf. A Physicochem. Eng. Asp. 364, 123–131.

Vladisavljević, G.T., Schubert, H., 2003. Preparation of emulsions with a narrow particle size distribution using microporous α-alumina membranes. J. Dispers. Sci. Technol. 24, 811–819.

Vladisavljević, G.T., Shimizu, M., Nakashima, T., 2004. Preparation of monodisperse multiple emulsions at high production rates by multi-stage premix membrane emulsification. J. Membr. Sci. 244, 97–106.

Vladisavljević, G.T., Shimizu, M., Nakashima, T., 2005. Permeability of hydrophilic and hydrophobic Shirasu-porous-glass (SPG) membranes to pure liquids and its microstructure. J. Membr. Sci. 250, 69–77.

Vladisavljević, G.T., Shimizu, M., Nakashima, T., 2006. Production of multiple emulsions for drug delivery systems by repeated SPG membrane homogenization: influence of mean pore size, interfacial tension and continuous phase viscosity. J. Membr. Sci. 284, 373–383.

Vladisavljević, G.T., Surh, J., McClements, D.J., 2006. Effect of emulsifier type on droplet disruption in repeated Shirasu porous glass membrane homogenization. Langmuir 22, 4526–4533.

Vladisavljević, G.T., Williams, R.A., 2006. Manufacture of large uniform droplets using rotating membrane emulsification. J. Colloid Interface Sci. 299, 396–402.

Vladisavljević, G.T., Williams, R.A., 2008. Recent developments in manufacturing particulate products from double-emulsion templates using membrane and microfluidic devices. In: Aserin, A. (Ed.), Multiple Emulsions: Technology and Applications. John Wiley & Sons, Inc., Hoboken.

Wang, L.Y., Ma, G.H., Su, Z.G., 2005. Preparation of uniform sized chitosan microspheres by membrane emulsification technique and application as a carrier of protein drug. J. Control. Release 106, 62–75.

Yanagishita, T., Fujimura, R., Nishio, K., Masuda, H., 2010. Fabrication of monodisperse polymer nanoparticles by membrane emulsification using ordered anodic porous alumina. Langmuir 26, 1516–1519.

Yanagishita, T., Kato, A., Masuda, H., 2017. Preparation of ideally ordered through-hole anodic porous alumina membranes by two-layer anodization. Jpn. J. Appl. Phys. 56.

Yasuda, M., Goda, T., Ogino, H., Glomm, W.R., Takayanagi, H., 2010. Preparation of uniform monomer droplets using packed column and continuous polymerization in tube reactor. J. Colloid Interface Sci. 349, 392–410.

Yobas, L., Martens, S., Ong, W.L., Ranganathan, N., 2006. High-performance flow-focusing geometry for spontaneous generation of monodispersed droplets. Lab Chip 6, 1073–1079.

Yuan, Q., Aryanti, N., Gutiérrez, G., Williams, R.A., 2009. Enhancing the throughput of membrane emulsification techniques to manufacture functional particles. Ind. Eng. Chem. Res 48, 8872–8880.

Zeng, W., Gomaa, H.G., Liu, J., Zhu, J., 2013. Intensification of production of O/W emulsions using oscillatory woven metal micro-screen (WMMS). Chem. Eng. Process. 73, 111–118.

Zhang, W., Wang, L., Liu, Y., Chen, X., Li, Q., Jia, J., Yang, T., Qiu, S., Ma, G., 2014a. Immune responses to vaccines involving a combined antigene-nanoparticle mixture and nanoparticle-encapsulated antigen formulation. Biomaterials 35, 6086–6097.

Zhang, W., Wang, L., Liu, Y., Chen, X., Li, J., Yang, T., An, W., Ma, X., Pan, R., Ma, G., 2014b. Comparison of PLA microparticles and alum as adjuvants for H5N1 influenza split vaccine: adjuvanticity evaluation and preliminary action mode analysis. Pharm. Res. 31, 1015–1031.

Zhou, Q.Z., Ma, G.H., Su, Z.G., 2009. Effect of membrane parameters on the size and uniformity in preparing agarose beads by premix membrane emulsification. J. Membr. Sci. 326, 694–700.

FURTHER READING

Kotta, S., Khan, A.W., Pramod, K., Ansari, S.H., Sharma, R.K., Ali, J., 2012. Exploring oral nanoemulsions for bioavailability enhancement of poorly water-soluble drugs. Expert Opin. Drug Deliv. 9, 585–598.

Vladisavljević, G.T., 2015. Structured microparticles with tailored properties produced by membrane emulsification. Adv. Colloid Interf. Sci. 225, 53–87.

Vladisavljević, G.T., Khalid, N., Neves, M.A., Kuroiwa, T., Nakajima, M., Uemura, K., Ichikawa, S., Kobayashi, I., 2013. Industrial lab on a chip: design, applications and scale up for drug discovery and delivery. Adv. Drug Deliv. Rev. 65, 1626–1663.

Part IV

Application of Nanoemulsions

Chapter 11

Applications of Nanoemulsions in Foods

Francesco Donsì
University of Salerno, Fisciano, Italy

Chapter Outline

11.1 INTRODUCTION

Emulsions are widely used in applications where the formation of a metastable dispersion of two immiscible fluids is required, as in cosmetic creams, lotions, hairsprays and sunscreens, pharmaceutical delivery systems for lipophilic drugs, paints, dry-cleaning formulations, agrochemical commodities, and the oil industry (Tadros, 2009). In food applications, emulsions can be found in many different products, such as butter and spreads (W/O emulsions), as well as mayonnaise, salad creams, desserts, and beverages (O/W emulsions), where they significantly contribute to the texture, rheology, and organoleptic properties of the food systems (McClements, 2015).

In the last decade, the introduction of more efficient emulsification technologies and highly surface-active emulsifiers made possible the large-scale production of nanoemulsions. In particular, O/W nanoemulsions exhibit unique properties, including the capability of efficiently crossing biological barriers to deliver active molecules, high-surface areas, high-surface reactivity, and unusual optical properties (Sagalowicz and Leser, 2010). The introduction of

Nanoemulsions. https://doi.org/10.1016/B978-0-12-811838-2.00011-4

nanoemulsions has enabled scientists to expand significantly the range of applications of emulsions in the pharmaceutical, cosmetic, and chemical fields. In contrast, the implementation of nanoemulsion technology in the food industry has been relatively slow due to a combination of different limiting factors, including the following:

- The capital costs of the homogenization systems required for high-throughput and generally low-added-value productions, typical of the food industry (Donsì et al., 2013a).
- The scarcity of naturally derived, food-compatible efficient emulsifiers, with controlled surface-active properties (McClements, 2017).
- Concerns of consumers and regulatory bodies about the use of nanotechnology in foods and the conservative position of the food industry compared with other industries (Silva et al., 2012).
- The still scarcely predictable interactions of nanoemulsions with complex food systems (Donsì et al., 2014; Donsì and Ferrari, 2016).
- Ensuring nanoemulsion stability over the expected shelf life of food products. In particular, Ostwald ripening mechanism is responsible for droplet growth, which becomes increasingly important as the mean droplet size decreases, and is prevented only when a dispersed phase with extremely low solubility in water is used, which is not the case for many food applications (Gutiérrez et al., 2008).

However, in the near future, the capital costs of emulsification equipment can be expected to quickly decrease, as a consequence of its increasing expansion in other fields. At the same time, different food-grade, naturally delivered surfactants, such as sugar esters and lecithins, or biopolymers, such as natural gums, vegetable or animal proteins, and modified starches, with enhanced emulsifying properties, are becoming available, which could lead to nanoemulsions with additional features, including tailored interfacial properties, enhanced loading capability, and specific responses to environmental triggers (Chen et al., 2006).

Therefore, the use of nanoemulsions in foods can be expected to exponentially increase, especially to impart augmented functionality in combination with enhanced sensorial attributes to food products.

From the analysis of the main driving forces of the most recent novel research activities on food nanoemulsions, the next generation of applications can be speculated to occur in the following main areas:

- The delivery of bioactive compounds with health-beneficial properties (Donsì et al., 2013b)
- The improved bioaccessibility and bioavailability during digestion of micronutrients (Troncoso et al., 2012)
- The product structuring/texturization or modification of texture (McClements, 2017)

- The food preservation through the delivery of natural antimicrobial agents (Donsì and Ferrari, 2016)

In this chapter, we will refer exclusively to O/W nanoemulsions, as W/O nanoemulsions have demonstrated, so far, significantly less potential in food applications.

11.2 NANOEMULSION FORMULATION FOR FOOD APPLICATIONS

In order to deliver a certain degree of functionality, nanoemulsions should be formulated and processed according to the principles of functional design, which relate their composition and size to the desired functionalities in the product (*in product*) and upon consumption (*in body*).

Additionally, another extremely important driving force in formulating nanoemulsions is the preservation of their physicochemical properties, which often translates into the need for efficient stabilization layers and of ripening inhibitors (Donsì and Ferrari, 2016).

11.2.1 Nanoemulsion Properties on Different Length Scales

Fig. 11.1 schematically depicts how the constitutive properties of nanoemulsions, as considered at the molecular level, affect the behavior in the nanoscale and microscale ranges, in terms of single-droplet characteristics and the

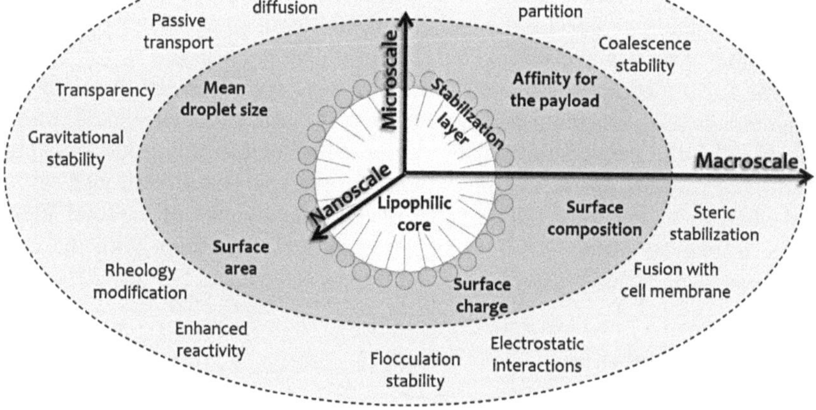

FIG. 11.1 Schematics of nanoemulsion properties, as they extend at different length scales.

interdroplet interactions, and are reflected on the macroscale as the behavior of the nanoemulsion droplets as a collective.

The structure of a nanoemulsion droplet is in principle extremely simple, consisting of a lipophilic core and a stabilization layer. However, unique functionalities can be developed by specifically formulating the lipophilic phase; by blending different oils, lipids, amphiphilic molecules, functional compounds, and stabilization layer; by combining one or more emulsifiers, with the eventual addition of one or more biopolymeric stabilization layers; and by controlling their respective concentrations (Sessa and Donsì, 2015). Together with the formulation, also the processing conditions enable to control the nano- and microscale properties of the nanoemulsion droplets and in particular their droplet size and surface area, the surface composition and charge, and the affinity for any encapsulated compounds.

The composition of the inner dispersed phase, through the regulation of its affinity for the encapsulated compounds, enables the protection of molecules sensitive to UV light or susceptible to chemical attack and the control of their release and of instability phenomena driven by Ostwald ripening, as a consequence of the partitioning in the aqueous phase (Donsì and Ferrari, 2016; Donsì et al., 2012a).

The mean droplet size of the nanoemulsions, together with their composition and concentration, significantly affects, at the macroscale, the kinetic stability, especially in terms of gravitational separation; the optical behavior, influencing the light scattering and absorption (McClements, 2002); and the transport properties, either through porous systems such as food matrices (Donsì et al., 2014) and biological membranes (Sessa et al., 2014).

Moreover, with the mean droplet size, also the surface area of the nanoemulsions is affected, with implications for their reactivity with biological cells and macromolecules and for all those phenomena involving surface forces, such as gelation (Helgeson, 2016).

As a result, lipid digestibility and bioavailability of encapsulated molecules are promoted by nanoemulsions (Salvia-Trujillo et al., 2017).

The surface composition and charge also have significant implications on the interaction of nanoemulsion droplets with the external environment. More specifically, the stabilization layer, composed of emulsifier(s) and possibly additional polymer layers, not only affects the physical stability through steric and electrostatic repulsion but also controls the interactions with external interfaces, driven by surface affinity, as in the case of the fusion with the cell membranes, or by electrostatic attractive forces (Donsì and Ferrari, 2016).

11.2.2 Formulation

As shown in Fig. 11.1, formulation plays a decisive role not only on the minimum achievable droplet size but also in determining several nanoemulsion properties, both at the microscale and at the macroscale. However, the selection of the materials and their concentration of use have to meet the severe constraints, coming from current regulations in place for food products, from the impact on the organoleptic properties, and from economic considerations.

The selection of a suitable emulsifier and of its concentration with respect to the oil phase is generally based not only on the required surface coverage for obtaining stable systems but also on the dynamics of surface adsorption and rearrangement at O/W interfaces, to ensure an efficient emulsification process (Donsì et al., 2012b; Donsì and Ferrari, 2016). However, the currently available naturally derived emulsifiers and stabilizers, which are food grade and can ensure the compatibility with the other food constituents, are still limited. This is an important issue to be considered, as the emulsifier properties control the final mean droplet size, more than the efficiency of the emulsification process itself. Typically, the hydrophilic-lipophilic balance (HLB) number gives some empirical indication on the selection of the emulsifier. In general, emulsifiers with HLB numbers between 8 and 18 are able to form monolayers with low interfacial tensions, promoting the conditions needed for the formation of nanometer-sized droplets (Donsì and Ferrari, 2016). However, it gives no clear prediction of the mean droplet size, which can be achieved upon emulsification, and it cannot easily extend to the use of biopolymers.

Emulsifier concentration also plays an important role on droplet surface coverage and interfacial tension, affecting droplet breakup efficiency (Walstra, 1993), droplet coalescence (Rao and McClements, 2012), and adsorption kinetics (Chanamai et al., 2002).

The efficiency of emulsification depends also on the relative value of the viscosity of the dispersed phase with respect to the viscosity of the continuous phase. In general, when it is close to unity, a finer mean droplet size can be achieved for a given energy input (Walstra, 1993).

Finally, the long-term stability of nanoemulsions can be controlled through adequate selection of the stabilization layer, in order to promote steric hindrance and electrostatic repulsion between the droplets.

In nanoemulsions, the finite solubility in water of many food oils and of many encapsulated compounds triggers Ostwald ripening, which is responsible for droplet growth. Its prevention is one of the main challenges associated with the formulation of physically stable food nanoemulsions (Chang and McClements, 2014; Donsì and Ferrari, 2016). Ostwald ripening consists of the growth of larger droplets at the expense of smaller ones, because of molecular diffusion of oil driven by the higher local solubility around smaller droplets than larger ones (Wooster et al., 2008). Long-chain triglycerides oils, which exhibit a negligible aqueous phase solubility, are therefore often used as ripening inhibitors when the formulation requires the encapsulation of components with an appreciable water solubility (Rao and McClements, 2012).

Examples of ripening inhibitors include corn oil (Chang et al., 2012; Ziani et al., 2011), sesame oil (Terjung et al., 2012), sunflower oil (Donsì et al., 2012a, 2014), and canola oil (Majeed et al., 2016). Alternatively, structured lipophilic cores can be designed, such as those based on organogels. The organogel core is obtained by adding a gelator agent, such as stearic acid, to the lipid phase, hence affecting the partition of the payload compounds into the aqueous phase (Zahi et al., 2014).

Ostwald ripening has also been reported to be inhibited by the use of macromolecules as stabilizers, such as soybean polysaccharide (Wu et al., 2014), sodium caseinate (Pan et al., 2014), whey protein-maltodextrin conjugates (Shah et al., 2013), or pectins (Guerra-Rosas et al., 2016). A layer of biopolymers on the droplet surfaces may inhibit Ostwald ripening by mechanically preventing droplet shrinkage or growth.

The formulation has a significant effect on other properties of food nanoemulsions, such as the surface charge of nanoemulsion droplets, which can contribute to their stability against aggregation (McClements and Rao, 2011). The electric charge can be induced either by using one or more charged emulsifiers or by adsorbing charged macromolecules on the droplet surfaces (Ziani et al., 2011).

Furthermore, the deposition of multiple layers of stabilizers and biopolymers around the lipid droplets, via the layer-by-layer electrostatic deposition technique, has been used as a tool to tailor the interfacial properties of nanoemulsions. For example, multiple chitosan layers were reported to delay the oxidation of omega-3 fatty acids, avoiding both the development of off-odor and the loss of nutritive value (Grigoriev and Miller, 2009). The layer-by-layer method consists in the fabrication of a nanoemulsion, followed by the deposition of biopolymer stabilizing layers at the surface of the oil droplets, driven by the electrostatic attraction between the templating interface and the forming monolayers. Each deposited layer not only fully compensates the charge of the previous templating layer but also imparts an uncompensated counter charge, generating a charge reversal, making possible the deposition of multiple biopolymer layers (Grigoriev and Miller, 2009). Multilayer emulsions exhibit improved stability to environmental stresses in comparison with conventional single-layer systems, without affecting the release in the gastrointestinal tract (Tokle et al., 2013).

11.2.3 *In Product* and *In Body* Behavior

The design of a food nanoemulsion should also be based on its desired *in-product* and *in-body* behavior.

In particular, the *in-product* behavior is affected by the following:

- The efficient dispersion in the food matrix and the compatibility with it
- The reactivity of any encapsulated compounds, which determines their stability and their interaction with other food components

- The complex environment of food matrices, with different interfaces between aqueous and lipid phases, where encapsulated compounds may adsorb
- The intense treatment conditions, to which food is subjected during processing, preservation, or preparation (high or low temperatures, pH extremes, intense shearing, etc.)

The *in-body* behavior is instead affected by the following:

- The release of the encapsulated compounds from the food matrix, preferably triggered by environmental changes, such as pH (chewing and gastrointestinal tract), temperature (body temperature and cooking), mechanical shear (chewing and mastication), enzymes (the gastrointestinal tract), and addition of moisture (dissolution and chewing)
- The fate of the encapsulated compounds during gastric and intestinal digestion
- The bioavailability of the encapsulated compounds, taking into account epithelial cell uptake, absorption in the bloodstream, and reaching the required target sites

In order to control and regulate the *in-product* and the *in-body* behavior, the encapsulated compounds need a suitable delivery system, which is able not only to promote their efficient dispersion and their protection during the most intense phases of processing, preservation, or preparation but also to control their release during mastication (taste masking or enhancing) and gastrointestinal digestion (bioaccessibility and bioavailability enhancing). In particular, the digestibility should also become part of the functional design of food nanoemulsions. Recent studies have shown that mean particle size and formulation significantly influence the rate of lipid digestion, which increases with decreasing droplet size (Troncoso et al., 2012; Majeed et al., 2016). In addition, also the composition of the lipophilic core affects the rate of fat digestion. For example, when it consists of medium-chain triglycerides, lipid digestion is faster than when it consists of canola oil (Majeed et al., 2016).

11.3 DELIVERY OF BIOACTIVE COMPOUNDS

In recent years, consumers' demand for health promotion and disease prevention through diet and nutrition increased exponentially, pushing the food industry toward the development of functional food products incorporating health-beneficial bioactive compounds. The number of bioactive compounds of interest for food functionalization includes several classes of molecules, which have been reported to positively affect health, and is continuously expanded not only because of clinical evidence of new interesting properties but also because of marketing reasons.

Table 11.1 reports a list of some classes of compounds that have been the object of investigation by food nanoemulsion scientists in recent years, including the main flavonoid and nonflavonoid phenolic compounds and carotenoids, together with their reported health-beneficial properties. It must be noted that, despite the copious number of research studies, which have demonstrated that the encapsulation of lipophilic active ingredients in nanoemulsions increases their bioavailability, the real benefits after incorporation in food products and after consumption (*in-product* and *in-body* behavior) have not been sufficiently elucidated to ensure that the potential benefits observed in vitro are preserved during the contact with other ingredients, food processing, storage, and digestion (Salvia-Trujillo et al., 2017).

The main challenge in developing functional foods is therefore associated with the incorporation of bioactive compounds in food matrices, using edible delivery systems, which are capable of encapsulating, protecting, and releasing the bioactive compounds and are suitable for the food industry (McClements and Rao, 2011). In this respect, O/W food nanoemulsions appear to be ideal candidates for delivery of lipophilic bioactive compounds. In particular, bioactive compounds are usually only required in low concentrations to exhibit their health-promoting activities. Nanoemulsions, hence, often contain an additional dispersing lipid phase, whose functional design can significantly improve the chemical stability, the release, and the biological activity of the bioactive compounds.

Table 11.1 reports also the main issues in formulating nanoemulsions for the delivery of different bioactive compounds and the main benefits associated with the use of nanoemulsions. In general, the use of nanoemulsions in the delivery of bioactive compounds has been reported to significantly improve their solubility in water dispersions, to increase their chemical stability by preventing or reducing their interaction with external agents (oxygen, UV light, heat, and food constituents), and to enhance their bioaccessibility, bioavailability, and bioactivity in the human body.

Depending on their polarity, encapsulated molecules may be localized primarily within the inner lipophilic core (bioactive-enriched core) or within the outer stabilization layer (bioactive-enriched shell), with significant implications on their *in-product* and *in-body* behavior (McClements and Rao, 2011). In the case of bioactive compounds with scarce solubility in the oil phase, as in the case of resveratrol (Sessa et al., 2011), curcumin (Yucel et al., 2015), quercetin (Dario et al., 2016), naringenin (Khan et al., 2015), and tangeretin (Wan et al., 2017), the lipophilic core of the nanoemulsions should be specifically designed to prevent their crystallization outside of the nanoemulsions, by means of the use of amphiphilic molecules, which change the microenvironment around the bioactive molecules. The specific selection of surfactants and cosurfactants enables also the stabilization at nanoemulsion interfaces of hydrophilic molecules, such as EGCG (Tsai and Chen, 2016).

TABLE 11.1 Main Classes and Categories of Bioactive Compounds Encapsulated in Nanoemulsions, With the Indication of the Main Challenges for Their Formulation, and the Added Value Deriving From the Use of Nanoemulsions

Class	Category	Compound	Desired Functionality	Challenges in Formulation	Added Value of Nanoemulsions
Flavonoids	Flavanols	EGCG, catechin, epicatechin	Free-radical scavenging, anticancer, decreasing cholesterol level in blood, preventing arterial sclerosis, thrombosis, heart attacks, reducing fat and sugar uptake	Catechins generally exhibit high-water solubility. Partition in the lipophilic core is improved by adding 1-dodecanol (Gadkari et al., 2017) or lecithin (Tsai and Chen, 2016)	Increase of the storage stability of easily degradable catechins (Gadkari et al., 2017)
	Flavonols	Quercetin, kaempferol, myricetin	Free-radical scavenging, fighting the effects of aging and inflammation, downregulating or suppressing inflammatory pathways and functions	The scarce solubility of most flavonols in oil phase requires the addition of amphiphilic molecules in the lipophilic core (Dario et al., 2016)	Improvement of thermal and photostability. Enhanced delivery (Dario et al., 2016)
	Flavones	Apigenin, luteolin, rutin, tangeretin	Antimutagenic, antiinvasive, and antiproliferative agent	Supersaturated flavones in oil phases easily form crystals, requiring the addition of compounds, such as soy protein isolates, to slow down the crystallization process (Wan et al., 2017)	Improve physical stability and bioaccessibility (Wan et al., 2017)

Continued

TABLE 11.1 Main Classes and Categories of Bioactive Compounds Encapsulated in Nanoemulsions, With the Indication of the Main Challenges for Their Formulation, and the Added Value Deriving From the Use of Nanoemulsions—cont'd

Class	Category	Compound	Desired Functionality	Challenges in Formulation	Added Value of Nanoemulsions
	Flavanones	Naringenin, hesperidin	Antiinflammatory, anticarcinogenic, hepatoprotective, and antilipid peroxidation	Flavanones exhibit poor water solubility. Oil and emulsifier are added to ensure maximum loading and stability in the nanoemulsion	Increase in release and bioavailability (Khan et al., 2015)
	Isoflavones	Daidzein, genistein, puerarin	Protection against hormone-related disorders, menopausal symptoms, heart disease and osteoporosis	Incorporation in oil droplets should be promoted by suitable emulsifiers (i.e., lecithin)	Improved permeation through epidermal layers (Silva et al., 2009)
Nonflavonoids	Hydroxybenzoic acids	Gallic acid, ellagic acid, *p*-hydroxybenzoic acid	Antiproliferative and antioxidant	Complexation with phospholipids is required to increase physical stability in nanoemulsions	Increased bioaccessibility, reduced interaction with the food matrix (Avachat and Patel, 2015)
	Hydroxycinnamic acids	Cinnamic acid, coumaric acid, ferulic acid, caffeic acid	Free-radical scavenging, preventing cell damage by ultraviolet light, antiaging, decreasing blood glucose	The solubility of some phenolic acids is very low in aqueous solutions, but they can easily be adsorbed at oil/water interface of nanoemulsions (Ebrahimi et al., 2013)	Improved biological activity, permeability, the absence of cytotoxic effect (Harwansh et al., 2015)

	Stilbenes	Trans-resveratrol	Anticancer, antiinflammation, lowering low-density lipoproteins (bad cholesterol), fighting the plaque buildup leading to Alzheimer's disease, preventing insulin resistance	Poor solubility in water and in oil requires either high amounts of surfactants or a combination of lipophilic and hydrophilic emulsifiers (Sessa et al., 2011)	Improved UV-light stability (Davidov-Pardo and McClements, 2015; Sessa et al., 2011). Enhanced cell permeation (Sessa et al., 2014)
	Curcuminoids	Curcumin	Antiinflammatory, antioxidant, boosting brain-derived neurotrophic factor, lowering the risk of heart disease, anticancer, fighting age-related chronic diseases	Poor solubility in water and in oil requires the use of amphiphilic molecules such as lecithin in the lipid phase (Donsì et al., 2011b; Vecchione et al., 2016)	Improved cell uptake (Yucel et al., 2015) and anticancer activity (Vecchione et al., 2016). Good food compatibility and reduced impact on sensory properties (Kumar et al., 2016)
Carotenoids	Carotenes	β-Carotene, lycopene	Antioxidant, free-radical scavenging, decreasing the risk of cardiovascular disease, oral cavity, and lung cancers	Light and oxidative instability requires the use of biopolymers in nanoemulsion formulation	Improved physical stability, control of lipolysis, and release (Fan et al., 2017)
	Xanthophylls	Lutein, zeaxanthin	Preventing macular degeneration, promoting normal visual functions, strengthening eye tissue, supporting visual acuity and brain function	Limited solubility and dissolution requires complex formulation (Yoo et al., 2010)	Improved bioavailability (Vishwanathan et al., 2009)

The encapsulation of bioactive molecules in nanoemulsions also contributes to their permeation by significantly enhancing the passive transport mechanisms through the cell membranes, as shown for resveratrol or curcumin permeating through Caco-2 cell monolayers (Sessa et al., 2014; Yucel et al., 2015). In particular, polyphenols, which represent a heterogeneous group of phytochemicals, with well-recognized beneficial effects on human health (Habauzit and Morand, 2012), are often characterized by extremely poor bioavailability and rapid metabolization by the human body, thus losing their potential beneficial effects (Donsì et al., 2011b; Sessa et al., 2014).

Moreover, the use of O/W nanoemulsions to encapsulate bioactive compounds is particularly suitable for food products, because of the easily scalable emulsification processes currently available (i.e., high-pressure homogenization), the possibility of using all natural ingredients with low concentrations of emulsifiers (Donsì et al., 2011c), and the high affinity for aqueous-based food matrices.

From the industrial point of view, only a limited number of food products containing nanoemulsions that deliver bioactive molecules are currently available in the food market (Salvia-Trujillo et al., 2017). In addition, there is currently a lack of strong scientific evidence of the advantages or disadvantages of such delivery systems. In fact, most of the research activities conducted on the functionality of food nanoemulsions are conducted in simple model systems. In particular, data about the possible interactions between the nanoemulsion droplets and other macromolecules present in food formulations, such as proteins, carbohydrates, or fibers, must be investigated to drive the implementation of such systems in food products (Salvia-Trujillo et al., 2017).

In addition, a more natural and more sustainable approach for the food industry would require the encapsulation of natural, unrefined extracts from plant tissues or agrifood by-products, such as grape marc, olive pomace, and tomato peels. On one side, this approach poses significant challenges for a successful nanoemulsion formulation, especially because of the higher number of different molecules being involved with specific properties. On the other side, it encounters the needs of the food industry for reduced costs, clean labels, and possibly environmentally benign processes. For example, we have recently investigated the use of grape marc nanoemulsions as a stabilizer against lipid oxidation of a hazelnut paste, with very encouraging results (Spigno et al., 2013). Another example is the encapsulation of water-soluble olive leaf phenolic compounds in double emulsions, upon encapsulation in a W/O nanoemulsion, followed by water dispersion and stabilization with a biopolymer (Mohammadi et al., 2016a). These double emulsions were developed and studied for the control of the oxidative stability of soybean oil (Mohammadi et al., 2016b). Similarly, also saffron extracts were encapsulated in double emulsions to improve the physical stability of the system (Esfanjani et al., 2015, 2017) and control the release rate (Mehrnia et al., 2016, 2017).

11.4 DELIVERY OF MICRONUTRITIVE COMPOUNDS

Besides bioactive compounds, which have health-beneficial properties but do not contribute to the nutritive profile of the food system, nanoemulsions have also been widely investigated to deliver some vital nutrients, which are metabolized by the organism but are required in small amounts. Among micronutritive compounds, fatty acids and in particular polyunsaturated fatty acids such as omega-3 and omega-6; liposoluble vitamins, such as vitamins A, D, and E; phytosterols; and dietary fibers have received particular interest. Table 11.2 summarizes the health benefits of some of the micronutrients, whose delivery to the

TABLE 11.2 Main Classes of Micronutritive Compounds Encapsulated in Nanoemulsions, With the Indication of the Added Value Deriving From the Use of Nanoemulsions

Class	Compound	Desired Functionality	Added Value of Nanoemulsions
Fatty acids	Omega-3, omega-6, and omega-9 polyunsaturated fatty acids	Fighting asthma, diabetes, arthritis, osteoporosis; reducing cholesterol, attention disorders, macular degeneration; contrasting digestive difficulties; anticancer	Protection against lipid oxidation, masking the unpleasant odors (Chen et al., 2016). Improved physical and oxidative stability (Walker et al., 2015), improved digestibility (Karthik and Anandharamakrishnan, 2016)
	Conjugated linoleic acid	Anticancer, antidiabetic, antilipogenic, and antiatherosclerotic effects, positively affecting immune function and bone modeling	Prevention of oxidation, minimization of impact on the food matrix (Fernandez-Avila et al., 2017), improved *in-product* bioavailability (Lane et al., 2014). Improved lipolysis and antiobesity activity (Kim et al., 2013a)
Phytosterols	Stigmasterol, β-sitosterol, campesterol	Lowering cholesterol	Improved storage stability and emulsification (Panpipat et al., 2013). Structuring the lipid phase of emulsion and improving the protection of other oils (Chen et al., 2016)

Continued

TABLE 11.2 Main Classes of Micronutritive Compounds Encapsulated in Nanoemulsions, With the Indication of the Added Value Deriving From the Use of Nanoemulsions—cont'd

Class	Compound	Desired Functionality	Added Value of Nanoemulsions
Vitamins	Vitamin A	Supporting the immune system, strengthening vision, antioxidant	Protection from chemical, enzymatic, and physical instability (Ricaurte et al., 2016). Sustained release and improved delivery through epidermal layers (Jenning et al., 2000)
	Vitamin E	Lowering cholesterol, free-radical scavenging, repairing damaged skin, balancing hormone disorders, improving vision, anticancer	Improved permeability through epidermal barriers, controlled release. Formulation with natural emulsifier is possible (Ozturk et al., 2014)
	Vitamin D	Regulating the absorption of calcium and phosphorous, facilitating the immune system function, growth and development of bones and teeth, reducing the risk of multiple sclerosis and of heart disease	Improved physical stability, especially against coalescence (Guttoff et al., 2015), improved bioaccessibility, controlled through the oil. Formulation with natural emulsifier is possible (Ozturk et al., 2015)
	Vitamin B9	Producing and maintaining new cells, preventing changes to DNA that may lead to cancer	Protection through preliminary encapsulation in W/O nanoemulsions, followed by dispersion in water as double emulsion (Assadpour et al., 2016a,b)
Dietary fibers	Arabinoxylans, cellulose, dextrins, inulin, lignin, chitins, pectins, beta-glucans, and oligosaccharides	Food bulking, metabolic effects, prebiotic activity	Fibers, prevalently located at the oil-water interface, can be exploited to form an edible coating with health-beneficial properties (Artiga-Artigas et al., 2017)

body has been investigated through the use of nanoemulsions, together with the added value of using nanoemulsions.

The encapsulation in nanoemulsions of liposoluble micronutrients is carried out mainly to improve their dispersion in aqueous-based products; to protect them from degradation during food processing, storage, and preparation or from interaction with other food ingredients; and to regulate their digestion rate and uptake in the gastrointestinal tract. Differently from the case of bioactive compounds, micronutrients are incorporated in nanoemulsions at higher concentrations, and not necessarily requiring the presence of a carrier lipid phase, as in the case of omega-3 and omega-6 oils (Jafari et al., 2008). For these compounds, also the issue associated with the masking of their unpleasant flavor represents a formidable challenge in the formulation of successful delivery systems. Moreover, recent studies have shown that their therapeutic efficacy depends also on nanoemulsion charge and size (García-Márquez et al., 2017). In contrast, vitamins A, D, and E are usually blended with a carrier oil, which also contributes to their preservation (Mehmood, 2015). In particular, the utilization of digestible triglycerides as carrier oils has been reported to increase their bioavailability (Salvia-Trujillo et al., 2017).

In the case of hydrophilic compounds, such as Vitamin B9, also known as folic acid, double emulsions were proposed to obtain stable water dispersions, where the folic acid is first encapsulated in W/O nanoemulsions and then dispersed as a double emulsion, with the advantage of an improved protection of the payload (Assadpour et al., 2016a,b).

Moreover, the deposition of multiple biopolymer layers has also been reported to positively affect the preservation of the encapsulated substances, without affecting the digestion rate (Tokle et al., 2013). In addition, the use of specific biopolymers might also enable the control of the digestion rate. For example, we have previously shown that a natural biopolymer such as zein is able to affect the rate of lipid digestion, as the rate-limiting step becomes the full or partial zein digestion by the protease trypsin (Donsì et al., 2017).

Recently, the use of natural or naturally derived emulsifiers for the delivery of micronutrients in nanoemulsions has gathered increasing interest from the food industry, but more research is needed to determine the oxidative stability of these systems (Uluata et al., 2015).

Dietary fibers, which include plant carbohydrates resistant to enzymatic hydrolysis in the gastrointestinal tract, are also important diet components because of their functional and metabolic effects (Satija and Hu, 2012). Their incorporation in nanoemulsions is reported to contribute in reducing their capacity to bind water and to alter organoleptic properties, by adsorbing them at O/W interfaces. The film-forming properties of the fibers can also be exploited to form an edible coating with health-beneficial properties (Artiga-Artigas et al., 2017).

11.5 DELIVERY OF FLAVORS AND COLORS

The delivery of flavors in food systems is a very challenging topic, which should take into account the preservation and persistence of highly volatile molecules over the entire shelf life of the product and their controlled release during food preparation and consumption, which in turn impacts how they are perceived by consumers. The use of essential oils in food products as natural antimicrobial agents has rapidly increased in the last years, as discussed in Section 11.7. However, the use of essential oils impacts the organoleptic properties of the food products where they are incorporated, due to their volatility and pungent taste. A possible approach is the spray-drying of the nanoemulsions containing flavoring molecules, such as D-limonene, using adequate macromolecules. Remarkably, the spray-drying process did not alter the nanoemulsion properties upon rehydration, when surface-active biopolymers were used as emulsifiers (Jafari et al., 2007). However, to date, the sensory impact of their addition to foods has been only speculatively considered, in relation to the capability of nanoemulsions to affect the release profile of essential oils (Majeed et al., 2016). Further studies are, therefore, needed on this topic.

In contrast, the problem of citral preservation upon addition to acidic beverages has solicited the interest of food scientists. Citral is an important lemonlike aroma compound, widely used in the food and drink industry. However, due to its molecular structure, citral is highly susceptible to degradation to off-flavors, primarily *p*-cymene, *p*-cresol, and *p*-methylacetophenone, during storage in a low pH or oxidative environment (Tian et al., 2017), typical of carbonated beverages.

The preservation of citral from fast degradation at low pH has greatly challenged the food industries in the last decades. Among the possible solutions, encapsulation in emulsion emerged as a simple but effective solution. In particular, the selection of the emulsifier (Choi et al., 2010; Djordjevic et al., 2008), the deposition of multiple biopolymer layers (Yang et al., 2012), and the addition of antioxidant molecules (Yang et al., 2011) appeared to efficiently contribute to citral preservation.

The optical behavior of emulsions is affected by their colloidal properties and specifically droplet size, refractive index, and concentration, which all influence the light scattering and absorption, and by the presence of any chromophoric materials (McClements, 2002). In particular, the lightness increases with increasing droplet concentration, decreasing droplet size, and decreasing dye concentration, whereas chromaticness exhibits an opposite behavior (Chanamai and McClements, 2001). When a light beam interacts with an emulsion, part of the light is absorbed or transmitted, and part is scattered or reflected. Light scattering determines the turbidity, opacity, or lightness of the system, where the emulsion is dispersed. In contrast, absorption determines the chromaticness, such as blueness, greenness, and redness (McClements, 2002).

In general, nanoemulsions with diameters below about 60 nm are optically transparent and therefore suitable for use in clear foods and beverages, whereas those with diameters from about 60 to 100 nm are turbid or opaque. Based on these considerations, the use of nanoemulsions to control the color of the food system is tricky and calls for the accurate control of different nanoemulsion properties, including mean droplet size and width of the droplet size distribution, refractive index, concentration, and stability of chromophoric components. Not many studies are available in the literature on these aspects. However, examples are reported where colloidal delivery systems, in the nanometric range, incorporating natural pigments, such as curcumin or indigo carmine, might be combined to develop tunable coloring systems for food applications (Patel et al., 2013). It is interesting from an industrial perspective to incorporate natural bioactive pigments, such as curcumin, indigo carmine, or β-carotene, into deliver systems since they contribute not only to the appearance but also to their potential health benefits.

11.6 PRODUCT STRUCTURING

Recently, the exploitation of nanoemulsions as a constitutive part of a food product, rather than an additive, emerged as a particularly interesting opportunity for the food industry. Therefore, an approach to the design of foods based on the collective behavior of colloidal particles, such as nanoemulsions, enables the development of exciting new properties. In particular, nanoemulsion droplets under many aspects behave like solid particulates and therefore are subject to gelation behavior, related to their concentration and surface properties. However, nanoemulsion droplets are characterized, differently from hard particles, by a soft structure and by the presence of emulsifiers and stabilizers at the O/W interface, which are in dynamic equilibrium with the aqueous phase (Vlassopoulos and Cloitre, 2014). Therefore, different mechanisms of gelation are induced, which are schematically depicted in Fig. 11.2:

1. Repulsive gelation, which is characteristic of nanoemulsion droplets with the same sign of surface charge. The repulsive charge cloud formed around

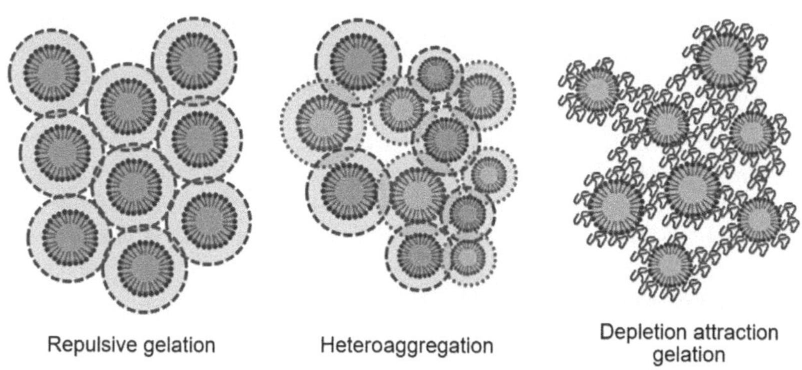

FIG. 11.2 Schematics of the possible gelation processes of nanoemulsion droplets.

the nanoemulsion droplets causes an increase in the effective oil volume fraction (ϕ_{eff}). When ϕ_{eff} becomes comparable with the volume fraction required for maximal random jamming, an elastic gel is formed, hampering the droplets mobility (Erramreddy and Ghosh, 2014).
2. Heteroaggregation or attractive gelation, which is observed when two populations of nanoemulsion droplets of opposite charge are mixed together. It is affected not only by the surface charge but also by the droplet size of the populations (McClements, 2017).
3. Depletion attraction gelation, which is caused by the presence of nonadsorbed polymers or micelles in the aqueous phase, which generates an osmotic attraction that promotes droplet aggregation (Helgeson, 2016). The depletion attraction generated by the polymers or micelles in the continuous phase causes extensive droplet aggregation, with the entrapment of the continuous phase into a strong gel (Erramreddy and Ghosh, 2014).

Nanoemulsion gels are a new class of soft materials, which exhibit a strong elastic behavior even at lower dispersed phase volume fraction. These gels exhibit great potential for use in low-fat foods, pharmaceuticals, and cosmetic products (Erramreddy and Ghosh, 2014).

Both electrostatic and polymer-induced interdroplet interactions are extremely powerful tools to enable the fine control of the macroscopic rheological behavior of the system at the molecular level, via changes of droplet microstructure and formulation (Helgeson, 2016). For example, surfactants with longer hydrophilic tails result in nanoemulsion gels with higher viscoelasticity (Wulff-Pérez et al., 2013).

However, to date, there are no examples on the market of the use of nanoemulsion gels as constitutive ingredients of food products. Further studies are required to elucidate the aspects related for the preparation of nanoemulsion gels with food-grade, naturally derived ingredients and their stability during food preservation and storage.

11.7 ANTIMICROBIAL AGENTS

In the last decades, the use of natural antimicrobial agents, such as essential oils, has attracted considerable attention for food preservation against food-borne pathogens and spoilage bacteria. However, due to their low solubility in water, essential oils need to be delivered through appropriate carriers to promote their efficiency (Donsì et al., 2011a). In the search for naturally derived alternatives to synthetic antimicrobial compounds, essential oils emerged because of their antimicrobial action involving multiple targets within the cell, rather than relying on one specific mechanism (Seow et al., 2014), thus helping to prevent the development of microbial resistance.

It is well established that the antimicrobial activity of essential oils derives from their molecular hydrophobicity, which causes an efficient interaction with the lipids of the cell membrane, increasing its permeability, disturbing the cell membrane structure, and causing the leakage of ions and cytoplasmic content (Salvia-Trujillo et al., 2015a; Donsì and Ferrari, 2016). The use of nanoemulsions, through the increase of dispersibility of essential oils in aqueous phase, where microorganisms proliferate, and the reduction of the resistances to mass transfer of essential oils to the cell membranes have been shown not only to enhance the antimicrobial activity but also to improve the physicochemical stability of essential oils (Donsì et al., 2011a, 2012a; Donsì and Ferrari, 2016).

Essential oil nanoemulsions have been widely investigated in vitro against different microorganisms, with extremely promising results. Nanoemulsions appear to be particularly effective against bacterial cells, being efficiently transported through their outer membranes, because of their small size and the exposition of the hydrophilic groups of the emulsifying molecules. For this reason, no significant difference is observed from the comparison of the antimicrobial activity of different essential oil nanoemulsions against gram+ and gram − bacteria (Wu et al., 2014; Shah et al., 2013). However, other works showed that *Lactobacillus delbrueckii* (gram+) exhibited a greater resistance than *E. coli* (gram −), suggesting that the nanoemulsions affect also other routes of essential oil bacterial inactivation (Donsì et al., 2011a, 2012a). In addition, the antimicrobial efficacy of essential oil nanoemulsions is also dependent on the specific essential oil components used, the microbial strains, and the nanoemulsion composition and droplet size (Donsì and Ferrari, 2016). In contrast, the inactivation of yeast cells by nanoemulsions was shown to be somewhat less effective compared with bacterial inactivation, both for *Saccharomyces cerevisiae* (Donsì et al., 2012a) and *Zygosaccharomyces bailii* (Zhang et al., 2014).

The incorporation of essential oil nanoemulsions in food products poses several challenges.

First of all, the antimicrobial system is required to exhibit a wide spectrum of activity, given the large number of spoilage and pathogenic microorganisms contaminating real foods. In addition, the degradation of the highly reactive molecules constituting the essential oils through the interaction with food ingredients, such as protein, carbohydrates, lipids, and minerals, should be prevented. Similarly, also the distribution of nanoemulsions at the multiple interfaces present in real foods should be controlled to ensure an even distribution (Gyawali and Ibrahim, 2014). Finally, a higher concentration of essential oils than in model systems might be required in real foods (Seow et al., 2014). Moreover, the high volatility and flavor of essential oils often has a significant impact on the sensory properties of food products (Kim et al., 2013b; Gyawali and Ibrahim, 2014). Therefore, the use of essential oils in the preservation of food products is often limited by their strong and distinct flavor profiles (Dima and Dima, 2015).

The encapsulation of essential oils in nanoemulsions can be expected not only to reduce the required concentration by improving their dispersion in food but also to partly mask their taste, via the alteration of their partition into the aqueous phase (Donsì et al., 2012a), and to reduce their interaction with other food components (Shah et al., 2012). Only a few studies have investigated issues related to the use of essential oil nanoemulsions in real food systems, and an even narrower number have considered the inactivation or inhibition of their endogenous flora.

Table 11.3 describes the use of essential oil nanoemulsions for the preservation of real foods. In particular, four main types of application are identified:

1. Direct mixing with liquid foods. Data for milk and fruit juices typically show that small droplet sizes enhance the *in-product* antimicrobial activity, because of the better distribution of the antimicrobial in the food matrix at concentrations above the solubility limit (Shah et al., 2013; Xue et al., 2015) and by minimizing the impact on product appearance (Donsì et al., 2011a).
2. Washing of food surfaces and in particular of vegetable leaves. Nanoemulsions are exploited to increase essential oil concentration in washing solutions and to improve the wettability of vegetable leaves. Nanoemulsions also help to minimize the interaction of the antimicrobial agents with the fresh produce (Bhargava et al., 2015).
3. Infusion in porous food matrices. The diffusion rate of essential oils through vegetable or animal tissues is significantly enhanced when they are embedded in a nanoemulsion, which facilitates the transport to the intercellular interstices of the food matrix (Donsì et al., 2014).
4. Coating with a biopolymer film loaded with nanoemulsions. Essential oils in nanoemulsions can be loaded in biopolymer films at a higher concentration than in the pure form, with minimal impact on their appearance (Otoni et al., 2014; Donsì et al., 2015; Severino et al., 2014a,b). This is a promising technology for the preservation of fresh-cut fruits and vegetables, because of the reduced concentration of essential oils for an effective antimicrobial action, the prolonged persistence of the biopolymer film, and the possible synergism of essential oils with the physical barrier of the coating.

11.8 CONCLUSIONS AND PERSPECTIVES

O/W nanoemulsions are simple but highly effective delivery systems that can be formulated from natural or naturally derived ingredients and produced at high throughput using easily scalable processes. They can be designed to protect functional *in-product* ingredients and to control their *in-body* behavior after ingestion. In addition, the capability of nanoemulsions to form gels can also be exploited in functional food design, where nanoemulsion gels become a constitutive part of the food product.

Despite the fact that nanoemulsions have only found limited application in commercial food products so far, they hold greater potential in several areas because of their ability to generate novel functionalities. In particular, the delivery of functional compounds, such as bioactive molecules, micronutrients,

TABLE 11.3 Application of Essential Oil in Real Food Products Against Microbial Contamination, Describing the Added Value of Encapsulating the Essential Oils in Nanoemulsions

Technique	Product	Essential Oil	Microbial Contamination	Added Value of Nanoemulsion
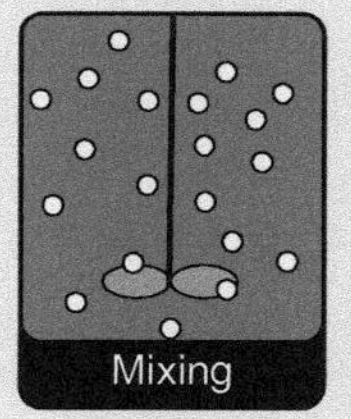	Milk	Eugenol, thyme oil	*E. coli*, *L. monocytogenes*, *S. enteritidis*	Higher antibacterial activity than free essential oil, due to the improved dispersion in the food matrix above the solubility limit (Shah et al., 2013; Xue et al., 2015)
	Orange juice	Eugenol	Native cultivable bacteria	Alteration of membrane permeability, with consequent increased antimicrobial activity (Ghosh et al., 2014)
	Pear and watermelon juice	Tea tree oil, cinnamaldehyde	*L. delbrueckii*, *E. coli*, *S. typhimurium*, *S. aureus*	Enhanced antimicrobial activity (Jo et al., 2015; Donsì et al., 2011a), minimized impact on appearance (Donsì et al., 2011a)
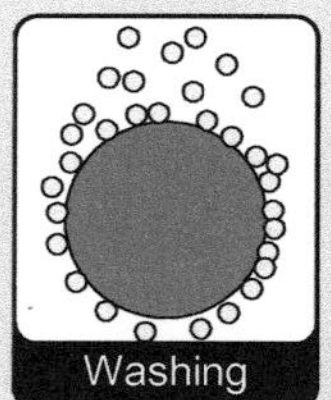	Fresh lettuce, spinach leaves	Carvacrol, eugenol, oregano oil	*E. coli*, *L. monocytogenes*, *S. typhimurium*, *S. enterica*	Easy application via spraying or immersion (Bhargava et al., 2015; Ruengvisesh et al., 2015)

Continued

TABLE 11.3 Application of Essential Oil in Real Food Products Against Microbial Contamination, Describing the Added Value of Encapsulating the Essential Oils in Nanoemulsions—cont'd

Technique	Product	Essential Oil	Microbial Contamination	Added Value of Nanoemulsion
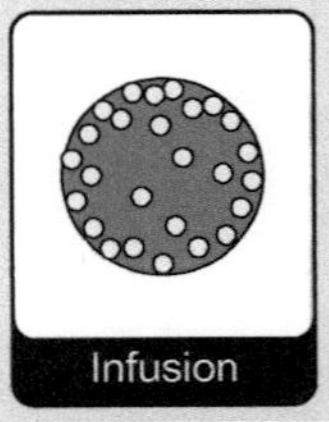	Pork sausages	Carvacrol	Endogenous flora	Increasing antimicrobial activity at decreasing droplet size (Donsì et al., 2014)
	Zucchini	Carvacrol	*E. coli*	Increasing permeation and antimicrobial activity at decreasing droplet size (Donsì et al., 2014)
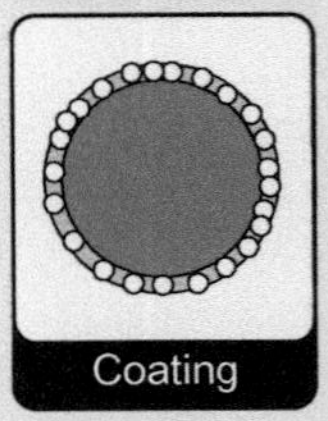	Apple pieces	Lemongrass	*E. coli*, Endogenous flora	Complete microbial inhibition while maintaining the product quality (Salvia-Trujillo et al., 2015b)
	Broccoli florets	Citrus oils	*L. monocytogenes*	Enhanced antimicrobial activity, in combination with other nonthermal processes (Severino et al., 2014b)
	Green beans	Mandarin oil	*L. innocua*	Synergistic antibacterial effect with chitosan and nonthermal processes (Donsì et al., 2015; Severino et al., 2014a)
	Rucola leaves	Lemon oil	Endogenous flora	Enhanced antimicrobial activity (Sessa et al., 2015)
	Sliced bread	Clove bud oil, oregano oil	Endogenous flora	Increased antibacterial and antifungal properties, with fewer preservative agents required (Otoni et al., 2014)

colorants, flavorings, or antimicrobial agents, into food products that naturally do not contain them, is highly desired to increase product value by enhancing health-beneficial properties, nutritive profiles, appearance, aroma, or shelf life. The main challenges for the successful application of nanoemulsions in foods include the development of nanoemulsion-based delivery systems that are formulated from natural or naturally derived ingredients that have an increased compatibility with food matrices and that can control the digestion, release, and absorption of functional compounds. Carefully designed nanoemulsions will increase the functionality of food products and lead to novel foods tailored for personalized nutrition of groups of the population, such as young or elderly people, with peculiar dietetic requirements. These novel nanoemulsion-based delivery systems will also have to gain approval from the regulatory bodies operating in different countries, such as the FDA in the United States and the EFSA in the EU, which are becoming increasingly severe about the introduction of novel foods with health claims and about the risks associated with the use of nanotechnology.

REFERENCES

Artiga-Artigas, M., Acevedo-Fani, A., Martín-Belloso, O., 2017. Improving the shelf life of low-fat cut cheese using nanoemulsion-based edible coatings containing oregano essential oil and mandarin fiber. Food Control 76, 1–12.

Assadpour, E., et al., 2016a. Evaluation of folic acid nano-encapsulation by double emulsions. Food Bioprocess Technol. 9 (12), 2024–2032.

Assadpour, E., et al., 2016b. Optimization of folic acid nano-emulsification and encapsulation by maltodextrin-whey protein double emulsions. Int. J. Biol. Macromol. 86, 197–207.

Avachat, A.M., Patel, V.G., 2015. Self nanoemulsifying drug delivery system of stabilized ellagic acid-phospholipid complex with improved dissolution and permeability. Saudi Pharm. J. 23 (3), 276–289.

Bhargava, K., et al., 2015. Application of an oregano oil nanoemulsion to the control of foodborne bacteria on fresh lettuce. Food Microbiol. 47, 69–73.

Chanamai, R., McClements, D.J., 2001. Prediction of emulsion color from droplet characteristics: dilute monodisperse oil-in-water emulsions. Food Hydrocoll. 15 (1), 83–91. Available at: http://www.sciencedirect.com/science/article/pii/S0268005X00000552. Accessed 12 April 2017.

Chanamai, R., Horn, G., McClements, D.J., 2002. Influence of oil polarity on droplet growth in oil-in-water emulsions stabilized by a weakly adsorbing biopolymer or a nonionic surfactant. J. Colloid Interface Sci. 247 (1), 167–176.

Chang, Y., McClements, D.J., 2014. Optimization of orange oil nanoemulsion formation by isothermal low-energy methods: influence of the oil phase, surfactant, and temperature. J. Agric. Food Chem. 62 (10), 2306–2312.

Chang, Y., McLandsborough, L., McClements, D.J., 2012. Physical properties and antimicrobial efficacy of thyme oil nanoemulsions: influence of ripening inhibitors. J. Agric. Food Chem. 60 (48), 12056–12063.

Chen, L., Remondetto, G.E., Subirade, M., 2006. Food protein-based materials as nutraceutical delivery systems. Trends Food Sci. Technol. 17 (5), 272–283.

Chen, X.-W., et al., 2016. Phytosterol structured algae oil nanoemulsions and powders: improving antioxidant and flavor properties. Food Funct. 7 (9), 3694–3702.

Choi, S.J., et al., 2010. Influence of droplet charge on the chemical stability of citral in oil-in-water emulsions. J. Food Sci. 75 (6), C536–40.

Dario, M.F., et al., 2016. Stability and safety of quercetin-loaded cationic nanoemulsion: in vitro and in vivo assessments. Colloids Surf. A Physicochem. Eng. Asp. 506, 591–599.

Davidov-Pardo, G., McClements, D.J., 2015. Nutraceutical delivery systems: resveratrol encapsulation in grape seed oil nanoemulsions formed by spontaneous emulsification. Food Chem. 167, 205–212.

Dima, C., Dima, S., 2015. Essential oils in foods: extraction, stabilization, and toxicity. Curr. Opin. Food Sci. 5, 29–35. Available at:http://www.sciencedirect.com/science/article/pii/S2214799315000958.

Djordjevic, D., et al., 2008. Stability of citral in protein- and gum arabic-stabilized oil-in-water emulsions. Food Chem. 106 (2), 698–705.

Donsì, F., Ferrari, G., 2016. Essential oil nanoemulsions as antimicrobial agents in food. J. Biotechnol. 233, 106–120. https://doi.org/10.1016/j.jbiotec.2016.07.005.

Donsì, F., Annunziata, M., et al., 2011a. Nanoencapsulation of essential oils to enhance their antimicrobial activity in foods. LWT Food Sci. Technol. 44 (9), 1908–1914. https://doi.org/10.1016/j.lwt.2011.03.003.

Donsì, F., Sessa, M., et al., 2011b. Encapsulation of bioactive compounds in nanoemulsion-based delivery systems. Procedia Food Sci. 1, 1666–1671. Available at:http://www.sciencedirect.com/science/article/pii/S2211601X11002471.

Donsì, F., Wang, Y., Huang, Q., 2011c. Freeze-thaw stability of lecithin and modified starch-based nanoemulsions. Food Hydrocoll. 25 (5), 1327–1336.

Donsì, F., Annunziata, M., et al., 2012a. Design of nanoemulsion-based delivery systems of natural antimicrobials: effect of the emulsifier. J. Biotechnol. 159 (4), 342–350. Available at:http://www.ncbi.nlm.nih.gov/pubmed/21763730.

Donsì, F., Sessa, M., Ferrari, G., 2012b. Effect of emulsifier type and disruption chamber geometry on the fabrication of food nanoemulsions by high pressure homogenization. Ind. Eng. Chem. Res. 51 (22), 7606–7618.

Donsì, F., Annunziata, M., Ferrari, G., 2013a. Microbial inactivation by high pressure homogenization: effect of the disruption valve geometry. J. Food Eng. 115 (3), 362–370.

Donsì, F., Sessa, M., Ferrari, G., 2013b. Nanometric-size delivery systems for bioactive compounds for the nutraceutical and food industries. In: Bagchi, D., et al. (Eds.), Bio-Nanotechnology: A Revolution in Food, Biomedical and Health Sciences. Blackwell Publishing Ltd., Oxford, pp. 619–666. https://doi.org/10.1002/9781118451915.ch37

Donsì, F., et al., 2014. Infusion of essential oils for food stabilization: unraveling the role of nanoemulsion-based delivery systems on mass transfer and antimicrobial activity. Innovative Food Sci. Emerg. Technol. 22, 212–220.

Donsì, F., et al., 2015. Green beans preservation by combination of a modified chitosan based-coating containing nanoemulsion of mandarin essential oil with high pressure or pulsed light processing. Postharvest Biol. Technol. 106, 21–32. https://doi.org/10.1016/j.postharvbio.2015.02.006.

Donsì, F., et al., 2017. Zein-based colloidal particles for encapsulation and delivery of epigallocatechin gallate. Food Hydrocoll. 63, 508–517.

Ebrahimi, P., et al., 2013. Ferulic acid lecithin-based nano-emulsions prepared by using spontaneous emulsification process. Iran. J. Chem. Chem. Eng. 32 (3), 17–25.

Erramreddy, V.V., Ghosh, S., 2014. Influence of emulsifier concentration on nanoemulsion gelation. Langmuir 30 (37), 11062–11074.

Esfanjani, A.F., et al., 2015. Nano-encapsulation of saffron extract through double-layered multiple emulsions of pectin and whey protein concentrate. J. Food Eng. 165, 149–155.

Esfanjani, A.F., Jafari, S.M., Assadpour, E., 2017. Preparation of a multiple emulsion based on pectin-whey protein complex for encapsulation of saffron extract nanodroplets. Food Chem. 221, 1962–1969.

Fan, Y., et al., 2017. Physicochemical stability and in vitro bioaccessibility of β-carotene nanoemulsions stabilized with whey protein-dextran conjugates. Food Hydrocoll. 63, 256–264.

Fernandez-Avila, C., Gutierrez-Merida, C., Trujillo, A.J., 2017. Physicochemical and sensory characteristics of a UHT milk-based product enriched with conjugated linoleic acid emulsified by ultra-high pressure homogenization. Innovative Food Sci. Emerg. Technol. 39, 275–283.

Gadkari, P.V., Shashidhar, M.G., Balaraman, M., 2017. Delivery of green tea catechins through oil-in-water (O/W) nanoemulsion and assessment of storage stability. J. Food Eng. 199, 65–76.

García-Márquez, E., Higuera-Ciapara, I., Espinosa-Andrews, H., 2017. Design of fish oil-in-water nanoemulsion by microfluidization. Innovative Food Sci. Emerg. Technol. 40, 87–91. Available at: http://www.sciencedirect.com/science/article/pii/S1466856416306683. Accessed 3 April 2017.

Ghosh, V., Mukherjee, A., Chandrasekaran, N., 2014. Eugenol-loaded antimicrobial nanoemulsion preserves fruit juice against, microbial spoilage. Colloids Surf. B: Biointerfaces 114, 392–397.

Grigoriev, D.O., Miller, R., 2009. Mono- and multilayer covered drops as carriers. Curr. Opin. Colloid Interface Sci. 14 (1), 48–59.

Guerra-Rosas, M.I., et al., 2016. Long-term stability of food-grade nanoemulsions from high methoxyl pectin containing essential oils. Food Hydrocoll. 52, 438–446.

Gutiérrez, J.M., et al., 2008. Nano-emulsions: new applications and optimization of their preparation. Curr. Opin. Colloid Interface Sci. 13 (4), 245–251.

Guttoff, M., Saberi, A.H., McClements, D.J., 2015. Formation of vitamin D nanoemulsion-based delivery systems by spontaneous emulsification: factors affecting particle size and stability. Food Chem. 171, 117–122. Available at: http://www.sciencedirect.com/science/article/pii/S0308814614013119. Accessed 3 April 2017.

Gyawali, R., Ibrahim, S.A., 2014. Natural products as antimicrobial agents. Food Control 46, 412–429.

Habauzit, V., Morand, C., 2012. Evidence for a protective effect of polyphenols-containing foods on cardiovascular health: an update for clinicians. Ther. Adv. Chronic Dis. 3 (2), 87–106.

Harwansh, R.K., et al., 2015. Enhanced permeability of ferulic acid loaded nanoemulsion based gel through skin against UVA mediated oxidative stress. Life Sci. 141, 202–211.

Helgeson, M.E., 2016. Colloidal behavior of nanoemulsions: interactions, structure, and rheology. Curr. Opin. Colloid Interface Sci. 25, 39–50.

Jafari, S.M., He, Y., Bhandari, B., 2007. Encapsulation of nanoparticles of D-limonene by spray drying: role of emulsifiers and emulsifying techniques. Dry. Technol. 25 (6), 1069–1079.

Jafari, S.M., et al., 2008. Nano-particle encapsulation of fish oil by spray drying. Food Res. Int. 41 (2), 172–183.

Jenning, V., Schäfer-Korting, M., Gohla, S., 2000. Vitamin A-loaded solid lipid nanoparticles for topical use: drug release properties. J. Control. Release 66 (2–3), 115–126.

Jo, Y.J., et al., 2015. Physical and antimicrobial properties of trans-cinnamaldehyde nanoemulsions in water melon juice. LWT Food Sci. Technol. 60 (1), 444–451.

Karthik, P., Anandharamakrishnan, C., 2016. Enhancing omega-3 fatty acids nanoemulsion stability and in-vitro digestibility through emulsifiers. J. Food Eng. 187, 92–105.

Khan, A.W., et al., 2015. Self-nanoemulsifying drug delivery system (SNEDDS) of the poorly water-soluble grapefruit flavonoid naringenin: design, characterization, in vitro and in vivo evaluation. Drug Deliv. 22 (4), 552–561.

Kim, D., et al., 2013a. Bioavailability of nanoemulsified conjugated linoleic acid for an antiobesity effect. Int. J. Nanomedicine 8, 451–459.

Kim, I.H., et al., 2013b. Plum coatings of lemongrass oil-incorporating carnauba wax-based nanoemulsion. J. Food Sci. 78 (10), E1551–9.

Kumar, D.D., et al., 2016. Formulation and characterization of nanoencapsulated curcumin using sodium caseinate and its incorporation in ice cream. Food Funct. 7 (1), 417–424.

Lane, K.E., et al., 2014. The bioavailability of an omega-3-rich algal oil is improved by nanoemulsion technology using yogurt as a food vehicle. Int. J. Food Sci. Technol. 49 (5), 1264–1271.

Majeed, H., et al., 2016. Influence of carrier oil type, particle size on in vitro lipid digestion and eugenol release in emulsion and nanoemulsions. Food Hydrocoll. 52, 415–422.

McClements, D.J., 2002. Colloidal basis of emulsion color. Curr. Opin. Colloid Interface Sci. 7 (5–6), 451–455.

McClements, D.J., 2015. Food emulsions: principles. In: Practice and Techniques, third ed. CRC Press, Boca Raton, FL.

McClements, D.J., 2017. The future of food colloids: next-generation nanoparticle delivery systems. Curr. Opin. Colloid Interface Sci. 28, 7–14.

McClements, D.J.D.J., Rao, J., 2011. Food-grade nanoemulsions: formulation, fabrication, properties, performance, biological fate, and potential toxicity. Crit. Rev. Food Sci. Nutr. 51 (4), 285–330.

Mehmood, T., 2015. Optimization of the canola oil based vitamin E nanoemulsions stabilized by food grade mixed surfactants using response surface methodology. Food Chem. 183, 1–7. Available at: http://www.sciencedirect.com/science/article/pii/S0308814615003726. Accessed 3 April 2017.

Mehrnia, M.A., et al., 2016. Crocin loaded nano-emulsions: factors affecting emulsion properties in spontaneous emulsification. Int. J. Biol. Macromol. 84, 261–267.

Mehrnia, M.-A., et al., 2017. Rheological and release properties of double nano-emulsions containing crocin prepared with Angum gum, Arabic gum and whey protein. Food Hydrocoll. 66, 259–267.

Mohammadi, A., Jafari, S.M., Assadpour, E., et al., 2016a. Nano-encapsulation of olive leaf phenolic compounds through WPC-pectin complexes and evaluating their release rate. Int. J. Biol. Macromol. 82, 816–822.

Mohammadi, A., Jafari, S.M., Esfanjani, A.F., et al., 2016b. Application of nano-encapsulated olive leaf extract in controlling the oxidative stability of soybean oil. Food Chem. 190, 513–519.

Otoni, C.G., et al., 2014. Edible films from methylcellulose and nanoemulsions of clove bud (Syzygium aromaticum) and oregano (Origanum vulgare) essential oils as shelf life extenders for sliced bread. J. Agric. Food Chem. 62 (22), 5214–5219.

Ozturk, B., et al., 2014. Formation and stabilization of nanoemulsion-based vitamin E delivery systems using natural surfactants: quillaja saponin and lecithin. J. Food Eng. 142, 57–63. Available at: http://www.sciencedirect.com/science/article/pii/S0260877414002568. Accessed 3 April 2017.

Ozturk, B., et al., 2015. Nanoemulsion delivery systems for oil-soluble vitamins: influence of carrier oil type on lipid digestion and vitamin D3 bioaccessibility. Food Chem. 187, 499–506.

Pan, K., et al., 2014. Thymol nanoencapsulated by sodium caseinate: physical and antilisterial properties. J. Agric. Food Chem. 62 (7), 1649–1657.

Panpipat, W., et al., 2013. Thermal properties and nanodispersion behavior of synthesized β-sitosteryl acyl esters: a structure-activity relationship study. J. Colloid Interface Sci. 407, 177–186.

Patel, A.R., et al., 2013. Colloidal approach to prepare colour blends from colourants with different solubility profiles. Food Chem. 141 (2), 1466–1471. Available at: http://www.ncbi.nlm.nih.gov/pubmed/23790940.

Rao, J., McClements, D.J., 2012. Impact of lemon oil composition on formation and stability of model food and beverage emulsions. Food Chem. 134 (2), 749–757. https://doi.org/10.1016/j.foodchem.2012.02.174.

Ricaurte, L., et al., 2016. Production of high-oleic palm oil nanoemulsions by high-shear homogenization (microfluidization). Innovative Food Sci. Emerg. Technol. 35, 75–85. Available at: http://www.sciencedirect.com/science/article/pii/S1466856416300649. Accessed 3 April 2017.

Ruengvisesh, S., et al., 2015. Inhibition of bacterial pathogens in medium and on spinach leaf surfaces using plant-derived antimicrobials loaded in surfactant micelles. J. Food Sci. 80 (11), M2522–M2529.

Sagalowicz, L., Leser, M.E., 2010. Delivery systems for liquid food products. Curr. Opin. Colloid Interface Sci. 15 (1–2), 61–72. https://doi.org/10.1016/j.cocis.2009.12.003.

Salvia-Trujillo, L., Rojas-Graü, A., et al., 2015a. Physicochemical characterization and antimicrobial activity of food-grade emulsions and nanoemulsions incorporating essential oils. Food Hydrocoll. 43, 547–556.

Salvia-Trujillo, L., Rojas-Grau, M.A., et al., 2015b. Use of antimicrobial nanoemulsions as edible coatings: impact on safety and quality attributes of fresh-cut fuji apples. Postharvest Biol. Technol. 105, 8–16.

Salvia-Trujillo, L., et al., 2017. Edible nanoemulsions as carriers of active ingredients: a review. Annu. Rev. Food Sci. Technol. 8, 439–466.

Satija, A., Hu, F.B., 2012. Cardiovascular benefits of dietary fiber. Curr. Atheroscler. Rep. 14 (6), 505–514.

Seow, Y.X., et al., 2014. Plant essential oils as active antimicrobial agents. Crit. Rev. Food Sci. Nutr. 54 (5), 625–644. Available at: http://www.ncbi.nlm.nih.gov/pubmed/24261536%5Cn. http://www.scopus.com/inward/record.url?eid=2-s2.0-84889025166&partnerID=tZOtx3y1.

Sessa, M., Donsì, F., 2015. Nanoemulsion-based delivery systems. In: Sagis, L.M.C. (Ed.), Microencapsulation and Microspheres for Food Applications. Elsevier, London, pp. 79–94. Available at: http://linkinghub.elsevier.com/retrieve/pii/B9780128003503000078.

Sessa, M., et al., 2011. Evaluation of the stability and antioxidant activity of nanoencapsulated resveratrol during in vitro digestion. J. Agric. Food Chem. 59 (23), 12352–12360.

Sessa, M., et al., 2014. Bioavailability of encapsulated resveratrol into nanoemulsion-based delivery systems. Food Chem. 147, 42–50.

Sessa, M., Ferrari, G., Donsì, F., 2015. Novel edible coating containing essential oil nanoemulsions to prolong the shelf life of vegetable products. Chem. Eng. Trans. 43, 55–60.

Severino, R., Vu, K.D., Donsì, F., Salmieri, S., et al., 2014a. Antibacterial and physical effects of modified chitosan based-coating containing nanoemulsion of mandarin essential oil and three non-thermal treatments against Listeria innocua in green beans. Int. J. Food Microbiol. 191, 82–88.

Severino, R., Vu, K.D., Donsì, F., Salmieri, S., et al., 2014b. Antimicrobial effects of different combined non-thermal treatments against Listeria monocytogenes in broccoli florets. J. Food Eng. 124, 1–10.

Shah, B., Davidson, P.M., Zhong, Q., 2012. Encapsulation of eugenol using Maillard-type conjugates to form transparent and heat stable nanoscale dispersions. LWT Food Sci. Technol. 49 (1), 139–148. https://doi.org/10.1016/j.lwt.2012.04.029.

Shah, B., Davidson, P.M., Zhong, Q., 2013. Nanodispersed eugenol has improved antimicrobial activity against Escherichia coli O157:H7 and Listeria monocytogenes in bovine milk. Int. J. Food Microbiol. 161 (1), 53–59.

Silva, A.P.C., et al., 2009. Development of topical nanoemulsions containing the isoflavone genistein. Pharmazie 64 (1), 32–35.

Silva, H.D., Cerqueira, M.A., Vicente A.A., 2012. Nanoemulsions for food applications: development and characterization. Food Bioprocess Technol. **5** (3), 854–867.

Spigno, G., et al., 2013. Nanoencapsulation systems to improve solubility and antioxidant efficiency of a grape marc extract into hazelnut paste. J. Food Eng. 114 (2), 207–214. Available at: http://www.sciencedirect.com/science/article/pii/S0260877412003974.

Tadros, T.F., 2009. Emulsion science and technology: a general introduction. In: Tadros, T.F. (Ed.), Emulsion Science and Technology. Wiley-VCH Verlag GmbH & Co. KGaA, Weinheim, pp. 1–56. Available at: http://doi.wiley.com/10.1002/9783527626564.ch1. Accessed 3 April 2017.

Terjung, N., et al., 2012. Influence of droplet size on the efficacy of oil-in-water emulsions loaded with phenolic antimicrobials. Food Funct. 3 (3), 290–301.

Tian, H., et al., 2017. Citral stabilization and characterization of nanoemulsions stabilized by a mixture of gelatin and tween 20 in an acidic system. J. Sci. Food Agric. 97 (9), 2991–2998.

Tokle, T., Mao, Y., McClements, D.J., 2013. Potential biological fate of emulsion-based delivery systems: lipid particles nanolaminated with lactoferrin and β-lactoglobulin coatings. Pharm. Res. 30 (12), 3200–3213.

Troncoso, E., Aguilera, J.M., McClements, D.J., 2012. Fabrication, characterization and lipase digestibility of food-grade nanoemulsions. Food Hydrocoll. 27 (2), 355–363.

Tsai, Y.-J., Chen, B.-H., 2016. Preparation of catechin extracts and nanoemulsions from green tea leaf waste and their inhibition effect on prostate cancer cell PC-3. Int. J. Nanomedicine 11, 1907–1926.

Uluata, S., McClements, D.J., Decker, E.A., 2015. Physical stability, autoxidation, and photosensitized oxidation of ω-3 oils in nanoemulsions prepared with natural and synthetic surfactants. J. Agric. Food Chem. 63 (42), 9333–9340.

Vecchione, R., et al., 2016. Curcumin bioavailability from oil in water nano-emulsions: in vitro and in vivo study on the dimensional, compositional and interactional dependence. J. Control. Release 233, 88–100.

Vishwanathan, R., Wilson, T.A., Nicolosi, R.J., 2009. Bioavailability of a nanoemulsion of lutein is greater than a lutein supplement. Nano Biomed. Eng. 1 (1), 38–49.

Vlassopoulos, D., Cloitre, M., 2014. Tunable rheology of dense soft deformable colloids. Curr. Opin. Colloid Interface Sci. 19 (6), 561–574.

Walker, R.M., Decker, E.A., McClements, D.J., 2015. Physical and oxidative stability of fish oil nanoemulsions produced by spontaneous emulsification: effect of surfactant concentration and particle size. J. Food Eng. 164, 10–20.

Walstra, P., 1993. Principles of emulsion formation. Chem. Eng. Sci. 48 (2), 333–349.

Wan, J., et al., 2017. Enhancement of physical stability and bioaccessibility of tangeretin by soy protein isolate addition. Food Chem. 221, 760–770.

Wooster, T.J., Golding, M., Sanguansri, P., 2008. Impact of oil type on nanoemulsion formation and ostwald ripening stability. Langmuir 24 (22), 12758–12765.

Wu, J.E., Lin, J., Zhong, Q., 2014. Physical and antimicrobial characteristics of thyme oil emulsified with soluble soybean polysaccharide. Food Hydrocoll. 39, 144–150.

Wulff-Pérez, M., et al., 2013. The effect of polymeric surfactants on the rheological properties of nanoemulsions. Colloid Polym. Sci. 291 (3), 709–716.

Xue, J., Michael Davidson, P., Zhong, Q., 2015. Antimicrobial activity of thyme oil co-nanoemulsified with sodium caseinate and lecithin. Int. J. Food Microbiol. 210, 1–8.

Yang, X., et al., 2011. Inhibition of citral degradation by oil-in-water nanoemulsions combined with antioxidants. J. Agric. Food Chem. 59 (11), 6113–6119.

Yang, X., et al., 2012. Stability of citral in emulsions coated with cationic biopolymer layers. J. Agric. Food Chem. 60 (1), 402–409.

Yoo, J.H., et al., 2010. Novel self-nanoemulsifying drug delivery system for enhanced solubility and dissolution of lutein. Arch. Pharm. Res. 33 (3), 417–426.

Yucel, C., et al., 2015. Submicron complex lipid carriers for curcumin delivery to intestinal epithelial cells: effect of different emulsifiers on bioaccessibility and cell uptake. Int. J. Pharm. 494 (1), 357–369.

Zahi, M.R., et al., 2014. Formation and stability of D-limonene organogel-based nanoemulsion prepared by a high-pressure homogenizer. J. Agric. Food Chem. 62 (52), 12563–12569.

Zhang, Z., et al., 2014. Effects of nisin on the antimicrobial activity of D-limonene and its nanoemulsion. Food Chem. 150, 307–312.

Ziani, K., et al., 2011. Influence of surfactant charge on antimicrobial efficacy of surfactant-stabilized thyme oil nanoemulsions. J. Agric. Food Chem. 59 (11), 6247–6255.

Chapter 12

Application of Nanoemulsions in Formulation of Pesticides

Jianguo Feng*, Qi Zhang*, Qi Liu*, Zhengxi Zhu*, David J. McClements[†] and Seid Mahdi Jafari[‡]
*Yangzhou University, Yangzhou, China, [†]University of Massachusetts, Amherst, MA, United States, [‡]Gorgan University of Agricultural Sciences and Natural Resources, Gorgan, Iran

Chapter Outline

12.1 INTRODUCTION

12.1.1 Background of Pesticides

Pesticides play a critical role in reducing yield loss and ensuring food safety (Cooper and Dobson, 2007; Zhang et al., 2015). In the future, the utilization of pesticides will become even more important to solve food security problems

Nanoemulsions. https://doi.org/10.1016/B978-0-12-811838-2.00012-6

associated with providing an adequate food supply for the growing world population (Carvalho, 2006; Delcour et al., 2015). Pesticides, which may contain synthetic and/or natural substances, are used to prevent, destroy, or control microorganisms, insects, weeds, and other pests that endanger agriculture and forestry and to regulate the growth of certain plant species.

Pesticides are classified based on the biological species that they are used to control: insecticides, fungicides, nematicides, herbicides, and plant growth regulators. They can be used for agricultural (Silva et al., 2012; Weaver et al., 2009), forestry (Faria et al., 2013), hygiene (Pavela, 2013), and nonagricultural purposes (Badawi et al., 2016). Commercially, pesticides are not typically used directly in their pure chemical form. Instead, they are usually incorporated into a suitable delivery system by adding an appropriate dosage of the pesticide into a suitable solvent or colloidal dispersion, which may then be used in a solid, liquid, or gaseous form (Rodham, 2000). Thus, commercial pesticide formulations can be divided into solid forms (such as granules or powders), liquid forms (such as solutions or suspensions), and gaseous forms (such as vapors). The liquid forms include solutions, emulsifiable concentrates, microemulsions, oil-in-water emulsions, and soluble concentrate (Wang et al., 2004). The development of suitable formulation technologies is therefore extremely important for the development of efficacious delivery systems for hydrophobic pesticides.

12.1.2 Current Problems in Application of Pesticides

There are currently a number of challenges that limit the efficacy of certain types of pesticides. As mentioned earlier, many pesticides are highly hydrophobic molecules that a have low solubility in water, and therefore, it is difficult to get them to the required site of action. Solid pesticides can be converted into a powdered form, but they often have an unsatisfactory performance because of their large particle size, poor adhesion properties, and low solubility (Mulqueen, 2003). For this reason, hydrophobic pesticides are often processed into liquid formulations, such as solutions, emulsifiable concentrates (Chin et al., 2012), microemulsions (Xu et al., 2010), and emulsions (Elshafei et al., 2010). Many of these liquid formulations can be prepared using relatively simple processing techniques and have excellent efficacy. At present, emulsifiable concentrates are one of the most widely used delivery systems for hydrophobic pesticides, accounting for ~40%–50% of the total formulation products. Annually, a total of around 300,000 tons of organic solvents, mainly volatile aromatic solvents (such as xylene), are used in the preparation of emulsifiable concentrate formulations. In addition, other solvents may also be used in their preparation such as *N,N*-dimethylformamide (DMF), methanol, and other cosolvents. There are numerous potential problems associated with utilizing these organic solvents, since they are flammable, explosive, phytotoxic to crops, harmful to humans and the environment and produce toxic wastes (Eddleston et al., 2012). Therefore, more environmentally friendly alternatives of delivering hydrophobic pesticides are required to reduce the amount of organic solvents (Wang et al., 2007b).

Microemulsions and emulsions are the main alternative pesticide formulations to emulsifiable concentrates. These colloidal delivery systems are typically prepared from water, surfactants, and oils, with or without organic solvents. In the cases where organic solvents are used, they are typically used at much lower levels than those used to prepare emulsifiable concentrates. Microemulsion and emulsion formulations typically have lower processing costs, cause less environmental pollution, and have higher efficiency than emulsifiable concentrate pesticides. However, there are advantages and disadvantages associated with these different kinds of colloidal delivery systems. It is difficult to incorporate high levels of hydrophobic pesticides in microemulsions due to the low loading capacity of this type of formulation (Huang et al., 2002). Moreover, typically, two to three times more surfactant is required to form microemulsions than emulsifiable concentrates or emulsions. Some microemulsions also require the use of a large amount of cosolvent (methanol and DMF, 10%–15%) and cosurfactant (*n*-propanol and *n*-butanol, 3%–5%) (Singla and Patanjali, 2013). These cosolvents and cosurfactants are hydrophilic polar compounds that can easily permeate into farmland and water resources but are difficult to be removed, therefore causing environmental pollution problems. Ingestion of these compounds may also lead to chronic toxicity in humans and animals. Therefore, in recent years, the development and application of microemulsion-based pesticides remain controversial and have limited their further development and application.

Compared with emulsifiable concentrates and microemulsions, emulsion-based pesticides contain substantially lower levels of organic solvents and surfactants. In addition, they can be produced at large scale using commonly utilized processing operations (such as homogenization) at relatively low costs. Typically, a conventional oil-in-water emulsion consists of oil droplets (diameter from around 200 nm to 200 μm) dispersed in an aqueous medium, with each droplet being surrounded by a layer of emulsifier molecules. In practice, there are some problems associated with utilizing conventional emulsions as delivery systems for hydrophobic pesticides that are associated with their relatively large droplet size. In particular, emulsions are thermodynamically unstable systems that tend to break down through a variable of mechanisms, including gravitational separation (creaming and sedimentation), droplet aggregation (flocculation and coalescence), and Ostwald ripening (Zhang et al., 2007). All these factors can adversely affect the efficiency and shelf life of the final product, thereby reducing its ability to control pests (Shi et al., 2016).

Many of the problems associated with emulsifiable concentrates, microemulsions, and emulsions can be overcome by using nanoemulsions. Like emulsions, nanoemulsions are thermodynamically unstable colloidal dispersions that consist of emulsifier-coated oil droplets dispersed in water, but the mean droplet diameter is smaller (<200 nm). The smaller size of the droplets in nanoemulsions increases their stability to gravitational separation and droplet aggregation and increases their deposition, diffusion, and permeability on plant foliage and insect body surfaces (Kah and Hofmann, 2014). Nanoemulsions provide an effective way to use pesticides efficiently, economically, and safely. In view of this, the composition, characteristics, mechanism of formation, and stability of pesticide nanoemulsions have important theoretical and practical significances on the promotion and application of pesticides.

12.2 TRADITIONAL PESTICIDE FORMULATIONS

Encapsulation of pesticides in well-designed delivery systems extends their shelf life, facilitates their handling and utilization, expands their range of usage, and reduces their toxicity. The design and fabrication of pesticide delivery systems that can be produced economically using large-scale production methods is therefore a critical stage in the development of commercial pesticides. Some of the most important criteria that should be considered when designing these delivery systems are as follows: (1) improves pesticide handling and application, (2) extends shelf life, (3) lowers pesticide toxicity, (4) controls the release rate and targeted application, (5) improves bioactivity, and (6) expands the control spectrum and delays development of resistance (Liu, 2012).

In this section, the traditional delivery systems used for hydrophobic pesticides are reviewed: solid formulations (such as wettable powders) and liquid formulations (such as emulsifiable concentrates, microemulsions, and emulsions).

12.2.1 Emulsifiable Concentrates

Emulsifiable concentrates are typically optically transparent oily liquid formulations that are prepared by dissolving a certain amount of pesticide in organic solvents (such as benzene, toluene, xylene, and solvent oil), which may also contain surfactants and other additives. These systems are then diluted with water prior to utilization, which leads to the spontaneous formation of an oil-in-water emulsion that contains pesticides inside the oil droplets. These systems are therefore similar to the self-emulsifying drug delivery systems (SEDDS) used in the pharmaceutical industry to encapsulate and deliver hydrophobic drugs. Emulsifiable concentrates have the following advantages for pesticide applications: (1) simple processing requirements, (2) low equipment requirements, (3) high levels of active ingredients, (4) excellent storage stability, (5) convenient usage, and (6) high-biological activity (Wu et al., 2013). The main disadvantage of using this form of delivery system is that they contain flammable organic solvents that can lead to high-production costs, increased combustibility, risks of explosion hazards, potential toxicity, and environmental damage.

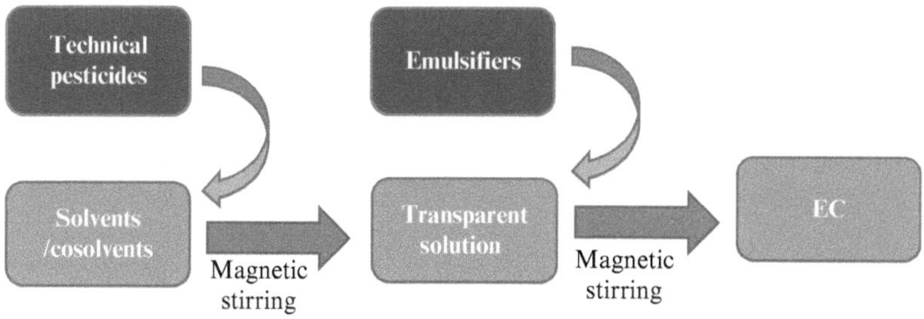

FIG. 12.1 Flow diagram for preparation of emulsifiable concentrate (EC) pesticides.

Emulsifiable concentrates are mainly composed of pesticides, solvents, and surfactants, but cosolvents, stabilizers, synergists, and other additives may also be utilized for some applications. A flow diagram for the preparation of emulsifiable concentrates containing pesticides is shown in Fig. 12.1.

- *Pesticides:* Pesticides with sufficient solubility and chemical stability in organic solvents can be processed into emulsifiable concentrates. Ideally, a high-loading capacity is desirable since this reduces the amount of solvent required and therefore reduces environmental problems. Organochlorine and organophosphorus insecticides are often prepared into emulsifiable concentrates.
- *Solvent*: The organic solvent used plays a key role in the dissolution and dilution of pesticides. A solvent should have a high-flash point, low volatility, good solubility characteristics, excellent compatibility with other components, low human toxicity, low cost, and abundant source.
- *Surfactant*: Emulsifiable concentrates mainly use mixed surfactants, which are made up of one or more nonionic and anionic surfactants. There are many different types of nonionic surfactants used to formulate this type of delivery system, whereas the main anionic surfactant used commercially is dodecyl benzene sulfonate.
- *Other additives*: Stabilizers, synergists, and other additives are often required depending on the nature of the pesticides used and the application requirements. Other constituents may also sometimes be utilized such as cosolvents and cosurfactants.

12.2.2 Microemulsions

Microemulsions are thermodynamically stable systems that are spontaneously formed when certain types and levels of surfactant, oil, and water are mixed together. Microemulsion formulations have a number of potential advantages for pesticide encapsulation and delivery: (1) decreased flammability and explosive risk because water is used as a solvent, (2) lower production costs since little or no organic solvents are needed, (3) ease of preparation since no

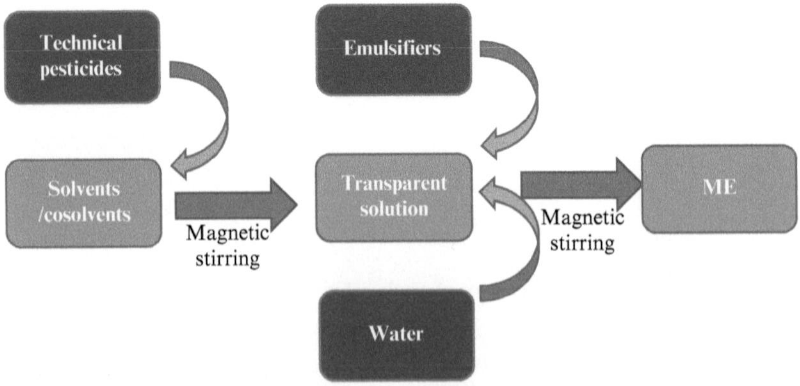

FIG. 12.2 Flow diagram for preparation of microemulsion (ME) pesticides.

specialized equipment is needed, (4) less environmental pollution, and (5) formation of fine oil droplets (10–100 nm) that have higher permeability and absorption rate (Chen and Yin, 2007). However, the main disadvantage of this system is that the surfactant level used is much higher than that required to form emulsifiable concentrates or emulsions.

A flow diagram for the preparation of microemulsions is shown in Fig. 12.2. The pesticides are dissolved in appropriate organic solvents and then stirred, and then, appropriate surfactants are added, which leads to the formation of a uniform and transparent viscous fluid. Finally, water is added to the pesticide-surfactant mixture with continuous stirring, which leads to the production of a transparent oil-in-water microemulsion.

- *Pesticides:* The stability of pesticides in water during microemulsion processing is an important factor that should be considered when developing these systems. For example, organophosphate insecticides easily hydrolyze and oxidize in the presence of water, which should be taken into account when designing this type of delivery system. The pesticide level in microemulsions intended for hygiene applications typically ranges from around 0.1% to 2% but is around 5%–50% for agricultural applications. Microemulsions are particularly suitable for hydrophobic pesticides that only need to be utilized at a low level because of their high bioactivity, such as pyrethroid insecticides.
- *Surfactants:* Surfactants are a key component in the formulation of microemulsions and can be used individually or in combination. The following points need to be considered when selecting an appropriate surfactant or surfactant blend for formulating microemulsion-based delivery systems:
  - *HLB value*: The hydrophilic-lipophilic balance (HLB) value of a surfactant plays a large role in determining its functionality. Surfactants with relatively low HLB values (<6) tend to be oil soluble and stabilize water-in-oil emulsions or microemulsions, whereas those with relatively

high-HLB values (>10) tend to be water soluble and stabilize oil-in-water emulsions or microemulsions. Using a mixture of low and high-HLB surfactants is often employed to facilitate microemulsion formation and improve their functionality. The HLB value may also be varied by using mixtures of nonionic and anionic surfactants, such as calcium dodecyl benzene sulfonate (Zhao et al., 2006).

- *Cloud point*: Nonionic surfactants are composed of a number of oxygen-containing groups, and their solubility in water decreases with increasing temperature due to dehydration of their polar head groups. A surfactant solution may become turbid when a certain temperature, called the cloud point, is reached (Batıgöç and Akbaş, 2017; Lindman et al., 2016). This phenomenon is reversible; the solution becomes transparent and homogeneous again after cooling.
- *Krafft point*: The solubility of the ionic surfactant rises sharply when the temperature increases above a certain value, which is known as the Krafft point (Bakshi et al., 2004).

- *Solvent*: Organic solvents are not typically required when the pesticides used are liquid at room temperature. However, for solid pesticides, one or more organic solvents are often required in microemulsion formulations to ensure that they are fully soluble. The most commonly used solvents are alcohols, ketones, or esters, but aromatic solvents are also sometimes used (Chen et al., 2000). The solvents used must have certain characteristics: they must solubilize high levels of the pesticides, be insoluble in water, be unreactive with other components, have low volatility and toxicity, and have low cost.

A number of analytic tests are normally carried out to ensure quality control of pesticide-loaded microemulsions:

- *Stability*: The stability of microemulsions at ambient temperature is determined by diluting 100 times with 342 mg/L standard hard water and then storing at 30°C for 1 h. The system is taken to be stable if it remains in a transparent state, with no oil slick or sediment being observed.
- *Low temperature stability*: The stability of microemulsions at low temperatures is determined by storing them at 0 ± 2°C for 7 days and then measuring any change in turbidity or overall appearance. The system is taken to be stable if it remains transparent with no aggregates or phase separation being observed.
- *High-temperature stability*: The stability of microemulsions to high temperatures is determined by storing them in ampoules at 54 ± 2°C for 14 days. The system is taken to be stable if it remains transparent, and there is less than 5%–10% reduction in the concentration of the active component due to degradation.
- *Transparent temperature range*: The range of temperatures at which a microemulsion remains physically stable typically spans from around 0°C to 40°C but may be as wide as −5°C to 60°C (Chen and Yin, 2007).

The thermal stability range of microemulsions can be determined by measuring the turbidity versus temperature profile. For example, 10mL of a microemulsion can be placed into a 25mL transparent test tube, and then, the system is cooled in an ice bath until it becomes turbid or the freezing point is reached, and this temperature is recorded (T_1). The same system can then be warmed using a hot bath, and the temperature at which it becomes turbid or phase separates is recorded (T_2). The transparent temperature range is then defined as $\Delta T = T_2 - T_1$.

12.2.3 Emulsions

As mentioned earlier, conventional oil-in-water emulsions can also be utilized to encapsulate and deliver hydrophobic pesticides. The oil phase of these emulsions typically consists of either a liquid pesticide or a solid pesticide dissolved in an organic solvent. This type of emulsion is usually produced commercially by homogenizing the oil phase with an aqueous phase containing a water-soluble emulsifier (such as a surfactant, phospholipid, or amphiphilic polymer). Initially, a coarse emulsion is usually formed using a high-shear mixer, and then, this may be passed through a mechanical homogenizer to further reduce to the droplet size. The most commonly used homogenizers are colloid mills, high-pressure valve homogenizers, microfluidizers, and sonicators. Emulsions are thermodynamically unstable systems but may be designed to have good kinetic stability by appropriate choice of stabilizers, such as emulsifiers, texture modifiers, weighting agents, and ripening inhibitors (Feng et al., 2015). Some of the major advantages of using emulsions as pesticide delivery systems are related to the fact that they can be formulated with lower levels of (or even without) organic solvents and surfactants: (1) They are nonflammable and not explosion hazards, and (2) they cause less environmental pollution and less human and animal toxicity (Li et al., 2012). However, they tend to be less stable during storage than both emulsifiable concentrates and microemulsions. A flow diagram for preparation of emulsions is shown in Fig. 12.3. The major components used to formulate emulsion-based delivery systems for pesticides are discussed below:

- *Pesticides*: Pesticides need to meet a number of conditions if they are going to be good candidates for encapsulation in emulsions: (1) They should be hydrophobic molecules with a water solubility <1000mg/L; (2) they should have good chemical stability when in contact with water, that is, not prone to hydrolysis; and (3) they should be liquid or solid with a low melting point (generally <60°C) in order to be suitable for preparing emulsions (Shi et al., 2016). Nevertheless, crystalline pesticides may be dissolved in an oil phase prior to emulsion formation. At present, pyrethroid insecticides and amide herbicides are commonly encapsulated in emulsion-based delivery systems.
- *Solvents*: Liquid pesticides do not usually require the use of an organic solvent to prepare stable emulsions; however, solid pesticides may need to be dissolved in an organic solvent prior to homogenization.

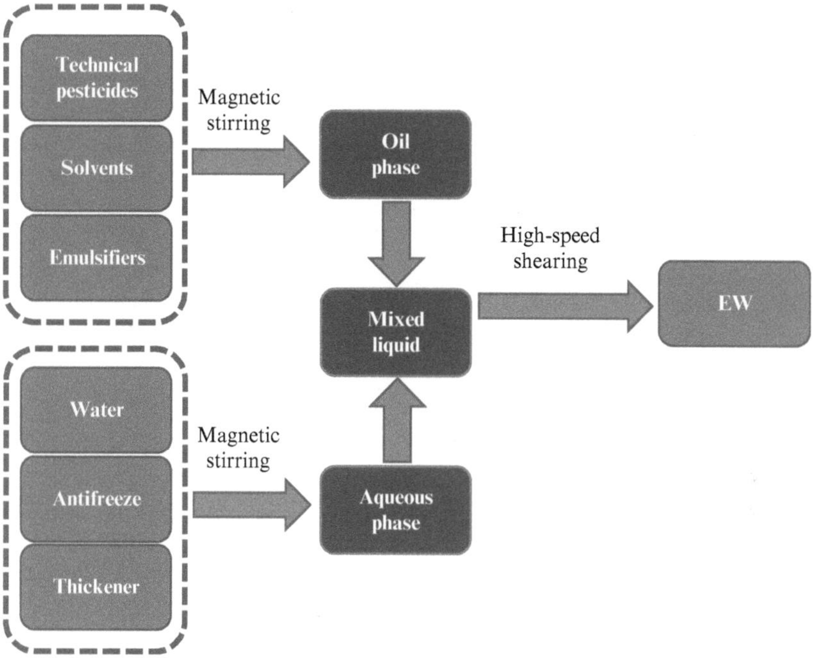

FIG. 12.3 Flow diagram for preparation of emulsion oil in water (EW) pesticides.

- *Emulsifiers*: The selection of an appropriate emulsifier or blend of emulsifiers is critical to the formulation of stable and effective emulsion-based delivery systems. A range of emulsifiers have been used to prepare pesticide emulsions, including phenol ether, phosphates, castor oil, polyoxyethylene ethers, block polyethers, and their mixtures (Feng et al., 2012b). However, a variety of other types of emulsifiers could also be used, including phospholipids, proteins, and polysaccharides.
- *Antifreeze agents*: The low temperature stability of emulsions can be improved by adding antifreeze agents, such as ethylene glycol, propylene glycol, glycerol, and urea.
- *Thickeners*: The stability of emulsions to gravitational separation can be improved by adding thickening or gelling agents, which may be natural polysaccharides (such as xanthan gum) or inorganic compounds (such as magnesium silicate).

12.3 DEVELOPMENTS OF PESTICIDE NANOEMULSIONS

In this section, the formulation and production of nanoemulsion-based delivery systems for pesticides is reviewed. Oil-in-water nanoemulsions are thermodynamically unstable colloidal dispersions that usually consist of emulsifier-coated oil droplets dispersed in water. In this respect, they are similar to conventional oil-in-water emulsions. However, nanoemulsions have been defined as systems containing droplets with diameters <200 nm, whereas conventional emulsions have been defined as systems containing droplets with diameters >200 nm (Gutiérrez et al., 2008; McClements, 2012). The small size of the droplets in nanoemulsions gives them some advantages over conventional emulsions: higher optical clarity; higher stability to droplet aggregation and gravitational separation; and higher bioactivity of encapsulated components (McClements, 2012). Nanoemulsions are distinguished from microemulsions based on their thermodynamic stability, rather than on their particle size characteristics. Nanoemulsions are thermodynamically unstable, whereas microemulsions are thermodynamically stable. Nevertheless, nanoemulsions can be formulated to have good long-term kinetic stability.

12.3.1 Composition of Pesticide Nanoemulsions

In this section, an overview of the main components used to formulate pesticide nanoemulsions is given. In Table 12.1, a comparison of different emulsions and nanoemulsions for preparation of pesticides has been presented.

Oil phase: The oil phase may consist entirely of the pesticide if it is liquid at ambient temperatures. However, additional oil-phase components may be required if a pesticide is solid at ambient temperature or if it has an appreciable water solubility. A crystalline pesticide can be dissolved in an appropriate solvent prior to homogenization. The solvent selected should have a number of characteristics: wide availability, low cost, low toxicity, good safety characteristics, high-pesticide solubility, low solubility in water, and good environmental compatibility. At present, the most commonly used solvents are (1) traditional aromatic solvents, such as toluene and xylene (Liu et al., 2011); (2) other solvents that have high solubility for the vast majority of pesticides but are expensive and are therefore usually mixed with aromatic solvents, such as common ketones (cyclohexanone and isophorone), alcohols (methanol and ethanol), and dimethylformamide and dimethyl sulfoxide; and (3) environmentally friendly solvents, such as methyl oleate (Feng et al., 2016) and methyl decanoate (Wang et al., 2007a). In principle, other environmentally friendly solvents could also be used, such as food-grade triglyceride oils (such as corn, soybean, sunflower, or rapeseed oils). It may also be necessary to mix a liquid pesticide that has an appreciable water solubility with a water-insoluble solvent (a "ripening inhibitor") to retard droplet growth due to Ostwald ripening, that is, diffusion of pesticide molecules from small to large droplets due to a difference in Laplace pressure (Chang et al., 2012).

Emulsifiers: A variety of different kinds of natural and synthetic emulsifiers can be used to formulate oil-in-water emulsions. Any emulsifier should meet the following characteristics:

- Excellent emulsifying performance, that is, can produce small oil droplets at low dosage levels

TABLE 12.1 Comparison Among ECs, MEs, EWs, and Nanoemulsions for Formulation of Pesticides

Parameter	Emulsifiable Concentrates (ECs)	Microemulsions (MEs)	Emulsions, Oil in Water (EWs)	Nanoemulsions
Appearance	Transparent	Transparent	Opaque	Semitransparent or transparent
Droplet size	Molecular dimension	10–100 nm	2–5 μm	20–500 nm
Thermodynamic stability	Stable	Stable	Instable	Instable
Water content	No water	High	High	High
Emulsifier dosage	High	Maximum	Low	Minimum
Pesticide loading capacity	High	Low	Low	Low
Production costs	Maximum	High	Low	Minimum
Application method	Diluted with water for spraying	Diluted with water for spraying	Diluted with water for spraying	Diluted with water for spraying

- Broad range of application for a variety of pesticides
- Good compatibility with pesticides, solvents, and other components
- Good stability to environmental conditions, such as water hardness and temperature changes
- Low toxicity and good environmental compatibility

The major functions of emulsifiers in pesticide nanoemulsions are briefly reviewed here:

(1) *Droplet formation:* An emulsifier should rapidly adsorb to the oil droplet surfaces created during homogenization and reduce the interfacial tension, thereby facilitating further droplet breakup (Matsumiya et al., 2011). In nanoemulsions, a highly curved interfacial layer is formed because of the small droplet size. As a result, there is a pressure difference, which is known as the Laplace pressure (ΔP), between the inner and outer sides of the interface; the pressure inside the droplet being greater than that outside it. Consequently, work (W) must be done by the homogenizer to overcome the Laplace pressure and promote droplet deformation and disruption. The Laplace pressure can be calculated using the following formulas (Sivastava et al., 2006):

$$W = \gamma_i \Delta A \tag{12.1}$$

$$\Delta P = \gamma_i (1/R_1 + 1/R_2) \tag{12.2}$$

Here, γ_i is the oil/water interfacial tension (mN/m), ΔA is the increase in interfacial area (m^2), and R_1 and R_2 are the concave radii of curvature of the interface (m).

Formulas (12.1), (12.2) indicate that more work is required to deform and disrupt the droplets when the interfacial tension increases or the droplet size decreases (Fan et al., 2010). Consequently, reducing the low interfacial tension will facilitate the formation of small droplets inside a homogenizer. However, it should be stressed that reduction of the interfacial tension is not the only factor that determines the formation of stable nanoemulsions. For example, low-carbon alcohols (such as pentanol) can reduce the interfacial tension to a very low level, but cannot form stable nanoemulsions because the droplets tend to rapidly coalesce with each other after formation. Conversely, some macromolecules (such as gelatin or gum arabic) do not to reduce the interfacial tension appreciably, but have a very excellent emulsifying stability because they form a protective coating around the droplets. Thus, lowering the interfacial tension can facilitate the formation of a nanoemulsion, but the reduction in interfacial tension alone is not sufficient to ensure the stability of the nanoemulsions.

(2) *Droplet stabilization:* An emulsifier should also form a protective coating around the oil droplets during homogenization (Jin et al., 2017). The emulsifier molecules may do this by generating a relatively strong steric and/or electrostatic repulsion between the oil droplets, or they may form a cohesive interfacial film that is difficult to disrupt when droplets collide together. The mechanical strength of the interfacial film is therefore a key factor affecting the stability of nanoemulsions (Liang et al., 2017). Typically, oil droplets must be almost fully covered with emulsifier molecules to prevent their coalescence. In practical applications, the use of mixed emulsifiers or the addition of small molecular alcohols can form a composite interfacial film that may improve the stability of nanoemulsions (Zeeb et al., 2015).

Co-emulsifiers: Emulsion formation and/or stability can sometimes be improved using coemulsifiers in addition to emulsifiers. Coemulsifiers are surface-active molecules that can adsorb to droplet surfaces but are unable to stabilize emulsions when used in isolation. The most commonly used coemulsifiers are shorter carbon-chain alcohols, which have a small polar head group and a hydrocarbon chain. These molecules may improve emulsion formation and stability by a number of physicochemical mechanisms, including decreasing the interfacial tension, increasing the flexibility of the interfacial layer, and altering the optimum curvature of the interfacial layer.

12.3.2 Advantages and Disadvantages of Pesticide Nanoemulsions

Nanoemulsions have a number of potential advantages over microemulsions and conventional emulsions: (1) Compared with microemulsions, the amount of surfactant required is significantly reduced (Rao and McClements, 2011), and (2) compared with conventional emulsions, nanoemulsions have better aggregation stability, exhibit less gravitational separation, have higher optical clarity, and give a higher bioavailability (Jafari et al., 2007a,b; Khot et al., 2012).

Nanoemulsions are thermodynamically unstable systems that may break down due to flocculation, coalescence, creaming, Ostwald ripening, and other phenomena (Robins et al., 2002; Rao and McClements, 2011; Jafari, et al., 2008; Assadpour et al., 2016a,b). According to Stoke's law, the upward movement of droplets in colloidal dispersions is proportional to the particle size squared, and so, the smaller particle size in nanoemulsions greatly slows down gravitational separation. Moreover, Brownian motion effects become more important for smaller particles, which may be sufficient to overcome gravity effects (Rao and McClements, 2011).

Ostwald ripening is one of most common forms of physical instability in nanoemulsions (Taylor, 1998). This phenomenon occurs due to a particle size dependence of the local solubility of the oil phase in the surrounding water phase; the water solubility of the oil phase in small droplets is greater than that in large droplets. Consequently, oil-phase molecules tend to move from the smaller droplets to the larger droplets, resulting in the disappearance of small droplets and an increase in the mean droplet diameter (Chebil et al., 2013). A theoretical description of Ostwald ripening in a two-phase system was independently developed by Lifshitz, Slezov, and Wagner and is referred to as the LSW theory (Chebil et al., 2013). One of their major results is that once the system has reached a steady state, the ripening rate (ω) is given by (Juárez and Whitby, 2012)

$$\omega = \frac{dr^3}{dt}\frac{8C(\infty)\gamma V_m D}{9\rho RT} \tag{12.3}$$

where r is the average droplet radius after time t, ρ is the density of the internal phase, D is the diffusion of the dispersed phase in the continuous phase, $C(\infty)$ is the bulk phase solubility (the solubility of an infinitely large droplet), V_m is the molar volume of the internal phase, R is the gas constant, and T is the absolute temperature. The cube of the mean droplet radius therefore has a linear relationship with time when Ostwald ripening occurs.

Ostwald ripening is affected by many factors such as the initial droplet diameter and particle size distribution, the types and levels of emulsifiers used, the oil-water ratio, the temperature, and the oil-phase composition (Singh et al., 2017; El-Din et al., 2017; Moghimi et al., 2016). Furthermore, the physical and chemical properties of the dispersed and continuous phases are also important factors for Ostwald ripening; good intermiscibility between the two phases tends to lead to Ostwald ripening (Zeeb et al., 2012).

12.3.3 Production of Pesticide Nanoemulsions

The methods used to prepare pesticide nanoemulsion can be divided into high-energy (Liu et al., 2011) and low-energy (Song et al., 2009) emulsification methods, which each have their own advantages and disadvantages.

12.3.3.1 High-Energy Processing Method

High-energy emulsification is a traditional method for the preparation of pesticide nanoemulsions, which typically requires the use of high-pressure homogenizers, microfluidizers, or sonicators (Jafari et al., 2006; Lee and Norton, 2013; McClements, 2012). These methods utilize a high-energy consumption resulting in high-production costs, and only ~0.1% of the energy is actually used for emulsification, with the rest being lost as heat.

Droplet breakup in high-energy emulsification is mainly due to the strong disruptive forces generated within mechanical homogenizers (Sivastava et al., 2006). These disruptive forces have to be large enough to overcome the Laplace pressure:

$$\Delta P = \frac{\gamma}{2r} \tag{12.4}$$

where γ is the interface tension and r is the droplet radius. The smaller the droplet, the greater the pressure gradient required for the deformation and disruption of the droplets and therefore the greater the mechanical forces and energy expenditure needed to form small droplets. Although high-energy emulsification methods require greater energy expenditure than low-energy ones, they typically require lower levels of emulsifiers, which can offset costs. Other advantages of high-energy emulsification for preparing nanoemulsions include applying an oil phase with high viscosity and high-molecular weight, such as silicone oil and fluorocarbon, and a greater flexibility in the types of emulsifiers that can be used.

A comparison was made between a high-energy method (microfluidization) and a low-energy method (spontaneous emulsification) for producing oil-in-water emulsions from food-grade ingredients (medium-chain triglycerides and Tweens) by Yang et al. (2012). The influence of surfactant type (Tween 80, Tween 85, and Tween 80/Tween 85) and surfactant-to-oil ratio (SOR = 0.1:5) on the formation of emulsions was examined. Both the low- and high-energy methods produced emulsions with ultrafine droplets ($r < 100$ nm). The microfluidization method required high-energy inputs and dedicated equipment but could produce ultrafine emulsions at much lower surfactant-to-oil ratio (SOR < 0.1). On the other hand, the spontaneous emulsification method only required simple mixing, but the method needed a much higher surfactant-to-oil ratio (SOR > 0.5) to produce droplets with $r < 100$ nm.

Different types of high-energy emulsification methods are discussed in more detail elsewhere in this book (Chapters 5–10).

12.3.3.2 *Low-Energy Processing Method*

Low-energy emulsification methods mainly utilize the chemical energy stored in the system components, and therefore require less energy input, utilize a simple preparation device, and have a relatively low cost. This method has received considerable attention in recent years (Liu et al., 2006; Mei et al., 2011a; Mehrnia et al., 2016). Low-energy emulsification methods can be conveniently divided into different categories (Mohammadi et al., 2016a,b):

(1) Phase inversion composition

In the phase inversion composition (PIC) method, a nanoemulsion is formed by changing the proportion of the different components in the system. In one approach, an aqueous phase is slowly added dropwise into an organic phase containing a hydrophilic surfactant at low speed. As the water content increases, W/O nanoemulsions or reverse micelles are formed, and then, they are gradually transformed into a double-continuous structure. A further increase in water content results in phase inversion, thereby resulting in the production of O/W nanoemulsions. Alternatively, an organic phase containing a hydrophilic surfactant can be added dropwise to an aqueous phase with mild stirring. This approach is sometimes also referred to as the spontaneous emulsification method.

Many factors influence the preparation of nanoemulsions by the PIC method. The size of the droplets produced depends on system composition, mixing order, agitation rate, dropping rate, temperature, and other environmental factors (Perazzo et al., 2015). Feng et al. (2016) systematically investigated the effects of addition sequence of organic/aqueous phases and initial location of emulsifiers on the stability of pesticide nanoemulsions. Nanoemulsions with similar formulations were prepared: beta-cypermethrin (hydrophobic pesticide); methyl oleate (organic solvent); polyoxyethylene castor oil ether (surfactant), and deionized water. The initial particle size distribution and emulsion stability depended strongly on the preparation method used. Stable nanoemulsions could be prepared by adding an aqueous phase into an organic phase containing surfactants or vice versa.

(2) Phase transition temperature

The phase transition temperature (PIT) method is based on a transition from an oil-in-water, to a bicontinuous microemulsion, to a water-in-oil system when certain surfactant-oil-water mixtures are heated (Saberi and McClements, 2015). Typically, only nonionic surfactants can be used in this approach since their head group becomes progressively dehydrated as they are heated. Usually, all components are mixed evenly, the temperature is constantly increased at low-speed stirring until the PIT is reached, and then the mixture is transferred to an ice bath for rapid cooling, which results in the formation of a nanoemulsion (Balaji et al., 2015).

(3) Microemulsion dilution

The microemulsion dilution method of preparing nanoemulsions involves rapidly diluting a microemulsion by addition of a large amount of continuous phase. O/W-type or double-continuous microemulsions are usually formed by mixing oil phase, surfactant, and a small amount of water and then adding a large amount of water and stirring rapidly to form a nanoemulsion (Wang et al., 2007a).

Different types of low-energy emulsification methods are discussed in more detail elsewhere in this book (Chapters 3 and 4).

12.4 INFLUENCING FACTORS FOR FORMATION AND STABILITY OF PESTICIDE NANOEMULSIONS

In recent years, research on pesticide nanoemulsions has mainly focused on emulsifier selection and formulation optimization, and there are few studies on the effects of pH, ionic strength, and temperature on their formation and stabilization. This lack of knowledge is limiting the application of nanoemulsions as pesticide delivery systems.

12.4.1 pH Stability

The pH stability of nanoemulsions is strongly dependent on the nature of the emulsifier coating the droplets. Nanoemulsions stabilized by nonionic surfactants are often stable across a wide range of pH values because the main repulsive force is steric rather than electrostatic. For ionic surfactants, the pH stability depends on the pK_a values of the charged groups. At pH values where these groups lose their charge, a nanoemulsion may become unstable to aggregation because of the reduction in electrostatic repulsion between the droplets. The stability of protein-coated oil droplets is highly dependent on pH because their charge goes from highly positive, to zero, to highly negative when the pH is increased from below to above their isoelectric point (p*I*) (Bai et al., 2016; Faridi Esfanjani et al., 2015; Faridi et al., 2017). The p*I* of a protein is related to its amino acid sequence and conformation and therefore varies from protein to protein. Research on pesticide nanoemulsions prepared using nonionic and ionic surfactants is fairly extensive, but there have been few studies on those prepared using protein emulsifiers.

12.4.2 Ionic Strength

Electrolyte ions also have significant effects on the formation and stability of nanoemulsions, particularly those that are stabilized by electrostatic repulsion. Electrolyte ions affect nanoemulsion properties mainly by altering the structure and function of the emulsifier. For example, electrolyte ions mainly have salting-in and salting-out effects, which can increase or decrease the cloud points of nonionic surfactants, respectively. Salting-in ions include Ca^{2+}, Mg^{2+}, H^+, and I^-, whereas salting-out ions include Na^+, K^+, Cl^-, and F^- (Mei et al., 2011b). Electrolyte ions may also alter nanoemulsion stability through electrostatic screening and ion-binding effects. Electrostatic screening occurs due to accumulation of counterions around the droplet surfaces, which partially neutralizes their charge. Ion binding occurs due to the strong association of an ion (usually multivalent) with an oppositely charged surface.

Feng et al. (2012a) investigated the effect of electrolyte ions on the stability of metolachlor nanoemulsions. They showed that electrolyte ions decreased the zeta-potential by compressing the electric double-layer thickness at the droplet surfaces, which led to droplet aggregation through an electrostatic screening effect. Higher electrolyte concentration and valence led to lower stability of the metolachlor emulsions.

12.4.3 Temperature

The effects of temperature on pesticide nanoemulsions mainly include two aspects: (i) emulsification temperature and (ii) storage temperature. Yu et al. (2012)

produced oil-in-water nanoemulsions with Span 80-Tween 80 as surfactants and paraffin oil as the oil phase via the PIC method at elevated temperatures. With increase of preparation temperature from 20°C to 70°C, the emulsion droplet diameter decreased from around 10.3 μm to 51 nm. These results showed that nanoemulsions cannot be successfully produced from highly viscous paraffin oil by the PIC method at 25°C, but they can at elevated temperatures. In practical applications, the ambient temperature of pesticide nanoemulsions may change from low (in winter) to high (in summer). Consequently, future work should focus on examining the impact of storage temperature on the physical stability and performance of pesticide nanoemulsions produced using different types of emulsifiers.

12.4.4 Oil-Water Ratio

In many commercial applications, nanoemulsions with high oil concentration are of interest because they enable higher loading capacities, reduce transport costs, and lead to interesting textural characteristics. Many types of pesticide nanoemulsions use solvents to dissolve the active ingredient, and the level of solvent utilized should be optimized; too little solvent may lead to nanoemulsions containing insufficient pesticides, whereas too much solvent may reduce nanoemulsion stability, increase production costs, and have negative environmental impacts.

12.4.5 Dilution Ratio

Many pesticide nanoemulsions are applied by spraying them after they have been diluted with water. The dilution ratio and the hardness of the water used may affect the stability and performance of the nanoemulsions; and therefore, they should be designed to have excellent dilution stability and dilution media adaptability.

Lim et al. (2012a) have exploited glyphosate-laden nanoemulsions for herbicide formulations. This study revealed that upon addition of nanoemulsion components (LFAMEs + LAPG + organosilicon + water) into 41% (w/w) glyphosate IPA, the preformulation formed a gel-like polymerized network. The nanoemulsion system occupied the junctions of the network of glyphosate IPA, thereby causing the network to become densely packed. Dilution of this preformulation with water led to the destruction of the polymerized network and dispersion of the nanoemulsions. The glyphosate IPA no longer appeared in needle-shaped structures, indicating that the nanoemulsion droplets had incorporated the glyphosate IPA structure and changed the morphology. This glyphosate-nanoemulsion associability can improve the pesticide delivery efficiency. In contrast, a less stable preformulation was diluted and analyzed.

The emulsion droplets were found to be unequally distributed and had low probability of being incorporated into the glyphosate IPA and thus remained as a needle-shaped structure. This deficiency of incorporation can impair the bioactivity of the herbicides.

12.5 APPLICATION PERFORMANCE OF PESTICIDE NANOEMULSIONS

Compared with nanoemulsions in other fields, there has been a fairly limited amount of research on the applications on pesticide nanoemulsions. The related literature on pesticide nanoemulsions (2007–17) is summarized in Table 12.2. In the future, the application of pesticide nanoemulsions will be used when people pay more attention to sustainable agricultural development.

12.5.1 Deposition, Diffusion, and Pervaporation of Pesticide Nanoemulsions

Unlike food and pharmaceutical nanoemulsions, pesticide nanoemulsions need to be diluted with water, and then applied using a spray device for uniform coverage (Bueno et al., 2017). Pesticides need to reach a certain concentration range in the target place (e.g., insect surface or plant leaves) in order to achieve a good control effect. Achieving an effective concentration range improves the deposition, diffusion, and permeability of the pesticides. Wetting is a process of where a liquid/solid interface replaces a gas/solid surface when a liquid is applied to a solid surface. Wettability is an important condition for the effective deposition of diluted pesticides in plant foliage and insect surfaces. Diluents with poor wetting ability are hardly stable on the solid surface and easily roll off. Conversely, when the wetting ability is too strong, the diluent forms a very thin liquid film that easily leaves the surface. These phenomena will reduce the amount of pesticide at the site of action. In general, solid surface wetting can be divided into bedewing, soaking, and spreading (Gharabaghi and Aghazadeh, 2014; Olivier, 2016; Cui, 2013):

12.5.1.1 Bedewing

Bedewing is a process wherein liquid replaces the gas of the solid surface, but cannot be fully expanded. The newly formed liquid/solid interface increases the free energy γ_{sl}, whereas the replaced gas/liquid and gas/solid interfaces reduce the free energy γ_{lg} and γ_{sg}, respectively. The overall change in the free energy of the system is related to the adhesion work and is calculated by formula (12.5)

$$-\Delta G_A = \gamma_{sg} + \gamma_{lg} - \gamma_{sl} = W_A \text{ (Adhesion work)} \qquad (12.5)$$

When the external work is done by the system $W_A > 0$, that is, $\Delta G_A < 0$, the wetting process will be thermodynamically spontaneous.

TABLE 12.2 A Brief Overview of Recent Studies in Pesticide Nanoemulsions (2007–17)

Pesticide Type	Processing Method	Oil Phase	Emulsifiers	Droplet Size	Types of Nano emulsions (O/W or W/O)	Charac terization Tools of Stability	Function and Applications	References
β-Cypermethrin	ME dilution	Methyl decanoate	Poly(oxyethylene) lauryl ether, with an average of 7 mol of ethylene oxide (EO)	40–200 nm	O/W	Polarizing light microscopy	The solubilization capacity of the concentrate was maintained on dilution, which shows this system is easy to scale up with less energy consumption and very suitable for agrochemical active delivery	Wang et al. (2007a)
Citronella oil	High-pressure homo genization	Citronella oil	Montanov@82 (mixture of cetearyl alcohol and cocoyl glucoside)	100–350 nm	O/W	Phase diagram	Nanoemulsions with optimal size of oil droplets and viscosity of medium should be recommended for stability during storage and ability to control the release of essential oil and the in vivo activity	Sakulku et al. (2009)

Triazophos	Phase transition component method	*N*-Octyl-2-pyrrolidone	Nonylphenol polyoxyethylene ether (an average number of ethylene oxide (EO) groups is 12)	<200 nm	O/W	Phase diagram, TEM	Triazophos can be protected from hydrolysis by incorporating into nanoemulsions	Song et al. (2009)
Avermectin	High-pressure process	Xylene	Polyoxyethylene (40) castor oil ether (EL-40), polyoxyethylene (20) castor oil ether (EL-20), and octyl phenyl polyoxyethylene phosphonate (OPP)	<200 nm	O/W	Zeta potential, rheological measurements	A more stable emulsion formulation of avermectin by the addition of AG and XG to relatively stable emulsions	Zhang and Liu (2011)
Bifenthrin	High-pressure process	Dimethyl benzene	Dipotassium monododecyl phosphate ($C_{12}H_{25}OPO_3K_2$) Polyoxyethylene 3-Lauryl ether ($C_{12}H_{25}O(CH_2CH_2O)_3H$)	<250 nm	O/W	Multiple light scatting	Nanoemulsions showed high-kinetic stability, without phase separation after 180 days of storage at 25°C, and the quality indicators were up to standard according to FAO standards	Liu et al. (2011)

Continued

TABLE 12.2 A Brief Overview of Recent Studies in Pesticide Nanoemulsions (2007–17)—cont'd

Pesticide Type	Processing Method	Oil Phase	Emulsifiers	Droplet Size	Types of Nano emulsions (O/W or W/O)	Charac terization Tools of Stability	Function and Applications	References
Glyphosate isopro pylamine (IPA)	Low-energy emulsification	Short-chain fatty acid methyl esters containing mixture of hexanoate, octanoate, decanoate, and laurate esters at 3.2, 50.8, 44.0, and 2.0 (w/w), respectively, and long-chain fatty acid methyl esters containing a mixture of octanoate, decanoate, laurate, myristate, palmitate, stearate, oleate, linoleate, linolenate, and arachidate esters at 0.1, 0.1, 52.1, 17.7, 8.9, 2.2, 15.8, 2.8, 0.2, and 0.1 (w/w)	Carbohydrate-derived alkyl polyglucoside (APG) surfactants: short-chain alkyl polyglucosides (SAPGs) with a mixture of alkyl-chain octyl and decyl at 45 and 55 (w/w), respectively, and long-chain alkyl polyglucosides (LAPGs) with a mixture of alkyl-chain dodecyl, tetradecyl, and hexadecyl at 68, 26, and 6 (w/w), respectively	<200 nm	O/W	Phase diagrams, Thermostability test, rheology	Nanoemulsions in glyphosate IPA formulations have yielded injury rates equivalent to those achieved with roundup	Lim et al. (2012a)

Glyphosate IPA	Low-energy emulsification	Stripped palm kernel fatty acid methyl octanoate/ decanoate/ laurate/ myristate/ palmitate/ stearate/ oleate/ linoleate/ linolenate/ arachidate (0.1;0.1; 52.1;17.7; 8.9;2.2; 15.8;2.8; 0.2;0.1, w/w)	Dodecyl/ tetradecyl/ hexadecyl (68;26;6, w/w) polyglucosides	<250 nm	O/W	TEM	The nanoemulsions incorporated with glyphosate IPA to form the well-dispersed nanoparticles	Lim et al. (2012b)
Glyphosate isopropylamine (IPA)	Low-energy emulsification	Fatty acid methyl hexanoate/ octanoate/ decanoate/ laurate (3.2;50.8; 44.0;2.0 (w/w))	Octyl/decyl (45;55 (w/w)) polyglucosides (short-chain APG), dodecyl/ tetradecyl/ hexadecyl (68:26:6 (w/w)) polyglucosides (long-chain APG)	<200 nm	O/W	Phase diagrams, thermo stability test	The particles of nanoemulsions were evenly distributed and promoted higher bioefficacy than the commercial formulation	Lim et al. (2013)

Continued

TABLE 12.2 A Brief Overview of Recent Studies in Pesticide Nanoemulsions (2007–17)—cont'd

Pesticide Type	Processing Method	Oil Phase	Emulsifiers	Droplet Size	Types of Nano emulsions (O/W or W/O)	Charac terization Tools of Stability	Function and Applications	References
Etofenprox	High-pressure homogenization	Mixture of xylene and butanol (m/m=7.80)	Mixed surfactants of polyoxyethylene (40) castor oil ether (EL-40) and polyoxyethylene (20) castor oil ether (EL-20). Block copolymer polyalkoxylated butyl ether (LQ) and anionic surfactant polyoxyethylene styrenated phenol ether phosphonate (P)	180–320 nm	O/W	Viscosity measurement, thermal analysis	Mixed surfactants can provide good storage and freeze-thaw stability for pesticide emulsions	Zhang et al. (2013)
Eucalyptus oil	Low-energy emulsification	Eucalyptus oil	Suprol T-60 (commercial Polysorbate 80 or Tween 80)	77–569 nm	O/W	Droplet size, ATR-FTIR analysis	Eucalyptus essential oil-based nanoemulsions with the aqueous filtrate can be used as toxicants for the control of stored-product beetles	Pant et al. (2014)

Diethyl phenyl acetamide (DEPA)	PIT emulsification	Dichloromethane	Tween 20 and soy lecithin	<200 nm	O/W	High-resolution TEM	Provide the template in attaining the monodispersed stable and bioefficacious nanometric form of several other poorly water-soluble organic compounds	Balaji et al. (2015)
Asteraceae essential oils	High-pressure homogenization	Asteraceae essential oils		<150 nm	O/W	TEM	Showed a considerable insecticidal activity against *C. maculatus*	Nenaah et al. (2015)
Fenpropathrin and cyflumetofen	Low-energy emulsification	Solvent naphtha, xylene = 4:1	Calcium alkyl aromatic sulfonate, polyoxyethylene aliphatic = 2:3	31–300 nm	O/W	Contact angle, adhesion work	Had the smaller contact angle and the larger adhesion work, which was conducive to the absorption of plant leaves to liquid medicine and thus improve efficacy	Zhao et al. (2016)
Beta-cypermethrin	Low-energy emulsification	Methyl oleate	Polyoxyethylene (40) castor oil ether (EL-40) and polyoxyethylene (20) castor oil ether (EL-20)	<200 nm	O/W	Multiple light scattering, thermal analysis	Successfully used the methyl oleate as an environmentally friendly alternative solvent for pesticide nanoemulsions, and a low-emulsifying process for highly stable pesticide nanoemulsions was developed	Feng et al. (2016)

Continued

TABLE 12.2 A Brief Overview of Recent Studies in Pesticide Nanoemulsions (2007–17)—cont'd

Pesticide Type	Processing Method	Oil Phase	Emulsifiers	Droplet Size	Types of Nano emulsions (O/W or W/O)	Charac terization Tools of Stability	Function and Applications	References
Deltamethrin	Ultrasonic emulsification	Toluene	Tween 80	<40 nm	O/W	Dynamic light scattering	Ensured the ability of deltamethrin to combat against the mosquito population even at lower concentrations, reducing the pesticide exposure load toward the environment and human population	Balaji et al. (2017)

12.5.1.2 Soaking

Soaking refers to the process of immersing a solid in a liquid; this process transforms a gas/solid interface into a liquid/solid interface. The liquid surface does not change during the soaking process. At constant temperature and constant pressure, the overall change in free energy of the system is related to the infiltration work and is calculated by the formula (12.6):

$$-\Delta G_I = \gamma_{sg} - \gamma_{sl} = W_I \text{ (Infiltration work)} \tag{12.6}$$

The infiltration work (W_I) represents the ability of the liquid to replace the gas at the solid surface. The spreading effect entails the spreading of force against the surface tension of the liquid. This parameter is also known as the adhesive tension and is commonly used and expressed by the formula (12.7)

$$W_I = A = -\Delta G_I = \gamma_{sg} - \gamma_{sl} \tag{12.7}$$

With constant temperature and pressure, the condition of the solid soaked in the liquid is $W_I > 0$, that is, $-\Delta G_I < 0$, the liquid will soak in the solid surface.

Spreading

Spreading refers to the process of replacing a liquid/solid interface with a gas/solid interface and expansion of the liquid surface. The system also adds the same area of gas/liquid interface. At constant temperature and pressure, the overall change in free energy that is related to the spreading factor is calculated by formula (12.8)

$$-\Delta G_s = \gamma_{gs} - \gamma_{lg} - \gamma_{sl} = S \text{(Spreading factor)} \tag{12.8}$$

When $S > 0$, the spreading process occurs spontaneously.

When the adhesion tension (A) between the liquid and the solid is greater than the surface tension of the liquid, the liquid will automatically spread on the solid surface.

For the same system, the occurrence conditions of three types of wetting are $W_A > W_I > S$. Therefore, when $S \geq 0$, W_A and W_I must be greater than zero. This result indicates that if the liquid can spread on the solid surface, the liquid becomes capable of bedewing and soaking the solid, and so, the spreading factor S is commonly used as an indicator of system wetting.

The occurrence conditions of the three types of wetting process indicate that the contribution of the gas/solid and solid/liquid interfaces to the three kinds of system wetting is consistent; all play a role in the form of adhesive tension A: the larger the γ_{sg}, the smaller the γ_{sl}, the greater the difference between them, and it is more conducive to wetting.

The contribution of liquid surface tension to the three processes is different; that is, higher γ_{lg} values are favorable for bedewing, whereas smaller γ_{lg} values are favorable for spreading, while wetting has nothing to do with γ_{lg}. Theoretically, after the wetting type is determined, the wetting effect can be obtained by adjusting the corresponding surface energy. The surface energy of a solid surface determines the wetting properties. The basic condition that the liquid can spontaneously spread on the solid surface is that the surface tension of the liquid is smaller than the solid surface energy, and the lower surface tension of the liquid is more favorable to the spread.

The empirical parameter characterizing the wettability of a solid surface is the critical (wetting) surface tension (γ_c), and the physical meaning of this parameter is that a liquid with a surface tension $<\gamma_c$ can spread on this low-energy surface. In fact, in the three kinds of interface energy, only γ_{lg} can be directly measured experimentally; therefore, the above criteria have only theoretical significance. Practical applications generally rely on the contact angle to determine the type of wetting, that is, the smaller the contact angle, the better the wettability (or the higher the surface hydrophilicity). On the contrary, the wettability worsens (or the higher the surface hydrophobicity). Many of the important targets for pesticide nanoemulsions have hydrophobic surfaces, such as insect wings (Aideo and Mohanta, 2016), and plant leaves (Tang et al., 2008).

In 1979, Young (1979) proposed the concept of "dose transfer" of pesticides and considered that the pesticide began its dose transfer once it was sprayed from the spraying machine. Although the goal of spraying pesticides is to control specific target organisms, a considerable fraction of a sprayed pesticide may go into the environment or onto nontarget organisms, resulting in loss of efficacy and potentially damaging the environment. The main forms of pesticide losses are evaporation, drift, and shedding from the plant surface. The diluent characteristics, droplet size, meteorological factors, and leaf characteristics of the target plant are the main factors influencing pesticide losses (Jensen and Olesen, 2014). The shape, size, growth direction, and surface wax of plant leaves will also affect the deposition of pesticides.

In addition, the crop temperature, humidity, wind direction, and wind power affect the formation and movement of the sprayed pesticide nanoemulsions, thereby influencing their deposition (Hilz and Vermeer, 2013). Finally, deposition of the pesticide solution on the target plant is affected by the physical and chemical properties of the pesticide, the surface structure of the plant, and environmental conditions.

12.5.2 Bioactivity of Pesticide Nanoemulsions

The biological activity of pesticides is an important criterion for the performance of pesticide nanoemulsions and has great significance for the control of diseases, pests, and weeds. Insect and plant leaf surfaces with waxy layers have poor hydrophilicity, and so, applied pesticides may easily fall off and have increased tendency to evaporate, seriously affecting their potency (Byun et al., 2009). The wetting and rainwash resistance of emulsifiable concentrates, wettable powders, and other traditional pesticide formulations is relatively poor, which is not conducive to the absorption and penetration of the pesticides, seriously restricting their biological activity. In practical applications, reducing the droplets size can significantly improve their adhesion and permeability on the target organisms, thereby enhancing the biological activity of pesticides and/or reducing their utilization levels. A nanoemulsion with smaller spray droplets, excellent wetting performance, rapid diffusion, and penetration at the target surface is conducive to improve the biological activity of pesticides. Neem oil nanoemulsions with different particle sizes (31.03–251.43 nm) were prepared by Karthikeyan et al. (2012). The results showed that the LC_{50} of Culex pipiens pallens was decreased with smaller droplet sizes.

12.6 CONCLUSION AND FURTHER REMARKS

Nanoemulsions are already widely used in the food, medical, cosmetic, and other fields and have great potential for pesticide applications. Pesticides are used to effectively control pests and weeds in agriculture and forestry production, thereby ensuring higher food production, less food waste, and greater farmers' income. On the other hand, excessive use of certain types of pesticides can cause environmental pollution and other problems. For this reason, there is considerable interest in the development of effective pesticide delivery systems that will enhance their efficacy, reduce their usage levels, limit their adverse effects on the environment, and reduce the risk to farm workers.

Nanoemulsion-based delivery systems for pesticides can be designed to be efficient, safe, and economical and to have good physicochemical stability, functional attributes (such as spreading and penetration), and enhanced bioavailability. Enhancement of pesticide bioavailability can reduce the usage levels required and therefore may have important ecological and environmental benefits and have likely to become one of the most important directions for future research. Finally, we highlight a number of areas where research is still needed to design effective nanoemulsions:

(1) Improvement in their environmental friendliness and toxicity profiles by using renewable and biodegradable solvents (such as biodiesels) and emulsifiers (such as vegetable oils).
(2) Increase in the content of active ingredients and multiple mixtures: increasing the content of active ingredients reduces the amount of solvent; multiple mixtures can expand the control spectrum, improve the control effect, and slow down the occurrence of pesticide resistance.
(3) Conduct more research on secondary dispersions and biological activities to expand the application scope and improve the performance (droplet size and penetration ability) of liquid formulations by careful selection of appropriate stabilizers (such as emulsifiers).
(4) Use more advanced instruments and scientific methods, and study the mechanisms of the formation, stability, and mode of action of pesticide nanoemulsions.
(5) Develop more complex colloidal delivery systems based on nanoemulsions that contain multiple components and that can control the release of pesticides.

REFERENCES

Aideo, S.N., Mohanta, D., 2016. Limiting hydrophobic behavior and reflectance response of dragonfly and damselfly wings. Appl. Surf. Sci. 387, 609–616.

Assadpour, E., Maghsoudlou, Y., Jafari, S.M., Ghorbani, M., Aalami, M., 2016a. Optimization of folic acid nano-emulsification and encapsulation by maltodextrin-whey protein double emulsions. Int. J. Biol. Macromol. 86, 197–207.

Assadpour, E., Maghsoudlou, Y., Jafari, S.M., Ghorbani, M., Aalami, M., 2016b. Evaluation of folic acid nano-encapsulation by double emulsions. Food Bioprocess Technol. 9, 2024–2032.

Badawi, N., Rosenbom, A.E., Jensen, A.M.D., Sørensen, S.R., 2016. Degradation and sorption of the fungicide tebuconazole in soils from golf greens. Environ. Pollut. 219, 368–378.

Bai, L., Huan, S.Q., Gu, J.Y., McClements, D.J., 2016. Fabrication of oil-in-water nanoemulsions by dual-channel microfluidization using natural emulsifiers: saponins, phospholipids, proteins, and polysaccharides. Food Hydrocoll. 61, 703–711.

Bakshi, M.S., Kaura, A., Miller, J.D., Paruchuri, V.K., 2004. Sodium dodecyl sulfate-poly (amidoamine) interactions studied by AFM imaging, conductivity, and Krafft temperature measurements. J. Colloid Interface Sci. 278, 472–477.

Balaji, A.P.B., Mishra, P., Kumar, R.S.S., Mukherjee, A., Chandrasekaran, N., 2015. Nanoformulation of poly(ethylene glycol) polymerized organic insect repellent by PIT emulsification method and its application for Japanese encephalitis vector control. Colloids Surf. B: Biointerfaces 128, 370–378.

Balaji, A.P.B., Sastry, T.P., Manigandan, S., Mukherjee, A., Chandrasekaran, N., 2017. Environmental benignity of a pesticide in soft colloidal hydrodispersive nanometric form with improved toxic precision towards the target organisms than non-target organisms. Sci. Total Environ. 579, 190–201.

Batıgöç, C., Akbaş, H., 2017. Thermodynamic parameters of clouding phenomenon in nonionic surfactants: the effect of the electrolytes. J. Mol. Liq. 231, 509–513.

Bueno, M.R., Cunha, J.P.A.R., Denise, G.S., 2017. Assessment of spray drift from pesticide applications in soybean crops. Biosyst. Eng. 154, 35–45.

Byun, D., Hong, J., Saputra, Ko, J.H., Lee, Y.J., Park, H.C., Byun, B.K., Lukes, J.R., 2009. Wetting characteristics of insect wing surfaces. J. Bionic Eng. 6, 63–70.

Carvalho, F.P., 2006. Agriculture, pesticides, food security and food safety. Environ. Sci. Pol. 9, 685–692.

Chang, T., McLandsborough, L., McClements, D.J., 2012. Physical properties and antimicrobial efficacy of thyme oil nanoemulsions: influence of ripening inhibitors. J. Agric. Food Chem. 60, 12056–12063.

Chebil, A., Desbrières, J., Nouvel, C., Six, J.L., Durand, A., 2013. Ostwald ripening of nanoemulsions stopped by combined interfacial adsorptions of molecular and macromolecular nonionic stabilizers. Colloids Surf. A Physicochem. Eng. Asp. 425, 24–30.

Chen, F.L., Yin, M.M., 2007. Discussion on pesticide microemulsion concept and its problems in production and application. Chin. J. Pestic. Sci. 9, 110–116.

Chen, F.L., Wang, Y., Zheng, F.N., Wu, Y.T., Liang, W.P., 2000. Studies on cloud point of agrochemical microemulsions. Colloids Surf. A Physicochem. Eng. Asp. 175, 257–262.

Chin, C.P., Lan, C.W., Wu, H.S., 2012. Application of biodiesel as carrier for insecticide emulsifiable concentrate formulation. J. Taiwan Inst. Chem. Eng. 43, 578–584.

Cooper, J., Dobson, H., 2007. The benefits of pesticides to mankind and the environment. Crop Prot. 26, 1337–1348.

Cui, Z.G., 2013. Fundamentals of Surfactants, Colloids, and Interface Chemistry. Chemical Industry Press, Beijing, pp. 11–12.

Delcour, I., Spanoghe, P., Uyttendaele, M., 2015. Literature review: impact of climate change on pesticide use. Food Res. Int. 68, 7–15.

Eddleston, M., Street, J.M., Self, I., Thompson, A., King, T., Williams, N., Naredo, G., Dissanayake, K., Yu, L.M., Worek, F., John, H., Smith, S., Thiermann, H., Harris, J.B., Eddie Clutton, R., 2012. A role for solvents in the toxicity of agricultural organophosphorus pesticides. Toxicology 294 (2), 94–103.

El-Din, M.N., Osman, D.I., Rashad, A.M., Mishrif, M.R., El-Sharaky, E.A., 2017. Physicochemical and rheological characterization of diesel fuel nanoemulsions at different water and surfactant contents. J. Mol. Liq. 231, 440–450.

Elshafei, G.M.S., El-Said, M.M., Attia, H.A.E., Mohammed, T.G.M., 2010. Environmentally friendly pesticides: essential oil-based w/o/w multiple emulsions for anti-fungal formulations. Ind. Crop. Prod. 31, 99–106.

Fan, Y.R., Simon, S., Sjöblom, J., 2010. Influence of nonionic surfactants on the surface and interfacial film properties of asphaltenes investigated by langmuir balance and brewster angle microscopy. Langmuir 26, 10497–10505.

Faria, J.M.S., Barbosa, P., Bennett, R.N., Mota, M., Figueiredo, C.A., 2013. Bioactivity against Bursaphelenchus xylophilus: nematotoxics from essential oils, essential oils fractions and decoction waters. Phytochemistry 94, 220–228.

Faridi Esfanjani, A., Jafari, S.M., Assadpoor, E., Mohammadi, A., 2015. Nano-encapsulation of saffron extract through double-layered multiple emulsions of pectin and whey protein concentrate. J. Food Eng. 165, 149–155.

Faridi, E.A., Jafari, S.M., Assadpour, E., 2017. Preparation of a multiple emulsion based on pectin-whey protein complex for encapsulation of saffron extract nanodroplets. Food Chem. 221, 1962–1969.

Feng, J.G., Zhang, X.J., Fan, T.F., Ma, C., Yu, C., Wu, X.M., 2012a. Effects of pH value, emulsification temperature and electrolytical ions on the stability of metolachlor emulsion in water. Chem. J. Chin. Univ. 33, 2521–2525.

Feng, J.G., Zhang, X.J., Zhao, Z.W., Chen, W.T., Fan, T.F., Wu, X.M., 2012b. The research application status of the emulsifiers for EW. Agrochemicals 51, 706–709.

Feng, J.G., Xiang, S., Qian, K., Zhu, F., Yu, Q.Y., Wu, X.M., 2015. Characterization methods for emulsion stability and their applications in the research and development for pesticide emulsions oil in water. Chin. J. Pestic. Sci. 17, 15–26.

Feng, J.G., Shi, Y.L., Yu, Q.Y., Sun, C.C., Yang, G.T., 2016. Effect of emulsifying process on stability of pesticide nanoemulsions. Colloids Surf. A Physicochem. Eng. Asp. 497, 286–292.

Gharabaghi, M., Aghazadeh, S., 2014. A review of the role of wetting and spreading phenomena on the flotation practice. Curr. Opin. Colloid Interface Sci. 19, 266–282.

Gutiérrez, J.M., González, C., Maestro, A., Solè, I., Pey, C.M., Nolla, J., 2008. Nano-emulsions: new applications and optimization of their preparation. Curr. Opin. Colloid Interface Sci. 13, 245–251.

Hilz, E., Vermeer, A.W.P., 2013. Spray drift review: the extent to which a formulation can contribute to spray drift reduction. Crop Prot. 44, 75–83.

Huang, Q.L., Yuan, H.Z., Yang, D.B., 2002. Studies on the microscopic mechanism of the thermodynamics stability of pesticide microemulsion. J. Integr. Agric. 35, 1483–1486.

Jafari, S.M., He, Y., Bhandari, B., 2006. Nano-emulsion production by sonication and microfluidization—a comparison. Int. J. Food Prop. 9, 475–485.

Jafari, S.M., He, Y., Bhandari, B., 2007a. Production of sub-micron emulsions by ultrasound and microfluidization techniques. J. Food Eng. 82, 478–488.

Jafari, S.M., He, Y., Bhandari, B., 2007b. Optimization of nano-emulsions production by microfluidization. Eur. Food Res. Technol. 225, 733–741.

Jafari, S.M., Assadpoor, E., He, Y., Bhandari, B., 2008. Re-coalescence of emulsion droplets during high-energy emulsification. Food Hydrocoll. 22, 1191–1202.

Jensen, P.K., Olesen, M.H., 2014. Spray mass balance in pesticide application: a review. Crop Prot. 61, 23–31.

Jin, Q.W., Li, X.B., Cai, Z.X., Zhang, F., Madhav, P.Y., Zhang, H.B., 2017. A comparison of corn fiber gum, hydrophobically modified starch, gum arabic and soybean soluble polysaccharide: interfacial dynamics, viscoelastic response at oil/water interfaces and emulsion stabilization mechanisms. Food Hydrocoll. 70, 329–344.

Juárez, J.A., Whitby, C.P., 2012. Oil-in-water pickering emulsion destabilisation at low particle concentrations. J. Colloid Interface Sci. 368, 319–325.

Kah, M., Hofmann, T., 2014. Nanopesticide research: current trends and future priorities. Environ. Int. 63, 224–235.

Karthikeyan, S., Jeeva, P.A., Jerobin, J., Mukherjee, A., Chandrasekaran, N., 2012. Formulation and characterization of nanoemulsion coatings from Azadirachta indica. Int. J. ChemTech Res. 4, 1566–1570.

Khot, L.R., Sankaran, S., Maja, J.M., Ehsani, R., Schuster, E.W., 2012. Applications of nanomaterials in agricultural production and crop protection: a review. Crop Prot. 35, 64–70.

Lee, L., Norton, I.T., 2013. Comparing droplet breakup for a high-pressure valve homogeniser and a Microfluidizer for the potential production of food-grade nanoemulsions. J. Food Eng. 114, 158–163.

Li, S.J., Guo, Y.F., Li, Y.F., Zhang, X.J., 2012. Research progress on stability mechanism of pesticide emulsion in water. Mod. Agrochem 11, 6–10.

Liang, H., Guo, D., Ma, L.R., Luo, J.B., 2017. Investigation of film formation mechanism of oil-in-water (O/W) emulsions at high speeds. Tribol. Int. 109, 428–434.

Lim, C.J., Basri, M., Omar, D., Rahman, M.B.A., Salleh, A.B., Rahman, R.N.Z.R.A., 2012a. Physicochemical characterization and formation of glyphosate-laden nano-emulsion for herbicide formulation. Ind. Crop. Prod. 36, 607–613.

Lim, C.J., Basri, M., Omar, D., Rahman, M.B.A., Salleh, A.B., Rahman, R.N.Z.R.A., Selamat, A., 2012b. Green nanoemulsion intervention for water-soluble glyphosate isopropylamine (IPA) formulations in controlling Eleusine indica (E. indica). Pestic. Biochem. Physiol. 102, 19–29.

Lim, C.J., Basri, M., Omar, D., Rahman, M.B.A., Salleh, A.B., Rahman, R.N.Z.R.A., 2013. Green nanoemulsion-laden glyphosate isopropylamine formulation in suppressing creeping foxglove (A. gangetica), slender button weed (D. ocimifolia) and buffalo grass (P. conjugatum). Pest Manag. Sci. 69, 104–111.

Lindman, B., Medronho, B., Karlström, G., 2016. Clouding of nonionic surfactants. Curr. Opin. Colloid Interface Sci. 22, 23–29.

Liu, G.W., 2012. Modern Pesticide Formulations Processing Technology. Chemical Industry Press, Beijing, pp. 1–7.

Liu, W.R., Sun, D.J., Li, C.F., Liu, Q., Xu, J., 2006. Formation and stability of paraffin oil-in-water nano-emulsions prepared by the emulsion inversion point method. J. Colloid Interface Sci. 303, 557–563.

Liu, Y., Wei, F.L., Wang, Y.Y., Zhu, G.N., 2011. Studies on the formation of bifenthrin oil-in-water nano-emulsions prepared with mixed surfactants. Colloids Surf. A Physicochem. Eng. Asp. 389, 90–96.

Matsumiya, K., Nakanishi, K., Matsumura, Y., 2011. Destabilization of protein-based emulsions by diglycerol esters of fatty acids. The importance of chain length similarity between dispersed oil molecules and fatty acid residues of the emulsifier. Food Hydrocoll. 25, 773–780.

McClements, D.J., 2012. Nanoemulsions versus microemulsions: terminology, differences, and similarities. Soft Matter 8, 1719–1729.

Mehrnia, M.A., Jafari, S.M., Makhmal-Zadeh, B.S., Maghsoudlou, Y., 2016. Crocin loaded nano-emulsions: factors affecting emulsion properties in spontaneous emulsification. Int. J. Biol. Macromol. 84, 261–267.

Mei, Z., Liu, S.Y., Wang, L., Jiang, J.J., Xu, J., Sun, D.J., 2011a. Preparation of positively charged oil/water nano-emulsions with a sub-PIT method. J. Colloid Interface Sci. 361, 565–572.

Mei, Z., Xu, J., Sun, D.J., 2011b. O/W nano-emulsions with tunable PIT induced by inorganic salts. Colloids Surf. A Physicochem. Eng. Asp. 375, 102–108.

Moghimi, R., Ghaderi, L., Rafati, H., Aliahmadi, A., McClements, D.J., 2016. Superior antibacterial activity of nanoemulsion of Thymus daenensis essential oil against E. coli. Food Chem. 194, 410–415.

Mohammadi, A., Jafari, S.M., Esfanjani, A.F., Akhavan, S., 2016a. Application of nano-encapsulated olive leaf extract in controlling the oxidative stability of soybean oil. Food Chem. 190, 513–519.

Mohammadi, A., Jafari, S.M., Assadpour, E., Esfanjani, A.F., 2016b. Nano-encapsulation of olive leaf phenolic compounds through WPC-pectin complexes and evaluating their release rate. Int. J. Biol. Macromol. 82, 816–822.

Mulqueen, P., 2003. Recent advances in agrochemical formulation. Adv. Colloid Interf. Sci. 106, 83–107.

Nenaah, G.E., Ibrahim, S.I.A., Al-Assiuty, B.A., 2015. Chemical composition, insecticidal activity and persistence of three asteraceae essential oils and their nanoemulsions against Callosobruchus maculatus (F.). J. Stored Prod. Res. 61, 9–16.

Olivier, P.L., 2016. Solid-state wetting at the nanoscale. Prog. Cryst. Growth Charact. Mater. 62, 177–202.

Pant, M., Dubey, S., Patanjali, P.K., Naik, S.N., Sharma, S., 2014. Insecticidal activity of eucalyptus oil nanoemulsion with karanja and jatropha aqueous filtrates. Int. Biodeterior. Biodegrad. 91, 119–127.

Pavela, R., 2013. Efficacy of naphthoquinones as insecticides against the house fly, Musca domestica L. Ind. Crop. Prod. 43, 745–750.

Perazzo, A., Preziosi, V., Guido, S., 2015. Phase inversion emulsification: current understanding and applications. Adv. Colloid Interf. Sci. 222, 581–599.

Rao, J.J., McClements, D.J., 2011. Formation of flavor oil microemulsions, nanoemulsions and emulsions: influence of composition and preparation method. J. Agric. Food Chem. 59, 5026–5035.

Robins, M.M., Watson, A.D., Wilde, P.J., 2002. Emulsions-creaming and rheology. Curr. Opin. Colloid Interface Sci. 7, 419–425.

Rodham, D.K., 2000. Colloid and interface science in formulation research for crop protection products. Curr. Opin. Colloid Interface Sci. 5, 280–287.

Saberi, A.H., McClements, D.J., 2015. Fabrication of protein nanoparticles and microparticles within water domains formed in surfactant-oil-water mixtures: phase inversion temperature method. Food Hydrocoll. 51, 441–448.

Sakulku, U., Nuchuchua, O., Uawongyart, N., Puttipipatkhachorn, S., Soottitantawat, A., Ruktanonchai, U., 2009. Characterization and mosquito repellent activity of citronella oil nanoemulsion. Int. J. Pharm. 372, 105–111.

Shi, Y.L., Yu, Q.Y., Sun, C.C., Dong, F., Yang, G.T., Feng, J.G., 2016. Review on research of physical stability of pesticides emulsion oil in water. China surfactant detergent. Cosmetics 46, 173–177.

Silva, G.N., Faroni, L.R.A., Sousa, A.H., Freitas, R.S., 2012. Bioactivity of Jatropha curcas L. to insect pests of stored products. J. Stored Prod. Res. 48, 111–113.

Singh, Y., Meher, J.G., Raval, K., Khan, F.A., Chaurasia, M., Jain, N.K., Chourasia, M.K., 2017. Nanoemulsion: concepts, development and applications in drug delivery. J. Control. Release 252, 28–49.

Singla, M., Patanjali, P.K., 2013. Phase behaviour of neem oil based microemulsion formulations. Ind. Crop. Prod. 44, 421–426.

Sivastava, V.K., Kini, G., Rout, D., 2006. Detergency in spontaneously formed emulsions. J. Colloid Interface Sci. 304, 214–221.

Song, S.L., Liu, X.H., Jiang, J.H., Qian, Y.H., Zhang, N., Wu, Q.H., 2009. Stability of triazophos in self-nanoemulsifying pesticide delivery system. Colloids Surf. A Physicochem. Eng. Asp. 350, 57–62.

Tang, X.L., Dong, J.F., Li, X.F., 2008. A comparison of spreading behaviors of Silwet L-77 on dry and wet lotus leaves. J. Colloid Interface Sci. 325, 223–227.

Taylor, P., 1998. Ostwald ripening in emulsions. Adv. Colloid Interf. Sci. 75, 107–163.

Wang, Y.Y., Liu, S.R., Zong, F.L., Ye, J.M., 2004. Pesticide formulations name and code instructions. Pestic. Sci. Adm. 25, 1–5.

Wang, L.J., Li, F., Zhang, G.Y., Dong, J.F., Eastoe, J., 2007a. Oil-in-water nanoemulsions for pesticide formulations. J. Colloid Interface Sci. 314, 230–235.

Wang, Y.H., Wang, M.H., Zhang, J.S., 2007b. Present situation on pesticide formulations. Agrochemicals 46, 300–304.

Weaver, M.A., Jin, X., Hoagland, R.E., Boyette, C.D., 2009. Improved bioherbicidal efficacy by Myrothecium verrucaria via spray adjuvants or herbicide mixtures. Biol. Control 50, 150–156.

Wu, X.M., Feng, J.G., Ma, C., 2013. Pesticide Preparation Processing Experiment. Chemical Industry Press, Beijing, pp. 15–18.

Xu, J., Fan, Q.J., Yin, Z.Q., Li, X.T., Du, Y.H., Jia, R.Y., Wang, K.Y., Lv, C., Ye, G., Geng, Y., Su, G., Zhao, L., Hu, T.X., Shi, F., Zhang, L., Wu, C.L., Tao, C., Zhang, Y.X., Shi, D.X., 2010. The preparation of neem oil microemulsion (*Azadirachta indica*) and the comparison of acaricidal time between neem oil microemulsion and other formulations in vitro. Vet. Parasitol. 169, 399–403.

Yang, Y., Marshallbreton, C., Leser, M.E., Sher, A.A., Mcclements, D.J., 2012. Fabrication of ultrafine edible emulsions: comparison of high-energy and low-energy homogenization methods. Food Hydrocoll. 29, 398–406.

Young, B.W., 1979. In: Studies on the retention and deposit characteristics of pesticide sprays on folliage[C].Application meeting, IXth CIGR congress. East Lancing, M Ipp. 1–4.

Yu, L.L., Li, C., Xu, J., Hao, J.C., Sun, D.J., 2012. Highly stable concentrated nanoemulsions by the phase inversion composition method at elevated temperature. Langmuir 28, 14547–14552.

Zeeb, B., Gibis, M., Fischer, L., Weiss, J., 2012. Influence of interfacial properties on Ostwald ripening in crosslinked multilayered oil-in-water emulsions. J. Colloid Interface Sci. 387, 65–73.

Zeeb, B., Herz, E., McClements, D.J., Weiss, J., 2015. Reprint of: impact of alcohols on the formation and stability of protein-stabilized nanoemulsions. J. Colloid Interface Sci. 449, 13–20.

Zhang, X.G., Liu, J.X., 2011. Effect of arabic gum and xanthan gum on the stability of pesticide in water emulsion. J. Agric. Food Chem. 59, 1308–1315.

Zhang, D.K., Wei, F.L., Zhu, G.N., Wu, X., Xu, Z.Y., 2007. Development status of pesticide emulsion oil in water in our country and research of stability mechanism. Mod. Agrochem. 6, 1–4.

Zhang, X.G., Liu, J.X., Fan, Z.J., 2013. Stability of etofenprox in water emulsion induced by block copolymer and surfactant. Colloids Surf. A Physicochem. Eng. Asp. 422, 191–198.
Zhang, C., Guanming, S., Shen, J., Hu, R.F., 2015. Productivity effect and overuse of pesticide in crop production in China. J. Integr. Agric. 14, 1903–1910.
Zhao, H., Lu, F.S., Li, P.Q., 2006. Effect of different factors on phase behavior in cyhalothrin microemulsion. Acta Phys. Chim. Sin. 22, 475–480.
Zhao, H.K., Lan, Y., Nan, C., Hu, Y., Rao, P., Tian, Y., Yan, W., Qian, K., He, L., 2016. Preparation of 8% fenpropathrin·cyflumetofen nano-emulsion and its performance. Sci. Agric. Sin. 49, 2700–2710.

Chapter 13

Application of Nanoemulsions in Drug Delivery

Kamla Pathak*, Satyanarayan Pattnaik* and Kalpana Swain†
*Uttar Pradesh University of Medical Sciences, Etawah, India, †Talla Padmavathi College of Pharmacy, Warangal, India

Chapter Outline

13.1 INTRODUCTION

Advances in technology have led to the evolution of pharmaceutical dosage forms from relatively simple systems to highly sophisticated ones, which are often referred to as novel drug delivery systems. Nanoemulsion-based delivery systems typically consist of a colloidal dispersion of oil and water phases, with mean droplet diameters typically ranging from about 50 to 500 nm in pharmaceutical applications (Assadpour et al., 2016a,b, 2017a,b). Nanoemulsions may exist as an oil-in-water (o/w) or water-in-oil (w/o) form, where the droplets are composed of either oil or water, respectively (McClements, 2011, 2012). Pharmaceutically acceptable surfactants that are generally regarded as safe (GRAS) are used as emulsifiers to stabilize nanoemulsions (Mehrnia et al., 2016, 2017). The capacity of nanoemulsions to dissolve large quantities of poorly water-soluble drugs, along with their ability to protect the drugs from degradation, makes them particularly effective drug delivery systems. The major advantages of nanoemulsions for this purpose include increased drug loading, enhanced drug solubility, improved bioavailability, controlled drug release, and enhanced protection from chemical or enzymatic degradation. Nanoemulsions also have

Nanoemulsions. https://doi.org/10.1016/B978-0-12-811838-2.00013-8

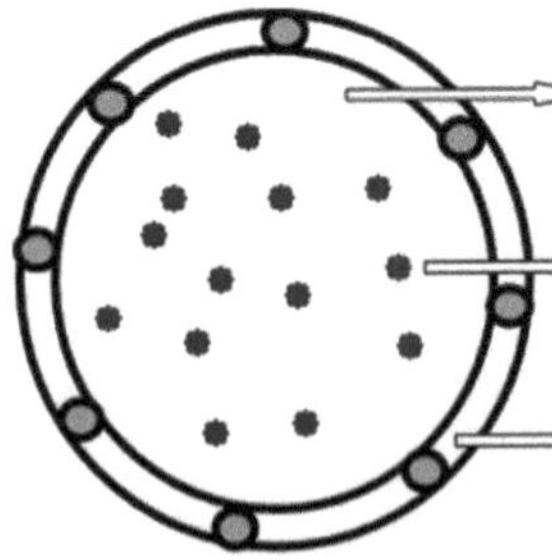

FIG. 13.1 Schematic representation of o/w nanoemulsion system.

better stability to droplet aggregation and gravitational separation than conventional emulsions because of their very small droplet size, which can lead to product's increased shelf life. Moreover, the small droplet size and large interfacial area may enhance targeted delivery of bioactive agents (Jafari et al., 2015). The relatively small size of the droplets in nanoemulsions also leads to other interesting physicochemical properties, such as high-optical clarity and unusual viscoelastic behavior. Nanoemulsions may have optical properties ranging from opaque to nearly transparent and rheological properties ranging from liquid to semisolid.

It should be noted that a number of different kinds of colloidal dispersions can be fabricated from oil, water, and surfactants that have some similar and some dissimilar characteristics (Fig. 13.1). Depending on the size of the dispersed droplets and their thermodynamic stability, they can be categorized as (i) *microemulsions* (thermodynamically stable isotropic systems with particle diameters between 20 and 100 nm), (ii) *nanoemulsions* (metastable stables with mean particle diameters below 200 nm), and (iii) *conventional emulsions* (metastable systems with mean particle diameters between 200 nm and several micrometers) (McClements, 2011, 2012). In this chapter, we will mainly focus on nanoemulsions.

Nanoemulsion can be formulated as delivery systems for the administration of drugs through various routes, including oral, parenteral, transdermal, intranasal, and ocular routes. This chapter focuses on the recent advances in the drug delivery applications of nanoemulsions. The formulation and production aspects of nanoemulsions are discussed elsewhere in this book.

13.2 DRUG DELIVERY APPLICATIONS

13.2.1 Oral Delivery

The most convenient, easiest, and cost-effective way for the noninvasive administration of drugs is via the oral route, and so, drug delivery systems based on this principle currently dominate the pharmaceutical market. Moreover, increased patient compliance makes it the optimal means for achieving therapeutic targets. However, this route of delivery has its limitations with respect to geriatric, pediatric, and possibly trauma epileptic patients where patient cooperation is a major constraint. In addition, certain drugs are inherently difficult to deliver through the oral route because of their nonconducive physiochemical properties. Oral delivery of poorly soluble drugs poses some serious problems with respect to drug solubility, stability, and absorption in the gastrointestinal tract (GIT) (Mallick et al., 2007). Peptide drugs are known to undergo hydrolysis and enzymatic degradation that limits their intestinal absorption and bioactivity. The limited ability of some drugs to permeate through the epithelium cell wall may also present additional drawbacks. Many approaches have been put forth to increase the overall bioavailability of drugs, including micronization/nanonization (Pattnaik et al., 2015a,b), solid dispersions (Senta-Loys et al., 2017; Huang and Dai, 2014), complexation with cyclodextrins (Mahapatra et al., 2013), amorphization (Mallick et al., 2008), and utilization of particulate delivery systems that are dispersible in aqueous environments (Pattnaik and Pathak, 2017; Pattnaik et al., 2015c). In the past decade, many research groups have reported significant oral bioavailability improvement of poorly soluble and hydrophobic drugs employing nanoemulsion technology (Table 13.1). Substantial improvements in C_{max} and the area under the curve (AUC) have been documented by many researchers when poorly soluble drugs were delivered using nanoemulsions. Fig. 13.2 shows the pharmacokinetic advantages of this platform over conventional approaches.

Peyer's patches are the most important structural units of the gut-associated lymphoid tissue, which are specialized for endocytosis and transport into intraepithelial spaces. Nanoparticles are engulfed by the M-cells, after which they are rapidly internalized and "shuttled" to the lymphocytes (Yoo et al., 2010; Rieux et al., 2006; Shakweh et al., 2004). The lymphatic absorption of a drug has an advantage over the portal blood route, because it avoids presystemic metabolism by the liver. In brief, lymphatic targeting of nanoemulsions can permit oral delivery of labile drugs protected by the carrier, oral delivery of poorly soluble drugs for improved bioavailability, oral delivery of vaccine antigens to gut-associated lymphoid tissue, translocation of antineoplastic drugs for treatment of lymphomas, sustained/controlled drug release, reduction of drug-related mucosal irritation, and avoidance of the hepatic first-pass effect.

In the GIT, lipids are effectively absorbed through various mechanisms. Thus, one of the feasible approaches to increase the absorption of drugs (especially protein drugs) is to load them inside lipids, such that there would be a significant increase in the level of absorption of drugs along with the lipids. The strategy is to use lipids as components of nanoemulsions to load the drugs, ultimately leading to increased absorption of drugs in the GIT. Use of nanoemulsions as oral drug delivery systems has been shown to increase the effectiveness of the drug at the target site (Khatri and Shao, 2017; Mamadou et al., 2017).

In vitro transport studies (PAMPA and Caco-2 models) have shown enhanced permeation of flavonoids and iridoids from Vitex agnus-castus extract-loaded nanoemulsions using triacetin as an oil phase and labrasol and cremophor EL

TABLE 13.1 Oral Bioavailability Enhancement With Nanoemulsion Formulations

Drug	Therapeutic Category	Remarks on Bioavailability Enhancement	Reference
Amlodipine besylate	Antihypertensive	Relative bioavailability compared with the drug suspension was 475%	Chhabra et al. (2011)
Arteether	Antimalarial	About threefold higher levels of AUC compared with drug in groundnut oil	Dwivedi et al. (2015)
Artemether	Antimalarial	2.6-fold higher than the plain drug	Laxmi et al. (2015)
Baicalin	Antipyretic, antiinflammatory bioactive	Sevenfold higher than free drug	Zhao et al. (2013)
Cefuroxime axetil	Antibiotic	About twofold higher AUC compared with drug suspension	Patel et al. (2012)
Curcumin	Multiple therapeutic applications	Ninefold higher compared with raw curcumin	Yu and Huang (2012)
Danazol	Synthetic steroid indicated in endometriosis	About 17-fold higher compared with raw drug	Devalapally et al. (2013)
Diclofenac	Antiinflammatory	Therapeutic availability was 2.89-fold higher compared with drug solution	Kandadi et al. (2012)
Diferuloylmethane	Anticancer	Relative bioavailability of nanoemulsion formulation to free drug was 855.02%	Sun et al. (2016)

TABLE 13.1 Oral Bioavailability Enhancement With Nanoemulsion Formulations—cont'd

Drug	Therapeutic Category	Remarks on Bioavailability Enhancement	Reference
Fisetin	Antitumor	24-fold higher compared with free fisetin	Ragelle et al. (2012)
Glyburide	Antidiabetic	1.5-fold increase of AUC compared with commercial tablet	Liu et al. (2014)
Mebudipine	Antihypertensive	2.6-, 2.0-, and 1.9-fold higher, respectively, compared with suspension, ethyl oleate solution, and micellar solution	Khani et al. (2016)
Megestrol acetate	Hormone replacement therapy, breast/ endometrial cancers	Fivefold higher compared with drug microcrystal suspension	Li et al. (2015)
Olmesartan medoxomil	Antihypertensive	2.8-fold higher compared with raw drug	Gorain et al. (2014)
Oxaliplatin and 5-fluorouracil	Antineoplastic	9.19- and 1.39-fold higher compared with respective drugs	Pangeni et al. (2016)
Paclitaxel	Anticancer	Absolute bioavailability improved up to 55.9%	Choudhury et al. (2014)
Probucol	Antihyperlipidemic	8.97-fold higher compared with control	Zhang et al. (2013)
Quercetin	Chemoprotective	Twofold higher AUC compared with control suspension	Tran et al. (2014)

Continued

TABLE 13.1 Oral Bioavailability Enhancement With Nanoemulsion Formulations—cont'd

Drug	Therapeutic Category	Remarks on Bioavailability Enhancement	Reference
Ramipril	Antihypertensive	The relative bioavailability when compared with conventional tablet and drug suspension was 176.3% and 428.76%, respectively	Shafiq et al. (2007)
Rosuvastatin	Antihyperlipidemic, anticancer	Relative bioavailability compared with the commercial tablets was 167%	Kamel and Mahmoud (2013)
Stiripentol	Anticonvulsant	Relative bioavailability was up to 206.2% compared with the conventional suspensions	Lu et al. (2015)
Vitamin E	Vitamin	Threefold higher compared with conventional emulsion	Parthasarathi et al. (2016)

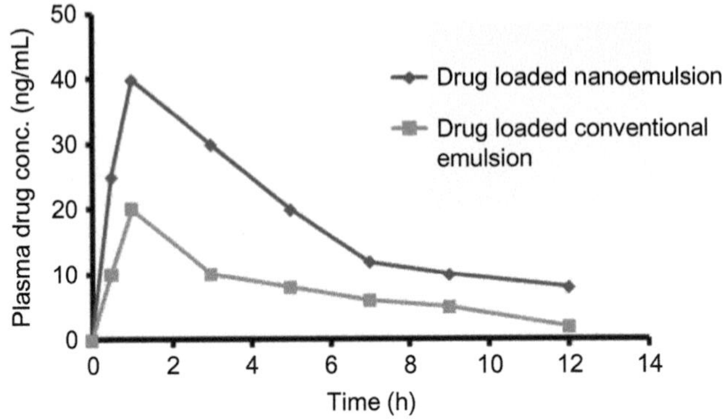

FIG. 13.2 Hypothetical drug plasma concentration versus time plots for orally administered drug-loaded nanoemulsion in comparison with conventional drug-loaded emulsion.

as surfactant and cosurfactant, respectively (Piazzini et al., 2017). In an interesting method to deliver nanoemulsions, Elkadi et al. (2017) developed self-nanoemulsifying tablets to improve the dissolution of simvastatin.

13.2.2 Parenteral Delivery

For drugs with low bioavailability and narrow therapeutic index, the parenteral route is considered to be one of the most common and effective routes (Gassmann et al., 1994; Kalepu and Nekkanti, 2015). However, for drugs with poor aqueous solubility, parenteral administration represents a major challenge in the area of drug delivery. Diverse strategies are described in the literature for this purpose, including modification of vehicle pH, addition of cosolvents, or formation of inclusion complexes with cyclodextrins (Loftsson and Brewster, 1996; Powell et al., 1998; Mallick et al., 2007). However, adjuvant biocompatibility, induction of pain, and possible precipitation of the drugs during the administration are considered as limiting factors for many of these approaches. In this context, nanostructured systems have shown promise. Lipid emulsions have been successfully used over the past few decades in patients requiring parenteral nutrition (Washington, 1996).The biocompatibility of the raw materials used for their production (oils of natural or semisynthetic origin and phospholipids) makes these systems a promising alternative for the administration of several types of pharmaceutical and nutritional molecules. The composition, structural properties, and preparation conditions of parenteral nanoemulsions should be strictly controlled, aiming at the parenteral administration and the stability of the system.

The stability of nanoemulsions used for parenteral delivery mainly depends upon their compositions, preparation techniques, and storage conditions (Cheng et al., 2016; Đorđević et al., 2015; Tan et al., 2015, 2016). However, the capacity of nanoemulsions to dissolve large quantities of hydrophobic bioactive substances, along with their biocompatibility, and the ability to protect drugs from chemical and enzymatic degradation make them ideal vehicles for parenteral delivery (Gue et al., 2016; Hormann and Zimmer, 2016).

Selection of suitable emulsifiers capable of forming mono- or multilayers around the oil droplets so as nanoemulsion formation and improve nanoemulsion stability is essential for this type of delivery system. The phospholipids most commonly used in the preparation of parenteral delivery systems originate from natural sources, such as lecithins, or are semisynthetic materials, such as dioleoylphosphatidylethanolamine (DOPE) and distearoylphosphatidylcholine (DSPC). Lecithins may be of animal or plant origin and hence are biocompatible and biodegradable. An appreciable fraction of lecithin phospholipids have charged polar head groups, such as phosphatidylserine, phosphatidylglycerol, and phosphatidic acid, which is of great importance to the long-term stability of the emulsions as they impart a relatively high-negative charge at the droplet surfaces that generates an electrostatic repulsion.

Although naturally occurring surfactants are preferred to synthetic surfactants, having suitable emulsification results, adjuvant emulsifying agents have been employed to impart greater emulsion stability. Nonionic surfactants from the poloxamers and polyoxyethylene sorbitans (Tweens) group have been shown to give promising result when used in combination with phospholipids because they lead to the formation of compact mixed films, giving greater stability to the formulation. However, hemolytic reactions and changes in the droplet diameter of nanoemulsions stabilized by Tween 80 during autoclaving limit their use. Polyoxyethylated castor oil derivatives (Cremophor EL) have also received considerable attention for parenteral use, but the occurrence of hemolysis and toxicity has been described for some of these compounds (Lu et al., 2008; Soderlind et al., 2003).

Most of the lipophilic drug candidates delivered as nanoemulsions, after parenteral administration, show a higher plasma concentration compared with the solution form. Moreover, an increase in the volume of distribution and reduction of the clearance leads to an increase in the half-life (Hwang et al., 2004). Further, the frequency and dosage of injections can be reduced throughout the drug therapy period as these emulsions guarantee the release of drugs in a sustained and controlled mode over long periods of time. Additionally, the lower tendency for droplet aggregation or gravitational separation to occur offers advantages over conventional emulsions for this route of administration.

13.2.3 Transdermal and Topical Delivery

Systemic drug delivery via the skin is particularly convenient for certain clinical conditions (Pattnaik et al., 2011, 2015d; Swain et al., 2010, 2011). In particular, the advantage of steady-state controlled drug delivery with self-administration also being possible, which may not be the case with the parenteral route (Swain et al., 2009), renders the transdermal delivery systems very attractive. The drug input can be terminated at any time by the patient just by removing the transdermal patch. However, the fundamental disadvantage that currently limits the use of this mode of administration for many applications is the barrier imposed by the skin to effective penetration of the bioactives. Transdermally applied drugs may penetrate the skin through hair follicles, sweat ducts, or stratum corneum and then reach the systemic circulation. However, the stratum corneum restricts their absorption to a large extent and therefore limits their bioavailability. For improved drug pharmacokinetics and targeting, the primary skin barriers need to be overcome. The utilization of delivery systems containing nanosized particles has been highly successful in overcoming this barrier (Sonneville-Aubrun et al., 2004; Pegoraro et al., 2012; Chaudhary et al., 2013). Nanoemulsion droplets are able to easily penetrate through the pores of the skin and reach the systemic circulation, thus getting channelized for effective delivery. There are various ways that the penetration of drugs through the skin can be enhanced by optimization of drug and vehicle properties, e.g., maximum penetration is usually observed when the drug is at its maximum thermodynamic activity like in the case of supersaturated solutions.

There has been a plethora of different drug molecules delivered to or through the skin using nanoformulations, including betamethasone (Sivaramakrishnan et al., 2004), clobetasol (Senyigit et al., 2010), corticosterone (Jensen et al., 2010), flufenamic acid (Luengo et al., 2006), flurbiprofen (Jain et al., 2005), tocopheryl acetate (Zhao et al., 2009), tolterodine tartrate (Elshafeey et al., 2009), and amlodipine (Kumar et al., 2009).

13.2.4 Intranasal Delivery

Intranasal drug delivery is another reliable route for the administration of certain types of drugs. Indeed, the nasal mucosa has emerged as a therapeutically viable channel for the administration of systemic drugs and also appears to be a favorable way to overcome the obstacles for the direct entry of drugs to the target site. This system has been accepted in the Ayurvedic system of Indian medicine since ancient times, and more recently, it has often been preferred over oral administration of drugs because it leads to better systemic bioavailability, presumably by avoiding the gastrointestinal metabolism of the drug (Rahisuddin et al., 2011). The intranasal route is also painless, noninvasive, and tolerated favorably. There are several problems associated with targeting drugs to the brain, especially hydrophilic ones and those of high-molecular weight (Bahadur and Pathak, 2012; Marianecci et al., 2015). This is because of the impervious nature of the endothelium that divides the systemic circulation from the brain, i.e., the blood-brain barrier (Pardridge, 1999). The olfactory region of the nasal mucosa provides a direct connection between the nose and the brain. Nanoemulsions loaded with drugs have been exploited for the treatment/management of conditions such as hormone replacement therapy (estradiol), osteoporosis (raloxifene), schizophrenia (olanzapine), smoking cessation (nicotine), enuresis (desmopressin), endometriosis (nafarelin), and motion sickness (metoclopramide) (Deshpande et al., 2014; Elsheikh et al., 2012; Kumar et al., 2008a,b). The possibilities for central nervous system (CNS) delivery via nasal administration have been investigated for the delivery of polar drugs to treat chronic CNS conditions such as Parkinson's or Alzheimer's disease (Kumar et al., 2008a,b; Mistry et al., 2009). The brain is considered to be one of the sanctuary sites for HIV, and hence, saquinavir has been targeted effectively to the brain as an intranasal nanoemulsion to treat this condition (Mahajan et al., 2014).Vaccines are also delivered as nanoemulsions through the intranasal route (discussed elsewhere in this chapter) and often have advantages over oral and parenteral routes. However, the major constraints on nasal drug delivery include a restricted capacity, difficulty in attaining dose accuracy, and reproducibility of delivery. The delivery device, formulation, and administration technique often influence drug distribution in the nasal cavity that in turn affects permeability, residence time, and metabolism in the nasal cavity.

13.2.5 Ocular Delivery

Most of the ocular therapeutics have been administered to the eye as simple aqueous solutions as eye drops by instillation to the lower conjunctival sac (Tatham et al., 2013). However, the major limitations of aqueous eye drops include their inability to deliver lipophilic and/or insoluble drugs, low ocular retention time, and limited resistance to the washout effect of blinking and tear turnover. Further, the less than 20% of the applied topical dose that remains in the ocular cavity is challenged by enzyme binding and metabolism, ocular permeation barriers, phagocytic activity, partial distribution to adjacent tissues, and systemic circulation (Gaudana et al., 2010). Hence, elimination is very significant in topical ocular drug administration leading to 1% or less of the administered dose actually penetrating the ocular surface. Moreover, when targeting a remote target tissue like the retina, the fraction of the administered dose that reaches the site of action will be much less due to further anatomical barriers, aqueous humor turnover, intraocular metabolic activity, binding to intraocular pigmented tissues, and phagocytosis by cell line other than the targeted tissue cells (Patel et al., 2013). Consequently, there are major challenges in the pharmaceutical industry to the successful development of effective formulations for topical ocular administration of drugs that can encapsulate many different active ingredients, remain longer on the ocular surface, and provide sustained therapeutic concentrations, in addition to meeting the regulatory criteria for approval.

Nanoemulsions could be employed to overcome some of these problems. Nanoemulsions are potent drug delivery vehicles for ophthalmic use due to their numerous advantages as sustained effect and high ability of drug penetration into the deeper layers of the ocular structure and the aqueous humor (Ammar et al., 2009; Daull et al., 2014). Compared to currently available approaches for administering eye drops, cationic nanoemulsions that have bioadhesive properties have been shown to be more efficient at delivering appropriate concentrations of bioactive molecules to the eye (Du Toit et al., 2011). The mechanism underlying the bioadhesiveness of these nanoemulsions is an electrostatic interaction that prolongs the residence time of the small oil droplets on the ocular surface. For instance, a commercial cationic nanoemulsion technology (Novasorb) uses positively charged droplets (Fig. 13.3) to create an electrostatic

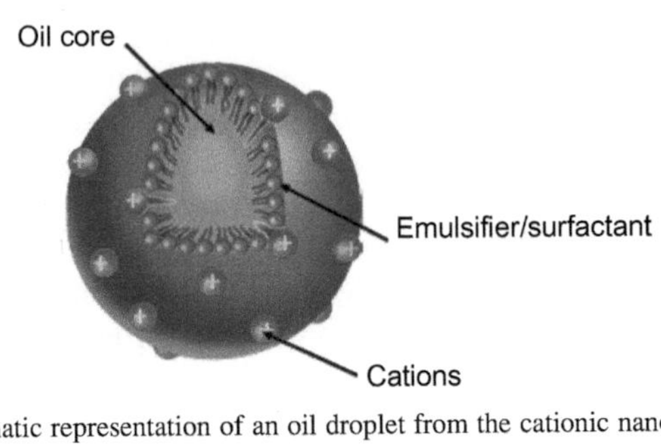

FIG. 13.3 Schematic representation of an oil droplet from the cationic nanoemulsions.

attraction with the negatively charged cells of the ocular surface (Lallemand et al., 2012). The corneal and conjunctival cells and the mucus layer of glycosyl amino glycans lining the ocular surface are negatively charged at a physiological pH, and upon application of a positively charged formulation to the eye, it is likely that an electrostatic attraction will occur prolonging the residence time of the formulation on the ocular surface. In addition, the small size of the oil droplets may facilitate enhanced absorption.

13.3 NANOEMULSIONS FOR VACCINE DELIVERY

Most of the research on nanoemulsions for vaccine delivery concentrates on intranasal mucosal methods (Das et al., 2012; Sun et al., 2015). As opposed to traditional vaccination routes, in the intranasal mucosal vaccination, the oil-based emulsion is administered into the nostrils. Intranasal vaccination is considered as the preferred immunization route due to the ease of administration and the lesser amount of antigen needed to induce an immune response. Though the exact mechanism of vaccine adjuvants is still to be explored, it is believed that improved antigen delivery and innate immune activation are possible mechanisms by which adjuvants enhance the immune response. Interestingly, nanoemulsion-based vaccine adjuvants act by enhancing antigen uptake by dendritic cells (DCs) and activating toll-like receptors (TLR) 2 and 4, which in turn enhance both humoral and cell-mediated Th1 and Th17 immune responses (Bielinska et al., 2010). Moreover, nanoemulsion-based mucosal adjuvants act without damaging the mucosal epithelium and have been demonstrated to be safe and well tolerated in early phase human clinical trials (Makidon et al., 2008). The delivery of nucleic acid-based vaccines is discussed elsewhere in this chapter.

Since, HIV can attack the mucosal immune system, developing mucosal immunity through the use of nanoemulsions has been considered as a useful strategy against HIV infection (Demberg and Robert-Guroff, 2009). Moreover, it is now demonstrated that genital mucosa immunity may be attained with vaccines that are administered into the nasal mucosa (Charles and Attama, 2011; Lycke, 2012). Nanoemulsions are now being used as adjuvants to transport attenuated organisms to a mucosal surface to produce the desired immune response. Nanoemulsion (with a 1:6 ratio of cationic-to-nonionic surfactants)-adjuvanted inactivated influenza vaccine delivered intranasally was found to elicit a strong mucosal influenza-specific IgA response, offering an important advantage over immune responses driven by parenteral vaccination (Das et al., 2012). Studies indicate that nanoemulsions appear to be an effective mucosal adjuvant for a recombinant *Bacillus anthracis* protective antigen anthrax vaccine (Bielinska et al., 2007). These developed vaccines were found to induce long-lasting, robust, and specific humoral and cellular responses and further appear to lack adverse effects and have the ability to stabilize the antigen.

13.4 NANOEMULSIONS FOR GENE DELIVERY

The concept of gene delivery is quite straightforward and involves the insertion of genes into somatic cells of patients for correction of an inborn genetic defect or creation of a new cellular function (Kaufmann et al., 2013; Wang and Gao, 2014). Since liposomes can easily bind to the cell surface, they have been used as nonviral vectors for gene delivery (Ropert, 1999). Unlike nonviral vectors, viral gene delivery systems have improved transfection capabilities. However, relative toxicities of these systems limit their applications. Some of the limitations of the use of viral vectors in gene therapy have stimulated research in nonviral strategies such as cationic nanoemulsions. The presence of cationic surfactants allows the complexation with negatively charged DNA via electrostatic interactions (Dai et al., 2011; Yue et al., 2011), resulting in DNA compaction and consequently nanocomplex formation. Stearylamine, a cationic lipid, is reported to be a suitable cationic surfactant for gene delivery. Stearylamine (a cationic lipid) containing cationic nanoemulsions with a droplet size smaller than 200 nm and a zeta potential higher than +30 mV was able to effectively complex the plasmid model PIRES2-EGFP, as confirmed by an agarose gel electrophoresis assay (Silva et al., 2016). Cationic nanoemulsion systems have been used to intranasally deliver TNFα siRNA to protect against neuroinflammation, as demonstrated using an LPS-induced model of neuroinflammation (Yadav et al., 2016). These results indicate that intranasal delivery of cationic nanoemulsions encapsulating TNFα siRNA offered an efficient means of gene knockdown, and this approach has significant potential in the prevention of neuroinflammation. Self-amplifying messenger RNA (mRNA) of positive-strand RNA viruses has now evolved as effective nonviral vectors for in situ expression of vaccine antigens (Bogers et al., 2015). For oral gene delivery, cell-penetrating peptides are found to amplify cellular uptake of plasmid DNA (pDNA)-loaded self-nanoemulsifying drug delivery systems (SNEDDS) by mucosal epithelial cells (Mahmood et al., 2016).

Liposomes are also widely exploited for delivery of genes. However, the distinct advantage of nanoemulsions over liposomes is their greater physical stability (Kim et al., 2000). It has been shown that most liposomal preparations aggregate on interaction with DNA, which can cause embolism upon systemic administration (Yi et al., 2000), limiting their applications in gene delivery. This is likely to be less of a problem with nanoemulsion-based delivery systems.

13.5 CONCLUSION AND FUTURE PROSPECTS

Nanoemulsions are being increasingly developed and utilized as pharmaceutical delivery systems for almost all routes of drug administration. Extensive research has been carried out with diverse applications in drug delivery, with the most studied area being the improved solubilization and delivery of poorly soluble actives. This technology platform may provide a solution to the delivery of "difficult-to-formulate" phytopharmaceuticals, where stability remains a major concern. However, cost-effective and reproducible manufacturing methods and effective, reliable, and safe functional components are required to ensure widespread industrial application of this technology.

REFERENCES

Assadpour, E., Maghsoudlou, Y., Jafari, S.-M., Ghorbani, M., Aalami, M., 2016a. Optimization of folic acid nano-emulsification and encapsulation by maltodextrin-whey protein double emulsions. Int. J. Biol. Macromol 86, 197–207.

Assadpour, E., Maghsoudlou, Y., Jafari, S.-M., Ghorbani, M., Aalami, M., 2016b. Evaluation of folic acid nano-encapsulation by double emulsions. Food Bioprocess Technol 9 (12), 2024–2032.

Assadpour, E., Jafari, S.-M., Maghsoudlou, Y., 2017a. Evaluation of folic acid release from spray dried powder particles of pectin-whey protein nano-capsules. Int. J. Biol. Macromol 95, 238–247.

Assadpour, E., Jafari, S.-M., 2017b. Spray drying of folic acid within nano-emulsions: optimization by Taguchi approach. Drying Technol 35 (9), 1152–1160.

Ammar, H.O., Salama, H.A., Ghorab, M., Mahmoud, A.A., 2009. Nanoemulsion as a potential ophthalmic delivery system for dorzolamide hydrochloride. AAPS PharmSciTech 10 (3), 808.

Bahadur, S., Pathak, K., 2012. Physicochemical and physiological considerations for efficient nose-to-brain targeting. Expert Opin. Drug Deliv. 9 (1), 19–31.

Bielinska, A.U., Janczak, K.W., Landers, J.J., Makidon, P., Sower, L.E., Peterson, J.W., Baker Jr., J.R., 2007. Mucosal immunization with a novel nanoemulsion-based recombinant anthrax protective antigen vaccine protects against Bacillus anthracis spore challenge. Infect. Immun. 75 (8), 4020–4029.

Bielinska, A.U., Gerber, M., Blanco, L.P., Makidon, P.E., Janczak, K.W., Beer, M., 2010. Induction of Th17 cellular immunity with a novel nanoemulsion adjuvant. Crit. Rev. Immunol. 30 (2), 189–199.

Bogers, W.M., Oostermeijer, H., Mooij, P., Koopman, G., Verschoor, E.J., Davis, D., Ulmer, J.B., Brito, L.A., Cu, Y., Banerjee, K., Otten, G.R., Burke, B., Dey, A., Heeney, J.L., Shen, X., Tomaras, G.D., Labranche, C., Montefiori, D.C., Liao, H.X., Haynes, B., Geall, A.J., Barnett, S.W., 2015. Potent immune responses in rhesus macaques induced by nonviral delivery of a self-amplifying RNA vaccine expressing HIV type 1 envelope with a cationic nanoemulsion. J Infect Dis 211 (6), 947–955.

Charles, L., Attama, A.A., 2011. Current state of nanoemulsions in drug delivery. J. Biomat. Nanobiotech. 2, 626–639.

Chaudhary, H., Kohli, K., Kumar, V., 2013. Nano-transfersomes as a novel carrier for transdermal delivery. Int. J. Pharm. 454 (1), 367–380.

Cheng, Y., Liu, M., Hu, H., Liu, D., Zhou, S., 2016. Development, optimization, and characterization of PEGylated nanoemulsion of prostaglandin E1 for long circulation. AAPS PharmSciTech 17 (2), 409–417.

Chhabra, G., Chuttani, K., Mishra, A.K., Pathak, K., 2011. Design and development of nanoemulsion drug delivery system of amlodipine besilate for improvement of oral bioavailability. Drug Dev. Ind. Pharm. 37 (8), 907–916.

Choudhury, H., Gorain, B., Karmakar, S., Biswas, E., Dey, G., Barik, R., Mandal, M., Pal, T.K., 2014. Improvement of cellular uptake, in vitro antitumor activity and sustained release profile with increased bioavailability from a nanoemulsion platform. Int. J. Pharm. 460 (1–2), 131–143.

Dai, Z., Gjetting, T., Mattebjerg, M.A., Wu, C., Andresen, T.L., 2011. Elucidating the interplay between DNA-condensing and free polycations in gene transfection through a mechanistic study of linear and branched pei. Biomaterials 32, 8626–8634.

Das, S.C., Hatta, M., Wilker, P.R., Myc, A., Hamouda, T., Neumann, G., Baker Jr., J.R., Kawaoka, Y., 2012. Nanoemulsion W_{80}5EC improves immune responses upon intranasal delivery of an inactivated pandemic H1N1 influenza vaccine. Vaccine 30 (48), 6871–6877.

Daull, P., Lallemand, F., Garrigue, J.-S., 2014. Benefits of cetalkonium chloride cationic oil-in-water nanoemulsions for topical ophthalmic drug delivery. J. Pharm. Pharmacol. 66 (4), 531–541.

Demberg, T., Robert-Guroff, M., 2009. Mucosal immunity and protection against HIV/SIV infection: strategies and challenges for vaccine design. Int. Rev. Immunol. 28 (1), 20–48.

Deshpande, D., Kethireddy, S., Gattacceca, F., Amiji, M., 2014. Comparative pharmacokinetics and tissue distribution analysis of systemically administered 17-β-estradiol and its metabolites in vivo delivered using a cationic nanoemulsion or a peptide-modified nanoemulsion system for targeting atherosclerosis. J. Control. Release 180, 117–124.

Devalapally, H., Silchenko, S., Zhou, F., McDade, J., Goloverda, G., Owen, A., Hidalgo, I.J., 2013. Evaluation of a nanoemulsion formulation strategy for oral bioavailability enhancement of danazol in rats and dogs. J. Pharm. Sci. 102 (10), 3808–3815.

Đorđević, S.M., Cekić, N.D., Savić, M.M., Isailović, T.M., Ranđelović, D.V., Marković, B.D., Savić, S.R., TimićStamenić, T., Daniels, R., Savić, S.D., 2015. Parenteral nanoemulsions as promising carriers for brain delivery of risperidone: design, characterization and in vivo pharmacokinetic evaluation. Int. J. Pharm. 493 (1–2), 40–54.

Du Toit, D.C., Pillay, V., Choonara, Y.E., Govender, T., Carmichael, T., 2011. Ocular drug delivery—a look towards nanobioadhesives. Expert Opin. Drug Deliv. 8 (1), 71–94.

Dwivedi, P., Khatik, R., Chaturvedi, P., Khandelwal, K., Taneja, I., Raju, K.S., Dwivedi, H., Singh, S.K., Gupta, P.K., Shukla, P., Tripathi, P., Singh, S., Tripathi, R., Wahajuddin, Paliwal, S.K., Dwivedi, A.K., Mishra, P.R., 2015. Arteether nanoemulsion for enhanced efficacy against Plasmodium yoelii nigeriensis malaria: an approach by enhanced bioavailability. Colloids Surf. B: Biointerfaces 126, 467–475.

Elkadi, S., Elsamaligy, S., Al-Suwayeh, S., Mahmoud, H., 2017. The development of self-nanoemulsifying liquisolid tablets to improve the dissolution of simvastatin. AAPS PharmSciTech. https://doi.org/10.1208/s12249-017-0743-z.

Elshafeey, A.H., Kamel, A.O., Fathallah, M.M., 2009. Utility of nanosized microemulsion for transdermal delivery of tolterodine tartrate: ex-vivo permeation and in-vivo pharmacokinetic studies. Pharm. Res. 26, 2446–2453.

Elsheikh, M.A., Elnaggar, Y.S., Gohar, E.Y., Abdallah, O.Y., 2012. Nanoemulsion liquid preconcentrates for raloxifene hydrochloride: optimization and in vivo appraisal. Int. J. Nanomedicine 7, 3787–3802.

Gassmann, P., List, M., Schweitzer, A., Sucker, H., 1994. Hydrosols: alternatives for the parenteral application of poorly water soluble drugs. Eur. J. Pharm. Biopharm. 40, 64–72.

Gaudana, R., Ananthula, H.K., Parenky, A., Mitra, A.K., 2010. Ocular drug delivery. AAPS J. 12 (3), 348–360.

Gorain, B., Choudhury, H., Kundu, A., Sarkar, L., Karmakar, S., Jaisankar, P., Pal, T.K., 2014. Nanoemulsion strategy for olmesartan medoxomil improves oral absorption and extended antihypertensive activity in hypertensive rats. Colloids Surf. B: Biointerfaces 115, 286–294.

Gue, E., Since, M., Ropars, S., Herbinet, R., Le Pluart, L., Malzert-Freon, A., 2016. Evaluation of the versatile character of a nanoemulsion formulation. Int. J. Pharm. 498 (1–2), 49–65.

Hormann, K., Zimmer, A., 2016. Drug delivery and drug targeting with parenteral lipid nanoemulsions—a review. J. Control. Release 223, 85–98.

Huang, Y., Dai, W.-G., 2014. Fundamental aspects of solid dispersion technology for poorly soluble drugs. Acta Pharm. Sin. B 4 (1), 18–25.

Hwang, S.R., Lim, S.-J., Park, J.-S., Kim, C.-K., 2004. Phospholipid-based microemulsion formulation of all-trans-retinoic acid for parenteral administration. Int. J. Pharm. 276 (1–2), 175–183.

Jafari, S.M., Fathi, M., Mandala, I., 2015. Emerging Product Formation Food Waste Recovery: Processing Technologies and Industrial Techniques. Elsevier, New York, pp. 293–317.

Jain, S.K., Chourasia, M.K., Masuriha, R., Soni, V., Jain, A., Jain, N.K., Gupta, Y., 2005. Solid lipid nanoparticles bearing flurbiprofen for transdermal delivery. Drug Deliv. 12, 207–215.

Jensen, L.B., Magnussson, E., Gunnarsson, L., Vermehren, C., Nielsen, H.M., Petersson, K., 2010. Corticosteroid solubility and lipid polarity control release from solid lipid nanoparticles. Int. J. Pharm. 390, 53–60.

Kalepu, S., Nekkanti, V., 2015. Insoluble drug delivery strategies: review of recent advances and business prospects. Acta Pharm. Sin. B 5 (5), 442–453.

Kamel, A.O., Mahmoud, A.A., 2013. Enhancement of human oral bioavailability and in vitro antitumor activity of rosuvastatin via spray dried self-nanoemulsifying drug delivery system. J. Biomed. Nanotechnol. 9 (1), 26–39.

Kandadi, P., Syed, M.A., Goparaboina, S., Veerabrahma, K., 2012. Albumin coupled lipid nanoemulsions of diclofenac for targeted delivery to inflammation. Nanomedicine 8 (7), 1162–1171.

Kaufmann, K.B., Büning, H., Galy, A., Schambach, A., Grez, M., 2013. Gene therapy on the move. EMBO Mol. Med. 5 (11), 1642–1661.

Khani, S., Keyhanfar, F., Amani, A., 2016. Design and evaluation of oral nanoemulsion drug delivery system of mebudipine. Drug Deliv. 23 (6), 2035–2043.

Khatri, P., Shao, J., 2017. Transport of lipid nano-droplets through MDCK epithelial cell monolayer. Colloids Surf. B: Biointerfaces 153, 237–243.

Kim, T.W., Chung, H., Kwon, I.C., Sung, H.C., Jeong, S.Y., 2000. In vivo gene transfer to the mouse nasal cavity mucosa using a stable cationic lipid emulsion. Mol. Cells 10 (2), 142–147.

Kumar, M., Misra, A., Babbar, A.K., Mishra, A.K., Mishra, P., Pathak, K., 2008a. Intranasal nanoemulsion based brain targeting drug delivery system of risperidone. Int. J. Pharm. 358 (1–2), 285–291.

Kumar, M., Misra, A., Mishra, A.K., Mishra, P., Pathak, K., 2008b. Mucoadhesive nanoemulsion-based intranasal drug delivery system of olanzapine for brain targeting. J. Drug Target. 16 (10), 806–814.

Kumar, D., Aqil, M., Rizwan, M., Sultana, Y., Ali, M., 2009. Investigation of a nanoemulsion as vehicle for transdermal delivery of amlodipine. Pharmazie 64, 80–85.

Lallemand, F., Daull, P., Benita, S., Buggage, R., Garrigue, J.-S., 2012. Successfully improving ocular drug delivery using the cationic nanoemulsion, novasorb. J. Drug Deliv. 2012.16 pages.

Laxmi, M., Bhardwaj, A., Mehta, S., Mehta, A., 2015. Development and characterization of nanoemulsion as carrier for the enhancement of bioavailability of artemether. Artif. Cells Nanomed. Biotechnol. 43 (5), 334–344.

Li, Y., Song, C.K., Kim, M.K., Lim, H., Shen, Q., Lee, D.H., Yang, S.G., 2015. Nanomemulsion of megestrol acetate for improved oral bioavailability and reduced food effect. Arch. Pharm. Res. 38 (10), 1850–1856.

Liu, H., Shang, K., Liu, W., Leng, D., Li, R., Kong, Y., Zhang, T., 2014. Improved oral bioavailability of glyburide by a self-nanoemulsifying drug delivery system. J. Microencapsul. 31 (3), 277–283.

Loftsson, T., Brewster, M.E., 1996. Pharmaceutical applications of cyclodextrins. 1. Drug solubilization and stabilization. J. Pharm. Sci. 85, 1017–1025.

Lu, Y., Wang, Y., Tang, X., 2008. Formulation and thermal sterile stability of a less painful intravenous clarithromycin emulsion containing vitamin E. Int. J. Pharm. 346 (1–2), 47–56.

Lu, R., Liu, S., Wang, Q., Li, X., 2015. Nanoemulsions as novel oral carriers of stiripentol: insights into the protective effect and absorption enhancement. Int. J. Nanomedicine 10, 4937–4946.

Luengo, J., Weiss, B., Schneider, M., Ehlers, A., Stracke, F., Konig, K., Kostka, K.H., Lehr, C.M., Schaefer, U.F., 2006. Influence of nanoencapsulation on human skin transport of flufenamic acid. Skin Pharmacol. Physiol. 19, 190–197.

Lycke, N., 2012. Recent progress in mucosal vaccine development: potential and limitations. Nat. Rev. Immunol. 12, 592–605.

Mahajan, M.S., Nerkar, P.P., Agrawal, A., 2014. Nanoemulsion-based intranasal drug delivery system of saquinavir mesylate for brain targeting. Drug Deliv. 21 (2), 148–154.

Mahapatra, A.K., Murthy, P.N., Patra, R.K., Pattnaik, S., 2013. Solubility enhancement of modafinil by complexation with β-cyclodextrin and hydroxypropyl β-cyclodextrin: a response surface modeling approach. Drug Deliv. Lett. 3, 210–219.

Mahmood, A., Prüfert, F., Efiana, N.A., Ashraf, M.I., Hermann, M., Hussain, S., Bernkop-Schnürch, A., 2016. Cell-penetrating self-nanoemulsifying drug delivery systems (SNEDDS) for oral gene delivery. Expert Opin. Drug Deliv. 13 (11), 1503–1512.

Makidon, P.E., Bielinska, A.U., Nigavekar, S.S., Janczak, K.W., Knowlton, J., Scott, A.J., 2008. Pre-clinical evaluation of a novel nanoemulsion-based hepatitis B mucosal vaccine. PLoS ONE 3(8).

Mallick, S., Pattnaik, S., Swain, K., De, P., 2007. Current perspectives of solubilization: potential for improved bioavailability. Drug Dev. Ind. Pharm. 33, 865–873.

Mallick, S., Pattnaik, S., Swain, K., De, P., Saha, A., Ghoshal, G., Mondal, A., 2008. Formation of physically stable amorphous phase of ibuprofen by solid state milling with kaolin. Eur. J. Pharm. Biopharm. 68, 346–351.

Mamadou, G., Charrueau, C., Dairou, J., LimasNzouzi, N., Eto, B., Ponchel, G., 2017. Increased intestinal permeation and modulation of presystemic metabolism of resveratrol formulated into self-emulsifying drug delivery systems. Int. J. Pharm. 521 (1–2), 150–155.

Marianecci, C., Rinaldi, F., Hanieh, P.N., Paolino, D., Marzio, L.D., Carafa, M., 2015. Nose to brain delivery: new trends in amphiphile-based "Soft" nanocarriers. Curr. Pharm. Des. 21 (36), 5225–5232.

McClements, D.J., 2011. Edible nanoemulsions: fabrication, properties, and functional performance. Soft Matter 7, 2297–2316.

McClements, D.J., 2012. Nanoemulsions versus microemulsions: terminology, differences, and similarities. Soft Matter 8, 1719–1729.

Mehrnia, M.A., Jafari, S.M., Makhmal-Zadeh, B.S., Maghsoudlou, Y., 2016. Crocin loaded nano-emulsions: factors affecting emulsion properties in spontaneous emulsification. Int. J. Biol. Macromol 84, 261–267.

Mehrnia, M.-A., Jafari, S.-M., Makhmal-Zadeh, B.S., Maghsoudlou, Y., 2017. Rheological and release properties of double nano-emulsions containing crocin prepared with Angum gum, Arabic gum and whey protein. Food Hydrocoll 66, 259–267.

Mistry, A., Stolnik, S., Illum, L., 2009. Nanoparticles for direct nose-to-brain delivery of drugs. Int. J. Pharm. 379 (1), 146–157.

Pangeni, R., Choi, S.W., Jeon, O.-C., Byun, Y., Park, J.W., 2016. Multiple nanoemulsion system for an oral combinational delivery of oxaliplatin and 5-fluorouracil: preparation and in vivo evaluation. Int. J. Nanomedicine 11, 6379–6399.

Pardridge, W.M., 1999. Blood-brain barrier biology and methodology. J. Neurovirol 5 (6), 556–569.

Parthasarathi, S., Muthukumar, S.P., Anandharamakrishnan, C., 2016. The influence of droplet size on the stability, in vivo digestion, and oral bioavailability of vitamin E emulsions. Food Funct. 7, 2294–2302.

Patel, Y., Poddar, A., Sawant, K., 2012. Formulation and characterization of Cefuroxime Axetil nanoemulsion for improved bioavailability. J. Pharm. Bioallied Sci. 4 (Suppl 1), S4–5.

Patel, A., Cholkar, K., Agrahari, V., Mitra, A.K., 2013. Ocular drug delivery systems: an overview. World J. Pharmacol. 2 (2), 47–64.

Pattnaik, S., Pathak, K., 2017. Mesoporous silica molecular sieve based nanocarriers: transpiring drug dissolution research. Curr. Pharm. Des. 23 (3), 467–480.

Pattnaik, S., Swain, K., Mallick, S., Lin, Z., 2011. Effect of casting solvent on crystallinity of ondansetron in transdermal films. Int. J. Pharm. 406, 106–110.

Pattnaik, S., Swain, K., Manaswini, P., Divyavani, E., Rao, J.V., Talla, V., Subudhi, S.K., 2015a. Fabrication of aceclofenac nanocrystals for improved dissolution: process optimization and physicochemical characterization. J Drug Deliv. Sci. Technol. 29, 199–209.

Pattnaik, S., Swain, K., Rao, J.V., Talla, V., Subudhi, S.K., 2015b. Aceclofenac nanocrystals for improved dissolution: influence of polymeric stabilizers. RSC Adv. 5 (112), 91960–91965.

Pattnaik, S., Swain, K., Rao, J.V., Talla, V., Prusty, K.B., Subudhi, S.K., 2015c. Polymer co-processing of ibuprofen through compaction for improved oral absorption. RSC Adv. 5 (91), 74720–74725.

Pattnaik, S., Swain, K., Rao, J.V., Talla, V., Mallick, S., 2015d. Temperature influencing permeation pattern of alfuzosin: an investigation using DoE. Medicina 51 (4), 253–261.

Pegoraro, C., MacNeil, S., Battaglia, G., 2012. Transdermal drug delivery: from micro to nano. Nanoscale 4 (6), 1881–1894.

Piazzini, V., Monteforte, E., Luceri, C., Bigagli, E., Bilia, A.R., Bergonzi, M.C., 2017. Nanoemulsion for improving solubility and permeability of Vitex agnus-castus extract: formulation and in vitro evaluation using PAMPA and Caco-2 approaches. Drug Deliv. 24 (1), 380–390.

Powell, M.F., Nguyen, T., Baloian, L., 1998. Compendium of excipients for parenteral formulations. PDA J. Pharm. Sci. Technol. 52, 238–311.

Ragelle, H., Crauste-Manciet, S., Seguin, J., Brossard, D., Scherman, D., Arnaud, P., Chabot, G.G., 2012. Nanoemulsion formulation of fisetin improves bioavailability and antitumour activity in mice. Int. J. Pharm. 427 (2), 452–459.

Rahisuddin, R., Sharma, P.K., Garg, G., Salim, M., 2011. Review on nasal drug delivery system with recent advancement. Int. J. Pharm. Pharm. Sci. 3 (Suppl. 2), 1–5.

Rieux, A., Fievez, V., Garinot, M., Schneider, Y.-J., Preat, V., 2006. Nanoparticles as potential oral delivery systems of proteins and vaccines: a mechanistic approach. J. Control. Release 116, 1–27.

Ropert, C., 1999. Liposomes as a gene delivery system. Braz. J. Med. Biol. Res. 32, 163–169.

Senta-Loys, Z., Bourgeois, S., Valour, J.-P., Briançon, S., Fessi, H., 2017. Orodispersible films based on amorphous solid dispersions of tetrabenazine. Int. J. Pharm. 518 (1–2), 242–252.

Senyigit, T., Sonvico, F., Barbieri, S., Ozer, O., Santi, P., Colombo, P., 2010. Lecithin/chitosan nanoparticles of clobetasol-17-propionate capable of accumulation in pig skin. J. Control. Release 142, 368–373.

Shafiq, S., Shakeel, F., Talegaonkar, S., Ahmad, F.J., Khar, R.K., Ali, M., 2007. Design and development of oral oil in water ramipril nanoemulsion formulation: in vitro and in vivo assessment. J. Biomed. Nanotechnol. 3 (1), 28–44.

Shakweh, M., Ponchel, G., Fattal, E., 2004. Particle uptake by Peyer's patches: a pathway for drug and vaccine delivery. Expert Opin. Drug Deliv. 1, 141–163.

Silva, A.L., Marcelino, H.R., Verissimo, L.M., Araujo, I.B., Agnez-Lima, L.F., do Egito, E.S., 2016. Stearylamine-containing cationic nanoemulsion as a promising carrier for gene delivery. J. Nanosci. Nanotechnol. 16 (2), 1339–1345.

Sivaramakrishnan, R., Nakamura, C., Mehnert, W., Korting, H.C., Kramer, K.D., Schafer-Korting, M., 2004. Glucocorticoid entrapment into lipid carriers characterization by parelectric spectroscopy and influence on dermal uptake. J. Control. Release 97, 493–502.

Soderlind, E., Wollbratt, M., von Corswant, C., 2003. The usefulness of sugar surfactants as solubilizing agents in parenteral formulations. Int. J. Pharm. 252 (1–2), 61–71.

Sonneville-Aubrun, O., Simonnet, J.T., L'Alloret, F., 2004. Nanoemulsions: a new vehicle for skincare products. Adv. Colloid Interf. Sci. 108–109, 145–149.

Sun, H., Wei, C., Liu, B., Jing, H., Feng, Q., Tong, Y., Yang, Y., Yang, L., Zuo, Q., Zhang, Y., Zou, Q., Zeng, H., 2015. Induction of systemic and mucosal immunity against methicillin-resistant Staphylococcus aureus infection by a novel nanoemulsion adjuvant vaccine. Int. J. Nanomedicine 10, 7275–7290.

Sun, L., Wan, K., Hu, X., Zhang, Y., Yan, Z., Feng, J., Zhang, J., 2016. Functional nanoemulsion-hybrid lipid nanocarriers enhance the bioavailability and anti-cancer activity of lipophilic diferuloylmethane. Nanotechnology 27(8).

Swain, K., Pattnaik, S., Sahu, S.C., Mallick, S., 2009. Feasibility assessment of ondansetron hydrochloride transdermal systems: physicochemical characterization and in vitro permeation studies. Lat. Am. J. Pharm. 28, 706–714.

Swain, K., Pattnaik, S., Sahu, S.C., Patnaik, K.K., Mallick, S., 2010. Drug in adhesive type transdermal matrix systems of ondansetron hydrochloride: optimization of permeation pattern using response surface methodology. J. Drug Target. 18, 106–114.

Swain, K., Pattnaik, S., Yeasmin, N., Mallick, S., 2011. Preclinical evaluation of drug in adhesive type ondansetron loaded transdermal therapeutic systems. Eur. J. Drug Metab. Pharmacokinet. 36 (4), 237–241.

Tan, S.L., Stanslas, J., Basri, M., AbediKarjiban, R.A., Kirby, B.P., Sani, D., Basri, H.B., 2015. Nanoemulsion-based parenteral drug delivery system of carbamazepine: preparation, characterization, stability evaluation and blood-brain pharmacokinetics. Curr. Drug Deliv. 12 (6), 795–804.

Tan, S.F., Masoumi, H.R., Karjiban, R.A., Stanslas, J., Kirby, B.P., Basri, M., Basri, H.B., 2016. Ultrasonic emulsification of parenteral valproic acid-loaded nanoemulsion with response surface methodology and evaluation of its stability. Ultrason. Sonochem. 29, 299–308.

Tatham, A.J., Sarodia, U., Gatrad, F., Awan, A., 2013. Eye drop instillation technique in patients with glaucoma. Eye (Lond) 27 (11), 1293–1298.

Tran, T.H., Guo, Y., Song, D., Bruno, R.S., Lu, X., 2014. Quercetin-containing self-nanoemulsifying drug delivery system for improving oral bioavailability. J. Pharm. Sci. 103 (3), 840–852.

Wang, D., Gao, G., 2014. State-of-the-art human gene therapy: part i. Gene delivery technologies. Discov. Med. 18 (97), 67–77.

Washington, C., 1996. Stability of lipid emulsions for drug delivery. Adv. Drug Deliv. Rev. 20, 131–145.

Yadav, S., Gandham, S.K., Panicucci, R., Amiji, M.M., 2016. Intranasal brain delivery of cationic nanoemulsion-encapsulated TNFα siRNA in prevention of experimental neuroinflammation. Nanomedicine 12 (4), 987–1002.

Yi, S.W., Yune, T.Y., Kim, T.W., Chung, H., Choi, Y.W., Kwon, I.C., Lee, E.B., Jeong, S.Y., 2000. A cationic emulsion/DNA complex as a physically stable and serum-resistant gene delivery system. Pharm. Res. 17 (3), 314–320.

Yoo, M.K., Kang, S.K., Choi, J.H., Park, I.K., Na, H.S., Lee, H.C., Kim, E.B., Lee, N.K., Nah, J.W., Choi, Y.J., Cho, C.S., 2010. Targeted delivery of chitosan nanoparticles to Peyer's patch using M cell-homing peptide selected by phage display technique. Biomaterials 31 (30), 7738–7747.

Yu, H., Huang, Q., 2012. Improving the oral bioavailability of curcumin using novel organogel-based nanoemulsions. J. Agric. Food Chem. 60 (21), 5373–5379.

Yue, Y.A., Jin, F., Deng, R., Cai, J.G., Chen, Y.C., Lin, M.C.M., Kung, H.F., Wu, C., 2011. Revisit complexation between DNA and polyethylenimine—effect of uncomplexed chains free in the solution mixture on gene transfection. J. Control. Release 155, 67–76.

Zhang, Z., Huang, J., Jiang, S., Liu, Z., Gu, W., Yu, H., Li, Y., 2013. A high-drug-loading self-assembled nanoemulsion enhances the oral absorption of probucol in rats. J. Pharm. Sci. 102 (4), 1301–1306.

Zhao, Y., Moddaresi, M., Jones, S.A., Brown, M.B., 2009. A dynamic topical hydrofluoroalkane foam to induce nanoparticle modification and drug release in situ. Eur. J. Pharm. Biopharm. 72, 521–528.

Zhao, L., Wei, Y., Huang, Y., He, B., Zhou, Y., Fu, J., 2013. Nanoemulsion improves the oral bioavailability of baicalin in rats: in vitro and in vivo evaluation. Int. J. Nanomedicine 8, 3769–3779.

FURTHER READING

Daull, P., Amrane, M., Garrigue, J.-S., 2017. Novasorb® cationic nanoemulsion and latanoprost: the ideal combination for glaucoma management? Glauc. Open Access 2 (1), 107.

Holmgren, J., Czerkinsky, C., 2005. Mucosal immunity and vaccines. Nat. Med. 11, S45–S53.

Chapter 14

Application of Nanoemulsions in Cosmetics

Odile Sonneville-Aubrun*, Megumi N. Yukuyama† and Aldo Pizzino‡

**L'Oréal, Research and Innovation, Chevilly-Larue, France, †University of Sao Paulo, Butanta, Brazil, ‡L'Oréal, Research and Innovation, St-Ouen, France*

Chapter Outline

14.1 INTRODUCTION

14.1.1 Generalities on Nanoemulsions

Addressing the consumers' growing demand for high-quality commercial products, which combine performance, safety, and sustainability, has been one of the greatest challenges for researchers in the cosmetics and personal-care field in recent years (Antignac et al., 2011; Philippe et al., 2012; Barbulova et al., 2015). The application of nanotechnology for formulating these types of product has gained considerable attention recently due to its potential benefits over conventional technologies.

Nanoemulsions are typically oil-in-water (O/W) or water-in-oil (W/O) colloidal dispersions that contain droplets that typically range from a few nanometer to 200 nm in diameter (Tadros et al., 2004; Solans et al., 2005; Mason et al., 2006),

Nanoemulsions. https://doi.org/10.1016/B978-0-12-811838-2.00014-X

although sometimes the upper particle-size range for this type of system is extended to 1000 nm (US FDA, 2014). The physicochemical properties and functional performance of this type of system depend on the droplet dimensions (Solans et al., 2005; Mason et al., 2006; Jafari et al., 2008). A nanoemulsion is a kinetically stable but a thermodynamically unstable system, in which Ostwald ripening is considered the main factor of its instability (Solans and Solé, 2012). Different visual appearances and textures can be obtained depending on system composition and structure, ranging from transparent to opaque, from low to high viscosity, and from "light" to "rich" touch when applied, enabling a wide range of products to be formulated (Ribeiro et al., 2015; Kong et al., 2011; Sonneville-Aubrun et al., 2004).

The growing industrial interest in nanoemulsions is demonstrated by an analysis of the increase in patent-protection activities. From 1994, about 900 patent families (pf) can be identified as "nanoemulsions" for cosmetic applications (Fig. 14.1), which represent more than 1900 granted patent applications during the past 20 years. About one-third of these patent families are focused on hair-care products and the rest on skin-, body-, or personal-care products. L'Oréal is the most active company in the industrial protection activities on nanoemulsions, followed by AmorePacific, Henkel, BASF, and P&G.

14.1.2 How Nanoemulsions Meet Cosmetics Needs

This section highlights the main applications of nanoemulsions in the cosmetics and personal-care field and their prospects and challenges. Nanoemulsions are a

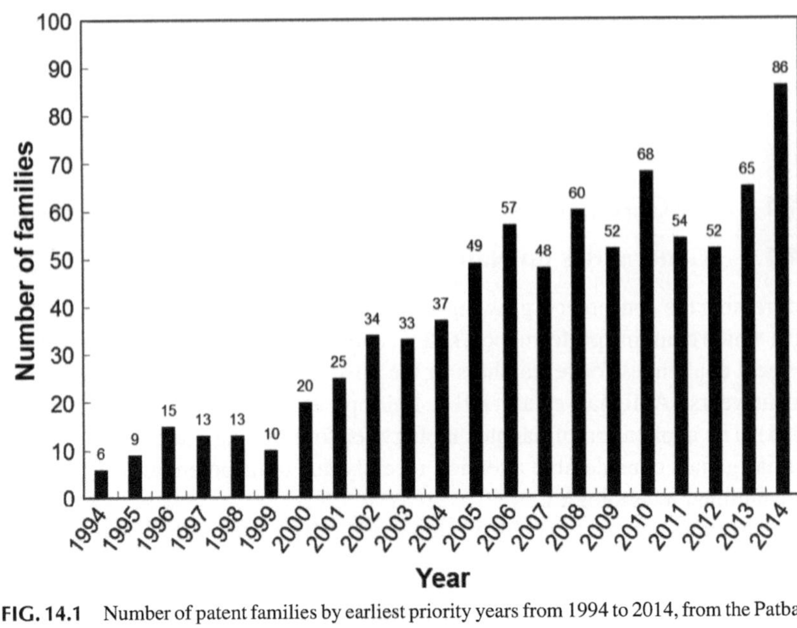

FIG. 14.1 Number of patent families by earliest priority years from 1994 to 2014, from the Patbase (general research strategy based on nanoemulsions).

promising system for cosmetic use due to several advantages over other technologies (such as emulsions, liposomes, and microemulsions), including ease of manufacturing, controllable particle sizes, high-kinetic stability, and relatively low surfactant levels. The versatility in formulation allows a wide range of delivery formulations to be developed, including sprays, fluids, gels, creams, and foams, with a variety of different functional attributes and sensorial benefits (Ribeiro et al., 2015; Samson et al., 2016; Silva et al., 2013).

The efficacy of cosmetic products depends on their textural properties and their ability to carry active and functional ingredients. Lipophilic, hydrophilic, and amphiphilic components can easily be incorporated into nanoemulsions because they contain oil, water, and interfacial regions. The small droplet size provides a large surface-to-volume ratio, which enhances the dispersed phase deposition on the hair and active-component ingredient penetration and permeability into the skin (Kong et al., 2011; Mahdi et al., 2011). Additionally, the topical application of nanoemulsions on the skin provides a uniform formation of a thin lipid film, thereby enabling higher performance and bioavailability. After several applications of cationic nanoemulsion on hair, a significant improvement was observed regarding dry-hair appearance over a prolonged period (Sonneville-Aubrun et al., 2004). The very small size of the droplets in nanoemulsions also offers the possibility to develop preservative-free products because they can be sterilized by passing through 0.2 μm filters.

14.2 CHALLENGES FOR COSMETICS NANOEMULSIONS

Developing safe, stable, efficient, and attractive products, with optimum cost benefit for consumers, is always a challenge for the cosmetic industry (Fig 14.2). Safety and environmental aspects and productivity and quality have to be considered very early in the development process of new products.

Nanoemulsions offer high versatility including manufacturing process selection and compounds selection such as surfactants, lipids, and active ingredients (Jafari et al., 2017). The investigation of the impact of process or formulation component variables on final products should be performed rationally (Yukuyama et al., 2016). The selection of an appropriate process for a particular system depends on the ease of scaling-up, reproducibility, and optimization of manufacturing time.

Understanding the behavior of surfactant-oil-water (SOW) phases during processing, storage, and contact with the skin or hair is also crucial. The solubilization and efficacy of lipophilic active components depend on the nature of the oil phase used to formulate O/W nanoemulsions (Yukuyama et al., 2017). The role of nanoemulsions as a carrier system must also guarantee the safety of the skin, avoiding penetration of undesirable external agents (Clares et al., 2014).

Ensuring long-term stability in both standard and stressed conditions, without altering the desirable sensory features of a product and its performance, requires good knowledge of the major destabilization phenomena

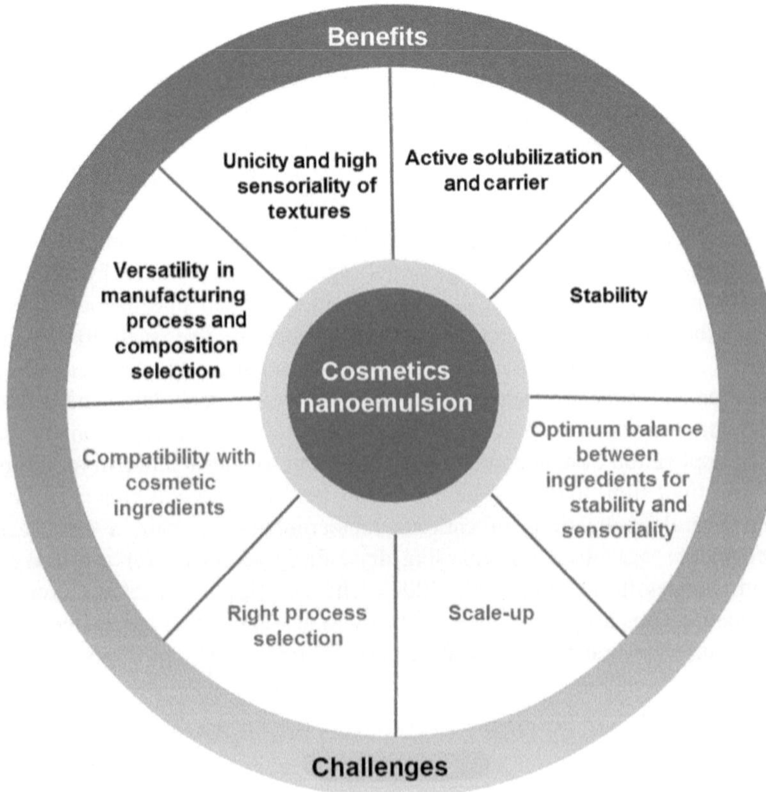

FIG. 14.2 Cosmetic nanoemulsions: benefits and challenges.

(Sonneville-Aubrun et al., 2004). Incompatibilities with some cosmetic ingredients, such as pigments, oxidants, multivalent salts, and polymers, restrict the use of nanoemulsions for certain applications and still have to be mastered.

The antimicrobial activity of preservatives in nanoemulsions is another concern to be considered. Parameters such as pH, water activity, and ethanol are known to affect the preservation of colloidal systems (Geis, 2006). However, improved knowledge of the impact of other parameters, such as particle size, on the preservation of cosmetic nanoemulsions is also needed (Fang et al., 2016).

14.3 FORMULATION PROCESSES

Nanoemulsions are thermodynamically unstable and need energy to be created, with a higher input energy being required to obtain smaller droplets, based on the Laplace equation (Mason et al., 2006). In cosmetics, nanoemulsions can be obtained either by high- or low-energy processes (respectively, external and internal energy input).

The choice of the most appropriate method mostly depends on system composition and/or scale-up requirements. The information reported in this section aims at facilitating the rational design of cosmetic-grade nanoemulsions using different fabrication methods. The most important advantages and drawbacks of each technique are also pointed out for cosmetic applications. A more detailed description of specific low- and high-energy methods is given elsewhere in this book.

14.3.1 High-Energy Process

In the cosmetic industry, high-pressure homogenization (HPH) and microfluidization are commonly used at the laboratory and industrial scale to obtain stable nanoemulsions over a large range of compositions (Sharma and Sarangdevot, 2012). Droplets are forced through a narrow channel under pressure, where they are submitted to large pressure drops, turbulent eddies, and high-shearing forces. The ultimate size of the nanoemulsion is determined by the balance between two opposing processes: droplet breakup and coalescence (Walstra, 1993), promoted by the intense shear forces that are generated within the homogenizer. The breakup of droplets occurs when the applied shear is greater than the Laplace pressure of the droplets.

The main operating parameters that affect the final droplet size are the pressure, the temperature, the number of passes, the valve and impingement design, and the flow rate. Other parameters are related to the compound selection: surfactant type, concentration, and interfacial properties; the presence of cosurfactants and cosolvents; the viscosity ratio between the dispersed and continuous phases; and the volume fraction of the dispersed phase (Walstra, 1993). In terms of process and formulation selection, technical requirements for nanoemulsions are stricter compared with emulsions to be able to obtain very small droplets.

14.3.1.1 Devices and Processes

A number of different homogenization devices are commonly used in the cosmetic industry: GEA Niro-Soavi Spa (GEA Niro-Soavi Spa, Parma, Italy) by L'Oréal (Ribier et al., 1995a,b) and O Boticário (Knapik et al., 2010) and Microfluidizer from Microfluidics Corp. (Newton, Massachusetts, the United States) by Shiseido (Okamoto et al., 2016). The processing temperature has to be above the melting temperature of the oil and surfactant phase.

Nanoemulsion fabrication generally proceeds in two steps and sometimes in three steps, with an additional dilution step for increasing the productivity or functionalizing the product (Fig. 14.3):

(1) First, a coarse emulsion is made by mixing the surfactant system with the oil and aqueous phases using a classical shearing tool (e.g., a rotor-stator or mixer).

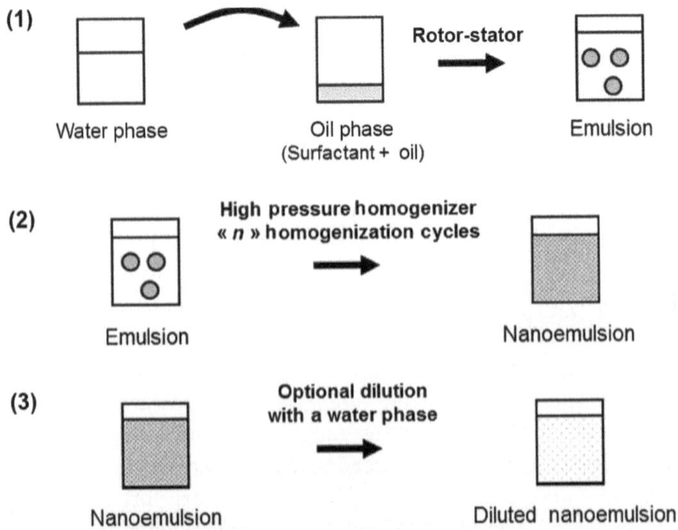

FIG. 14.3 Schematic representation of the homogenization step process.

(2) Second, the coarse emulsion is recirculated several times through a high-pressure homogenizer to obtain a small droplet size.
(3) Third, the nanoemulsion may be diluted with additional aqueous phase or incorporated as a preparation into another composition.

The number of recirculation cycles to obtain an optimum droplet size depends on the homogenizer, the applied pressure, and the system composition. On the one hand, three homogenization cycles has been reported to be sufficient to yield a stationary size at 60 and 120 MPa (Knapik et al., 2010; Liedtke et al., 2000). On the other hand, a nanoemulsion with a stationary size was only obtained after at least six homogenization cycles in other studies (Kentish et al., 2008; Qian and McClements, 2011). The number of recirculation steps has to be checked experimentally and adapted to the composition and the homogenization conditions.

An example is presented on Fig. 14.4 comparing data from Sonneville-Aubrun et al. with a Niro-Soavi Panda at 120 MPa (isostearate of PEO-8, stearoyl glutamate/isopropyl myristate/isocetyl stearate/dipropylene glycol/glycerin/water) with data from Knapik et al. (2010) with a Niro-Soavi Spa model NS100 at 120 MPa (steareth-2/stearyl alcohol/cetearyl alcohol/almond oil/capric-caprylic triglycerides/tocopherol acetate/mineral oil/water). In the first case, the size evolution can be fitted by an exponential function to the number of recirculation cycles as presented in Eq. (14.1), where D_F describes the emulsion final diameter and N is a characteristic decay number, as has been utilized with sonication (Delmas et al., 2011):

$$D(n) = D_F + A \cdot e^{-n/N} \tag{14.1}$$

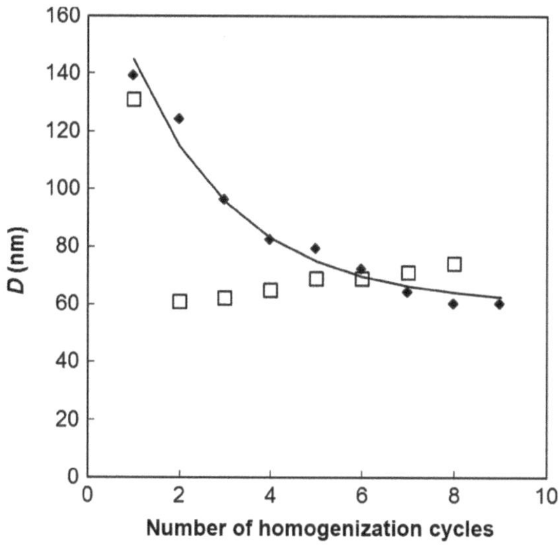

FIG. 14.4 Nanoemulsion droplet diameter as a function of number of homogenization cycles. *Diamond symbols* correspond to experimental data from Sonneville-Aubrun et al. and *square symbols* from Knapik et al. (2010).

Forming nanoemulsions generally requires applied pressures above 60 MPa. For example, nanoemulsions containing 170 and 190 nm median droplet diameters were obtained with using ethoxylated surfactant and an oligomeric surfactant at 60 MPa, respectively (Knapik et al., 2010). With the same surfactants, increasing the pressure to 120 MPa led to nanoemulsions with median diameters of 61 and 82 nm size, respectively.

14.3.1.2 Formulation Parameters

Besides process parameters, the composition of the system has to be carefully adapted to obtain stable nanoemulsions, such as the surfactant/oil weight ratio, surfactant type, oil- and aqueous-phase viscosities, and aqueous-phase additives.

Surfactant/oil ratio: The droplet size of a nanoemulsion formed by a high-pressure homogenizer is a result of a succession of fragmentation and coalescence events, where the surfactant plays a key role. On one hand, by lowering the interfacial tension, the surfactant facilitates droplet breakup. On the other hand, surfactant adsorption to oil/water interfaces prevents droplet coalescence (Walstra, 1993).

Two emulsification regimes can be identified depending on the surfactant/oil ratio and the energy input (Taisne et al., 1996). First, in the "surfactant-poor" regime, the surfactant is limited, and droplet coalescence occurs until the surfactant coverage is high enough to stabilize the droplets under the prevailing homogenizing conditions. In this case, the droplet size is determined by the surfactant/oil ratio. Second, in the "surfactant-rich" regime, the surfactant is in

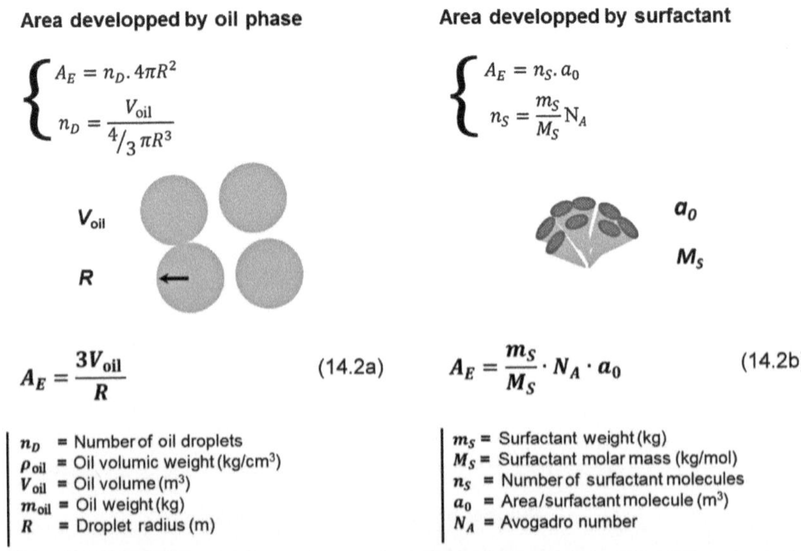

FIG. 14.5 Theoretical estimation of area developed by oil droplets and area developed by surfactant at interface of oil droplets.

excess, and coalescence is rare. The droplet size is then mostly determined by the applied pressure generated by the homogenizer.

Assuming that all the surfactant is located at oil/water interface, simple calculations allow estimating the minimum amount of surfactant (m_s) needed to form a nanoemulsion of a given radius (R) for a given oil volume, knowing the average area per surfactant molecule at an interface and the surfactant molecular weight, as presented in Fig. 14.5.

The equality between the two equations, Eqs. (14.2a), (14.2b), gives a simple relation (Eq. 14.3) between surfactant weight and droplet size to estimate either the amount of surfactant needed for a given droplet size or the theoretical droplet size for a given oil/surfactant ratio:

$$m_S = 3 \cdot \frac{M_S}{a_0} \cdot \frac{m_{oil}}{\rho_{oil} \cdot R} \cdot \frac{1}{N_A} \tag{14.3}$$

As an example, the surfactant needed to obtain a nanoemulsion with a 40 nm droplet diameter using 15% hydrocarbon oil is estimated at about 4.7% for a nonionic surfactant ($M_{TA} \sim 600$ g/mol and $a_0 \sim 60$ Å^2/mol), corresponding to an oil/surfactant weight ratio of about 3.2.

To determine the transition between the two regimes, nanoemulsions were produced with 5% surfactant (PEO8-isostearate/stearoyl glutamate, 9:1) and increasing weight fractions of oil (avocado oil/cyclopentasiloxane, 2:1) between 5% and 50% with 15% ethanol using a Niro-Soavi OBL20 at 1200 bar (Sonneville-Aubrun, unpublished).

As presented in Fig. 14.6, two regimes can be distinguished in the results obtained. For low oil/surfactant weight ratio (<4) corresponding to the "surfactant-rich" regime, the mean droplet diameter is almost constant around 45 nm. Surfactant excess in the water phase is expected under these conditions. For higher oil/surfactant weight ratios (4–10), corresponding to the "surfactant-poor" regime, the droplet size increased linearly with the oil/surfactant weight ratio in good agreement with the estimated droplet diameter based on Eqs. (14.2a), (14.2b).

Determination of the transition between these two regimes allows one to optimize the surfactant concentration for a given targeted size and oil content.

Surfactants: In cosmetics, a wide variety of surfactants can be used to obtain nanoemulsions through homogenization (Table 14.1). These include (i) classical nonionic surfactants such as ethoxylated fatty alcohols or fatty acids, sucrose fatty esters, alkyl polyglucosides, and polyglycerol esters; (ii) amphiphilic oligomers; and (iii) mixtures of fatty alcohols and surfactants or phospholipids that form crystalline lamellar phases.

A comparison between classical surfactants has been made by keeping the other parameters constant (Fig. 14.7), 4.5% surfactant/0.5% stearoyl glutamate/10% isocetyl stearate/5% isopropyl myristate/10% dipropylene glycol/5% glycerin/water with a Niro-Soavi OBL20 device at 120 MPa (Sonneville-Aubrun et al., unpublished). Droplet diameters between 40 and 60 nm could be obtained over a wide (5–15) HLB range.

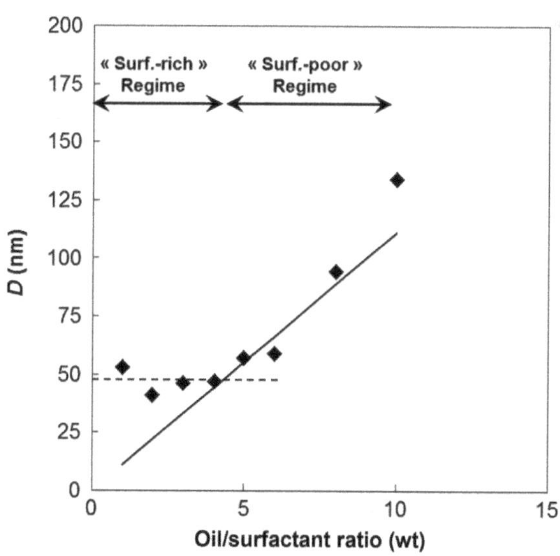

FIG. 14.6 Evolution of the nanoemulsion droplet diameter as a function of oil/surfactant weight ratio. The *straight line* represents the estimated diameter ($m_{TA} \sim 600$ g/mol and $a_0 \sim 60$ Å^2/mol), and the *dashed line* represents the saturated diameter.

TABLE 14.1 Examples of Surfactants Used to Stabilize Cosmetic Nanoemulsions

Surfactant Type	Examples	References
PEO alkyl ether PEO alkyl ester	PEO-8 isostearate PEO-20 stearate Steareth-10	Ribier et al. (1995b) Simonnet et al. (1998e) Franco and Heine (1997) Quemin (2001) Gesztesi et al. (2007) Knapik et al. (2010)
Alkyl polyglucoside Sucrose alkyl ester	Cetearyl glucoside Sucrose distearate	Simonnet et al. (1998a)
Sorbitan alkyl ester PEO sorbitan alkyl ester	Polysorbate 61	Simonnet et al. (1998d)
Polyglycerol alkyl ester	Decaglycerol monostearate	Simonnet et al. (1998c)
Phosphoric alkyl Citric alkyl ether	K cetyl phosphate Trilaureth-9 citrate	Knapik et al. (2010) Simonnet et al. (1998b)
Lecithin Gemini	Soybean lecithin Dilauramidoglutamide lysine	Zhou et al. (2010) Klang and Valenta (2011)
Alkyl trimethyl ammonium Alkyl amido propyl trimethyl ammonium	Behenyl trimethyl ammonium chloride	Dubief et al. (2009)
Amphiphilic oligomer	Poloxamer 231 Inulin lauryl carbamate	Simonnet et al. (1999a,b) Knapik et al. (2010)
Alpha-gel with fatty alcohol and surfactant	Stearyl alcohol/behenyl alcohol/K stearate/K behenate	Ribier et al. (1995a) Okamoto et al. (2016)

Many systems described in the literature contain both nonionic and ionic surfactants to ensure good stability over the long term. Some authors perform zeta potential measurements to verify the droplet surface charge and to anticipate aggregation phenomena (Tian et al., 2016). Some cosmetics may contain functional amphiphilic lipids, which bring skin or hair benefits, such as phytosphingosine (Yilmaz and Borchert, 2006). These will be discussed in the following sections of this chapter.

Polyols: Polyols are classical ingredients in cosmetics, where they are used as humectants, penetration enhancers for active ingredients, solvents, etc.

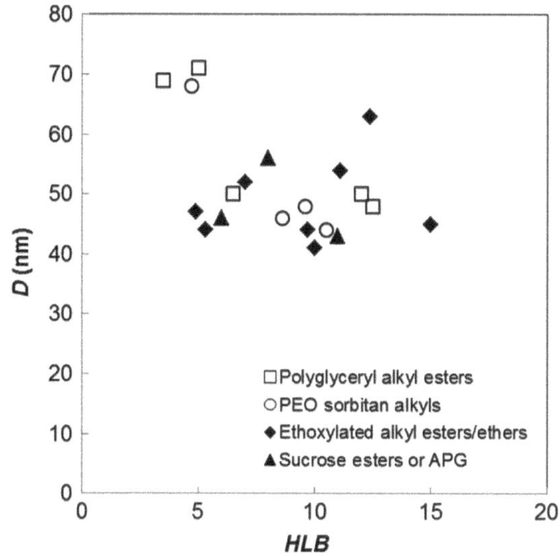

FIG. 14.7 Nanoemulsion droplet diameter as a function of surfactant HLB. For surfactants with HLB < 5, slightly larger diameters were obtained with very poor emulsion stability, even in the presence of an anionic surfactant at a low concentration.

They can also be seen as formulation aids to reduce droplet size and the formula-product transparency during the homogenization process.

An experiment has been conducted with formulas by increasing the amount of dipropylene glycol from 0% to 10% (Sonneville-Aubrun et al., unpublished). The nanoemulsions contained 5% surfactant (PEO-8 isostearate/stearoyl glutamate, 9:1) and 25% oil (isocetyl stearate/isopropyl myristate, 2:1). A high-pressure homogenizer (Niro-Soavi OBL20) operated at 120 MPa was used to perform the experiments. As shown in Fig. 14.8, the nanoemulsion droplet size and turbidity decreased by increasing the concentration of dipropylene glycol (Hach 2100).

Different factors may explain the positive impact of dipropylene glycol on nanoemulsion transparency: (i) a decrease in refractive index difference between the oil and water phase; (ii) an increase in the aqueous-phase viscosity from 1.8 cP without dipropylene glycol to 4 cP with 10% dipropylene glycol, with a theoretical reduction in droplet size by a factor of 0.8 according to the power law found in the literature (Walstra, 1993); and (iii) a decrease in interfacial tension from 43 to 25 mN/m following addition of 10% dipropylene glycol (for isocetyl stearate/isopropyl myristate oil phase). A positive impact on droplet size and transparency has also been observed for lecithin nanoemulsions upon addition of glycerin (Zhou et al., 2010).

Glycols can easily be added to tune the properties of nanoemulsions, being aware of their potential negative impacts on sensoriality (stickiness) at higher dosages.

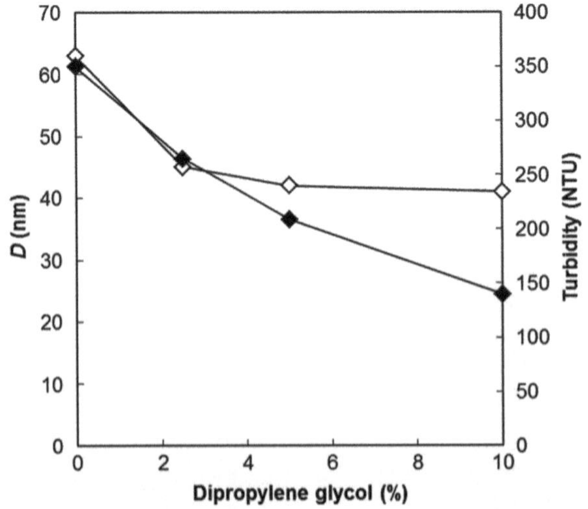

FIG. 14.8 Nanoemulsion droplet diameter *(empty diamond symbols)* and turbidity *(full diamond symbols)* as a function of dipropylene concentration.

TABLE 14.2 Current Oils Used in Cosmetics

Family	Oils	Chain Lengths	*Mw* (g/mol)	η at 25°C (cP)
Triglyceride	Caprylic/capric triglyceride Avocado oil Apricot kernel oil	$C_{8\text{–}10}$ $C_{16}C_{18:1}$ 20/60 $C_{18:1}C_{18:2}$ 66/28	542 930 930	27 71 69
Esters	Isopropyl myristate Isopropyl palmitate Isostearyl neopentanoate Isocetyl stearate	$C_{17}C_{34}O_2$ $C_{19}C_{38}O_2$ $C_{23}C_{46}O_2$ $C_{34}C_{68}O_2$	270 298 354 508	5.7 12 18 30
Alkanes	Isododecane Isohexadecane Vaseline Parleam	$C_{12}C_{26}$ $C_{16}C_{34}$ – $C_{30}C_{62}$	170 226 345 422	1.9 4.5 26 40

Oils: A large variety of oils can be used in cosmetics depending on the targeted benefits: emolliency and rich feeling, active solubilization and bioavailability, makeup remover, volatility, light feeling, shininess, etc. Table 14.2 assesses some oil families used to prepare cosmetic nanoemulsions.

According to Walstra (1993), the droplet size increases with oil-phase viscosity, following a power law, $D \sim \eta^{0.37}$. The experimental data presented in

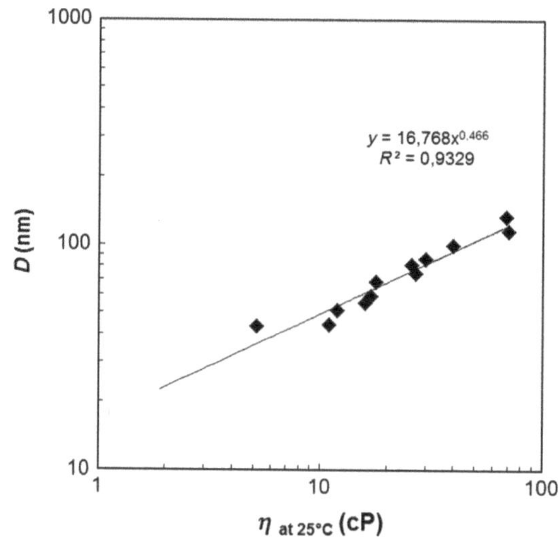

FIG. 14.9 Nanoemulsion droplet diameter *D (full diamond symbols)* as a function of oil-phase viscosity η at 25°C.

Fig. 14.9 show a power law $D \sim \eta^{0.47}$, with a slightly higher exponent. Long-chain triglycerides tend to give larger droplets compared with esters. The viscosity of the oil phase plays a key role in producing nanoemulsions and can be optimized by mixing different oils. HPH remains limited to oils with low to moderate viscosities, which covers most cosmetic applications.

In conclusion, high-energy homogenization may yield to a wide range of compositions and offers to formulators the opportunity to create new and attractive products that respond to technical, toxicological, and environmental requirements of cosmetic products. However, high-energy homogenization typically requires costly and specific equipment. The following section focuses on an interesting alternative provided by low-energy processes.

14.3.2 Low Energy Process

Nanoemulsions can be obtained by using the internal-system energy during the emulsification process (phase inversion and/or phase transition). For industrial applications, these methods require a good understanding of the equilibrium-phase behavior of the surfactant-oil-water system used and the evolution of the physicochemical properties of the system during emulsification. Nanoemulsions can be obtained by spontaneous emulsification, which is mainly driven by the surfactant chemical potentials and mass transfer (Solans and Solé, 2012). These methods are known as low-energy methods and require relatively simple process devices for the scale-up.

Phase inversion temperature (*PIT*) *method*: One of the most widely known low-energy methods to obtain nanoemulsions in cosmetics is based on the PIT principle introduced by Shinoda in the 1960s (Shinoda and Saito, 1968; Shinoda and Arai, 1964, 1967). The PIT method utilizes a transitional phase inversion (Salager et al., 2004) induced by specific temperature-sensitive nonionic polyethoxylated surfactants. O/W nanoemulsions can be obtained due to changes in the surfactant spontaneous curvature at the oil-water interface and at some degrees below the PIT, that is, not so far from the minimum interfacial tension locus. Salager and coworkers showed that the optimum domain to get the smallest droplet size corresponds to the best compromise between droplet breakup and coalescence in SOW dispersions. This is favored by the extremely low interfacial tension near and on both sides of the optimal formulation where hydrophilic lipophilic deviation (HLD) is zero (Forgiarini et al., 2015; Salager et al., 1996, 2000, 2010; Tolosa et al., 2006; Shinoda and Saito, 1969). These physicochemical phenomena are well described and explained in the literature (Solans et al., 2003).

In the 1990s, Förster used low-energy methods to obtain nanoemulsions composed of typical cosmetic ingredients, such as ceteareth-12 or polyoxyethylene eicosyl/docosyl ether. Förster examined the influence of oil polarity, using decyl oleate, 2-octyl dodecanol, or isopropyl myristate and their mixtures with mineral oil (Förster, 1997; Förster et al., 1990, 1995). From these results, some formulation rules can be identified to reach stable nanoemulsion at room temperature, that is, surfactant concentration and hydrophilicity, and the oil nature should be adapted to fix the PIT system between 70°C and 90°C. Within this range, nanoemulsions are stable according to standardized long-time stability tests at 45°C (Förster et al., 1992).

The following equation obtained from the HLD expression shows some physicochemical variables that influence the *PIT* value (Salager et al., 2001), notably the surfactant and oil nature and the presence of the alcohols that can be considered as cosurfactants:

$$PIT = 25 + \frac{[EON - \alpha - b \cdot S + k \cdot EACN - a \cdot A]}{t} \tag{14.4}$$

Here, *EON* corresponds to the number of ethylene oxide groups on the surfactant, *EACN* to the equivalent alkane carbon number linked to the polarity and carbon number of the oil, *S* to the salt concentration in the water phase, and *A* to the alcohol or cosolvent concentration in the water phase; *α*, *k*, and *t* are surfactant characteristic parameters; "*a*" and "*b*" represent, respectively, the specific constants of the alcohol and salt (Ontiveros et al., 2013).

During emulsification, to ensure nanoemulsion stability, the gap between *PIT* and storage temperature and the cooling rate to ensure a narrow drop-size distribution and a translucent appearance in the final nanoemulsion are important (Förster et al., 1990; Shinoda and Saito, 1969). Obtaining monodisperse drop-size distributions also helps to prevent Oswald ripening.

Surfactant concentration and water/oil ratio are important parameters to optimize when formulating nanoemulsions using this method. Tadros and coworkers showed how increasing the surfactant concentration causes the *PIT* value of industrial-grade nonionic surfactants to decrease (Tadros et al., 2004). This behavior allows one to ensure that there is a sufficient level of surfactant to stabilize the droplets formed and eventually cross the liquid-crystal (LC) zones [in the Winsor IV (WIV) behavior]. It has been shown that higher surfactant concentration and dynamic emulsification conditions allow the droplet lifetime to be increased, thereby reaching translucent nanoemulsions. Furthermore, better results can be obtained if the formulation is performed near to the X point in the limit of the WIII and WIV systems in the gamma-cut phase behavior. This point is currently associated with the surfactant efficacy. Pizzino et al. show the evolution of the gamma-cut phase behavior for nonionic polyethoxylated surfactants at different water/oil ratio (Pizzino et al., 2009). From these results, the water/oil ratio can be used to slightly increase the surfactant efficacy and shift the X point and the WIV-LC zone toward lower concentrations.

The sub-PIT process was introduced by Roger et al. as a new process to obtain nanoemulsions (Roger et al., 2009). The latter work signals the existence of a clear boundary and the advantages offered by this zone. These results correspond to the same observations pointed out by Shinoda 50 years before (Shinoda and Saito, 1969) and were extensively explained during the past 30 years by many authors (Tolosa et al., 2006; Solans et al., 2003, 2005; Salager et al., 1996). From Shinoda's original work, "better, stable and fine emulsion can be obtained if a system is shaken at first close to the PIT (about 2–4°C below the PIT is the best) to fine dispersion and then rapidly cooled down to storage temperature, at which the coalescence rate is slow. We designate this process as emulsification by the PIT-method (Shinoda and Saito, 1969)." The scale-up of this method however is not easy to transpose as explained later.

Recent works in the cosmetic domain show how the PIT method allows one to produce nanoemulsions suitable for application in skin and hair products. Dario et al. present a stable quercetin-loaded cationic nanoemulsion composed mainly of 5.6% PEO-20 oleyl and 3.4% PEO-3 oleyl, 5% caprylic-capric triglyceride, and 1% cetyltrimethyl ammonium chloride, a cationic surfactant (Dario et al., 2016a,b). In these works, the authors claim that nanoemulsions were obtained by a sub-PIT process.

At an industrial scale, the PIT method consists in heating the chosen surfactant/oil/water combination above the PIT under constant stirring. The system is then cooled down by adding cold water or by circulation of ice water in the double-wall reactor. Working at a temperature higher than the PIT allows one to ensure high-cooling kinetics during the phase transition that favors nanoemulsion monodispersity and stability. However, if the cooling process is started at or near the PIT, the thermal inertia due to the large reactor volume can be high enough to quickly lose the small droplet size and increase polydispersity (Tolosa et al., 2006; Solans et al., 2003). The same results are obtained for lower cooling rates (Ozawa et al., 1997; Kunieda et al., 1996). Under these process conditions, opaque emulsions are formed rather than clear nanoemulsions.

On one hand, adding ice water into the reactor mass helps to homogenize the temperature of the system; however, it also promotes strong composition changes, principally by decreasing the surfactant concentration or by changing the water/oil ratio. On the other hand, cooling down the system by external ice water circulation allows the system composition to remain constant even if the heat transfer into the reactor mass is low and with the same consequences as discussed in the previous paragraph.

For cosmetic purposes, the PIT method is specific to thermal-sensitive surfactants such as PEO derivatives, thereby strongly restricting the choice in surfactants. To overcome this constraint, other low-energy methods can be used, as described in the following section.

PIC-Method: Another low-energy process takes advantage, at constant temperature, of the phase transition that occurs during a dilution process. This method studied by Forgiarini et al. is known as phase inversion composition (PIC) method (Solans et al., 2003; Forgiarini et al., 2001). In general, the PIC method consists in the dilution by water (or oil) of an initial binary surfactant/oil (or surfactant/water) mixture under stirring. The dilution provokes an inversion of surfactant curvature as in a PIT process. The way by which the emulsion is produced is still unclear as pointed out by Sonneville-Aubrun et al., even if detailed small-angle neutron scattering (SANS) and NMR studies during dilution steps help to understand the changes in evolution of the system structure and the link with the SOW phase behavior (Solè et al., 2012; Sonneville-Aubrun et al., 2009; Forgiarini et al., 2001). Obtaining nanoemulsions by the PIC method cannot be explained solely by the system equilibrium properties. Transient mesomorphic phases, lamellar or cubic, have been observed during the inversion process (Sonneville-Aubrun et al., 2009; Maestro et al., 2008). Low interfacial tension and inversion kinetics may also be necessary conditions to form nanodroplets.

This method allows one to use a wide range of surfactant types that are possible with PIT methods, as long as the amphiphilic/oil system has the ability to spontaneously form lamellar or cubic phase in contact with the dilution phase; thus, ionic surfactant and nonionic surfactants other than the polyethoxylated ones can also be used (Yu et al., 2012; Solè et al., 2006a,b, 2010).

Formulation and composition variables are not the only factors that need to be taken into account. The vessel geometry and other process parameters, such as addition rates and stirring energy, are also key factors to reach translucent, fine, and monodispersed nanoemulsions (Solè et al., 2006b, 2010; Solans et al., 2003).

Solè and coworkers studied the PIC method scale-up from 100 to 600 mL vessel using an experimental design approach (Solè et al., 2010). The main process factors identified for scaling-up are the vessel geometry, the dilution phase rate addition, and the energy of stirring during emulsification. Translucent nanoemulsions could be produced using various systems. For industrial devices, these results are promising but should nevertheless be considered as laboratory-scale trials. More complex process parameters need to be considered when developing devices for scaling up to 1 ton. As previously mentioned, a lower addition rate of the dilution phase and stirring geometries limit the possibility to obtain translucent, fine, monodisperse nanoemulsions. At large scales, lower addition rates and higher shearing rates may perturb the phase transition and the emulsification kinetics. Other industrial factors that need to be considered are processing times and costs.

In cosmetics, the PIC technology has been applied to develop O/W nanoemulsions for hair care and topical applications as reported by Dubief et al. (2009) and Aubrun-Sonneville and Guiramand (2014). Other compositions were also studied and reported in the literature (Yu et al., 2012; Solè et al., 2006a, 2010).

Spontaneous Emulsification: Spontaneous emulsification may occur when two immiscible liquids with different free-energy levels are brought into contact. This phenomenon is triggered by gradients of the chemical potential between the phases, which under certain conditions lead to negative values of the free energy of emulsification (Solans et al., 2016). This method is generally based on solvent, cosurfactant, or surfactant diffusion from the initial oil or water phase to the water or oil dilution phase. Considerable efforts have been made to better understand the mechanisms for spontaneous emulsification, as summarized by Solans et al. (2016). Even though some questions still remain (Saberi et al., 2015), we can assume that the phenomenon is driven by diffusion followed by nucleation, with the potential contribution of other factors such as interfacial turbulences (Rayleigh instabilities) and local low interfacial tensions (Solans et al., 2016; Forgiarini et al., 2015; Anton et al., 2008; Miller, 2006; Ganachaud and Katz, 2005; Lopez-Montilla et al., 2002). The diffusion path can be plotted on an equilibrium ternary-phase diagram (Anton et al., 2008) that constitutes an important tool to understand and control these systems. Without surfactants, this phenomenon was also studied and known as the Ouzo effect (Ganachaud and Katz, 2005).

Nonionic surfactants, such as Span 80 and polyglycerol polyricinoleate, have been used to obtain nanoemulsions using the spontaneous emulsification method (Mehrnia et al., 2016). Vitamin E-enriched nanoemulsions have also been obtained using this low-energy method (Yukuyama et al., 2016; Saberi et al., 2013a,b,c, 2014). These studies show that the oil composition has a major impact on the nanoemulsion droplet size. Other variables, such as the surfactant concentration and type, the mixing temperature, and the stirring energy when the organic phase is added to the aqueous phase, also impact the mean droplet size.

In cosmetics, this method is very promising to obtain nanoemulsions by simply mixing ingredients. However, the fine control of the diffusion path may represent a challenge to obtain homogeneous and reproducible nanoemulsions. In terms of formulation parameters, spontaneous emulsification allows low internal-phase nanoemulsions to be prepared, typically below 5% after solvent evaporation (Ganachaud and Katz, 2005; Solans et al., 2005). Many parameters such as the oil viscosity, the surfactant structure, and the water solubility of the organic solvent are important in determining the quality of the nanoemulsions obtained by this method (Yukuyama et al., 2016). Use of solvents and the low concentration in oil volume fraction may represent a limit for some applications.

14.4 CONTROLLING NANOEMULSION STABILITY AND TEXTURE

Most legal frames, worldwide, specify that cosmetic products should remain stable for at least 3 years, irrespective of local climates, ranging from tropical to cold conditions. Accelerated stability tests are therefore often performed to guarantee a constant quality of the products to consumers, including temperature cycles and long-term cold (4°C) and hot (45–55°C) storage. As with all other formulations, nanoemulsions have to be designed to comply with these stability requirements.

Furthermore, consumers do not use the same products, according to climate, skin/hair types, product function, and cultural or fashion habits. For instance, for face products, Japanese women apply different layers of very fluid textures from cosmetic waters to milky lotions, whereas European women tend to use light to rich creams. Being able to fine-tune the texture, visual appearance, sensory feel, and the functional benefits of nanoemulsions used in cosmetic products is therefore important.

14.4.1 Stability Control

Emulsions are by nature metastable systems that encounter well-known instability phenomena such as aggregation, sedimentation/creaming, coalescence, and Ostwald ripening (Chapter 2). However, the expression of instability phenomena strongly depends on the type of nanoemulsion. For instance, in the case in W/O nanoemulsions, Ostwald ripening and coalescence are generally favored by short distances between droplets and a significant water solubilization in the oil phase.

In nanoemulsions, very small droplet sizes and rather low volume fractions are unfavorable to droplet coalescence and sedimentation/creaming. However, numerous examples show the dramatic effects of Ostwald ripening and flocculation phenomena in cosmetic compositions that appear to be the dominant destabilization mechanisms (Sonneville-Aubrun et al., 2004; Wooster et al., 2008).

Ostwald Ripening: Ostwald ripening is characterized by a coarsening of emulsion droplets over time due to molecular diffusion of dispersed phase between droplets through the continuous phase from smaller droplets to larger ones (Lifshitz and Slyosov, 1961; Wagner, 1961). For an oil-in-water emulsion,

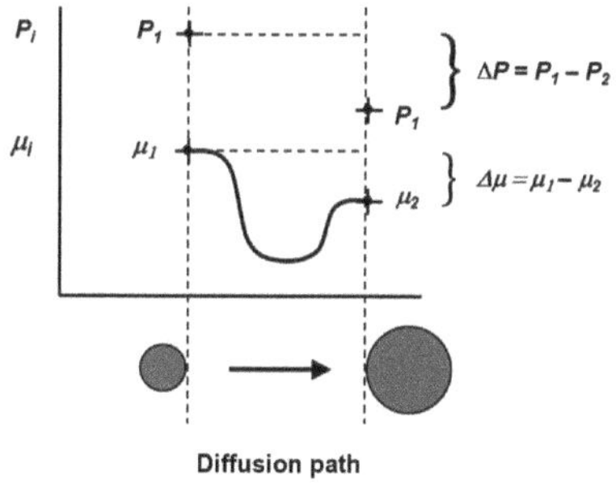

$$\mu_1 = \mu^\circ + \frac{2\gamma}{r_1} \cdot V^m > \mu_2 = \mu^\circ + \frac{2\gamma}{r_2} \cdot V^m \qquad (14.5)$$

μ_i = Chemical potential of oil in droplet i
μ° = Reference chemical potential of oil
γ = Interfacial tension (mN/m)
r_i = Radius of droplet i
V^m = Molar volume of oil drop (m^3/mol)
P = Pressure of drop i

FIG. 14.10 Expression of differences in oil chemical potential according to emulsion droplet size.

the process is driven by the difference of the oil chemical potential between oil droplets of different sizes: the lower the droplet size, the higher the Laplace pressure and the oil chemical potential (Fig. 14.10).

Ostwald ripening rate "ω" is proportional to the oil solubility "s" and the oil diffusion coefficient "D_f" in the continuous phase according to the LSW theory (Lifshitz and Slyosov, 1961; Wagner, 1961):

$$\omega = \frac{32}{27}\pi \frac{\gamma \cdot V^m \cdot D_f \cdot s}{R \cdot T} \qquad (14.6)$$

Depending on oil polarity, oil molecular weight, and aqueous-phase composition, the Ostwald ripening rate can vary by several orders of magnitude. Table 14.3 exhibits the calculated Ostwald ripening rate for three alkanes with increasing molecular weight, taking 10 mN/m as interfacial-tension value. The time needed to double the droplet size of an emulsion for two initial droplet diameters was also calculated.

Dodecane nanoemulsions are very unstable to Ostwald ripening due to the relatively high solubility of this oil in the aqueous phase. The increase in droplet

TABLE 14.3 Solubility Data for Alkanes and Calculated Ostwald Ripening Rates

					Time Needed to Double the Initial Droplet Size	
Oil	C_i	*Mw* (g/mol)	Solubility/ Water (mL/mL)	Ripening Rate (nm^3/h)	$D_{init.}$= *50 nm*	$D_{init.}$= *1 μm*
Dodecane	12	170	3.40×10^{-9}	2.1×10^4	22 h	19 years
Tetradecane	14	198	2.10×10^{-10}	1.4×10^3	14 days	300 years
Hexadecane	16	226	1.30×10^{-11}	9.0×10^1	7 months	4000 years

diameter is directly linked to a given transferred volume of oil, which is relatively higher for smaller droplets. A few hours or days should suffice to double the initial droplet size for nanoemulsions, as compared with years for micron-size emulsions.

Ostwald ripening rates for cosmetic oils were monitored by the measurement of droplet size over time by dynamic quasielastic light scattering (Sonneville-Aubrun et al., 2004). Fig. 14.11 presents data for three nanoemulsions containing 4.5% PEO-8 isostearate, 0.5% stearoyl glutamate, 15% oil, 10% dipropylene glycol, 5% glycerin, and water made with a Niro-Soavi OBL20 device at 120 MPa. Three different oils were used: isopropyl myristate, isohexadecane, and isopropyl palmitate. For isohexadecane, the measured Ostwald ripening rate was around 1000 times larger than the calculated rates for hexadecane in a model emulsion. This discrepancy may come from the glycols present in the aqueous phase that increase the oil solubility in the continuous phase.

Fig. 14.12 presents the Ostwald ripening rate for different oils (Sonneville-Aubrun et al., 2004). As obtained by different authors, the Ostwald ripening rate decreases with oil carbon number or oil molecular weight, following a semilogarithmic law (Taylor, 2003). Furthermore, for comparable molecular weight or carbon number, the Ostwald ripening rate increases with oil polarity ($V_{alkane} < V_{ester} < V_{triglyceride}$).

Cosmetic formulations must be stable at different temperatures encountered in real life. For this reason, stability tests are usually performed at refrigerator (4°C), room temperature (~25°C), and elevated temperature (45°C) for several months.

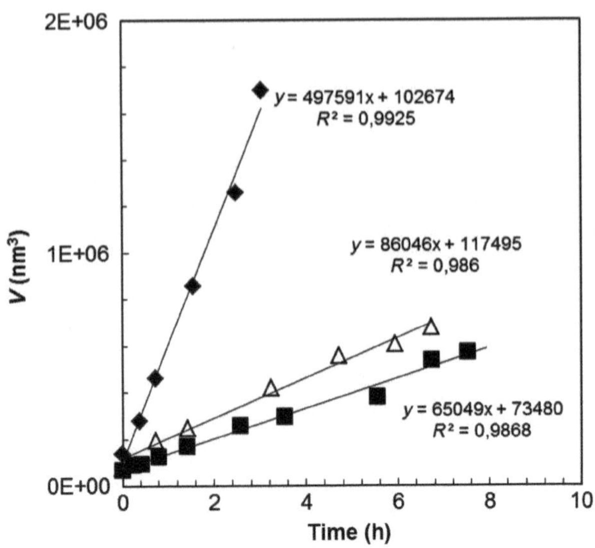

FIG. 14.11 Evolution of droplet volume upon time for different cosmetic oils: *full diamond symbols*, isopropyl myristate; *empty triangle symbols*, isododecane; *full square symbols*, isopropyl palmitate.

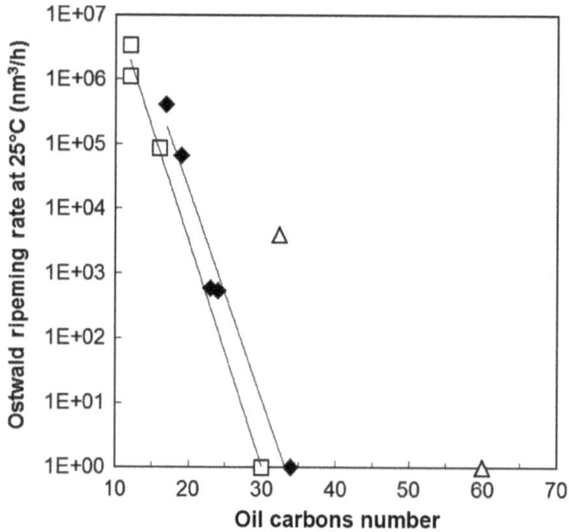

FIG. 14.12 Evolution of droplet volume over time for different cosmetic oils: *empty square symbols*, alkanes; *full diamond symbols*, fatty esters; *empty triangle and diamond symbols*, triglycerides.

According to previous studies, the Ostwald ripening rate should follow an Arrhenius law with temperature (Delmas et al., 2011). Fig. 14.13 presents experimental data on nanoemulsions containing 4.5% PEO-8 isostearate, 0.5% stearoyl glutamate, 15% oil, 10% dipropylene glycol, 5% glycerin, and water made with a Niro-Soavi OBL20 device at 120 MPa (Sonneville-Aubrun

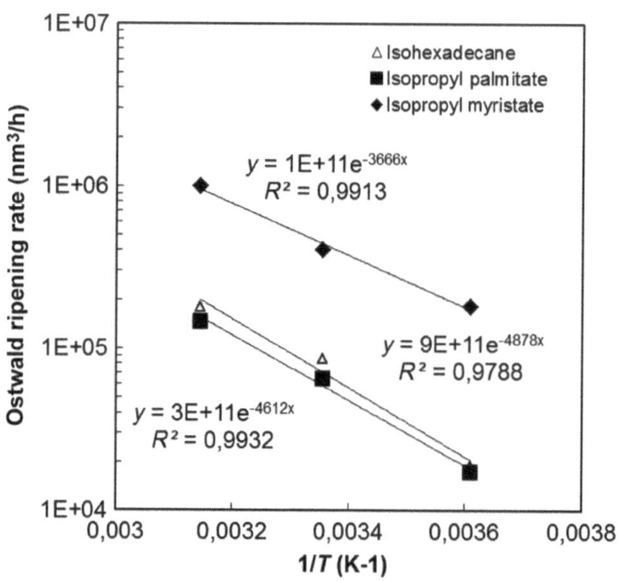

FIG. 14.13 Evolution of Ostwald ripening rate with $1/T$ for three different oils: *full diamond symbols*, isopropyl myristate; *empty triangle symbols*, isohexadecane; *full square symbols*, isopropyl palmitate.

et al., unpublished). The graph shows an exponential decrease in Ostwald ripening rate with 1/T, confirming the former results. From 4°C to 45°C, the Ostwald ripening rate was found to increase by about one order of magnitude.

Among each oil family, it is still possible to find oils with no Ostwald ripening phenomenon to achieve nanoemulsion stability. However, to better optimize the sensoriality and performance of the products, it is important to find solutions that stabilize nanoemulsions with low-molecular-weight oils and polar oils. Higushi and Misra (1962) first proposed guidelines to stabilize emulsions sensitive to Ostwald ripening by using mixtures of oils. According to theory, two scenarios may occur for an emulsion prepared with two oils of different nature, oil_1 "soluble" and oil_2 "insoluble." In Eq. (14.7), an additional term of entropy of mixing can compensate for the Laplace pressure term to obtain a stable system:

$$\mu_1 = \mu_1^{\circ} + \frac{4 \cdot \gamma \cdot V_1^m}{D} - R \cdot T \cdot x_2 \cdot \left(\frac{D^i}{D}\right)^3 \tag{14.7}$$

The stability condition is linked to the molar fraction of the "insoluble" oil in a droplet that has to be above a critical value depending on the interfacial tension, oil molar volume, droplet diameter, and temperature as expressed by Eqs. (14.8a), (14.8b) presented in Fig. 14.14.

If the stability condition is not satisfied, the droplet-size distribution evolves toward a bimodal nanoemulsion with smaller droplets enriched in "insoluble" oil and larger droplets enriched in "soluble" oil. The critical molar fraction of "insoluble" oil is all the more important that the initial droplet size is small.

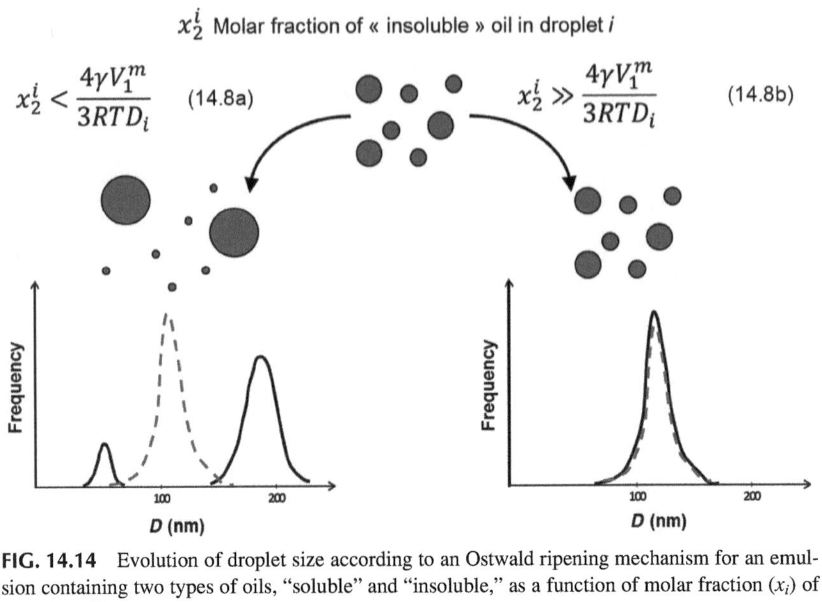

FIG. 14.14 Evolution of droplet size according to an Ostwald ripening mechanism for an emulsion containing two types of oils, "soluble" and "insoluble," as a function of molar fraction (x_i) of insoluble oil.

The critical weight fraction of "insoluble oil" has been determined theoretically and compared with experimental data for nanoemulsions containing mixtures of "insoluble oil" and "soluble oil." Theoretically, 10% of isocetyl stearate in weight relative to oil phase could stabilize a nanoemulsion with 40 nm diameter containing isododecane, taking a value of 10 mN/m for the interfacial tension. Experimentally, 50% isocetyl stearate in weight relative to oil phase was necessary to stabilize a 40 nm nanoemulsion containing isododecane (Sonneville-Aubrun et al., 2004).

14.4.2 Textures: From Lotions to Gels

The texture of cosmetic products is a complex and multidimensional notion that drives consumers' liking/acceptance by conveying an immediate hedonic perception, uniqueness, a surprise, or an excitement. Nanoemulsions have a wide range of textures, starting from lotions to stiff gels. Three major factors determine product consistency: the dispersed phase volume fraction, the presence of polymeric thickener or gelling agents, and the presence of crystalline lamellar phases.

Dispersed fraction: By increasing the dispersed phase volume fraction, an emulsion can be changed from a simple viscous liquid to an elastic solid of a substantial shear modulus (Mason et al., 2006). For classical emulsions, the droplets become transiently caged by their neighbors at a volume fraction of about 0.58 (hard-sphere glass transition) and become compressed with deformed interfaces above a volume fraction of about 0.64 (random hard-sphere close packing).

Nanoemulsions were made with increasing volume fraction using 5% surfactant (PEO-8 isostearate/stearoyl glutamate, 9:1), oil (isocetyl stearate/isopropyl myristate, 2:1) from 10% to 25%, dipropylene glycol, glycerol, and water, with a Niro-Soavi OBL20 at 120 MPa. Oscillatory measurements have been performed to characterize the change in viscoelasticity with volume disperse fraction with a Haake RS150 (stress sweep at 1 Hz, 25°C) (Arnaud-Roux and Sonneville-Aubrun, unpublished). Complex shear modulus and loss angle as a function of dispersed phase fraction are presented on Fig. 14.15. Comparable with classical emulsions, the consistency of nanoemulsions sharply increases with an increase of the dispersed phase volume fraction. The liquid-solid transition happens at lower volume fractions for repulsive nanoemulsions. Stiff gels have been observed between 0.25 and 0.3 volume fraction in dispersed phase in accordance with Kawada (2010).

Fig. 14.16 represents the complex shear modulus and loss angle as a function of shear stress for a 25% oil nanoemulsion. It shows a large linear domain with an abrupt transition around 1000 Pa in relation to the sensorial characteristics of the product, a stiff gel that melts upon shearing.

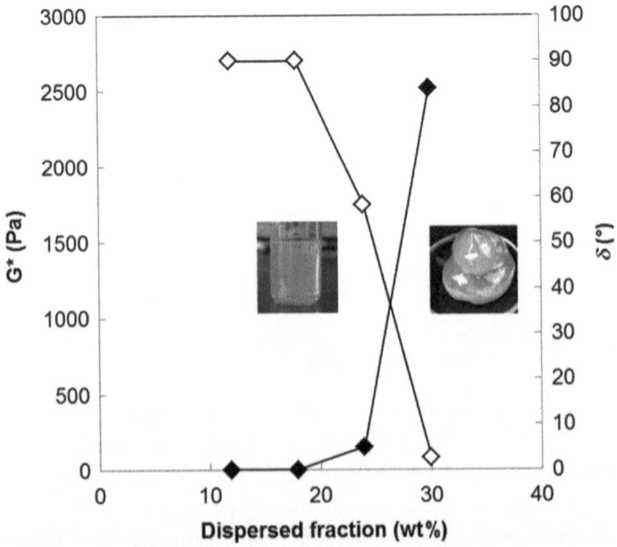

FIG. 14.15 Complex modulus G* *(full diamond symbols)* and loss angle δ *(empty diamond symbols)* as a function of dispersed phase weight fraction, in the linear domain.

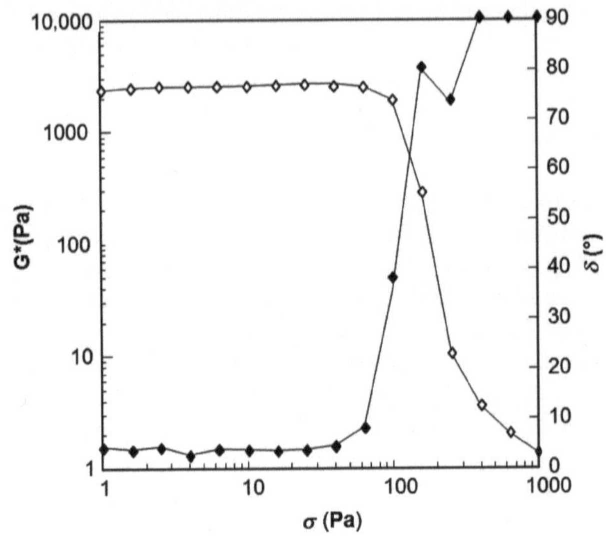

FIG. 14.16 Stress sweep for a concentrated O/W nanoemulsion of 25% oil.

This gel texture is ascribed to the crystal-like lattice structure of nanodroplets, as observed by transmission electron microscopy of a freeze-fractured surface of the specimen (Kawada, 2010). SANS revealed the presence of an ordered structure in addition to spherical domains with a radius of 17 nm. This long-range order is, in principle, due to electrostatic repulsion between charged nanodroplets.

TABLE 14.4 Droplet Interdistances for Different Droplet Sizes and Oil Volume Fractions

Oil Fraction (wt%)	Oil + Surfactant Fraction (v%)	$D = 40$ nm d (nm)	$D = 60$ nm d (nm)	$D = 300$ nm d (nm)	$D = 1$ μm d (nm)
25	34.2	8.5	14.3	84.9	290.6
23	31.9	9.6	16.1	93.8	320.3
21	29.7	11	18.1	103.6	353.2
19	27.4	12.4	20.3	114.7	390.0

A simple calculation of interdroplet distances "*d*," assuming a centered cubic lattice, shows distances smaller than 10 nm that can be reached by nanoemulsions for 25% oil in weight (Table 14.4). This distance will therefore be in the range of repulsive distances between droplets:

$$d = \left(\left(\frac{1}{2}\right)^{1/6} \cdot \left(\frac{\pi}{3 \cdot \varnothing_{\text{oil}}}\right)^{1/3} - 1 \right) \cdot D \tag{14.9}$$

The gel textures appear extremely shear-thinning, leaving a melting feeling upon application. This rheological behavior might be an issue in picking up the product from a jar. In that case, the texture is linked to the oil volume fraction. Obtaining gel textures with lower oil content for lighter products is often desirable. Thickeners or gelling agents can give access to other textures and sensory feelings.

Thickeners or gelling agents: Cosmetic O/W emulsions are classically thickened by polymers that offer a wide range of structures and cosmetic benefits. Usually, the addition of polymers is performed at the end of the emulsification process, to avoid disturbing droplet fragmentation and degrading the polymer during homogenization. A pregel can be prepared and added to a concentrated nanoemulsion. In some cases, it might be favorable to add the polymer during the emulsification process (i.e., low-molecular-weight polymers with low sensitivity to shear).

Microgels such as carbomers are very commonly used in cosmetics and especially in skin-care products for their texturing and sensorial properties. Some authors have used them to thicken nanoemulsions at low oil volume fractions (Oliveira et al., 2011). However, they are sensitive to formulation ingredients such as ionic species (salts, actives, and preservative), with a dramatic decrease in consistency. In the case of nanoemulsions with medium to high-volume fractions, as soon as they are introduced, the system turns white, and micron-size irregular structures are visible through microscopic observations (Fig. 14.17).

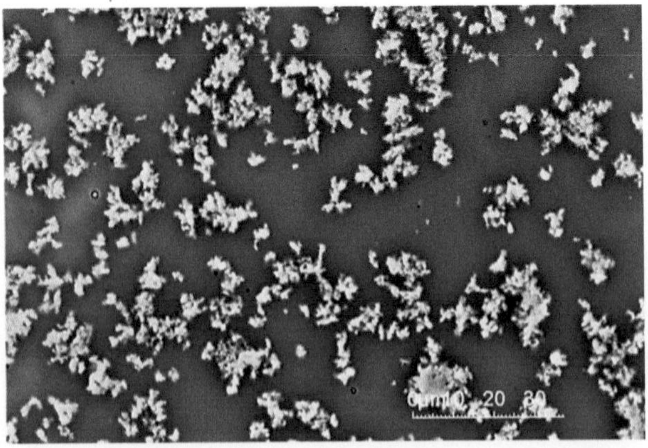

FIG. 14.17 Optical microscopy image from a thickened nanoemulsion with carbomer (Sonneville-Aubrun et al., 2004).

Associative polymers such as hydrophobically modified ethoxylated urethane polymers (HEUR) can easily thicken nanoemulsions even at low oil dispersed volume fractions. The droplets participate in the three-dimensional network formed by the polymer in water. A thickened nanoemulsion may present a fluid character at low frequency, turning to solid at high frequency as shown in Fig. 14.18 for a nanoemulsion containing 5% surfactant and 15% oil and 0.5% HEUR (Arnaud-Roux and Sonneville-Aubrun, unpublished).

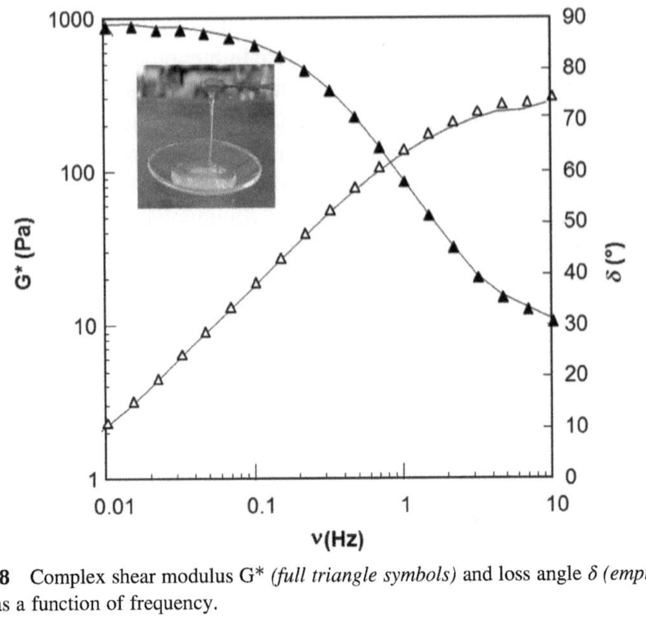

FIG. 14.18 Complex shear modulus G* *(full triangle symbols)* and loss angle δ *(empty triangle symbols)* as a function of frequency.

Linear and nonassociative polymers can also be used to thicken nanoemulsions, such as hydroxyethyl cellulose and xanthan gum (Samson et al., 2016; Ribeiro et al., 2015). However, depletion flocculation mechanisms may happen even at rather low concentrations in the dispersed phase and polymer, with a dramatic effect on stability and transparency of the nanoemulsion.

Crystalline lamellar phase: Fatty alcohol/surfactant associations have been used for a long time to texture emulsions without polymer, through the formation of a crystalline lamellar phase in the bulk and at the droplet surface (Barry, 1968; Okamoto et al., 2016). These structures also play a role in nanoemulsion efficacy by creating an occlusive and protective layer onto the skin for hydration. Okamoto et al. (2016) have demonstrated the ability of a decreased droplet size to get new milky lotions, lighter upon application without compromising hydration benefits. The nanoemulsion is broken down during its application onto the skin, thereby releasing the fatty alcohol and surfactant molecules into the aqueous phase.

14.5 EXAMPLES OF COSMETIC APPLICATIONS

Besides the noticeable interest in nanoemulsions for their unique texture, other benefits of nanoemulsions have been valued in many cosmetic applications, especially in skin care for the solubilization and delivery of active ingredients. In the hair-care domain, they play an important role in the silicon deposition onto hair fiber or scalp treatment. The following section gives an overview of different cosmetic applications of nanoemulsions.

14.5.1 Skin Care

Skin structure and penetration pathways: The human skin is the largest organ in the body, consisting of distinct layers, providing, as its main function, a protection to the body from environmental harms (Ribeiro et al., 2015; Wu and Guy, 2009). Starting from the outermost layer of skin, the *stratum corneum* (SC) of the epidermis is composed of dead cells, that is, corneocytes (Fig. 14.19). Below the SC, the viable epidermis comprises keratinocytes, melanocytes, and Langerhans cells. Keratinocyte cells are continuously regenerated from the basal cell layer of the epidermis and then differentiate upward toward the surface. Underlying the epidermis, the dermis contains fibroblasts, collagen, elastin, hair follicles, sebaceous glands, sweat glands, nerve, blood, and lymphatic vessels and is considered the major support and feeding component of the skin. The subcutaneous fat layer or hypodermis is mainly comprised of hypertrophic adipocytes and provides a thermal barrier to the body (Wu and Guy, 2009; Ammala, 2013; Kong et al., 2011; Alzorqi et al., 2016).

At the skin surface, the flat corneocytes are assembled in parallel and overlap each other like "bricks" and surrounded by a "cement-like" structure of lipid matrix composed of ceramides, fatty acids, and cholesterol. Despite its

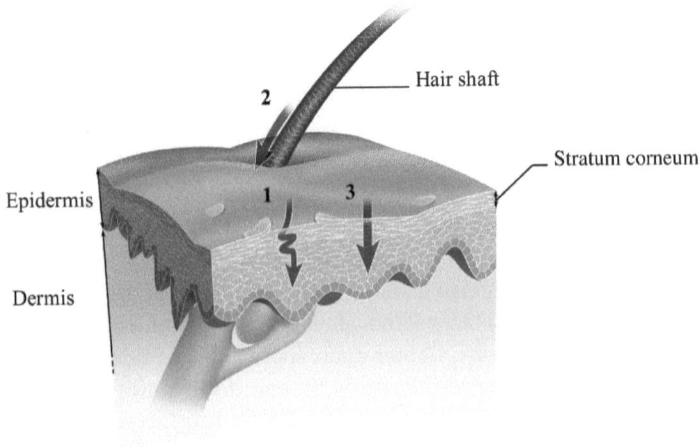

FIG. 14.19 Schematic illustration of skin structure and active component penetration pathways. 1 = intercellular, 2 = follicular, and 3 = intracellular.

permanent losses in corneocytes through the regular process of desquamation, this structure acts as an excellent barrier against penetration of external agents by topical administration or exposure and the component loss from internal to external part of the skin (Kong et al., 2011; Wu and Guy, 2009).

There are a number of possible pathways for molecules to penetrate through the skin barrier, and three main routes are generally considered. First, the intercellular pathway that is most well-known occurs by molecular diffusion via lipid layers among the interface of corneocytes. Second, the follicular pathway with its dense network of blood capillaries also acts as an active component "reservoir." Third, the transcellular pathway involves molecular transportation directly through the corneocytes and lipid layers (Morrow et al., 2007; Mihranyan et al., 2012).

Skin disorders and needs: Disturbance of the SC barrier function caused by a disorder in the corneocyte arrangement or delipidation may increase transepidermal water loss, leading to a decrease in skin hydration and a dry-skin syndrome, a common complaint among the population, particularly among older subjects (Bernardi et al., 2011; Yilmaz and Borchert, 2006; Ribeiro et al., 2015). Another undesirable phenomenon for consumers is brought about by the global skin-aging process of two types, intrinsic and extrinsic aging. Intrinsic (chronological) aging is a natural process caused by genetic factors. It is an irreversible degeneration process into which a progressive alteration of dermal extracellular-matrix network leads to the loss of skin firmness and elasticity. Extrinsic aging manifests in deep lines and wrinkles and rough and hyperpigmented skin, as a consequence of exposure to sun/ultraviolet (UV) rays or overexposure to oxidative stresses (Samson et al., 2016). Several chromophores in the skin (such as melanin, DNA, RNA, lipid, and aromatic amino acids) absorb UV, leading to photochemical reactions and free radicals, otherwise known as reactive oxygen species (ROS). The excess of ROS induces the oxidative-stress state caused by the deterioration of the natural antioxidant defense mechanisms, leading to protein denaturation and lipid peroxidation and finally to cell and tissue death (Alzorqi et al., 2016; Mahdi et al., 2011). These oxidative damages contribute (Krutmann et al., 2017) to several skin disorders such as photoaging, inflammation, and photocarcinogenesis. These incidences of skin disorders spur the development of high-quality products, targeting high performance, safety, and cost benefits for consumers.

Ceramides solubilization and delivery: Ceramides are major SC lipids, representing 40%–50% of its intercellular lipids. They play crucial roles in the skin barrier function, cell adhesion, epidermal differentiation, and induction of tumor-cell apoptosis. A ceramide comprises a fatty acid and a sphingoid base, and nine subclasses of ceramides have been identified in the SC, according to the acid chains and sphingoid base. Ceramide compositions vary according to skin conditions such as aging and dry skin. Previous studies reported correlations between a decrease in ceramide content and dry skin or skin with an atopic dermatitis, indicating the relevance of topical application of lipid mixtures for these conditions' treatments (Choi and Maibach, 2005).

Nonetheless, ceramide application is limited due to their difficult solubilization into water-based systems. Research has been carried out to reduce or overcome these limitations by using nanoemulsions. Aiming at restoring these disturbed skin barriers, different nanoemulsions containing specific lipids were prepared and compared (Yilmaz and Borchert, 2006): positively charged oil/water nanoemulsion (PN); positively charged oil/water nanoemulsion containing ceramide 3B, naturally found SC lipids such as ceramide 3, cholesterol, and palmitic acid (PNSC); and negatively charged O/W nanoemulsion containing SC lipids (NNSC). The positive charge was achieved by incorporating phytosphingosine (PS), an amphiphilic sphingoid base naturally found in the human body. Carbopol 940 was further added after nanoemulsion formation by high-HPH at 500 bar. The droplet diameter of the nanoemulsions produced in these studies was around 200 nm.

An in vivo study of skin hydration, elasticity, and erythema was performed on 14 healthy female volunteers, using instrumental measurements, such as Corneometer 825 (SC water content), Cutometer SEM 575 (skin firmness), and Mexameter® 18 (skin darkness). Results showed that all formulations increased skin hydration and elasticity. However, PNSC demonstrated significantly higher efficacy for skin hydration and elasticity, as compared with PN and NNSC. These results indicate that phytosphingosine plays an important role in delivering SC lipids to the skin. However, the impact of nanoemulsion droplet size on the efficacy still has to be investigated.

Plant extracts for skin protection and moisturizing: Plant extracts are widely used in cosmetic preparations due to several beneficial properties, such as antioxidant activities and moisturizing properties (Bernardi et al., 2011;

Zhong et al., 2010). Generally speaking, natural antioxidants comprise three important groups: phenolic acids, flavonoids, and high-molecular-weight polyphenols (Svobodová et al., 2003). Numerous in vitro studies have demonstrated that antioxidant compounds can directly scavenge ROS and reactive nitrogen species, chelate potentially prooxidant metal ions (Fe^{3+}, Al^{3+}, and Cu^{2+}), and complex with prooxidant proteins (Mahdi et al., 2011; Munin and Edwards-Lévy, 2011; Svobodová et al., 2003). Other categories of natural components such as water-soluble polysaccharides are also of great interest.

One of the main constraints of the natural compounds' application, mainly the polyphenols, lies on their instabilities upon exposure to external factors during processing and storage and their limited solubility. The stability problem may result in changes in the color and odor of the final products and considerable activity loss (Munin and Edwards-Lévy, 2011). A number of studies have shown the feasibility of fabricating nanoemulsions loaded with plant-based polyphenols using the PIC method and/or high-shear device (Mahdi et al., 2011; Ribeiro et al., 2015). Small droplet diameters (30 nm) could easily be obtained using classical surfactants such as Tween 80/Span 80 mixtures. Comparison with regular emulsions would be necessary to demonstrate the advantage of nanoemulsions either for delivery or hydration benefits.

Another study was performed using a quercetin-loaded nanoemulsion (flavonoid) prepared by the sub-PIT method with a PEO-oleyl ether mixture (Dario et al., 2016b). The active was found to act as a cosurfactant with a strong positive impact on droplet size and nanoemulsion stability. In nanoemulsions, the quercetin was mainly located at the oil/water interface, whereas in coarse emulsions, it was mainly dispersed in the aqueous phase in a crystallized form due to its low solubility in water. In this case, nanoemulsions appear superior to deliver poorly soluble amphiphilic antioxidant with a good photostability.

Rice bran oil nanoemulsions have been developed using the emulsion phase inversion method (Bernardi et al., 2011) to provide a stable, low-irritant, and high-moisturizing product. The unsaponifiable fractions of rice (*Oryza sativa*) bran oil are rich in tocopherols/tocotrienols and gamma-oryzanol, which are well-known for their antioxidant properties. The obtained nanoemulsions composed of 10% rice bran oil and 10% surfactant sorbitan oleate/PEO-30 castor oil had a mean droplet diameter of around 70 nm. in vivo assessment of these nanoemulsions demonstrated good tolerance and increased skin hydration with satisfactory results, considered as superior to commercial products according to the authors.

UV filters dispersions for sun care: Preventing and protecting the skin from the harmful effects of UV rays require the development of efficient topical products, which cover a wide range of UV-radiation levels. ultraviolet A (UVA) waves range from 320 to 400 nm, whereas ultraviolet B (UVB) rays range from 290 to 320 nm. UVAs are mostly responsible for extrinsic aging processes due to their penetration into the dermis, whereas UVBs cause sunburns and promote skin malignant processes (Pillai et al., 2005). Photoprotective agents are usually classified into organic and inorganic/mineral absorbers. Organic UV filters act by chemical absorption and comprise para-aminobenzoic acid derivatives, salicylates, cinnamates and camphor derivatives as main UVB sunscreens and avobenzone, ecamsule, and drometrizole trisiloxane as UVA sunscreens. Inorganic absorbers prevent the radiation by physical blockage (reflection) and include titanium dioxide as UVB absorber and zinc oxide as UVA absorber (Coutinho et al., 2015; Puglia et al., 2014). The development of an efficient colloidal system that overcomes the limitations of current sunscreen formulations, such as photostability, water resistance, risk of penetration, and high-sun protection factor, is a challenge (Puglia et al., 2014).

O/W nanoemulsions containing both chemical and physical UV filters (octyl methoxycinnamate and lipophilic nano-TiO_2) were processed by ultrasound with classical ethoxylated alkyl ethers (Silva et al., 2013). Without physical UV filters, very fine and stable nanoemulsions (3–10 nm) could be obtained with limited quantity in chemical UV filter providing in vitro SPF as low as 3. Incorporation of physical UV filters led to a huge increase in mean droplet diameter to 400–500 nm, due to the presence of hydrophobic TiO_2 particles in the oil phase. This example illustrates the limit in using nanoemulsions for mineral UV filtration.

Other authors studied the impact of structuring the oil phase of nanoemulsions containing chemical UV filters. Nanoemulsions and nanostructured lipid carriers (NLC) were formulated with different UV filters in the oil phase (Puglia et al., 2014). Nanoemulsions only contained liquid oil in the dispersed phase, whereas NLCs contained a mixture of liquid oil and waxy material. They were stabilized by Pluronic F68 and prepared using an ultrasonication method. Mean droplet diameters between 120 and 320 nm were successfully obtained. No significant differences in the spectral profile were observed between nanoemulsions and NLCs. In vitro percutaneous absorption studies showed reduced permeation of sun filters for NLC compared with nanoemulsions. However, the epidermal penetration values were low and below the accepted minimum safety limit, for both nanoemulsions and NLC.

Skin cell targeting with amphiphilic ligands: Direct ligand incorporation into nanoemulsions was explored to target specific skin cells such as keratinocytes and fibroblasts (Atrux-Tallau et al., 2013). The authors studied two different amphiphilic ligands: a polypeptide lysine-threonine-threonine-lysine-serine (KTTKS) lipophilized with palmitic acid (palmitoyl-KTTKS) and a trisaccharide chain glucose-glucose-rhamnose, asiaticoside. The peptide KTTKS is a subfragment of collagen I with potent activity on fibroblasts, improving extracellular matrix synthesis and inhibiting collagenase activity. Asiaticoside may link to lecithin-type receptors on skin cells. Direct emulsification of all components (the oil/wax mixture, surfactants, and targeting ligand at 1% w/w of dispersed phase) using ultrasonication enabled formation of 50 and 120 nm nanoemulsions. An in vitro cell-absorption assay using HaCaT cell line showed a preferential binding of asiaticoside-targeted nanoemulsion droplets to the keratinocytes. When tested on fibroblasts, palmitoyl KTTKS-targeted nanoemulsions demonstrated specific interaction with fibroblasts. This study shows that different skin-cell types might be targeted with selected ligands by increasing their adhesion onto cells.

Further studies have been performed to test ligand-targeted nanoemulsions for whitening active delivery (licorice) to melanocytes (Atrux-Tallau et al., 2014). Four ligand candidates were tested: a palmitoyl-peptide (palmitoyl-GQPR), a lipidized hyaluronic acid (caproyl-HA), and two amphiphilic actives (polydatin and isopilosine). The use of caproyl-HA significantly improved the ability of licorice-loaded nanoemulsion to reduce melanin concentration on melanocytes in vitro. In vivo efficacy has to be demonstrated.

14.5.2 Hair Fiber and Scalp

Hair structure: Hair is formed by dynamic, complex processes of structural alignment and keratinization of proliferated cells originating from the hair matrix at the base of the hair follicle and soon entering a multistep differentiation process upward (Franbourg and Leroy, 2005). As shown in Fig. 14.20, the outermost layer of the hair is composed of a cuticle, a protective layer that tightly surrounds the cortex, the major component of the hair.

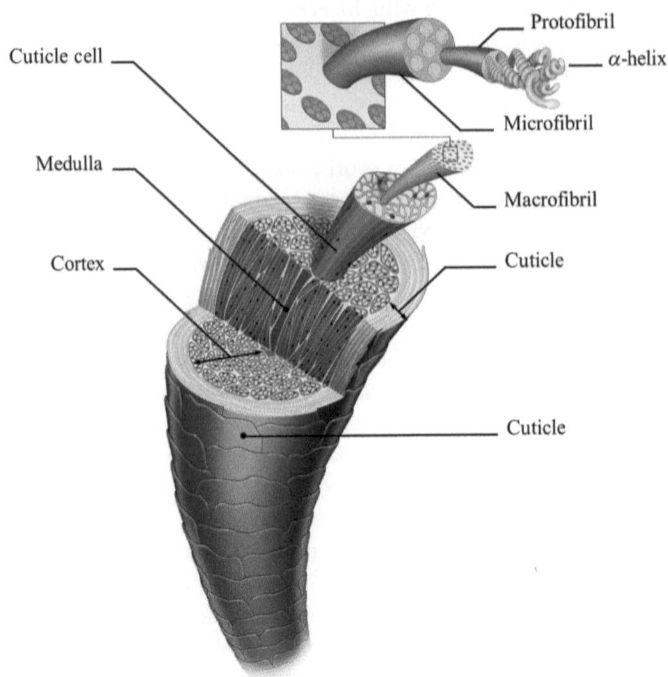

FIG. 14.20 Schematic illustration of hair structure.

The continuous phase named the cell membrane complex (CMC) acts as cement between the cells. Cuticle CMC contains polysaccharides and a lipid, 18-methyleicosanoic acid (18-MEA), that confer hydrophobic properties to the hair surface. The cortex mainly consists of highly arranged keratin filaments called microfibrils, which are packed into larger units called macrofibrils. The fiber is made of 60% keratins and 40% keratin-associated proteins, composing the matrix that stabilizes the microfibrils in macrofibrils. Among the structure of cortex, melanin pigments are dispersed, providing color characteristics to hair. Finally, located in the center of the fiber are the medulla cells.

Hair care and needs: The condition of the hair surface is daily altered by weathering, which may encompass a number of factors, aerial pollution included. But the main damage inducer is exposure to UV rays that causes photooxidative cleavage of cystine linkages and free-radical production, thereby generating cysteic acid end groups that confer a high-anionic character to the hair surface and increased porosity. Among the 18 amino acids that compose hair keratins, cystine amounts to about 16% and even 20%–30% in the outer cuticle. Hence, the chemical behavior of the hair surface may be highly impacted.

Another highly damaging process is daily handling, namely, frequent or excessive brushing of hair for dry untangling, mainly in women with midlong or long hair. The cuticle scales of the hair from root to distal end are eroded, uplifted, scratched, and torn out, becoming increasingly thinner up to being erased in split ends. Chemical processes such as hair straightening alter cuticles in order to achieve the desired visual effect (Lohani et al., 2014). These external events lead to brittleness and loss of shine, impairing smoothness and high-friction level of hair.

Thus, the development of efficient hair-conditioning products aiming at an optimal repair and softness of the hair surface; easy combing and brushing; improved control, gloss, manageability, body and volume, or bounce balance between rinsing; active compound deposition and penetration; safety and delivery of active compounds to the right destination without systemic circulation absorption, presents paramount challenges and opportunities in the cosmetics and personal-care field.

Another common hair concern is hair loss, either caused by *androgenetic alopecia* or *alopecia areata*, where hair follicles on a small scalp region are affected by a chronic inflammatory process (Aljuffali et al., 2014).

Silicon deposition for hair conditioning: Silicon oil-in-water nanoemulsion, stabilized by nonionic surfactants (Span 80/Tween 80) and a cosurfactant (*n*-butanol), was developed and compared with classical emulsions regarding silicon deposition on hair (Hu et al., 2012). Nanoemulsions of mean droplet diameter of 300 nm were obtained by the phase inversion composition method and dispersed in a shampoo. They were compared with micron-size emulsions. Experimental results demonstrated a significant increase in silicon-oil deposition on hair treated with a direct correlation of droplet size: the smaller the size, the greater the adsorption onto the hair. Nanoemulsions provide potential formulations with effective combination of washing and hair-care properties into a one-step process, with a higher performance compared with classical emulsions.

Scalp treatment for hair loss: Two lipid nanosystems, named squarticles and composed of squalene (sebum lipid), were developed by Aljuffali et al. (2014) to deliver active material against hair loss to the hair follicles, NLC and nanoemulsions. Diphencyprone (DPCP) is a topical drug for *alopecia areata* treatment that induces hair regrowth through local immune response. Minoxidil is another drug (OTC at 2% in some countries) for treating *androgenic alopecia* by promoting vasodilation and hair regrowth. These drugs may cause side effects such as skin irritation and risk of systemic absorption. Drug encapsulation in the nanosystems possibly reduces these drawbacks, by drug targeting into the hair follicles, controlling drug release, and increasing drug stability. As sebum is secreted by the sebaceous duct, which connects to the hair follicles, good interaction of squarticles with the follicles is expected. Mean particle diameters of 177 nm for NLC and 194 nm for nanoemulsions were obtained using a sonication method, with concentration of 0.15% w/v (7.3 mM) and 0.8% w/v (38.2 mM) of DPCP and minoxidil, respectively. in vitro and in vivo tests demonstrated a reduced systemic absorption of drug and a higher follicular uptake for squarticles versus control with free active.

14.5.3 Preservative System for Cosmetic Nanoemulsions

Aiming at investigating the relationship between the nanoemulsion droplet size and the antimicrobial activity, a new alternative combination of preservative system composed of caprylyl glycol, phenethyl alcohol, and glyceryl caprylate was developed (Fang et al., 2016). The antimicrobial activities were analyzed through minimum inhibitory concentration (MIC)/minimum bacteria concentration (MBC)/minimum fungicidal concentration (MFC) methods, and the results demonstrated a synergic effect of the combination of these preservative ingredients in five common microorganisms, as compared with their use alone. A broad-spectrum antimicrobial activity was observed with this blend, meeting the PCPC (Personal Care Products Council) and European Pharmacopoeia requirements at a concentration of 1.4%. The effect of the size of the emulsion droplets on preservative activity was observed, indicating the gradual enhancement of this activity with the increase of emulsion droplets in the range of 100–900 nm. According to this study, nanoemulsion droplet size is a factor that influences the preservation of cosmetic products. As a consequence, preservative dose has to be adapted to nanoemulsions to guarantee a good preservation.

14.6 CONCLUSIONS

This chapter has shown that nanoemulsions have a potential to be applied in a wide range of cosmetic and personal-care products for various purposes.

The small size of the droplets in nanoemulsions provides desirable optical, stability, rheological, and delivery properties that cannot be achieved with conventional emulsions. Research is continuing to develop cost-effective methods of producing these nanoemulsions from consumer-friendly ingredients and in testing their properties using in vivo models.

REFERENCES

Aljuffali, I.A., Sung, C.S., Shen, F.M., Huang, C.T., Fang, J.Y., 2014. Squarticles as a lipid nanocarrier for delivering diphencyprone and minoxidil to hair follicles and human dermal papilla cells. AAPS J. 16 (1), 140–150.

Alzorqi, I., Ketabchi, M.R., Sudheer, S., Manickam, S., 2016. Optimization of ultrasound induced emulsification on the formulation of palm-olein based nanoemulsions for the incorporation of antioxidant b-D-glucan polysaccharides. Ultrason. Sonochem. 31, 71–84.

Ammala, A., 2013. Biodegradable polymers as encapsulation materials for cosmetics and personal care markets. Int. J. Cosmet. Sci. 35, 113–124.

Antignac, E., Nohynek, G.J., Re, T., Clouzeau, J., Toutain, H., 2011. Safety of botanical ingredients in personal care products/cosmetics. Food Chem. Toxicol. 49, 324–341.

Anton, N., Benoit, J.P., Saulnier, P., 2008. Design and production of nanoparticles formulated from nano-emulsion templates—a review. J. Control. Release 128, 185–199.

Atrux-Tallau, N., Delmas, T., Han, S.H., Kim, J.W., Bibette, J., 2013. Skin cell targeting with self-assembled ligand addressed nanoemulsion droplets. Int. J. Cosmet. Sci. 35, 310–318.

Atrux-Tallau, N., Lasselin, J., Han, S.H., Delmas, T., Bibette, J., 2014. Quantitative analysis of ligand effects on bioefficacy of nanoemulsion encapsulating depigmenting active. Colloids Surf. B: Biointerfaces 122, 390–395.

Aubrun-Sonneville, O., Guiramand, C., 2014. Oil-in-Water Emulsion. US 8,741,322 B2.

Barry, B.W., 1968. Crystallization in polyhedral emulsion particles. J. Pharm. Pharmacol. 20 (6), 483–484.

Barbulova, A., Colucci, G., Apone, F., 2015. New trends in cosmetics: by-products of plant origin and their potential use as cosmetic active ingredients. Cosmetics 2, 82–92.

Bernardi, D.S., Pereira, T.A., Maciel, N.R., Bortoloto, J., Vieira, G.S., Oliveira, G.C., Rocha-Filho, P.A., 2011. Formation and stability of oil-in-water nanoemulsions containing rice bran oil: in vitro and in vivo assessments. J. Nanobiotechnol. 9, 44.

Choi, M.J., Maibach, H.I., 2005. Role of ceramides in barrier function of healthy and diseased skin. Am. J. Clin. Dermatol. 6 (4), 215–223.

Clares, B., Calpena, A.C., Parra, A., Abrego, G., Alvarado, H., Fangueiro, J.F., Souto, E.B., 2014. Nanoemulsions (NEs), liposomes (LPs) and solid lipid nanoparticles (SLNs) for retinyl palmitate: effect on skin permeation. Int. J. Pharm. 473, 591–598.

Coutinho, C.S.C., Santos, E.P., Mansur, C.R.E., 2015. Nanosystems in photoprotection. J. Nanosci. Nanotechnol. 15 (12), 9679–9688.

Dario, M.F., Santos, M.S.C.S., Viana, A.S., Arêas, E.P.G., Bou-Chacra, N.A., Oliveira, M.C., Minas da Piedade, M.E., Baby, A.R., Velasco, M.V.R., 2016a. A high loaded cationic nanoemulsion for quercetin delivery obtained by sub-PIT method. Colloids Surf. A Physicochem. Eng. Asp. 489, 256–264.

Dario, M.F., Oliveira, C.A., Cordeiro, L.R.G., Rosado, C., Mariz, I.F.A., Maçôas, E., Santos, M.S.C.S., Minas da Piedade, M.E., Baby, A.R., Velasco, M.V.R., 2016b. Stability and safety of quercetin-loaded cationic nanoemulsion: in vitro and in vivo assessments. Colloids Surf. A Physicochem. Eng. Asp. 506, 591–599.

Delmas, T., Piraux, H., Couffin, A.C., Texier, I., Vinet, F., Poulin, P., Cates, M.E., Bibette, J., 2011. How to prepare and stabilize very small nanoemulsions. Langmuir 27 (5), 1683–1692.

Dubief, C., Fack, G., Nicolas-Morgantini, L., Restle, S., 2009. Process for the Preparation of a Cationic Nanoemulsion, and Cosmetic Composition. US 7,476,393 B2.

Fang, B., Yu, M., Zhang, W., Wang, F., 2016. A new alternative to cosmetics preservation and the effect of the particle size of the emulsion droplets on preservation efficacy. Int. J. Cosmet. Sci. 38, 496–503.

Forgiarini, A., Esquena, J., Gonzalez, C., Solans, C., 2001. Formation of nanoemulsions by low-energy emulsification methods at constant temperature. Langmuir 17, 2076–2083.

Forgiarini, A., Marquez, L., Celis, M.T., Salager, J.L., 2015. Nanoemulsiones—formación con baja energía. In: Larez-Velasquez, C.L., Koteich, S., Lopez, F. (Eds.), Nanopartícula: Fundamentos y Aplicaciones. Universidad de Los Andes, Mérida, pp. 273–293 (Chapter 16).

Förster, T., 1997. Principles of emulsion formulation. In: Rieger, M.M., Rhein, L.D. (Eds.), Surfactants in Cosmetics, second ed. In: Surfactants Science Series, vol. 68. Marcel Dekker, New York, pp. 105–125.

Förster, T., Schambil, F., Tesmann, H., 1990. Emulsification by the phase inversion temperature method: the role of self-bodying agents and the influence of oil polarity. Int. J. Cosmet. Sci. 12, 217–227.

Förster, T., Schambil, F., Von Rybinski, W., 1992. Production of fine disperse and long-term stable oil-in-water emulsions by the phase inversion temperature method. J. Dispers. Sci. Technol. 13, 183–193.

Förster, T., Von Rybinski, W., Wadle, A., 1995. Influence of microemulsion phases on the preparation of fine-disperse emulsions. Adv. Colloid Interf. Sci. 58, 119–149.

Franbourg, A., Leroy, F., 2005. Hair structure, function and physicochemical properties. In: Bouillon, C., Wilkinson, J. (Eds.), The Science of Hair Care. CRC Press Informa Healthcare, New-York, London, pp. 1–30 (Chapter 1).

Franco, P., Heine, J., 1997. Water-Thin Emulsion Formed by High Pressure Homogenization Process. US 5,994,414.

Ganachaud, F., Katz, J., 2005. Nanoparticles and nanocapsules created using the Ouzo effect: spontaneous emulsification as an alternative to ultrasonic and high-shear devices. ChemPhysChem 6, 209–216.

Geis, P.A., 2006. Preservation strategies. In: Geis, P.A. (Ed.), Cosmetic Microbiology: a Practical Approach. Taylor & Francis Group, LLC, New York, London, pp. 163–180 (Chapter 7).

Gesztesi, J.L., Moraes Santos, L., De Tarso Hennies, P., Araujo Macian, K., 2007. Oil-in-Water Nanoemulsion, a Cosmetic Composition and a Cosmetic Product Comprising It, a Process for Preparing Said Nanoemulsion. US 8,956,597.

Higushi, W.I., Misra, J., 1962. Physical degradation of emulsions via the molecular diffusion route and the possible prevention thereof. J. Pharm. Sci. 51, 459.

Hu, Z., Liao, M., Chen, Y., Cai, Y., Meng, L., Liu, Y., Lv, N., Liu, Z., Yuan, W., 2012. A novel preparation method for silicone oil nanoemulsions and its application for coating hair with silicone. Int. J. Nanomedicine 7, 5719–5724.

Jafari, S.M., Assadpoor, E., He, Y., Bhandari, B., 2008. Re-coalescence of emulsion droplets during high-energy emulsification. Food Hydrocoll. 22 (7), 1191–1202.

Jafari, S.M., Paximada, P., Mandala, I., Assadpour, E., Mehrnia, M.A., 2017. 2—Encapsulation by nanoemulsions. In: Nanoencapsulation Technologies for the Food and Nutraceutical Industries. Academic Press, New York, pp. 36–73.

Kawada, H., Kume, T., Matsunaga, T., Iwai, H., Sano, T., Shibayama, M., 2010. Structure and rheology of a self-standing nanoemulsion. Langmuir 26 (4), 2430–2437.

Kentish, S., Wooster, T.J., Ashokkumar, M., Balachandran, S., Mawson, R., Simons, L., 2008. The use of ultrasonics for nanoemulsion preparation. Innovative Food Sci. Emerg. Technol. 9, 170–175.

Klang, V., Valenta, C., 2011. Lecithin-based nanoemulsions. J. Drug Delivery Sci. Technol. 21 (1), 55–76.

Knapik, R., Rocha dos Santos Taques, M.C., Praes, C., Lima de Oliveira, L., 2010. Characterizing nanoemulsions prepared by high pressure homogenization under various emulsifying conditions. Cosmet. Toilet. 125 (3), 72–78.

Kong, M., Chen, X.G., Kweon, D.K., Park, H.J., 2011. Investigations on skin permeation of hyaluronic acid based nanoemulsion as transdermal carrier. Carbohydr. Polym. 86, 837–843.

Krutmann, J., Bouloc, A., Sore, G., Bernard, B.A., Passeron, T., 2017. The skin aging exposome. J. Dermatol. Sci. 85, 152–161.

Kunieda, H., Fukui, Y., Uchiyama, H., Solans, C., 1996. Spontaneous formation of highly concentrated water-in-oil emulsions (gel-emulsions). Langmuir 12, 2136–2140.

Liedtke, S., Wissing, S., Müller, R.H., Mäder, K., 2000. Influence of high pressure homogenisation equipment on nanodispersions characteristics. Int. J. Pharm. 196, 183–185.

Lifshitz, I.M., Slyosov, V.V., 1961. The kinetics of precipitation from supersaturated solid solutions. J. Phys. Chem. Solids 19, 35–50.

Lohani, A., Verma, A., Joshi, H., Yadav, N., Karki, N., 2014. Nanotechnology-based cosmeceuticals. ISRN Dermatol. 2014. 14 p.

Lopez-Montilla, J.C., Herrera-Morales, P.E., Shah, D.O., 2002. New method to quantitatively determine the spontaneity of the emulsification process. Langmuir 18, 4258–4262.

Maestro, A., Sole, I., Gonzalez, C., Solans, C., Gutierrez, J.M., 2008. Influence of the phase behavior on the properties of ionic nanoemulsions prepared by the phase inversion composition method. J. Colloid Interface Sci. 327, 433–439.

Mahdi, E.S., Noor, A.M., Sakeena, M.H., Abdullah, G.Z., Abdulkarim, M.F., Sattar, M.A., 2011. Formulation and in vitro release evaluation of newly synthesized palm kernel oil esters-based nanoemulsion delivery system for 30% ethanolic dried extract derived from local Phyllanthus urinaria for skin antiaging. Int. J. Nanomedicine 6, 2499–2512.

Mason, T.J., Wilking, J.N., Meleson, K., Chang, C.B., Graves, S.M., 2006. Nanoemulsions: formation, structure, and physical properties. J. Phys. Condens. Matter 18, 635–666.

Mehrnia, M.A., Jafari, S.M., Makhmal-Zadeh, B.S., Maghsoudlou, Y., 2016. Crocin loaded nanoemulsions: factors affecting emulsion properties in spontaneous emulsification. Int. J. Biol. Macromol. 84, 261–267.

Mihranyan, A., Ferraz, N., Stromme, M., 2012. Current status and future prospects of nanotechnology in cosmetics. Prog. Mater. Sci. 57, 875–910.

Miller, C., 2006. Spontaneous emulsification: recent developments with emphasis on self-emulsification. In: Sjöblom, J. (Ed.), Emulsions and Emulsion Stability. In: Surfactant Science Series, vol. 132. Taylor and Francis, Boca Raton, FL, pp. 107–126.

Morrow, D., McCarron, P., Woolfson, A., Donnelly, R., 2007. Innovative strategies for enhancing topical and transdermal drug delivery. Open Drug Delivery J. 1, 36–59.

Munin, A., Edwards-Lévy, F., 2011. Encapsulation of natural polyphenolic compounds; a review. Pharmaceutics 3, 793–829.

Okamoto, T., Tomomasa, S., Nakajima, H., 2016. Preparation and thermal properties of fatty alcohol/surfactant/oil/water nanoemulsions and their cosmetic, applications. J. Oleo Sci. 65 (1), 27–36.

Oliveira, J.S., Aguiar, T.A., Mezadri, H., Santos, O.D.H., 2011. Attainment of hydrogel-thickened nanoemulsions with tea tree oil (*Melaleuca alternifolia*) and retinyl palmitate. Afr. J. Biotechnol. 10 (60), 13014–13018.

Ontiveros, J.F., Pierlot, C., Catte, M., Molinier, V., Pizzino, A., Salager, J.L., Aubry, J.M., 2013. Classification of ester oils according to their equivalent alkane carbon number (EACN) and asymmetry of fish diagrams of $C_{10}E_4$/ester oil/water systems. J. Colloid Interface Sci. 403, 67–76.

Ozawa, K., Solans, C., Kunieda, H., 1997. Spontaneous formation of highly concentrated oil-in-water emulsions. J. Colloid Interface Sci. 188, 275–281.

Philippe, M., Didillon, B., Gilbert, L., 2012. Industrial commitment to green and sustainable chemistry: using renewable materials & developing eco-friendly processes and ingredients in cosmetics. Green Chem. (4), 952–956.

Pillai, S., Oresajo, C., Hayward, J., 2005. Ultraviolet radiation and skin aging: roles of reactive oxygen species, inflammation and protease activation, and strategies for prevention of inflammation-induced matrix degradation—a review. Int. J. Cosmet. Sci. 27, 17–34.

Pizzino, A., Molinier, V., Catte, M., Salager, J.L., Aubry, J.M., 2009. Bidimensional analysis of the phase behavior of a well-defined surfactant ($C_{10}E_4$)/oil (n-Octane)/water—temperature system. J. Phys. Chem. B 113, 16142–16150.

Puglia, C., Damiani, E., Offerta, A., Rizza, L., Tirendi, G.G., Tarico, M.S., Curreri, S., Bonina, F., Perrotta, R.E., 2014. Evaluation of nanostructured lipid carriers (NLC) and nanoemulsions as carriers for UV-filters: characterization, in vitro penetration and photostability studies. Eur. J. Pharm. Sci. 51, 211–217.

Qian, C., McClements, D., 2011. Formation of nanoemulsions stabilized by model food-grade emulsifiers using high-pressure homogenization: factors affecting particle size. Food Hydrocoll. 25, 1000–1008.

Quemin, E., 2001. Translucent Nanoemulsion, Production Method, and Uses Thereof in the Cosmetic, Dermatological and/or Ophthalmological Fields. EP2002711954.

Ribeiro, R.C., Barreto, S.M., Ostrosky, E.A., Da Rocha-Filho, P.A., Veríssimo, L.M., Ferrari, M., 2015. Production and characterization of cosmetic nanoemulsions containing Opuntia ficus-indica (L.) mill extract as moisturizing agent. Molecules 20, 2492–2509.

Ribier, A., Simonnet, J.T., Michelet, J., 1995a. Dermatologic or Cosmetic Composition Made Up by an Oil-in-Water Emulsion Based on Oily Globules Coated With a Lamellar Liquid Crystal Coating. EP0705593A1.

Ribier, A., Simonnet, J.T., Legret, S., 1995b. Transparent Nanoemulsion Less Than 100 nm Based on Fluid Non-Ionic Amphiphilic Lipids and Use in Cosmetics or in Dermopharmaceuticals. US 5,753,241.

Roger, K., Cabane, B., Olsson, U., 2009. Formation of 10–100 nm size-controlled emulsions through a sub-PIT cycle. Langmuir 26, 3860–3867.

Saberi, A.H., Fang, Y., McClements, D.J., 2013a. Fabrication of vitamin E-enriched nanoemulsions by spontaneous emulsification: effect of propylene glycol and ethanol on formation, stability, and properties. Food Res. Int. 54, 812–820.

Saberi, A.H., Fang, Y., McClements, D.J., 2013b. Fabrication of vitamin E-enriched nanoemulsions: factors affecting particle size using spontaneous emulsification. J. Colloid Interface Sci. 391, 95–102.

Saberi, A.H., Fang, Y., McClements, D.J., 2013c. Effect of glycerol on formation, stability, and properties of vitamin-E enriched nanoemulsions produced using spontaneous emulsification. J. Colloid Interface Sci. 411, 105–113.

Saberi, A.H., Fang, Y., McClements, D.J., 2014. Stabilization of vitamin E-enriched mini-emulsions: influence of organic and aqueous phase compositions. Colloids Surf. A Physicochem. Eng. Asp. 449, 65–73.

Saberi, A.H., Fang, Y., McClements, D.J., 2015. Formation of thermally reversible optically transparent emulsion-based delivery systems using spontaneous emulsification. Soft Matter 11, 9321–9329.

Salager, J.L., Pérez-Sánchez, M., Garcia, Y., 1996. Physicochemical parameters influencing the emulsion drop size. Colloid Polym. Sci. 274, 81–84.

Salager, J.L., Márquez, N., Graciaa, A., Lachaise, J., 2000. Partitioning of ethoxylated octylphenol surfactants in microemulsion-oil-water systems. Influence of temperature and relation between partitioning coefficient and physicochemical formulation. Langmuir 16, 5534–5539.

Salager, J.L., Anton, R.E., Andérez, J.M., Aubry, J.M., 2001. Formulation des micro-émulsions par la méthode du HLD. © Techniques de l'Ingénieur, traité Génie des procédés J2 157, pp. 1–20.

Salager, J.L., Forgiarini, A., Marquez, L., Peña, A., Pizzino, A., Rodriguez, M.P., Rondon-Gonzalez, M., 2004. Using emulsion inversion in industrial processes. Adv. Colloid Interf. Sci. 108 –109, 259–272.

Salager, J.L., Bullon, J., Pizzino, A., Rondon-Gonzalez, M., Tolosa, L., 2010. Emulsion formulation engineering for the practitioner. In: Somasundarad, P. (Ed.), Encyclopedia of Surface and Colloid Science. In: 1:1. Taylor & Francis Group, London, pp. 1–16.

Samson, S., Basri, M., Masoumi, H.R.F., Malek, E.A., Karjiban, R.A., 2016. An artificial neural network based analysis of factors controlling particle size in a virgin coconut oil-based nanoemulsion system containing copper peptide. Plos One. https://doi.org/10.1371/journal.pone.0157737.

Sharma, S., Sarangdevot, K., 2012. Nanoemulsions for cosmetics. IJARPB 2 (3), 408–415.

Shinoda, K., Arai, H., 1964. The correlation between phase inversion temperature in emulsion and cloud point in solution of nonionic emulsifier. J. Phys. Chem. 68, 3485–3490.

Shinoda, K., Arai, H., 1967. The effect of phase volume on the phase inversion temperature of emulsions stabilized with nonionic surfactants. J. Colloid Interface Sci. 25, 429–431.

Shinoda, K., Saito, H., 1968. The effect of temperature on the phase equilibria and the types of dispersions of the ternary system composed of water, cyclohexane, and nonionic surfactant. J. Colloid Interface Sci. 26, 70–74.

Shinoda, K., Saito, H., 1969. The stability of O/W type emulsions as a function of temperature and the HLB of emulsifiers: the emulsification by PIT-method. J. Colloid Interface Sci. 30, 258–263.

Silva, F.F., Ricci-Junior, E., Mansur, C.R., 2013. Nanoemulsions containing octyl methoxycinnamate and solid particles of TiO2: preparation, characterization and in vitro evaluation of the solar protection factor. Drug Dev. Ind. Pharm. 39, 1378–1388.

Simonnet, J.T., Sonneville, O., Legret, S., 1998a. Nanoemulsion Based on Sugar Fatty Esters or on Sugar Fatty Ethers and its Uses in the Cosmetics, Dermatological and/or Ophthalmological Fields. US 6,689,371.

Simonnet, J.T., Sonneville, O., Legret, S., 1998b. Nanoemulsion Based on Phosphoric Acid Fatty Acid Esters and Its Uses in the Cosmetics, Dermatological, Pharmaceutical, and/or Ophthalmological Fields. US 6,274,150.

Simonnet, J.T., Sonneville, O., Legret, S., 1998c. Nanoemulsion Based on Glycerol Fatty Esters, and Its Uses in the Cosmetics, Dermatological and/or Ophthalmological Fields. US 6,541,018.

Simonnet, J.T., Sonneville, O., Legret, S., 1998d. Nanoemulsion Based on Oxyethylenated or Non-Oxyethylenated Sorbitan Fatty Esters, and Its Uses in the Cosmetics, Dermatological and/or Ophthalmological Fields. US 6,335,022.

Simonnet, J.T., Sonneville, O., Legret, S., 1998e. Nanoemulsion Based on Ethoxylated Fatty Ethers or on Ethoxylated Fatty Esters and Its Uses in the Cosmetics, Dermatological and/or Ophthalmological Fields. US 6,375,960.

Simonnet, J.T., Sonneville, O., Legret, S., 1999a. Nanoemulsion Based on Ethylene Oxide and Propylene Oxide Block Copolymers and Its Uses in the Cosmetics, Dermatological and/or Ophthalmological Fields. US 6,464,990.

Simonnet, J.T., Sonneville, O., Legret, S., 1999b. Nanoemulsion Based on Alkyl Ether Citrates and Its Uses in the Cosmetics, Dermatological, Pharmacological and/or Ophthalmological Fields. US 6,413,527.

Solans, C., Solé, I., 2012. Nano-emulsions: formation by low-energy methods. Curr. Opin. Colloid Interface Sci. 17 (5), 246–254.

Solans, C., Esquena, J., Forgiarini, A.M., Uson, N., Morales, D., Izquierdo, P., Azemar, N., Garcia-Celma, J.M., 2003. Nano-emulsions: formation, properties, and applications. In: Mittal, K.L., Shah, D.O. (Eds.), Adsorption and Aggregation of Surfactants in Solutions. vol. 109, pp. 525–554.

Solans, C., Izquierdo, P., Nolla, J., Azemar, N., Garcia-Celma, J.M., 2005. Nano-emulsions. Curr. Opin. Colloid Interface Sci. 10, 102–110.

Solans, C., Morales, D., Homs, M., 2016. Spontaneous emulsification. Curr. Opin. Colloid Interface Sci. 22, 88–93.

Solè, I., Maestro, A., Pey, C.M., Gonzales, C., Solans, C., Gutierrez, J.M., 2006a. Nano-emulsions preparation by low energy methods in an ionic surfactant system. J. Colloid Interface Sci. 344, 417–423.

Solè, I., Maestro, A., Gonzales, C., Solans, C., Gutierrez, J.M., 2006b. Optimization of nano-emulsion preparation by low-energy methods in an Ionic surfactant system. Langmuir 22, 8326–8332.

Solè, I., Pey, C.M., Maestro, A., Gonzales, C., Porras, M., Solans, C., Gutierrez, J.M., 2010. Nano-emulsions prepared by the phase inversion composition method: preparation variables and scale up. Colloids Surf. A Physicochem. Eng. Asp. 288, 138–143.

Solè, I., Solans, C., Maestro, A., Gonzales, C., Gutierrez, J.M., 2012. Study of nano-emulsion formation by dilution of microemulsions. J. Colloid Interface Sci. 376, 133–139.

Sonneville-Aubrun, O., Simonnet, J.T., L'Alloret, F., 2004. Nanoemulsions: a new vehicle for skincare products. Adv. Colloid Interf. Sci. 108–109, 145–149.

Sonneville-Aubrun, O., Babayan, D., Bordeaux, D., Lindner, P., Rata, G., Cabane, B., 2009. Phase transition pathways for the production of 100 nm oil-in-water emulsions. Phys. Chem. Chem. Phys. 11, 101–110.

Svobodová, A., Psotová, J., Walterová, D., 2003. Natural phenolics in the prevention of UV-induced skin damage. A review. Biomed. Pap. Med. Fac. Univ. Palacky Olomouc Czech Repub. 147 (2), 137–145.

Tadros, T., Izquierdo, P., Esquena, J., Solans, C., 2004. Formation and stability of nano-emulsions. Adv. Colloid Interf. Sci. 108–109, 303–318.

Taisne, L., Walstra, P., Cabane, B., 1996. Transfer of oil between emulsion droplets. J. Colloid Interface Sci. 184, 378–390.

Taylor, P., 2003. Ostwald ripening in emulsions: estimation of solution thermodynamics of the disperse phase. Adv. Colloid Interf. Sci. 106, 261–285.

Tian, Y., Guo, Y., Zhang, W., 2016. Effect of oil type, aliphatic alcohol, and ionic surfactants on the formation and stability of ceramide-2 enriched nanoemulsions. J. Dispers. Sci. Technol. 37 (8), 1115–1122. https://doi.org/10.1080/01932691.2015.1083440.

Tolosa, L.I., Forgiarini, A., Moreno, P., Salager, J.L., 2006. Combined effects of formulation and stirring on emulsion drop size in the vicinity of three-phase behavior of surfactant-oil-water systems. Ind. Eng. Chem. Res. 45, 3810–3814.

United States Food and Drug Administration, 2014. Guidance for Industry Considering Whether an FDA-Regulated Product Involves the Application of Nanotechnology. http://www.fda.gov/RegulatoryInformation/Guidances/ucm257698.htm. Accessed 25 October 2016.

Wagner, Z., 1961. Elektrochem. Ber. Bunsenges. Phys. Chem. 65, 581.

Walstra, P., 1993. Principles of emulsion formation. Chem. Eng. Sci. 48, 333–349.

Wooster, T.J., Golding, M., Sanguansi, P., 2008. Impact of oil type on nanoemulsion formation and ostwald ripening stability. Langmuir 24, 12758–12765.

Wu, X., Guy, R.H., 2009. Applications of nanoparticles in topical drug delivery and in cosmetics. J. Drug Delivery Sci. Technol. 19 (6), 371–384.

Yilmaz, E., Borchert, H.H., 2006. Effect of lipid-containing, positively charged nanoemulsions on skin hydration, elasticity and erythema-an in vivo study. Int. J. Pharm. 307, 232–238.

Yu, L., Li, C., Xu, J., Hao, J., Sun, D., 2012. Highly stable concentrated nanoemulsions by the phase inversion composition method at elevated temperature. Langmuir 28, 14547–14552.

Yukuyama, M.N., Ghisleni, D.D., Pinto, T.J., Bou-Chacra, N.A., 2016. Nanoemulsion: process selection and application in cosmetics—a review. Int. J. Cosmet. Sci. 38, 13–24.

Yukuyama, M.N., Kato, E.T., Löbenberg, R., Bou-Chacra, N.A., 2017. Challenges and future prospects of nanoemulsion as a drug delivery system. Curr. Pharm. Des. 23 (3), 495–508.

Zhong, X.K., Jin, X., Lai, F.Y., Lin, Q.S., Jiang, J.G., 2010. Chemical analysis and antioxidant activities in vitro of polysaccharide extracted from Opuntia ficus-indica Mill. cultivated in China. Carbohydr. Polym. 82, 722–727.

Zhou, H., Yue, Y., Liu, G., Li, Y., Zhang, J., Gong, Q., Yan, Z., Duan, M., 2010. Preparation and characterization of a lecithin nanoemulsion as a topical delivery system. Nanoscale Res. Lett. 5, 224–230. https://doi.org/10.1007/s11671-009-9469-5.

FURTHER READING

Dhankhar, P., 2014. Homogenization fundamentals. IOSR J. Eng. 04 (5), 1–4.

Franbourg, A., Hallegot, P., Baltenneck, F., Toutain, C., Leroy, F., 2003. Current research on ethnic hair. J. Am. Acad. Dermatol. 48, 115–119.

Kabalnov, A.S., Pertzov, A.V., Schchukin, E.D., 1987. Ostwald ripening in emulsions: I. Direct observations of Ostwald ripening in emulsions. J. Colloid Interface Sci. 118, 590–597.

Kabalnov, A.S., Makarov, K.N., Pertzov, A.V., Schchukin, E.D., 1990. Ostwald ripening in emulsions: 2. Ostwald ripening in hydrocarbon emulsions: experimental verification of equation for absolute rates. J. Colloid Interface Sci. 138, 98–104.

Kabalnov, A.S., Makarov, K.N., Schchukin, E.D., 1992. Stability of perfluoroalkyl halide emulsions. Colloids Surf. 62, 101–104.

Meleson, K., Graves, S., Mason, T.G., 2004. Formation of concentrated nanoemulsions by extreme shear. Soft Mater. 2 (2–3), 109–123.

Shinoda, K., 1968. Proc. 5th International Congress of Surface Active Substances, Barcelona vol. 3, 275.

US Food and Drug Administration, 2015. Is It a Cosmetic, a Drug, or Both? (Or Is It Soap?). http://www.fda.gov/cosmetics/guidancecomplianceregulatoryinformation/ucm074201.htm. Accessed 31 January 2017.

Walstra, P., 1996. Emulsion stability. In: Becher, P. (Ed.), Encyclopedia of Emulsion Technology. In: vol. 4. Marcel Dekker, New York, Basel, Hong-Kong, pp. 1–62 (Chapter 1).

Wolfram, L.J., 2003. Human hair: a unique physicochemical composite. J. Am. Acad. Dermatol. 48 (6), 106–114.

Chapter 15

Application of Nanoemulsions in the Synthesis of Nanoparticles

Rafael Muñoz-Espí* and Olaia Álvarez-Bermúdez*,†
*Institut de Ciència dels Materials (ICMUV), Universitat de València, Paterna, Spain, †Max Planck Institute for Polymer Research, Mainz, Germany

Chapter Outline

Nanoemulsions. https://doi.org/10.1016/B978-0-12-811838-2.00015-1

15.1 INTRODUCTION

Nanodroplets of a liquid in another immiscible liquid are able to provide confined spaces in which chemical reactions can take place, so that they can be considered as "nanoreactors." This strategy has been widely exploited for synthesis of nanoparticles of different nature, which will be the focus of this chapter.

Although the application of nanoemulsions for synthesis is most commonly exemplified by polymer nanoparticles, also inorganic and organic-inorganic materials have been prepared. Herein, we will review the different types of particles obtained from nanoemulsions and the state of the art in the corresponding synthetic strategies. Before that, however, we should pay some attention to the definition of nanoemulsion and the variety in nomenclature, which is often ambiguous and not free of controversy.

15.1.1 Definitions and Naming Problems

While the concept of *nanodroplet* is quite straightforward and it refers obviously to a droplet with nanometric size, how to name a liquid dispersion of nanodroplets can be more problematic. In the context of particle synthesis, and in particular among polymer scientists, the historical term to refer to an emulsion with droplet sizes between 50 and 500 nm has been *miniemulsion*, typically linked with the polymerization process referred to as *miniemulsion polymerization*.

Based on the stability and on the droplet size, Sudol and El-Aasser (1997) proposed a differentiation between *macro*emulsions, *mini*emulsions, and *micro*emulsions. Emulsions with large droplet size (1–100 μm) and relatively short stability (typically in the minute scale) are referred to as *macroemulsions* (or simply *emulsions*). In the middle range of this classification, *miniemulsions* are kinetically stabilized emulsions, stable from days to months and with droplet sizes between 50 to 500 nm, as mentioned earlier. Finally, *microemulsions* are thermodynamically stable systems—in contrast to macro- and miniemulsions—with sizes below 50 nm. Due to the very small droplet size, microemulsions are translucent or even transparent to the naked eye.

When a polymerization process is involved, the term "miniemulsion" has also been used with mechanistic implications. According to Landfester (2001),"in macroemulsion polymerization, the latex particle does not correspond to the primary emulsion droplet, and the size is established by kinetic processes," whereas in miniemulsion polymerization, the size of the droplets "is essentially given by dispersion process and droplet stability, but not by polymerization parameters" and "the latexes are essentially a polymerized copy of the original droplets." However, other authors have questioned whether this "one-to-one copy" is an intrinsic feature of miniemulsions. Asua (2002), for instance, affirms that "the one-to-one copy is a desirable feature of miniemulsion polymerization, but the conditions required to achieve this goal seem to be very difficult to attain in practice." For him, it is a "possibility that might be reached" rather than something intrinsic, and he recommends to "be cautious about using this concept to analyze miniemulsion data." With time, the "one-to-one copy" seems to have been widely accepted as linked to the definition of miniemulsion, and in the IUPAC Recommendations of 2011 on "Terminology of polymers and polymerization processes in dispersed systems" (Slomkowski et al., 2011), *miniemulsion polymerization* is defined as the "[p]olymerization of a mini-emulsion of monomer in which all of the polymerization occurs within preexisting monomer particles without the formation of new particles."[1]

Promoted by scientists probably oblivious to these "controversies" of the polymer community, in recent years, the younger term *nanoemulsion* has became popular to refer more or less to the same thing as miniemulsion—at least in most of the cases—not only in food sciences (McClements, 2012), but also in colloid science in general (Solans et al., 2005; Solans and Solé, 2012). We only need to think on the title of this book to realize its wide acceptance. The name *nanoemulsion* seems to be purely based on size appreciations and should be, in principle, free of any mechanistic connotation. However, the fact that microemulsion droplets are also nanometric (in spite of the prefix *micro-*, which could wrongly make us to think of micrometers) generates another ambiguity: are microemulsions also nanoemulsions or not? If the stability is the decisive argument, the answer should be clearly negative, since microemulsions are thermodynamically stable and nanoemulsions are—according to most authors—not. But it seems to be a matter of definition, and different authors may use different nomenclatures. As a side comment, but related to the naming variety, it should also be mentioned that some authors (e.g., Solans and Solé, 2012) are defenders of the hyphenation after the prefix, writing *nano-emulsion* instead of *nanoemulsion*. From a linguistic point of view, though, there is no solid argument to justify the hyphen. The prefixes *macro-*, *micro-*, *mini-*, and *nano-* are typically written without hyphen nowadays, so why should *nano-emulsion* be written with hyphen, but *nanoparticle* or *nanotechnology* without? Accordingly, we discourage the hyphenation.

From all stated earlier, it should become clear that names used for nanometric emulsions are not trivial. Miniemulsion and nanoemulsion are mostly overlapping terms, with ambiguous limits, very especially in the context of applications for nanoparticle synthesis. Therefore, in the following sections, we will use both words quite indistinguishably. Nevertheless, in the case of "miniemulsion polymerization," we will use this term, which is the historically established one.

1. We keep in this definition the hyphen in *mini-emulsion* originally used by the authors.

15.2 POLYMER NANOPARTICLES FROM NANOEMULSIONS

The synthesis of polymer nanoparticles from nanoemulsions is typically linked, as already said, to the technique known as miniemulsion polymerization. To understand the difference between this technique and conventional emulsion polymerization, we should first explain briefly how the latter works. For further details on mechanistic and kinetic aspects of all emulsion polymerization techniques, the reader is addressed to the already "classical" book edited by Lovell and El-Aasser (1997), *Emulsion Polymerization and Emulsion Polymers*.

In the most common case of conventional oil-in-water *emulsion polymerization*, droplets of a hydrophobic monomer (e.g., styrene or methyl methacrylate) are dispersed in an aqueous solution of a water-soluble surfactant (e.g., sodium dodecyl sulfate, SDS), obtaining relatively large (several micrometers) monomer droplets, which act as "monomer reservoir." The excess of surfactant (above the critical micelle concentration) originates micelles in the continuous phase. Diffusion of small amounts of monomer from the reservoir droplets to the micelles will follow. The addition of a water-soluble initiator (e.g., potassium persulfate) will lead to a fast polymerization of the monomer inside the micelles, which will turn into polymer particles. Until full consumption, monomer will further diffuse from the reservoir droplets to the growing polymer particles, reacting with initiator molecules. The result is a so-called *latex*, which is not anymore an emulsion, but a suspension of polymer particles in the continuous phase (water in this described characteristic example).

Conventional emulsion (or *macro*emulsion) polymerization is well-established and has been widely applied in industry for decades in the preparation of synthetic rubbers, certain plastics, and many polymer dispersions. Unfortunately, though, the number of polymers that can be produced in this way is rather restricted, being mostly applied to radical homopolymerization of apolar solvents. This limitation is caused by the polymerization mechanism in macroemulsion polymerization, according to which "the polymer particles are the product of a kinetically controlled growth and are built from the center to the surface, where all the monomer has to be transported by diffusion through the water phase," quoting Antonietti and Landfester (2002). Thus, this kinetic control restricts the accessible composition and leads to lack of homogeneity even for radical *co*polymerization, which are handicaps of the technique. It is in this context, and taking into account the "one-to-one copy" concept referred to Section 15.1, that *mini*emulsion polymerization appears as an advantageous alternative.

In miniemulsions, Ostwald ripening (i.e., the diffusion of material from smaller to bigger droplets due to difference in Laplace pressure) is significantly decreased—suppressed in an ideal case—by the addition of an osmotic agent, which should have low molar mass, very low solubility in the continuous phase, and very high solubility in the monomer (i.e., in the disperse phase). Hexadecane and cetyl alcohol (1-hexadecanol) are examples of very common and effective hydrophobic osmotic pressure agents used in direct (oil-in-water) miniemulsions. In inverse (water-in-oil) miniemulsions, salts or sugars are often used with the same role (Antonietti and Landfester, 2002). On the other hand, the coalescence between droplets is hindered by the use of effective surfactants. Finally, high shear forces (e.g., ultrasounds, high-pressure homogenization) are used to kinetically stabilize miniemulsion systems. Under these idealized metastable conditions, with homogeneous droplet sizes and suppressed Ostwald ripening and coalescence, each droplet acts as an independent "nanoreactor" and turns into a polymer nanoparticle after polymerization, without diffusion from monomer reservoir droplets to growing particles.

15.2.1 Polymers and Copolymers by Miniemulsion (Co) Polymerization

Most of the work about miniemulsion polymerization deals with radical polymerization and, more particularly, with free radical polymerization. However, this is not the only type of polymerization reactions that can be conducted in miniemulsions. Reversible addition-fragmentation chain-transfer (RAFT) polymerization, atom-transfer radical polymerization (ATRP), polyaddition, polycondensation, anionic polymerization, and even oxidative polymerization have been reported for both direct and inverse miniemulsions, depending on the monomer and the solvents used. The possibilities of miniemulsion polymerization for the synthesis of an amazingly large list of polymers and copolymers have been extensively reviewed over the last two decades (Landfester, 2001, 2003, 2009; Antonietti and Landfester, 2002; Asua, 2002, 2004; Crespy and Landfester, 2010; Landfester et al., 2010). Table 15.1 contains a selected list of polymers and copolymers synthesized by miniemulsion polymerization. The list does not pretend to be exhaustive, but to give an idea of the versatility of the technique. Some of the copolymers in this list were prepared by inclusion of functional comonomers, which yields functionalized nanoparticles, which are briefly discussed later.

Free radical miniemulsion polymerization works, in general, extremely well and is a very effective technique to prepare polymer nanoparticles with defined size. The polydispersity of the polymer chains is, as expected for free radical polymerization, rather uncontrolled, but this is less relevant for many applications in which the focus is on the particles themselves and not necessarily on the length of the constituting polymer chains. A "standard" representative case, which we describe here for the sake of illustration, is the miniemulsion polymerization of styrene in a direct system (Tiarks et al., 2001b). The organic disperse phase is comprised of the monomer and an osmotic pressure agent (e.g., hexadecane, as indicated earlier). Small quantities of functionalizing comonomers can be added, if required for the concrete application. The continuous phase is an aqueous solution of a surfactant with a high hydrophilic-lypophilic balance (e.g., SDS). Most commonly, a very hydrophobic initiator

TABLE 15.1 Polymerization Reaction Conducted in Miniemulsion With Corresponding Selected Examples

Polymerization Reaction	Miniemulsion Type	(Co)Polymer	Reference(s)
Free radical polymerization	Direct	Polystyrene	Alduncin et al. (1994)
	Direct	PMMA	Mouran et al. (1996)
	Direct	Poly(vinyl acetate)	Wang and Schork (1994)
	Direct	Polyacrylonitrile	Landfester and Antonietti (2000)
	Direct	PVC	Dewald et al. (1984)
	Direct	Poly(S/MMA)	Rodríguez et al. (1989)
	Direct	Poly(S/AA)	Tiarks et al. (2001b) and Holzapfel et al. (2005)
	Direct	Poly(S/AEMA)	Holzapfel et al. (2005)
	Inverse	Poly(acrylic acid)	Landfester et al. (2000)
	Inverse	Poly(HEMA-*co*-MAA)	Klinger and Landfester (2011)
	Inverse	Polyacrylamide	Landfester et al. (2000)
ATRP	Direct	Poly(*n*-butyl methacrylate)	Matyjaszewski et al. (2000) and Qiu et al. (2000)
	Inverse	PEO-*block*-poly (OEOMA)	Oh et al. (2006)
RAFT	Direct	Polystyrene	de Brouwer et al. (2000)
	Direct	PMMA	de Brouwer et al. (2000)
	Inverse	Polyacrylamide	Qi et al. (2007)
Metal-catalyzed	Direct	Polyethylene	Soula et al. (2001)

TABLE 15.1 Polymerization Reaction Conducted in Miniemulsion With Corresponding Selected Examples—cont'd

Polymerization Reaction	Miniemulsion Type	(Co)Polymer	Reference(s)
Polyaddition	Direct	Polyurethane	Tiarks et al. (2001a)
	Inverse	Polyurethane	Paiphansiri et al. (2009)
Polycondensation	Direct	Polyesters	Barrére and Landfester (2003)
	Inverse	Polyurea	Crespy et al. (2007)
Anionic polymerization	Direct	PTFPMS	Barrér et al. (2001)
	Inverse	Poly(ε-polycaprolactam)	Crespy and Landfester (2005)
Oxidative polymerization	Direct	Polyaniline	Marie et al. (2003) and Lv et al. (2013)
	Direct	Polypyrrole	Lv et al. (2013)
	Direct	PEDOT	Sanchis et al. (2016)
	Inverse	Polyaniline	Marie et al. (2003)

AA, acrylic acid; *AEMA*, 2-aminoethyl methacrylate; *HEMA*, hydroxyethyl methacrylate; *MAA*, methacrylic acid; *MMA*, methyl methacrylate; *OEOMA*, oligo(ethylene glycol) monomethyl ether methacrylate; *PEDOT*, poly(3,4-ethylenedioxythiophene); *PEO*, poly(ethylene oxide); *PMMA*, poly(methyl methacrylate); *PTFPMS*, 1,3,5-tris(trifluoropropylmethyl)cyclotrisiloxane; *PVC*, poly(vinyl chloride); *S*, styrene.

(e.g., azobisisobutyronitrile or 2,2′-azodi(2-methylbutyronitrile)), highly soluble in the monomer, is used to ensure the initiation inside the droplets. However, water-soluble initiators can also be applied (Sudol and El-Aasser, 1997). The choice of the initiator can be a key factor depending on the monomers and comonomers introduced in the system. Fig. 15.1 shows the classical scheme of a direct miniemulsion polymerization.

To illustrate the versatility of the strategy and the variety of monomers that can be polymerized by free radical miniemulsion polymerization, Fig. 15.2 presents scanning electron microscopy (SEM) images of different polyacrylate polymers with amino acids as pending groups. Since amino acids have an intrinsic chirality, chiral nanoparticles can be produced in this manner (Preiss et al., 2015).

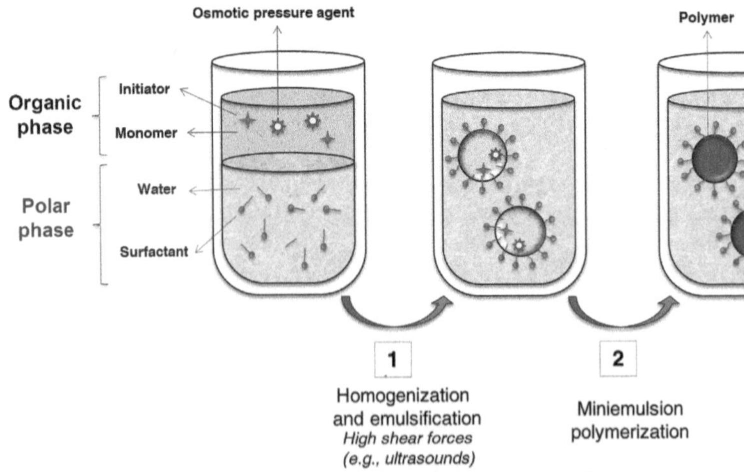

FIG. 15.1 Schematic representation of a typical direct miniemulsion polymerization.

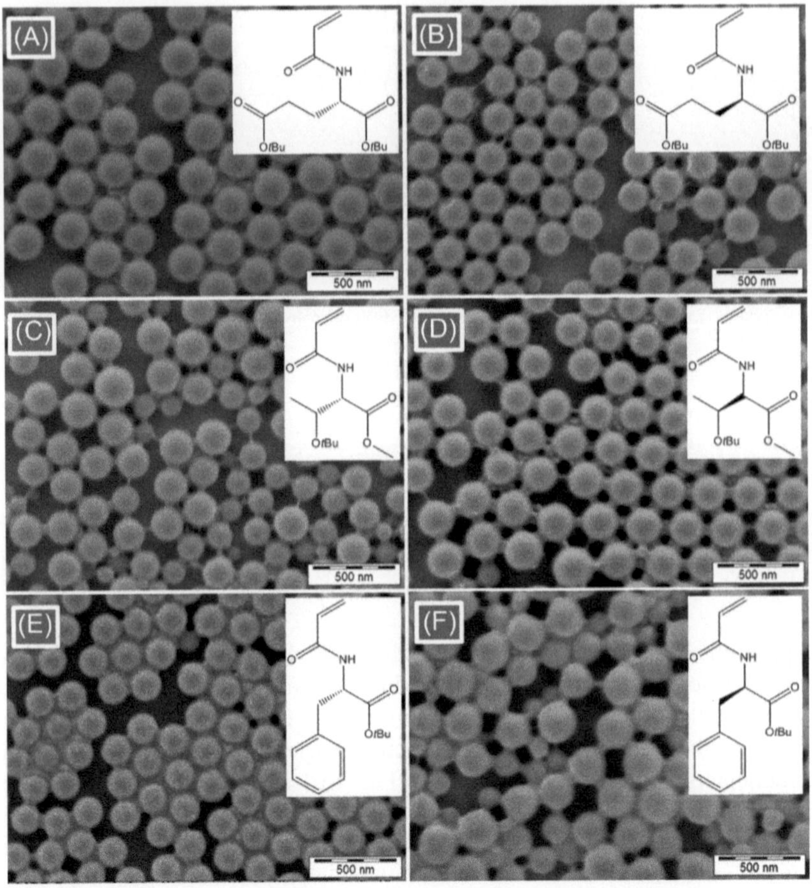

FIG. 15.2 SEM images of nanoparticles prepared by miniemulsion polymerization of chiral monomers: (A) *N*-acryl-di-*tert*-butyl-L-glutamic acid, (B) *N*-acryl-di-*tert*-butyl-D-glutamic acid, (C) *N*-acryl-*O*-*tert*-butyl methyl ester-L-threonine, (D) *N*-acryl-*O*-*tert*-butyl methyl ester-D-threonine, (E) *N*-acryl-*tert*-butyl ester-L-phenylalanine, and (F) *N*-acryl-*tert*-butyl ester-D-phenylalanine. The chemical structure of the monomers is represented in the insets. *(Adapted with permission from Preiss, L.C., Werber, L., Fischer, V., Hanif, S., Landfester, K., Mastai, Y., Muñoz-Espí, R., 2015. Amino-acid-based chiral nanoparticles for enantioselective crystallization. Adv. Mater. 27, 2728–2732. Copyright 2015 by John Wiley & Sons Ltd.)*

When the monomers to be polymerized are hydrophilic, inverse miniemulsions are used for the polymerization. The inverse miniemulsion polymerization of water-soluble monomers was reviewed a few years ago by Capek (2010), and the reader is addressed to this work for details. On the other hand, as an interesting feature of miniemulsions, if one of the reacting monomers is hydrophilic and the other is hydrophobic, the reaction will take place at the interface (i.e., interfacial polymerization), which can lead under certain conditions to the formation of capsules. This strategy is especially used in polyaddition and polycondensation reactions (see, for instance, Crespy et al., 2007).

15.2.2 Surface-Functionalized Nanoparticles

The addition in the miniemulsion formulation of a functional comonomer with a certain affinity to the continuous phase allows the generation of copolymers with the functional comonomer preferably placed at the surface, yielding surface-functionalized polymer nanoparticles. Functionalities such as alcohol, carboxylate, amino, phosphate, phosphonate, sulfonate, or ethylene oxide can be easily introduced on the surface of particles of a nonfunctional supporting polymer (e.g., polystyrene, poly(methyl methacrylate)). This approach has been reviewed in several articles (Crespy and Landfester, 2010; Froimowicz et al., 2013).

If the functional monomer used possesses amphiphilicity and is surface active, it can simultaneously act as a surfactant. Such a *surf*ace-active mono*mer* is named with the acronym *surfmer*. If the stabilization provided by this polymerizable surfactant is sufficient, no further surfactant is needed, which is a great advantage for many applications, because only the desired copolymer is present and no removal of surfactant is required. The stability of the system may be also positively affected under conditions at which conventional surfactants could de-attach, since the surfmer is covalently bound to the particle after copolymerization. Surfmers with different functional groups have been synthesized and used in miniemulsion polymerization:

- Sulfate (Matahwa et al., 2006)
- Sulfosuccinate (Boisson et al., 2001)
- Phosphate (Crespy et al., 2006; Fischer et al., 2013)
- Phosphonate (Sauer et al., 2012a; Fischer et al., 2013)
- Carboxylate (Guyot et al., 2003; Preiss et al., 2016)
- Quaternary ammonium (Matahwa et al., 2006; Taniguchi et al., 2008)
- BODIPY (fluorescent dye) (Sauer et al., 2012b)

Fig. 15.3 presents SEM images of surface-functionalized nanoparticles prepared by miniemulsion polymerization of styrene and surfmers containing phosphonate, phosphonate, and carboxylate groups (Fischer et al., 2013; Preiss et al., 2016). The chemical structure of the used surfmers is also depicted.

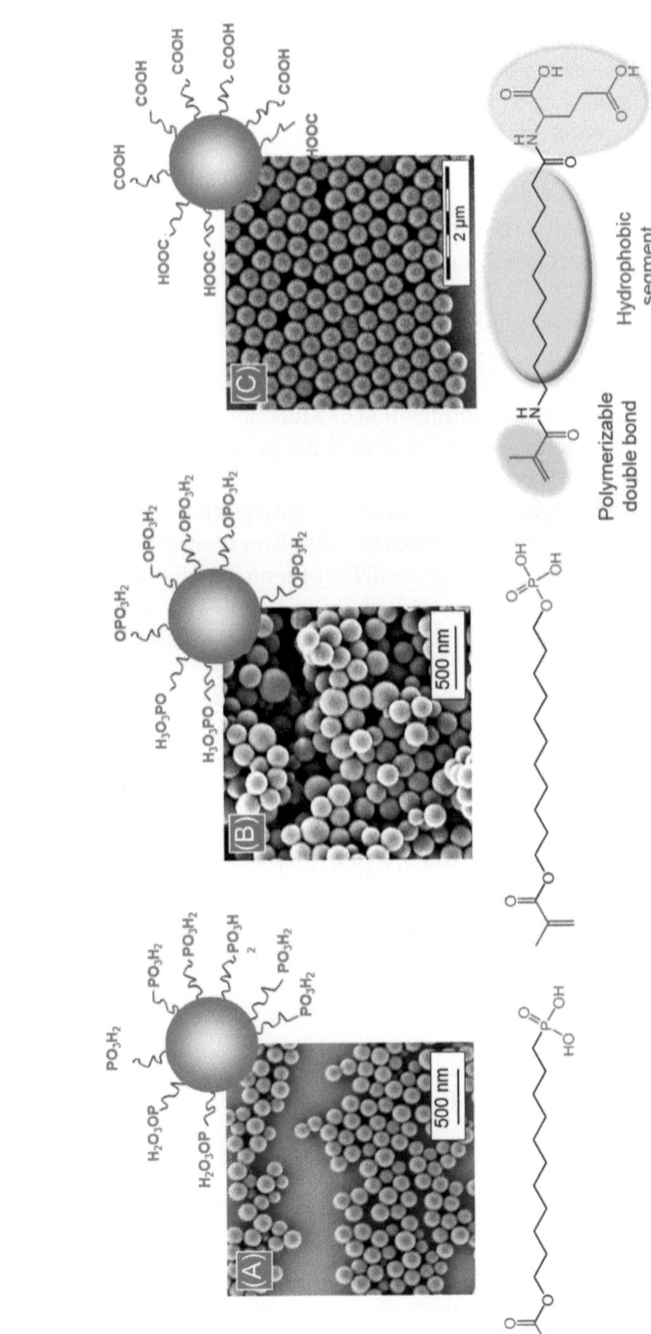

FIG. 15.3 Surface-functionalized polystyrene nanoparticles prepared by miniemulsion polymerization in the presence of surfmers (Fischer et al., 2013; Preiss et al., 2016).

15.2.3 Polymer Nanoparticles by Emulsion-Solvent Evaporation and by Ouzo Effect

Polymer nanoparticles can also be prepared in miniemulsions without miniemulsion polymerization. The so-called (mini)emulsion-solvent evaporation process involves the dissolution of a polymer in an appropriate solvent. This solution acts a disperse phase and is emulsified with an immiscible continuous phase of a second solvent. After evaporation of the solvent in which the polymer is dissolved, the polymer precipitates in the droplet and a nanoparticle is formed (Crespy and Landfester, 2007; Staff et al., 2013b).

The emulsion-solvent evaporation technique has been typically used in pharmaceutical applications for microsized particles and mainly for encapsulation of drugs in biodegradable polymers, but it has been somehow "rediscovered" in the last years for the preparation of nanoparticles in nanoemulsion systems. The method has been nicely reviewed by Staff et al. (2013a).

A kind of related process, but involving self-emulsification, is the so-called "Ouzo effect." It needs to be said that the spontaneous emulsions formed by Ouzo effect should be probably called microemulsions, which strictly speaking are not nanoemulsions, even if they also lead to nanoparticles. However, the term nanoemulsion appears sometimes in literature when referring to emulsions resulting from the Ouzo effect, so we are briefly mentioning the process here. The Ouzo effect is named after the famous Greek anise-flavored liquor. Ouzo—and similar anise liquors as raki, arak, pastis, or cazalla—contains anethol and some other oils dissolved in ethanol (and some water). At room temperature, the oils are in solution and the liquid appears as transparent. The addition of water to the system, a nonsolvent for the oils but miscible in the ethanol, leads to a spontaneous emulsification. The ethanol and the water mix and the oil phase separate in the form of nanodroplets, so that the system turns milky. The same strategy has been used for preparation of polymer particles (Ganachaud and Katz, 2005; Aschenbrenner et al., 2013). First the polymer is dissolved in a good organic solvent that is miscible in water (which is a bad solvent for the polymer). Afterward, the solution is added to an aqueous phase, so that there is a fast solvent diffusion and a spontaneous phase separation and emulsification. Finally, the organic solvent can be evaporated from the system. Spontaneous emulsification and the Ouzo effect have been very recently reviewed by Solans et al. (2016).

15.2.4 Polymer Nanocapsules From Nanoemulsions

As already mentioned, nanoemulsion systems are also suited for generating polymer nanocapsules, that is, a polymer shell occluding a liquid core. There are two possibilities to generate such structures: (i) phase separation of the components of the system and (ii) interfacial polymerization.

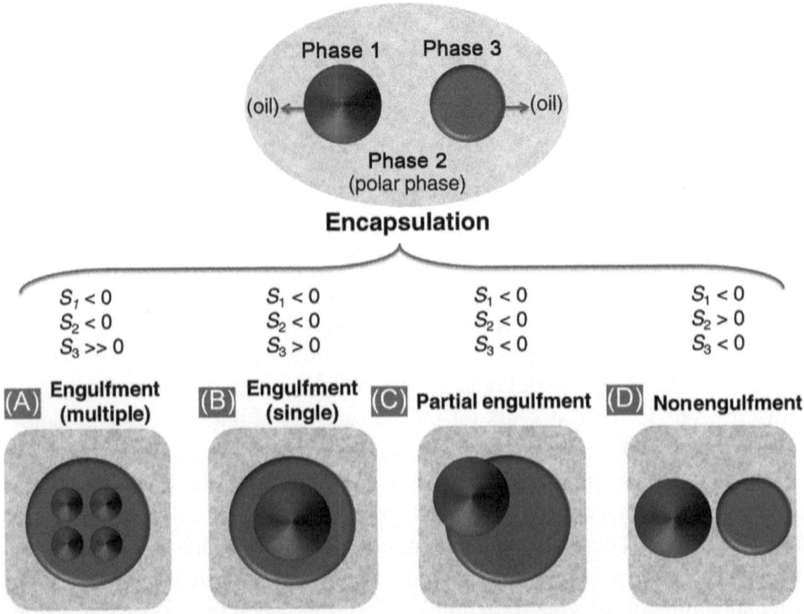

FIG. 15.4 Morphologies resulting from a three-component system comprised of two oil phases (1 and 3) and a polar continuous phase (2), depending on the spreading coefficients, S_i. *(Based on Torza, S., Mason, S., 1970. Three-phase interactions in shear and electrical fields. J. Colloid Interface Sci. 33 (1), 67–83. https://doi.org/10.1016/0021-9797(70)90073-1. Available from http://www.sciencedirect.com/science/article/pii/0021979770900731, and Trongsatitkul, T., Budhlall, B.M., 2011. Multicore–Shell PNIPAm-co-PEGMa microcapsules for cell encapsulation. Langmuir 27 (22), 13468–13480. https://doi.org/10.1021/la203030j.)*

For explaining the phase separation case, the model proposed by Torza and Mason (1970) can be used. In this model, shown in Fig. 15.4, the morphology of a system comprised of three immiscible liquid phases—with two of them combined in the disperse phase—is described by the spreading coefficient, S_i, which depends on the interfacial tension between the phases. For a phase i, the spreading coefficient S_i is given by

$$S_i = \gamma_{jk} - (\gamma_{ij} + \gamma_{ik}) \tag{15.1}$$

where γ is the interfacial tension and i, j, and k denote the phases. If the phase to be encapsulated is labeled as 1, the continuous polar phase as 2, and the second oil as 3, depending on the values of the spreading coefficients corresponding to the three phases, different situations can occur:

(A) Engulfment of phase 1 in phase 3 with formation of multiple droplets: $S_1 < 0$, $S_2 < 0$, $S_3 \gg 0$
(B) Engulfment of phase 1 in phase 3 with a core-shell morphology: $S_1 < 0$, $S_2 < 0$, $S_3 > 0$
(C) Partial engulfment: $S_1 < 0$, $S_2 < 0$, $S_3 < 0$
(D) Nonengulfment: $S_1 < 0$, $S_2 > 0$, $S_3 < 0$

If the second case occurs, the formation of a capsule is possible. This model is useful to explain phase separation in three different cases of formation of polymer capsules in nanoemulsions:

(i) Capsules obtained by direct miniemulsion polymerization, increasing the amount of hydrophobic agent (e.g., hexadecane) with respect to the monomer (Wohnhaas et al., 2013; Katta et al., 2015).
(ii) Capsules obtained by emulsion-solvent evaporation (e.g., Crespy and Landfester, 2007).
(iii) Capsules obtained by Ouzo effect (Ganachaud and Katz, 2005).

The other possibility to produce capsules in nanoemulsions is the use of interfacial polymerization, commonly in inverse systems. One of the most typical examples is the formation of polyurethane/polyurea by polyaddition or polycondensation of diisocyanates (Crespy et al., 2007; Rosenbauer et al., 2009). In addition to low-molecular diamines and diols, which are common reagents in such reactions, also biopolymers like cross-linked hydroxyethyl starch can be placed in the aqueous disperse phase as a water-soluble reagent (Freichels et al., 2013).

15.3 INORGANIC NANOPARTICLES FROM NANOEMULSIONS

15.3.1 Nanodroplets as Templates for Inorganic Synthesis

In the same way that nanodroplets containing organic monomers can act as nanoreactors for polymerization, nanodroplets containing precursors of inorganic materials can also be nanoreactors for precipitation/crystallization processes.

The specific topic of inorganic synthesis in nanoemulsions was reviewed in certain detail a few years ago (Muñoz-Espí et al., 2012). In the following years, parts of other works have also addressed it (Muñoz-Espí et al., 2013; Diodati et al., 2015). The synthesis of inorganic materials in nanoemulsions continues to develop very rapidly, and significant contributions have been made in very recent time. While avoiding to repeat unnecessarily what it has already been said elsewhere, in this section we will briefly discuss the different synthetic strategies for the preparation of inorganic nanoparticles from nanoemulsions, and we will update the state of the art in the field, with emphasis in the most recent achievements.

Pileni's research group was one of the pioneers in the application and investigation of droplets and colloidal soft templates (such as micelles and surfactant structures) to the synthesis and morphological control of inorganic nanoparticles, mostly in macro- and microemulsions (Pileni, 1993, 1997, 2003).

The most common pathway to synthesize inorganic nanoparticles from nanoemulsions, schematically represented in Fig. 15.5A, is the combination of two emulsions, the first containing the main precursor (typically a metal precursor) and the second containing a reducing or precipitating agent. After

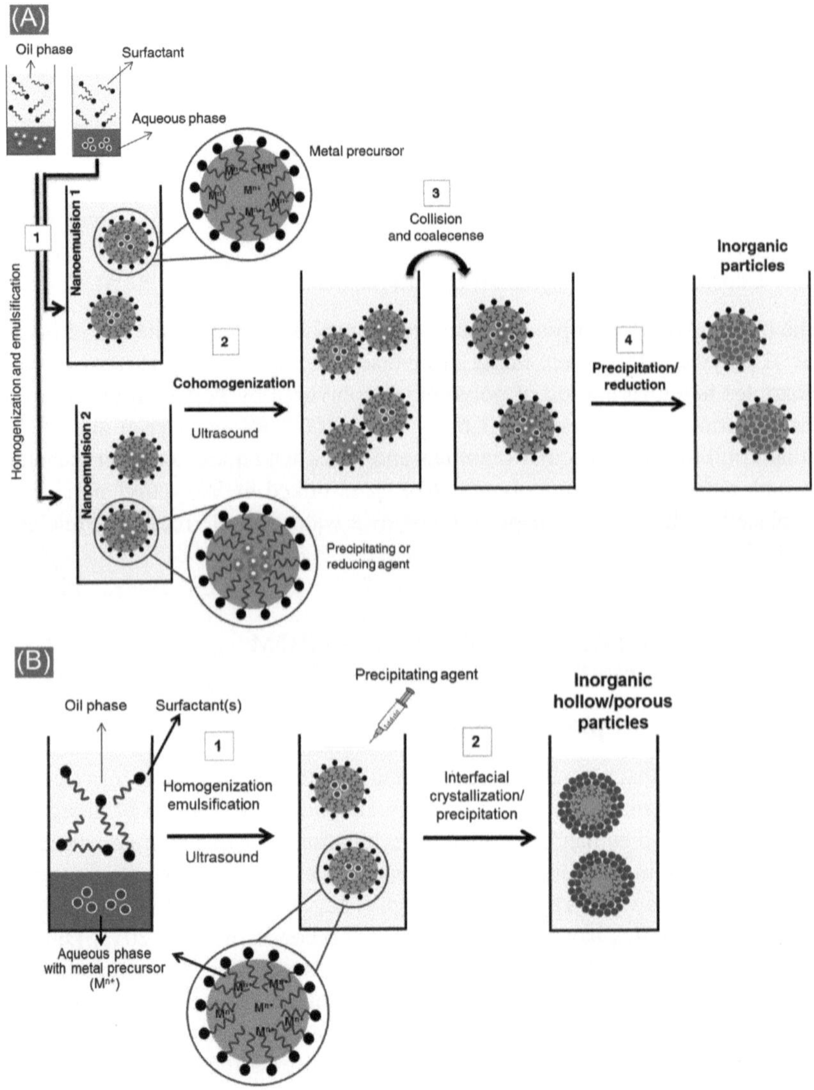

FIG. 15.5 Schematic representation of two methods for the formation of inorganic nanoparticles from nanoemulsions: (A) combination of two nanoemulsions containing the metal precursor and the precipitating/reducing agent; (B) addition of a precipitating/reducing agent to a nanoemulsion.

mixing both emulsions, the collision and coalescence of the droplets with both reactants drives the precipitation reaction (e.g., the reduction of a metal by addition of a reducing agent or the formation of a metal oxide or hydroxide by addition of a base). Another possibility, shown in Fig. 15.5B, is the use of one single nanoemulsion, adding the precipitating agent to the continuous phase. In this way, the precipitation may occur at the liquid-liquid interface and result in hollow particles, as described below. The formation of solid or hollow particles is determined by the kinetic of the precipitation and by the solubility in the disperse phase of the precipitating agent used.

Although in this chapter we mainly focus on the systems referred to as "miniemulsions," in the case of synthesis of inorganic nanoparticles, microemulsion is by far the most commonly used system involving nanosized droplets. It is more than questionable, as previously discussed, to include microemulsions in the nanoemulsion definition, but in the daily laboratory practice the limits are sometimes less well defined. Therefore, we will dedicate a few lines here to the use of microemulsions in nanoparticle synthesis. For further details, earlier reviews on the topic (Holmberg, 2004; Eastoe et al., 2006; Boutonnet et al., 2008) can be looked up.

A large variety of metal oxide, metal hydroxide, metal chalcogenide, and metal nanoparticles have been reported in microemulsion. Although much of the work is conducted in inverse systems (Buchold and Feldmann, 2007; Gyger et al., 2010; Leidinger et al., 2010), which is the obvious result of the water-soluble character of typical metal salts used as precursors, also oil-soluble precursors have been used in direct microemulsions (Ge et al., 2006; Sánchez-Domínguez et al., 2009, 2010, 2011; Okoli et al., 2012). Other inorganic systems, such as zirconium phosphates (Bellezza et al., 2006) and hydrotalcites (Bellezza et al., 2009), have also been reported.

For miniemulsion systems, their versatility in the inorganic synthesis was already demonstrated in the early 2000s by Willert et al. (2001), who reported the crystallization of calcium carbonate and iron oxide, as well as the recrystallization of molten salts (zirconium oxychloride) and metals (gallium and alloys). In the following years, the same group also reported the preparation of complex lanthanide phosphors (Taden et al., 2004), TiO_2 (Rossmanith et al., 2008), CeO_2 (Nabih et al., 2011), and $Ti_{1-x}Zr_xO_2$ (Schiller et al., 2010). Especially active in recent years have been Gross and coworkers, who have reported the synthesis in nanoemulsions of ZnO (Dolcet et al., 2012), Au-TiO_2 (Heutz et al., 2013), lanthanide-doped $Ca(OH)_2$ and $Mg(OH)_2$ (Butturini et al., 2014), lanthanide-doped CaF_2 (Dolcet et al., 2015a), and ZnS (Dolcet et al., 2015b).

Very interesting results have been lately obtained by our research team through the combination of the nanoemulsion technique with the hydrothermal route, that is, by submitting nanoemulsions containing the inorganic precursors to nonstandard conditions of temperature and pressure. With this method, we have been able to obtain different spinel ferrites at much lower temperatures (i.e., 80°C without any postsynthesis thermal treatment) than usually reported in literature for these systems, which is the result of the confinement exerted by the nanodroplets (Antonello et al., 2017).

15.3.2 Interfacial Precipitation and Crystallization in Nanoemulsions: Formation of Capsules

By applying a very similar pathway to the one used to produce polymer nanocapsules by interfacial miniemulsion polymerization, inorganic capsules can be obtained by sol-gel processes taking place at the interface. Wu's group (Peng et al., 2008) reported the preparation of silica capsules from oil-in-water nanoemulsions containing the precursor tetraethyl orthosilicate (TEOS) in octane as a disperse phase and an aqueous solution of surfactant as a continuous phase. TEOS is hydrolized at the liquid-liquid interface, which leads to the formation of hollow silica spheres that trap inside the hydrophobic core. The inclusion of functionalized silanes (e.g., 3-aminopropyltriethoxysilane or 3-mercaptopropyltrimethoxysilane) during the synthesis allows for the functionalization of the capsule surface (Fickert et al., 2012). Silica capsules with a hydrophilic liquid core can also be analogously obtained from inverse nanoemulsions (Hood et al., 2015). In this case, the disperse phase contains water, and the silica precursor (e.g., TEOS) is added to the continuous phase of an water-in-oil nanoemulsion. The hydrophobicity of the surface of the resulting capsules can be controlled by introducing silanes of varying polarity during the synthesis (Hood et al., 2017).

Sol-gel processes of metals of group 4 were also conducted at the droplet interface in inverse nanoemulsion systems (Hajir et al., 2012). Hydrous zirconia and hafnia ($MO(OH)_2 \cdot nH_2O$, M = Zr, Hf) could be obtained in this way from metal oxychloride precursors ($ZrOCl_2 \cdot 8\ H_2O$ and $HfOCl_2 \cdot 8\ H_2O$). Triethylamine, an organic base soluble in the organic continuous phase, but also partially soluble in the aqueous disperse phase, was found to be a suitable precipitating agent to drive the sol-gel process at the liquid-liquid interface. According to the proposed mechanism, depicted in Fig. 15.6 and extensible to similar processes, triethylamine (Et_3N) generates hydroxide ions at the interface and within the droplets:

$$Et_3N + H_2O \rightleftharpoons Et_3NH^+OH^- \tag{15.2}$$

The oxychloride precursor, which most likely form oligomeric clusters, is hydrolyzed to the metal hydroxide, with triethylammonium chloride as a byproduct:

$$2\,Et_3NH^+OH^- + MOCl_2 \rightleftharpoons MO(OH)_2 + 2\,Et_3NH^+Cl^- \tag{15.3}$$

The presence of OH^- ions catalyzes the condensation reaction of the metal hydroxo species at the interface.

A very similar technique, also with triethylamine as a precipitating agent, was reported for the preparation of $YCrO_3$ hollow nanoparticles (Singh et al., 2013), which were found to display an anomalous magnetic behavior when compared with similar materials prepared in the bulk solution and without droplet confinement.

The materials obtained under these synthetic conditions for typical sol-gel systems (silica, zirconia, hafnia, titania, etc.) are amorphous and crystallize only upon calcination. However, for other transition metal oxide and lanthanide oxide systems, such as CuO, γ-Fe_2O_3, and CeO_2, the crystallization can occur

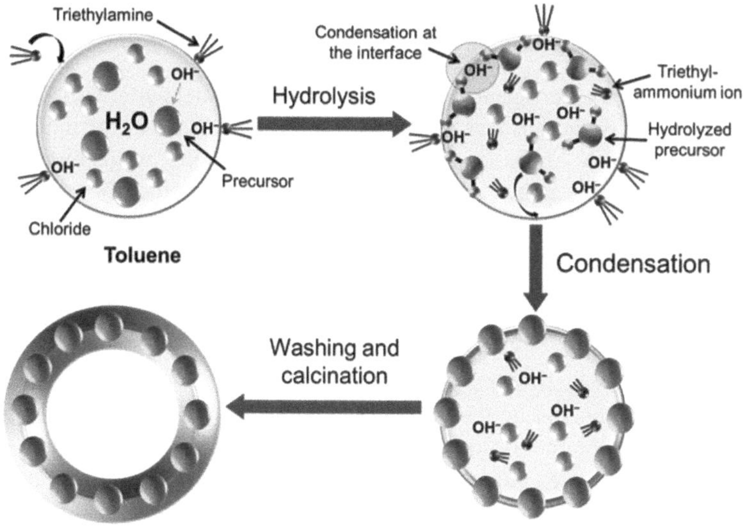

FIG. 15.6 Scheme of the formation of zirconia and hafnia capsules by interfacial precipitation in the presence of triethylamine. *(Reproduced from Hajir, M., Dolcet, P., Fischer, V., Holzinger, J., Landfester, K., Muñoz-Espí, R., 2012. Sol–gel processes at the droplet interface: hydrous zirconia and hafnia nanocapsules by interfacial inorganic polycondensation. J. Mater. Chem. 22 (12), 5622–5628. https://doi.org/10.1039/c2jm15353j, with permission from the Royal Society of Chemistry.)*

already at room temperature or at relatively mild temperatures below 100°C, without being necessary any thermal treatment after synthesis (Singh et al., 2015; Varol et al., 2016). By employing the required stoichiometry of precursors in the nanoemulsions, ternary oxide systems can also be prepared, as it has been shown for $Ce_{1-x}Cu_xO_2$ (Singh et al., 2017). Examples of hollow nanoparticles prepared in inverse nanoemulsion systems by interfacial precipitation are shown in Fig. 15.7.

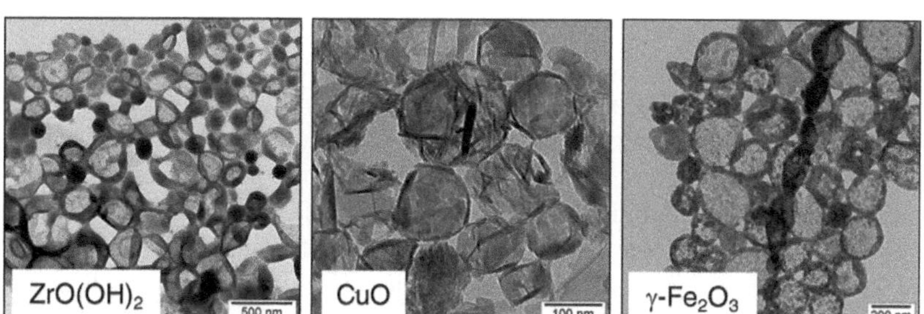

FIG. 15.7 TEM images of amorphous hydrated zirconia (Hajir et al., 2012), CuO (Singh et al., 2015), and γ-Fe_2O_3 (Varol et al., 2016) hollow nanoparticles prepared by interfacial precipitation in inverse nanoemulsion systems.

15.4 POLYMER/INORGANIC HYBRID NANOPARTICLES FROM NANOEMULSIONS

The incorporation of inorganic components within polymer particles has gained a great interest in materials research during the last two decades. The integration in the same material of the easy processability, the stability, or the biocompatibility of polymer materials and the functional features specifically provided by inorganic components (e.g., optical, magnetic, electrical, catalytic functions) offers the possibility of developing new complex and multifunctional materials with applicability in a wide range of sectors.

Strictly speaking, the inorganic nanoparticles described in the previous section are never truly "inorganic"—unless calcined—but organic-inorganic hybrid particles, which is the result of the presence of surfactants and other organic components during the synthesis in the nanoemulsions. These organic components are commonly incorporated into the inorganic material and may not be easy to remove. The percentage of organic material is, in general, not negligible. In this section, however, we will not refer to such hybrid materials, but to systems in which an organic polymeric material is one of the main components and not just residual. In a previous work, we reviewed in detail the preparation methods of hybrid nanoparticles (Hood et al., 2014). Here, we aim to focus exclusively on those strategies involving nanoemulsions.

15.4.1 Encapsulation or Integration of Inorganic Components Within Polymer Particles Prepared in Nanoemulsions

The combination at the nanoscale of components with different chemical nature (organic and inorganic) can result in a synergy that goes beyond the simple addition of the individual properties of the forming elements.

Polymer/inorganic hybrid nanoparticles can be designed to tune and combine specific functionalities with controlled accessibility according to their final application. The relative disposition of the inorganic components within the polymer matrix has a key role in the properties of the resulting material. Despite the great advances in this area, a general understanding of the precise morphological control of polymer/inorganic nanoparticles as a function of the components and the synthetic parameters remains an open challenge. Among other deciding parameters, the different polarity of the polymer and the inorganic components may hinder their integration within a hybrid system. In this framework, nanoemulsions are a versatile platform of synthesis that overcomes the intrinsic morphological limitations of other colloidal methods, such as dispersion, suspension, or seeded emulsion polymerization. While other techniques are restricted to the encapsulation of a single particle surrounded by a polymer shell or to the disposition of the inorganic components on the polymer surface, the use of nanoemulsions allows for a selective migration and location of the inorganic components within the polymer matrix. Thus, not only the

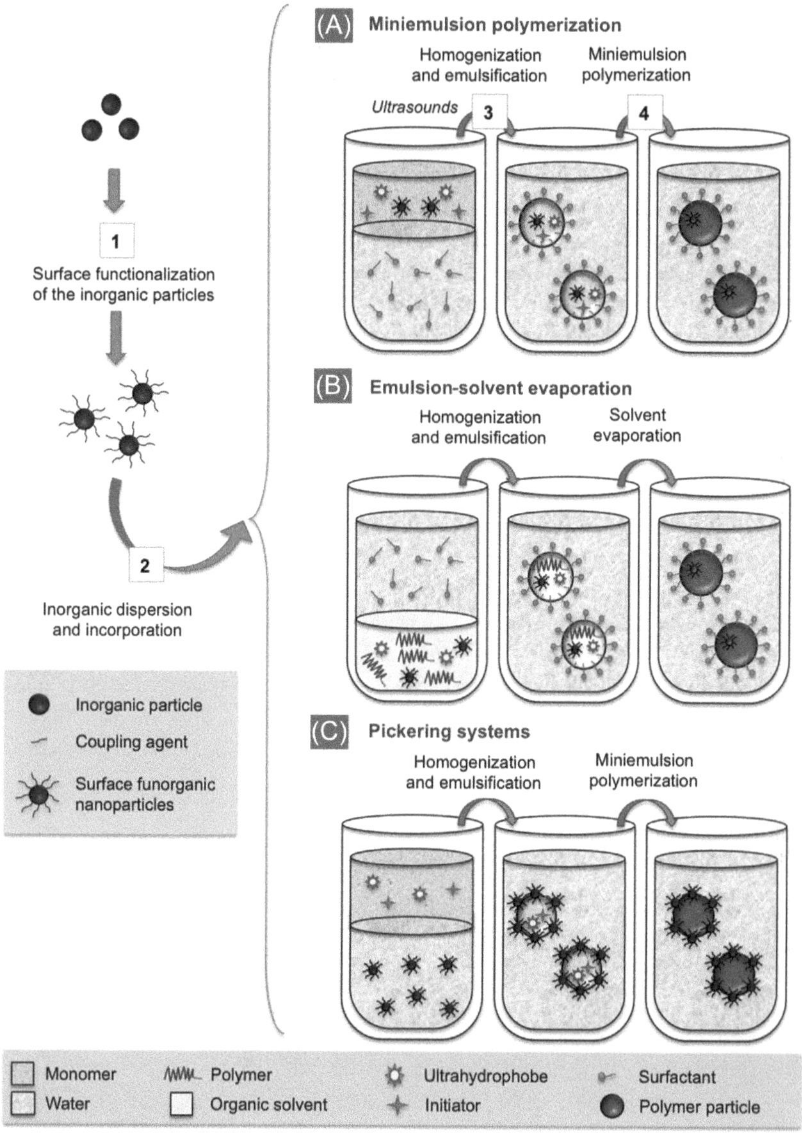

FIG. 15.8 Different strategies described in this section for the encapsulation or integration of inorganic components in polymer particles prepared in nanoemulsions: (A) miniemulsion polymerization in the presence of functionalized inorganic particles; (B) emulsion-solvent evaporation; and (C) Pickering nanoemulsions.

homogeneous encapsulation of inorganic particles, but also Janus-like, core-shell, and raspberry-like morphologies can be reached.

Fig. 15.8 introduces three main synthetic strategies commonly used in nanoemulsions for the integration and/or encapsulation of preformed inorganic nanoparticles within a polymer matrix. The three strategies follow a common principle: the preformed inorganic nanoparticles (mostly hydrophilic as prepared) are hydrophobized to a certain extent to compatibilize their surface with the monomer/polymer or interface in/at which they are expected to be incorporated. Typically, surface functionalization of the inorganic components is achieved by using appropriate functionalizing agents (e.g., carboxylates, thiols, or silane components).

Fig. 15.8A shows a typical case of direct *miniemulsion polymerization* in which the inorganic particles, previously synthesized and hydrophobized, are further dispersed in the monomer to form the organic disperse phase. The emulsification is achieved by using high shear forces (e.g., ultrasounds, high-pressure homogenization, microfluidizer), followed by the formation of the polymer matrix by free radical polymerization. *Emulsion-solvent evaporation* (Fig. 15.8B), whose principles have been described earlier, presents an alternative approach. During the evaporation of the solvent and the precipitation of the polymer, the inorganic components become incorporated within the resulting matrix. Finally, in *Pickering nanoemulsions*, schematically represented in Fig. 15.8C, the role of the surfactant is fulfilled by inorganic nanoparticles, which are first dispersed in the continuous phase, being further assembled at the droplet interface during the homogenization/emulsification step. After polymerization, the particles remain at the polymer surface creating an inorganic outer shell. In the following, we will review a few representative examples of each strategy, and we will discuss the key role of the specific functionalization of the inorganic component.

15.4.1.1 Miniemulsion Polymerization

Miniemulsion polymerization is one of the most common techniques for the synthesis of hybrid polymer/inorganic nanoparticles in nanoemulsions. The versatility of the technique has been demonstrated with numerous inorganic and polymer systems. Styrene (Bourgeat-Lami et al., 2012), methyl methacrylate (Schoth et al., 2014), and L-lactide (Urban et al., 2009), for instance, have been some of the reported monomers, often copolymerized with comonomers such as 4-vinyl pyridine (Schoth et al., 2014; Haase et al., 2012), divinylbenzene (DVB) (van Berkel and Hawker, 2010), butyl acrylate (Bourgeat-Lami et al., 2012; Zhou et al., 2006), or butyl methacrylate (Schoth et al., 2016a). Regarding the inorganic component, silica (Haase et al., 2012; Schoth et al., 2014) has been extensively studied as a model system, but other functional inorganic nanoparticles have also been reported, including magnetite (Schoth et al., 2016b; Mori and Kawaguchi, 2007; Feuser et al., 2015; Gong et al., 2009), nickel (Staudt et al., 2013), ceria (Aguirre et al., 2015), titania (Lv et al., 2014; Yang et al., 2015), zinc oxide (Hecht et al., 2012b), and gold or silver (van Berkel and Hawker, 2010).

Functionalized nanoparticles are not the only type of inorganic components that can be integrated in polymers produced in miniemulsion polymerization, but also molecular well-defined entities, such as metal oxoclusters, can be copolymerized with organic monomers (Pablico et al., 2011; Benedetti et al., 2016).

Following the same principle, the miniemulsion polymerization strategy can also be carried out in inverse systems, in which hydrophilic inorganic salts (e.g., zinc nitrate) are encapsulated in a polymer matrix resulting from the polymerization of water-soluble monomers, such as acrylamide (Kobitskaya et al., 2010).

15.4.1.2 Emulsion-Solvent Evaporation

In the emulsion-solvent evaporation technique, the inorganic incorporation results from a self-assembly process based on physicochemical interactions driven by the minimization of the interfacial energy toward the thermodynamic equilibrium. In contrast to the miniemulsion polymerization technique, no chemical reactions take place within the nanodroplets, so that no covalent incorporation of the inorganic components can occur, and the characteristics of the polymer (e.g., molecular weight, composition) and/or the inorganic elements remain essentially unchanged. This alternative synthetic route allows for the integration of the same polymer/inorganic components as miniemulsion polymerization, but there may be significant differences in the resulting hybrid morphology (Schoth et al., 2014).

Fluorescent and magnetic hybrid nanoparticles were prepared by encapsulation of iron oxide nanoparticles and *N*-(2,6-diisopropylphenyl)-perylene-3,4-dicarbonacid-imide within a poly(L-lactide) matrix via an emulsion-solvent evaporation technique in miniemulsion (Urban et al., 2009). In the same fashion polystyrene/magnetite Janus nanoparticles were prepared by combination of miniemulsion polymerization and emulsion-solvent evaporation (Kaewsaneha et al., 2013). An elegant case of Janus nanoparticles with titanium dioxide and polyvinyl formal was reported by Lv et al. (2014). In this case, the complexity of the system is further increased by the introduction of hexadecane in the formulation of the organic phase. This latest component is nonsoluble in the polymer and remains as a liquid independent compartment after the precipitation of the polymer. The substitution of hexadecane by dodecyl methacrylate and DVB allowed for conduction of a miniemulsion polymerization process within the monomer compartment, which resulted from a previous step of emulsion-solvent evaporation. Hybrid Janus nanoparticles with two polymeric phases (one elastomer and one amorphous), separated by an inorganic interface of TiO_2 nanoparticles, were obtained.

15.4.1.3 Pickering Nanoemulsions

In so-called stems (Pickering, 1907) colloidal stabilization of the nanoemulsion is promoted by a layer of inorganic nanoparticles irreversibly adsorbed at the liquid-liquid interface, which acts reducing the interfacial area between disperse and continuous phases (Binks, 2002). Pickering stabilization has been commonly used in the synthesis of hybrid nanoparticles by miniemulsion polymerization. The technique is widely spread in direct systems. In this case, the chemical environment determined by the pH and the ionic force of the aqueous phase has a great influence in the characteristics of the surface and the stabilizing ability of the particles.

Among the inorganic systems used in this strategy, clay platelets (Cauvin et al., 2005) and surface-functionalized nanoparticles of silica (Haase et al., 2012; Fortuna et al., 2009), titanium dioxide (Gonzàlez et al., 2015; Bonnefond et al., 2016), and cerium(IV) oxide (Zgheib et al., 2012) have been reported. A detailed review on Pickering systems in heterophase polymerization was published by Schrade et al. (2013).

Pickering stabilization in inverse systems is also possible, but in contrast to direct nanoemulsions only a few examples have been reported. For instance, silanized silica nanoparticles were used to promote Pickering stabilization of a water-in-oil nanoemulsion, allowing an interfacial polymerization process that resulted in hybrid polyurethane/silica nanocapsules with an aqueous core (Schoth et al., 2015).

15.4.1.4 Role of Functionalization in Structure Control

The morphological development of hybrid polymer/inorganic nanoparticles relies on the original disposition of the inorganic components within the nanoemulsion droplets and its evolution during the particle formation process, which is at the same time determined by a self-assembly mechanism driven by the minimization of the overall interfacial energy of the system.

The interfacial energy is expressed as the product of the contact area (A_{ij}) and the interfacial tension (γ_{ij}) between two components. In the case of hybrid nanoparticles dispersed in an aqueous phase, the overall interfacial energy of the system is understood as the sum of the interfacial energies involved between the polymer (P), the inorganic nanoparticles (I), and the aqueous phase (W). In the case of hybrid nanoparticles including a single inorganic species, the very low interfacial tension between the inorganic components with identical surface chemistry can be neglected. As a result, the preferential morphology of the hybrid nanostructure is determined by the interfacial tensions between the different phases: γ_{PW}, γ_{IP}, and γ_{IW}:

$$E=\sum_{ij} A\gamma = A_{PW}\gamma_{PW}+A_{IW}\gamma_{IW}+A_{IP}\gamma_{IP}+A_{II}\gamma_{II}=A_{PW}\gamma_{PW}+A_{IW}\gamma_{IW}+A_{IP}\gamma_{IP} \tag{15.4}$$

The morphological development has been studied by Asua (2014), who proposed a model that allows a structural mapping adapted for each hybrid system. According to this model, the dominance of the effect of an individual interfacial tension over the others will result in a specific hybrid morphology, ranging from the homogeneous encapsulation to Janus-like, core-shell structures, or the complete segregation of the inorganic particles respect to the polymer matrix. Thus, the transition between the available morphologies can be promoted by modifying the interfacial tensions involved in the process. The interfacial tensions of the overall system can be tuned with the introduction of changes in the nature of initiator used in the polymerization reaction (Mori and Kawaguchi, 2007; Staudt et al., 2013), the concentration of the surfactant (Zhou et al., 2006; Gong et al., 2009; Hecht et al., 2012a), or the relative hydrophobicity between the polymer and the inorganic particles (i.e., by changing the monomer (Staudt et al., 2013) or via the surface functionalization of the inorganic particles (Schoth et al., 2014, 2016b)).

A specific hydrophobization of the inorganic components is required to ensure a preferential migration within the polymer matrix, depending on the desired final location, the type of monomer, the polymerization environment, and the size, morphology, and nature of the inorganic nanoparticles. Different compounds (e.g., monomers, carboxylic acids, alkoxysilanes, thiols, or ionic surfactants) can be used as coupling agents to promote the inorganic surface hydrophobization.

Silane chemistry offers a broad spectrum of chain lengths and accessible functional groups. Fig. 15.9 shows the structure of poly(methyl methacrylate)/silica

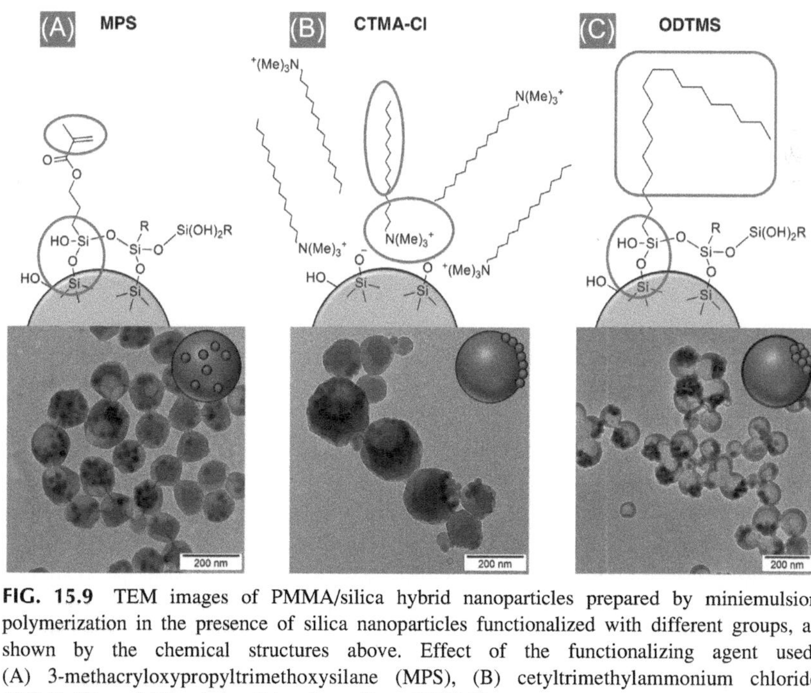

FIG. 15.9 TEM images of PMMA/silica hybrid nanoparticles prepared by miniemulsion polymerization in the presence of silica nanoparticles functionalized with different groups, as shown by the chemical structures above. Effect of the functionalizing agent used: (A) 3-methacryloxypropyltrimethoxysilane (MPS), (B) cetyltrimethylammonium chloride (CTMA-Cl), and (C) octadecyltrimethoxysilane (ODTMS).

nanoparticles prepared by in situ miniemulsion polymerization and reached with different surface functionalizations. The impact of the ionic interaction between a positively charged ammonium salt (cetyltrimethylammonium chloride) and the negatively charged silica surface is compared with the condensation of two alkoxysilanes with different structure, 3-methacryloxypropyltrimethoxysilane (MPS) and octadecyltrimethoxysilane. The introduction of short alkyl chains with polymerizable units (such as MPS) results into an homogeneous distribution of the inorganic nanoparticles within the polymer matrix, while the presence of inert and long alkyl chains drives to the location of the inorganic nanoparticles on the polymer surface or even to Janus-like morphologies (Schoth et al., 2014). Similar structures were observed in the incorporation of magnetite nanoparticles modified with alkoxysilanes and oleic acid (Schoth et al., 2016b) within a polymer matrix. The essential role of the surface modification in the morphological control of the hybrid polymer/inorganic nanoparticles has been clearly demonstrated.

15.4.2 Polymer Nanoparticles Formed in Nanoemulsions as Templates for Inorganic Synthesis

Polymer nanoparticles resulting from nanoemulsion strategies can be used as supports or templates for the synthesis of smaller inorganic nanoparticles, to lead to hierarchical hybrid nanoparticles. Since nanoemulsions, which are the focus of this chapter, are involved in the synthesis of the polymer nanoparticles themselves and not on the preparation of the resulting hybrid particles, we will only refer very briefly to this strategy.

The main idea, as illustrated in Fig. 15.10, is the addition of a metal precursor (typically a metal salt) to a dispersion of surface-functionalized nanoparticles. The surface of the particle should have the ability to complex, or at least to "retain," the metal cations. A precipitating agent—which could be a reducing agent if the aim is to obtain metallic nanoparticles, a base for the case of metal oxides, or a chalcogenide salt for metal chalcogenides—is added to the particle dispersion at a certain dropping rate. The addition rate can be decisive for the effective crystallization on the surface and a minimized precipitation in the bulk phase (Fischer et al., 2013). An example of polystyrene/CeO_2 nanoparticles obtained by this strategy (Mari et al., 2015) is presented in Fig. 15.11. Polystyrene nanoparticles had been functionalized at the surface via miniemulsion copolymerization with acrylic acid (AA) or ethylene glycol methacrylate phosphate (EGMP).

It should be mentioned that if functionalized microgel particles are used, the crystallization of the inorganic matter can take place within the particles, which act as "scaffolds" for the crystallization. For a detailed list of different inorganic materials synthesized *on* and *within* polymer nanoparticles of different materials, a previous work of our team can be consulted (Muñoz-Espí et al., 2013).

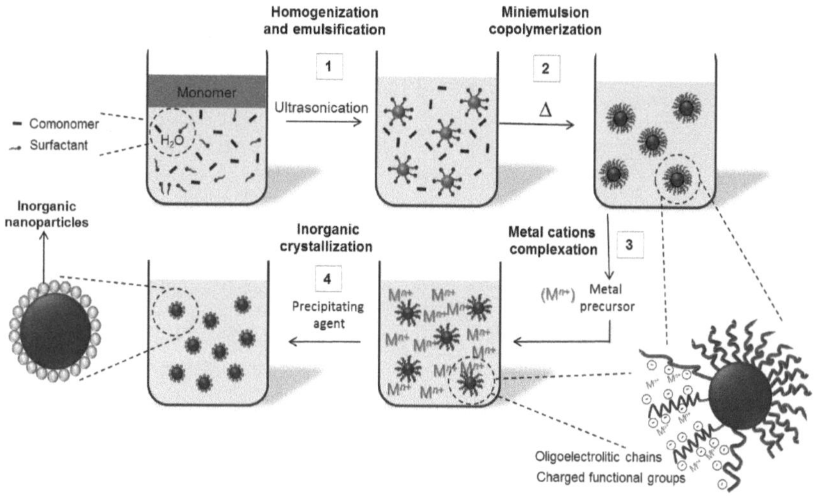

FIG. 15.10 Formation of metal oxide nanoparticles on the surface of functionalized polymer nanoparticles. *(Adapted with permission from Mari, M., Müller, B., Landfester, K., Muñoz-Espí, R., 2015. Ceria/polymer hybrid nanoparticles as efficient catalysts for the hydration of nitriles to amides. ACS Appl. Mater. Interfaces 7 (20), 10727–10733. https://doi.org/10.1021/acsami.5b01847. Copyright (2015) American Chemical Society.)*

15.4.3 Polymer/Inorganic Hybrid Capsules

The strategies previously described for the preparation of polymer nanocapsules, based on phase separation and interfacial polymerization processes, are similarly applied in the synthesis of hybrid polymer/inorganic capsules.

The phase separation process between two immiscible liquids that are integrated within the disperse phase of a nanoemulsion was used, for instance, in the preparation of raspberry-like poly(styrene-*co*-4-vinylpiridine)/silica nanocapsules. Hexadecane was introduced as the hydrophilic liquid core to be engulfed by a mixture of monomers, which was afterwards subjected to a Pickering miniemulsion copolymerization stabilized with silica nanoparticles (Cao et al., 2011).

In inverse miniemulsion interfacial polymerization was applied to prepare polyurea magnetic nanocapsules by reaction of 1,6-diaminohexane with toluene diisocyanate (TDI) in the presence of magnetite nanoparticles included within the aqueous phase (Rosenbauer et al., 2009). The same polymerization principle was applied in the preparation of polyurethane capsules, using hydrophobized silica nanoparticles to promote Pickering stabilization (Schoth et al., 2015). In another example Hood et al. (2015) reported the preparation of poly(urethane/urea)/silica hybrid nanocapsules with an interpenetrating structure of the silica and the polymer regions. Interestingly, in this strategy the polymerization at the droplet interface takes simultaneously place with the sol-gel process to form silica.

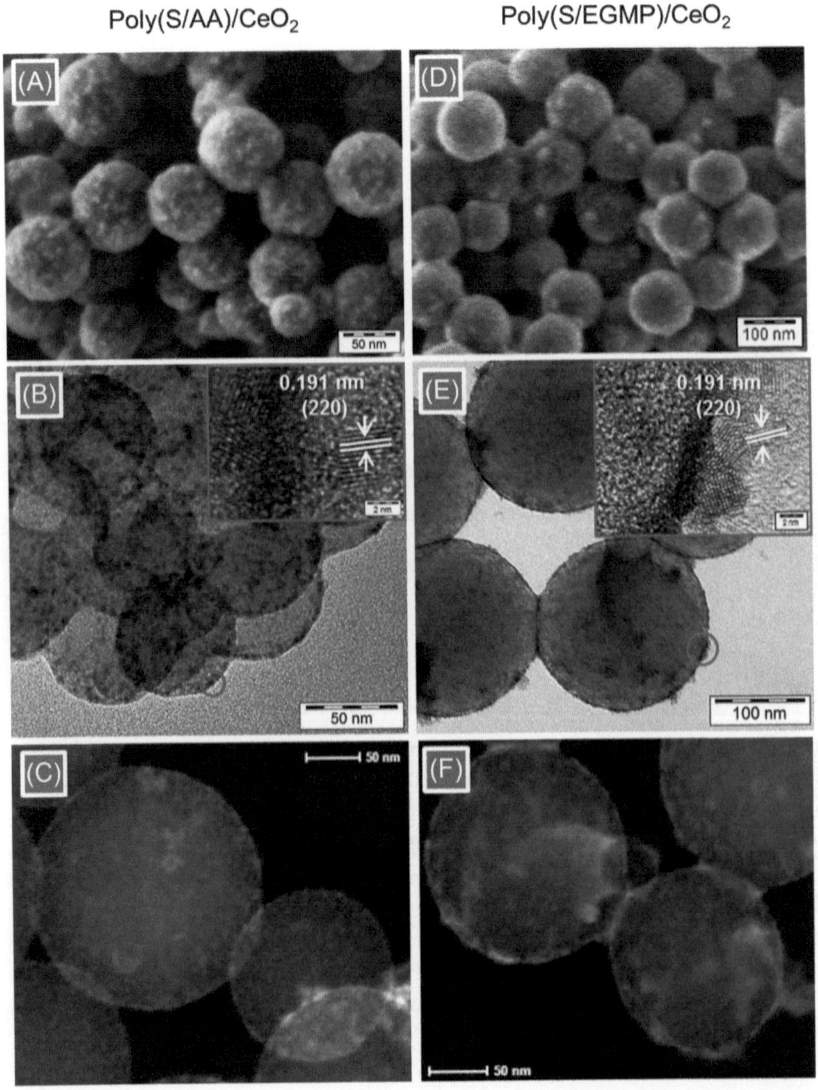

FIG. 15.11 (A) SEM images, (B) TEM images, and (C) dark-field TEM images of CeO_2/poly(S/AA) hybrid nanoparticles prepared with SDS as a surfactant. (D–F) Analogous micrographs for poly(S/EGMP) prepared also with SDS. Insets in panels (B) and (E) show high-resolution TEM images of selected areas. *S*, styrene; *AA*, acrylic acid; *EGMP*, ethylene glycol methacrylate phosphate. *(Reproduced with permission from Mari, M., Müller, B., Landfester, K., Muñoz-Espí, R., 2015. Ceria/polymer hybrid nanoparticles as efficient catalysts for the hydration of nitriles to amides. ACS Appl. Mater. Interfaces 7 (20), 10727–10733. https://doi.org/10.1021/acsami.5b01847. Copyright (2015), American Chemical Society.)*

15.5 FURTHER APPLICATIONS IN SYNTHETIC PROCESSES OF NANOPARTICLES PREPARED IN NANOEMULSIONS

In this last section, we will address a completely different strategy implying not only the use of nanoemulsions for the synthesis of nanoparticles, but also the application of such nanoparticles as additives or controlling agents for the synthesis of further nanostructures. Although these nanostructures are mostly not nanoparticles, the strategy is clearly related to synthesis, and we could not finish a chapter on applications of nanoemulsion on synthesis without mentioning it briefly. In particular we will discuss how polymer nanoparticles resulting from nanoemulsions have been used to control the growth of crystalline nanostructures.

The presence of any type of additives, including also nanoparticles, can affect the formation of crystals by influencing solubility-precipitation equilibria, nucleation, growth, Ostwald ripening, aggregation, etc., which can result in a control of size, morphology, or even the stabilization of polymorphs.

Since miniemulsion polymerization offers a great tool for surface functionalization, as seen in previous sections, the resulting nanoparticles can be very active as additives in crystallization processes. Initiated by Wegner and coworkers in the early 2000s (Wegner et al., 2001), this idea was followed up in the subsequent years for a variety of systems. Polymer nanoparticles prepared by miniemulsion polymerization have been used as controlling agents in the crystallization of zinc oxide (Muñoz-Espí et al., 2006; Gorna et al., 2007), calcium carbonate (Wegner et al., 2007; Lu et al., 2005; Kim et al., 2010), and calcium oxalate (Fischer et al., 2011). The latex nanoparticles become incorporated in the inorganic growing matrix, leading to hybrid materials. Although the practical applicability of such materials is, in principle, rather limited, functionalized nanoparticles are very convenient model systems for studying (bio)mineralization processes, because they can be very easily visualized by electron microscopy. As an illustrative paradigmatic case, studies with zinc oxide showed clearly how carboxyl-functionalized nanoparticles adsorb specifically on the {001} facets of growing zinc oxide crystals, retarding the growth in the [100] direction, which leads shorter and wider crystals (Muñoz-Espí et al., 2006). This effect of latex particles on inorganic growth is depicted in Fig. 15.12. Although spectroscopic studies indicated that the physical properties of zinc oxide are affected by the polymer nanoparticles (Muñoz-Espí et al., 2007b), the crystal perfection remains essentially undisturbed even at contents of polymer of up to 10 wt%, as proven by X-ray diffraction (Muñoz-Espí et al., 2007a).

Finally, another application in crystallization processes of polymer nanoparticles prepared in nanoemulsions is their use as specific chiral seeds for enantioselective crystallization of so-called conglomerate systems. Amino-acid-based chiral nanoparticles as those shown in Fig. 15.2 were used in the crystallization of the conglomerate *rac*-asparagine. Interestingly, enantiomeric

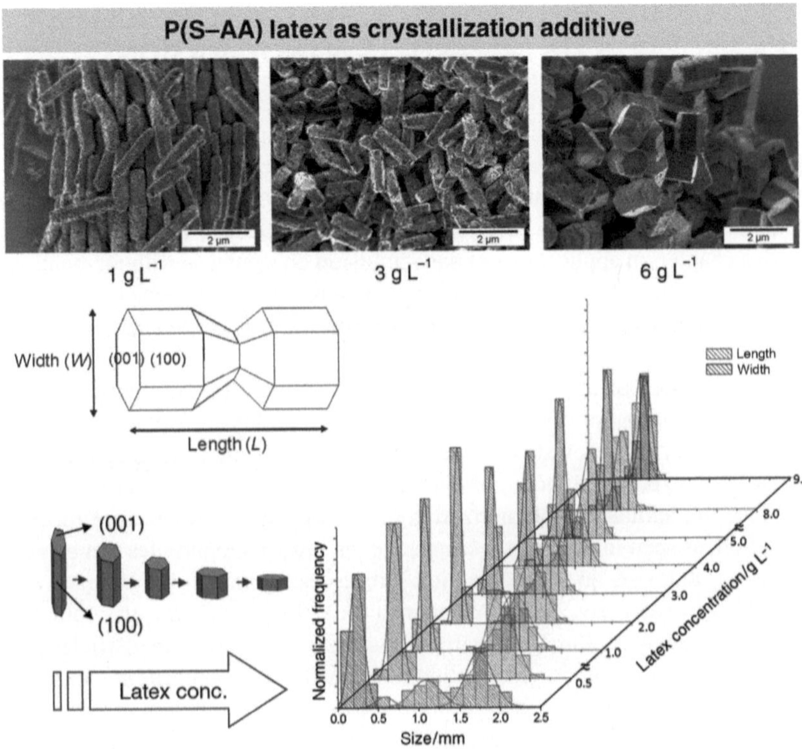

FIG. 15.12 Change in the aspect ratio (length/width) of zinc oxide crystals prepared in the presence of different concentrations of poly(styrene/acrylic acid) nanoparticles synthesized by miniemulsion polymerization. SEM images and histograms of the lengths and widths of the prepared crystals. *(Part of the figure from Muñoz-Espí, R., Qi, Y., Lieberwirth, I., Gómez, C.M., Wegner, G., 2006. Surface-functionalized latex particles as controlling agents for the mineralization of zinc oxide in aqueous medium. Chem. Eur. J. 12(1), 118–129. https://doi.org/10.1002/chem.200500860. Reproduced with permission. Copyright 2015 by John Wiley & Sons Ltd.)*

excesses up to 23% of either D- or L-asparagine were observed when the crystallization was carried out in the presence of D- and L-particles, respectively (Preiss et al., 2015). Crystallization of the same system in the presence of polystyrene nanoparticles prepared with amino-acid-based surfmers also showed a preferential enantioselective crystallization (Preiss et al., 2016).

15.6 SUMMARY AND PERSPECTIVES

This chapter has reviewed the versatility of nanoemulsions for the synthesis of nanoparticles. One of the most common applications of nanoemulsions in synthesis, typically linked to the so-called miniemulsion polymerization, is the preparation of polymer nanoparticles. The nanodroplets containing the monomer act as "nanoreactors" for the polymerization. Although radical polymerization is very typical for nanoemulsions and works in general very well, other polymerization reactions, such as polyaddition, polycondensation, anionic, metal-catalyzed, or oxidative polymerization, are also possible. Miniemulsion polymerization allows the synthesis of homopolymers and copolymers from both hydrophilic and hydrophobic monomers. The preparation of capsules is also possible, by controlling the ratios of the oil phases present in the system, by interfacial polymerization, or by the so-called emulsion-solvent evaporation technique (in this latter case there is no polymerization, but a precipitation of a preformed polymer). In addition, by introducing functional monomers in the formulation, surface-functionalized nanoparticles can also be readily obtained by miniemulsion copolymerization.

The synthesis of inorganic nanoparticles is other of the important applications of nanoemulsions. In a very similar way to polymerization, the droplets contain the precursor—in this case, an inorganic precursor—which is converted to particles by driving a fast precipitation or reduction reaction. Capsules can be obtained by interfacial precipitation/crystallization. Very especially in the field of nanoemulsions for inorganic synthesis, the combination of the colloidal confinement of the droplets with hydrothermal/solvothermal routes opens new possibilities, which we expect that will be exploited in upcoming years. The synergy of nanoemulsions with nonstandard temperature and pressure conditions allows for obtaining phases at lower temperatures than typically required for synthesis in bulk conditions.

Nanoemulsions have been shown to be also an excellent platform for the synthesis of polymerization of polymer/inorganic hybrid nanoparticles. The inorganic component can be incorporated into the polymer during the formation of the polymer particles (from monomers in the case of miniemulsion polymerization or by precipitation of a solution of the polymer in the emulsion-solvent evaporation technique), but can also be attached to or deposited on preformed particles, and even be formed in situ, using the polymer nanoparticles as supports or templates for crystallization. In recent time, much effort has been dedicated to the combination of more than one function in one single particle. Moving beyond the state of the art in the synthesis of multifunctional nanoparticles will be most probably an important goal of the community in the field in a near future. Low polydispersity may be relatively easy to achieve for simple systems, but it remains a challenge when nanoparticles become more complex in composition and morphology. The systematic understanding of the key features determining the structure in multicomponent systems will also remain an active area of research.

Finally, in the last section of the chapter, we have briefly discussed the use of nanoparticles prepared in nanoemulsions for further applications in chemical synthesis, namely as controlling agents or specific seeds for crystallization processes of organic and inorganic materials.

ACKNOWLEDGMENTS

The authors thank the financial support by the Spanish Ministry of Economy, Industry and Competitiveness (Ramón y Cajal grant to RME, grant no. RYC-2013-13451) and the Max Planck Society through the funding of the Max Planck Partner Group on Colloidal Methods for Multifunctional Materials (CM3) at the University of Valencia, headed by R. Muñoz-Espí. Prof. Katharina Landfester (Mainz) is gratefully acknowledged for the regular inspiring discussions.

REFERENCES

Aguirre, M., Paulis, M., Barrado, M., Iturrondobeitia, M., Okariz, A., Guraya, T., Ibarretxe, J., Leiza, J.R., 2015. Evolution of particle morphology during the synthesis of hybrid acrylic/ CeO_2 nanocomposites by miniemulsion polymerization. J. Polym. Sci. A Polym. Chem. 53 (6), 792–799. https://doi.org/10.1002/pola.27504.

Alduncin, J.A., Forcada, J., Asua, J.M., 1994. Miniemulsion polymerization using oil-soluble initiators. Macromolecules 27 (8), 2256–2261. https://doi.org/10.1021/ma00086a041.

Antonello, A., Jakob, G., Dolcet, P., Momper, R., Kokkinopoulou, M., Landfester, K., Muñoz-Espí, R., Gross, S., 2017. Synergy of miniemulsion and solvothermal conditions for the low-temperature crystallization of magnetic nanostructured transition-metal ferrites. Chem. Mater. 29 (3), 985–997. https://doi.org/10.1021/acs.chemmater.6b03467.

Antonietti, M., Landfester, K., 2002. Polyreactions in miniemulsions. Prog. Polym. Sci. 27 (4), 689–757. https://doi.org/10.1016/S0079-6700(01)00051-X.

Aschenbrenner, E., Bley, K., Koynov, K., Makowski, M., Kappl, M., Landfester, K., Weiss, C.K., 2013. Using the polymeric ouzo effect for the preparation of polysaccharide-based nanoparticles. Langmuir 29 (28), 8845–8855. https://doi.org/10.1021/la4017867.

Asua, J.M., 2002. Miniemulsion polymerization. Prog. Polym. Sci. 27 (7), 1283–1346.

Asua, J.M., 2004. Emulsion polymerization: from fundamental mechanisms to process developments. J. Polym. Sci. A Polym. Chem. 42 (5), 1025–1041.

Asua, J.M., 2014. Mapping the morphology of polymer-inorganic nanocomposites synthesized by miniemulsion polymerization. Macromol. Chem. Phys. 215 (5), 458–464. https://doi.org/10.1002/macp.201300696.

Barrér, M., Maitre, C., Dourges, M.A., Hémery, P., 2001. Anionic polymerization of 1,3,5-tris(trifluoropropylmethyl)cyclotrisiloxane (f3) in miniemulsion. Macromolecules 34 (21), 7276–7280. https://doi.org/10.1021/ma010559z.

Barrére, M., Landfester, K., 2003. Polyester synthesis in aqueous miniemulsion. Polymer 44 (10), 2833–2841. https://doi.org/10.1016/S0032-3861(03)00151-4.

Bellezza, F., Cipiciani, A., Costantino, U., Marmottini, F., Quotadamo, M.A., 2006. Zirconium phosphate nanoparticles from water-in-oil microemulsions. Colloid Polym. Sci. 285 (1), 19–25. https://doi.org/10.1007/s00396-006-1525-4.

Bellezza, F., Cipiciani, A., Costantino, U., Nocchetti, M., Posati, T., 2009. Hydrotalcite-like nanocrystals from water-in-oil microemulsions. Eur. J. Inorg. Chem. 18, 2603–2611. https://doi.org/10.1002/ejic.200801114.

Benedetti, C., Cazzolaro, A., Carraro, M., Graf, R., Landfester, K., Gross, S., Muñoz-Espí, R., 2016. Dual role of zirconium oxoclusters in hybrid nanoparticles: cross-linkers and catalytic sites. ACS Appl. Mater. Interfaces 8 (39), 26275–26284. https://doi.org/10.1021/acsami.6b07023.

Binks, B.P., 2002. Particles as surfactants-similarities and differences. Curr. Opin. Colloid Interface Sci. 7 (1), 21–41. https://doi.org/10.1016/S1359-0294(02)00008-0.

Boisson, F., Uzulina, I., Guyot, A., 2001. Polymerisable surfactants in miniemulsion polymerisation. Macromol. Rapid Commun. 22 (14), 1135–1142. https://doi.org/10.1002/1521-3927(20011001)22:14⟨1135::AID-MARC1135⟩3.0.CO;2-X.

Bonnefond, A., Ibarra, M., Gonzalez, E., Barrado, M., Chuvilin, A., Asua, J.M., Leiza, J.R., 2016. Photocatalytic and magnetic titanium dioxide/polystyrene/magnetite composite hybrid polymer particles. J. Polym. Sci. A Polym. Chem. 54 (20), 3350–3356. https://doi.org/10.1002/pola.28224.

Bourgeat-Lami, E., Farzi, G.A., David, L., Putaux, J.L., McKenna, T.F., 2012. Silica encapsulation by miniemulsion polymerization: distribution and localization of the silica particles in droplets and latex particles. Langmuir 28 (14), 6021–6031. https://doi.org/10.1021/la300587b.

Boutonnet, M., Lögdberg, S., Elm Svensson, E., 2008. Recent developments in the application of nanoparticles prepared from w/o microemulsions in heterogeneous catalysis. Curr. Opin. Colloid Interface Sci. 13 (4), 270–286. https://doi.org/10.1016/j.cocis.2007.10.001.

Buchold, D.H.M., Feldmann, C., 2007. Nanoscale γ-AiO(OH) hollow spheres: synthesis and container-type functionality. Nano Lett. 7 (11), 3489–3492. https://doi.org/10.1021/nl072074f.

Butturini, E., Dolcet, P., Casarin, M., Speghini, A., Pedroni, M., Benetti, F., Motta, A., Badocco, D., Pastore, P., Diodati, S., Pandolfo, L., Gross, S., 2014. Simple, common but functional: biocompatible and luminescent rare-earth doped magnesium and calcium hydroxides from miniemulsion. J. Mater. Chem. B 2 (38), 6639–6651. https://doi.org/10.1039/c4tb01206b.

Cao, Z., Schrade, A., Landfester, K., Ziener, U., 2011. Synthesis of raspberry-like organic-inorganic hybrid nanocapsules via Pickering miniemulsion polymerization: colloidal stability and morphology. J. Polym. Sci. A Polym. Chem. 49 (11), 2382–2394. https://doi.org/10.1002/pola.24668.

Capek, I., 2010. On inverse miniemulsion polymerization of conventional water-soluble monomers. Adv. Colloid Interface Sci. 156 (1–2), 35–61. https://doi.org/10.1016/j.cis.2010.02.006.

Cauvin, S., Colver, P.J., Bon, S.A.F., 2005. Pickering stabilized miniemulsion polymerization: preparation of clay armored latexes. Macromolecules 38 (19), 7887–7889. https://doi.org/10.1021/ma051070z.

Crespy, D., Landfester, K., 2005. Anionic polymerization of ε-caprolactam in miniemulsion: synthesis and characterization of polyamide-6 nanoparticles. Macromolecules 38 (16), 6882–6887. https://doi.org/10.1021/ma050616e.

Crespy, D., Landfester, K., 2007. Preparation of nylon 6 nanoparticles and nanocapsules by two novel miniemulsion/solvent displacement hybrid techniques. Macromol. Chem. Phys. 208 (5), 457–466. https://doi.org/10.1002/macp.200600487.

Crespy, D., Landfester, K., 2010. Miniemulsion polymerization as a versatile tool for the synthesis of functionalized polymers. Beilstein J. Org. Chem. https://doi.org/10.3762/bjoc.6.130.

Crespy, D., Musyanovych, A., Landfester, K., 2006. Synthesis of polymer particles and nanocapsules stabilized with PEO/PPO containing polymerizable surfactants in miniemulsion. Colloid Polym. Sci. 284 (7), 780–787. https://doi.org/10.1007/s00396-005-1446-7.

Crespy, D., Stark, M., Hoffmann-Richter, C., Ziener, U., Landfester, K., 2007. Polymeric nanoreactors for hydrophilic reagents synthesized by interfacial polycondensation on miniemulsion droplets. Macromolecules 40 (9), 3122–3135. https://doi.org/10.1021/ma0621932.

de Brouwer, H., Tsavalas, J.G., Schork, F.J., 2000. Living radical polymerization in miniemulsion using reversible addition fragmentation chain transfer. Macromolecules 33 (25), 9239–9246. https://doi.org/10.1021/ma001205v.

Dewald, R.C., Hart, L.H., Carroll, W.F., 1984. PVC miniemulsion polymerization. I. Origin of droplet families. J. Polym. Sci. Polym. Chem. Ed. 22 (11), 2923–2930. https://doi.org/10.1002/pol.1984.170221115.

Diodati, S., Dolcet, P., Casarin, M., Gross, S., 2015. Pursuing the crystallization of mono- and polymetallic nanosized crystalline inorganic compounds by low-temperature wet-chemistry and colloidal routes. Chem. Rev. 115 (20), 11449–11502. https://doi.org/10.1021/acs.chemrev.5b00275.

Dolcet, P., Casarin, M., Maccato, C., Bovo, L., Ischia, G., Gialanella, S., Mancin, F., Tondello, E., Gross, S., 2012. Miniemulsions as chemical nanoreactors for the room temperature synthesis of inorganic crystalline nanostructures: ZnO colloids. J. Mater. Chem. 22 (4), 1620–1626. https://doi.org/10.1039/c1jm13301b.

Dolcet, P., Mambrini, A., Pedroni, M., Speghini, A., Gialanella, S., Casarin, M., Gross, S., 2015. Room temperature crystallization of highly luminescent lanthanide-doped CaF_2 in nanosized droplets: first example of the synthesis of metal halogenide in miniemulsion with effective doping and size control. RSC Adv. 5 (21), 16302–16310. https://doi.org/10.1039/c4ra12006j.

Dolcet, P., Maurizio, C., Casarin, M., Pandolfo, L., Gialanella, S., Badocco, D., Pastore, P., Speghini, A., Gross, S., 2015. An effective two-emulsion approach to the synthesis of doped ZnS crystalline nanostructures. Eur. J. Inorg. Chem. (4), 706–714. https://doi.org/10.1002/ejic.201403004.

Eastoe, J., Hollamby, M.J., Hudson, L., 2006. Recent advances in nanoparticle synthesis with reversed micelles. Adv. Colloid Interface Sci. 128, 5–15. https://doi.org/10.1016/j.cis.2006.11.009.

Feuser, P.E., Bubniak, L.d.S., Silva, M.C.d.S., Viegas, A.d.C., Castilho Fernandes, A., Ricci-Junior, E., Nele, M., Tedesco, A.A.C., Sayer, C., de Araújo, P.H.H., 2015. Encapsulation of magnetic nanoparticles in poly(methyl methacrylate) by miniemulsion and evaluation of hyperthermia in U87MG cells. Eur. Polym. J. 68, 355–365. https://doi.org/10.1016/j.eurpolymj.2015.04.029.

Fickert, J., Rupper, P., Graf, R., Landfester, K., Crespy, D., 2012. Design and characterization of functionalized silica nanocontainers for self-healing materials. J. Mater. Chem. 22, 2286–2291. https://doi.org/10.1039/C2JM15151K.

Fischer, V., Landfester, K., Muñoz-Espí, R., 2011. Stabilization of calcium oxalate metastable phases by oligo(L-glutamic acid): effect of peptide chain length. Cryst. Growth Des. 11 (5), 1880–1890. https://doi.org/10.1021/cg200058d.

Fischer, V., Lieberwirth, I., Jakob, G., Landfester, K., Muñoz-Espí, R., 2013. Metal oxide/polymer hybrid nanoparticles with versatile functionality prepared by controlled surface crystallization. Adv. Funct. Mater. 23 (4), 451–466. https://doi.org/10.1002/adfm.201201839.

Fortuna, S., Colard, C.A., Troisi, A., Bon, S.A., 2009. Packing patterns of silica nanoparticles on surfaces of armored polystyrene latex particles. Langmuir 25 (21), 12399–12403. https://doi.org/10.1021/la9010289.

Freichels, H., Wagner, M., Okwieka, P., Meyer, R.G., Mailänder, V., Landfester, K., Musyanovych, A., 2013. (Oligo)mannose functionalized hydroxyethyl starch nanocapsules: en route to drug delivery systems with targeting properties. J. Mater. Chem. B 1 (34), 4338. https://doi.org/10.1039/c3tb20138d.

Froimowicz, P., Muñoz-Espí, R., Landfester, K., Musyanovych, A., Crespy, D., 2013. Surface-functionalized particles: from their design and synthesis to materials science and bio-applications. Curr. Org. Chem. 17 (9), 900–912. https://doi.org/10.2174/1385272811317090004.

Ganachaud, F., Katz, J.L., 2005. Nanoparticles and nanocapsules created using the ouzo effect: spontaneous emulsification as an alternative to ultrasonic and high-shear devices. ChemPhysChem 6 (2), 209–216. https://doi.org/10.1002/cphc.200400527.

Ge, J.P., Chen, W., Liu, L.P., Li, Y.D., 2006. Formation of disperse nanoparticles at the oil/water interface in normal microemulsions. Chem. Eur. J. 12 (25), 6552–6558. https://doi.org/10.1002/chem.200600454.

Gong, T., Yang, D., Hu, J., Yang, W., Wang, C., Lu, J.Q., 2009. Preparation of monodispersed hybrid nanospheres with high magnetite content from uniform Fe_3O_4 clusters. Colloids Surf. A 339 (1–3), 232–239. https://doi.org/10.1016/j.colsurfa.2009.02.034.

Gonzàlez, E., Bonnefond, A., Barrado, M., Casado Barrasa, A.M., Asua, J.M., Leiza, J.R., 2015. Photoactive self-cleaning polymer coatings by TiO_2 nanoparticle Pickering miniemulsion polymerization. Chem. Eng. J. 281, 209–217. https://doi.org/10.1016/j.cej.2015.06.074.

Gorna, K., Muñoz-Espí, R., Gröhn, F., Wegner, G., 2007. Bioinspired mineralization of inorganics from aqueous media controlled by synthetic polymers. Macromol. Biosci. 7 (2), 163–173. https://doi.org/10.1002/mabi.200600209.

Guyot, A., Graillat, C., Favero, C., 2003. Anionic surfmers in mini-emulsion polymerisation. C. R. Chim. 6 (11-2), 1319–1327. https://doi.org/10.1016/j.crci.2003.07.018.

Gyger, F., Hubner, M., Feldmann, C., Barsan, N., Weimar, U., 2010. Nanoscale S_nO_2 hollow spheres and their application as a gas-sensing material. Chem. Mater. 22 (16), 4821–4827. https://doi.org/10.1021/cm1011235.

Haase, M.F., Grigoriev, D.O., Mohwald, H., Shchukin, D.G., 2012. Development of nanoparticle stabilized polymer nanocontainers with high content of the encapsulated active agent and their application in water-borne anticorrosive coatings. Adv. Mater. 24 (18), 2429–2435. https://doi.org/10.1002/adma.201104687.

Hajir, M., Dolcet, P., Fischer, V., Holzinger, J., Landfester, K., Muñoz-Espí, R., 2012. Sol–gel processes at the droplet interface: hydrous zirconia and hafnia nanocapsules by interfacial inorganic polycondensation. J. Mater. Chem. 22 (12), 5622–5628. https://doi.org/10.1039/c2jm15353j.

Hecht, L.L., Wagner, C., Ozcan, O., Eisenbart, F., Koehler, K., Landfester, K., Schuchmann, H.P., 2012. Influence of the surfactant concentration on miniemulsion polymerization for the preparation of hybrid nanoparticles. Macromol. Chem. Phys. 213 (20), 2165–2173. https://doi.org/10.1002/macp.201200219.

Hecht, L.L., Winkelmann, M., Wagner, C., Landfester, K., Gerlinger, W., Sachweh, B., Schuchmann, H.P., 2012. Miniemulsions for the production of nanostructured particles. Chem. Eng. Technol. 35 (9), 1670–1676. https://doi.org/10.1002/ceat.201200196.

Heutz, N.A., Dolcet, P., Birkner, A., Casarin, M., Merz, K., Gialanella, S., Gross, S., 2013. Inorganic chemistry in a nanoreactor: Au/TiO_2 nanocomposites by photolysis of a single-source precursor in miniemulsion. Nanoscale 5 (21), 10534–10541. https://doi.org/10.1039/c3nr03977c.

Holmberg, K., 2004. Surfactant-templated nanomaterials synthesis. J. Colloid Interface Sci. 274 (2), 355–364. https://doi.org/10.1016/j.jcis.2004.04.006.

Holzapfel, V., Musyanovych, A., Landfester, K., Lorenz, M.R., Mailänder, V., 2005. Preparation of fluorescent carboxyl and amino functionalized polystyrene particles by miniemulsion polymerization as markers for cells. Macromol. Chem. Phys. 206 (24), 2440–2449. https://doi.org/10.1002/macp.200500372.

Hood, M., Mari, M., Muñoz-Espí, R., 2014. Synthetic strategies in the preparation of polymer/inorganic hybrid nanoparticles. Materials 7 (5), 4057–4087. https://doi.org/10.3390/ma7054057.

Hood, M.A., Paiphansiri, U., Schaeffel, D., Koynov, K., Kappl, M., Landfester, K., Muñoz-Espí, R., 2015. Hybrid poly(urethane–urea)/silica nanocapsules with pH-sensitive gateways. Chem. Mater. 27 (12), 4311–4318. https://doi.org/10.1021/acs.chemmater.5b00690.

Hood, M.A., Encinas, N., Vollmer, D., Graf, R., Landfester, K., Muñoz-Espí, R.M., 2017. Controlling hydrophobicity of silica nanocapsules prepared from organosilanes. Colloids Surf. A. https://doi.org/10.1016/j.colsurfa.2017.05.047.

Kaewsaneha, C., Tangboriboonrat, P., Polpanich, D., Eissa, M., Elaissari, A., 2013. Facile method for preparation of anisotropic submicron magnetic janus particles using miniemulsion. J. Colloid Interface Sci. 409, 66–71. https://doi.org/10.1016/j.jcis.2013.07.067.

Katta, K., Busko, D., Avlasevich, Y., Muñoz-Espí, R., Baluschev, S., Landfester, K., 2015. Synthesis of triplet-triplet annihilation upconversion nanocapsules under protective conditions. Macromol. Rapid Commun. 36 (11), 1084–1088. https://doi.org/10.1002/marc.201400670.

Kim, Y.Y., Ribeiro, L., Maillot, F., Ward, O., Eichhorn, S.J., Meldrum, F.C., 2010. Bio-inspired synthesis and mechanical properties of calcite-polymer particle composites. Adv. Mater. 22 (18), 2082–2086. https://doi.org/10.1002/adma.200903743.

Klinger, D., Landfester, K., 2011. Dual stimuli-responsive poly(2-hydroxyethyl methacrylate-*co*-methacrylic acid) microgels based on photo-cleavable cross-linkers: pH-dependent swelling and light-induced degradation. Macromolecules 44 (24), 9758–9772. https://doi.org/10.1021/ma201706r.

Kobitskaya, E., Ekinci, D., Manzke, A., Plettl, A., Wiedwald, U., Ziemann, P., Biskupek, J., Kaiser, U., Ziener, U., Landfester, K., 2010. Narrowly size distributed zinc-containing poly(acrylamide) latexes via inverse miniemulsion polymerization. Macromolecules 43 (7), 3294–3305. https://doi.org/10.1021/ma902553a.

Landfester, K., 2001. Polyreactions in miniemulsions. Macromol. Rapid Commun. 22 (12), 896–936. https://doi.org/10.1002/1521-3927(20010801)22:12⟨896::AID-MARC896⟩3.0.CO;2-R.

Landfester, K., 2003. Miniemulsions for Nanoparticle Synthesis. In: vol. 227. Springer, Berlin, Heidelberg, pp. 75–123. https://doi.org/10.1007/3-540-36412-9_4 Chapter 4 ISBN 978-3-540-00418-9.

Landfester, K., 2009. Miniemulsion polymerization and the structure of polymer and hybrid nanoparticles. Angew. Chem. Int. Ed. Engl. 48 (25), 4488–4507. https://doi.org/10.1002/anie.200900723.

Landfester, K., Antonietti, M., 2000. The polymerization of acrylonitrile in miniemulsions: "crumpled latex particles" or polymer nanocrystals. Macromol. Rapid Commun. 21 (12), 820–824. https://doi.org/10.1002/1521-3927(20000801)21:12⟨820::AID-MARC820⟩3.0.CO;2-9.

Landfester, K., Willert, M., Antonietti, M., 2000. Preparation of polymer particles in nonaqueous direct and inverse miniemulsions. Macromolecules 33 (7), 2370–2376. https://doi.org/10.1021/ma991782n.

Landfester, K., Musyanovych, A., Mailänder, V., 2010. From polymeric particles to multifunctional nanocapsules for biomedical applications using the miniemulsion process. J. Polym. Sci. A Polym. Chem. 48 (3), 493–515. https://doi.org/10.1002/pola.23786.

Leidinger, P., Popescu, R., Gerthsen, D., Feldmann, C., 2010. Nanoscale La(OH)(3) hollow spheres and fine-tuning of its outer diameter and cavity size. Small 6 (17), 1886–1891. https://doi.org/10.1002/smll.201000575.

Lovell, P.A., El-Aasser, M.S., 1997. Emulsion Polymerization and Emulsion Polymers. John Wiley, Chichester.

Lu, C.H., Qi, L.M., Cong, H.L., Wang, X.Y., Yang, J.H., Yang, L.L., Zhang, D.Y., Ma, J.M., Cao, W.X., 2005. Synthesis of calcite single crystals with porous surface by templating of polymer latex particles. Chem. Mater. 17 (20), 5218–5224. https://doi.org/10.1021/cm0513029.

Lv, L.P., Zhao, Y., Vilbrandt, N., Gallei, M., Vimalanandan, A., Rohwerder, M., Landfester, K., Crespy, D., 2013. Redox responsive release of hydrophobic self-healing agents from polyaniline capsules. J. Am. Chem. Soc. 135 (38), 14198–14205. https://doi.org/10.1021/ja405279t.

Lv, L.P., Zhao, Y., Zhou, H.X., Landfester, K., Crespy, D., 2014. From core–shell and janus structures to tricompartment submicron particles. Polymer 55 (3), 715–720. https://doi.org/10.1016/j.polymer.2013.12.061.

Mari, M., Müller, B., Landfester, K., Muñoz-Espí, R., 2015. Ceria/polymer hybrid nanoparticles as efficient catalysts for the hydration of nitriles to amides. ACS Appl. Mater. Interfaces 7 (20), 10727–10733. https://doi.org/10.1021/acsami.5b01847.

Marie, E., Rothe, R., Antonietti, M., Landfester, K., 2003. Synthesis of polyaniline particles via inverse and direct miniemulsion. Macromolecules 36 (11), 3967–3973. https://doi.org/10.1021/ma0257550.

Matahwa, H., McLeary, J.B., Sanderson, R.D., 2006. Comparative study of classical surfactants and polymerizable surfactants (surfmers) in the reversible addition-fragmentation chain transfer mediated miniemulsion polymerization of styrene and methyl methacrylate. J. Polym. Sci. A Polym. Chem. 44 (1), 427–442. https://doi.org/10.1002/pola.21071.

Matyjaszewski, K., Qiu, J., Tsarevsky, N.V., Charleux, B., 2000. Atom transfer radical polymerization of *n*-butyl methacrylate in an aqueous dispersed system: a miniemulsion approach. J. Polym. Sci. A Polym. Chem. 38 (S1), 4724–4734. https://doi.org/10.1002/1099-0518(200012)38:1+⟨4724::AID-POLA120⟩3.0.CO;2-Q.

McClements, D.J., 2012. Crystals and crystallization in oil-in-water emulsions: implications for emulsion-based delivery systems. Adv. Colloid. Interface Sci. 174, 1–30. https://doi.org/10.1016/j.cis.2012.03.002.

Mori, Y., Kawaguchi, H., 2007. Impact of initiators in preparing magnetic polymer particles by miniemulsion polymerization. Colloids Surf. B 56 (1–2), 246–254. https://doi.org/10.1016/j.colsurfb.2006.11.023.

Mouran, D., Reimers, J., Schork, F.J., 1996. Miniemulsion polymerization of methyl methacrylate with dodecyl mercaptan as cosurfactant. J. Polym. Sci. A Polym. Chem. 34 (6), 1073–1081. https://doi.org/10.1002/(SICI)1099-0518(19960430)34:6⟨1073::AID-POLA16⟩3.0.CO;2-4.

Muñoz-Espí, R., Qi, Y., Lieberwirth, I., Gómez, C.M., Wegner, G., 2006. Surface-functionalized latex particles as controlling agents for the mineralization of zinc oxide in aqueous medium. Chem. Eur. J. 12 (1), 118–129. https://doi.org/10.1002/chem.200500860.

Muñoz-Espí, R., Chandra, A., Wegner, G., 2007. Crystal perfection in zinc oxide with occluded carboxyl-functionalized latex particles. Cryst. Growth Des. 7 (9), 1584–1589. https://doi.org/10.1021/cg060858l.

Muñoz-Espí, R., Jeschke, G., Lieberwirth, I., Gómez, C.M., Wegner, G., 2007. ZnO–latex hybrids obtained by polymer-controlled crystallization: a spectroscopic investigation. J. Phys. Chem. B 111 (4), 697–707. https://doi.org/10.1021/jp066380d.

Muñoz-Espí, R., Weiss, C.K., Landfester, K., 2012. Inorganic nanoparticles prepared in miniemulsion. Curr. Opin. Colloid Interface Sci. 17 (4), 212–224. https://doi.org/10.1016/j.cocis.2012.04.002.

Muñoz-Espí, R., Mastai, Y., Gross, S., Landfester, K., 2013. Colloidal systems for crystallization processes from liquid phase. CrystEngComm 15 (12), 2175. https://doi.org/10.1039/c3ce26657e.

Nabih, N., Schiller, R., Lieberwirth, I., Kockrick, E., Frind, R., Kaskel, S., Weiss, C.K., Landfester, K., 2011. Mesoporous CeO_2 nanoparticles synthesized by an inverse miniemulsion technique and their catalytic properties in methane oxidation. Nanotechnology. 22(13). https://doi.org/10.1088/0957-4484/22/13/135606.

Oh, J.K., Tang, C., Gao, H., Tsarevsky, N.V., Matyjaszewski, K., 2006. Inverse miniemulsion ATRP: a new method for synthesis and functionalization of well-defined water-soluble/

cross-linked polymeric particles. J. Am. Chem. Soc. 128 (16), 5578–5584. https://doi.org/10.1021/ja060586a.

Okoli, C., Sánchez-Domínguez, M., Boutonnet, M., Järås, S., Civera, C., Solans, C., Kuttuva, G.R., 2012. Comparison and functionalization study of microemulsion-prepared magnetic iron oxide nanoparticles. Langmuir 28 (22), 8479–8485. https://doi.org/10.1021/la300599q.

Pablico, M.H., Mertzman, J.E., Japp, E.A., Boncher, W.L., Nishida, M., Van Keuren, E., Lofland, S.E., Dollahon, N., Rubinson, J.F., Holman, K.T., Stoll, S.L., 2011. Miniemulsion synthesis of metal-oxo cluster containing copolymer nanobeads. Langmuir 27 (20), 12575–12584. https://doi.org/10.1021/la2029774.

Paiphansiri, U., Dausend, J., Musyanovych, A., Mailänder, V., Landfester, K., 2009. Fluorescent polyurethane nanocapsules prepared via inverse miniemulsion: surface functionalization for use as biocarriers. Macromol. Biosci. 9 (6), 575–584. https://doi.org/10.1002/mabi.200800293.

Peng, B., Chen, M., Zhou, S., Wu, L., Ma, X., 2008. Fabrication of hollow silica spheres using droplet templates derived from a miniemulsion technique. J. Colloid Interface Sci. 321 (1), 67–73. https://doi.org/10.1016/j.jcis.2007.12.044.

Pickering, S.U., 1907. CXCVI—emulsions. J. Chem. Soc. Trans. 91, 2001–2021. https://doi.org/10.1039/CT9079102001.

Pileni, M.P., 1993. Water-in-oil colloidal droplets used as microreactors. Adv. Colloid Interface Sci. 46, 139–163. https://doi.org/10.1016/0001-8686(93)80039-e.

Pileni, M.P., 1997. Nanosized particles made in colloidal assemblies. Langmuir 13 (13), 3266–3276. https://doi.org/10.1021/la960319q.

Pileni, M.P., 2003. The role of soft colloidal templates in controlling the size and shape of inorganic nanocrystals. Nat. Mater. 2 (3), 145–150. https://doi.org/10.1038/nmat817.

Preiss, L.C., Werber, L., Fischer, V., Hanif, S., Landfester, K., Mastai, Y., Muñoz-Espí, R., 2015. Amino-acid-based chiral nanoparticles for enantioselective crystallization. Adv. Mater. 27, 2728–2732.

Preiss, L.C., Wagner, M., Mastai, Y., Landfester, K., Muñoz-Espí, R., 2016. Amino-acid-based polymerizable surfactants for the synthesis of chiral nanoparticles. Macromol. Rapid Commun. 37 (17), 1421–1426. https://doi.org/10.1002/marc.201600210.

Qi, G., Jones, C.W., Schork, F.J., 2007. RAFT inverse miniemulsion polymerization of acrylamide. Macromol. Rapid Commun. 28 (9), 1010–1016. https://doi.org/10.1002/marc.200700026.

Qiu, J., Pintauer, T., Gaynor, S.G., Matyjaszewski, K., Charleux, B., Vairon, J.P., 2000. Mechanistic aspect of reverse atom transfer radical polymerization of *n*-butyl methacrylate in aqueous dispersed system. Macromolecules 33 (20), 7310–7320. https://doi.org/10.1021/ma000720q.

Rodríguez, V.S., Delgado, J., Silebi, C.A., El-Aasser, M.S., 1989. Interparticle monomer transport in miniemulsions. Ind. Eng. Chem. Res. 28 (1), 65–74. https://doi.org/10.1021/ie00085a013.

Rosenbauer, E.M., Landfester, K., Musyanovych, A., 2009. Surface-active monomer as a stabilizer for polyurea nanocapsules synthesized via interfacial polyaddition in inverse miniemulsion. Langmuir 25 (20), 12084–12091. https://doi.org/10.1021/la9017097.

Rossmanith, R., Weiss, C.K., Geserick, J., Hüsing, N., Hormann, U., Kaiser, U., Landfester, K., 2008. Porous anatase nanoparticles with high specific surface area prepared by miniemulsion technique. Chem. Mater. 20 (18), 5768–5780. https://doi.org/10.1021/cm800533a.

Sánchez-Domínguez, M., Boutonnet, M., Solans, C., 2009. A novel approach to metal and metal oxide nanoparticle synthesis: the oil-in-water microemulsion reaction method. J. Nanopart. Res. 11 (7), 1823–1829. https://doi.org/10.1007/s11051-009-9660-8.

Sánchez-Domínguez, M., Liotta, L.F., Di Carlo, G., Pantaleo, G., Venezia, A.M., Solans, C., Boutonnet, M., 2010. Synthesis of CeO_2, ZrO_2, $Ce_{0.5}Zr_{0.5}O_2$, and TiO_2 nanoparticles by a novel oil-in-water microemulsion reaction method and their use as catalyst support for co-oxidation. Catal. Today 158 (1–2), 35–43. https://doi.org/10.1016/j.cattod.2010.05.026.

Sánchez-Domínguez, M., Koleilat, H., Boutonnet, M., Solans, C., 2011. Synthesis of Pt nanoparticles in oil-in-water microemulsions: phase behavior and effect of formulation parameters on nanoparticle characteristics. J. Dispersion Sci. Technol. 32 (12), 1765–1770. https://doi.org/10.1080/01932691.2011.616170.

Sanchis, M.J., Redondo-Foj, B., Carsí, M., Ortiz-Serna, P., Culebras, M., Gómez, C.M., Cantarero, A., Muñoz-Espí, R., 2016. Controlling dielectrical properties of polymer blends through defined PEDOT nanostructures. RSC Adv. 6 (66), 62024–62030. https://doi.org/10.1039/c6ra05597d.

Sauer, R., Froimowicz, P., Scholler, K., Cramer, J.M., Ritz, S., Mailänder, V., Landfester, K., 2012. Design, synthesis, and miniemulsion polymerization of new phosphonate surfmers and application studies of the resulting nanoparticles as model systems for biomimetic mineralization and cellular uptake. Chem. Eur. J. 18 (17), 5201–5212. https://doi.org/10.1002/chem.201103256.

Sauer, R., Turshatov, A., Baluschev, S., Landfester, K., 2012. One-pot production of fluorescent surface-labeled polymeric nanoparticles via miniemulsion polymerization with Bodipy surfmers. Macromolecules 45 (9), 3787–3796. https://doi.org/10.1021/ma300090a.

Schiller, R., Weiss, C.K., Landfester, K., 2010. Phase stability and photocatalytic activity of Zr-doped anatase synthesized in miniemulsion. Nanotechnology. 21(40). https://doi.org/10.1088/0957-4484/21/40/405603.

Schoth, A., Wagner, C., Hecht, L.L., Winzen, S., Muñoz-Espí, R., Schuchmann, H.P., Landfester, K., 2014. Structure control in PMMA/silica hybrid nanoparticles by surface functionalization. Colloid Polym. Sci. 292 (10), 2427–2437. https://doi.org/10.1007/s00396-014-3316-7.

Schoth, A., Landfester, K., Muñoz-Espí, R., 2015. Surfactant-free polyurethane nanocapsules via inverse Pickering miniemulsion. Langmuir 31 (13), 3784–3788. https://doi.org/10.1021/acs.langmuir.5b00442.

Schoth, A., Adurahim, E.S., Bahattab, M.A., Landfester, K., Muñoz-Espí, R., 2016. Waterborne polymer/silica hybrid nanoparticles and their structure in coatings. Macromol. React. Eng. 10, 47–54. https://doi.org/10.1002/mren.201500029.

Schoth, A., Keith, A.D., Landfester, K., Muñoz-Espí, R., 2016. Silanization as a versatile functionalization method for the synthesis of polymer/magnetite hybrid nanoparticles with controlled structure. RSC Adv. 6, 53903–53911.

Schrade, A., Landfester, K., Ziener, U., 2013. Pickering-type stabilized nanoparticles by heterophase polymerization. Chem. Soc. Rev. 42 (16), 6823–6839. https://doi.org/10.1039/c3cs60100e.

Singh, I., Nigam, A.K., Landfester, K., Muñoz-Espí, R., Chandra, A., 2013. Anomalous magnetic behavior below 10 K in $YCrO_3$ nanoparticles obtained under droplet confinement. Appl. Phys. Lett. 103 (18), 1077–3118. https://doi.org/10.1063/1.4826503.

Singh, I., Landfester, K., Chandra, A., Muñoz-Espí, R., 2015. A new approach for crystallization of copper(II) oxide hollow nanostructures with superior catalytic and magnetic response. Nanoscale 7 (45), 19250–19258. https://doi.org/10.1039/c5nr05579b.

Singh, I., Landfester, K., Muñoz-Espí, R., 2017. Evolution of hollow nanostructures in hybrid Ce 1 $-\times$ Cu $\times$ O2 under droplet confinement leading to synergetic effects on the physical properties. Nanotechnology 28 (7), 075601. https://doi.org/10.1088/1361-6528/aa5376.

Slomkowski, S., Alemán, J.V., Gilbert, R.G., Hess, M., Horie, K., Jones, R.G., Kubisa, P., Meisel, I., Mormann, W., Stanislaw, P., Stepto, R.F.T., 2011. Terminology of polymers and polymerization and processes in dispersed systems and (IUPAC Recommendations and 2011). Pure Appl. Chem. 83, 2229–2259. https://doi.org/10.1351/PAC-REC-10-06-03.

Solans, C., Solé, I., 2012. Nano-emulsions: formation by low-energy methods. Curr. Opin. Colloid Interface Sci. 17 (5), 246–254. https://doi.org/10.1016/j.cocis.2012.07.003.

Solans, C., Izquierdo, P., Nolla, J., Azemar, N., Garcia-Celma, M.J., 2005. Nano-emulsions. Curr. Opin. Colloid Interface Sci. 10 (3–4), 102–110. https://doi.org/10.1016/j.cocis.2005.06.004.

Solans, C., Morales, D., Homs, M., 2016. Spontaneous emulsification. Curr. Opin. Colloid Interface Sci. 22, 88–93. https://doi.org/10.1016/j.cocis.2016.03.002.

Soula, R., Novat, C., Tomov, A., Spitz, R., Claverie, J., Drujon, X., Malinge, J., Saudemont, T., 2001. Catalytic polymerization of ethylene in emulsion. Macromolecules 34 (7), 2022–2026. https://doi.org/10.1021/ma0017135.

Staff, R.H., Landfester, K., Crespy, D., 2013. Recent advances in the emulsion solvent evaporation technique for the preparation of nanoparticles and nanocapsules. Adv. Polym. Sci. 262, 329–344. https://doi.org/10.1007/12_2013_233.

Staff, R.H., Schaeffel, D., Turshatov, A., Donadio, D., Butt, H.J., Landfester, K., Koynov, K., Crespy, D., 2013. Particle formation in the emulsion–solvent evaporation process. Small 9 (20), 3514–3522. https://doi.org/10.1002/smll.201300372.

Staudt, T., Machado, T.O., Vogel, N., Weiss, C.K., Araujo, P.H.H., Sayer, C., Landfester, K., 2013. Magnetic polymer/nickel hybrid nanoparticles via miniemulsion polymerization. Macromol. Chem. Phys. 214, 2213–2222. https://doi.org/10.1002/macp.201300329.

Sudol, E.D., El-Aasser, M.S., 1997. Miniemulsion Polymerization. In: John Wiley, Chichester, pp. 699–722.

Taden, A., Antonietti, M., Heilig, A., Landfester, K., 2004. Inorganic films from three different phosphors via a liquid coating route from inverse miniemulsions. Chem. Mater. 16 (24), 5081–5087. https://doi.org/10.1021/cm049958+.

Taniguchi, T., Takeuchi, N., Kobaru, S., Nakahira, T., 2008. Preparation of highly monodisperse fluorescent polymer particles by miniemulsion polymerization of styrene with a polymerizable surfactant. J. Colloid Interface Sci. 327 (1), 58–62. https://doi.org/10.1016/j.jcis.2008.08.003.

Tiarks, F., Landfester, K., Antonietti, M., 2001. One-step preparation of polyurethane dispersions by miniemulsion polyaddition. J. Polym. Sci. A Polym. Chem. 39 (14), 2520–2524. https://doi.org/10.1002/pola.1228.

Tiarks, F., Landfester, K., Antonietti, M., 2001. Preparation of polymeric nanocapsules by miniemulsion polymerization. Langmuir 17 (3), 908–918.

Torza, S., Mason, S., 1970. Three-phase interactions in shear and electrical fields. J. Colloid Interface Sci. 33 (1), 67–83. https://doi.org/10.1016/0021-9797(70)90073-1.

Urban, M., Musyanovych, A., Landfester, K., 2009. Fluorescent superparamagnetic polylactide nanoparticles by combination of miniemulsion and emulsion/solvent evaporation techniques. Macromol. Chem. Phys. 210 (11), 961–970. https://doi.org/10.1002/macp.200900071.

van Berkel, K.Y., Hawker, C.J., 2010. Tailored composite polymer-metal nanoparticles by miniemulsion polymerization and thiolene functionalization. J. Polym. Sci. A Polym. Chem. 48 (7), 1594–1606. https://doi.org/10.1002/pola.23917.

Varol, H., Álvarez Bermúdez, O., Dolcet, P., Kuerbanjiang, B., Gross, S., Landfester, K., Muñoz-Espí, R., 2016. Crystallization at nanodroplet interfaces in emulsion systems: a soft-template strategy for preparing porous and hollow nanoparticles. Langmuir 32 (49), 13116–13123. https://doi.org/10.1021/acs.langmuir.6b02954.

Wang, S., Schork, F.J., 1994. Miniemulsion polymerization of vinyl acetate with nonionic surfactant. J. Appl. Polym. Sci. 54 (13), 2157–2164. https://doi.org/10.1002/app.1994.070541318.

Wegner, G., Baum, P., Müller, M., Norwig, J., Landfester, K., 2001. Polymers designed to control nucleation and growth of inorganic crystals from aqueous media. Macromol. Symp. 175, 349–355.

Wegner, G., Demir, M.M., Faatz, M., Gorna, K., Muñoz-Espí, R., Guillemet, B., Gröhn, F., 2007. Polymers and inorganics: a happy marriage? Macromol. Res. 15 (2), 95–99.

Willert, M., Rothe, R., Landfester, K., Antonietti, M., 2001. Synthesis of inorganic and metallic nanoparticles by miniemulsification of molten salts and metals. Chem. Mater. 13 (12), 4681–4685. https://doi.org/10.1021/cm011121g.

Wohnhaas, C., Mailänder, V., Dröge, M., Filatov, M.A., Busko, D., Avlasevich, Y., Baluschev, S., Miteva, T., Landfester, K., Turshatov, A., 2013. Triplet-triplet annihilation upconversion based nanocapsules for bioimaging under excitation by red and deep-red light. Macromol. Biosci. 13 (10), 1422–1430. https://doi.org/10.1002/mabi.201300149.

Yang, L., Zhou, S., Wu, L., 2015. Preparation of waterborne self-cleaning nanocomposite coatings based on TiO_2/PMMA latex. Prog. Org. Coat. 85, 208–215. https://doi.org/10.1016/j.porgcoat.2015.04.012.

Zgheib, N., Putaux, J.L., Thill, A., D'Agosto, F., Lansalot, M., Bourgeat-Lami, E., 2012. Stabilization of miniemulsion droplets by cerium oxide nanoparticles: a step toward the elaboration of armored composite latexes. Langmuir 28 (14), 6163–6174. https://doi.org/10.1021/la300494g.

Zhou, J., Zhang, S., Qiao, X., Li, X., Wu, L., 2006. Synthesis of SiO_2/poly(styrene-*co*-butyl acrylate) nanocomposite microspheres via miniemulsion polymerization. J. Polym. Sci. A Polym. Chem. 44 (10), 3202–3209. https://doi.org/10.1002/pola.21434.

Part V

Characterization and Analysis of Nanoemulsions

Chapter 16

Characterization of Particle Properties in Nanoemulsions

Ana I. Bourbon*, Raquel F.S. Gonçalves*, António A. Vicente* and Ana C. Pinheiro*,†
*University of Minho, Braga, Portugal, †Institute of Experimental Biology and Technology, Oeiras, Portugal

Chapter Outline

16.1 INTRODUCTION

The functional attributes of emulsion-based products depend on the selection of the most appropriate types and concentrations of raw material (e.g., oil, water, emulsifiers, active compounds, and thickening agents) and the most adequate processing conditions (e.g., homogenization, mixing, storage temperature, and pasteurization) (McClements, 2004). These factors influence the particle properties of nanoemulsions, such as their size, concentration, charge, physical state, interfacial properties, and final applications (e.g., food, cosmetic, or pharmaceutical products).

Nanoemulsions. https://doi.org/10.1016/B978-0-12-811838-2.00016-3

The small droplet size of nanoemulsions, that is, mean radii between 10 and 500 nm, is responsible for their better stability to gravitational separation because the relatively small particle size means that Brownian motion effects dominate gravitational forces (McClements, 2015a). They also have good stability against droplet aggregation because the range of attractive forces acting between the droplets decreases with decreasing particle size, while the range of steric repulsion is less dependent on particle size (Silva et al., 2012). In addition, the relatively small droplet size compared with the wavelength of light ($r \ll \lambda$) means that they tend to be transparent or only slightly turbid, and therefore, they may be incorporated into optically transparent products without adversely affecting their clarity, for example, fortified soft drinks and waters (Mason et al., 2006; McClements, 2011; Mehrnia et al., 2016). Another important particle property of nanoemulsions is its spherical shape, which is the result of the contribution of a high interfacial tension and the small droplet size resulting in a high Laplace pressure that favors the minimization of the oil-water interfacial area (McClements, 2011). The design of nanoemulsions with specific properties depends on how particle properties can be rationally controlled.

In this chapter, the effect of parameters such as processing conditions and ingredients on particle properties (e.g., size, physical state of the droplets, and interfacial characteristics) will be analyzed. Moreover, different techniques will be discussed that can be used to evaluate the particle properties of nanoemulsions. The formulation and production of nanoemulsions have been described in other chapters in this book.

Different analytic techniques can be used to evaluate the particle properties in nanoemulsions (Jafari and Esfanjani, 2017; Jafari et al., 2017). In order to obtain reliable data, the choice of the most appropriate technique must be based on the samples' characteristics and on knowledge of the operation principles and advantages/drawbacks of each technique.

16.2 PARTICLE SIZE

It is very important to control and measure the particle size, since this property plays a major role in nanoemulsion characteristics such as appearance, stability, optical properties, rheology, release profile, and bioavailability (Jafari et al., 2008; Assadpour et al., 2016).

Nanoemulsion particle size can be measured using various analytic methods, and the nature of the particle-size data collected depends on the method used. For example, an average particle size or a particle-size distribution may be measured and reported. Also, depending on the method chosen, these data can be obtained by direct observation of the sample (e.g., microscopy) or could be inferred from measurements of some particle-size-dependent physical property of the nanoemulsions by fitting the data with an appropriate mathematical model (McClements, 2015c). The most widely used methods for measuring particle-size data are discussed below.

16.2.1 Microscopy

Microscopy can be used as a direct imaging technique for nanoemulsions. Besides providing important information about nanoemulsions' morphology and structure, microscopy techniques can also provide valuable information about their size, being often used as a complement to other sizing techniques (Hosseini et al., 2015). Various microscopy techniques are available, with the most popular ones used to analyze nanoemulsions being electron microscopy (Li et al., 2017), laser scanning confocal microscopy (LSCM) (Su and Zhong, 2016), and atomic force microscopy (AFM) (Ghasemi et al., 2017, 2018). These techniques typically have better resolution and sensitivity than conventional optical microscopy.

In fact, electron microscopy is widely used to examine the structure of nanoemulsions, providing information about the concentration, dimensions, and location of components within the sample. Two types of electron microscopy are commonly used: transmission electron microscopy, which gives a two-dimensional cross section of the sample, and scanning electron microscopy, which gives a three-dimensional image (Klang et al., 2012). LSCM can also provide extremely valuable information about the structure of nanoemulsions and has the advantage of allowing the observation of the microstructure of individual components of a system (and consequently the determination of each component's size) by using the natural fluorescence of certain components or by using fluorescent dyes that bind selectively to specific components. This method also allows the generation of three-dimensional images of samples. Also, AFM has the ability to provide valuable information about nanoemulsions' structure at the atomic and molecular level, since high resolution (± 0.1 nm) can be achieved, providing a high-resolution three-dimensional profile of the surface under study (Silva et al., 2012).

However, these microscopy techniques present some drawbacks, with the major ones being the fact that sample preparation often alters sample structure and is time-consuming, since it is often necessary to analyze different regions within a sample to obtain statistically reliable data. More details on these microscopy techniques can be found in Chapter 17.

16.2.2 Light Scattering

Light scattering is a well-established technique to investigate the properties of particles in solutions, such as size, molecular weight, diffusion, and interaction strength.

16.2.2.1 Static Light Scattering

Static light scattering (SLS), also called laser diffraction, is a widely used technique to determine the droplet-size distributions in nanoemulsions. SLS is based on the principle that a beam of light that irradiates a sample is scattered by the droplets. By measuring the intensity of the scattered light (i.e., mean intensity) at several angles, it is possible to determine the particle-size distribution using a suitable theoretical model, for example, Mie theory. Because small particles scatter light at wide angles, the lower measurement limit of SLS is typically 30–100 nm (Egerton and Tooley, 2014), and the upper limit is around 1000 μm. The advantages of this method are that it is a fast and nondestructive technique (Clementi et al., 2012); however, to obtain reliable data, before measurement, samples must be diluted to avoid multiple-scattering effects and stirred to ensure homogeneity. Also, the complex theoretical model used to interpret light scattering data is usually simplified making some a priori assumptions: in most cases, it is assumed that the scattering particles are spherical, homogeneous, and noninteracting. However, care should be taken once these assumptions may not always be valid and may lead to errors under certain circumstances (McClements, 2015b).

16.2.2.2 Dynamic Light Scattering

Dynamic light scattering (DLS), also known as photon correlation spectroscopy or quasielastic light scattering (Bhattacharjee, 2016), is one of the most popular methods used to determine the particle size of nanoemulsions. During DLS measurements, the sample is irradiated by a light beam, and the intensity-time fluctuations of the reflected or transmitted light beam are measured (Lim et al., 2013). It is possible to determine their particle sizes from the frequency shift of the incident light beam when scattered by particles in motion inside a colloidal suspension. The scattered intensity is monitored as a function of time and as opposed to SLS, and the measurement is normally carried out at a single scattering angle (typically 90 degree). For spherical particles in dilute suspensions (where there are no particle-particle interactions), the particle hydrodynamic radius, R_H, can be calculated from its translational diffusion coefficient by the Stokes-Einstein equation:

$$D_\mathrm{f} = \frac{k_\mathrm{B} T}{6\pi\eta R_H} \tag{16.1}$$

where k_B is the Boltzmann constant, T is the absolute temperature, and η is the viscosity of the continuous phase.

The particle-size data are usually presented as the intensity-weighted mean of the hydrodynamic diameter and the polydispersity index (PDI). The PDI characterizes the width of the particle-size distribution and thus the homogeneity of the formulation, with a PDI <0.2 being indicative of a narrow droplet-size distribution, whereas values close to 1 are found for samples that exhibit a very high particle-size heterogeneity (Klang and Valenta, 2011).

DLS is capable of analyzing particles with diameters between about 3 nm and 3 μm, and it presents several advantages over other particle-size methodologies, such as (i) short measurement times, (ii) simple and mostly automated process, (iii) no preexperimental calibration required, (iv) noninvasive, and (v) highly sensitive. Also, DLS allows analysis of particles that are smaller than the lower detection limit of SLS. One of the major limitations of DLS is that the viscosity of the aqueous phase is assumed to be Newtonian, which is not always the case for emulsions (McClements, 2015b), and the need to use low concentrations (i.e., diluted samples) to avoid multiple scattering (Egerton and Tooley, 2014).

16.2.3 Electric Pulse Counting

Electric pulse counting, also known as the Coulter counter, is based on measurement of emulsion droplet size and distribution by detecting the magnitude of electric conductive variation when passing a dilute emulsion through a small orifice in an electrode (Hu et al., 2017). Due to the fact that oil has a much lower electric conductivity than water, the electric current decreases each time an oil droplet passes through the orifice. This decrease in current is converted into an electric pulse, and when the droplets are small compared with the diameter of the orifice, the droplet size can be related to the height of the pulse through

$$d^3 = kP \tag{16.2}$$

where d is the droplet diameter, P is the pulse height, and k is an instrument constant determined by measuring the pulse height of a suspension of monodisperse particles of known diameter.

This method is suitable for measuring particles with diameters ranging between about 0.4 and 1200 μm; however, a number of tubes with different-sized orifices are required to cover this whole range (Lines, 1996). In order to obtain accurate measurements, the diameter of the emulsion droplets should be between 2% and 60% of the diameter of the orifice.

Moreover, there are a number of practical issues associated with this technique that must be taken into consideration (McClements, 2015b): (i) samples should be diluted in a way that only one particle passes through the orifice at a time; (ii) the emulsion droplets must be suspended in an electrolyte solution, to obtain an aqueous phase with sufficiently large electric conductivity, and an electrolyte solution that does not promote droplet aggregation should be chosen; (iii) weakly flocculated droplets may be disrupted as they pass through the small orifice; (iv) for emulsions containing a wide droplet-size distribution, glass tubes with different orifice sizes are needed to cover the full-size distribution; (v) emulsion droplets may cream or sediment before being pulled through the orifice, if there is a large density contrast between the dispersed and continuous phase; and (vi) the lower size limit is 0.4 μm. For measuring particle sizes in nanoemulsions, electric pulse counting is often considered to be more reliable and accurate than light scattering.

16.2.4 Sedimentation

This technique is based on the principle that large particles sediment more quickly than small ones (Allen, 1992). Therefore, the size of nanoemulsions in suspension may be inferred by measuring the velocity at which droplets cream or sediment in a gravitational or centrifugal field. Using different instruments, a wide range of sizes (between 1 nm and 1 mm) can be covered.

For gravitational sedimentation, Stokes' Law (Eq. 16.3) can be used to determine the droplet size by measuring the velocity (v) at which spherical particles move through a Newtonian liquid:

$$v = \frac{2(\rho_2 - \rho_1)gr^2}{9\eta_1} \tag{16.3}$$

where ρ_2 is the density of the droplets, ρ_1 is the density of the surrounding fluid, g is the acceleration due to gravity, r is the droplet radius, and η_1 is the viscosity of the surrounding fluid.

The gravitational sedimentation is typically limited to large particles since the rate of sedimentation of small particles is too low, which results in a very high analysis time, and also due to the fact that Brownian motion effects of small particles become too large to allow effective settling. In fact, Stokes' equation cannot be used to reliably estimate the size of droplets with radii less than about 0.1 μm because of Brownian motion effects.

Centrifugal sedimentation extends the range of particle sizes that could be analyzed by this approach. At sufficiently high centrifugal forces, the sedimentation rate is sufficiently high that perturbations due to Brownian motion may be neglected (Egerton and Tooley, 2014). In this case, the droplet motion can be characterized by a sedimentation coefficient (S) that depends on the velocity ($v(x)$) at which the droplets move through the surrounding liquid and on the angular velocity (ω) at which the tube is centrifuged and their distance from the center of the rotor (x):

$$S = \frac{v(x)}{\omega^2 x} \tag{16.4}$$

Being the radius of an isolated spherical particle in a fluid is related to S by the following equation:

$$r = \sqrt{\frac{9\eta_1 S}{2(\rho_2 - \rho_2)}} \tag{16.5}$$

This method allows fractionating the droplets in an emulsion into different size classes, which could be an advantage for further analyzing each size class (McClements, 2015b).

16.2.5 Ultrasonic Spectrometry

Ultrasonic spectrometry is based on the measurement of the change in ultrasonic velocity and/or attenuation coefficient with frequency, which is strongly affected by the size distribution of the particles (Shukla et al., 2010). Basically, there are two steps involved: the measurement of the ultrasonic velocity and/or attenuation coefficient of a nanoemulsion over a wide range of frequencies and then the conversion of these values into a droplet-size distribution using an appropriate mathematical theory, that is, finding the droplet size and concentration that gives the best fit between the experimental data and the theory (McClements, 1996).

As an ultrasonic wave propagates through an emulsion, its velocity and attenuation are altered due to its interaction with the droplets, which may involve both scattering and absorption mechanisms. The scattering of ultrasound by a droplet can be divided into three categories according to the relationship between the droplet size (i.e., radius, r) and the ultrasonic wavelength (λ) (McClements, 2015b): (1) long-wavelength regime ($r < \lambda/20$), (2) intermediate-wavelength regime ($\lambda/20 < r < 20\lambda$), and (3) short-wavelength regime ($r > 20\lambda$). Most ultrasonic measurements on nanoemulsions ($r = 10$–100 nm) using typical ultrasonic frequencies (0.1–100 MHz) fall into the long-wavelength regime. In this case, the ultrasonic properties of fairly dilute emulsions ($\phi < 15\%$) can be related to their physicochemical characteristics using the following equation:

$$\frac{K^2}{k_1} = \left(1 - \frac{3\varnothing i A_0}{k_1^3 r^3}\right)\left(1 - \frac{9\varnothing i A_1}{k_1^3 r^3}\right) \tag{16.6}$$

where K is the complex propagation constant of the emulsion, k_1 is the complex propagation constant of the continuous phase, r is the droplet radius, and A_0 and A_1 terms are the monopole and dipole scattering coefficients of the individual droplets.

This technique can be used to measure droplet sizes in range between about 0.1 nm and 1000 μm and has as main advantages the fact of being nondestructive, is capable of rapid and precise measurements, is applied to samples that are optically opaque, does not require sample preparation (e.g., dilution), and can be used in situ (Forrester et al., 2016). The main disadvantages of this technique include the need of a large amount of information about the physicochemical properties of the component phases (often not readily available in the literature) to interpret the measurements and the possible interference of air bubbles with the signal from the emulsion droplets (McClements, 1996). In addition, care must be taken to ensure that the theory used to interpret the ultrasonic spectra obtained from an emulsion is appropriate to obtain reliable measurements.

16.2.6 Nuclear Magnetic Resonance

Nuclear magnetic resonance (NMR) is based on measurements of the interactions between radio waves and the nuclei of hydrogen atoms to obtain information about materials' properties. Within the arsenal of NMR techniques, pulsed field gradient (PFG)-based methods are one of the most powerful for assessment of particle-size distributions (Voda and Van Duynhoven, 2009). PFG NMR is based on the restriction of the self-diffusion of the molecules of the discrete phase (i.e., droplets) in an emulsion by the boundary between the phases (i.e., interior droplet surfaces) (Hollingsworth and Johns, 2003). Basically, a series of radio frequency pulses are applied to the sample, causing the excitation of some of their hydrogen nuclei to higher energy levels and leading to the generation of a detectable NMR signal. The amplitude of this signal depends on the movement of the nuclei in the sample, and by measuring its attenuation at different times, it is possible to identify when the diffusion becomes restricted, thereby allowing estimation of the droplet size (Lee et al., 1998).

This technique can be used to measure particle sizes between about 100 nm and 100 μm and would therefore not be suitable to application to most nanoemulsions. However, it can provide valuable information about the structure of flocculated emulsions or can monitor changes in nanoemulsions' droplet size due to coalescence or Ostwald ripening. NMR is a nondestructive method, which can be used to analyze concentrated and optically opaque emulsions in situ. However, the widespread use of NMR for size measurement is limited due to the fact that the equipment needed is relatively expensive and requires skilled operators (McClements, 2015b).

As shown, a wide variety of analytic methods are currently available for the measurement of the particle size of nanoemulsions. Each method has its own advantages and disadvantages, and the choice of a particular technique depends on the material being tested and the information required. Table 16.1 shows a comparison of the most common techniques used to measure nanoemulsions' droplet size.

16.3 PARTICLE CONCENTRATION

Particle concentration in a nanoemulsion plays a major role in its cost, appearance, rheology, taste, stability, and nutritional attributes (McClements, 2015c). For example, an emulsion tends to become more turbid and more viscous as the concentration of oil droplets increases. This property determines the relative proportion of oil and water in a product and can be controlled by altering the initial amount of oil used to prepare the nanoemulsions or by diluting or concentrating the system after its production.

The particle concentration is often known because the ingredients' concentrations are carefully controlled during nanoemulsion production. However,

TABLE 16.1 Comparison of the Techniques to Measure Droplet Size

Method	Droplet-Size Range	Concentration Range	Advantages	Disadvantages	Cost
Microscopy	>5 nm	0.1%–50%	Allows structure visualization	Destructive Need sample preparation Time-consuming Expensive	$$$$
SLS	30 nm–1000 μm	<0.1%	Easy to operate Fast	Need sample preparation	$$
DLS	1 nm–6 μm	From <0.01% to >10%	Easy to operate Fast	Need sample preparation	$$
Coulter	0.4–1200 μm	<0.1%	Fast Reliable and accurate	Need sample preparation Influence of measurement conditions Need various glass tubes	$$
Sedimentation	1 nm and 1 mm	Depends on the method	Allows droplets' fractionation	Need sample preparation Destructive	$
Ultrasonic spectroscopy	0.1–1000 μm	<50%	Fast Precise No sample preparation Nondestructive Applied to optically opaque samples In situ use	Knowledge of physicochemical properties Interference of air bubbles	$$$
NMR	0.2 and 100 μm	<80%	No sample preparation Nondestructive Applied to optically opaque samples	Expensive	$$$$

$, inexpensive; $$$$, very expensive.

alterations in particle concentration could occur due to, for example, creaming or sedimentation during storage. Particle concentration is often described in terms of the dispersed-phase volume fraction (∅):

$$\varnothing = \frac{V_D}{V_E} \tag{16.7}$$

where V_D is the volume of emulsion droplets and V_E is the total volume of the emulsion.

Many of the techniques used to determine droplet size (i.e., light scattering, electric pulse counting, ultrasonic spectroscopy, and NMR) can also be used to measure the dispersed-phase volume fraction. Other techniques are briefly discussed below.

16.3.1 Proximate Analysis

Proximate analysis techniques can often be used to provide information about the oil or water content of emulsions. For example, solvent extraction can be used to determine the oil content, while oven drying can be used to determine the moisture content. However, many of these techniques are destructive and often labor-intensive and time-consuming and therefore are not suitable for rapid quality control or online measurements (McClements, 2008).

16.3.2 Electrical Conductivity

Dispersed-phase volume fraction of a nanoemulsion can also be determined by measuring its electric conductivity, ε_e, due to the differences in the electric conductivity of water and oil (Clausse, 1983). In fact, ε_e decreases as the concentration of oil within an emulsion increases. The dispersed-phase volume fraction can be related to the electric conductivity by the following equation, which is valid for diluted emulsions:

$$\varnothing = \left(\frac{\varepsilon_e - \varepsilon_1}{\varepsilon_e - 2\varepsilon_1}\right)\left(\frac{\varepsilon_2 - 2\varepsilon_1}{\varepsilon_2 - \varepsilon_1}\right) \tag{16.8}$$

where ε_1 and ε_2 are the electric conductivity of the continuous phase and dispersed phase, respectively.

The electric conductivity of a nanoemulsion can be measured using a simple conductivity cell, and this technique can be used with concentrated and optically opaque emulsions without the need of any sample preparation. Although the measurements are independent of the size of the nanoemulsion droplets, they depend on the physical state of its constituents, which means that it may be necessary to heat an emulsion to melt the crystals before the measurement (McClements, 2015b).

16.3.3 Density Measurements

Density measurements are one of the simplest means of determining the dispersed-phase volume fraction of nanoemulsions (Pal, 1994). The density (ρ_e) of a nanoemulsion can be expressed as

$$\rho_e = \varnothing\rho_2 + (1 - \varnothing)\rho_1 \tag{16.9}$$

Here, ρ_1 and ρ_2 are the densities of the continuous and dispersed phases, respectively. From Eq. (16.9), it is clear that the dispersed-phase volume fraction of a nanoemulsion can be determined by measuring its density and knowing the density of the oil and aqueous phases. The density of nanoemulsions can be measured using different methods, such as using a pycnometer (Ribes et al., 2017), a hydrometer (Buffo et al., 2001), or an oscillating U-tube densitometer (Pal, 1994).

The main advantages of this approach are that typically, inexpensive equipment is required, it is a nondestructive technique, and it can be used to analyze nanoemulsions that are concentrated and optically opaque. A major limitation of density measurements is the fact that the accuracy of the determination may be influenced by the physical state of the emulsion constituents (McClements, 2015b).

16.4 PARTICLE CHARGE

The particle charge is a relevant property for the stability and functional activity of nanoemulsions. Usually, the droplets in nanoemulsions have an electric charge due to adsorption of mineral ions, ionized emulsifiers, or biopolymers to their surfaces (Rao and McClements, 2011).

The particle charge can be characterized by different electric properties such as surface-charge density (σ), surface potential (ψ_0), and ζ-potential (ζ). The surface-charge density is the number of electric charges per unit of surface area. The surface electric potential is the free energy required to increase the surface-charge density from zero to σ by taking charges from an infinite distance to the surface. The ionic strength influences the surface potential of the droplets due to electrostatic screening effects (Rao and McClements, 2011). The ζ-potential is the most commonly used parameter to determine the electric properties of the droplets in nanoemulsions due to its good representation in terms of electric properties and easier determination (Rao and McClements, 2011). It consists in the surface potential at the shear plane, that is, the limit between mobile and stationary phases (Honary and Zahir, 2013). For electrostatically stabilized systems, the ζ-potential can be correlated to the stability of the nanoemulsions. Values of ζ-potential above 60 mV typically provide excellent stability and good stability above 30 mV; when between 5 and 15 mV, ζ-potential is in the region of limited flocculation, and between 3 and 5 mV, it corresponds to the maximum flocculation (Heurtault, 2003; Silva et al., 2011). Many factors can influence the nanoemulsions' ζ-potential, such as the medium pH, the electrolyte concentration, the particle morphology, the state of hydration, the sources of the particles, and the treatment with different surfactants (Simunkova et al., 2009). The surface potential of a droplet is counterbalanced by the excess charge of the counterions (ions with opposite charge) in the surrounding electrolyte medium, which is referred to as the electric double layer (McClements, 2015e; Norde, 2011). Once the distribution of counterions and coions around the particle surface occurs in this layer, any movement between rigid (Stern layer) and mobile (diffuse layer) phases results in the production of an electrokinetic potential (please see Fig. 16.1) (Fairhurst and Lee, 2011a). The electrokinetic potential can be qualified based on the driving force and the resultant movement. The driving force can be an applied electric potential gradient, that is, electric field, or a mechanical potential gradient, that is, pressure difference, gravitational, or centrifugal field. Relative to the resultant particle movement, it is divided between the mobile and the stationary phase (Norde, 2011). The ζ-potential is calculated from measurements of the electrodynamic properties of the particles using suitable mathematical models and can be determined from particle electrophoresis, sedimentation potential, streaming potential, and electroosmosis measurements (Fairhurst and Lee, 2011a; Norde, 2011).

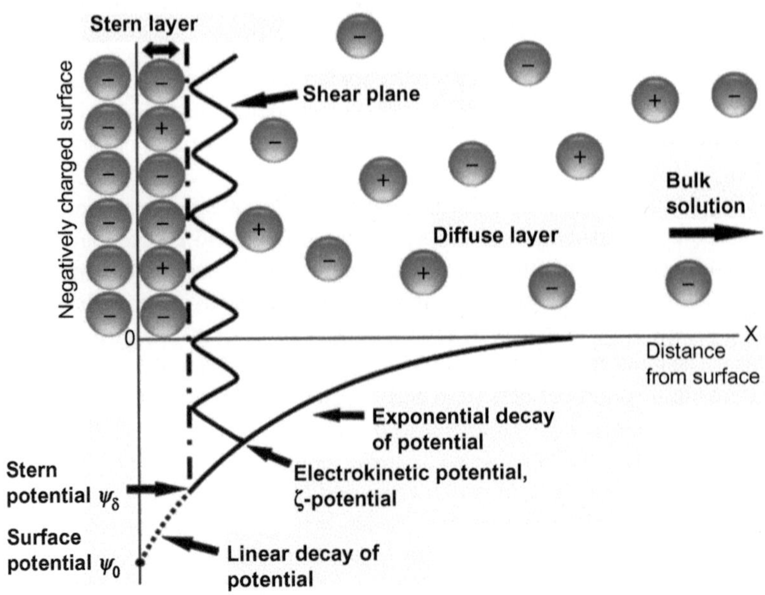

FIG. 16.1 Schematic representation of the electric double layer at a charged interface in aqueous solution. *(Modified from Fairhurst, D., Lee, R.W., 2011. The zeta potential & its use in pharmaceutical applications—part 1: charged interfaces in polar & non-polar media & the concept of the zeta potential. Drug Dev. Deliv. 11, 60–64.)*

16.4.1 Electroosmosis

In the electroosmosis method, an electric field is applied to a colloidal dispersion, and the dispersed phase is stationary, while the continuous phase moves. Considering a capillary with the charged inner walls containing an electrolyte solution, the motion of the ions in the solution is caused by the application of a potential difference between the two edges of the capillary. This movement will result in a flow of the liquid in the same direction as the ions. The ζ-potential is determined by measurement of the electric current (I) and the electroosmotic volume flux (J_{eo}) through the capillary using an electrolyte solution of known viscosity (η), specific conductivity (κ_{sp}), and permittivity (ε) (Norde, 2011):

$$J_{eo} = \frac{\zeta \varepsilon \varepsilon_0 I}{\eta \kappa_{sp}} \tag{16.10}$$

16.4.2 Electrophoresis

The electrophoresis is promoted through the application of an electric field between two electrodes where the charged particles move in a stationary liquid phase to the oppositely charged electrode (Uskokovic, 2012; McClements, 2016). The sign of the charge may be derived from the direction of their movement. Electrophoretic mobility (μ) of charged particles is measured through changes in the intensity of the scattered light produced by particles that move in the alternating electric field. This parameter allows calculation of the ζ-potential of the particles using the Smoluchowski equation (Uskokovic, 2012):

$$\zeta = \frac{\eta \mu}{\varepsilon} \tag{16.11}$$

This mathematical model is dependent on the size and charge of the particles, the thickness of Debye length (κ^{-1}), the viscosity of the solution, and the strength of the applied electric field. Debye length is related to the thickness of the electric double layer, that is, the distance that it is possible to detect the electric properties within the surrounding solution (Rao and McClements, 2011; McClements, 2016).

The most common technique that uses this approach is electrophoresis light scattering (ELS), which is also known as laser-Doppler electrophoresis. Another technique that uses electrophoresis is phase analysis light scattering (PALS). The principal differences between these two techniques are the fact that ELS produces a histogram of the ζ-potential measured in the suspensions and can also determine the particle-size distribution, whereas PALS can only provide an average value for the ζ-potential. The main limitation of instruments that use ELS or PALS techniques is the need to dilute the nanoemulsion before the measurement, which leads to possible instability of the nanoemulsion and the subsequent extrapolation of ζ-potential from the dilute suspensions to the practical concentration of the nanoemulsion (Fairhurst and Lee, 2011a).

16.4.3 Streaming Current

Streaming current techniques can be considered to be the opposite of electro-osmotic ones. In other words, a pressure difference (Δp) is applied between the two ends of a capillary with charged inner walls containing an electrolyte solution, causing the solution to flow. Thus, an electric current is produced by the mechanical stress (Fairhurst and Lee, 2011a; Norde, 2011). The streaming current is measured when the two electrodes are connected to a low resistance (Norde, 2011).

Consider the situation where the thickness of the electric double layer is much smaller than the radius of the capillary ($\kappa a \gg 1$). Applying the boundary conditions $\psi = \zeta$ at $x = 0$ and $\psi = 0$ at $d\psi(x)/dx = 0$ at $x = a$, it is possible to obtain the relation between the ζ-potential and streaming current (I_s) through the following equation:

$$\frac{I_s}{\Delta p} = \frac{\varepsilon \varepsilon_0 \pi a^2}{\eta l} \zeta \tag{16.12}$$

in which a is the radius of the capillary and l is the length of the capillary.

16.4.4 Sedimentation Potential

In the sedimentation potential technique, the dispersed phase moves through the stationary liquid under the influence of gravity, which produces an electric potential. However, when the gravitational force is replaced by an acoustic field, it is named electroacoustic attenuation. There are two ways to do electroacoustic measurements: (1) electrosonic amplitude (ESA) and (2) colloid vibration potential (CVP) (Fairhurst and Lee, 2011a).

ESA is a method in which an electric signal is applied to an emulsion and the acoustic signal produced by the oscillating particles is obtained. An alternating electric field is applied to a nanoemulsion causing a rapid movement of the charged droplets (Dukhin et al., 2000). This oscillation of the droplets creates a wave pressure with the same frequency as the alternating electric field that is detected by an ultrasonic transducer (Dukhin and Goetz, 2001). The amplitude of the signal received by the transducer (i.e., ESA) is proportional to the dynamic mobility (μ_d) of the particles in the emulsion. The following equation shows the relation of the dynamic mobility with the ζ-potential and the size of the particles (McClements, 2016):

$$\mu_d = \left(\frac{\varepsilon_0 \varepsilon_R \zeta}{\eta}\right) G \left(\frac{\omega r^2 \rho}{\eta}\right) \tag{16.13}$$

where ε_R is the relative permittivity of the continuous phase; ω is the angular frequency; r is the radius droplet; and ρ are the viscosity and density of the continuous phase, respectively; and G is a function that varies between 1 at low frequencies and 0 at high frequencies. The ζ-potential can be determined at low frequencies, when it is independent of the particle size.

The CVP technique is the opposite of the ESA technique. It consists in the application of an ultrasonic wave to the nanoemulsion, which causes an oscillation of the droplets due to the density difference between themselves and the surrounding medium. The oscillation of the charged droplets produces an alternating electric field that is detected by electrodes. The amplitude of the signal detected by the electrodes (known as CVP) is used to calculate the ζ-potential and the size of the droplet (Dukhin and Goetz, 2001; Dukhin et al., 2000).

The principal advantage of the electroacoustic methods is the possibility to measure the ζ-potential of the droplets in concentrated nanoemulsions without dilution. However, they may only be used in nanoemulsions with charged droplets and with a significant density difference to the surrounding medium (McClements, 2016).

16.5 PARTICLE PHYSICAL STATE

Most nanoemulsions are composed of liquid oils, but nowadays, it is also possible to form particles with oils that are fully or partly crystallized at the application temperature (McClements and Xiao, 2012). Solid lipid nanoparticles (SLN) and nanostructured lipid carriers are two types of nanoemulsions where the lipids are fully or partly crystallized. Generally, during the preparation of the nanoemulsion, the lipid phase is maintained in a liquid state by keeping the temperature above the melting point. The nanoemulsion is then cooled to cause the lipid phase to fully or partly crystallize. The state of the lipid particles in the resultant nanoemulsion can be controlled through their composition and preparation conditions such as temperature and oil type (McClements, 2011). The physical state of the lipid phase influences texture, stability, mouthfeel, and appearance of the nanoemulsions. Moreover, its crystallinity also influences the release rates and the encapsulation efficiency of the compounds (McClements, 2011, 2015d). The droplets are crystalized at a controlled rate to prevent their instability. Moreover, the concentration, type, and location of the crystals within the droplets can be controlled through the selection of the oil type, the thermal history, the emulsifier type, and the droplet size (McClements, 2011). Due to supercooling and curvature effects, the lipids incorporated within small oil droplets can crystallize at a much lower temperature than in the bulk phase, and the nature of the crystals formed in the droplets may be different from those formed in the bulk phase (Gibout et al., 2007; Wissing et al., 2004). There are several techniques to measure or to analyze the physical state of the particles in nanoemulsions.

16.5.1 Thermal Analysis

16.5.1.1 Differential Scanning Calorimetry

Differential scanning calorimetry (DSC) is a thermal technique that can be used to analyze the physical properties of nanoemulsions over a range of temperatures or times (Silva et al., 2011). This technique can provide information about properties such as melting point, heat of fusion, phase transitions (glass transition endotherm, crystallization exotherm, and fusion endotherm), percentage of crystallinity, crystallization kinetics, and oxidative stability (Gill et al., 2010; Heurtault, 2003). In general, DSC quantifies the heat that is radiated or absorbed by the sample when its temperature is changed in a controlled fashion. The operation mechanism of DSC can be divided into two types (Gill et al., 2010):

(1) *Heat-flux DSC*: This method consists of heating the sample and an empty reference in the same furnace at a linear heating rate, and the temperature difference between them is measured using thermocouples. The resultant heat flow (q) is determined by the thermal equivalent of Ohm's law:

$$q = \frac{\Delta T}{R} \tag{16.14}$$

where ΔT is the temperature difference between the sample material and the reference and R is the resistance of thermoelectric disc.

(2) *Power-compensated DSC*: This method consists of heating the sample material and the reference in different furnaces, being both maintained at the same temperature; the difference of the thermal power required to keep both samples at the same temperature is measured and plotted as a function of temperature or time.

The crystallinity percentage can be determined by DSC using two methods. The first one consists in the elaboration of calibration curves of 100% amorphous and crystalline standards and different mixtures of them (Gill et al., 2010). The crystallinity percentage (X_c) of a determined sample material is calculated through the following equation:

$$X_c = \frac{\Delta H}{\Delta H_0} 100 \tag{16.15}$$

where ΔH is the enthalpy of fusion of the sample and ΔH_0 is the enthalpy of fusion of 100% crystalline standard. The second one consists in the subtraction of the heat of crystallization from the heat of fusion of the sample. These result in the heat of primary crystallinity of the sample; thus, the crystallinity percentage can be calculated through the following equation:

$$X_c = \frac{\Delta H_f - \Delta H_c}{\Delta H_{cry}^{f}} m \tag{16.16}$$

16.5.1.2 Differential Thermal Analysis

Differential thermal analysis (DTA) methods are very similar to DSC methods and can also be used to provide similar information about the physical properties of nanoemulsions. DTA measures the temperature difference (ΔT) between a sample and reference over a temperature range using thermocouples below each sample. The results are plotted as ΔT versus temperature and provide valuable information about physical and chemical changes in the sample (Clausse, 2010; McClements, 2016).

16.5.1.3 Ultrasonic Spectrometry

Ultrasonic spectrometry can also be used to analyze the phase transitions that occur in nanoemulsions (Buckin et al., 2003). The ultrasonic velocity is one of the parameters that can be measured through ultrasonic spectrometry, and it is dependent on the density and elasticity of the medium and temperature (Smyth et al., 2005). The ultrasonic velocity is higher when the fat is solid than when it is liquid (Buckin et al., 2003; McClements, 2016). There are two ways to do these measurements: isothermally, where the temperature is kept constant and the ultrasonic velocity is measured as a function of time, or through temperature scanning, where the ultrasonic velocity is determined as a function of the increase or decrease of the temperature at a controlled rate (McClements, 2016). Furthermore, ultrasonic spectroscopy allows the measurement of the crystal growth through time and temperature and the evaluation of the quantity of the crystals in the sample. The principal advantage of this technique is the possibility to analyze the sample directly without having to make any dilution (Smyth et al., 2005).

16.5.1.4 X-Ray Diffraction

X-ray diffraction (XRD) is a nondestructive technique, and it is capable of providing information about chemical composition, physical properties, and the crystallographic structure of samples. The principle of the XRD is the observation of the scattered intensity of an X-ray beam into a sample as a function of the wavelength or energy, incident and scattered angles, and the polarization (Silva et al., 2011).

16.5.1.5 Dilatometry

Dilatometry is used to analyze the crystallinity of the continuous and dispersed phases of a nanoemulsion. This method measures the density changes when a material melts or crystallizes and determines the fraction of crystalline droplets (ϕ_c) using the following equation (McClements, 2016):

$$\phi_{\mathrm{c}} = \frac{\rho_{\mathrm{e}} - \rho_{\mathrm{eL}}}{\rho_{\mathrm{eS}} - \rho_{\mathrm{eL}}} \tag{16.17}$$

in which ρ_e is the density of a nanoemulsion with droplets partially crystallized and ρ_{eS} and ρ_{eL} are the densities of the same nanoemulsion when the droplets are completely solid or liquid, respectively.

Dilatometry measurements can be performed through the determination of the changes on the height of the mercury or colored water in a capillary tube that is connected to a glass blub with a known mass sample. The changes observed in the height of mercury are promoted by changes in the volume of the sample when it crystallizes or melts. The results can be presented as a density change or a volume change. It can be carried out at constant temperature over time or through a temperature slope (McClements, 2016).

16.5.1.6 Nuclear Magnetic Resonance

NMR is an analytic technique that can be used to analyze concentrated and opaque nanoemulsions without sample dilution. NMR has largely replaced the traditional dilatometry technique for monitoring phase changes in materials. The solid content is determined by monitoring the interactions of radio waves with the nuclei of the hydrogen atoms in the sample. The ratio of the solid-to-liquid droplets present in a nanoemulsion depends on measuring the amplitude and decay time of the NMR signal produced by the excitation state of the hydrogen nuclei: solids lead to a more rapid decay of the NMR signal than liquids (McClements, 2016).

16.6 INTERFACIAL CHARACTERISTICS

Interfacial properties are particularly important parameters for the design of stable nanoemulsion systems. The principal interfacial properties that can be determined and controlled are the charge, composition, structure, tension, and rheology. All are related with several nanoemulsions' functionalities such as stability, interfacial permeability, environmental responsiveness, rheology, and interactions (Rao and McClements, 2011). Table 16.2 presents several techniques used for the characterization of the interfacial properties of nanoemulsions.

The interfacial charge determines the electrostatic interactions of nanoemulsion droplets and therefore influences their stability and physicochemical properties. The charge of the droplet is dependent on pH, ionic strength, and the nature of the emulsifiers adsorbed to the surfaces (McClements, 2015e). There are several techniques to determine and monitor the interfacial charge (Table 16.2).

Interfacial composition also influences nanoemulsion stability, rheology, and formation. Nanoemulsions can be formed by one or more emulsifiers, and their adsorption is dependent on the initial emulsifier concentration, ionic strength, pH, temperature, and homogenization conditions. The interfacial composition is determined by the concentration of different surface-active

TABLE 16.2 Description of Techniques Used to Measure the Different Interfacial Properties of Nanoemulsions

Properties	Techniques	Description	Examples	Articles
Interfacial charge	Electrokinetics techniques	Based on the movement of a charged particle or a charge surface in relation to an electrolytic solution	Electrophoresis, electroosmosis, sedimentation, potential, streaming current and potential	Pinheiro et al., (2016), Addai-Mensah and Ralston (2004), and Duval et al. (2011)
Interfacial composition	Depletion technique	Consists in the determination of the concentration of the adsorbed emulsifier in the interface through the difference between the initial concentration and the nonadsorbed emulsifier concentration	Chromatography, electrophoresis	Robson and Dalgleish (1987), Ye (2008), Singh et al. (2003), and McClements (2015e)
Interfacial structure	Scattering techniques	The principle of these techniques is the application of incident beam of radiation (X-rays, light, or neutrons) to the sample and measuring the intensity and angle of the scattered beam	Dynamic light scattering, X-ray scattering, neutron scattering, small-angle X-ray scattering, small-angle neutron scattering	Ostertag et al. (2012), Liu et al. (2017), Acharya and Hartley (2012), Shukat et al. (2011), Wang et al. (2009), and Hoffmann et al. (2012)
	Microscopic techniques	Provide a direct visualization of internal and surface of nanostructures	Fluorescent microscopy, Raman microscopy, atomic force microscopy, electron microscopy, Brewster angle microscopy	Preetz et al. (2010), Nik et al. (2012), Liu et al. (2017), and Acharya and Hartley (2012)
	Spectroscopy techniques	Can work with transmission mode, in which the electromagnetic wave is propagated through the sample or reflection mode, where the electromagnetic wave is reflected from the interface	Infrared, ultraviolet-visible, fluorescence, nuclear magnetic resonance, circular dichroism, X-ray	Nik et al. (2012), Fairhurst and Lee (2011b), Orte et al. (2011), Mattos et al. (2016), and McClements (2015e)

Continued

TABLE 16.2 Description of Techniques Used to Measure the Different Interfacial Properties of Nanoemulsions—cont'd

Properties	Techniques	Description	Examples	Articles
	Reflection techniques	Provide information about thickness, refractive index, and concentration profile through the phase and amplitude of reflected wave from the surface	Ellipsometry, surface plasmon resonance, neutron reflexion	Wong et al. (2014) and Reynolds et al. (2002)
	Calorimetry techniques	Can provide information about the conformation of the proteins that are adsorbed in the surface of the droplets through the comparison with values of nonadsorbed proteins	Differential scanning calorimetry	Silva et al. (2014), Teeranachaideekul et al. (2017), Mattos et al. (2016), and Gao and McClements (2016)
Interfacial tension	Direct measurement by microbalance	The interfacial tension is related to the restoring force, and it can be determined directly through the microbalance	Du Nouy ring method, Wilhelmy plate method	Bhuiyan et al. (2015), Huang et al. (2001), Sajjadi (2007), and Drelich et al. (2002)
	Measurement of capillary pressure	Based on the maximum pressure needed to push out a gas bubble from a capillary tube into the liquid	Maximum bubble pressure	Piotrowski et al. (2012), Wojciechowski et al. (2014), and McClements (2015e)
	Analysis of capillary-gravity forces	The interfacial tension is determined through the drop volume in which it separates from the capillary tube	Drop volume/weight method	van der Graaf et al. (2004), Hong and Kim (2014), and Sünder et al. (2001)
	Gravity-distorted drops	Determines the static interfacial tension through the measurement of the drop shape using microscopic, image software and mathematical equations	Sessile and pendant drop methods	Böttcher et al. (2017) and Sugiura et al. (2004)

	Reinforced distortion of drop	Used to measure the static interfacial tensions of liquids. The interfacial tension determined is a function of angular frequency	Spinning drop method	Forgiarini et al. (2001) and McClements (2015e)
Interfacial rheology	Interfacial shear rheology	Only changes the shape of the interface and its area size keeps constant. Provides information about the network formation	Concentric cylinder technique	Lippacher et al. (2004), Böttcher et al. (2017), and Karbaschi et al. (2014)
	Interfacial dilational rheology	Consists in the measurement of the area changes along the time when the droplet interface is subjected to compression/expansion movements	Capillary pressure tensiometry, drop/bubble profile tensiometry	Ravera et al. (2008), Chebil et al. (2013), Böttcher et al. (2017), and Nash and Erk (2017)

components present, the method used in nanoemulsion formation, the relative affinity of the emulsifiers for the interface, and the solution conditions (Bai and McClements, 2016).

Interfacial structure is determined by the structural organization of the surface-active molecules at the droplet surfaces. Although the surfaces of nanoemulsion droplets are often assumed to be made of homogeneous layers, in practice, they are heterogeneous and structurally complex. The emulsifier organization and composition can vary horizontally and vertically relative to the interface, thus leading to changes in physicochemical and sensory properties of nanoemulsions (McClements, 2015e).

Interfacial tension is a measure of the free energy required to increase the interfacial area by a unit amount. This property has an important influence on the droplet size during homogenization, on the stability of the droplets, and in the packaging patterns of large droplets in concentrated emulsions (Bai et al., 2017).

Interfacial rheology depends on the viscoelastic behavior of the droplet interface. This is important due to the influence of rheological properties on the texture, formation, and stability of nanoemulsions. Nanoemulsions are subjected to several mechanical stresses, which result in different types of deformations of the droplet interface such as expansion or contraction of the surface area and the movement of different regions of the interface relative to each other without altering the surface area (Karbaschi et al., 2014).

16.7 CONCLUSIONS

Particle properties such as particle size, concentration, charge, physical state, and interfacial characteristics play a major role in determining nanoemulsion characteristics such as appearance, stability, rheology, release profile, and bioavailability. These properties can be controlled by the careful selection of formulation parameters and production conditions.

A wide variety of techniques are available to provide information about particle properties in nanoemulsions, each one with its own particular advantages and disadvantages. Knowing their underlying principles is the first step to choose the most appropriate technique in order to extract reliable and significant information.

ACKNOWLEDGMENTS

Ana C. Pinheiro acknowledges the Foundation for Science and Technology (FCT) for her fellowship (SFRH/BPD/101181/2014). This work was supported by the Portuguese Foundation for Science and Technology (FCT) under the scope of the project PTDC/AGR-TEC/5215/2014, of the strategic funding of UID/BIO/04469/2013 unit and COMPETE 2020 (POCI-01-0145-FEDER-006684), and BioTecNorte operation (NORTE-01-0145-FEDER-000004) funded by the European Regional Development Fund under the scope of Norte 2020—Programa Operacional Regional do Norte.

REFERENCES

Acharya, D.P., Hartley, P.G., 2012. Progress in microemulsion characterization. Curr. Opin. Colloid Interface Sci. 17, 274–280.

Addai-Mensah, J., Ralston, J., 2004. Interfacial chemistry and particle interactions and their impact upon the dewatering behaviour of iron oxide dispersions. Hydrometall. 74, 221–231.

Allen, T., 1992. A review of sedimentation methods of particle size analysis. In: Particle Size Analysis. The Royal Society of Chemistry, Cambridge.

Assadpour, E., Maghsoudlou, Y., Jafari, S.-M., Ghorbani, M., Aalami, M., 2016. Optimization of folic acid nano-emulsification and encapsulation by maltodextrin-whey protein double emulsions. Int. J. Biol. Macromol. 86, 197–207.

Bai, L., McClements, D.J., 2016. Formation and stabilization of nanoemulsions using biosurfactants: Rhamnolipids. J. Colloid Interface Sci. 479, 71–79.

Bai, L., Huan, S., Li, Z., McClements, D.J., 2017. Comparison of emulsifying properties of food-grade polysaccharides in oil-in-water emulsions: gum Arabic, beet pectin, and corn fiber gum. Food Hydrocoll. 66, 144–153.

Bhattacharjee, S., 2016. DLS and zeta potential—what they are and what they are not? J. Control. Release 235, 337–351.

Bhuiyan, M.H.U., Saidur, R., Mostafizur, R.M., Mahbubul, I.M., Amalina, M.A., 2015. Experimental investigation on surface tension of metal oxide–water nanofluids. Int. Commun. Heat Mass Transfer 65, 82–88.

Böttcher, S., Keppler, J.K., Drusch, S., 2017. Mixtures of Quillaja saponin and beta-lactoglobulin at the oil/water-interface: adsorption, interfacial rheology and emulsion properties. Colloids Surf. A Physicochem. Eng. Asp. 518, 46–56.

Buckin, V.A., O'Driscoll, B., Smyth, C., 2003. Ultrasonic spectroscopy for material analysis. Recent advances. Spectrosc. Eur. 15, 20–25.

Buffo, R.A., Reineccius, G.A., Oehlert, G.W., 2001. Factors affecting the emulsifying and rheological properties of gum acacia in beverage emulsions. Food Hydrocoll. 15, 53–66.

Chebil, A., Desbrières, J., Nouvel, C., Six, J.-L., Durand, A., 2013. Ostwald ripening of nanoemulsions stopped by combined interfacial adsorptions of molecular and macromolecular nonionic stabilizers. Colloids Surf. A Physicochem. Eng. Asp. 425, 24–30.

Clausse, D., 1983. Dielectric properties of emulsions and related systems. In: Becher, P. (Ed.), Encyclopedia of Emulsion Technology: Vol. 1. Basic Theory. Marcel Dekker, New York.

Clausse, D., 2010. Differential thermal analysis, differential scanning calorimetry, and emulsions. J. Therm. Anal. Calorim. 101, 1071–1077.

Clementi, L.A., Vega, J.R., Gugliotta, L.M., Quirantes, A., 2012. Characterization of spherical core–shell particles by static light scattering. Estimation of the core- and particle-size distributions. J. Quant. Spectrosc. Radiat. Transf. 113, 2255–2264.

Drelich, J., Fang, C., White, C., 2002. Measurement of interfacial tension in fluid-fluid systems. In: Hubbard, A.T. (Ed.), Encyclopedia of Surface and Colloid Science. Marcel Dekker, New York.

Dukhin, A.S., Goetz, P.J., 2001. Acoustic and electroacoustic spectroscopy for characterizing concentrated dispersions and emulsions. Adv. Colloid Interf. Sci. 92, 73–132.

Dukhin, A.S., Goetz, P.J., Wines, T.H., Somasundaran, P., 2000. Acoustic and electroacoustic spectroscopy. Colloids Surf. A Physicochem. Eng. Asp. 173, 127–158.

Duval, J.F., Kuttner, D., Nitschke, M., Werner, C., Zimmermann, R., 2011. Interrelations between charging, structure and electrokinetics of nanometric polyelectrolyte films. J. Colloid Interface Sci. 362, 439–449.

Egerton, T.A., Tooley, I.R., 2014. Physical characterization of titanium dioxide nanoparticles. Int. J. Cosmet. Sci. 36, 195–206.

Fairhurst, D., Lee, R.W., 2011a. The zeta potential & its use in pharmaceutical applications—part 1: charged interfaces in polar & non-polar media & the concept of the zeta potential. Drug Dev. Deliv. 11, 60–64.

Fairhurst, D., Lee, R.W., 2011b. The zeta potential & its use in pharmaceutical applications—part 2: measurement techniques & uses. Drug Dev. Deliv. 11, 50–55.

Forgiarini, A., Esquena, J., González, C., Solans, C., 2001. Formation of nano-emulsions by low-energy emulsification methods at constant temperature. Langmuir 17, 2076–2083.

Forrester, D.M., Huang, J., Pinfield, V.J., 2016. Characterisation of colloidal dispersions using ultrasound spectroscopy and multiple-scattering theory inclusive of shear-wave effects. Chem. Eng. Res. Des. 114, 69–78.

Gao, S., McClements, D.J., 2016. Formation and stability of solid lipid nanoparticles fabricated using phase inversion temperature method. Colloids Surf. A Physicochem. Eng. Asp. 499, 79–87.

Ghasemi, S., Jafari, S.M., Assadpour, E., Khomeiri, M., 2017. Production of pectin-whey protein nano-complexes as carriers of orange peel oil. Carbohydr. Polym 177 (Suppl. C), 369–377.

Ghasemi, S., Jafari, S.M., Assadpour, E., Khomeiri, M., 2018. Nanoencapsulation of d-limonene within nanocarriers produced by pectin-whey protein complexes. Food Hydrocoll. https://doi.org/10.1016/j.foodhyd.2017.09.030.

Gibout, S., Jamil, A., Kousksou, T., Zéraouli, Y., Castaing-Lasvignottes, J., 2007. Experimental determination of the nucleation probability in emulsions. Thermochim. Acta 454, 57–63.

Gill, P., Moghadam, T.T., Ranjbar, B., 2010. Differential scanning calorimetry techniques: applications in biology and nanoscience. J. Biomol. Tech. 21, 167–193.

Heurtault, B., 2003. Physico-chemical stability of colloidal lipid particles. Biomaterials 24, 4283–4300.

Hoffmann, I., Heunemann, P., Farago, B., Grillo, I., Holderer, O., Pach, M., Gradzielski, M., 2012. Structure and dynamics of nanoemulsions: insights from combining dynamic and static neutron scattering. Phys. Rev. E Stat. Nonlinear Soft Matter Phys. 86.

Hollingsworth, K.G., Johns, M.L., 2003. Measurement of emulsion droplet sizes using PFG NMR and regularization methods. J. Colloid Interface Sci. 258, 383–389.

Honary, S., Zahir, F., 2013. Effect of zeta potential on the properties of nano-drug delivery systems—a review (part 1). Trop. J. Pharm. Res. 12, 255–264.

Hong, J.S., Kim, J.G., 2014. Variation of interfacial tension by nanoclay particles in oil-in-water emulsions. Compos. Interfaces 21, 703–713.

Hosseini, A., Jafari, S.M., Mirzaei, H., Asghari, A., Akhavan, S., 2015. Application of image processing to assess emulsion stability and emulsification properties of Arabic gum. Carbohydr. Polym. 126, 1–8.

Hu, Y.-T., Ting, Y., Hu, J.-Y., Hsieh, S.-C., 2017. Techniques and methods to study functional characteristics of emulsion systems. J. Food Drug Anal. 25, 16–26.

Huang, X., Kakuda, Y., Cui, W., 2001. Hydrocolloids in emulsions: particle size distribution and interfacial activity. Food Hydrocoll. 15, 533–542.

Jafari, S.M., Esfanjani, A.F., 2017. 14: Instrumental analysis and characterization of nanocapsules. Nanoencapsulation Technologies for the Food and Nutraceutical Industries. Academic Press, New York, pp. 524–544.

Jafari, S.M., Esfanjani, A.F., Katouzian, I., Assadpour, E., 2017. Chapter 10—Release, Characterization, and Safety of Nanoencapsulated Food Ingredients Nanoencapsulation of Food Bioactive Ingredients. Academic Press, New York, pp. 401–453.

Jafari, S.M., Assadpoor, E., HE, Y., Bhandari, B., 2008. Re-coalescence of emulsion droplets during high-energy emulsification. Food Hydrocoll. 22, 1191–1202.

Karbaschi, M., Lotfi, M., Krägel, J., Javadi, A., Bastani, D., Miller, R., 2014. Rheology of interfacial layers. Curr. Opin. Colloid Interface Sci. 19, 514–519.

Klang, V., Valenta, C., 2011. Lecithin-based nanoemulsions. J. Drug Delivery Sci. Technol. 21, 55–76.

Klang, V., Matsko, N.B., Valenta, C., Hofer, F., 2012. Electron microscopy of nanoemulsions: an essential tool for characterisation and stability assessment. Micron 43, 85–103.

Lee, H.-Y., McCarthy, M.J., Dungan, S.R., 1998. Experimental characterization of emulsion formation and coalescence by nuclear magnetic resonance restricted diffusion techniques. J. Am. Oil Chem. Soc. 75, 463–475.

Li, X., Wang, L., Wang, B., 2017. Optimization of encapsulation efficiency and average particle size of Hohenbuehelia serotina polysaccharides nanoemulsions using response surface methodology. Food Chem. 229, 479–486.

Lim, J., Yeap, S.P., Che, H.X., Low, S.C., 2013. Characterization of magnetic nanoparticle by dynamic light scattering. Nanoscale Res. Lett. 8, 381.

Lines, R.W., 1996. The electrical sensing zone method (the coulter principle). In: Knapp, J.Z. (Ed.), Liquid and Surface Borne Particle Measurement Handbook. Marcel Dekker, New York.

Lippacher, A., Muller, R.H., Mader, K., 2004. Liquid and semisolid SLN dispersions for topical application: rheological characterization. Eur. J. Pharm. Biopharm. 58, 561–567.

Liu, F., Ma, C., Zhang, R., Gao, Y., Julian McClements, D., 2017. Controlling the potential gastrointestinal fate of beta-carotene emulsions using interfacial engineering: impact of coating lipid droplets with polyphenol-protein-carbohydrate conjugate. Food Chem. 221, 395–403.

Mason, T.G., Wilking, J.N., Meleson, K., Chang, C.B., Graves, S.M., 2006. Nanoemulsions: formation, structure, and physical properties. J. Phys. Condens. Matter 18, R635–R666.

Mattos, C.B., Rodrigues, M.R., Cordeiro, M., Nunes, R.J., Teixeira, H.F., Lima, V.R., Koester, L.S., 2016. Nanoemulsions containing a synthetic chalcone: photodegradation, in vitro release, and interaction studies. J. Photochem. Photobiol. A Chem. 328, 42–49.

McClements, D.J., 1996. Principles of ultrasonic droplet size determination in emulsions. Langmuir 12, 3454–3461.

McClements, D.J., 2004. Context and background. In: Food Emulsions. third ed. CRC Press, Taylor and Francis Group, Boca Raton, FL.

McClements, D.J., 2008. Lipid-based emulsions and emulsifiers. In: Casimir, C., Akoh, D.B.M. (Eds.), Food Lipids: Chemistry, Nutrition, and Biotechnology. CRC Press, Taylor and Francis Group, Boca Raton, FL.

McClements, D.J., 2011. Edible nanoemulsions: fabrication, properties, and functional performance. Soft Matter 7, 2297–2316.

McClements, D., 2015a. Food Emulsions: Principles, Practice, and Techniques. CRC Press, Boca Raton, FL, p. 14.

McClements, D.J., 2015b. Characterization of emulsion properties. In: McClements, D.J. (Ed.), Food Emulsions: Principles, Practices and Techniques, third ed. CRC Press, Taylor and Francis Group, Boca Raton, FL.

McClements, D.J., 2015c. Context and background. In: McClements, D.J. (Ed.), Food Emulsions: Principles, Practices and Techniques, third ed. CRC Press, Taylor and Francis Group, Boca Raton, FL.

McClements, D.J., 2015d. Emulsion ingredients. In: McClements, D.J. (Ed.), Food Emulsions: Principles, Practices, and Techniques, third ed. CRC Press, Taylor and Francis Group, Boca Raton, FL.

McClements, D.J., 2015e. Interfacial properties and their characterization. In: McClements, D.J. (Ed.), Food Emulsions: Principles, Practices, and Techniques. third ed. CRC Press, Taylor and Francis Group, Boca Raton, FL.

McClements, D.J., 2016. Characterization of emulsion properties. In: McClements, D.J. (Ed.), Food Emulsions: Principles, Practices, and Techniques. third ed. CRC Press, Taylor and Francis Group, Boca Raton, FL.

McClements, D.J., Xiao, H., 2012. Potential biological fate of ingested nanoemulsions: influence of particle characteristics. Food Funct. 3, 202–220.

Mehrnia, M.A., Jafari, S.M., Makhmal-Zadeh, B.S., Maghsoudlou, Y., 2016. Crocin loaded nano-emulsions: factors affecting emulsion properties in spontaneous emulsification. Int. J. Biol. Macromol. 84, 261–267.

Nash, J.J., Erk, K.A., 2017. Stability and interfacial viscoelasticity of oil-water nanoemulsions stabilized by soy lecithin and Tween 20 for the encapsulation of bioactive carvacrol. Colloids Surf. A Physicochem. Eng. Asp. 517, 1–11.

Nik, A.M., Langmaid, S., Wright, A.J., 2012. Nonionic surfactant and interfacial structure impact crystallinity and stability of beta-carotene loaded lipid nanodispersions. J. Agric. Food Chem. 60, 4126–4135.

Norde, W., 2011. Electrochemistry of interfaces. In: Norde, W. (Ed.), Colloids and Interfaces in Life Sciences and Bionanotechnology. second ed. CRC Press, Taylor and Francis Group, Boca Raton, FL.

Orte, A., Ruedas-Rama, M.J., Paredes, J.M., Crovetto, L., Alvarez-Pez, J.M., 2011. Dynamics of water-in-oil nanoemulsions revealed by fluorescence lifetime correlation spectroscopy. Langmuir 27, 12792–12799.

Ostertag, F., Weiss, J., McClements, D.J., 2012. Low-energy formation of edible nanoemulsions: factors influencing droplet size produced by emulsion phase inversion. J. Colloid Interface Sci. 388, 95–102.

Pal, R., 1994. Techniques for measuring the composition (oil and water content) of emulsions—a state of the art review. Colloids Surf. A Physicochem. Eng. Asp. 84, 141–193.

Pinheiro, A.C., Coimbra, M.A., Vicente, A.A., 2016. In vitro behaviour of curcumin nanoemulsions stabilized by biopolymer emulsifiers—effect of interfacial composition. Food Hydrocoll. 52, 460–467.

Piotrowski, M., Lewandowska, J., Wojciechowski, K., 2012. Biosurfactant-protein mixtures: Quillaja Bark Saponin at water/air and water/oil interfaces in presence of beta-lactoglobulin. J. Phys. Chem. B 116, 4843–4850.

Preetz, C., Hauser, A., Hause, G., Kramer, A., Mader, K., 2010. Application of atomic force microscopy and ultrasonic resonator technology on nanoscale: distinction of nanoemulsions from nanocapsules. Eur. J. Pharm. Sci. 39, 141–151.

Rao, J., McClements, D.J., 2011. Formation of flavor oil microemulsions, nanoemulsions and emulsions: influence of composition and preparation method. J. Agric. Food Chem. 59, 5026–5035.

Ravera, F., Ferrari, M., Liggieri, L., Loglio, G., Santini, E., Zanobini, A., 2008. Liquid–liquid interfacial properties of mixed nanoparticle–surfactant systems. Colloids Surf. A Physicochem. Eng. Asp. 323, 99–108.

Reynolds, P.A., Henderson, M.J., Holt, S.A., White, J.W., 2002. Interfacial structure of a high internal phase emulsion near a solid surface. Langmuir 18, 9153–9157.

Ribes, S., Fuentes, A., Talens, P., Barat, J.M., 2017. Application of cinnamon bark emulsions to protect strawberry jam from fungi. LWT Food Sci. Technol. 78, 265–272.

Robson, E.W., Dalgleish, D.G., 1987. Interfacial composition of sodium caseinate emulsions. J. Food Sci. 52, 1694–1698.

Sajjadi, S., 2007. Formation of fine emulsions by emulsification at high viscosity or low interfacial tension; a comparative study. Colloids Surf. A Physicochem. Eng. Asp. 299, 73–78.

Shukat, R., Bourgaux, C., Relkin, P., 2011. Crystallisation behaviour of palm oil nanoemulsions carrying vitamin E. J. Therm. Anal. Calorim. 108, 153–161.

Shukla, A., Prakash, A., Rohani, S., 2010. Online measurement of particle size distribution during crystallization using ultrasonic spectroscopy. Chem. Eng. Sci. 65, 3072–3079.

Silva, H.D., Cerqueira, M.Â., Vicente, A.A., 2011. Nanoemulsions for food applications: development and characterization. Food Bioprocess Technol. 5, 854–867.

Silva, H.D., Cerqueira, M.Â., Vicente, A.A., 2012. Nanoemulsions for food applications: development and characterization. Food Bioprocess Technol. 5, 854–867.

Silva, J.C.D., Borrin, T.R., Ruy, P., Brito, T.C., Pinheiro, A.C., Vicente, A.A., Pinho, S.C.D., 2014. Characterization, physicochemical stability, and evaluation of in vitro digestibility of solid lipid microparticles produced with palm kernel oil and tristearin. Food Sci. Technol. 34, 532–538.

Simunkova, H., Pessenda-Garcia, P., Wosik, J., Angerer, P., Kronberger, H., Nauer, G.E., 2009. The fundamentals of nano- and submicro-scaled ceramic particles incorporation into electrodeposited nickel layers: zeta potential measurements. Surf. Coat. Technol. 203, 1806–1814.

Singh, H., Tamehana, M., Hemar, Y., Munro, P.A., 2003. Interfacial compositions, microstuctures and properties of oil-in-water emulsions formed with mixtures of milk proteins and κ-carrageenan: 1. sodium caseinate. Food Hydrocoll. 17, 539–548.

Smyth, C., O'Driscoll, B., Lawrence, M.J., Hickey, S., O'Regan, T., Buckin, V., 2005. High-resolution ultrasonic spectroscopy: analysis of microemulsions. Spectroscopy 20, 44–49.

Su, D., Zhong, Q., 2016. Lemon oil nanoemulsions fabricated with sodium caseinate and Tween 20 using phase inversion temperature method. J. Food Eng. 171, 214–221.

Sugiura, S., Nakajima, M., Oda, T., Satake, M., Seki, M., 2004. Effect of interfacial tension on the dynamic behavior of droplet formation during microchannel emulsification. J. Colloid Interface Sci. 269, 178–185.

Sünder, A., Scherze, I., Muschiolik, G., 2001. Physico-chemical characteristics of oil-in-water emulsions based on whey protein–phospholipid mixtures. Colloids Surf. B: Biointerfaces 21, 75–85.

Teeranachaideekul, V., Chantaburanan, T., Junyaprasert, V.B., 2017. Influence of state and crystallinity of lipid matrix on physicochemical properties and permeation of capsaicin-loaded lipid nanoparticles for topical delivery. J. Drug Delivery Sci. Technol. 39, 300–307.

Uskokovic, V., 2012. Dynamic light scattering based microelectrophoresis: main prospects and limitations. J. Dispers. Sci. Technol. 33, 1762–1786.

van der Graaf, S., Schroen, C.G., Van der sman, R.G., Boom, R.M., 2004. Influence of dynamic interfacial tension on droplet formation during membrane emulsification. J. Colloid Interface Sci. 277, 456–463.

Voda, M.A., Van Duynhoven, J., 2009. Characterization of food emulsions by PFG NMR. Trends Food Sci. Technol. 20, 533–543.

Wang, L., Tabor, R., Eastoe, J., Li, X., Heenan, R.K., Dong, J., 2009. Formation and stability of nanoemulsions with mixed ionic-nonionic surfactants. Phys. Chem. Chem. Phys. 11, 9772–9778.

Wissing, S.A., Kayser, O., Muller, R.H., 2004. Solid lipid nanoparticles for parenteral drug delivery. Adv. Drug Deliv. Rev. 56, 1257–1272.

Wojciechowski, K., Kezwon, A., Lewandowska, J., Marcinkowski, K., 2014. Effect of β-casein on surface activity of Quillaja bark saponin at fluid/fluid interfaces. Food Hydrocoll. 34, 208–216.

Wong, P.T., Wang, S.H., Ciotti, S., Makidon, P.E., Smith, D.M., Fan, Y., Schuler, C.F.T., Baker, J.R., 2014. Formulation and characterization of nanoemulsion intranasal adjuvants: effects of surfactant composition on mucoadhesion and immunogenicity. Mol. Pharm. 11, 531–544.

Ye, A., 2008. Interfacial composition and stability of emulsions made with mixtures of commercial sodium caseinate and whey protein concentrate. Food Chem. 110, 946–952.

Chapter 17

Characterization of Physicochemical Properties of Nanoemulsions: Appearance, Stability, and Rheology

Cheryl Chung and David J. McClements
University of Massachusetts, Amherst, MA, United States

Chapter Outline

17.1 INTRODUCTION

As highlighted in earlier chapters, nanoemulsions are being increasingly utilized in a diverse range of industries because of their characteristic physicochemical properties and functional attributes, such as their high optical clarity, good long-term stability, novel rheological properties, and ability to increase the activity of encapsulated components (Donsi and Ferrari, 2016; Hormann and Zimmer, 2016; Mason et al., 2006; McClements, 2011; Shams and Sahari, 2016; Yukuyama et al., 2016). The design of nanoemulsions for particular applications depends on understanding the relationship between their composition/structure on the one hand and their physicochemical properties/

Nanoemulsions. https://doi.org/10.1016/B978-0-12-811838-2.00017-5

functional attributes on the other hand. The analytic tools required to characterize the droplets in nanoemulsions are covered in Chapter 16 and have been reviewed extensively elsewhere (McClements and McClements, 2016). The current chapter focuses upon analytic techniques commonly used to characterize the bulk physicochemical attributes of nanoemulsions, such as their optical, stability, and rheological properties.

17.2 APPEARANCE

The appearance of a commercial product is the first sensory attribute that many consumers experience, and so, it often has a strong impact on their decision to purchase and like a product (Helgason et al., 2015; McClements, 2002a,b). The appearance of a nanoemulsion can be separated into a number of contributing factors: homogeneity, opacity, and color. A kinetically stable nanoemulsion appears homogeneous because the droplets are evenly distributed throughout the system. However, nanoemulsions are thermodynamically unstable systems and therefore have a tendency to breakdown over time, due to mechanisms such as gravitational separation, flocculation, coalescence, and Ostwald ripening (Mason et al., 2006). These instability mechanisms can lead to visible phase separation of a nanoemulsion-based product so that it no longer appears homogeneous. For example, there may be a white "cream" layer at the top (oil-droplet rich) and a transparent "serum" layer at the bottom (oil-droplet depleted), which would be visibly unappealing in many commercial products, including beverages, foods, personal care products, and cosmetics. This problem may be overcome by ensuring that the nanoemulsion remains physically stable within the product throughout storage, transport, and utilization. For example, the system should be designed so that the droplets do not aggregate or cream in the product.

The *opacity* of a nanoemulsion is largely determined by the scattering of light by the oil droplets, whereas the *color* is mainly determined by the selective absorption of light waves by any chromophores it contains. The final appearance of a nanoemulsion is therefore determined by the number, size, properties, and location of the droplets it contains, as well as any other materials that scatter or absorb light waves (McClements, 2015; Zhang and Reineccius, 2016).

17.2.1 Optical Properties of Nanoemulsions

Overall, the perceived appearance—homogeneity, opacity, and color—of a nanoemulsion is the result of its interactions with light waves (McClements, 2002a,b). When a light source is incident upon a nanoemulsion, the electromagnetic radiation may be *transmitted*, *reflected*, *absorbed*, *scattered*, and *refracted* before reaching the photoreceptors in the eyes or the optical detector within an analytic instrument. These nanoemulsion-light interactions determine the overall appearance of a material. Moreover, the nature of the light source itself (intensity versus wavelength profile) also impacts the perceived appearance of nanoemulsions and should therefore be carefully controlled when analyzing their

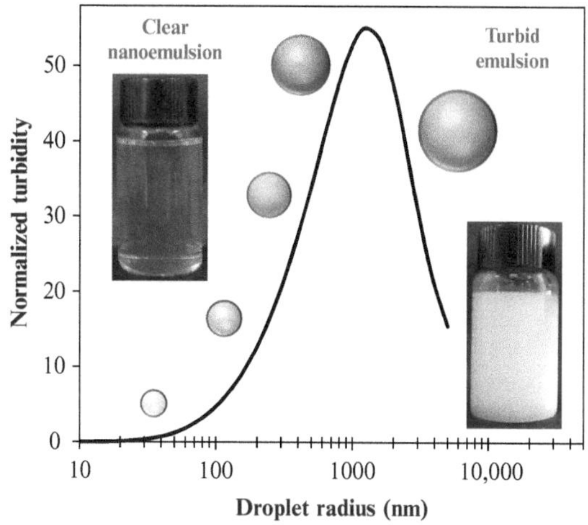

FIG. 17.1 Turbidity of emulsions influenced by particle size of particles. Small droplets lead to clear nanoemulsions, while large droplets lead to turbid emulsions.

optical properties. The diameter of the droplets in nanoemulsions may vary from around 20 to 200 nm, which means that their appearance may vary from clear (smaller droplets) to opaque (larger droplets) (Wooster et al., 2008). This effect is due to the fact that the intensity of light scattering from a nanoemulsion depends on the size of the droplets relative to the wavelength of light, with the most intense scattering occurring when the droplet diameter is fairly similar to the wavelength (Fig. 17.1; McClements, 2002a,b). Nanoemulsion-based products often contain chromophores that preferentially absorb light waves at specific wavelength ranges, and so, their colors may vary considerably, for example, red, blue, green, and yellow. The selection of an appropriate analytic technique to quantify the appearance of a nanoemulsion depends on understanding their interactions with light waves, and so, we begin by describing these interactions.

17.2.1.1 Transmission and Reflectance of Light

When light waves impinge upon nanoemulsions, they are partly transmitted and partly reflected (Fig. 17.2), with the transmitted fraction depending on the refractive indexes of air, wall materials, and nanoemulsions and the angle of the light source (Chantrapornchai et al., 1999; Kortüm, 1969). In general, the light waves reflected from a material may be either *specular* or *diffuse* depending on the surface topography, that is, the surface roughness relative to the wavelength of light. For smooth surfaces, the angle of reflection is equal to the angle of incidence (specular), whereas for rough surfaces, the light is reflected over a wide range of angles (diffuse) (Fig. 17.2; Kortüm, 1969). For nanoemulsions, one would expect their surfaces to be relatively smooth, and so, the reflection should mainly be specular.

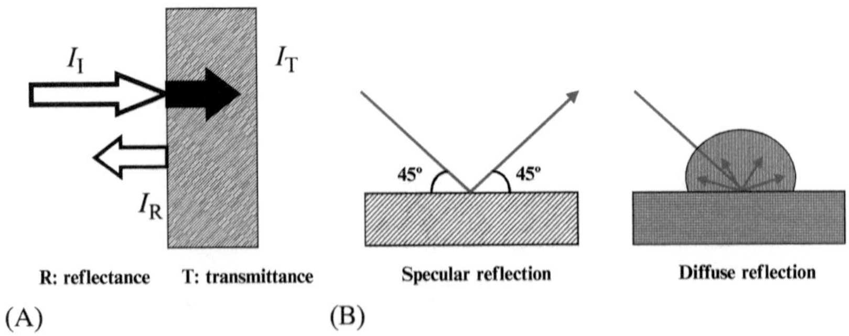

FIG. 17.2 (A) Transmittance and reflectance of light upon impingement on a sample. (B) Specular and diffuse reflection of light upon impingement on sample.

17.2.1.2 Absorption of Light

When light waves travel through a material, some of the photons may be absorbed by molecular species with particular electronic structures, so that some of the energy in the light wave is transferred to the molecules within the material. This process only occurs when the differences in the energy levels of the outer-shell electrons are similar to the energy associated with photons in the visible region of the electromagnetic spectrum (380–750nm). Substances that absorb in the visible region of the electromagnetic spectrum are referred to as *colorants* (dyes or pigments) (Fabian and Hartmann, 1980; Zollinger, 2003). Different substances have different electronic outer-shell structures and therefore absorb light waves in different regions of the spectrum, thereby producing different colors. Nanoemulsions may contain various types of chromophores in the oil, water, and interfacial regions that may absorb light and therefore alter their color.

17.2.1.3 Scattering of Light

Light scattering occurs when a light wave encounters an inhomogeneity (such as a nanoemulsion droplet) and is directed into directions that are different from that of the incident wave (Bohren and Huffman, 1998; Chantrapornchai et al., 1999). The extent of light scattering by a nanoemulsion is determined by the characteristics of the droplets it contains, such as size, concentration, and refractive index contrast (McClements, 2002a,b; Zhang and Reineccius, 2016; Zhang et al., 2013). As mentioned earlier, a nanoemulsion may go from transparent to opaque as the droplet size goes from well below the wavelength of light ($d \ll \lambda$) similar to the wavelength of light ($d \approx \lambda$) (Fig. 17.1). The opacity or lightness of a nanoemulsion typically increases as the droplet concentration increases because there are more scattering events. Moreover, the type of surfactants or emulsifiers used may also affect the scattering and resulting turbidity of a

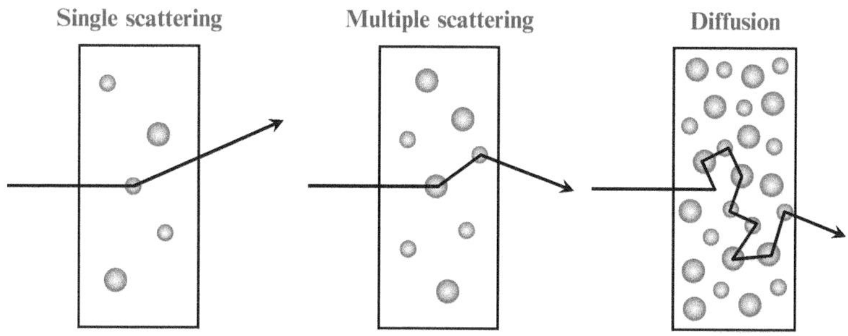

FIG. 17.3 Schematic diagram of the dependence of the light scattering behavior of nanoemulsions with different droplet concentrations.

nanoemulsion by altering the droplet size, aggregation state, or refractive index contrast (Zhang et al., 2015; Zhang and Reineccius, 2016). For example, a study showed that nanoemulsions prepared with whey proteins had a higher turbidity than those prepared using modified starch or quillaja saponin (Zhang and Reineccius, 2016). Nevertheless, the effect of emulsifier type on the turbidity of a nanoemulsion diminished when the droplets became less than about 100 nm in diameter. The physical state of the droplets also affects the scattering of light by nanoemulsions. For example, studies have shown that crystallization of the lipid droplets in nanoemulsions leads to an increase in turbidity due to the concomitant changes in refractive index (Helgason et al., 2015). In a highly dilute system, a light wave typically only encounters a single droplet as it propagates through a nanoemulsion (*single scattering*), but in a concentrated system, it may encounter numerous droplets (*multiple scattering*). In very highly concentrated systems, light waves may propagate through nanoemulsions by a diffusion process (Fig. 17.3). An understanding of the nature of the scattering in a nanoemulsion is often important because the selection of an appropriate analytic tool to characterize their appearance may depend on it.

17.2.2 Quantitative Characterization of Appearance (Instrumental Analysis)

Humans find it difficult to objectively characterize and describe the visual appearance of objects. For example, two objects may be described as "blue" but actually have very different color characteristics. Therefore, there is a need to quantify the appearance of nanoemulsions using standardized methods and techniques. In general, the appearance of nanoemulsions can be characterized using two types of instruments that rely on measurements of the interactions of light waves with nanoemulsions: spectrophotometers and trichromatic colorimeters. The type of instrument used depends on their availability, the nature of the sample to be tested (e.g., whether it is transparent or opaque), and the type of information required (e.g., color coordinates or spectra).

17.2.2.1 Spectrophotometric Colorimeters

The transmittance or reflectance spectrum of a nanoemulsion can be quantified using a UV-visible spectrophotometer that measures the transmission or reflection of light as a function of wavelength in the visible region of the electromagnetic spectrum. The color coordinates of nanoemulsions can be calculated from the transmittance or reflectance spectrum using appropriate theories (Kortüm, 1969; McClements, 2002a,b). In general, a spectrophotometer consists of a light source, wavelength selector, sample holder, and light detector. Spectrophotometers can be used for transmission and/or reflectance measurements depending on their design.

Transmission Spectrophotometry

In transmission spectrophotometers, a white light beam, which covers the whole visible spectrum, is passed through a wavelength selector to isolate light waves of a specific wavelength. The monochromatic wave then passes through a measurement cell containing the sample, and the intensity of the transmitted wave is quantified using a light detector (Chantrapornchai et al., 1999). The fraction of light transmitted can be measured across a certain wavelength range or at a specific wavelength. The transmittance (T) is the ratio of the intensity of the light transmitted through the sample (I_S) compared with the intensity of the light transmitted through a reference material containing only the continuous phase (I_R), that is, $T = I_S/I_R$. Transmission measurements are most suitable for the analysis of optically transparent samples that allow an appreciable amount of light to pass through them, such as dilute nanoemulsions or nanoemulsions containing very small droplets.

Reflectance Spectrophotometry

Reflectance spectrophotometry is more suitable for analysis of optically opaque samples that reflect a large amount of light, such as concentrated nanoemulsions that contain relatively large droplets (Kortüm, 1969). Similar to transmission spectrophotometers, a beam of white light is passed through a wavelength selector to isolate radiation of a specific wavelength (Kortüm, 1969; McClements, 2015). The monochromatic wave is then directed upon a measurement cell containing the sample, and then, the intensity of the light reflected from the surface of the sample is measured using a light detector positioned at a particular angle. The reflectance (R) is the ratio of the intensity of the light reflected from the sample (I_S) compared with the intensity of the light reflected from a reference material (I_R), that is, $R = I_S/I_R$. Typically, the reference material is a white plate that reflects almost 100% of the light waves incident upon it. As mentioned earlier, the reflected light can be either *specular* or *diffuse*, which will determine the nature of the detector that can be used to measure it. To accurately measure the reflectance of different materials, the reflected light should be measured at a right angle, for instance, 90 degree to the incident light source for specular reflectance, and over a range of different angles for diffuse reflectance (Kortüm, 1969; McClements, 2015). Detailed information about the optical arrangements of reflectance spectrophotometers has been covered elsewhere (Kortüm, 1969; McClements, 2015).

17.2.2.2 Trichromatic Colorimeters

Trichromatic colorimeters are specifically designed to measure the lightness and color of nanoemulsions based on the tristimulus coordinate concept, such as the L^*, a^*, and b^* systems (Cretu et al., 2009; McClements, 2002a,b, 2015; Schanda, 2007). Fig. 17.4 shows the graphic representation of these coordinates: L^* represents lightness/darkness with the scale ranging from 0 for black to 100 for whiteness, a^* represents the red-green axis running from negative ($-a^*$) for green to positive ($+a^*$) for red, and b^* represents the blue-yellow axis running from negative ($-b^*$) for blue to positive ($+b^*$) for yellow. Thus, the color of a nanoemulsion can be defined by specifying its L^*, a^*, and b^* values.

Instrumental colorimeters are commercially available to quantify the tristimulus color coordinates of the materials. These instruments are usually relatively inexpensive to purchase, simple to use, and provide results in less than a minute. Trichromatic colorimeters are designed to mimic the photoreceptors (cones) in the human eye, which are sensitive to red, blue, and green colors. Traditionally, colorimeters consisted of a light source, a set of three filters designed to simulate the human cones (red, blue, and green), and a light detector. The intensity of the light transmitted or reflected by a sample was then measured for each of the three filters, which led to tristimulus coordinates

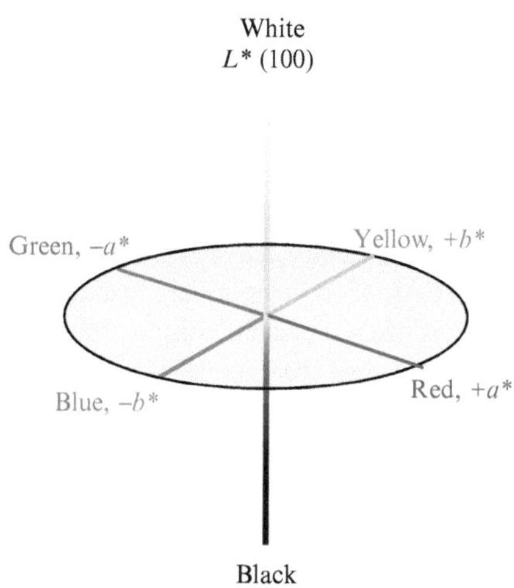

FIG. 17.4 Representation of color of a material in three-dimensional space using L^*, a^*, and b^* tristimulus coordinates.

(R, B, and G) that could be used to quantify the color of the sample. More modern colorimeters measure the transmission or reflectance spectrum of a sample and then convert it into tristimulus color coordinates using suitable mathematical models (Schanda, 2007).

17.2.2.3 *Impact of Measurement Cells*

The measuring cells that contain samples can significantly affect the analysis of the optical properties of nanoemulsions (Chantrapornchai et al., 1999). The refractive index and thickness of the measuring cell wall influence the transmission and reflectance of light from the sample being analyzed, which needs to be taken into account for accurate measurements. This can be done either by carrying out an appropriate theoretical analysis of the optical system used or by using carefully designed sample cells and blanks.

17.2.2.4 *Image Analysis of Color*

An alternative instrumental method to determine the color of nanoemulsion-based materials is to use imagine acquisition and analysis methods (Hosseini et al., 2015). Image acquisition is performed using a digital or video camera under standardized illumination conditions, for example, a light source with a well-defined intensity versus wavelength spectrum and a sample background with well-defined light reflection properties. Image analysis is then performed using a suitable software package (such as ImageJ, NIH) on a personnel computer. Consequently, lightness factors (L^*) can be obtained from photographs of the nanoemulsions. This approach has been used to study the impact of natural emulsifier concentration (gum arabic) on the creaming stability of oil-in-water emulsions (Hosseini et al., 2015).

17.2.3 Qualitative Characterization of Appearance (Sensory Analysis)

The appearance of a nanoemulsion can also be characterized by sensory analysis using human panelists (Granato and Masson, 2010; Linke and Drusch, 2016; Schäfer et al., 2011). Indeed, sensory analysis of the optical properties of a commercial product is often a critical component in the product development stage. In general, there are three commonly used methods in sensory analysis: *discrimination*, *descriptive*, and *preference* (hedonic) tests. *Discriminant* (difference) tests are used to determine whether subjects can discriminate two or more samples that may have different properties (compositions or structures). For example, in a triangle test, panelists are given three samples (two test samples + one control sample or vice versa), and then, the panelists are instructed to indicate which one is different. *Descriptive* tests provide quantitative information about some specific properties of a sample, such as the color intensity or turbidity of a nanoemulsion-based product. For example, panelists may be instructed to rank the turbidity of a nanoemulsion on a scale ranging from 0 (least turbid) to 9 (most turbid). *Preference tests* are used to determine which sample from a group of test samples that consumers prefer. For example, panelists may be instructed to state their preference for a particular sample when they are presented with three different nanoemulsion samples of varying compositions or structures. Information about the different sensory analysis methods has been described and applied in the cited literature earlier in this section. Much of this information can be directly applied to testing the sensory attributes of nanoemulsions.

17.3 STABILITY

Nanoemulsions are thermodynamically unstable systems because the two separated phases (oil and water) have a lower free energy than the nanoemulsion itself, which is primarily due to the hydrophobic effect (the tendency to reduce the unfavorable contact area between oil and water) (Wooster et al., 2016). Over a sufficiently long storage time, nanoemulsions therefore have a tendency to undergo physical changes that lead to their eventual phase separation. The physical destabilization of nanoemulsions may occur through a number of mechanisms, with the most important being *gravitational separation*, *flocculation*, *coalescence*, *Ostwald ripening*, and *phase inversion* (Fig. 2.2). These different destabilization mechanisms are discussed in detail in Chapter 2 of this book, and so, only a brief overview is given here because this information is needed to select the most appropriate analytic tool to characterize their properties (McClements, 2015; Tadros et al., 2004). Nanoemulsions may also breakdown over time due to chemical degradation of certain components as a result of chemical reactions such as oxidation, reduction, or hydrolysis. For example, nanoemulsions containing polyunsaturated lipids may undergo chemical degradation due to lipid oxidation reactions, whereas some types of biopolymers used in nanoemulsions as emulsifiers or thickening agents may become depolymerized due to hydrolysis reactions.

To monitor and identify the destabilization mechanisms, it is important to characterize the physicochemical properties of the nanoemulsions in the initial stage and throughout storage. Typically, storage is carried out under the range of conditions that simulate those a commercial product might experience in practice. This knowledge can then be used to improve the shelf life of the nanoemulsions by optimizing their formulation, processing conditions, packaging, handling, and/or storage conditions.

17.3.1 Gravitational Separation

17.3.1.1 Principles

Over time, nanoemulsions can destabilize by forming a cream layer at the top or a sediment layer at the bottom (Fig. 2.3). These phenomena, which are collectively known as *gravitational separation*, are mainly due to the density difference between the dispersed and continuous phases in a nanoemulsion. Creaming is the upward movement of the droplets that occurs when they have a lower density than the surrounding liquid, whereas sedimentation is the downward movement of the droplets that occurs when they are a higher density than the surrounding liquid. Oil typically has a lower density than water, and so, the droplets in oil-in-water nanoemulsions tend to cream, whereas those in water-in-oil nanoemulsions tend to sediment. However, sedimentation can be observed in oil-in-water nanoemulsions when the oil droplets are coated by a relatively thick and dense layer of emulsifier (McClements, 2011).

The creaming velocity can be predicted using Stokes' law (Eq. 17.1; McClements, 2007):

$$v_{\text{Stokes}} = -\frac{2gr^2(\rho_2 - \rho_1)}{9\eta_1} \tag{17.1}$$

This equation indicates that the creaming velocity is influenced by the radius of the droplets (r), the acceleration due to gravity (g), the density of the dispersed and continuous phases (ρ_2 and ρ_1), and the shear viscosity of the continuous phase (η_1). Thus, a nanoemulsion becomes more susceptible to gravitational separation as the droplet size increases, density contrast increases, and continuous phase viscosity decreases. Stokes' law ignores Brownian motion effects, which are particularly important for the small droplets in nanoemulsions. Brownian motion favors a more random distribution of droplets throughout the volume of a nanoemulsion and therefore tends to oppose gravitational separation. Indeed, gravitational separation may be almost completely avoided in nanoemulsions containing very small droplets, $d < 50$ nm (McClements, 2011; Solans et al., 2005; Tadros et al., 2004).

17.3.1.2 Characterization

To a first approximation, the stability of nanoemulsions to gravitational separation can be predicted using Stokes' law and knowledge of the droplet size, density contrast, and continuous phase viscosity. However, this approach may not give reliable predictions for many types of nanoemulsions due to numerous reasons: (i) The Brownian motion effects are ignored (which are particularly important for small droplets), (ii) the particle size may change during storage (e.g., due to flocculation, coalescence, or Ostwald ripening), (iii) the droplet-droplet interactions are ignored (which are important for concentrated systems), and (iv) the adsorbed emulsifier layer may make an appreciable contribution to the size and density of the droplets. Consequently, Stokes' law should only be used to give a rough idea of the potential creaming instability of a nanoemulsion.

A simple method to determine the stability of a nanoemulsion to gravitational separation is to use a long-term storage study. The nanoemulsion is contained in a transparent container (e.g., a clear test tube or bottle), and then, the heights of any separated layers formed during storage are measured over time, for example, a serum layer at the bottom or a cream layer at the top

(Fig. 2.3; Guerra-Rosas et al., 2016; Mirhosseini et al., 2008). This information can then be used to calculate the rate of creaming in the nanoemulsions over time, d*H*/d*t*. There are also commercial instruments available that have been specifically designed to automatically monitor the gravitational separation of colloidal suspensions (such as nanoemulsions) during storage. These instruments have a laser beam that measures the transmission and/or backscattering of light from the sample as a function of height (Balsamo et al., 2014; Huck-Iriart et al., 2016; Mengual et al., 1999). The resulting light intensity versus height profiles can then be used to monitor the gravitational separation of the sample. For example, when creaming occurs, there is an increase in the concentration of oil droplets at the top of the sample, which leads to a higher reflectance and lower transmittance of the light waves. Conversely, there is a decrease in the concentration of oil droplets in the lower part of the sample, which leads to a lower reflectance and higher transmittance. This type of instrument typically utilizes a monochromatic beam of near-infrared light that vertically scans the test tube at small distance increments to provide a detailed profile of the percentage of transmittance and backscattered light as a function of sample height (Fig. 17.5; Mengual et al., 1999). The changes in transmittance or backscattering of a sample can be measured over time (days, weeks, or months) to determine the rate of droplet movement (creaming or sedimentation). Modern analytic instruments based on this principle allow multiple samples to be automatically scanned over a specified storage period.

A similar type of analytic instrument is commercially available that is designed to reduce the observation time required to detect droplet movement by applying a centrifugal force. In this type of accelerated test, samples are

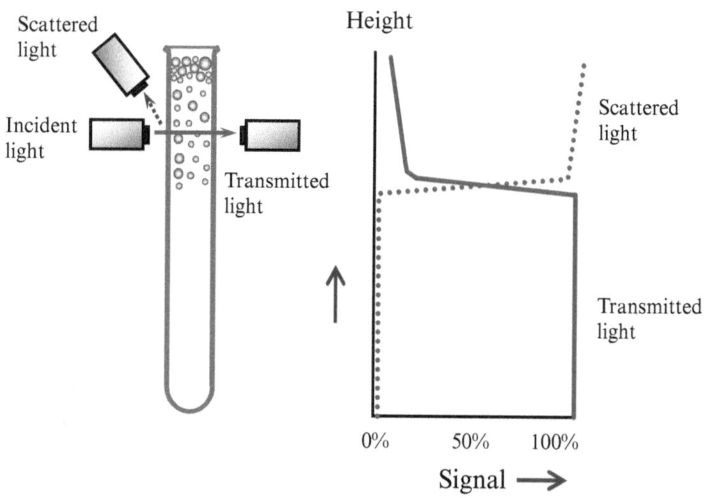

FIG. 17.5 Schematic diagram of light scattering device use for monitoring instability (creaming or sedimentation) in nanoemulsions. A light source and detectors scan the nanoemulsion in the vertical direction to record the intensity of scattered and transmitted light.

placed in centrifugation tubes that are subjected to a fixed centrifugation force over a certain period, and changes in the transmittance and/or reflectance are measured over time. The minimum centrifugation force and time that the nanoemulsions can tolerate before destabilizing can provide valuable quantitative information about the physical stability of the systems.

Other types of instrument are also available to measure the creaming stability of nanoemulsions, including ultrasonic imaging, magnetic resonance imaging (MRI), and electric conductivity imaging. Ultrasonic imaging has been used to monitor the changes in droplet concentration with sample height based on measuring the ultrasonic velocity and attenuation coefficient in the vertical direction (Basaran et al., 1998). MRI has also been used to evaluate the stability of emulsions to creaming by constructing T_2 relaxation time maps from magnetic resonance images (Onuki et al., 2015, 2016). Destructive methods have also been used to measure the concentration of droplets in different layers by physically removing the separated layers. However, these methods require a high number of samples for long-term storage studies.

17.3.2 Droplet Aggregation

17.3.2.1 Principles

The droplets in a nanoemulsion continuously collide with each other due to Brownian motion, gravitational forces, or applied mechanical forces (McClements, 2015; Solans et al., 2005; Tadros et al., 2004). These collisions may cause the droplets to aggregate with one another due to *flocculation* or *coalescence*. Flocculation is the process whereby a number of droplets associate with each other, but each droplet maintains its original size. Conversely, coalescence is the process where a number of droplets encounter each other and fuse together to form a single larger droplet. In most commercial applications, it is desirable to prevent droplet aggregation, which can be achieved by ensuring that the repulsive forces acting between the droplets (such as steric and electrostatic) outweigh the attractive forces (such as van der Waals, hydrophobic, or depletion). Analytic techniques are therefore required to characterize the aggregation state of the droplets in nanoemulsions, so as to optimize their formulation.

17.3.2.2 Characterization

A wide range of analytic techniques can be used to provide information about the aggregation state of the droplets in nanoemulsions, including particle sizing instruments, microscopes, and bulk physicochemical measurements. Some of these analytic techniques can distinguish between flocculation and coalescence mechanisms in nanoemulsions, whereas others only indicate that droplet aggregation has occurred.

Flocculation

The formation of *flocs* in a nanoemulsion can be determined either directly or indirectly using a variety of analytical methods, including microscopy, particle sizing, rheology, and gravitational stability methods (McClements and McClements, 2016). Typically, the presence and properties of the flocs is measured over time or after the nanoemulsion is exposed to particular environmental conditions (such as pH, ionic strength, or temperature).

Microscopy methods are one of the most powerful means of characterizing the properties of the flocs formed within a nanoemulsion. The smallest dimensions of the flocs within a nanoemulsion that can be discriminated depend on the limit of resolution of the microscope used, which is determined by its principles of operation. Typically, optical microscopes cannot distinguish features below about 500 nm and are therefore unsuitable for providing information about the dimensions of the individual droplets in nanoemulsions, but they are suitable for providing information about the formation of large flocs containing many droplets. Fluorescent microscopes can provide additional information about floc properties because different components (such as oils, starches, or proteins) within a nanoemulsion can be selectively stained using appropriate fluorescent dyes; consequently, their relative locations in the flocs can be determined (Mun et al., 2017). Higher resolution images and detailed information about floc structure can be obtained using transmission electron microscopy (TEM) or scanning electron microscopy (SEM; Carvalho Silva et al., 2014; Dario et al., 2016; Grapentin et al., 2015; Khan et al., 2015; Uzun et al., 2016). As an example, a TEM image of a flocculated nanoemulsion consisting of a mixture of negatively charged and positively charged droplets is shown in Fig. 17.6.

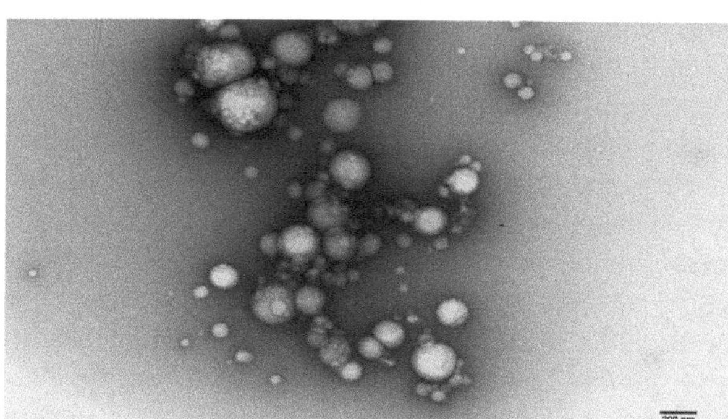

FIG. 17.6 TEM image of a flocculated 20% oil-in-water nanoemulsion containing a mixture of negatively charged and positively charged droplets *(Courtesy of Erica Mao, University of Massachusetts.)*

Cryo-TEM is capable to image a particle of 0.2 nm and focus a selected area to reveal the detailed structure of the spot (e.g., crystallinity). However, extensive sample preparation is needed to produce a thin layer of sample, which may alter the structure of the system being examined. SEM has an order of magnitude worse resolution than TEM, but less extensive sample preparation is typically needed. For example, samples often have to be dehydrated and sputter-coated with gold prior to analysis. SEM relies on the scattering of electrons from the surface of a sample, whereas TEM relies on the transmission of electrons through a sample. SEM is able to provide a good representation of the sample surface with a 3D-like structure (Silva et al., 2012). Microscopy methods are sometimes more sensitive than light scattering methods for detecting changes in particle size over time. For example, a recent study showed that dynamic light scattering was unable to accurately detect small changes in particle size in nanoemulsions during storage, whereas TEM was (Grapentin et al., 2015). Finally, molecular level detail can be obtained using atomic force microscopy (AFM), which has a very high resolution (0.1 nm) in some of the more sophisticated instruments available. AFM is also capable of viewing samples under real-time conditions, for example, under physiological conditions or during biological processes (Dario et al., 2016; Preetz et al., 2010; Ruozi et al., 2005; Salvia-Trujillo et al., 2013; Silva et al., 2012) and in a topology style. It is important to note that careful sample preparation should be carried out for all types of microscopy to avoid having artifacts within the images.

Information about floc properties within nanoemulsions can often be obtained by measuring their particle size distributions using commercial instruments specifically designed for this purpose (McClements and McClements, 2016). There is an increase in the particle size of a nanoemulsion when the droplets associate with each other and form flocs. The most widely used instruments for characterizing the presence of flocs in nanoemulsions are static and dynamic light scattering. Static light scattering instruments operate by directing a laser beam through a test sample contained within an optical chamber and then measuring the scattering pattern (intensity versus angle). The particle size distribution is then calculated from the scattering pattern using a suitable mathematical model (Mie theory) and knowledge of the optical properties of the component phases (refractive index and absorption). Dynamic light scattering instruments operate by measuring the translational diffusion coefficient (D) of the droplets by measuring the intensity versus time signal from a laser beam directed at the sample. The particle size distribution is then calculated from the diffusion coefficient using a suitable theory (e.g., Stokes-Einstein theory; Pecora, 1985). Other types of particle sizing instrument can also be used to detect changes in particle size due to flocculation, such as single particle counting, ultrasonic spectroscopy, electroacoustics, and neutron scattering (McClements, 2015). However, these instruments are much less commonly used than static and dynamic light scattering methods.

It is important to carry out particle sizing measurements carefully to obtain reliable results. Nanoemulsions often have to be diluted appreciably prior to measurement using light scattering methods to avoid multiple scattering effects. This process must be carried out cautiously so as not to change the characteristics of the flocs within the system. Nanoemulsions should be diluted with a fluid that has similar properties to that of the continuous phase surrounding the droplets, for example, a similar pH, ionic strength, and temperature. Samples often have to be stirred during the course of sample preparation and analysis, which may also cause changes in the structure of any flocs in the system. Stirring may either promote droplet aggregation by increasing the collision frequency, or it may disrupt flocs by mechanically breaking them down. Consequently, dilution and stirring conditions must be investigated and standardized to obtain reliable measurements.

The bulk properties of nanoemulsions can also be analyzed to monitor the flocculation process, such as changes in appearance or rheology (McClements and McClements, 2016). As mentioned earlier, the rate of gravitational separation in a nanoemulsion increases as the particle size increases. Consequently, flocculation may lead to an increase in the creaming velocity of the particles in a nanoemulsion (provided that the droplet concentration is not so high that a particle network is formed that inhibits creaming). The gravitational separation of a nanoemulsion can be monitored using the analytic techniques described in Section 17.3.1. Flocculation can also be detected indirectly by measuring the changes in the rheology of a nanoemulsion. The shear viscosity of a nanoemulsion tends to increase after flocculation because of the increase in the effective volume fraction of the dispersed phase associated with the continuous phase trapped inside the flocs (Helgeson, 2016). The extent of this increase depends on the fraction of droplets that have flocculated and the structure of the flocs formed and the strength of the bonds holding the droplets together. The shear viscosity of a nanoemulsion can be measured using viscometers or dynamic shear rheometers (Section 17.4.2). The flow profile of a nanoemulsion obtained from shear viscosity or oscillation measurements may provide valuable information about the interactions between the droplets in the flocs. The shear stress required to breakdown the flocs and reduce the viscosity tends to increase as the strength of the attractive forces between the droplets increases (Helgeson, 2016).

Coalescence

The occurrence of coalescence in a nanoemulsion can often be characterized using the same methods used to characterize flocculation (McClements and McClements, 2016). Coalescence leads to an increase in the size of the individual droplets in a nanoemulsion, which can be monitored by measuring changes in the particle size distribution using light scattering methods. However, it is often difficult to discriminate between particle size increases caused by coalescence and flocculation. In this case, it is advisable to utilize microscopy

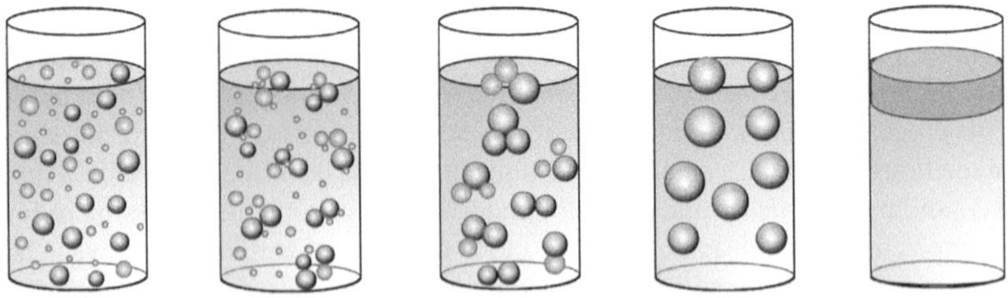

FIG. 17.7 Schematic diagram of progressive droplet coalescence mechanism that leads to eventual phase separation.

methods to directly observe the changes in the structure of the system, for example, optical, electron, or AFM. Alternatively, it may be possible to distinguish between flocculation and coalescence by measuring the particle size distribution before and after adding a deflocculant (such as a small molecule surfactant). If the particle size distribution remains the same, then the sample is probably coalesced, but if it changes back to that of the original droplets, then it was probably flocculated. Coalescence may also be detected by measuring the extent of creaming in a nanoemulsion by measuring the heights of the separated layers or the light scattering profile (transmittance/reflectance) of the samples in a vertical test tube. Coalescence leads to an increase in particle size and should therefore lead to more rapid creaming.

Nanoemulsions that are highly susceptible to droplet coalescence may eventually exhibit a phenomenon known as "oiling off," where a distinct layer of oil (rather than cream) is visible at the top of the sample. This process occurs because the large droplets merge with each other until they form a separate phase (Fig. 17.7). The amount of free oil released can be measured to determine the extent of oiling off that has occurred. This can be achieved by measuring the height of the oil layer or by using an organic solvent to extract the oil (Karim et al., 2016; Rusli et al., 2006). Alternatively, the amount of free oil can be quantified by adding an aliquot of dyed oil to the top of the nanoemulsion and measuring the reduction in its absorbance due to dilution effects: the greater the amount of free oil present, the greater the reduction in absorbance (Palanuwech et al., 2003).

A number of accelerated tests can also be utilized to determine the susceptibility of a nanoemulsion to coalescence under different mechanical stresses, such as centrifugation, shearing, whipping, or homogenization (McClements, 2015). In these tests, the nanoemulsion is subjected to a well-defined mechanical stress for a certain period, and then, the changes in particle size distribution or oiling off are measured.

17.3.3 Ostwald Ripening

17.3.3.1 Principles

Ostwald ripening occurs in nanoemulsions when large droplets grow and small droplets shrink due to diffusion of oil molecules from the small droplets to the large droplets through the intervening continuous phase (Fig. 2.6). This phenomenon occurs because the water solubility of the oil phase in the immediate vicinity of an oil droplet increases as its dimensions decrease (McClements, 2015). Thus, there is a higher concentration of oil phase in the water surrounding small droplets than surrounding large droplets, which leads to a concentration gradient that drives oil molecules from the small droplets toward the larger ones. Ostwald ripening is an important instability mechanism in nanoemulsions formulated using high water-soluble oils, for example, short-chain triglycerides, flavor oils, and essential oils (Chang and McClements, 2014; Rao and McClements, 2012; Wooster et al., 2008). Conversely, it is much less prevalent in nanoemulsions containing long-chain triglycerides because of their very low water solubility (Sakai et al., 2002). The stability of nanoemulsions to Ostwald ripening can be improved by adding ripening inhibitors to the oil phase (Chebil et al., 2013; Wooster et al., 2008).

17.3.3.2 Characterization

The stability of nanoemulsions to Ostwald ripening can be characterized using similar analytic methods as used to monitor coalescence, such as light scattering, microscopy, and creaming stability. The most widely used approach is to measure the changes in the particle size distribution during storage under controlled conditions using a suitable light scattering technique and then to calculate the change in the mean particle diameter (d_N) over time. It is often difficult to distinguish between Ostwald ripening and coalescence because they both lead to an increase in droplet size over time. However, they can sometimes be distinguished because of differences in the way that the particle size distribution changes over time. For Ostwald ripening, after an initial nonsteady-state period, the particle size distribution reaches a certain shape, which then moves up the size axis with time as the larger droplets grow at the expense of the smaller ones. Moreover, there is often a linear relationship between the cube of the mean particle radius and time, which can be analyzed using the Lifshitz-Slyozov-Wagner (LSW) theory to determine the Ostwald ripening rate (McClements, 2015; Rahn-Chique and Urbina-Villalba, 2015). Some researchers have reported that the expected linear increase in R^3 with time (t) due to Ostwald ripening may not occur if flocculation and coalescence are also prevalent (Rahn-Chique and Urbina-Villalba, 2015; Urbina-Villalba et al., 2009). In these more complex situations, particle growth in oil-in-water emulsions can be predicted using emulsion stability simulations that include flocculation, coalescence, and Ostwald ripening (Urbina-Villalba et al., 2009).

17.3.4 Chemical Destabilization

Many nanoemulsion-based products are susceptible to chemical degradation because they contain chemically labile constituents. Some examples of degradation of labile constituents include *oxidation* of polyunsaturated lipids, hydrolysis of biopolymers, fading of pigments, and loss of flavors (Bou et al., 2011; Chen et al., 2013; Karthik and Anandharamakrishnan, 2016; Qian et al., 2012; Shen et al., 2010). The rate of chemical reactions is often different in nanoemulsions than in conventional emulsions. When the chemical reactions occur at an oil-water interface, then a smaller droplet size will lead to a higher droplet surface area, thereby increasing the reaction rate. Chemical reactions may also occur more rapidly in nanoemulsions that contain small droplets if they are accelerated by exposure to light, since a greater fraction of light waves can penetrate into an optically transparent system. Each chemical degradation process depends on the nature of the reactants involved, and so, they will not be discussed in detail here. The analytic techniques required to monitor the chemical destabilization of a nanoemulsion are often highly system specific but typically involve chemical, spectroscopic, electrophoresis, chromatographic, and/or mass spectrometry methods. More details about the chemical degradation of polyunsaturated lipids in nanoemulsions are provided in a recent review article (Walker et al., 2015).

17.4 RHEOLOGY

The rheological properties of nanoemulsions have a significant influence on the physicochemical properties and sensory attributes of commercial products (Gupta et al., 2016). In particular, emulsion rheology impacts their processing (flow through pipes, mixing efficiency, and packaging), stability, textural properties (flowability and pourability), and sensory attributes (mouthfeel) (Chung et al., 2013; Kaltsa et al., 2016; Kalyon and Aktas, 2014). Consequently, it is important to have analytic instruments to characterize their rheological properties either online or offline (Derkach, 2009; Kalyon and Aktas, 2014). Moreover, measurements of the rheology of nanoemulsions can provide valuable information about their composition or structural properties and are therefore a useful analytic tool for understanding these systems. In this section, a brief introduction to the major factors impacting the rheology of nanoemulsions is given, and then, analytic techniques that can be used to characterize their rheological properties are discussed.

17.4.1 Rheological Properties of Nanoemulsions

Nanoemulsions can exhibit a broad range of rheological behaviors depending on their compositions and structures and the test conditions used (Derkach, 2009; Gupta et al., 2016; Helgeson, 2016). For example, they may exhibit liquid, viscoelastic, plastic, or elastic behaviors depending on droplet concentration, size, and interfacial properties. Dilute nanoemulsions containing noninteracting droplets often behave as ideal (Newtonian) liquids, whose rheology can be characterized in terms of their shear viscosity. More concentrated nanoemulsions and those containing strongly interacting droplets may behave as nonideal liquids, whose shear viscosity depends on the magnitude of the applied shear stress. Commonly, shear-thinning behavior is observed in this type of system where the viscosity decreases as the shear stress increases, and so, more complex mathematical models are required to describe their nonideal flow behavior. Highly concentrated nanoemulsions, where the droplets are jammed closely together, often exhibit plastic and/or viscoelastic behavior. A plastic material behaves as a solid-like material when at rest or below a certain applied shear stress but behaves as a liquid-like material above a certain stress threshold, known as the yield stress. A viscoelastic material simultaneously exhibits both liquid-like (viscous) and solid-like (elastic) behavior when a stress is applied (Derkach, 2009; Malkin and Kulichikhin, 2015; McClements, 2015). All of these rheological types can be categorized as exhibiting *ideal* or *nonideal* behavior, depending on their response to test conditions. For example, water is an ideal liquid (Newtonian liquid), and the measured shear viscosity is independent of shear stress or time (Malkin and Kulichikhin, 2015; McClements, 2015). Conversely, many concentrated nanoemulsions are nonideal liquids, whose shear viscosity decreases with increasing shear stress or time.

The rheological properties of a nanoemulsion can be characterized in terms of the stress-strain (solids) or stress-rate of strain (liquids) relationship (Jafari et al., 2012; Mehrnia et al., 2017). For a solid material, a simple stress (τ) can be applied to the sample, and the resultant strain (γ) can be determined:

$$\text{Stress}\,(\tau) = \text{Constant} \times \text{Strain}\,(\gamma) \tag{17.2}$$

The stress applied (force per unit area) on the material is proportional to the strain (relative deformation) below a critical stress value, with the constant of proportionality being a fundamental characteristic of the material known as the *elastic modulus*. When the stress is applied, the material rapidly changes its shape until it reaches a particular level of deformation, and when the stress is removed, the material returns back to its original shape. Above the critical stress value, the material may flow or fracture, and so, the stress-strain relationship is no longer proportional, and the material may not return back to its original shape once the stress is removed (McClements, 2015; Tabilo-Munizaga and Barbosa-Canovas, 2005). The rheology of nanoemulsions with solid-like characteristics can be determined using compression or shear rheometers (Kalyon and Aktas, 2014).

For a liquid material, a simple stress (τ) is applied to the sample, and then, the resultant rate of strain or shear rate ($d\gamma/dt$) is determined:

$$\text{Stress}\,(\tau) = \text{Constant} \times \text{Rate of Strain}\,(d\gamma/dt) \tag{17.3}$$

The rate of strain is a measure of how rapidly the material deforms in response to the applied stress, while the constant of proportionality is the *shear viscosity*, which is a fundamental characteristic of the liquid being tested related to its resistance to flow. For liquids, a shear stress has to be applied to initiate flow, but once the stress is removed, the system will continue to flow until the kinetic energy stored within the material is dissipated due to friction (McClements, 2015).

The factors that influence the rheological properties of nanoemulsions were described in Chapter 2 and so will not be discussed in detail here. Typically, the viscosity of a nanoemulsion tends to increase as the droplet concentration increases, the strength of the particle interactions increases, and the droplet size decreases (Derkach, 2009; Helgeson, 2016; Malkin and Kulichikhin, 2015).

17.4.2 Measurement of Rheological Properties

The rheological properties of a material can be determined using various approaches depending on the nature of the material being tested and the type of information required. The instruments used to characterize nanoemulsion rheology vary from simple and inexpensive (such as glass capillary tubes) to complex and expensive (such as dynamic shear rheometers). Three commonly applied force/stress relationships used during rheological measurements are *compression*, *elongation*, and *shear* (Fig. 17.8). The universal testing machine and texture analyzer are two of the most commonly used instruments that can be used to characterize solid or semisolid nanoemulsions based on compression and elongation measurements (Bähler et al., 2016; Gilbert et al., 2013a,b; Tai et al., 2014). In these tests, a sample is placed between two plates, with one plate fitted with a probe, and the sample is compressed at a controlled rate. The stress and strain experienced by the sample can then be derived from the force and deformation recorded by the probe.

17.4.2.1 Shear Rheology Measurements

Simple or dynamic shear rheology instruments are available for testing different types of materials. Simple shear rheology instruments (viscometers) apply a constant force and measure the resulting strain (solids) or rate of strain (liquids). This type of instrument is used to measure the shear modulus versus deformation (solids) or the apparent viscosity versus strain rate (liquids). Dynamic shear

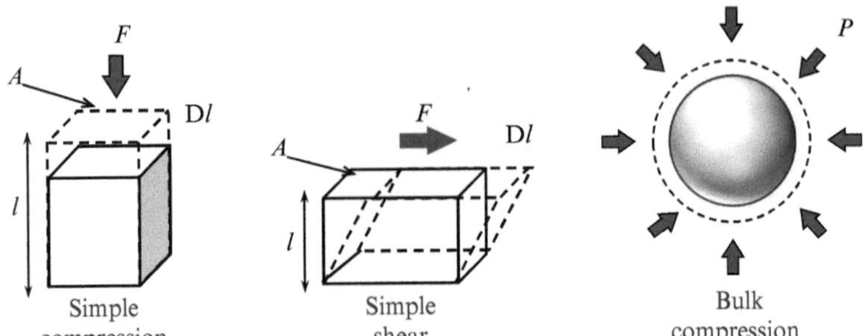

FIG. 17.8 Deformation of elastic solid materials under different nature of stress applied.

rheology instruments can do these types of measurement, but they can also apply an oscillating force and measure the resulting oscillating strain, which allows them to determine the viscoelastic properties of semisolid materials. These instruments are usually attached to a personal computer that can control the experiment and then store, analyze, and present the resulting data. In some instruments, the changes in rheological properties with temperature, time, or frequency can be measured. A number of measurement cells are available to hold different types of samples, with the most common being (i) concentric cylinders, (ii) cone and plate, (iii) plate and plate, (iv) vane blade, and (v) serrated surfaces. The selection of a measurement cell is dependent on the type of materials and test conditions used; for example, concentric cylinder cells are often the most suitable for analyzing liquids like nanoemulsions (Barnes and Carnali, 1990; Fischer and Windhab, 2011; Tabilo-Munizaga and Barbosa-Canovas, 2005).

Two types of shear rheology measurements are typically used to characterize the rheological properties of nanoemulsions: (i) S*mall deformation* tests are used to nondestructively determine the fundamental rheological properties of solid or semisolid samples without altering their properties; (ii) *large deformation* tests are used to determine the shear viscosity of fluids or the breaking stress or strain of semisolids and solids (McClements, 2015; Tabilo-Munizaga and Barbosa-Canovas, 2005; Tadros, 2015). They usually involve applying a stress that is large enough to cause the material to flow or become permanently deformed or disrupted.

Small Deformation

Creep and *stress-relaxation* measurements are *transient* tests with the solid or semisolid materials being placed between two plates (e.g., cone and plate or plate and plate measurement cells). In *creep* tests, a constant stress is applied to a material over a certain time, and the resulting strain is recorded. After the stress is removed, the strain is measured over time to obtain *creep recovery* data, that is, *compliance* (J) against time (Fig. 17.9). The compliance is defined

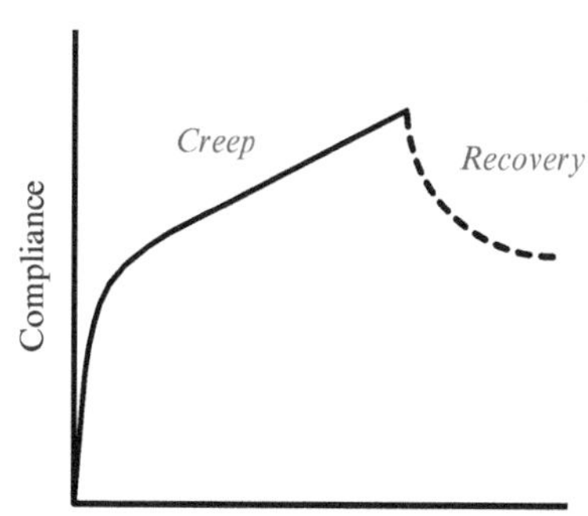

FIG. 17.9 A representative diagram of compliance versus time curve of viscoelastic materials.

as the ratio of the strain to the applied stress. This kind of measurement can give valuable insights into the strength of the bonds in the material being tested (Gilbert et al., 2013a,b; McClements, 2015; Tadros, 2015). In *stress-relaxation* tests, a constant strain is applied to a material, and the stress required to keep the material at the specified strain over a period of time is measured (Tabilo-Munizaga and Barbosa-Canovas, 2005; Tadros, 2015). Similar information as the creep tests can be obtained from stress-relaxation tests.

Besides this type of transient test, *dynamic* tests can also be carried out to determine the rheological properties of viscoelastic or solid materials. From these tests, the *storage modulus*, *loss modulus*, *complex modulus*, and *phase angle* of a material can be determined by applying a sinusoidal stress and measuring the resultant sinusoidal strain. The frequency and strain of the applied stress can be varied to obtain valuable information about material properties. For nondestructive analysis of material properties, the strain should be so small that it does not cause permanent changes in material properties, that is, it is within the *linear viscoelastic region* of the material, where stress is proportional to strain. The data obtained can be presented as storage modulus (G') and loss modulus (G'') versus relevant experimental parameters (such as time, temperature, or frequency). The storage modulus provides information about the elastic properties of a material (the stored energy), while the loss modulus provides information about the viscous properties (the energy dissipated as heat). If G' is greater than G'', the nanoemulsion is predominately solid-like, whereas if G'' is greater than G', then the nanoemulsion is predominantly liquid-like (Derkach, 2009; Kalyon and Aktas, 2014; Tabilo-Munizaga and Barbosa-Canovas, 2005). Alternatively, the same data can be presented as a complex modulus (G^*) and phase angle (δ). The phase angle provides insights into the viscoelastic properties of materials, where an elastic material has $\delta = 0$ degree, a viscous fluid has $\delta = 90$ degree, and a viscoelastic material is somewhere in between (Malkin and Kulichikhin, 2015; Tabilo-Munizaga and Barbosa-Canovas, 2005; Tadros, 2015).

Large Deformation

Large deformation tests are often carried out to simulate processes that a nanoemulsion may experience during production, storage, or utilization, such as flow through a pipe, mixing, pouring, or mastication. In this case, the rheological properties are tested using relatively high stresses that cause flow or large deformations. This type of test is commonly used for nanoemulsions that are either fluids or semisolids. For fluids, the viscosity versus applied shear stress is usually measured, whereas for semisolids, the shear modulus versus applied stress is measured (Fig. 17.10). The viscosity (η) of a nanoemulsion is related to the shear stress (τ) and shear rate (γ) by the following expression, $\tau = \eta\gamma$ (Brown and Jaeger, 2014). In practice, the viscosity actually varies with the applied shear stress, and so, an *apparent* viscosity is reported at a particular shear stress.

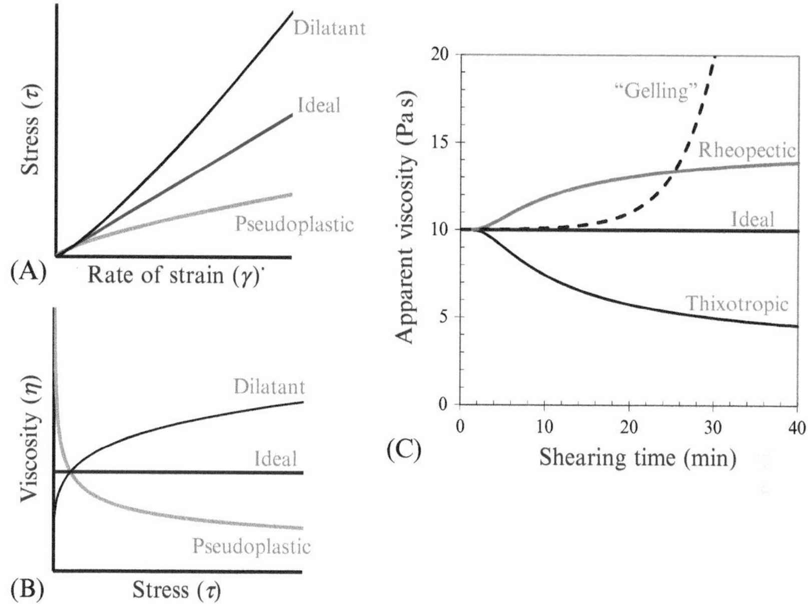

FIG. 17.10 Viscosity of different ideal and nonideal liquid materials: (A) shear stress versus rate of strain, (B) viscosity versus shear stress, and (C) viscosity versus shearing time.

Plots of the apparent viscosity versus shear stress can be used to determine whether a nanoemulsion is shear thinning, ideal, or shear thickening (Brown and Jaeger, 2014; Tabilo-Munizaga and Barbosa-Canovas, 2005). Relatively concentrated nanoemulsions may be viscoelastic or pseudoplastic systems due to strong attractive or repulsive interactions acting between the droplets. In this case, a plot of applied shear stress versus resulting shear strain (or rate of strain) can be used to determine the elastic modulus, yield stress, and plastic viscosity. The range of shear stresses selected to test a material is often designed to simulate the mechanical forces it would experience in practice (e.g., pumping, mixing, or mastication; Malkin, 2013; Tabilo-Munizaga and Barbosa-Canovas, 2005).

Shear viscosity and shear modulus measurements can be fitted to appropriate mathematical models to provide information about the rheological properties of the test materials. Some of the commonly used models that are suitable for describing the viscosity of nanoemulsions are the power law model, Ellis model, and Herschel-Bulkley model (Karataş and Arslan, 2016; Pal, 2015; Park and Song, 2010; Tadros, 2015).

Experimental Errors

As with any instrumental technique, rheological measurements made on nanoemulsions must be carried out carefully to avoid experimental errors. In the case of shear rheology tests using a rheometer, the experimental errors include instrument sensitivity, sample loading effects, wall slip effects, gravitational separation, and rheometer gap effects (McClements, 2015). One of the most serious potential sources of error when measuring concentrated nanoemulsion is the *wall slip* effect (Jesinghausen et al., 2016; Kalyon and Aktas, 2014; Paredes et al., 2015). This error arises when a sample containing polymers or particle phase separates from the liquid continuous phase and lags behind it during shearing. The thin liquid layer may act as a lubricant and cause slip at the measurement cell wall. Consequently, the measurement obtained is mainly determined by the viscosity of the thin liquid layer near the rheometer surface, rather than the bulk rheology of the overall sample. To minimize wall slip effects, serrated (roughened) or vane measurement cells can be used instead of smooth surface cells.

17.4.2.2 Advanced Measurement Methods

More advanced measurement methods have been developed in recent years to obtain deeper insights into the rheological behavior of nanoemulsions. Many of these methods have been developed to understand the rheological properties of nanoemulsions during mastication. The *tribology* method has been developed to simulate oral processing and to correlate sensory perception. This method uses *thin-film rheology* based on the principle that food is compressed between the tongue and palate during mastication to form thin layer of material (Chen and Stokes, 2012; Prakash et al., 2013; Stokes et al., 2013). A *tribometer* is an instrument where a thin layer of test sample is placed between a spinning disc and spinning ball to measure its coefficient of friction. The rate of strain can be varied on the rotating ball, while the rotational speed of the disc can also be varied. The information obtained from tribology measurements includes friction, lubrication, and adhesion properties of a sample during mastication. More detailed information about tribology is available from the cited literature.

Another method that has also been developed to mimic the deformation encountered during mastication (compression, elongation, and shearing) is *squeezing flow* rheology. This method uses compression-elongation forces to apply a combination of elongation, compression, and shear flows to a test material that is being squeezed between two plates. The resulting force-deformation profiles can be used to obtain important rheological parameters such as apparent maximum stress, apparent yield stress, consistency, and flow index (Campanella and Peleg, 2002; Corradini and Peleg, 2005; Corradini et al., 2000). This method can be carried out using a universal texture analyzer. In recent years, a dynamic shear-compression rheometer has been utilized to combine squeezing flow measurements with conventional shear rheology measurements (Chung et al., 2012, 2013). A sequence of compression, shearing, and decompression is applied to the sample to mimic mastication in the human mouth (Fig. 17.11). The compression-shearing-decompression cycle is repeated a number of times to simulate mastication of different types of material.

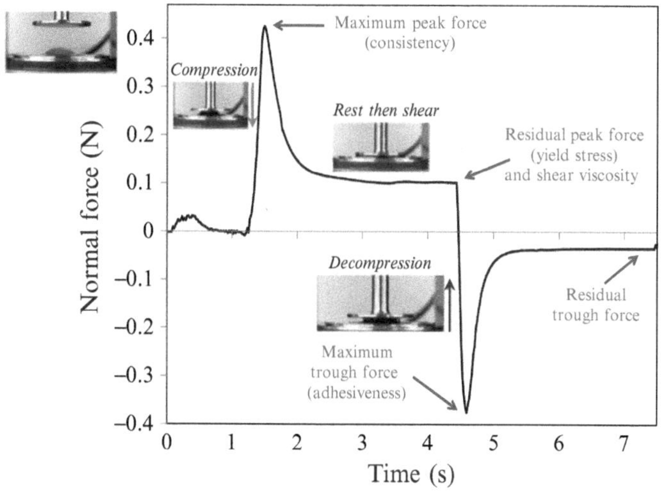

FIG. 17.11 A representative profile of force versus time during compression-shear-decompression of a semisolid emulsion sample between two parallel plates. Different rheological properties can be obtained from the squeezing-shearing method.

Information such as consistency, yield stress, adhesiveness, and shear viscosity can be obtained from these profiles.

17.4.2.3 Empirical Measurement Methods

Simple empirical test methods are available to measure the viscosity of materials that are inexpensive, rapid, and more fitting for certain measurements. For example, a *penetrometer* may be used to measure the hardness of semisolid samples, a *funnel* may be used to measure the consistency of viscous liquids flowing through a tube, and a *ball-cylinder* device may be used for measuring the time it takes for a dense spherical ball to fall through a sample. These empirical measurements can be rapidly carried out to make quick comparisons between samples especially during the production stage. However, these measurements do not usually provide fundamental information about a product.

17.5 CONCLUSION

It is important to characterize the physicochemical properties of nanoemulsions using appropriate tools and instruments so as to accurately describe their functional attributes (appearance, stability, and rheology). After choosing a suitable method, one must carry out the experiments carefully to avoid producing inaccurate measurements, for example, wall slip during viscosity measurements or alterations in microstructure during sample preparation for light scattering or microscopy measurements. A good understanding of the principles underlying different analytic methods is therefore essential for correctly characterizing nanoemulsion systems.

REFERENCES

Bähler, B., Nägele, M., Weiss, J., Hinrichs, J., 2016. Temperature and time-dependent relaxation of compressed cheese curd cubes: effect on structuring of pasta-filata cheese. J. Texture Stud. 47, 58–67.

Balsamo, V., Nguyen, D., Phan, J., 2014. Non-conventional techniques to characterize complex SAGD emulsions and dilution effects on emulsion stabilization. J. Pet. Sci. Eng. 122, 331–345.

Barnes, H., Carnali, J., 1990. The vane-in-cup as a novel rheometer geometry for shear thinning and thixotropic materials. J. Rheol. 34, 841–866.

Basaran, T.K., Demetriades, K., McClements, D.J., 1998. Ultrasonic imaging of gravitational separation in emulsions. Colloids Surf. A Physicochem. Eng. Asp. 136, 169–181.

Bohren, C.F., Huffman, D.R., 1998. Absorption and Scattering of Light by Small Particles. John Wiley & Sons, New York.

Bou, R., Boon, C., Kweku, A., Hidalgo, D., Decker, E.A., 2011. Effect of different antioxidants on lycopene degradation in oil-in-water emulsions. Eur. J. Lipid Sci. Technol. 113, 724–729.

Brown, E., Jaeger, H.M., 2014. Shear thickening in concentrated suspensions: phenomenology, mechanisms and relations to jamming. Rep. Prog. Phys. 77.

Campanella, O.H., Peleg, M., 2002. Squeezing flow viscometry for nonelastic semiliquid foods—theory and applications. Crit. Rev. Food Sci. Nutr. 42, 241–264.

Carvalho Silva, R., Alexandre Muehlmann, L., Rodrigues Da Silva, J., de Bentes Azevedo, R., Madeira Lucci, C., 2014. Influence of nanostructure composition on its morphometric characterization by different techniques. Microsc. Res. Tech. 77, 691–696.

Chang, Y.H., McClements, D.J., 2014. Optimization of orange oil nanoemulsion formation by isothermal low-energy methods: influence of the oil phase, surfactant, and temperature. J. Agric. Food Chem. 62, 2306–2312.

Chantrapornchai, W., Clydesdale, F., McClements, D.J., 1999. Theoretical and experimental study of spectral reflectance and color of concentrated oil-in-water emulsions. J. Colloid Interface Sci. 218, 324–330.

Chebil, A., Desbrieres, J., Nouvel, C., Six, J.L., Durand, A., 2013. Ostwald ripening of nanoemulsions stopped by combined interfacial adsorptions of molecular and macromolecular nonionic stabilizers. Colloids Surf. A Physicochem. Eng. Asp. 425, 24–30.

Chen, J.S., Stokes, J.R., 2012. Rheology and tribology: two distinctive regimes of food texture sensation. Trends Food Sci. Technol. 25, 4–12.

Chen, Q., Zhong, F., Wen, J., McGillivray, D., Quek, S.Y., 2013. Properties and stability of spray-dried and freeze-dried microcapsules co-encapsulated with fish oil, phytosterol esters, and limonene. Dry. Technol. 31, 707–716.

Chung, C., Degner, B., McClements, D.J., 2012. Instrumental mastication assay for texture assessment of semi-solid foods: combined cyclic squeezing flow and shear viscometry. Food Res. Int. 49, 161–169.

Chung, C., Olson, K., Degner, B., McClements, D.J., 2013. Textural properties of model food sauces: correlation between simulated mastication and sensory evaluation methods. Food Res. Int. 51, 310–320.

Corradini, M.G., Peleg, M., 2005. Consistency of dispersed food systems and its evaluation by squeezing flow viscometry. J. Texture Stud. 36, 605–629.

Corradini, M.G., Stern, V., Suwonsichon, T., Peleg, M., 2000. Squeezing flow of semi liquid foods between parallel Teflon coated plates. Rheol. Acta 39, 452–460.

Cretu, R., Dima, S., Olteanu, M., 2009. Stability and colour evolution of emulsions with sodium caseinate. Rev. Chim. 60, 683–688.

Dario, M.F., Santos, M.S.C., Viana, A.S., Arêas, E.P., Bou-Chacra, N.A., et al., 2016. A high loaded cationic nanoemulsion for quercetin delivery obtained by sub-PIT method. Colloids Surf. A Physicochem. Eng. Asp. 489, 256–264.

Derkach, S.R., 2009. Rheology of emulsions. Adv. Colloid Interf. Sci. 151, 1–23.

Donsi, F., Ferrari, G., 2016. Essential oil nanoemulsions as antimicrobial agents in food. J. Biotechnol. 233, 106–120.

Fabian, J., Hartmann, H., 1980. Light Absorption of Organic Colorants: Theoretical Treatment and Empirical Rules. Springer-Verlag, Berlin Available online 2013.

Fischer, P., Windhab, E.J., 2011. Rheology of food materials. Curr. Opin. Colloid Interface Sci. 16, 36–40.

Gilbert, L., Picard, C., Savary, G., Grisel, M., 2013a. Rheological and textural characterization of cosmetic emulsions containing natural and synthetic polymers: relationships between both data. Colloids Surf. A Physicochem. Eng. Asp. 421, 150–163.

Gilbert, L., Savary, G., Grisel, M., Picard, C., 2013b. Predicting sensory texture properties of cosmetic emulsions by physical measurements. Chemom. Intell. Lab. Syst. 124, 21–31.

Granato, D., Masson, M.L., 2010. Instrumental color and sensory acceptance of soy-based emulsions: a response surface approach. Food Sci. Technol. 30, 1090–1096.

Grapentin, C., Barnert, S., Schubert, R., 2015. Monitoring the stability of perfluorocarbon nanoemulsions by Cryo-TEM image analysis and dynamic light scattering. PLoS One 10.

Guerra-Rosas, M.I., Morales-Castro, J., Ochoa-Martinez, L.A., Salvia-Trujillo, L., Martin-Belloso, O., 2016. Long-term stability of food-grade nanoemulsions from high methoxyl pectin containing essential oils. Food Hydrocoll. 52, 438–446.

Gupta, A., Eral, H.B., Hatton, T.A., Doyle, P.S., 2016. Nanoemulsions: formation, properties and applications. Soft Matter 12, 2826–2841.

Helgason, T., Salminen, H., Kristbergsson, K., McClements, D.J., Weiss, J., 2015. Formation of transparent solid lipid nanoparticles by microfluidization: influence of lipid physical state on appearance. J. Colloid Interface Sci. 448, 114–122.

Helgeson, M.E., 2016. Colloidal behavior of nanoemulsions: interactions, structure, and rheology. Curr. Opin. Colloid Interface Sci. 25, 39–50.

Hormann, K., Zimmer, A., 2016. Drug delivery and drug targeting with parenteral lipid nanoemulsions—a review. J. Control. Release 223, 85–98.

Hosseini, A., Jafari, S.M., Mirzaei, H., Asghari, A., Akhavan, S., 2015. Application of image processing to assess emulsion stability and emulsification properties of Arabic gum. Carbohydr. Polym. 126, 1–8.

Huck-Iriart, C., Montes-de-Oca-Ávalos, J., Herrera, M.L., Candal, R.J., Pinto-de-Oliveira, C.L., Linares-Torriani, I., 2016. New insights about flocculation process in sodium caseinate-stabilized emulsions. Food Res. Int. 89, 338–346.

Jafari, S.M., Beheshti, P., Assadpoor, E., 2012. Rheological behavior and stability of d-limonene emulsions made by a novel hydrocolloid (Angum gum) compared with Arabic gum. J. Food Eng. 109 (1), 1–8.

Jesinghausen, S., Weiffen, R., Schmid, H.-J., 2016. Direct measurement of wall slip and slip layer thickness of non-Brownian hard-sphere suspensions in rectangular channel flows. Exp. Fluids 57, 153.

Kaltsa, O., Spiliopoulou, N., Yanniotis, S., Mandala, I., 2016. Stability and physical properties of model macro- and nano/submicron emulsions containing fenugreek gum. Food Hydrocoll. 61, 625–632.

Kalyon, D.M., Aktas, S., 2014. Factors affecting the rheology and processability of highly filled suspensions. In: Prausnitz, J.M., Doherty, M.F., Segalman, R.A. (Eds.), Annual Review of Chemical and Biomolecular Engineering. In: vol. 5. Annual Reviews, Palo Alto, p. 229.

Karataş, M., Arslan, N., 2016. Flow behaviours of cellulose and carboxymethyl cellulose from grapefruit peel. Food Hydrocoll. 58, 235–245.

Karim, F.T., Sarker, Z.M., Ghafoor, K., Al-Juhaimi, F.Y., Jalil, R., et al., 2016. Microencapsulation of fish oil using hydroxypropyl methylcellulose as a carrier material by spray drying. J. Food Process. Preserv. 40, 140–153.

Karthik, P., Anandharamakrishnan, C., 2016. Enhancing omega-3 fatty acids nanoemulsion stability and in-vitro digestibility through emulsifiers. J. Food Eng. 187, 92–105.

Khan, I.U., Serra, C.A., Anton, N., Er-Rafik, M., Blanck, C., et al., 2015. Microfluidic conceived Trojan microcarriers for oral delivery of nanoparticles. Int. J. Pharm. 493, 7–15.

Kortüm, G., 1969. Reflectance Spectroscopy: Principles, Methods, Applications. Spinger-Verlag, New York.

Linke, C., Drusch, S., 2016. Turbidity in oil-in-water-emulsions—key factors and visual perception. Food Res. Int. 89, 202–210.

Malkin, A.Y., 2013. Non-newtonian viscosity in steady-state shear flows. J. Non-Newtonian Fluid Mech. 192, 48–65.

Malkin, A.Y., Kulichikhin, V.G., 2015. Structure and rheology of highly concentrated emulsions: a modern look. Russ. Chem. Rev. 84, 803.

Mason, T.G., Wilking, J.N., Meleson, K., Chang, C.B., Graves, S.M., 2006. Nanoemulsions: formation, structure, and physical properties. J. Phys. Condens. Matter 18, R635–R666.

McClements, D.J., 2002a. Colloidal basis of emulsion color. Curr. Opin. Colloid Interface Sci. 7, 451–455.

McClements, D.J., 2002b. Theoretical prediction of emulsion color. Adv. Colloid Interf. Sci. 97, 63–89.

McClements, D.J., 2007. Critical review of techniques and methodologies for characterization of emulsion stability. Crit. Rev. Food Sci. Nutr. 47, 611–649.

McClements, D.J., 2011. Edible nanoemulsions: fabrication, properties, and functional performance. Soft Matter 7, 2297–2316.

McClements, D.J., 2015. Food Emulsions: Principles, Practices, and Techniques. CRC Press, Washington, DC.

McClements, J., McClements, D.J., 2016. Standardization of nanoparticle characterization: methods for testing properties, stability, and functionality of edible nanoparticles. Crit. Rev. Food Sci. Nutr. 56, 1334–1362.

Mehrnia, M.-A., Jafari, S.-M., Makhmal-Zadeh, B.S., Maghsoudlou, Y., 2017. Rheological and release properties of double nano-emulsions containing crocin prepared with Angum gum, Arabic gum and whey protein. Food Hydrocoll. 66, 259–267.

Mengual, O., Meunier, G., Cayré, I., Puech, K., Snabre, P., 1999. TURBISCAN MA 2000: multiple light scattering measurement for concentrated emulsion and suspension instability analysis. Talanta 50, 445–456.

Mirhosseini, H., Tan, C.P., Aghlara, A., Hamid, N.S.A., Yusof, S., Chern, B.H., 2008. Influence of pectin and CMC on physical stability, turbidity loss rate, cloudiness and flavor release of orange beverage emulsion during storage. Carbohydr. Polym. 73, 83–91.

Mun, S., Kim, J., McClements, D.J., Kim, Y.R., Choi, Y., 2017. Fluorescence imaging of spatial location of lipids and proteins during digestion of protein-stabilized oil-in-water emulsions: a simulated gastrointestinal tract study. Food Chem. 219, 297–303.

Onuki, Y., Funatani, C., Yokawa, T., Yamamoto, Y., Fukami, T., et al., 2015. Magnetic resonance imaging of the phase separation in mixed preparations of moisturizing cream and steroid ointment after centrifugation. Chem. Pharm. Bull. 63, 377–383.

Onuki, Y., Kida, C., Funatani, C., Hayashi, Y., Takayama, K., 2016. MRI as a promising tool for evaluation of the stability of cosmetic emulsions. Int. J. Cosmet. Sci. 38, 272–278.

Pal, R., 2015. Rheology of suspensions of solid particles in power-law fluids. Can. J. Chem. Eng. 93, 166–173.

Palanuwech, J., Potineni, R., Roberts, R.F., Coupland, J.N., 2003. A method to determine free fat in emulsions. Food Hydrocoll. 17, 55–62.

Paredes, J., Shahidzadeh, N., Bonn, D., 2015. Wall slip and fluidity in emulsion flow. Phys. Rev. E. 92.

Park, E.-K., Song, K.-W., 2010. Rheological evaluation of petroleum jelly as a base material in ointment and cream formulations: steady shear flow behavior. Arch. Pharm. Res. 33, 141–150.

Pecora, R., 1985. Dynamic Light Scattering: Applications of Photon Correlation Spectroscopy. Plenum Press, New York Available online 2013.

Prakash, S., Tan, D.D.Y., Chen, J.S., 2013. Applications of tribology in studying food oral processing and texture perception. Food Res. Int. 54, 1627–1635.

Preetz, C., Hauser, A., Hause, G., Kramer, A., Mäder, K., 2010. Application of atomic force microscopy and ultrasonic resonator technology on nanoscale: distinction of nanoemulsions from nanocapsules. Eur. J. Pharm. Sci. 39, 141–151.

Qian, C., Decker, E.A., Xiao, H., McClements, D.J., 2012. Inhibition of beta-carotene degradation in oil-in-water nanoemulsions: influence of oil-soluble and water-soluble antioxidants. Food Chem. 135, 1036–1043.

Rahn-Chique, K., Urbina-Villalba, G., 2015. Use of the Lifshitz-Slyosov-Wagner theory for the prediction of the drop size of an emulsion subject to flocculation and coalescence. Interciencia 40, 847–853.

Rao, J., McClements, D.J., 2012. Food-grade microemulsions and nanoemulsions: role of oil phase composition on formation and stability. Food Hydrocoll. 29, 326–334.

Ruozi, B., Tosi, G., Forni, F., Fresta, M., Vandelli, M.A., 2005. Atomic force microscopy and photon correlation spectroscopy: two techniques for rapid characterization of liposomes. Eur. J. Pharm. Sci. 25, 81–89.

Rusli, J.K., Sanguansri, L., Augustin, M.A., 2006. Stabilization of oils by microencapsulation with heated protein-glucose syrup mixtures. J. Am. Oil Chem. Soc. 83, 965–972.

Sakai, T., Kamogawa, K., Nishiyama, K., Sakai, H., Abe, M., 2002. Molecular diffusion of oil/water emulsions in surfactant-free conditions. Langmuir 18, 1985–1990.

Salvia-Trujillo, L., Rojas-Graü, M.A., Soliva-Fortuny, R., Martín-Belloso, O., 2013. Effect of processing parameters on physicochemical characteristics of microfluidized lemongrass essential oil-alginate nanoemulsions. Food Hydrocoll. 30, 401–407.

Schäfer, C., Neidhart, S., Carle, R., 2011. Application and sensory evaluation of enzymatically texturised vegetable proteins in food models. Eur. Food Res. Technol. 232, 1043–1056.

Schanda, J., 2007. Colorimetry: Understanding the CIE System. John Wiley & Sons, Hoboken, NJ.

Shams, N., Sahari, M.A., 2016. Nanoemulsions: preparation, structure, functional properties and their antimicrobial effects. Appl. Food Biotechnol. 3, 138–149.

Shen, Z., Augustin, M.A., Sanguansri, L., Cheng, L.J., 2010. Oxidative stability of microencapsulated fish oil powders stabilized by blends of chitosan, modified starch, and glucose. J. Agric. Food Chem. 58, 4487–4493.

Silva, H.D., Cerqueira, M.Â., Vicente, A.A., 2012. Nanoemulsions for food applications: development and characterization. Food Bioprocess Technol. 5, 854–867.

Solans, C., Izquierdo, P., Nolla, J., Azemar, N., Garcia-Celma, M., 2005. Nano-emulsions. Curr. Opin. Colloid Interface Sci. 10, 102–110.

Stokes, J.R., Boehm, M.W., Baier, S.K., 2013. Oral processing, texture and mouthfeel: from rheology to tribology and beyond. Curr. Opin. Colloid Interface Sci. 18, 349–359.

Tabilo-Munizaga, G., Barbosa-Canovas, G.V., 2005. Rheology for the food industry. J. Food Eng. 67, 147–156.

Tadros, T., 2015. Viscoelastic properties of sterically stabilised emulsions and their stability. Adv. Colloid Interf. Sci. 222, 692–708.

Tadros, T., Izquierdo, P., Esquena, J., Solans, C., 2004. Formation and stability of nano-emulsions. Adv. Colloid Interf. Sci. 108, 303–318.

Tai, A., Bianchini, R., Jachowicz, J., 2014. Texture analysis of cosmetic/pharmaceutical raw materials and formulations. Int. J. Cosmet. Sci. 36, 291–304.

Urbina-Villalba, G., Forgiarini, A., Rahn, K., Lozsán, A., 2009. Influence of flocculation and coalescence on the evolution of the average radius of an O/W emulsion. Is a linear slope of R [combining macron] 3 vs. t an unmistakable signature of Ostwald ripening? Phys. Chem. Chem. Phys. 11, 11184–11195.

Uzun, S., Kim, H., Leal, C., Padua, G.W., 2016. Ethanol-induced whey protein gels as carriers for lutein droplets. Food Hydrocoll. 61, 426–432.

Walker, R., Decker, E.A., McClements, D.J., 2015. Development of food-grade nanoemulsions and emulsions for delivery of omega-3 fatty acids: opportunities and obstacles in the food industry. Food Funct. 6, 42–55.

Wooster, T.J., Golding, M., Sanguansri, P., 2008. Impact of oil type on nanoemulsion formation and ostwald ripening stability. Langmuir 24, 12758–12765.

Wooster, T.J., Labbett, D., Sanguansri, P., Andrews, H., 2016. Impact of microemulsion inspired approaches on the formation and destabilisation mechanisms of triglyceride nanoemulsions. Soft Matter 12, 1425–1435.

Yukuyama, M.N., Ghisleni, D.D.M., Pinto, T.J.A., Bou-Chacra, N.A., 2016. Nanoemulsion: process selection and application in cosmetics—a review. Int. J. Cosmet. Sci. 38, 13–24.

Zhang, J., Reineccius, G.A., 2016. Factors controlling the turbidity of submicron emulsions stabilized by food biopolymers and natural surfactant. LWT Food Sci. Technol. 71, 162–168.

Zhang, L., Hayes, D.G., Chen, G., Zhong, Q., 2013. Transparent dispersions of milk-fat-based nanostructured lipid carriers for delivery of β-carotene. J. Agric. Food Chem. 61, 9435–9443.

Zhang, J., Peppard, T.L., Reineccius, G.A., 2015. Preparation and characterization of nanoemulsions stabilized by food biopolymers using microfluidization. Flavour Fragr. J. 30, 288–294.

Zollinger, H., 2003. Color Chemistry: Syntheses, Properties, and Applications of Organic Dyes and Pigments. John Wiley & Sons, Zürich.

Chapter 18

Characterization of Gastrointestinal Fate of Nanoemulsions

Ruojie Zhang and David J. McClements
University of Massachusetts, Amherst, MA, United States

Chapter Outline

18.1 INTRODUCTION

There is a growing interest in understanding and controlling the gastrointestinal fate of nanoemulsions within the human body (McClements and Li, 2010a; Cerqueira et al., 2014; Mei et al., 2013). This information can be used to design nanoemulsion-based delivery systems for controlled or targeted delivery of encapsulated bioactive components in different regions of the gastrointestinal tract (GIT), e.g., pharmaceuticals, nutraceuticals, or vitamins (van Aken, 2010; McClements, 2012a; Date et al., 2016; Frohlich and Roblegg, 2016). Alternatively, it can be used to create functional food products designed to

Nanoemulsions. https://doi.org/10.1016/B978-0-12-811838-2.00018-7

improve human health and well-being, e.g., foods that have reduced calories or that control hormonal responses, such as hunger, satiety, and satiation (McClements, 2015).

In this chapter, an overview of the conditions that nanoemulsions experience as they pass through the various stages of the human GIT (the mouth, stomach, small intestine, and colon) is discussed. The impact of these conditions on the composition and structure of nanoemulsions within the GIT is then considered, as well as the impact of various factors on the GIT fate of nanoemulsions. Analytic tools that can be used to study the physicochemical properties of nanoemulsions and the bioavailability of bioactive compounds under GIT conditions are highlighted, including in vitro simulated GIT models, cell culture models, and animal and human studies.

18.2 OVERVIEW OF GASTROINTESTINAL FATE OF NANOEMULSIONS

After ingestion, nanoemulsions are exposed to a complex physiochemical-physiological environment as they pass through the GIT (McClements et al., 2009; Singh et al., 2009). Understanding the nature of this environment is useful for designing nanoemulsion-based products that perform in a particular way after ingestion. A brief overview of the environmental conditions occurring within the various regions of the GIT is therefore given in this section (Fig. 18.1), and then, their potential influence on the gastrointestinal fate of

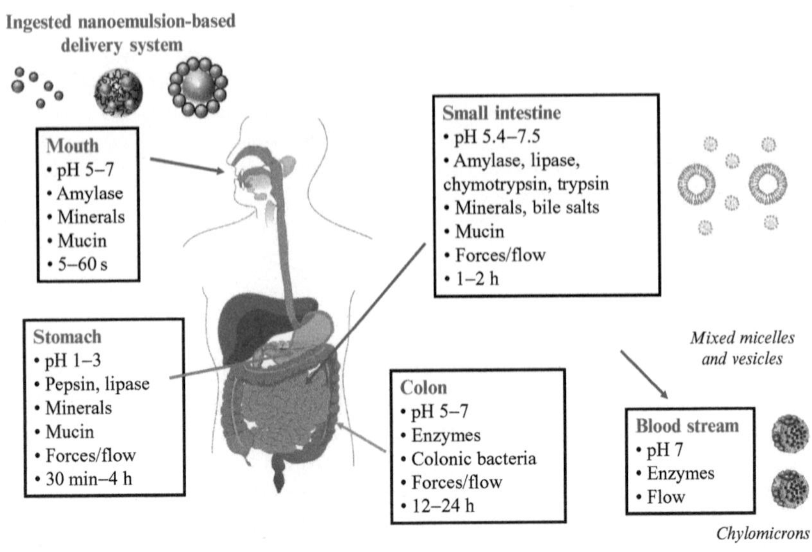

FIG. 18.1 Schematic diagram of the physicochemical and physiological conditions in different regions of the human gastrointestinal tract. *(The diagram of the human body was taken from http://en.wikipedia.org/wiki/Digestive_tract (copyright-free).)*

nanoemulsions is discussed (McClements et al., 2009; Singh et al., 2009). The different GIT regions vary in their pH, ionic compositions, enzyme activities, biopolymer levels, biosurfactant levels, and flow/stress patterns (McClements and Li, 2010a; Joyce et al., 2016; Gharibzahedi and Jafari, 2017). Exposure to these different regions may cause appreciable alterations in the properties of nanoemulsions, which should be taken into account when designing nanoemulsion-based delivery systems for specific applications (McClements and Xiao, 2012; Jafari and McClements, 2017).

Ingested nanoemulsion droplets may be present in a free or trapped form. The droplets may be freely suspended in an aqueous medium (such as a beverage), or they may be trapped within hydrogel matrices (as in some desserts and yogurts) or solid matrices (as in some dried spices, flavors, and colors). The gastrointestinal fate of freely suspended droplets is largely determined by their initial composition and structure, whereas the gastrointestinal fate of trapped droplets also depends on the nature of the matrix surrounding them (Zhang et al., 2015a,b; Golding and Wooster, 2010; Salvia-Trujillo et al., 2013; Beysseriat et al., 2006). For instance, the matrix may have to be dissociated or digested before the oil droplets are released, which will therefore determine the time and location that the droplets come into direct contact with the gastrointestinal fluids. This phenomenon can be used to develop nanoemulsion-based delivery systems that control the release of encapsulated components within the GIT, such as hydrogel beads (McClements et al., 2008; McClements and Li, 2010b; Matalanis et al., 2011; Yeung et al., 2016; Zhang et al., 2016b).

18.2.1 Mouth

The oral cavity is the first environment that nanoemulsions experience after ingestion (Fig. 18.1) (McClements et al., 2008, 2009; Singh et al., 2009; van Aken, 2010; Singh and Sarkar, 2011). The length of time that a nanoemulsion remains in the mouth depends on the nature of the delivery system utilized (e.g., a drink, solid food, capsule, or chewing gum) and may vary from a few seconds to minutes. Even though the time spent in the mouth is quite short, exposure to this region of the GIT may cause appreciable alterations in the characteristics of nanoemulsion droplets that alter their subsequent gastrointestinal fate. An ingested nanoemulsion will undergo both compositional and structural changes within the mouth, including the disruption of any solid material to small pieces by masticating, mixing with saliva, temperature changes, alterations in solution conditions (pH and mineral composition), and interactions with macromolecular components in saliva. Human saliva is a watery fluid that contains various biopolymers (such as mucin), enzymes (such as amylase), and salts (such as sodium and calcium) and has a fairly neutral pH (Kalantzi et al., 2006; Amado et al., 2005; Vitorino et al., 2004b). Mucin may promote the aggregation of nanoemulsion droplets in the mouth due to depletion or bridging flocculation (Silletti et al., 2007; Vingerhoeds et al., 2005). The susceptibility of ingested nanoemulsions to this kind of flocculation depends on the size and interfacial characteristics of the droplets they contain. The digestive enzyme amylase may hydrolyze any starch that is present within a nanoemulsion formulation, which typically causes an appreciable decrease in the viscosity of oral fluids (Chung et al., 2012). The oil droplets in nanoemulsions may interact with the inner surfaces of the oral cavity (such as the teeth, tongue and palette) depending on their interfacial properties (van Aken, 2010; Camacho et al., 2015a,b; Stokes et al., 2013). These interactions will influence the perceived mouthfeel of the product, which is an important sensory component of many nanoemulsion-based products.

There may also be appreciable changes in the characteristics of nanoemulsion-based delivery systems containing trapped oil droplets in the mouth. Environmental conditions within the mouth, such as pH, ionic composition, temperature, and enzyme activity, may trigger the breakdown of the matrix surrounding the oil droplets and thereby promote their release (Buettner, 2002; Hansson et al., 2003; van Ruth et al., 2001; Joyce et al., 2016). Nanoemulsion-based delivery systems can be fabricated to release oil droplets within the oral environment based on a various number of trigger mechanisms, including temperature, pH, ionic strength, and enzyme activity (Zhang et al., 2015d). For example, nanoemulsion droplets trapped in starch-based hydrogels can be released due to the hydrolysis of the starch molecules by amylase in the saliva (Mun et al., 2015a,b). There are numerous potential commercial applications where release of a bioactive agent within the mouth is desirable, such as lipophilic flavors that generate a desirable flavor profile (Madene et al., 2006; Sørensen et al., 2003), lipophilic antimicrobials to combat tooth decay and bad breath (Huang et al., 2010a; Loesche, 1999), and certain oil-soluble drugs (Porter et al., 2008; Pouton, 2006). After exposure to oral conditions, the characteristics of the oil droplets in nanoemulsions in the mouth may be quite different from those in the initial product. Moreover, changes in nanoemulsion properties in the mouth will alter the subsequent gastrointestinal fate of the oil droplets. For instance, the rate and extent of lipid digestion depends on their surface characteristics and aggregation state of the oil droplets, which may be altered in the mouth. The material resulting from oral processing of a food within the mouth is referred to as the "bolus."

18.2.2 Stomach

The bolus resulting from the oral phase travels down the esophagus and enters the gastric chamber. In the stomach, the nanoemulsion is further processed by the human GIT so that it is converted into a form that is suitable for subsequent absorption in the small intestine. Under fasted conditions, the gastric fluids have a pH value around 1–3 and an ionic strength around 100 mM (with the main salt ions being sodium, calcium, and potassium) and contain various digestive enzymes (especially gastric proteases and lipases) and surface-active substances (such as secreted phospholipids and proteins) (McClements and Li, 2010a; Kalantzi et al., 2006; Ekmekcioglu, 2002; Lindahl et al., 1997). Under fed conditions, the pH of the stomach may increase appreciably immediately after the food enters the stomach and then progressively decrease back to fasted condition as the food is processed and passes into the small intestine. The pH-time profile of the stomach under fed conditions depends on the type and quantity of food consumed and especially on the pH and buffering capacity of the food (Kalantzi et al., 2006). Under fed conditions, the ionic strength of the gastric fluids may also vary considerably due to the presence of different levels of mineral ions in ingested foods and due to dilution effects caused by consumption of liquid foods (Kalantzi et al., 2006). Alterations in the pH and ionic composition of the gastric fluids may have a major impact on the stability of nanoemulsions within the stomach because of their influence on the electrostatic interactions between the oil droplets and other charged components in the system. The pH determines the magnitude and sign of the electric charge of proteins, polysaccharides, lipid droplets, mineral ions, and other food components. The stability of a nanoemulsion to gastric conditions is therefore highly sensitive to the nature of the emulsifier molecules that coat the droplets (Fig. 18.2). For instance, protein-coated lipid droplets often aggregate under stomach conditions due to electrostatic screening, ion-binding or polyelectrolyte bridging effects, whereas nonionic surfactant-coated ones exhibit little aggregation (van Aken et al., 2011; Mun et al., 2007). The highly acidic environment of the gastric fluids may also promote the chemical degradation of some components present within nanoemulsions, e.g., the hydrolysis of surfactants or the oxidation of polyunsaturated lipids.

The most important digestion enzymes that are present within the stomach are gastric pepsin and lipase (Sams et al., 2016). Protein hydrolysis by pepsin in

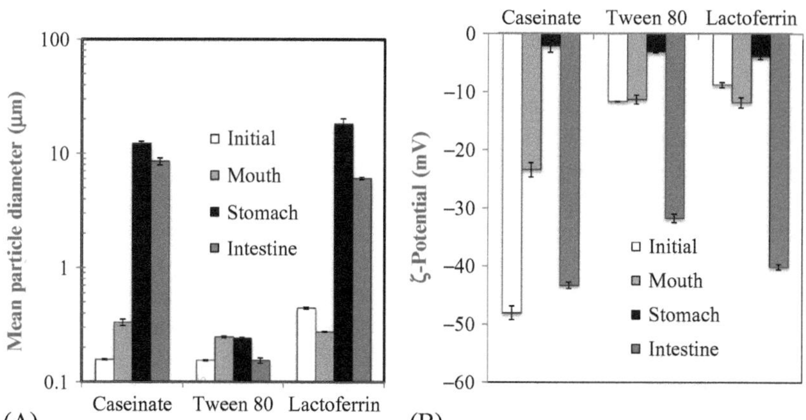

FIG. 18.2 Changes of colloidal dispersions initially containing lipid nanoparticles as they pass through different stages of the GI tract: (A) particle size and (B) surface charge.

the stomach may alter the gastrointestinal fate of both free and trapped lipid droplets in nanoemulsion-based delivery systems. If the lipid droplets in a nanoemulsion are coated by proteins, then hydrolysis of the adsorbed protein layer by pepsin may lead to instability because the droplets are no longer protected from aggregation. If the lipid droplets are trapped within a protein microgel, then hydrolysis of the protein matrix may lead to release of the lipid droplets. The rate and extent of protein hydrolysis by pepsin under gastric conditions depends on protein type, with flexible open proteins (such as casein) being digested more rapidly than compact globular proteins (such as native β-lactoglobulin) (Mandalari et al., 2009).

Gastric lipase is a highly glycosylated globular protein that initiates lipid digestion within the stomach (Mu and Hoy, 2004; Sams et al., 2016). The enzyme activity of gastric lipase is lower than that of pancreatic lipase, with only about 4%–40% of lipid digestion occurring in the stomach (Fave et al., 2004; Michalski, 2009). The degree of lipid digestion in the stomach tends to increase with decreasing droplet size, because there is a larger surface area of oil exposed to the enzymes (McClements and Li, 2010a; Pafumi et al., 2002). There are a variety of surface-active substances in the gastric fluids, which may come from any ingested foods or from GIT secretions, including phospholipids, proteins, peptides, and bile salts. These substances may also alter the composition, structure, and interfacial properties of the lipid droplets in nanoemulsions (Wenner et al., 2000; Macierzanka et al., 2009; Lindahl et al., 1997). For instance, these surface-active substances may adsorb to the surfaces of the droplets and displace the original surfactants, thereby altering interfacial composition and possibly aggregation state. The nanoemulsion droplets dispersed in the gastric fluids are exposed to mechanical forces (flow-force patterns) arising from stomach movements responsible for disrupting, mixing, and transporting digested food materials (Schulze, 2006; Schwizer et al., 2003; Shafik et al., 2006).

A number of studies have investigated the fate of nanoemulsions in the GIT using in vitro or in vivo gastrointestinal models (Singh et al., 2009; Singh and Sarkar, 2011; Golding et al., 2011; McClements and Xiao, 2012). The composition, dimensions, charge, and aggregation state of the lipid droplets in nanoemulsions often change considerably after exposure to stomach conditions. The nature of these changes depends on the properties of the lipid droplets entering the stomach. For example, lipid droplets coated with proteins tend to be much more prone to flocculation than those coated with nonionic surfactants (Fig. 18.2). Controlling the stability of lipid droplets to aggregation and creaming in the stomach impacts their subsequent gastrointestinal fate (e.g., rate of digestion in the small intestine) and the physiological response of the human body (e.g., the release of hormones that influence satiation, satiety, or hunger) (Lundin et al., 2008; Halford and Harrold, 2012; Chung et al., 2016). Nanoemulsions that are stable to stomach conditions may lead to slower gastric emptying and a higher satiety response that those that are unstable (Marciani et al., 2006, 2007, 2009). Consequently, there is interest in the design of gastric stable

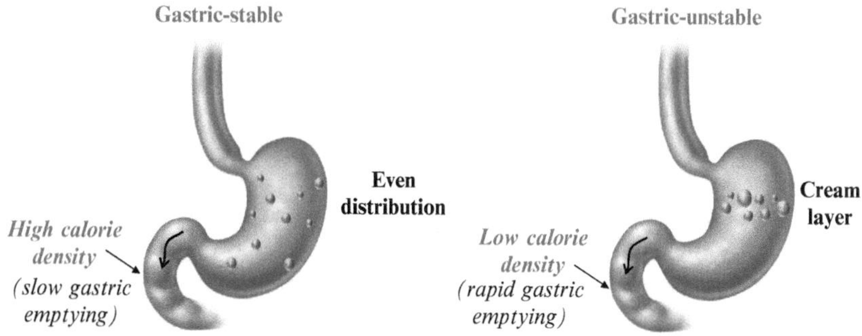

FIG. 18.3 Stability of nanoemulsions within gastric condition: gastric stable emulsions that remain evenly distributed throughout the stomach lead to slower gastric emptying and therefore a higher feeling of satiety than the gastric unstable emulsions.

nanoemulsions that do not aggregate or cream within the stomach (Fig. 18.3). The stability of nanoemulsions in the stomach also depends on the presence of other constituents in the gastric fluids. For example, the aggregation stability of protein-coated lipids within the stomach may be either increased or decreased in the presence of polysaccharides, such as chitosan, pectin, methyl cellulose, and fucoidan. Droplets may become more unstable due to the ability of polysaccharides to promote bridging or depletion flocculation, or they may become more stable due to the ability of the polysaccharides to form a protective coating around them (Espinal-Ruiz et al., 2014; Tokle et al., 2012; Zhang et al., 2015a; Chang and McClements, 2015). The type and concentration of polysaccharides present must therefore be carefully controlled in order to control the gastrointestinal fate of the nanoemulsions.

18.2.3 Small Intestine

The fluids arising from the stomach (chyme) pass through the pylorus sphincter and enter the small intestine, which is the region within the human GIT where most of macronutrients and micronutrients are digested and/or absorbed (McClements and Xiao, 2012; Singh et al., 2009). The fluids in the small intestine contain a complex mixture of bile salts, phospholipids, digestive enzymes, and coenzymes (such as pancreatic lipases, colipases, phospholipases, proteases, and amylases), bicarbonate salts, and other minerals that facilitate the digestion, transport, and absorption of nutrients (McClements and Li, 2010a). The small intestine can be conveniently divided into three sequential regions (duodenum, jejunum, and ileum), with lipid digestion normally occurring primarily in the proximal region (duodenum), which has the highest lipase activity (McClements and Li, 2010a). After entering the duodenum, the acidic chyme is mixed with alkaline small intestinal secretions, causing the pH of the mixture to become close to neutral. These pH conditions are optimum for the pancreatic digestive enzymes to operate. The surface-active bile salts and phospholipids may facilitate the emulsification of any bulk lipids or large lipid droplets by adsorbing to the lipid droplet surfaces (McClements and Li, 2010a; Tso and Crissinger, 2000). Moreover, they may compete with and displace any surface-active substances already present at the lipid droplet surfaces, thereby altering the interfacial composition and aggregation state of the droplets. Pancreatic lipase/colipase complexes can then adsorb to the surfaces of the lipid droplets and convert the emulsified triacylglycerols (TAGs) into free fatty acids (FFAs) and monoacylglycerols (MAGs) (Mukherjee, 2003). The FFAs and MAGs generated by this process interact with bile salts and phospholipids in the small intestinal fluids to form mixed micelles, which consist of small micelles, large vesicles, and liquid crystals (Figs. 18.4 and 18.5). These mixed micelles may also solubilize any other lipophilic substances that were present inside the lipid droplets, such as cholesterol, vitamins, and nutraceuticals (Porter et al., 2007). The mixed micelles then carry the solubilized components across the mucus layer and to the surfaces of the epithelium cells where they can be absorbed. The solubilization capacity of the mixed micelle phase depends on the nature of the ingested delivery system and foods. The mucus layer can be considered to be a thin semipermeable hydrogel with pore dimensions of around 400 nm (Cone, 2009). Small particles that are not strongly attracted to the mucus network will easily pass through, whereas large particles or particles that are strongly attracted to the mucus network will be prevented from passing through. The ability of mixed micelles to enhance the bioavailability of ingested

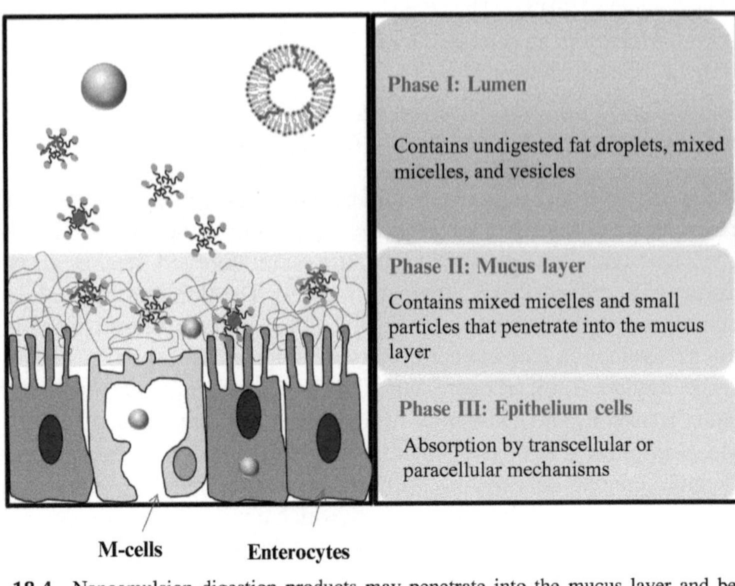

FIG. 18.4 Nanoemulsion digestion products may penetrate into the mucus layer and become trapped (mucoadhesion) or transported across it. Nanoemulsion that are not digested and that travel through the mucus layer may be directly adsorbed by the epithelium cells.

Mixed micelles: Vesicles and micelles

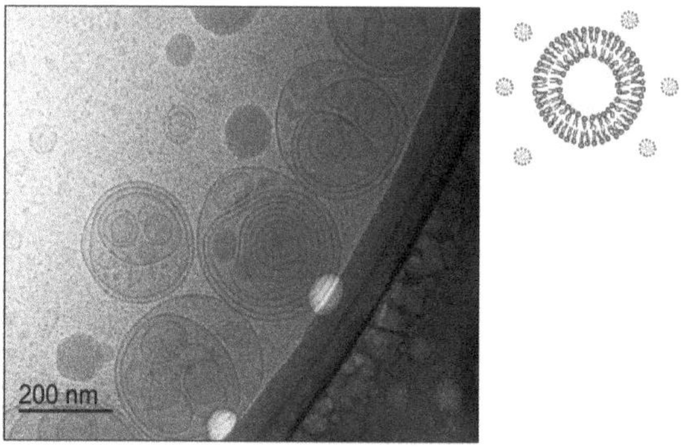

FIG. 18.5 Morphology of mixed micelles formed from digestible lipids under simulated GIT conditions as determined by cryo-TEM. *(Figure kindly supplied by Mingfei Yao, University of Massachusetts.)*

hydrophobic bioactives depends strongly on the nature of the fatty acids present in the mixed micelles, e.g., level, chain length, and unsaturation (Yao et al., 2017). The charge status of FFAs also plays an important role in their absorption. It has been proposed that the surface of the epithelium cells has a lower pH than that of the intestinal lumen, which causes fatty acids to become protonated after they encounter the epithelium cells (Stillhart et al., 2014). The protonation of the FFAs increases their hydrophobicity and therefore facilitates their incorporation into the phospholipid bilayer that makes up a large part of the epithelium cell membrane.

In general, the factors that influence the digestion, solubilization, and absorption of lipid droplets within small intestine fall into two categories: (i) the nature of the ingested lipid droplets and (ii) the properties of the small intestinal fluids. The influence of the nature of the ingested lipid droplets will be discussed in Section 18.3. In this section, we mainly focus on the impact of the composition of the small intestinal fluids on lipid digestion and absorption (McClements and Li, 2010a). Bile salts may promote lipid digestion when they are present at fairly low levels because they can solubilize lipid digestion products (such as FFAs and MAGs), thereby pulling them away from the droplet surfaces and allowing the lipase to access the underlying TAGs (McClements and Li, 2010a). Conversely, they may inhibit lipid digestion when present at relatively high levels because they adsorb strongly to the lipid droplet surfaces and prevent the lipase/colipase complex from adsorbing. Relatively low levels of multivalent cations (such as Ca^{2+} and Mg^{2+}) in the small intestinal fluids may promote lipid digestion due to their ability to form insoluble salts with long-chain saturated fatty acids, thereby pulling them away from the lipid droplet surfaces and allowing the lipase/colipase complex to access the underlying lipids (Li et al., 2011; Reid, 2004; Vaskonen, 2003; Karupaiah and Sundram, 2007). Conversely, high levels of multivalent cations may inhibit lipid digestion by promoting droplet flocculation, thereby reducing the access of the lipase to the lipid droplet surfaces (McClements, 2015; Li et al., 2011). Multivalent ions may also inhibit lipid digestion by promoting the gelation of certain types of ionic biopolymers (such as alginate) that form hydrogels that trap the lipid droplets, thereby preventing the lipase from reaching their surfaces. In general, the levels of the different constituents within the small intestinal fluids may vary considerably depending on the nature of the food consumed and the status of the individual (Tso and Crissinger, 2000), which could alter the gastrointestinal fate of nanoemulsions.

18.2.4 Colon

The large intestine is the final part of the GIT and can be conveniently divided into three main regions: cecum, colon, and, rectum. The colon has a major function in fermenting any ingestible components (such as dietary fibers) that have passed through the upper GIT (Basit, 2005). Typically, most nanoemulsions would be digested and absorbed in the stomach and small intestine and would therefore not reach the large intestine. However, they may be formulated from some ingestible components that do reach the colon. For example, this may occur for nanoemulsions fabricated from indigestible lipids (such as mineral oils or waxes) or indigestible polysaccharides (such as dietary fibers) or for nanoemulsions that are encapsulated within microgels made up of indigestible polysaccharides (such as alginate beads) (McClements and Li, 2010b).

In some circumstances, it may be desirable to develop nanoemulsion-based systems that can deliver specific components to the colon (Lovelyn and Attama, 2011; Gupta et al., 2009; Cai et al., 2010; McClements and Li, 2010b). For instance, the delivery of tributyrin (a short-chain triglyceride) or curcumin (a hydrophobic phytochemical) may inhibit the viability of colon cancer cells (Li et al., 2009; Gou et al., 2011; Yallapu et al., 2012; Huang et al., 2010b).

The targeted delivery of hydrophobic bioactives to the colon can be achieved by encapsulating nanoemulsion droplets inside microgels (Agnihotri et al., 2004; Brannon-Peppas and Blanchette, 2012; Yang et al., 2002; Chen et al., 2005; Hamidi et al., 2008). The microgels are designed so that they remain stable within the upper GIT but are disrupted in the lower GIT, thereby releasing the bioactive agents to the colon. This type of microgel can be formed by assembling indigestible biopolymers, such as dietary fibers, into small particles that trap the nanoemulsion droplets (Hamidi et al., 2008). The microgels can be designed to release the bioactive agents in response to various types of environmental triggers, including pH, time, pressure, enzyme activity, and bacterial activity (Philip and Philip, 2010; Chourasia and Jain, 2003; Yang et al., 2002). The dietary fibers commonly used to assemble colon-targeting microgels include alginate, pectin, carrageenan, cellulose, chitosan, galactomannan, guar gum, inulin, and dextran (Chourasia and Jain, 2004; Sinha and Kumria, 2001; Kosaraju, 2005; Park et al., 2010). Another promising aspect of utilizing dietary fibers to fabricate these colon-targeting delivery systems is that they produce fermentation products (such as short-chain fatty acids) that stimulate the preferential growth of colonic bacteria that may be beneficial to human health (Tan et al., 2014; Campbell et al., 1997; Topping and Clifton, 2001; Wong et al., 2006).

It is always important when developing colon-targeting delivery systems to be sue that they do not have any adverse effects on human health. For example, excess maltodextrin has been shown to impair cellular antibacterial responses and suppress intestinal antimicrobial defense mechanisms in the colon, and unabsorbed iron supplements can increase pathogen abundance and cause intestinal inflammation in infants (Nickerson et al., 2014, 2015; Jaeggi et al., 2015).

18.3 CHANGES IN NANOEMULSION PROPERTIES DURING GIT TRAVEL

18.3.1 Particle Composition and Structure

The composition and structure of lipid droplets may undergo substantial changes as nanoemulsions travel through the GIT. Prior to ingestion, nanoemulsion-based delivery systems may have a variety of different structures, including free droplets, droplet clusters, colloidosomes, multiple emulsions, and droplet-filled microgels (Fig. 18.6). These initial structures may be more or less altered as the delivery systems pass through the different stages of the GIT (the mouth, stomach, small intestine, and colon), due to changes in

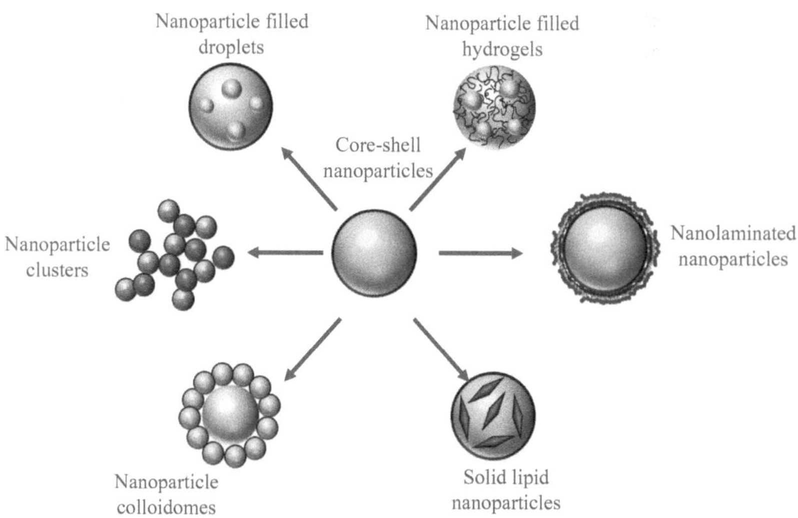

FIG. 18.6 A variety of structures can be assembled using lipid nanoparticles as building blocks, including nanoparticle-filled droplets and hydrogels, nanoparticle clusters, nanoparticle colloidosomes, solid lipid nanoparticles, and nanolaminated nanoparticles.

temperature, pH, ionic strength, dilution, surfactant levels, biopolymer levels, and enzyme activities. For example, the lipid droplets in a microgel may be released when the hydrogel matrix is digested. The composition of the delivery systems may also be changed appreciably due to their exposure to these varying environmental conditions, e.g., due to exchange of constituents with the surrounding gastrointestinal fluids or due to chemical or enzymatic reactions (Zhang et al., 2015c,d). For example, there may be competitive adsorption processes at the particle surfaces, or there may be digestion of any lipids, proteins, or starches present. It is therefore important to understand how the composition and structure of an ingested delivery systems change within the different stages of the GIT.

18.3.2 Particle Dimensions

The size of the particles in a nanoemulsion-based delivery system may change appreciably as it passes through the GIT (Fig. 18.7). This change might occur due to numerous different physicochemical processes (McClements and Xiao, 2012; Singh et al., 2009): (i) *flocculation*, two or more particles may associate with each other without altering of their individual integrities; (ii) *coalescence*, a number of liquid particles may fuse together and form a larger particle; (iii) *partial coalescence*, a number of partially crystalline particles may fuse

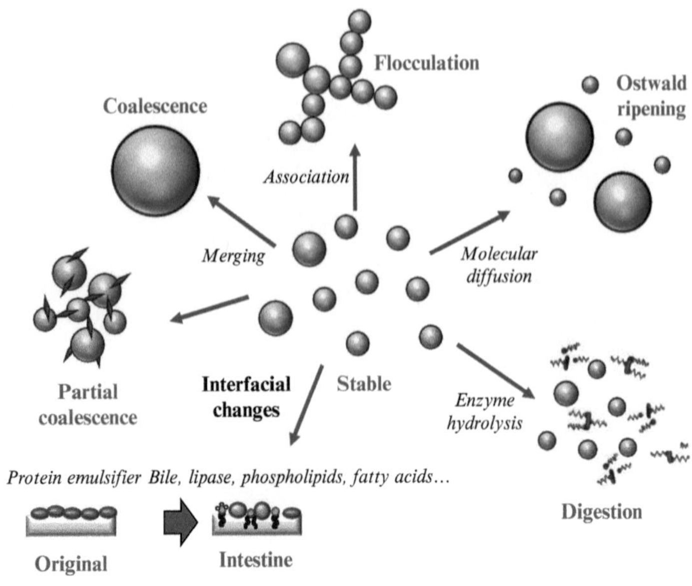

FIG. 18.7 The properties of nanoemulsion may occur in a number of different ways as they pass through the GI tract. Some potential changes in particle dimensions and interfacial properties are illustrated below.

together to form a clump; (iv) *Ostwald ripening*, large particles may grow at the expense of smaller ones due to molecular diffusion of oil molecules through the aqueous phase; (v) *digestion*, a particle may shrink due to enzymatic degradation of the core and/or shell components; and (vi) *solubilization*, a particle may shrink due to transfer of some of the lipid molecules into the surrounding aqueous phase. For these reasons, it is important to follow changes in the dimensions and structural organization of nanoemulsion-based delivery systems throughout the GIT. A number of researchers have reported substantial changes in the dimensions of ingested nanoemulsions as they pass through the GIT using both in vivo and in vitro models (van Aken, 2010; Singh and Sarkar, 2011; Tokle et al., 2012; Li et al., 2012a; Augustin et al., 2011; Liu et al., 2012; Zhang et al., 2015b).

An example of the microstructural changes that occur in protein-stabilized oil-in-water nanoemulsions as they pass through a simulated GIT is shown in Fig. 18.8. Relatively small lipid droplets (radius = 150 nm) that were stable

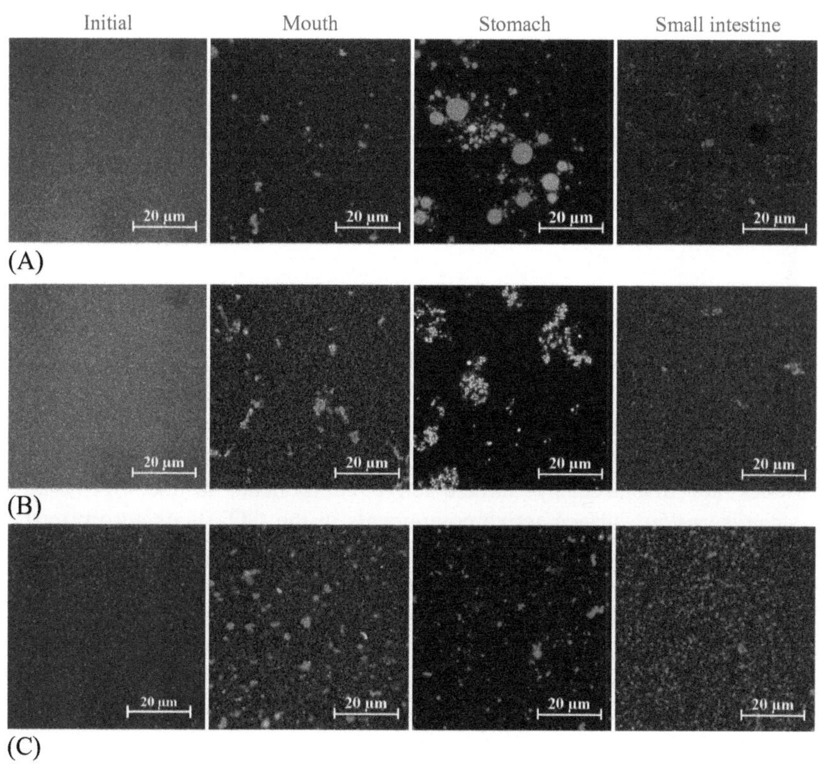

FIG. 18.8 Changes in the microstructure of protein-coated lipid nanoemulsions as they pass through different stages of a simulated GIT. Different gastrointestinal fate can be observed for nanoemulsions fabricated from different lipid phases: (A) MCT, (B) krill oil, and (C) orange oil.

to aggregation were present in the initial nanoemulsions prior to ingestion. A substantial amount of droplet aggregation occurred when the nanoemulsions were exposed to simulated oral conditions, which can be attributed to bridging and depletion flocculation induced by mucin in the simulated saliva. Extensive droplet flocculation and coalescence were observed when these nanoemulsions were then exposed to stomach conditions, which can be attributed to the high ionic strength, high acidity, and enzyme activity of the gastric fluids. Large lipid droplets were not observed in the samples after exposure to small intestine conditions due to digestion by lipase. Interestingly, nanoemulsions made using different lipid phases behaved differently under simulated GIT conditions (Zhang et al., 2015b). The nanoemulsions containing MCT contained relatively large oil droplets in the stomach phase, suggesting that droplet coalescence had occurred. Conversely, the nanoemulsions containing krill oil appeared to contain large nebulous clumps after exposure to the stomach stage, which was attributed to the presence of relatively high levels of phospholipids in these samples. In addition, the nature of the emulsifier initially coating the lipid particles has also been shown to have an appreciable impact on their gastrointestinal fate (Mun et al., 2007; Nik et al., 2011). This effect is demonstrated in Fig. 18.2, which shows the particle size and charge of the lipid droplets in nanoemulsions stabilized by different kinds of emulsifier when they were exposed to different regions of the GIT. Lipid droplets coated by casein underwent extensive aggregation in the stomach phase, whereas those coated by Tween 20 were relatively stable. In addition, it has been reported that nanoemulsions coated by protein-polysaccharide conjugates have higher gastrointestinal stability compared with those coated by protein alone (Yang et al., 2015; Xu et al., 2014).

18.3.3 Interfacial Properties

The interfacial properties of the lipid droplets in nanoemulsions may also be altered appreciably as they travel through the GIT, which can be attributed to various physicochemical and physiological processes (Fig. 18.7): (i) *competitive adsorption*, the original emulsifier molecules on the lipid droplet surfaces are displaced by surface-active components in the GIT fluids; (ii) *coadsorption*, surface-active components in the GIT fluids adsorb to the lipid droplet surfaces and penetrate between the original emulsifier molecules; (iii) *multilayer formation*. surface-active components in the GIT fluids may adsorb on top of the original interfacial layer; and (iv) *digestion or degradation*, some of the original emulsifier molecules on the lipid droplet surfaces may be digested by digestive enzymes or acids in the GIT fluids (Mackie and Macierzanka, 2010). The interfacial properties of the lipid droplets in a nanoemulsion may therefore undergo appreciable changes before they reach the point where they are digested and absorbed (Liu et al., 2012; Mun et al., 2007; Nik et al., 2012; Gallier et al., 2012; Troncoso et al., 2012; Wilde and Chu, 2011; Ye et al., 2011).

The change in interfacial properties of lipid droplets as they pass through the GIT depends on their initial interfacial properties. For example, Fig. 18.2 shows that the ζ-potential of the droplets in nanoemulsions in different regions of the GIT depends on initial emulsifier type. It may therefore be possible to regulate the gastrointestinal fate of nanoemulsions by carefully designing the initial interfacial properties.

18.3.4 Physical State

The physical state of the lipid phase (solid/liquid) or biopolymer phase (gel/sol) in a nanoemulsion may also change as it passes through the GIT. For example, the lipids in a solid lipid nanoparticle may be solid at ambient or refrigerator temperatures within a commercial product, but it may melt when it is exposed to a warmer body temperature after ingestion (Zhang et al., 2015d; Gao and McClements, 2016). Conversely, the lipids in a nanoemulsion may be liquid at elevated temperatures within a commercial product (e.g., a cooked food), but they may crystallize when they are exposed to a colder body temperature (Qian et al., 2012; Zheng et al., 2013; How et al., 2013). Similarly, certain types of bioactive components may be soluble within a commercial product, but they may precipitate in the GIT after ingestion (McClements, 2012b). Conversely, a bioactive agent may be crystalline in a commercial product but dissolve after exposure to GIT conditions. Changes in the physical state of the constituents that make up nanoemulsions have been shown to alter their gastrointestinal fate and should therefore be taken into account when designing nanoemulsion-based delivery systems (Nik et al., 2012; Bonnaire et al., 2008; Li et al., 2012b; Golding et al., 2011).

18.4 IN VITRO AND IN VIVO GIT MODELS FOR NANOEMULSIONS

Knowledge of the GIT fate of nanoemulsions is important for designing nanoemulsion-based delivery systems with specific features, such as the ability to control hormone responses (hunger, satiety, and satiation), to enhance the oral bioavailability of bioactive agents, and to target the release of bioactive agents into specific sites within the GIT. Consequently, analytic tools are required to determine the GIT fate of nanoemulsions with different compositions and structures, so that formulations with the required attributes can be developed (Abuhelwa et al., 2017; Arranz et al., 2016). In general, the analytic tools available for studying the GIT fate of nanoemulsions can be divided into two classes: (i) in vivo approaches and (ii) in vitro approaches. In vivo approaches involve feeding or gavaging an animal or human with a nanoemulsion and then measuring changes in its properties as it passes through the GIT and/or changes in the blood, urine, breath, or specific tissues of appropriate biomarkers. In vivo studies are usually considered to be the "gold standard" for determining the GIT fate of nanoemulsions because they mimic the conditions of consumers ingesting nanoemulsions. However, experiments with animals and humans are expensive, time-consuming, labor-intensive, and highly variable and often have ethical concerns. Moreover, it is often difficult to obtain samples of the nanoemulsion from different regions of the GIT to test their behavior. For this reason, in vitro methods that can be carried out in the laboratory using analytic equipment and chemical supplies are often favored in the initial stages of nanoemulsion development. These methods are designed to simulate the different regions of the GIT (the mouth, stomach, small intestine, and colon) in terms of the solution conditions, incubation times, and flow profiles (Fig. 18.9). In vitro methods are cheaper and quicker to carry out than in vivo methods, typically produce more consistent data, and allow samples to be collected and analyzed from different regions of the GIT (McClements and Li, 2010a; Minekus et al., 2014; Guerra et al., 2012; Arranz et al., 2016). Thus, in vitro methods have a number of characteristics that mean that they are widely used when developing nanoemulsion-based systems:

(1) They can be used to rapidly screen a large number of nanoemulsions with different formulations during the early development stages, thereby enabling optimum compositions and structures to be identified for further testing using more representative in vivo methods.

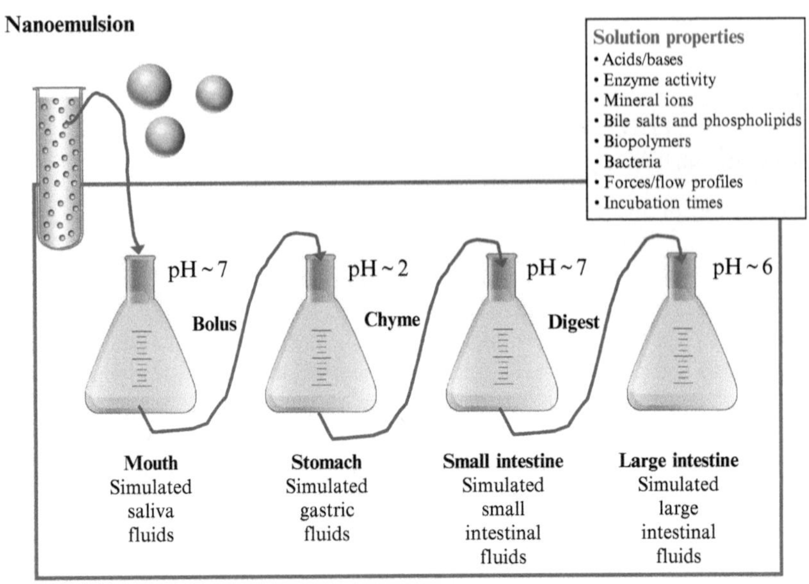

FIG. 18.9 Multiple-stage static gastrointestinal model (the mouth, stomach, small intestine, and large intestine) for simulating the gastrointestinal fate of nanoemulsions.

(2) They can be used to establish the relationship between the structure and composition of a nanoemulsion and its physicochemical behavior within the GIT.
(3) They can be used to determine the bioaccessibility and transformation of bioactive agents within the GIT.
(4) They can be used to determine the absorption of bioactive agents when used in conjunction with cell culture models.
(5) They can be used to establish the release mechanisms of nanoemulsion-based delivery systems in different regions of the GIT (such as the mouth, stomach, small intestine, and colon).

Numerous in vitro models have been introduced over the years to study the GIT fate of nanoemulsions and other ingested materials, which vary in their accuracy in simulating actual GIT conditions (McClements and Li, 2010a; Minekus et al., 2014). Some models focus on a specific part of the GIT (such as the mouth, stomach, small intestine, or colon), whereas others subject test samples to the full GIT (the mouth, stomach, small intestine, and colon) (Chen, 2006). The most sophisticated in vitro models use complex mechanical apparatus and solution compositions to subject test samples to conditions that closely mimic human conditions, such as incubation times, flow/force conditions (such as shear, compression, and elongation), variable pH profiles, and sequential addition of digestive secretions (such as acids, bases, minerals, enzymes, and bile salts) (Minekus et al., 1995; Vardakou et al., 2011; Dickinson et al., 2012; Shani-Levi et al., 2013). These sophisticated methods are relatively time-consuming, laborious, and expensive to carry out, but they provide more accurate data that can often be correlated to the results from in vivo studies. For this reason, many researchers use much simpler in vitro models that are faster, easier, and cheaper to perform but less accurately mimic real GIT conditions. One of the most common simple methods is the three-stage static GIT model that simulates the mouth, stomach, and small intestine (McClements and Li, 2010a; Guerra et al., 2012). In this case, the test sample is sequentially mixed with simulated oral, gastric, and intestinal fluids that have specified compositions (pH, ionic strength, enzyme activity, bile salts, and mucin) for incubation periods and mixing conditions that mimic GIT conditions.

The properties of particles in nanoemulsion-based products may change appreciably when they are exposed to GIT environments. Moreover, bioactive agents encapsulated inside the oil droplets may be retained, released, or chemically/enzymatically transformed in different regions of the GIT. Appropriate analytic techniques are therefore required to monitor the gastrointestinal fate of the oil droplets in nanoemulsions and any coingested bioactive agents.

18.4.1 Static In Vitro Gastrointestinal Model

The three-stage static in vitro gastrointestinal model is commonly used to study the fate of nanoemulsions in the GIT (Arranz et al., 2016; Li et al., 2011;

Minekus et al., 2014). In this case, the nanoemulsions are sequentially exposed to a series of simulated gastrointestinal fluids with different compositions at body temperature, i.e., ≈37°C (Fig. 18.9). The in vitro model described below has been widely used in our laboratory and is specifically designed for following the GIT fate of nanoemulsions.

Mouth: The nanoemulsions are mixed with simulated saliva fluid (SSF) for a defined incubation time (2 min) at a defined rotational speed (100 rpm). In the case of semisolid or solid nanoemulsion formulations, it may be necessary to have some kind of mechanical devices to simulate the mastication process. The SSF is a watery solution with pH around 6.8 that contains numerous components to simulate human saliva, such as water, acids, buffers, minerals, biopolymers (mucin), and enzymes (amylase) (Kalantzi et al., 2006; Amado et al., 2005; Vitorino et al., 2004a,b; Humphrey and Williamson, 2001). There are numerous recipes for simulating the SSF composition in the literature (Sarkar et al., 2009; Versantvoort et al., 2005). Mucins, which are large-molecular-weight glycoproteins present in saliva, should be included in the SSF since they may alter the properties of nanoemulsions, e.g., by binding to oil droplet surfaces or by inducing droplet aggregation (Sanz and Luyten, 2006; Kalantzi et al., 2006). It is recommended to include around 0.3 wt% mucins in SSF based on literature reports (Ionta et al., 2014; Kapsimalis et al., 1966). Salivary α-amylase is important to include in the SSF when a test sample contains starch, since certain types of starch molecules may be hydrolyzed by this enzyme in the mouth. An α-amylase activity concentration of $100\,U\,mL^{-1}$ is recommended since this is the average concentration reported during mastication (Yamaguchi et al., 2004; Tangsrianugul et al., 2015). A volume ratio of 1:1 is suggested for mixing test samples with SSF to simulate the dilution of ingested materials within the mouth by saliva. Although the oral chewing and/or swallowing time varies from few seconds to minutes depending on the nature of ingested material (e.g., liquid, semisolid, or solid), it is suggested to carry out the simulated oral phase for 2 min to ensure good reproducibility from batch to batch. The mixture collected at the completion of the mouth stage is referred to as the "bolus."

Stomach: The bolus collected from the mouth phase is mixed with simulated gastric fluid (SGF) and is then incubated for a defined time (2 h) at certain stirring conditions (100 rpm). For samples that remain semisolid or solid when reaching the stomach phase, it is advisable to utilize a mechanical device that simulates the mechanical forces/fluid flows associated with stomach motility (Ferrua et al., 2011; Kong and Singh, 2010). The SGF is a highly acidic watery solution whose composition is designed to simulate stomach conditions, e.g., acids, buffers, minerals, biopolymers, and digestive enzymes (proteases and lipases). Gastric pepsin is the major protease in the stomach fluids and hydrolyzes some proteins into peptides and amino acids. Gastric lipase is a highly glycosylated globular protein present in the stomach fluids that hydrolyzes TAGs into diacylglycerols (DAGs), MAGs, and FFAs. Typically, only a small fraction of the TAGs are hydrolyzed by gastric lipase, while the rest are hydrolyzed by pancreatic lipase in the small intestine. The US Pharmacopeia (USP) has recommended a recipe for the SGF that is relatively simple to prepare and has a pH, ionic strength, and enzyme activity (3.2 mg pepsin/mL) that represents human gastric fluids (Fu et al., 2002). For the study of nanoemulsions, it is useful to add 0.3% mucin to the SGF because this substance is known to be present in human gastric fluids and may alter droplet properties, e.g., by adsorbing to their surfaces or promoting their aggregation (Versantvoort et al., 2005). Even though gastric lipase is known to be present in human gastric fluids, it is commonly omitted from SGF because there is currently no commercially available reliable and economic source of human gastric lipase or alternatives with similar performance (Minekus et al., 2014). A volume ratio of 1:1 is suggested for mixing the bolus sample with the SGF to simulate dilution in the stomach. In this static in vitro model, the pH is adjusted to 2.5 after mixing the bolus sample with the SGF to mimic the highly acidic gastric fluids (but in practice, the pH of the stomach many change considerably depending on the nature of the food consumed). The material collected at the completion of the stomach stage is referred to as "chyme."

Small intestine: The chyme collected from the stomach stage is mixed with simulated small intestinal fluids (SSIF) for a defined time (2 h) using defined stirring conditions (shearing at 4 s^{-1}). The SSIF is a neutral watery solution that contains a mixture of constituents designed to mimic human small intestine conditions, such as bases, bile salts, phospholipids, buffers, mineral salts, and digestive enzymes (amylases, proteases, and lipases). The key factors that impact lipid digestion in the small intestine phase are ionic composition, bile salt concentration, and lipase activity. Screening studies suggest that a concentration of 150 mM NaCl and 10 mM $CaCl_2$ in the final samples after mixing with SSIF can be used to mimic the ionic composition of the small intestine (Li et al., 2011). There are two options available for mimicking the enzyme activity of the small intestine phase: (i) pancreatic lipase and (ii) pancreatin. In the first option, purified enzyme (pancreatic lipase) is used because it is chemically more well defined, and there is less batch-to-batch variation (McClements and Li, 2010a). In this case, a pancreatic lipase concentration of 1.6 mg/mL in the final reaction vessel is recommended. A bile salt concentration of 5 mg/mL is then used to mimic human intestinal conditions and to avoid inhibition of the pancreatic lipase by bile salts (Mun et al., 2006; Zangenberg et al., 2001). If the lipase used does not contain colipase, then this cofactor should be added to ensure good lipase activity. In the second option, pancreatin is used to simulate the intestinal enzyme activity, which is a complex mixture of different digestive enzymes (including lipase, protease, and amylase). This approach may be preferred for test samples that contain a mixture of different components, such as lipid, protein, and starch. In this case, a pancreatin concentration of 2.4 mg/mL is recommended and a bile salt concentration of 20 mg/mL (Li et al., 2011). However, due to the complex composition of pancreatin, it is more difficult to define its precise enzyme activity. This problem can be overcome by separately adding individual pancreatic enzymes to SSIF (McClements and Li, 2010a). The rate of macronutrient digestion can often be monitored using a pH-stat method in relatively simple systems (i.e., those that contain a single major macronutrient). In nanoemulsions, this method is often used to follow the rate and extent of lipid digestion. After lipase is added to the reaction chamber, the amount of alkaline solution (OH^-) required to neutralize the FFAs generated when the TAGs are hydrolyzed is determined and then converted into the fraction of lipids that have been digested at a particular incubation time (McClements and Li, 2010a). To compare the lipid digestion profile of different samples, it is suggested to use a fixed initial lipid content (300 mg) in the reaction chamber (Li et al., 2011). The material collected at the completion of the small intestine stage is referred to as the "digest." The digest can either be passed to a simulated colon stage, or it can be used to analyze the bioaccessibility, absorption, and transformation of bioactive agents (Fig. 18.10).

Large intestine: The digest sample resulting from the small intestine phase is mixed with simulated large intestinal fluids (SLIF) for a defined time using defined agitation conditions. The colon region itself may be divided into different regions that can be simulated using different conditions. One of the most important features of the human colon is that it contains a complex mixture of anaerobic bacteria that may interact with and transform components in the digest. Typically, the bacterial populations utilized in in vitro methods are

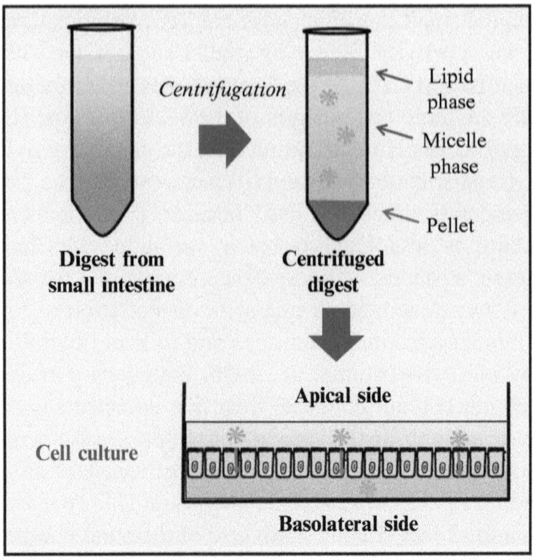

FIG. 18.10 The bioaccessibility, transformation, and absorption of bioactive agents can be determined by centrifuging the digest from the small intestine phase, collecting the micelle phase, and placing on a cell culture model.

cultivated from animal or human feces. The in vitro methods developed to mimic the large intestine of humans vary from single-stage static batch methods to multiple-stage dynamic methods that use a cocktail of anaerobic bacteria cultures (Rumney and Rowland, 1992; Macfarlane and Macfarlane, 2007; Ouwehand et al., 2009). Typically, a test sample is incubated in a single or multiple SLIFs for a certain time, and then, changes in its properties and/or the properties of the bacterial population are determined. A challenge to simulating the behavior of nanoemulsions within the large intestine is the differences in the nature of the bacterial populations between individuals. For this reason, SLIFs containing a well-defined enzyme cocktail designed to mimic the enzyme activity of colonic bacteria may be used (Yang, 2008; Souto-Maior et al., 2010). The inherent challenges associated with establishing, maintaining and performing experiments using a simulated large intestine means that many researchers prefer to use in vivo animal models (Macfarlane and Macfarlane, 2007). Moreover, if a nanoemulsion is expected to be completely digested and absorbed within the upper GIT, then this step can be omitted.

18.4.2 Characterization of Changes in Nanoemulsion Properties in GIT

As discussed in Section 18.3, the composition and structure of nanoemulsions may be changed appreciably as they travel through the GIT, and knowledge of these changes can be used to design nanoemulsion-based systems that behave in a specific desirable way (McClements and Xiao, 2012; McClements, 2013). In this section, we discuss analytic methods that can be used to characterize changes in the properties of nanoemulsions in the GIT.

Particle composition: Initially, the oil droplets in a nanoemulsion may be freely dispersed within a watery phase, or they may be embedded in larger particles, such as hydrogels or food fragments. The composition of the particles in these nanoemulsion-based systems will be altered as they pass through the different regions of the GIT due to a number of physicochemical and biochemical processes, including enzymatic degradation of particular constituents, dissociation of physical structures, and molecular exchange with the surrounding gastrointestinal fluids (Fig. 18.7). In certain situations, the particles can be isolated from the gastrointestinal fluids using filtration or centrifugation processes, and then, their composition can be determined using suitable analytic techniques, such as chemical, chromatography, electrophoresis, or spectroscopy methods. Alternatively, changes in the composition of the particles may be inferred from measurements of changes in the composition of the surrounding fluids using same analytic techniques. For some systems, information about the composition of the particles in gastrointestinal fluids can be determined in situ using appropriate imaging methods. For instance, fluorescence microscopy can be used to determine the concentration of specific components in a particular region within a sample by measuring its fluorescence spectra (Fig. 18.8). Indirect information about changes in the composition of the interfacial region of particles can often be inferred from measurements of the particle charge (ζ-potential) using electrophoresis methods.

Particle dimensions and aggregation state: The dimensions and structural organization of the particles within a nanoemulsion-based system may change appreciably as they pass through the GIT due to flocculation, coalescence, partial coalescence, and digestion (Fig. 18.7). Particle aggregation may occur due to alterations in the composition of the gastrointestinal fluids surrounding the particles (such as pH, ionic strength, enzyme activity, or biopolymer content) that alter the colloidal interactions acting between the particles. Particle digestion may occur as the result of different kinds of enzymes in the gastrointestinal fluids (such as lipases and proteases). Changes in the dimensions and aggregation state of the particles in the GIT may be characterized using particle sizing techniques, such as dynamic light scattering, static light scattering, optical microscopy, or electron microscopy. In some cases, the particles may have to be isolated from the surrounding gastrointestinal fluids prior to analysis so as to prevent interference from other particulate matter that might be present (such as fragments of partially digested food). Confocal fluorescence microscopy is widely used to monitor the gastrointestinal fate of nanoemulsions because different constituents (such as fats or proteins) can be stained to elucidate their location within the sample (Fig. 18.8).

Physical state: The physical state (solid or liquid) of the constituents within a nanoemulsion-based system may change during passage through the GIT depending on their characteristic transition temperatures relative to body temperature. For example, lipids may melt or crystallize. Biopolymer matrices may undergo glass-rubbery transitions. Partially soluble materials, such as sugars or salts, may dissolve or precipitate. Changes in the physical state of the components within nanoemulsions may be monitored using a number of analytic methods (Nielsen, 2010). Optical microscopy combined with a polarized light filter can be utilized to detect the location of crystals or liquid crystals within materials. X-ray diffraction and infrared instruments can provide information about molecular packing in crystalline materials. Differential scanning calorimetry (DSC) can be used to monitor the crystallization, melting, and polymorphic changes of materials.

Release of bioactive agents: The release of bioactive agents from nanoemulsions in the GIT can be monitored using various analytic tools. Chemical imaging techniques (such as those based on fluorescence, infrared, or Raman spectroscopy) can be used to follow changes in the concentration of labeled or nonlabeled substances in different locations of a sample over time. Alternatively, the concentration of a released component in the particles or in the GIT fluids surrounding the particles can be measured using a suitable analytic technique once the particles have been separated, e.g., by dialysis, filtration, or centrifugation. A number of the most commonly used analytic techniques include chemical analysis, chromatography, electrophoresis, mass spectrometry, and spectroscopy, with the most appropriate technique depending on the type of component that is being analyzed (Nielsen, 2010).

18.4.3 Bioaccessibility and Absorption of Nutrients and Bioactive Agents in GIT

The digested nutrients and released bioactive agents from a nanoemulsion are usually absorbed by the layer of epithelium cells that makes up the walls of the GIT, with the majority of the absorption typically occurring in the small intestine (Arranz et al., 2016; Li et al., 2011; Minekus et al., 2014; Jafari et al., 2017a,b). The absorption of nutrients and bioactive agents can be modeled using in vitro, ex vivo, or cell culture models depending on the purpose of the study.

In vitro models: The bioaccessibility of a substance is the fraction of the material that is in a suitable form for absorption. For lipophilic nutrients and bioactive agents, this can be taken to be the fraction that is solubilized by the mixed micelles formed in the small intestine. The mixed micelles transport these lipophilic substances through the mucus layer and into the epithelium cells where they can be absorbed. Thus, the bioaccessibility can be taken to be the concentration of the lipophilic substances in the mixed micelles divided by their overall concentration in the small intestine. In some cases, the bioaccessibility of a substance can be determined directly without carrying out any isolation procedures; e.g., the FFAs arising from lipid digestion can be measured by acid titration or enzymatic methods (Mun et al., 2005, 2006; Brogård et al., 2007). In other cases, the mixed micelles may have to be separated from the rest of the digest by dialysis, filtration, or centrifugation prior to analysis (Christensen et al., 2004; Do Thi et al., 2009). For example, the bioaccessibility of lipophilic bioactive agents (such as carotenoids or curcumin) released after nanoemulsion digestion can be separated by centrifugation and then analyzed using appropriate spectroscopic or chromatographic methods (Zou et al., 2015; Zhang et al., 2016a). In another approach, the test samples are placed inside a dialysis bag, and the amount of lipophilic substance transported through the semipermeable membrane in the mixed micelles is measured using a suitable method (Mouécoucou et al., 2003; Villaume et al., 2004).

Ex vivo models: The absorption of substances from the lumen of the GIT can also be modeled using ex vivo models (Jain et al., 2016). In this case, a piece of the small intestine from an animal is used as a semipermeable membrane to more closely mimic in vivo absorption conditions (Luo et al., 2013). One of the most common methods used to conduct this type of research utilizes an Ussing chamber (Dahan and Hoffman, 2007; Hess et al., 2005). In this approach, a section of the small intestine is removed from an animal after it has been sacrificed, carefully washed to remove any residual material, and then clamped between two chambers. The first chamber contains the test sample (donor chamber), while the second chamber contains buffer solution (receiver chamber) (Dahan and Hoffman, 2007). The transport of bioactive agents or nutrients through the small intestine is then measured over time using appropriate analytic techniques (Bhaskar et al., 2009; Gannu et al., 2010; Gavini et al., 2005; Gupta et al., 2011). In another approach, the test sample is placed inside the section of small intestine, which is then clamped at both ends, and then placed in a buffer solution. The fraction of bioactive agents or nutrients that are transported through the small intestine and into the buffer solution is then measured using a suitable method (Luo et al., 2013).

Cell culture methods: One of the most widely used methods of determining the absorption of bioactive agents and nutrients is to use cell culture models, which have many features that mimic those of the epithelium cells lining the small intestine. A variety of different kinds of cell culture model have been developed, with the most common being the Caco-2 model (Versantvoort et al., 2002; Reboul et al., 2006; Dhuique-Mayer et al., 2007). A layer of Caco-2 cells is usually grown on the surface of a petri dish or on a semipermeable membrane that is placed in a petri dish (Fig. 18.10). After being exposed to simulated GIT conditions, the test sample resulting from the small intestine phase is placed in contact with the Caco-2 cells, and then, the fraction of the substance of interest that is absorbed by or passes through the cells is determined using an appropriate analytic technique. In more sophisticated models, cocultures are used consisting of two or more different types of cells to mimic different aspects of human gastrointestinal walls, such as M cells, mucus-producing cells, and Caco-2 cells.

18.4.4 Animal and Human Studies for GIT Fate of Nanoemulsions

18.4.4.1 In Vivo Approaches

The in vitro and ex vivo methods described above are particularly suited for screening samples with different compositions and structures, and for identifying key physicochemical events that occur with the GIT (McClements, 2010; Minekus et al., 2014). However, they cannot simulate the extremely complicated processes occurring within a human GIT. For this reason, in vivo studies using animals and/or humans are sometimes carried out to provide a more realistic representation of the actual GIT fate of nanoemulsions (McClements et al., 2008). Because they are time-consuming, laborious, and expensive to carry out, in vivo studies are usually only performed on a limited number of test samples that have first been identified using in vitro studies. The nanoemulsions to be tested may be fed alone, as part of a meal or as part of a diet over a certain period. A number of different parameters may then be measured depending on the purpose of the study. Noninvasive measurements of changes in the behavior, activity, or body weight of animals or humans can be made over time. Invasive measurements of changes in the weight, composition, or function of specific target organs (such as the kidneys, liver, heart, spleen, or brain) of animals can be measured after they have been sacrificed (Khandavilli and

Panchagnula, 2007; Murillo et al., 2016; Yadav et al., 2016). The absorption, metabolism, and/or excretion of a particular component within an ingested nanoemulsion can be monitored by analyzing changes in the composition of the blood, breath feces, or urine of an animal or human (Kim et al., 2014; Cho et al., 2014). Invasive tubes can be used to collect samples from different regions of the GIT (such as the stomach) of animals or humans. More recently, imaging methods have been used to visualize changes in the location, composition, and structure of foods within the GIT, e.g., magnetic resonance, ultrasound, X-ray, or fluorescence imaging methods (Xing et al., 2012; Xiao et al., 2013; Yao et al., 2014).

18.4.4.2 In Vitro-In Vivo Correlations

As mentioned earlier, in vitro test methods have certain advantages over in vivo ones in the initial stages of nanoemulsion developed, because they are quicker, cheaper, and more reproducible, and they can provide information about the physicochemical processes occurring within the GIT (Porter and Charman, 2001; Chen, 2006; Dahan and Hoffman, 2007). On the other hand, they are too simplistic to accurately simulate the complex conditions occurring inside the human GIT, and so the results obtained should be treated with caution. For this reason, the gastrointestinal fate of nanoemulsions is often evaluated using both in vitro and in vivo studies. In some cases, animal or human studies can be minimized if a strong correlation between in vitro and in vivo methods can be established (Porter and Charman, 2001; Dahan and Hoffman, 2007; Pouton and Porter, 2008). In vitro-in vivo correlations (IVIVCs) can be established by measuring the bioavailability of a specific component in a nanoemulsion using both in vitro and in vivo methods and then comparing the results obtained. Alternatively, computer simulation methods can be utilized to model the processes occurring within the GIT and then predict the gastrointestinal fate of nanoemulsions based on knowledge of their initial composition and structure (Dressman et al., 2011; Mathias and Crison, 2012).

18.5 CONCLUSIONS

This chapter has provided an overview of the GIT fate of nanoemulsions and nanoemulsion-based delivery systems after ingestion. Research in this area has grown rapidly in recent years as nanoemulsions are being increasingly utilized to control the biological activity of foods, supplements, and drugs. Indeed, the main driving force for this research endeavor is the rational design of nanoemulsions that can improve human health and wellness. Nanoemulsion-based products may be designed to control hormone release in the GIT and therefore modulate feelings of hunger, satiety, and satiation. This type of product would be useful for controlling calorie consumption and thus be suitable for tackling obesity and overweight. Alternatively, nanoemulsion-based products may be fortified with nutraceuticals, vitamins, or drugs that prevent or treat chronic diseases, including cardiovascular disease, diabetes, or cancer. In this case, the nanoemulsions would be designed to protect the bioactive components during storage but release them in the GIT at a suitable location. The design and fabrication of nanoemulsion-based products to meet these requirements depends on a thorough understanding of their behavior within the human GIT. Finally, the potential toxicity of ingested nanoemulsions will also depend on their GIT fate, which should be considered when designing products with novel properties.

REFERENCES

Abuhelwa, A.Y., Williams, D.B., Upton, R.N., Foster, D.J.R., 2017. Food, gastrointestinal pH, and models of oral drug absorption. Eur. J. Pharm. Biopharm. 112, 234–248.

Agnihotri, S.A., Mallikarjuna, N.N., Aminabhavi, T.M., 2004. Recent advances on chitosan-based micro-and nanoparticles in drug delivery. J. Control. Release 100, 5–28.

Amado, F.M.L., Vitorino, R.M.P., Domingues, P.M.D.N., Lobo, M.J.C., Duarte, J.A.R., 2005. Analysis of the human saliva proteome. Expert Rev. Proteomics 2, 521–539.

Arranz, E., Corredig, M., Guri, A., 2016. Designing food delivery systems: challenges related to the in vitro methods employed to determine the fate of bioactives in the gut. Food Funct. 7, 3319–3336.

Augustin, M.A., Patten, G., De Luca, A., Abeywardena, M., Lockett, T., Head, R., Sanguansri, L., 2011. Intestinal passage of microencapsulated fish oil in rats following oral administration. Food Funct. 2, 684–696.

Basit, A.W., 2005. Advances in colonic drug delivery. Drugs 65, 1991–2007.

Beysseriat, M., Decker, E.A., McClements, D.J., 2006. Preliminary study of the influence of dietary fiber on the properties of oil-in-water emulsions passing through an in vitro human digestion model. Food Hydrocoll. 20, 800–809.

Bhaskar, K., Anbu, J., Ravichandiran, V., Venkateswarlu, V., Rao, Y.M., 2009. Lipid nanoparticles for transdermal delivery of flurbiprofen: formulation, in vitro, ex vivo and in vivo studies. Lipids Health Dis. 8, 6.

Bonnaire, L., Sandra, S., Helgason, T., Decker, E.A., Weiss, J., McClements, D.J., 2008. Influence of lipid physical state on the in vitro digestibility of emulsified lipids. J. Agric. Food Chem. 56, 3791–3797.

Brannon-Peppas, L., Blanchette, J.O., 2012. Nanoparticle and targeted systems for cancer therapy. Adv. Drug Deliv. Rev. 64, 206–212.

Brogård, M., Troedsson, E., Thuresson, K., Ljusberg-Wahren, H., 2007. A new standardized lipolysis approach for characterization of emulsions and dispersions. J. Colloid Interface Sci. 308, 500–507.

Buettner, A., 2002. Influence of human salivary enzymes on odorant concentration changes occurring in vivo. 1. Esters and thiols. J. Agric. Food Chem. 50, 3283–3289.

Cai, Z., Wang, Y., Zhu, L.-J., Liu, Z.-Q., 2010. Nanocarriers: a general strategy for enhancement of oral bioavailability of poorly absorbed or pre-systemically metabolized drugs. Curr. Drug Metab. 11, 197–207.

Camacho, S., den Hollander, E., van de Velde, F., Stieger, M., 2015a. Properties of oil/water emulsions affecting the deposition, clearance, and after-feel sensory perception of oral coatings. J. Agric. Food Chem. 63, 2145–2153.

Camacho, S., Liu, K., van der Linden, A., Stieger, M., van de Velde, F., 2015b. Formation, clearance and mouthfeel perception of oral coatings formed by emulsion-filled gels. J. Texture Stud. 46, 399–410.

Campbell, J.M., Fahey, G.C., Wolf, B.W., 1997. Selected indigestible oligosaccharides affect large bowel mass, cecal and fecal short-chain fatty acids, pH and microflora in rats. J. Nutr. 127, 130–136.

Cerqueira, M.A., Pinheiro, A.C., Silva, H.D., Ramos, P.E., Azevedo, M.A., Flores-Lopez, M.L., Rivera, M.C., Bourbon, A.I., Ramos, O.L., Vicente, A.A., 2014. Design of bio-nanosystems for oral delivery of functional compounds. Food Eng. Rev. 6, 1–19.

Chang, Y., McClements, D.J., 2015. Interfacial deposition of an anionic polysaccharide (fucoidan) on protein-coated lipid droplets: impact on the stability of fish oil-in-water emulsions. Food Hydrocoll. 51, 252–260.

Chen, X.D., 2006. GIT physicochemical modeling—a critical. Crit. Rev. 2, 4.

Chen, L.-G., Liu, Z.-L., Zhuo, R.-X., 2005. Synthesis and properties of degradable hydrogels of konjac glucomannan grafted acrylic acid for colon-specific drug delivery. Polymer 46, 6274–6281.

Cho, H.T., Salvia-Trujillo, L., Kim, J., Park, Y., Xiao, H., McClements, D.J., 2014. Droplet size and composition of nutraceutical nanoemulsions influences bioavailability of long chain fatty acids and coenzyme Q10. Food Chem. 156, 117–122.

Chourasia, M.K., Jain, S.K., 2003. Pharmaceutical approaches to colon targeted drug delivery systems. J. Pharm. Pharm. Sci. 6, 33–66.

Chourasia, M.K., Jain, S.K., 2004. Polysaccharides for colon targeted drug delivery. Drug Deliv. 11, 129–148.

Christensen, J.Ø., Schultz, K., Mollgaard, B., Kristensen, H.G., Mullertz, A., 2004. Solubilisation of poorly water-soluble drugs during in vitro lipolysis of medium-and long-chain triacylglycerols. Eur. J. Pharm. Sci. 23, 287–296.

Chung, C., Degner, B., McClements, D.J., 2012. Instrumental mastication assay for texture assessment of semi-solid foods: combined cyclic squeezing flow and shear viscometry. Food Res. Int. 49, 161–169.

Chung, C., Smith, G., Degner, B., McClements, D.J., 2016. Reduced fat food emulsions: physicochemical, sensory, and biological aspects. Crit. Rev. Food Sci. Nutr. 56, 650–685.

Cone, R.A., 2009. Barrier properties of mucus. Adv. Drug Deliv. Rev. 61, 75–85.

Dahan, A., Hoffman, A., 2007. The effect of different lipid based formulations on the oral absorption of lipophilic drugs: the ability of in vitro lipolysis and consecutive ex vivo intestinal permeability data to predict in vivo bioavailability in rats. Eur. J. Pharm. Biopharm. 67, 96–105.

Date, A.A., Hanes, J., Ensign, L.M., 2016. Nanoparticles for oral delivery: design, evaluation and state-of-the-art. J. Control. Release 240, 504–526.

Dhuique-Mayer, C., Borel, P., Reboul, E., Caporiccio, B., Besancon, P., Amiot, M.-J., 2007. β-cryptoxanthin from citrus juices: assessment of bioaccessibility using an in vitro digestion/caco-2 cell culture model. Br. J. Nutr. 97, 883–890.

Dickinson, P.A., Rmaileh, R.A., Ashworth, L., Barker, R.A., Burke, W.M., Patterson, C.M., Stainforth, N., Yasin, M., 2012. An investigation into the utility of a multi-compartmental, dynamic, system of the upper gastrointestinal tract to support formulation development and establish bioequivalence of poorly soluble drugs. AAPS J. 14, 196–205.

do Thi, T., van Speybroeck, M., Barillaro, V., Martens, J., Annaert, P., Augustijns, P., van Humbeeck, J., Vermant, J., van den Mooter, G., 2009. Formulate-ability of ten compounds with different physicochemical profiles in smedds. Eur. J. Pharm. Sci. 38, 479–488.

Dressman, J.B., Thelen, K., Willmann, S., 2011. An update on computational oral absorption simulation. Expert Opin. Drug Metab. Toxicol. 7, 1345–1364.

Ekmekcioglu, C., 2002. A physiological approach for preparing and conducting intestinal bioavailability studies using experimental systems. Food Chem. 76, 225–230.

Espinal-Ruiz, M., Parada-Alfonso, F., Restrepo-Sánchez, L.-P., Narváez-Cuenca, C.-E., McClements, D.J., 2014. Impact of dietary fibers [methyl cellulose, chitosan, and pectin] on digestion of lipids under simulated gastrointestinal conditions. Food Funct. 5, 3083–3095.

Fave, G., Coste, T.C., Armand, M., 2004. Physicochemical properties of lipids: new strategies to manage fatty acid bioavailability. Cell. Mol. Biol. 50, 815–832.

Ferrua, M.J., Kong, F., Singh, R.P., 2011. Computational modeling of gastric digestion and the role of food material properties. Trends Food Sci. Technol. 22, 480–491.

Frohlich, E., Roblegg, E., 2016. Oral uptake of nanoparticles: human relevance and the role of in vitro systems. Arch. Toxicol. 90, 2297–2314.

Fu, T.-J., Abbott, U.R., Hatzos, C., 2002. Digestibility of food allergens and nonallergenic proteins in simulated gastric fluid and simulated intestinal fluid a comparative study. J. Agric. Food Chem. 50, 7154–7160.

Gallier, S., Ye, A., Singh, H., 2012. Structural changes of bovine milk fat globules during in vitro digestion. J. Dairy Sci. 95, 3579–3592.

Gannu, R., Palem, C.R., Yamsani, V.V., Yamsani, S.K., Yamsani, M.R., 2010. Enhanced bioavailability of lacidipine via microemulsion based transdermal gels: formulation optimization, ex vivo and in vivo characterization. Int. J. Pharm. 388, 231–241.

Gao, S., McClements, D.J., 2016. Formation and stability of solid lipid nanoparticles fabricated using phase inversion temperature method. Colloids Surf. A Physicochem. Eng. Asp. 499, 79–87.

Gavini, E., Rassu, G., Sanna, V., Cossu, M., Giunchedi, P., 2005. Mucoadhesive microspheres for nasal administration of an antiemetic drug, metoclopramide: in-vitro/ex-vivo studies. J. Pharm. Pharmacol. 57, 287–294.

Gharibzahedi, S.M.T., Jafari, S.M., 2017. The importance of minerals in human nutrition: bioavailability, food fortification, processing effects and nanoencapsulation. Trends Food Sci. Technol. 62, 119–132.

Golding, M., Wooster, T.J., 2010. The influence of emulsion structure and stability on lipid digestion. Curr. Opin. Colloid Interface Sci. 15, 90–101.

Golding, M., Wooster, T.J., Day, L., Xu, M., Lundin, L., Keogh, J., Clifton, P., 2011. Impact of gastric structuring on the lipolysis of emulsified lipids. Soft Matter 7, 3513–3523.

Gou, M., Men, K., Shi, H., Xiang, M., Zhang, J., Song, J., Long, J., Wan, Y., Luo, F., Zhao, X., 2011. Curcumin-loaded biodegradable polymeric micelles for colon cancer therapy in vitro and in vivo. Nanoscale 3, 1558–1567.

Guerra, A., Etienne-Mesmin, L., Livrelli, V., Denis, S., Blanquet-Diot, S., Alric, M., 2012. Relevance and challenges in modeling human gastric and small intestinal digestion. Trends Biotechnol. 30, 591–600.

Gupta, H., Bhandari, D., Sharma, A., 2009. Recent trends in oral drug delivery: a review. Recent Pat. Drug Deliv. Formul. 3, 162–173.

Gupta, S., Chavhan, S., Sawant, K.K., 2011. Self-nanoemulsifying drug delivery system for adefovir dipivoxil: design, characterization, in vitro and ex vivo evaluation. Colloids Surf. A Physicochem. Eng. Asp. 392, 145–155.

Halford, J.C.G., Harrold, J.A., 2012. Satiety-enhancing products for appetite control: science and regulation of functional foods for weight management. Proc. Nutr. Soc. 71, 350–362.

Hamidi, M., Azadi, A., Rafiei, P., 2008. Hydrogel nanoparticles in drug delivery. Adv. Drug Deliv. Rev. 60, 1638–1649.

Hansson, A., Leufvén, A., van Ruth, S., 2003. Partition and release of 21 aroma compounds during storage of a pectin gel system. J. Agric. Food Chem. 51, 2000–2005.

Hess, S., Rotshild, V., Hoffman, A., 2005. Investigation of the enhancing mechanism of sodium N-[8-(2-Hydroxybenzoyl) Amino] caprylate effect on the intestinal permeability of polar molecules utilizing a voltage clamp method. Eur. J. Pharm. Sci. 25, 307–312.

How, C.W., Rasedee, A., Abbasalipourkabir, R., 2013. Characterization and cytotoxicity of nanostructured lipid carriers formulated with olive oil, hydrogenated palm oil, and polysorbate 80. IEEE Trans. Nanobiosci. 12, 72–78.

Huang, C.B., George, B., Ebersole, J.L., 2010a. Antimicrobial activity of N-6, N-7 and N-9 fatty acids and their esters for oral microorganisms. Arch. Oral Biol. 55, 555–560.

Huang, Q., Yu, H., Ru, Q., 2010b. Bioavailability and delivery of nutraceuticals using nanotechnology. J. Food Sci. 75, R50–R57.

Humphrey, S.P., Williamson, R.T., 2001. A review of saliva: normal composition, flow, and function. J. Prosthet. Dent. 85, 162–169.

Ionta, F.Q., Mendonça, F.L., de Oliveira, G.C., de Alencar, C.R.B., Honório, H.M., Magalhães, A.C., Rios, D., 2014. In vitro assessment of artificial saliva formulations on initial enamel erosion remineralization. J. Dent. 42, 175–179.

Jaeggi, T., Kortman, G.A.M., Moretti, D., Chassard, C., Holding, P., Dostal, A., Boekhorst, J., Timmerman, H.M., Swinkels, D.W., Tjalsma, H., 2015. Iron fortification adversely affects the gut microbiome, increases pathogen abundance and induces intestinal inflammation in kenyan infants. Gut 64, 731–742.

Jafari, S.M., McClements, D.J., 2017. Nanotechnology approaches for increasing nutrient bioavailability. In: Advances in Food and Nutrition Research. Academic Press, New York.

Jafari, S.M., Katouzian, I., Rajabi, H., Ganje, M., 2017a. 13—Bioavailability and release of bioactive components from nanocapsules. In: Nanoencapsulation Technologies for the Food and Nutraceutical Industries. Academic Press, New York, pp. 494–523.

Jafari, S.M., Esfanjani, A.F., Katouzian, I., Assadpour, E., 2017b. Chapter 10—Release, characterization, and safety of nanoencapsulated food ingredients. In: Nanoencapsulation of Food Bioactive Ingredients. Academic Press, New York, pp. 401–453.

Jain, A.S., Dhawan, V.V., Sarmento, B., Nagarsenker, M.S., 2016. In vitro and ex vivo evaluations of lipid anti-cancer nanoformulations: insights and assessment of bioavailability enhancement. AAPS PharmSciTech 17, 553–571.

Joyce, P., Whitby, C.P., Prestidge, C.A., 2016. Nanostructuring biomaterials with specific activities towards digestive enzymes for controlled gastrointestinal absorption of lipophilic bioactive molecules. Adv. Colloid Interf. Sci. 237, 52–75.

Kalantzi, L., Goumas, K., Kalioras, V., Abrahamsson, B., Dressman, J.B., Reppas, C., 2006. Characterization of the human upper gastrointestinal contents under conditions simulating bioavailability/bioequivalence studies. Pharm. Res. 23, 165–176.

Kapsimalis, P., Rosenthal, S.L., Updegrave, W., Evans, R., Hurley, B., Cobe, H.M., 1966. The relationship between caries activity, flow rate, total nitrogen and the mucin content of saliva. J. Oral Med. 21, 107–110.

Karupaiah, T., Sundram, K., 2007. Effects of stereospecific positioning of fatty acids in triacylglycerol structures in native and randomized fats: a review of their nutritional implications. Nutr. Metab. 4, 16.

Khandavilli, S., Panchagnula, R., 2007. Nanoemulsions as versatile formulations for paclitaxel delivery: peroral and dermal delivery studies in rats. J. Investig. Dermatol. 127, 154–162.

Kim, J.-Y., Lee, M.-S., Jung, S., Joo, H., Kim, C.-T., Kim, I.-H., Seo, S., Oh, S., Kim, Y., 2014. Anti-obesity efficacy of nanoemulsion oleoresin capsicum in obese rats fed a high-fat diet. Int. J. Nanomedicine 9, 301.

Kong, F., Singh, R.P., 2010. A human gastric simulator (hgs) to study food digestion in human stomach. J. Food Sci. 75, E627–E635.

Kosaraju, S.L., 2005. Colon targeted delivery systems: review of polysaccharides for encapsulation and delivery. Crit. Rev. Food Sci. Nutr. 45, 251–258.

Li, Y., Le Maux, S., Xiao, H., McClements, D.J., 2009. Emulsion-based delivery systems for tributyrin, a potential colon cancer preventative agent. J. Agric. Food Chem. 57, 9243–9249.

Li, Y., Hu, M., McClements, D.J., 2011. Factors affecting lipase digestibility of emulsified lipids using an in vitro digestion model: proposal for a standardised pH-stat method. Food Chem. 126, 498–505.

Li, Y., Kim, J., Park, Y., McClements, D.J., 2012a. Modulation of lipid digestibility using structured emulsion-based delivery systems: comparison of in vivo and in vitro measurements. Food Funct. 3, 528–536.

Li, Y., Zheng, J., Xiao, H., McClements, D.J., 2012b. Nanoemulsion-based delivery systems for poorly water-soluble bioactive compounds: influence of formulation parameters on polymethoxyflavone crystallization. Food Hydrocoll. 27, 517–528.

Lindahl, A., Ungell, A.-L., Knutson, L., Lennernäs, H., 1997. Characterization of fluids from the stomach and proximal jejunum in men and women. Pharm. Res. 14, 497–502.

Liu, Y., Hou, Z., Lei, F., Chang, Y., Gao, Y., 2012. Investigation into the bioaccessibility and microstructure changes of β-carotene emulsions during in vitro digestion. Innovative Food Sci. Emerg. Technol. 15, 86–95.

Loesche, W.J., 1999. The antimicrobial treatment of periodontal disease: changing the treatment paradigm. Crit. Rev. Oral Biol. Med. 10, 245–275.

Lovelyn, C., Attama, A.A., 2011. Current state of nanoemulsions in drug delivery. J. Biomed. Nanotechnol. 2, 626.

Lundin, L., Golding, M., Wooster, T.J., 2008. Understanding food structure and function in developing food for appetite control. Nutr. Diet. 65, S79–S85.

Luo, Z., Liu, Y., Zhao, B., Tang, M., Dong, H., Zhang, L., Lv, B., Wei, L., 2013. Ex vivo and in situ approaches used to study intestinal absorption. J. Pharmacol. Toxicol. Methods 68, 208–216.

Macfarlane, G.T., Macfarlane, S., 2007. Models for intestinal fermentation: association between food components, delivery systems, bioavailability and functional interactions in the gut. Curr. Opin. Biotechnol. 18, 156–162.

Macierzanka, A., Sancho, A.I., Mills, E.N.C., Rigby, N.M., Mackie, A.R., 2009. Emulsification alters simulated gastrointestinal proteolysis of B-casein and B-lactoglobulin. Soft Matter 5, 538–550.

Mackie, A., Macierzanka, A., 2010. Colloidal aspects of protein digestion. Curr. Opin. Colloid Interface Sci. 15, 102–108.

Madene, A., Jacquot, M., Scher, J., Desobry, S., 2006. Flavour encapsulation and controlled release–a review. Int. J. Food Sci. Technol. 41, 1–21.

Mandalari, G., Adel-Patient, K., Barkholt, V., Baro, C., Bennett, L., Bublin, M., Gaier, S., Graser, G., Ladics, G.S., Mierzejewska, D., 2009. In vitro digestibility of B-casein and B-lactoglobulin under simulated human gastric and duodenal conditions: a multi-laboratory evaluation. Regul. Toxicol. Pharmacol. 55, 372–381.

Marciani, L., Wickham, M., Singh, G., Bush, D., Pick, B., Cox, E., Fillery-Travis, A., Faulks, R., Marsden, C., Gowland, P.A., 2006. Delaying gastric emptying and enhancing cholecystokinin release and satiety by using acid stable fat emulsions. Gastroenterology A227. Wb Saunders Co-Elsevier Inc Independence Square West Curtis Center, Ste 300, Philadelphia, PA 19106-3399 USA.

Marciani, L., Wickham, M., Singh, G., Bush, D., Pick, B., Cox, E., Fillery-Travis, A., Faulks, R., Marsden, C., Gowland, P.A., 2007. Enhancement of intragastric acid stability of a fat emulsion meal delays gastric emptying and increases cholecystokinin release and gallbladder contraction. Am. J. Physiol. Gastrointest. Liver Physiol. 292, G1607–G1613.

Marciani, L., Faulks, R., Wickham, M.S.J., Bush, D., Pick, B., Wright, J., Cox, E.F., Fillery-Travis, A., Gowland, P.A., Spiller, R.C., 2009. Effect of intragastric acid stability of fat emulsions on gastric emptying, plasma lipid profile and postprandial satiety. Br. J. Nutr. 101, 919–928.

Matalanis, A., Jones, O.G., McClements, D.J., 2011. Structured biopolymer-based delivery systems for encapsulation, protection, and release of lipophilic compounds. Food Hydrocoll. 25, 1865–1880.

Mathias, N.R., Crison, J., 2012. The use of modeling tools to drive efficient oral product design. AAPS J. 14, 591–600.

McClements, D.J., 2010. Emulsion design to improve the delivery of functional lipophilic components. Annu. Rev. Food Sci. Technol. 1, 241–269.

McClements, D.J., 2012a. Advances in fabrication of emulsions with enhanced functionality using structural design principles. Curr. Opin. Colloid Interface Sci. 17, 235–245.

McClements, D.J., 2012b. Crystals and crystallization in oil-in-water emulsions: implications for emulsion-based delivery systems. Adv. Colloid Interf. Sci. 174, 1–30.

McClements, D.J., 2013. Edible lipid nanoparticles: digestion, absorption, and potential toxicity. Prog. Lipid Res. 52, 409–423.

McClements, D.J., 2015. Food Emulsions: Principles, Practices, And Techniques. CRC Press, Boca Raton, FL.

McClements, D.J., Li, Y., 2010a. Review of in vitro digestion models for rapid screening of emulsion-based systems. Food Funct. 1, 32–59.

McClements, D.J., Li, Y., 2010b. Structured emulsion-based delivery systems: controlling the digestion and release of lipophilic food components. Adv. Colloid Interf. Sci. 159, 213–228.

McClements, D.J., Xiao, H., 2012. Potential biological fate of ingested nanoemulsions: influence of particle characteristics. Food Funct. 3, 202–220.

McClements, D.J., Decker, E.A., Park, Y., 2008. Controlling lipid bioavailability through physicochemical and structural approaches. Crit. Rev. Food Sci. Nutr. 49, 48–67.

McClements, D.J., Decker, E.A., Park, Y., Weiss, J., 2009. Structural design principles for delivery of bioactive components in nutraceuticals and functional foods. Crit. Rev. Food Sci. Nutr. 49, 577–606.

Mei, L., Zhang, Z.P., Zhao, L.Y., Huang, L.Q., Yang, X.L., Tang, J.T., Feng, S.S., 2013. Pharmaceutical nanotechnology for oral delivery of anticancer drugs. Adv. Drug Deliv. Rev. 65, 880–890.

Michalski, M.C., 2009. Specific molecular and colloidal structures of milk fat affecting lipolysis, absorption and postprandial lipemia. Eur. J. Lipid Sci. Technol. 111, 413–431.

Minekus, M., Marteau, P., Havenaar, R., 1995. Multicompartmental dynamic computer-controlled model simulating the stomach and small intestine. Alternatives to Laboratory Animals: ATLA 23, 197–209.

Minekus, M., Alminger, M., Alvito, P., Ballance, S., Bohn, T., Bourlieu, C., Carriere, F., Boutrou, R., Corredig, M., Dupont, D., 2014. A standardised static in vitro digestion method suitable for food–an international consensus. Food Funct. 5, 1113–1124.

Mouécoucou, J., Sanchez, C., Villaume, C., Marrion, O., Frémont, S., Laurent, F., Méjean, L., 2003. Effects of different levels of gum arabic, low methylated pectin and xylan on in vitro digestibility of B-lactoglobulin. J. Dairy Sci. 86, 3857–3865.

Mu, H.L., Hoy, C.E., 2004. The digestion of dietary triacylglycerols. Prog. Lipid Res. 43, 105–133.

Mukherjee, M., 2003. Human digestive and metabolic lipases—a brief review. J. Mol. Catal. B Enzym. 22, 369–376.

Mun, S., Decker, E.A., McClements, D.J., 2005. Influence of droplet characteristics on the formation of oil-in-water emulsions stabilized by surfactant—chitosan layers. Langmuir 21, 6228–6234.

Mun, S., Decker, E.A., Park, Y., Weiss, J., McClements, D.J., 2006. Influence of interfacial composition on in vitro digestibility of emulsified lipids: potential mechanism for chitosan's ability to inhibit fat digestion. Food Biophys. 1, 21–29.

Mun, S., Decker, E.A., McClements, D.J., 2007. Influence of emulsifier type on in vitro digestibility of lipid droplets by pancreatic lipase. Food Res. Int. 40, 770–781.

Mun, S., Kim, Y.R., McClements, D.J., 2015a. Control of beta-carotene bioaccessibility using starch-based filled hydrogels. Food Chem. 173, 454–461.

Mun, S., Kim, Y.R., Shin, M., McClements, D.J., 2015b. Control of lipid digestion and nutraceutical bioaccessibility using starch-based filled hydrogels: influence of starch and surfactant type. Food Hydrocoll. 44, 380–389.

Murillo, A.G., Aguilar, D., Norris, G.H., Dimarco, D.M., Missimer, A., Hu, S., Smyth, J.A., Gannon, S., Blesso, C.N., Luo, Y., 2016. Compared with powdered lutein, a lutein nanoemulsion increases plasma and liver lutein, protects against hepatic steatosis, and affects lipoprotein metabolism in guinea pigs. J. Nutr. 146, 1961–1969.

Nickerson, K.P., Homer, C.R., Kessler, S.P., Dixon, L.J., Kabi, A., Gordon, I.O., Johnson, E.E., Carol, A., Mcdonald, C., 2014. The dietary polysaccharide maltodextrin promotes salmonella survival and mucosal colonization in mice. PLoS One. 9.

Nickerson, K.P., Chanin, R., Mcdonald, C., 2015. Deregulation of intestinal anti-microbial defense by the dietary additive, maltodextrin. Gut Microbes 6, 78–83.

Nielsen, S.S., 2010. Food Analysis. Springer, New York, NY.

Nik, A.M., Wright, A.J., Corredig, M., 2011. Impact of interfacial composition on emulsion digestion and rate of lipid hydrolysis using different in vitro digestion models. Colloids Surf. B: Biointerfaces 83, 321–330.

Nik, A.M., Langmaid, S., Wright, A.J., 2012. Digestibility and B-carotene release from lipid nanodispersions depend on dispersed phase crystallinity and interfacial properties. Food Funct. 3, 234–245.

Ouwehand, A.C., Tiihonen, K., Mäkeläinen, H., Rautonen, N., Hasselwander, O., Sworn, G., 2009. Non-starch polysaccharides in the gastrointestinal tract. In: Designing Functional Foods: Measuring and Controlling Food Structure Breakdown and Nutrient Absorption. Woodhead Publishing Ltd., Cambridge, pp. 126–147

Pafumi, Y., Lairon, D., De La Porte, P.L., Juhel, C., Storch, J., Hamosh, M., Armand, M., 2002. Mechanisms of inhibition of triacylglycerol hydrolysis by human gastric lipase. J. Biol. Chem. 277, 28070–28079.

Park, J.H., Saravanakumar, G., Kim, K., Kwon, I.C., 2010. Targeted delivery of low molecular drugs using chitosan and its derivatives. Adv. Drug Deliv. Rev. 62, 28–41.

Philip, A.K., Philip, B., 2010. Colon targeted drug delivery systems: a review on primary and novel approaches. Oman Med. J. 25, 79–87.

Porter, C.J.H., Charman, W.N., 2001. In vitro assessment of oral lipid based formulations. Adv. Drug Deliv. Rev. 50, S127–S147.

Porter, C.J.H., Trevaskis, N.L., Charman, W.N., 2007. Lipids and lipid-based formulations: optimizing the oral delivery of lipophilic drugs. Nat. Rev. Drug Discov. 6, 231–248.

Porter, C.J.H., Wasan, K.M., Constantinides, P., 2008. Lipid-Based Systems for the Enhanced Delivery of Poorly Water Soluble Drugs. Elsevier, Atlanta, GA.

Pouton, C.W., 2006. Formulation of poorly water-soluble drugs for oral administration: physicochemical and physiological issues and the lipid formulation classification system. Eur. J. Pharm. Sci. 29, 278–287.

Pouton, C.W., Porter, C.J.H., 2008. Formulation of lipid-based delivery systems for oral administration: materials, methods and strategies. Adv. Drug Deliv. Rev. 60, 625–637.

Qian, C., Decker, E.A., Xiao, H., McClements, D.J., 2012. Solid lipid nanoparticles: effect of carrier oil and emulsifier type on phase behavior and physical stability. J. Am. Oil Chem. Soc. 89, 17–28.

Reboul, E., Richelle, M., Perrot, E., Desmoulins-Malezet, C., Pirisi, V., Borel, P., 2006. Bioaccessibility of carotenoids and vitamin E from their main dietary sources. J. Agric. Food Chem. 54, 8749–8755.

Reid, I.R., 2004. Effects of calcium supplementation on circulating lipids. Drugs Aging 21, 7–17.

Rumney, C.J., Rowland, I.R., 1992. In vivo and in vitro models of the human colonic flora. Crit. Rev. Food Sci. Nutr. 31, 299–331.

Salvia-Trujillo, L., Qian, C., Martín-Belloso, O., McClements, D.J., 2013. Influence of particle size on lipid digestion and β-carotene bioaccessibility in emulsions and nanoemulsions. Food Chem. 141, 1472–1480.

Sams, L., Paume, J., Giallo, J., Carriere, F., 2016. Relevant pH and lipase for in vitro models of gastric digestion. Food Funct. 7, 30–45.

Sanz, T., Luyten, H., 2006. Effect of thickening agent in the in vitro mouth, stomach and intestine release of tyrosol from enriched custards. Food Hydrocoll. 20, 703–711.

Sarkar, A., Goh, K.K.T., Singh, H., 2009. Colloidal stability and interactions of milk-protein-stabilized emulsions in an artificial saliva. Food Hydrocoll. 23, 1270–1278.

Schulze, K., 2006. Imaging and modelling of digestion in the stomach and the duodenum. Neurogastroenterol. Motil. 18, 172–183.

Schwizer, W., Fox, M., Steingötter, A., 2003. Non-invasive investigation of gastrointestinal functions with magnetic resonance imaging: towards an "ideal" investigation of gastrointestinal function. Gut 52, Iv34–Iv39.

Shafik, A., El Sibai, O. & Shafik, A. A. 2006. Mechanism of gastric emptying through the pyloric sphincter: a human study. Med. Sci. Monit., 13, Cr24-Cr29.

Shani-Levi, C., Levi-Tal, S., Lesmes, U., 2013. Comparative performance of milk proteins and their emulsions under dynamic in vitro adult and infant gastric digestion. Food Hydrocoll. 32, 349–357.

Silletti, E., Vingerhoeds, M.H., Norde, W., van Aken, G.A., 2007. The role of electrostatics in saliva-induced emulsion flocculation. Food Hydrocoll. 21, 596–606.

Singh, H., Sarkar, A., 2011. Behaviour of protein-stabilised emulsions under various physiological conditions. Adv. Colloid Interf. Sci. 165, 47–57.

Singh, H., Ye, A., Horne, D., 2009. Structuring food emulsions in the gastrointestinal tract to modify lipid digestion. Prog. Lipid Res. 48, 92–100.

Sinha, V.R., Kumria, R., 2001. Polysaccharides in colon-specific drug delivery. Int. J. Pharm. 224, 19–38.

Sørensen, L.B., Møller, P., Flint, A., Martens, M., Raben, A., 2003. Effect of sensory perception of foods on appetite and food intake: a review of studies on humans. Int. J. Obes. 27, 1152–1166.

Souto-Maior, J.F.A., Reis, A.V., Pedreiro, L.N., Cavalcanti, O.A., 2010. Phosphated crosslinked pectin as a potential excipient for specific drug delivery: preparation and physicochemical characterization. Polym. Int. 59, 127–135.

Stillhart, C., Dürr, D., Kuentz, M., 2014. Toward an improved understanding of the precipitation behavior of weakly basic drugs from oral lipid-based formulations. J. Pharm. Sci. 103, 1194–1203.

Stokes, J.R., Boehm, M.W., Baier, S.K., 2013. Oral processing, texture and mouthfeel: from rheology to tribology and beyond. Curr. Opin. Colloid Interface Sci. 18, 349–359.

Tan, J., Mckenzie, C., Potamitis, M., Thorburn, A.N., Mackay, C.R., Macia, L., 2014. The role of short-chain fatty acids in health and disease. Adv. Immunol. 121.

Tangsrianugul, N., Suphantharika, M., McClements, D.J., 2015. Simulated gastrointestinal fate of lipids encapsulated in starch hydrogels: impact of normal and high amylose corn starch. Food Res. Int. 78, 79–87.

Tokle, T., Lesmes, U., Decker, E.A., McClements, D.J., 2012. Impact of dietary fiber coatings on behavior of protein-stabilized lipid droplets under simulated gastrointestinal conditions. Food Funct. 3, 58–66.

Topping, D.L., Clifton, P.M., 2001. Short-chain fatty acids and human colonic function: roles of resistant starch and nonstarch polysaccharides. Physiol. Rev. 81, 1031–1064.

Troncoso, E., Aguilera, J.M., McClements, D.J., 2012. Fabrication, characterization and lipase digestibility of food-grade nanoemulsions. Food Hydrocoll. 27, 355–363.

Tso, P., Crissinger, K., 2000. Overview of digestion and absorption. In: Biochemical and Physiological Aspects of Human Nutrition. W.B. Saunders, Philadelphia, PA, pp. 75–90.

van Aken, G.A., 2010. Relating food emulsion structure and composition to the way it is processed in the gastrointestinal tract and physiological responses: what are the opportunities? Food Biophys. 5, 258–283.

van Aken, G.A., Bomhof, E., Zoet, F.D., Verbeek, M., Oosterveld, A., 2011. Differences in vitro gastric behaviour between homogenized milk and emulsions stabilised by tween 80, whey protein, or whey protein and caseinate. Food Hydrocoll. 25, 781–788.

van Ruth, S.M., Grossmann, I., Geary, M., Delahunty, C.M., 2001. Interactions between artificial saliva and 20 aroma compounds in water and oil model systems. J. Agric. Food Chem. 49, 2409–2413.

Vardakou, M., Mercuri, A., Barker, S.A., Craig, D.Q.M., Faulks, R.M., Wickham, M.S.J., 2011. Achieving antral grinding forces in biorelevant in vitro models: comparing the USP dissolution apparatus II and the dynamic gastric model with human in vivo data. AAPS PharmSciTech 12, 620–626.

Vaskonen, T., 2003. Dietary minerals and modification of cardiovascular risk factors. J. Nutr. Biochem. 14, 492–506.

Versantvoort, C.H.M., Ondrewater, R.C.A., Duizer, E., van de Sandt, J.J.M., Gilde, A.J., Groten, J.P., 2002. Monolayers of IEC-18 cells as an in vitro model for screening the passive transcellular and paracellular transport across the intestinal barrier: comparison of active and passive transport with the human colon carcinoma CACO-2 cell line. Environ. Toxicol. Pharmacol. 11, 335–344.

Versantvoort, C.H.M., Oomen, A.G., van de Kamp, E., Rompelberg, C.J.M., Sips, A.J.A.M., 2005. Applicability of an in vitro digestion model in assessing the bioaccessibility of mycotoxins from food. Food Chem. Toxicol. 43, 31–40.

Villaume, C., Sanchez, C., Méjean, L., 2004. β-Lactoglobulin/polysaccharide interactions during in vitro gastric and pancreatic hydrolysis assessed in dialysis bags of different molecular weight cut-offs. Biochim. Biophys. Acta Gen. Subj. 1670, 105–112.

Vingerhoeds, M.H., Blijdenstein, T.B.J., Zoet, F.D., van Aken, G.A., 2005. Emulsion flocculation induced by saliva and mucin. Food Hydrocoll. 19, 915–922.

Vitorino, R., Lobo, M.J.C., Duarte, J.A.R., Ferrer-Correia, A.J., Domingues, P.M., Amado, F.M.L., 2004a. Analysis of salivary peptides using hplc—electrospray mass spectrometry. Biomed. Chromatogr. 18, 570–575.

Vitorino, R., Lobo, M.J.C., Ferrer-Correira, A.J., Dubin, J.R., Tomer, K.B., Domingues, P.M., Amado, F.M.L., 2004b. Identification of human whole saliva protein components using proteomics. Proteomics 4, 1109–1115.

Wenner, J., Gunnarsson, T., Graffner, H., Lindell, G., 2000. Influence of smoking and helicobacter pylori on gastric phospholipids. Dig. Dis. Sci. 45, 1648–1652.

Wilde, P.J., Chu, B.S., 2011. Interfacial & colloidal aspects of lipid digestion. Adv. Colloid Interf. Sci. 165, 14–22.

Wong, J.M.W., De Souza, R., Kendall, C.W.C., Emam, A., Jenkins, D.J.A., 2006. Colonic health: fermentation and short chain fatty acids. J. Clin. Gastroenterol. 40, 235–243.

Xiao, Q., Zheng, X., Bu, W., Ge, W., Zhang, S., Chen, F., Xing, H., Ren, Q., Fan, W., Zhao, K., 2013. A core/satellite multifunctional nanotheranostic for in vivo imaging and tumor eradication by radiation/photothermal synergistic therapy. J. Am. Chem. Soc. 135, 13041–13048.

Xing, H., Bu, W., Zhang, S., Zheng, X., Li, M., Chen, F., He, Q., Zhou, L., Peng, W., Hua, Y., 2012. Multifunctional nanoprobes for upconversion fluorescence, MR and CT trimodal imaging. Biomaterials 33, 1079–1089.

Xu, D., Yuan, F., Gao, Y., Panya, A., McClements, D.J., Decker, E.A., 2014. Influence of whey protein–beet pectin conjugate on the properties and digestibility of β-carotene emulsion during in vitro digestion. Food Chem. 156, 374–379.

Yadav, S., Gandham, S.K., Panicucci, R., Amiji, M.M., 2016. Intranasal brain delivery of cationic nanoemulsion-encapsulated tnfα sirna in prevention of experimental neuroinflammation. Nanomedicine 12, 987–1002.

Yallapu, M.M., Jaggi, M., Chauhan, S.C., 2012. Curcumin nanoformulations: a future nanomedicine for cancer. Drug Discov. Today 17, 71–80.

Yamaguchi, M., Kanemori, T., Kanemaru, M., Takai, N., Mizuno, Y., Yoshida, H., 2004. Performance evaluation of salivary amylase activity monitor. Biosens. Bioelectron. 20, 491–497.

Yang, L., 2008. Biorelevant dissolution testing of colon-specific delivery systems activated by colonic microflora. J. Control. Release 125, 77–86.

Yang, L., Chu, J.S., Fix, J.A., 2002. Colon-specific drug delivery: new approaches and in vitro/in vivo evaluation. Int. J. Pharm. 235, 1–15.

Yang, Y., Cui, S., Gong, J., Miller, S.S., Wang, Q., Hua, Y., 2015. Stability of citral in oil-in-water emulsions protected by a soy protein–polysaccharide maillard reaction product. Food Res. Int. 69, 357–363.

Yao, J., Yang, M., Duan, Y., 2014. Chemistry, biology, and medicine of fluorescent nanomaterials and related systems: new insights into biosensing, bioimaging, genomics, diagnostics, and therapy. Chem. Rev. 114, 6130–6178.

Yao, M., McClements, D.J., Zhao, F., Craig, R.W., Xiao, H., 2017. Controlling the gastrointestinal fate of nutraceutical and pharmaceutical-enriched lipid nanoparticles: from mixed micelles to chylomicrons. NanoImpact 5, 13–21.

Ye, A., Cui, J., Singh, H., 2011. Proteolysis of milk fat globule membrane proteins during in vitro gastric digestion of milk. J. Dairy Sci. 94, 2762–2770.

Yeung, T.W., Üçok, E.F., Tiani, K.A., McClements, D.J., Sela, D.A., 2016. Microencapsulation in alginate and chitosan microgels to enhance viability of bifidobacterium longum for oral delivery. Front. Microbiol. 7, 494.

Zangenberg, N.H., Müllertz, A., Kristensen, H.G., Hovgaard, L., 2001. A dynamic in vitro lipolysis model: I. Controlling the rate of lipolysis by continuous addition of calcium. Eur. J. Pharm. Sci. 14, 115–122.

Zhang, R., Zhang, Z., Zhang, H., Decker, E.A., McClements, D.J., 2015a. Influence of emulsifier type on gastrointestinal fate of oil-in-water emulsions containing anionic dietary fiber (pectin). Food Hydrocoll. 45, 175–185.

Zhang, R., Zhang, Z., Zhang, H., Decker, E.A., McClements, D.J., 2015b. Influence of lipid type on gastrointestinal fate of oil-in-water emulsions: in vitro digestion study. Food Res. Int. 75, 71–78.

Zhang, Z., Zhang, R., Chen, L., Tong, Q., McClements, D.J., 2015c. Designing hydrogel particles for controlled or targeted release of lipophilic bioactive agents in the gastrointestinal tract. Eur. Polym. J. 72, 698–716.

Zhang, Z., Zhang, R., Tong, Q., Decker, E.A., McClements, D.J., 2015d. Food-grade filled hydrogels for oral delivery of lipophilic active ingredients: temperature-triggered release microgels. Food Res. Int. 69, 274–280.

Zhang, R., Zhang, Z., Zou, L., Xiao, H., Zhang, G., Decker, E.A., McClements, D.J., 2016a. Enhancement of carotenoid bioaccessibility from carrots using excipient emulsions: influence of particle size of digestible lipid droplets. Food Funct. 7, 93–103.

Zhang, Z., Zhang, R., Zou, L., McClements, D.J., 2016b. Protein encapsulation in alginate hydrogel beads: effect of pH on microgel stability, protein retention and protein release. Food Hydrocoll. 58, 308–315.

Zheng, M., Falkeborg, M., Zheng, Y., Yang, T., Xu, X., 2013. Formulation and characterization of nanostructured lipid carriers containing a mixed lipids core. Colloids Surf. A Physicochem. Eng. Asp. 430, 76–84.

Zou, L., Liu, W., Liu, C., Xiao, H., McClements, D.J., 2015. Utilizing food matrix effects to enhance nutraceutical bioavailability: increase of curcumin bioaccessibility using excipient emulsions. J. Agric. Food Chem. 63, 2052–2062.

Chapter 19

Safety of Nanoemulsions and Their Regulatory Status

Touseef A. Wani*, Farooq A. Masoodi*, Seid Mahdi Jafari[†] and David J. McClements[‡]

*University of Kashmir, Srinagar, India, [†]Gorgan University of Agricultural Sciences and Natural Resources, Gorgan, Iran, [‡]University of Massachusetts, Amherst, MA, United States

Chapter Outline

19.1 INTRODUCTION

In general, engineered nanoscale materials (ENMs), including both organic and inorganic nanoparticles, may be intentionally incorporated into foods, supplements, or pharmaceuticals for their desirable functional attributes, or they may inadvertently find their way into these products, for example, from the ingredients, packaging, or processes used in their preparation (McClements and Xiao, 2017). As highlighted earlier in this book, nanoscale materials (such as nanoemulsions) may behave differently from their macroscale counterparts (such as emulsions). Consequently, it is important to ensure that the nanoemulsions intentionally or unintentionally present in commercial products intended for oral consumption are safe.

Nanoemulsions. https://doi.org/10.1016/B978-0-12-811838-2.00019-9

19.2 SAFETY OF NANOEMULSIONS

There is no simple way to determine the safety of nanoscale products in general and of nanoemulsions in particular. A great deal of research has been carried out to establish the potential toxicity of inorganic nanoemulsions, but much less research has been carried out to assess the safety and toxicity of nanoemulsions. Consequently, much of our current understanding of the potential toxicity of nanoemulsions comes from the studies on other types of nanoscale materials. There are numerous factors that may impact the potential toxicological impact of nanoemulsions on biological systems, which can be divided into a number of broad categories:

- Nanoemulsion composition
- Nanoemulsion structure
- Administration route of nanoemulsions

19.2.1 Nanoemulsion Composition

Nanoemulsions may be fabricated from a variety of natural and/or synthetic ingredients, including oils, surfactants, cosurfactants, weighting agents, ripening inhibitors, thickeners, and gelling agents (Wani et al., 2016; Jafari et al., 2017a,b; Assadpour and Jafari, 2017; Mehrnia et al., 2017). Some of these components are not utilized in conventional emulsions, or are used at much lower levels than those found in nanoemulsions (Jafari et al., 2015; Mohammadi et al., 2016; Faridi Esfanjani et al., 2017). Consequently, it is possible that nanoemulsions may have different toxicity profiles than emulsions due to differences in the nature of the ingredients used to formulate them. Many of the ingredients used to formulate nanoemulsions would not be expected to promote toxicity, because they are already widely used at high levels in other products, including protein-, polysaccharide-, and phospholipid-based emulsifiers. However, some synthetic surfactants may promote irritation or toxicity when used at high levels, such as anionic surfactants (e.g., lactic acid esters), cationic surfactants (e.g., lauric arginate), and nonionic surfactants (e.g., Tweens and Spans) (McClements and Rao, 2011). There have been relatively few studies of the impact of nanoemulsion composition on their toxicity. However, a Caco-2 cell culture study reported no apparent difference in toxicity of nanoemulsions and emulsions stabilized by different kinds of emulsifiers (modified starch, Tween 20, or whey protein), suggesting that decreasing the droplet size did not have a major effect on their toxicity (Yu and Huang, 2013). In order to produce small droplets, nanoemulsions are often formulated using cosurfactants, organic solvents, weighting agents, or ripening inhibitors, which may also contribute to their toxicity. For instance, some components may increase cell membrane permeability, open tight junctions, or interfere with active absorption or efflux mechanisms, thereby indirectly altering the uptake of bioactive components.

19.2.2 Nanoemulsion Structure

The structural characteristics of nanoemulsions (Chapter 2) may also impact their potential toxicity, for example, particle size or charge. The small size of the droplets in nanoemulsions often leads to a higher bioavailability of ingested bioactive components for a number of reasons: (1) The equilibrium solubility of an encapsulated bioactive component increases as the size of the droplets containing it decreases leading to a greater concentration gradient for absorption; (2) the rate of lipid digestion and bioactive release increases as the droplet size increases (droplet surface area decreases), leading to more rapid absorption; (3) small droplets can penetrate through the pores in the mucus layer, also leading to faster absorption (Acosta, 2009; McClements, 2013; Jafari and McClements, 2017; Jafari, 2017a,b). Consequently, there may be a greater absorption of a bioactive component from a nanoemulsion than from a conventional emulsion. In many cases, this may be desirable, but in some cases, it may be undesirable. For instance, some nutraceuticals may have beneficial health effects when present at low to intermediate blood serum levels but have toxic effects when present at higher levels (such as vitamin E). Alternatively, an increased absorption of some bioactive agents (such as pesticides) may be undesirable.

Particle charge (ζ-potential) may also impact the safety and toxicity of nanoemulsions. The sign and magnitude of the ζ-potential on nanoemulsion droplets may alter their physical stability, penetration through mucus layers, or interactions with cellular membranes. For instance, studies have shown that the ability of nanoemulsion droplets to penetrate mucus layers depends on the magnitude of the ζ-potential (Suchaoin et al., 2016). In addition, the magnitude and sign of the ζ-potential on inorganic nanoparticles have been shown to influence their interactions with biological cells (Schwegmann et al., 2010).

19.2.3 Interaction of Nanoemulsions With the Biological Systems

The safety and toxicity of nanoemulsions depends on their interactions with biological systems. Over the last few years, there have been a great number of systematic studies on the interactions of inorganic nanoparticles with biological systems. However, there have been relatively few studies on the interactions of nanoemulsion droplets with biological systems. A brief overview of the influence of nanoemulsions on biological systems is provided in Table 19.1.

Nanoemulsions loaded with vitamin E altered the expression of various genes that resulted in alteration of energy and xenobiotic metabolism (CYP1A1 and glutathione S-transferase), formation of ketone bodies (B-hydroxybutyrate and palmitoleic acid), and elevation of stearoyl-CoA desaturase (Park et al., 2017). Olmesartan medoxomil is an angiotensin II receptor blocker that has been converted into a nanoemulsion form to increase its water solubility and oral bioavailability. Studies involving oral administration of olmesartan medoxomil

TABLE 19.1 Impact of Nanoemulsions on the Biological Systems

Study	Bioactive	Surfactant	Size	Subject/Cell Line	Result	Reference
Hepatic gene expression and serum metabolites	Vitamin E	–	112 nm	Male Sprague Dawley rats	Alteration of energy and xenobiotic metabolism, formation of ketone bodies and elevation of stearoyl-CoA desaturase	Park et al. (2017)
Safety profiling	Olmesartan medoxomil	Solutol HS 15	<50 nm	Wistar rats	No potential hematologic, biochemical, or structural toxicity	Gorain et al. (2016)
Cytotoxicity and antiangiogenic efficacy	Resveratrol	Transcutol HP	103 nm	Breast cancer cells (MCF-7). Chick embryo chorioallantoic membrane	Cytotoxic to MCF-7 cells and antiangiogenic	Pund et al. (2014)
Toxicity	*Eucalyptus staigeriana* essential oil	Tween 80	274.3 nm	Female Wistar albino rats	Ovicidal and larvicidal effects on *H. contortus*. No hematologic alterations in rats	Ribeiro et al. (2015)

Intranasal delivery toxicity	Curcumin	Tween 80	14.4–118.9 nm	Human neuroblastoma cell line (SK-N-SH)	Nontoxic and safe	Sood et al. (2014)
Cytotoxicity of tocopheryl polyethylene glycol 1000 succinate	D-α-tocopherol	Tween 80	25–35 nm	Human liver cancer cell line (HepG2)	No toxicity	Kaur et al. (2017)
Cytotoxicity	Curcumin	Lecithin	29.95 and 80.02 nm	Hepatocellular cell lines (HepG2). Human embryonic kidney cells (HEK293)	No toxicity	Lin et al. (2014)
Safety and immunogenicity	Mucosal adjuvant $W_{80}5EC$	Tween 80	–	Humans	No toxicity	Stanberry et al. (2012)
Safety against nephrotoxicity	Paclitaxel	Tween 20	63.42 nm	Swiss albino mice	No hematologic, biochemical, or structural toxicity at 6.5 and 3 mg/kg oral doses. Decrease in RBC, hemoglobin, and neutrophil counts at 12.8 mg/kg	Choudhury et al. (2017)

Continued

TABLE 19.1 Impact of Nanoemulsions on the Biological Systems—cont'd

Study	Bioactive	Surfactant	Size	Subject/Cell Line	Result	Reference
Antiglioma, cytotoxicity, and genotoxicity	Pomegranate seed oil	Tween 80	<250 nm	Peripheral human blood. Rat glioma cell line (C6)	Low toxicity against human blood cells at 0.05 mg/mL	Ferreira et al. (2016)
Dyslipidemia	Garlic oil	Tween 80	24.9 nm	Wistar rats	No hematologic or histological changes at 0.46 mL/kg dosage. Immediate mortality at 18.63 mL/kg	Ragavan et al. (2017)
Safety	Quercetin	Polyoxyethylene (20) oleyl ether, polyoxyethylene (3) oleyl ether, and cetyl trimethyl ammonium chloride	–	Chorioallantoic membrane	Causes hemorrhagic phenomena	Dario et al. (2016)

nanoemulsions to Wistar rats for 28 days at subchronic toxicity doses revealed a significant concentration of this bioactive substance in the animal's brains but with low toxicity as revealed by hematologic, biochemical, and structural analysis (Gorain et al., 2016). While assessing the antiangiogenic activity of resveratrol-loaded nanoemulsions in chick embryos, no toxicity was observed (Pund et al., 2014). A comparison of acute toxicity in mice between *Eucalyptus staigeriana* essential oil and its nanoemulsion carriers revealed LD_{50} doses of 1604 and 3496 mg/mL, respectively. *E. staigeriana* nanoemulsion poses ovicidal and larvicidal effects on *Haemonchus contortus* without altering the hematologic parameters after treatment (Ribeiro et al., 2015). In another study, mucoadhesive nanoemulsions loaded with curcumin fabricated using chitosan did not show any toxicity while targeting it for the brain through intranasal delivery (Sood et al., 2014). Nanoemulsions of D-α-tocopherol made from lemon oil, Tween 80, and water and stabilized by tocopheryl polyethylene glycol 1000 succinate were found to be nontoxic to HepG2 cells with cell viability as high as 100% (Kaur et al., 2017).

Anticolorectal cancer effects of lecithin and hydroxypropyl methylcellulose in puerarin nanosuspensions were analyzed using HT-29 cells. The cytotoxicity tests revealed that puerarin nanosuspensions were able to cease the proliferation of HT-29 cells (Wang et al., 2013). Nanoemulsion mucosal adjuvant $W_{80}5EC$ combined with approved seasonal influenza antigens showed no toxicity and was well tolerated in healthy adult human volunteers (Stanberry et al., 2012). Nanoemulsions of paclitaxel, an anticancer drug, in a 28-day subchronic toxicity study (6.5 and 3 mg/kg oral doses) showed no prominent hematologic, biochemical, or structural alterations. However, a prominent decrease in RBCs, hemoglobin, and neutrophils was observed at 12.8 mg/kg body weight dose (Choudhury et al., 2017). Nanoemulsions of pomegranate seed oil showed a hemolytic effect at high concentrations (Ferreira et al., 2016). An acute toxicity study of garlic oil nanoemulsions (18.63 mL/kg) administered orally to female rats resulted in immediate death. However, no hematologic or histological changes were observed at a dosage of 0.46 mL/kg garlic oil nanoemulsion and 0.5 mL/kg Tween 80 (Ragavan et al., 2017). Quercetin nanoemulsions have been shown to have good biocompatibility with the skin and induce low eye irritation (Dario et al., 2016). A recent study showed that curcumin-loaded nanoemulsions had no cytotoxic effects on normal fibroblasts, had good anti-inflammatory properties, and were able to promote the death of colon cancer cells (Vecchione et al., 2016).

Most of the recent literature on the safety and toxicity of nanoemulsions in biological systems suggest that their application is safe. There are few studies where researchers have found that nanoemulsions are toxic probably because of the very high doses employed. Hence, it could be concluded here that most of the nanoemulsions are safe when used at low concentrations. McClements and Rao (2011) have proposed a potential approach to evaluate the toxicity of nanoemulsions as shown in Fig. 19.1.

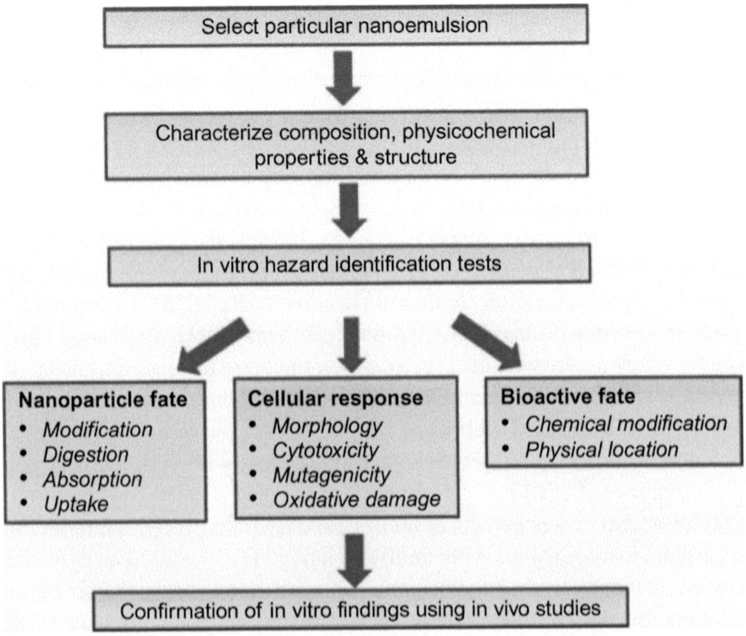

FIG. 19.1 A proposed protocol for the evaluation of potential toxicity of food-grade nanoemulsions. *(Adapted from McClements, D.J., Rao, J., 2011. Food-grade nanoemulsions: formulation, fabrication, properties, performance, biological fate, and potential toxicity. Crit. Rev. Food Sci. Nutr. 51, 285–330.)*

19.2.4 Administration Route of Nanoemulsions

The access route of nanoemulsions into biological systems is critical in determining their potential safety and toxicity. Nanoemulsions aimed for food applications are generally administered orally. However, pulmonary, intravenous, intramuscular, and ocular administrations of nanoemulsions are not uncommon (Wang et al., 2017). Orally ingested nanoemulsions comprised of digestible lipids are usually rapidly digested within the gastrointestinal tract (GIT), leading to the formation of mixed micelles consisting of monoacylglycerides and free fatty acids that are then absorbed (Chapter 18). Consequently, the end products of the digestion of nanoemulsions may be similar to those of conventional emulsions, although they are produced at a faster rate, which could alter blood serum levels (McClements, 2013). Typically, only a very small portion of the ingested nanoemulsion droplets will therefore remain intact and be able to cross the mucous layer and be absorbed by the epithelial cells (Müller et al., 2011; McClements, 2013). The potential toxicity of the small fraction of nanoparticles that are absorbed directly in the gut without digestion is not understood. When indigestible lipids are used to formulate nanoemulsions, there may be a much larger fraction of nanoparticles that are directly absorbed. The properties of the nanoemulsion droplets (such as particle size, charge, and composition) may then impact their absorption, distribution, metabolism, and excretion (Bouwmeester et al., 2009). Our understanding of this important area is currently limited, and there is clearly much scope for further research in the future.

In medical applications, nanoemulsions may also be delivered by intravenous administration because it can lead to enhanced permeability and retention of bioactive components (Wang et al., 2017). It is recommended that the mean particle diameter of nanoemulsions intended for these applications should be in the range of around 100–300nm (Sun and Yeo, 2012). If the droplets are too large, they may block the blood vessels, but if they are too small, they may be rapidly and indiscriminately absorbed. Dosage of the injection is one of the prime factors that decides the safety and toxicity of the injected component.

Intramuscular administration is another route for sustained release of bioactive components in biological systems (Wei et al., 2013). Controlling the size of the droplets in nanoemulsions intended for this type of application is again important for controlling drug pharmacokinetics. Intramuscular administration of curcumin and curcumin didecanoate nanosuspensions was shown to produce no toxicity in the spleen, heart, lung, liver, and kidney (Wei et al., 2013). Therefore, there is scope for exploration of nanoemulsions intended for intravenous and intramuscular administration.

For cosmetic-loaded nanoemulsions, risk concerning the skin permeation pattern of active ingredients should be considered since it has been demonstrated that nanoemulsions can enhance their permeation profiles (Chapter 14). In general, cosmetic ingredients can penetrate the upper layer of the human skin, but they should not permeate in depth through the skin. Therefore, the risk that their permeation profile could be enhanced by using nanoemulsions should be considered as a critical attribute for assuring consumer safety (Musazzi et al., 2017). As a matter of fact, there is room for reconsidering the regulatory framework on the basis of the risk assessment of nanoemulsion-based cosmetic products.

19.3 REGULATORY STATUS OF NANOEMULSIONS

19.3.1 Definitions and Current Status

Although there are typically no separate regulatory definitions covering nanoemulsions per se, there have been various definitions of nanomaterials by global regulating agencies and researchers that are applicable to nanomaterials in general. One of the generally used definitions of nanofoods was put forward by Joseph and Morrison (2006) as "food that has been cultivated, produced, processed, or packaged using nanotechnology techniques or tools, or to which manufactured nanomaterials have been added." However, the European Commission (2011) considers the size, distribution, and physical state of material particles while defining a nanomaterial as "a natural, incidental or manufactured material containing particles, in an unbound state or as an aggregate or as an agglomerate and where, for 50% or more of the particles in the number size distribution, one or more external dimensions is in the size range 1–100 nm. In specific cases and where warranted by concerns for the environment, health, safety or competitiveness the number size distribution threshold of 50% may be replaced by a threshold between 1 and 50%." Moreover, a material having a specific surface area by volume $>60\,m^2/cm^3$ is also considered "nano" by the European Commission (2011). As many as 24 definitions of "nanomaterial" have been provided by the FDA. These are based on the considerations that a material contains particles within the nanoscale range (1–100 nm) or the properties of the material are attributable to its dimension even if the dimensions fall out of the nanoscale range (up to 1000 nm) (FDA, 2014).

To the best of the authors' knowledge, there is currently no single internationally accepted body that regulates definitions, safety, toxicity, preparation, and characterization of nanoemulsions. The European Union and Switzerland are the only two regions of the world where nanospecific regulations have been incorporated in the legislation for agriculture, feed, and food (Amenta et al., 2015). Nanoregulations in the US Food and Drug Administration (FDA) are rather stratified and are building mainly upon guidance from industry (FDA, 2015). The European regulatory framework for cosmetic products is almost complete. However, due to the complexity in the field of medicine, the assessment of nanomaterials is based on a case-by-case evaluation (Musazzi et al., 2017). In order to ensure enough protection to humans and the environment, different countries in the world should collaborate with each other and exchange information without compromising the development of new beneficial products and their global marketing (Amenta et al., 2015).

19.3.2 Scientific Suggestions for Nano-Regulations

According to Sodano et al. (2016), acknowledging the political issues is necessary for the benefits of new technologies to outweigh the costs. Moreover, the enforcement of direct forms of regulation such as mandatory labeling and the establishment of a public register of products and producers could enhance the regulation of food nanotechnology. Table 19.2 provides some areas for the industry, legislators, and government to deliberate the regulations specific for the use of nanotechnology in foods.

Müller et al. (2011) divided nanoparticles on the basis of size and biodegradability into four classes and associated with them their persistency in the human body. The particles in the high-risk class are biopersistent nanoparticles, while those in the low-risk class are biodegradable nanoparticles (Fig. 19.2). According to these authors, it is important that the nanoparticles are degraded within the human body or at least eliminated from it. Nondegradable nanoparticles being too large for renal filtration are not eliminated, and the cells act as a

TABLE 19.2 A List of Recommendations on the Development of Regulations for Nanofoods

1. Considering the criteria including particle size range, measurement methods, processing methods, physical and chemical properties, and safety concerns
2. Regulating nanofoods and food products containing nanoparticles, of which their larger forms may be already permitted materials, as novel foods, ingredients, or additives
3. Several terms such as "micronized food," "ultrafine food," or "nanofood" can be considered. For the micronized food, a wider range of particle size (i.e., up to a few micron) may be considered
4. Upper particle size limit for nanofoods and ultrafine foods may be loosened to a value above 100 nm depending on the product types
5. Not only focusing on sizes but also concerning the changes in the bioactivity, physicochemical property, and functions of nanoscale materials in relation to their size reduction
6. Dividing nanofood products into different categories such as liquid, powder, aerosol, suspension, emulsion, and liposome for proper classification, management, and analysis
7. Requirement of food labeling to identify the presence of nanomaterials in the product and provide possible particle size range and relevant safety information
8. Carrying out societal and ethical research for food nanotechnology
9. Nanotoxicity screening should include the physicochemical characterization of nanomaterials, in vitro assays (cellular and noncellular), and in vivo animal studies
10. Safety assessment or nanotoxicity study should be required for the nanoscale materials, which are chemically modified
11. The nanoscale materials, which do not exist in nature, should be regarded as new substances and handled with more attention (i.e., control measures, nanotoxicity study, personal protective equipment, and health monitoring)
12. Potential toxicity of each individual nanoparticulate material needs to be considered on a case-by-case basis
13. Safety assessments and nanotoxicity study should be required for the nanoscale materials prepared by bottom-up synthetic approach, especially for those near or smaller than 100 nm even if their larger forms are proved to be safe for consumption
14. Risk assessment should include the potential release of manufactured nanoparticles from the packaging materials into food materials
15. Addressing any regulatory gaps in the existing food- and health-related regulatory frameworks in order to control the use of ingredients in the form of manufactured, free nanoparticles in consumer products
16. Laboratory protocols for handling nanoscale materials and dusts are needed to protect nanotechnology workers and scientists from exposure to occupational health hazards. It is recommended to start with standard hygiene procedures, including gloving, protective clothing, and highly efficient respirators capable of removing nanoparticles, and to move ahead as new information becomes available

Adapted from Chau, C.F., Wu, S.H., Yen, G.C., 2007. The development of regulations for food nanotechnology. Trends Food Sci. Technol. 18, 269–280.

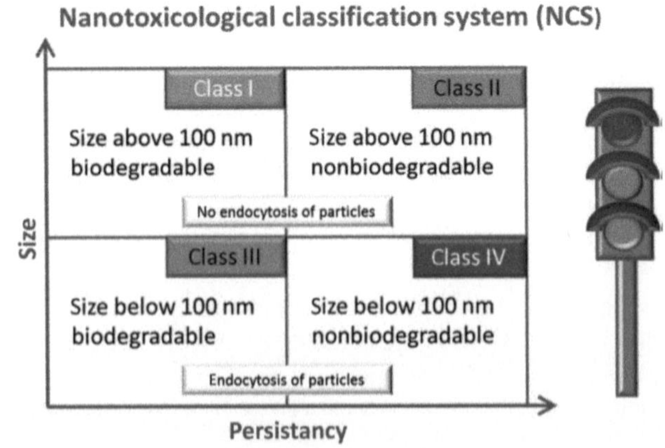

FIG. 19.2 Suggested nanotoxicological classification system (NCS) for a precise differentiation of nanotoxicological risks derived from nanoparticulate carriers in the future. *(Adapted from Müller, R.H., Gohla, S., Keck, C.M., 2011. State of the art of nanocrystals—special features, production, nanotoxicology aspects and intracellular delivery. Eur. J. Pharm. Biopharm. 78, 1–9.)*

waste deposit site for these nanoparticles. Therefore, materials that are not biodegradable have little for acceptance by regulatory authorities for food and pharmaceutical applications. This has already led the European Union to regulate the labeling of cosmetic products as "nano" when they contain particles that are smaller than 100nm and biopersistent (class IV of NCS). As such, products containing biodegradable particles of size <100nm (class III of NCS) are not considered as "nano" products.

We have recently proposed a universal nomenclature system of encapsulated products based on capsule size, wall materials, active ingredients, and production techniques (Wani et al., 2017). The nomenclature system is applicable to encapsulated products of any type. To get the name for the encapsulated products, the four aspects are arranged as follows: size (wall material, active ingredient) technology employed, for example,

i. 65–106 μm (alginate, insulin) emulsification
ii. 108 nm (β-cyclodextrin-pluronic polymer, curcumin) precipitation

The labeling of such type gives straightforward information regarding particle size (nano or not) and wall material (biodegradable or nonbiodegradable). A similar kind of labeling scheme may be useful for regulating the use of nanoemulsions and other nanoparticles in commercial products.

A comprehensive review in the field of regulations and legislations for nanofood products has been presented by Jafari et al. (2017a,b) titled "Safety and regulatory issues of nanocapsules" in different countries and organizations.

19.4 CONCLUSION AND PERSPECTIVES

Nanoemulsions are one of the simplest and most effective delivery systems for the encapsulation and release of hydrophobic bioactive agents. They may have different compositions and structures depending on the nature of the ingredients and processing methods used to fabricate them. They may also be administered through different routes (such as the mouth, skin, nose, eyes, or intravenously) depending on their intended application. The additional ingredients or higher ingredient levels often required to produce the very small droplets in nanoemulsions may affect their toxicity due to the change in system composition compared with conventional nanoemulsions. However, this problem can often be overcome by formulating nanoemulsions using only those ingredients known to be safe when used at the required levels. The small size of the droplets in nanoemulsions may also influence their biological fate, by altering the stability, solubility, or absorption of encapsulated bioactive agents. In many cases, increasing the bioavailability of bioactive agents is desirable as it leads to increased bioactivity. However, in some cases, this may be undesirable because the bioactive agent itself is toxic (such as pesticides) or because the bioactive agent is beneficial when absorbed at relatively low to intermediate levels but toxic at higher levels (such as some oil-soluble vitamins). There is still a relatively poor understanding of the physicochemical basis of the potential toxicity of nanoemulsions, and it is clear that more in vitro and in vivo studies are required in this area.

The regulation of nanoscale products (including nanoemulsions) is still in its early stages, and there is no global regulatory authority in this area. However, the European Union and Switzerland are the main regions where nanoregulations have been developed and enforced. There is still a pressing need for regulatory authorities around the globe to develop robust standardized procedures for analyzing nanoscale products and establishing their safety. The development of these regulations will ensure that the benefits of nanotechnology can be realized, without adversely affecting the health of the consumer and the environment.

ACKNOWLEDGMENTS

Touseef Ahmed Wani acknowledges the Council of Scientific & Industrial Research (CSIR), New Delhi, India, for the award of Senior Research Fellowship (SRF) grant (09/251/(0070)/2K17). We also thank the US Department of Agriculture, Australian Government NIFA Program (2016-08782) for partial support of this research.

REFERENCES

Acosta, E., 2009. Bioavailability of nanoparticles in nutrient and nutraceutical delivery. Curr. Opin. Colloid Interface Sci. 14, 3–15.

Amenta, V., Aschberger, K., Arena, M., Bouwmeester, H., Moniz, F.B., Brandhoff, P., Gottardo, S., Marvin, H.J.P., Mech, A., Pesudo, L.Q., Rauscher, H., Schoonjans, R., Vettori, M.V., Weigel, S., Peters, R.J., 2015. Regulatory aspects of nanotechnology in the agri/feed/food sector in EU and non-EU countries. Regul. Toxicol. Pharmacol. 73, 463–476.

Assadpour, E., Jafari, S.M., 2017. Spray drying of folic acid within nano-emulsions: optimization by Taguchi approach. Dry. Technol. 35 (9), 1152–1160.

Bouwmeester, H., Dekkers, S., Noordam, M.Y., Hagens, W.I., Bulder, A.S., de Heer, C., ten Voorde, S.E.C.G., Wijnhoven, S.W.P., Marvin, H.J.P., Sips, A.J.A.M., 2009. Review of health safety aspects of nanotechnologies in food production. Regul. Toxicol. Pharmacol. 53, 52–62.

Choudhury, H., Gorain, B., Tekade, R.K., Pandey, M., Karmakar, S., Pal, T.K., 2017. Safety against nephrotoxicity in paclitaxel treatment: Oral nanocarrier as an effective tool in preclinical evaluation with marked in vivo antitumor activity. Regul. Toxicol. Pharmacol. 91, 179–189.

Dario, M.F., Oliveira, C.A., Cordeiro, L.R.G., Rosado, C., Marizd, I.F.A., Macôasd, E., Santose, M.S.C.S., da Piedadee, M.E.M., Baby, A.R., Velasco, M.V.R., 2016. Stability and safety of quercetin-loaded cationic nanoemulsion: in vitro and in vivo assessments. Colloids Surf. A Physicochem. Eng. Asp. 506, 591–599.

European Commission, 2011. Commission Recommendation of 18 October 2011 on the definition of nanomaterial (2011/696/EU). Off. J. Eur. Union 275, 38–40.

Faridi Esfanjani, A., Jafari, S.M., Assadpour, E., 2017. Preparation of a multiple emulsion based on pectin-whey protein complex for encapsulation of saffron extract nanodroplets. Food Chem. 221, 1962–1969.

FDA, 2014. Guidance for Industry—Considering whether an FDA-Regulated Product Involves the Application of Nanotechnology. FDA.

FDA, 2015. Guidance for Industry Use of Nanomaterials in Food for Animals. https://www.fda.gov/downloads/AnimalVeterinary/GuidanceComplianceEnforcement/GuidanceforIndustry/UCM401508.pdf.

Ferreira, L.M., Gehrcke, M., Cervi, V.F., Bitencourt, P.E.R., da Silveira, E.F., Azambuja, J.H., Ramos, A.P., Nascimento, K., Moretto, M.B., Braganhol, E., Sagrillo, M.R., Cruz, L., 2016. Pomegranate seed oil nanoemulsions with selective antiglioma activity: optimization and evaluation of cytotoxicity, genotoxicity and oxidative effects on mononuclear cells. Pharm. Biol. 54, 2968–2977.

Gorain, B., Choudhury, H., Tekade, R.K., Karan, S., Jaisankar, P., Pal, T.K., 2016. Comparative biodistribution and safety profiling of olmesartan medoxomil oil-in-water oral nanoemulsion. Regul. Toxicol. Pharmacol. 82, 20–31.

Jafari, S.M., 2017a. 1—An overview of nanoencapsulation techniques and their classification. In: Jafari, S.M. (Ed.), Nanoencapsulation Technologies for the Food and Nutraceutical Industries. Academic Press, New York, pp. 1–34.

Jafari, S.M., 2017b. Chapter 1—An introduction to Nanoencapsulation techniques for the food bioactive ingredients. In: Jafari, S.M. (Ed.), Nanoencapsulation of Food Bioactive Ingredients. Academic Press, New York, pp. 1–62.

Jafari, S.M., Fathi, M., Mandala, I., 2015. Emerging product formation. In: Galanakis, C. (Ed.), Food Waste Recovery: Processing Technologies and Industrial Techniques. Elsevier Inc., New York, pp. 293–317.

Jafari, S.M., Katouzian, I., Akhavan, S., 2017a. 15—Safety and regulatory issues of nanocapsules. In: Jafari, S.M. (Ed.), Nanoencapsulation Technologies for the Food and Nutraceutical Industries. Academic Press, New York, pp. 545–590.

Jafari, S.M., McClements, D.J., 2017. Nanotechnology approaches for increasing nutrient bioavailability. In: Toldra, F. (Ed.), Advances in Food and Nutrition Research. Academic Press, New York.

Jafari, S.M., Paximada, P., Mandala, I., Assadpour, E., Mehrnia, M.A., 2017b. 2—Encapsulation by nanoemulsions. In: Jafari, S.M. (Ed.), Nanoencapsulation Technologies for the Food and Nutraceutical Industries. Academic Press, New York, pp. 36–73.

Joseph, T., Morrison, M., 2006. Nanoforum report: Nanotechnology in agriculture and food. ftp://ftp.cordis.europa.eu/pub/nanotechnology/doc/nanotechnology_in_agriculture_and_food.pdf/.

Kaur, K., Kumar, R., Arpita, G., Uppal, S., Bhatia, A., Mehta, S.K., 2017. Physiochemical and cytotoxicity study of TPGS stabilized nanoemulsion designed by ultrasonication method. Ultrason. Sonochem. 34, 173–182.

Lin, C.C., Lin, H.Y., Chi, M.H., Shen, C.M., Chen, H.W., Yang, W.J., Lee, M.H., 2014. Preparation of curcumin microemulsions with food–grade soybean oil/lecithin and their cytotoxicity on the HepG2 cell line. Food Chem. 154, 282–290.

McClements, D.J., 2013. Edible lipid nanoparticles: digestion, absorption, and potential toxicity. Prog. Lipid Res. 52, 409–423.

McClements, D.J., Rao, J., 2011. Food–grade nanoemulsions: formulation, fabrication, properties, performance, biological fate, and potential toxicity. Crit. Rev. Food Sci. Nutr. 51, 285–330.

McClements, D.J., Xiao, H., 2017. Is nano safe in foods? Establishing the factors impacting the gastrointestinal fate and toxicity of organic and inorganic food-grade nanoparticles. npj Sci. Food 1 (1), 6.

Mehrnia, M.-A., Jafari, S.M., Makhmal-Zadeh, B.S., Maghsoudlou, Y., 2017. Rheological and release properties of double nano-emulsions containing crocin prepared with Angum gum, Arabic gum and whey protein. Food Hydrocoll. 66, 259–267.

Mohammadi, A., Jafari, S.M., Esfanjani, A.F., Akhavan, S., 2016. Application of nano-encapsulated olive leaf extract in controlling the oxidative stability of soybean oil. Food Chem. 190, 513–519.

Müller, R.H., Gohla, S., Keck, C.M., 2011. State of the art of nanocrystals—special features, production, nanotoxicology aspects and intracellular delivery. Eur. J. Pharm. Biopharm. 78, 1–9.

Musazzi, U.M., Franzè, S., Minghetti, P., Casiraghi, A., 2017. Emulsion versus nanoemulsion: how much is the formulative shift critical for a cosmetic product? Drug Deliv. Transl. Res. https://doi.org/10.1007/s13346-017-0390-7.

Musazzi, U.M., Marini, V., Casiraghi, A., Minghetti, P., 2017. Is the European regulatory framework sufficient to assure the safety of citizens using health products containing nanomaterials? Drug Discov. Today 22, 870 – 882.

Park, C.Y., Jang, C.H., Lee, D.Y., Cho, H.T., Kim, Y.J., Park, Y.H., Imm, J.Y., 2017. Changes in hepatic gene expression and serum metabolites after oral administration of overdosed vitamin-E-loaded nanoemulsion in rats. Food Chem. Toxicol. 109, 421–427.

Pund, S., Thakur, R., More, U., Joshi, A., 2014. Lipid based nanoemulsifying resveratrol for improved physicochemical characteristics, in vitro cytotoxicity and in vivo antiangiogenic efficacy. Colloids Surf. B: Biointerfaces 120, 110–117.

Ragavan, G., Muralidaran, Y., Sridharan, B., Ganesh, R.N., Viswanathan, P., 2017. Evaluation of garlic oil in nano–emulsified form: Optimization and its efficacy in high–fat diet induced dyslipidemia in Wistar rats. Food Chem. Toxicol. 105, 203–213.

Ribeiro, W.L.C., Camurca-Vasconcelos, A.L.F., Macedo, I.T.F., dos Santos, J.M.L., de Araújo-Filho, J.V., de Carvalho Ribeiro, J., de Abreu Pereira, V., de Araújo Viana, D., de Paula, H.C.B.,

Bevilaqu, C.M.L., 2015. In vitro effects of *Eucalyptus staigeriana* nanoemulsion on *Haemonchus contortus* and toxicity in rodents. Vet. Parasitol. 212, 444–447.

Schwegmann, H., Feitz, A.J., Frimmel, F.H., 2010. Influence of the zeta potential on the sorption and toxicity of iron oxide nanoparticles on *S. cerevisiae* and *E. coli*. J. Colloid Interface Sci. 347, 43–48.

Sodano, V., Gorgitano, M.T., Quaglietta, M., Verneau, F., 2016. Regulating food nanotechnologies in the European Union: open issues and political challenges. Trends Food Sci. Technol. 54, 216–226.

Sood, S., Jain, K., Gowthamarajan, K., 2014. Optimization of curcumin nanoemulsion for intranasal delivery using design of experiment and its toxicity assessment. Colloids Surf. B: Biointerfaces 113, 330–337.

Stanberry, L.R., Simon, J.K., Johnson, C., Robinson, P.L., Morry, J., Flack, M.R., Gracon, S., Myc, A., Hamouda, T., Baker Jr., J.R., 2012. Safety and immunogenicity of a novel nanoemulsion mucosal adjuvant W_{80}5EC combined with approved seasonal influenza antigens. Vaccine 30, 307–316.

Suchaoin, W., de Sousa, I.P., Netsomboon, K., Lam, H.T., Laffleur, F., Bernkop-Schnurch, A., 2016. Development and in vitro evaluation of zeta potential changing self-emulsifying drug delivery systems for enhanced mucus permeation. Int. J. Pharmacol. 510, 255–262.

Sun, B., Yeo, Y., 2012. Nanocrystals for the parenteral delivery of poorly water-soluble drugs. Curr. Opin. Solid State Mater. Sci. 16, 295–301.

Vecchione, R., Quagliariello, V., Calabria, D., Calcagno, V., De Luca, E., Iaffaioli, R.V., Netti, P.A., 2016. Curcumin bioavailability from oil in water nano-emulsions: in vitro and in vivo study on the dimensional, compositional and interactional dependence. J. Control. Release 233, 88–100.

Wang, L., Du, J., Zhou, Y., Wang, Y., 2017. Safety of nanosuspensions in drug delivery. Nanomedicine 13, 455–469.

Wang, Y., Ma, Y., Zheng, Y., Song, J., Yang, X., Bi, C., Zhang, D., Zhang, Q., 2013. In vitro and in vivo anticancer activity of a novel puerarin nanosuspension against colon cancer, with high efficacy and low toxicity. Int. J. Pharm. 441, 728–735.

Wani, T.A., Gani, A., Wani, S.M., Wani, I.A., Masoodi, F.A., Nissar, N., Shagoo, M.A., 2016. Suitability of different food grade materials for the encapsulation of some functional foods well reported for their advantages and susceptibility. Crit. Rev. Food Sci. Nutr. 56, 2431–2454.

Wani, T.A., Masoodi, F.A., Wani, I.A., 2017. The possible nomenclature of encapsulated products. Food Chem. 234, 119–120.

Wei, X.L., Han, Y.R., Quan, L.H., Liu, C.Y., Liao, Y.H., 2013. Oily nanosuspension for long-acting intramuscular delivery of curcumin didecanoate prodrug: preparation, characterization and in vivo evaluation. Eur. J. Pharm. Sci. 49, 286–293.

Yu, H., Huang, Q., 2013. Investigation of the cytotoxicity of food-grade nanoemulsions in Caco-2 cell monolayers and HepG2 cells. Food Chem. 141, 29–33.

Index

Note: Page numbers followed by *f* indicate figures, and *t* indicate tables.

D

E

F

G

H

I

K

L

N

O

P

T

U

V

W

X

Y

Z

FSC
www.fsc.org
MIX
Papier | Fördert
gute Waldnutzung
FSC® C083411

Kolibri 360 GmbH
Ferdinand-Jühlke-Straße 7
99095 Erfurt, Deutschland
produktsicherheit@kolibri360.de